Physiological Ecology

Physiological Ecology

How Animals Process Energy, Nutrients, and Toxins

William H. Karasov and Carlos Martínez del Rio

PRINCETON UNIVERSITY PRESSS — PRINCETON AND OXFORD

LIBRARY OF CONGRESS CATALOGING-IN-PUBLICATION DATA

Karasov, William H., 1953-
Physiological ecology: how animals process energy, nutrients, and
toxins/William H. Karasov and Carlos Martínez del Rio.
p. cm.
Includes index.
ISBN-13: 978-0-691-07453-5 (clothbound: alk. paper)
ISBN-10: 0-691-07453-4 (clothbound : alk. paper) 1. Animal ecophysiology.
I. Martínez del Rio, Carlos, 1956- II. Title.
QP82.K34 2006
591.7–DC22
2006037344

British Library Cataloging-in-Publication Data is available

This book has been composed in Adobe Caslon

Printed on acid-free paper.

press.princeton.edu

Printed in the United States of America

1 3 5 7 9 10 8 6 4 2

CONTENTS

SECTION V: LIMITING NUTRIENTS

SECTION VI: PRODUCTION IN BUDGETS OF MASS AND ENERGY

Physiological ecology should have two primary goals: to understand how the interaction of organism and the environment determines characteristics that are relevant to ecology (e.g. age specific fecundity and mortality, movement patterns, foraging, . . . etc.); and to understand how these individual characteristics affect population and interspecific dynamics.

—*Kingsolver 1989*

PREFACE

Resource acquisition and allocation is a basic task of all animals. Among physiological processes, those that influence resource use are the most likely to influence an animal's behavior, its population dynamics, its role in biotic communities, and its contribution to ecosystem fluxes. But graduate and undergraduate students with ecological inclinations are often hard pressed to find user-friendly introductions to the physiological and biochemical disciplines that underpin these conceptual and practical applications. A behavioral ecologist interested in the physiological mechanisms that shape feeding behavior must wade through a diverse and fairly technical literature on animal energetics, nutrition, and digestive physiology. The same comment can be made about students or professional ecologists interested in understanding the effect of plant secondary compounds or environmental toxicants on their ecological systems. We wrote this book for advanced undergraduates, graduate students, and their mentors as an accessible introduction to the physiological principles that are important to understanding how animals use resources. Each chapter makes explicit links between the physiological mechanisms of resource utilization and ecological phenomena such as diet selection, ecotoxicology, ecological fluxes, trophic ecology, and the annual cycles of animals.

Joel Kingsolver's (1989) quote at the opening of this preface nicely characterizes a perspective of this book that we think is in harmony with one that has been gaining force in ecology over the past decade. This perspective recognizes the importance of advances in the understanding of basic ecological principles to solving many urgent environmental problems (Lubchenco et al. 1991), that to succeed ecologists should favor (but not exclusively) questions that lead to predictive rigor (such as "how much," "how many," "when," and "where") rather than solely explanation ("why" or "how come"; Peters 1991), and that many (but not all) ecological processes, from the level of individuals to the biosphere, are determined by the rates at which organisms take up, transform, and expend energy and materials (Sterner and Elser 2002; Brown et al. 2004). Thus, we wrote this book with the belief that ecologists and physiologists undertaking mechanistic studies of resource acquisition and allocation can make important contributions to academic ecology and its ability to help solve societal problems.

The perspective of this book is more similar to that adopted by plant physiological ecologists than to that adopted by animal physiologists. Traditionally, animal physiologists have emphasized evolution and the fit between animals and their environment, whereas plant physiological ecologists have focused on the ecological consequences of plant physiological mechanisms (Feder et al. 2000). Some of our colleagues choose their study animals because they are the best model systems to investigate a particular physiological mechanism. They follow August Krogh's dictum that there will be some animal on which a mechanism can be most conveniently studied (Krogh 1929). Although this comparative perspective is rich and remains fertile, it is one that we will not follow. Physiological traits are astoundingly diverse, and physiologists are often tempted to describe the features of their systems in exquisite detail. As much as possible, we tried to avoid this temptation. We wrote a book that includes sufficient mechanistic detail to understand what we believe are fundamental physiological processes with ecological relevance, but which is not comprehensive in its coverage of the immense diversity of function. Although our emphasis is primarily ecological, we did not ignore evolution. Evolutionary and ecological physiology are inextricably linked. Throughout the book we use evolutionary theory and the comparative method as tools to understand the diversity of physiological traits that influence resource use. The results of evolutionary physiology, a rich and enormously successful discipline, inform our book, but do not constitute its core.

Our ecological emphasis led us to incorporate sections that would not typically be included in a conventional animal physiology book. We include chapters on ecological stoichiometry, isotopic ecology, nutritional symbioses, and ecotoxicology. Lamentably, some topics firmly established within the domain of ecological physiology, such as heat exchange, ecological endocrinology, and ecological immunology, we could handle only fleetingly and primarily in relation to our main focus on resource utilization.

One of our purposes in writing this book was not only to provide a conceptual introduction to ecological physiology, but also to describe its methods. Methods will be described, often in boxes, in sufficient detail so that readers can evaluate their utility for their own work. We have found that nonphysiologists are often intimidated by the apparent need to use complicated analytical methods to answer ecological questions from a physiological perspective. Often, however, the methods used by ecological physiologists are relatively simple, and even when they are not, the broad principles on which they are based can be described and understood easily. Thus, an important reason to include methods in our book is to provide students and researchers with a set of potential tools to answer ecological questions. Although our book is not a manual of ecophysiological techniques, it can be used as a guide to methods and as an introduction to the literature on them.

The book is broadly divided into six sections. Sections I–III emphasize nutrient and toxin assimilation construed very broadly, whereas sections IV–VI emphasize the fate of materials after they have been assimilated. After a chapter that introduces themes common throughout the book (section I "Overview"), we devote a chapter to the chemical ecology of food (section II "Chemical Ecology of Food"). The following sections roughly trace the pathway of an ingested nutrient or toxin. First, the substance must be assimilated. It must be digested (broken down) and transported across the wall of the gastrointestinal tract (section III "Digestive Ecology"). Three of the four chapters that constitute this section are a survey of digestive ecology, which is the study of the behavioral and ecological consequences of the mechanisms used by animals to assimilate food. One of the chapters in section III (chapter 5 "Photosynthetic Animals and Gas-Powered Mussels: The Physiological Ecology of Nutritional Symbioses") adopts an expansive view of digestion to include the assimilation of organic materials produced by animals' symbionts. We have dwelt on the topic of nutritional symbioses and described some of them in significant detail, because we find that this is an area that nutritional ecologists have neglected. After assimilation, the substance is transformed by a variety of metabolic pathways and catabolized, excreted, or incorporated into tissues (section IV "Postabsorptive Processing of Food"). In chapter 7 we focus on the postabsorptive processing of nutrients, and in chapter 9 on toxins, because their processing constitutes an important component in the field of ecotoxicology. Isotopes have been a key tool to study these processes, and in chapter 8 we introduce relevant principles and important ecological implications of the new field of isotopic ecology. Section V ("Key Nutrients") is dedicated to water and several elements with fundamental ecological importance, and it introduces another relatively new field, ecological stoichiometry. In the final section (section VI "Organismic Budgets and Ecology") we take up the issue of links between rates of mass and energy gain and life history features such as growth and reproduction.

Physiological ecology is a growing field with many unanswered questions. Throughout the book we have identified areas that we believe call for investigation and that we deem ripe for fertile research. Our book attempts not only to describe what we know about the mechanisms that shape how animals use resources, but also to identify new questions, and to justify the value of answering these questions. We hope that some of these questions, and other material in the book, will stimulate and guide work on the mechanisms of resource use in ecology. Improved mechanistic understanding can increase the explanatory and predictive power of ecological theory. The ability of ecologists to contribute to solutions to societal problems such as consequences of global climate change, the biodiversity crisis, and environmental pollution depends partly on their understanding of the principles of energy and material flow in animals.

References

Brown, J. H., J. F. Gillooly, A. P. Allen, V. M. Savage, and G. B. West. 2004. Toward a metabolic theory of ecology. *Ecology* 85:1771–1789.

Feder, M. E., A. F. Bennett, and R. B. Huey. 2000. Evolutionary physiology. *Annual Review of Ecology and Systematics* 31:315–341.

Kingsolver, J. G. 1989. Weather and the population dynamics of insects: Integrating physiology and population ecology. *Physiological Zoology* 62:314–334.

Krogh, A. 1929. Progress in physiology. *American Journal of Physiology.* 90:243–251.

Lubchenco, J., A. M. Olson, L. B. Brubaker, S. R. Carpenter, M. M. Holland, S. P. Hubbell, S. A. Levin, J. A. MacMahon, P. A. Matson, J. M. Melillo, H. A. Mooney, C. R. Peterson, H. R. Pulliam, L. A. Real, P. J. Regal, and P. G. Risser. 1991. The sustainable biosphere initiative: An ecological research agenda. *Ecology* 72:371–412.

Peters, R. H. 1991. *A critique for ecology.* Cambridge University Press, Cambridge.

Sterner, R. W., and J. J. Elser. 2002. *Ecological stoichiometry: The biology of the elements from molecules to the atmosphere.* Princeton University Press, Princeton, N.J.

ACKNOWLEDGMENTS

W<small>E DEDICATE THIS</small> book to our colleagues who made the discoveries that we chronicle. There is a lot of important research that, unfortunately, we could not include. We regret these omissions, and hope that our colleagues will understand. We also dedicate the book to the students who have decided to quest for discoveries in physiological ecology. We hope that this book will incite them to ask new questions and to challenge old ideas—ours included.

We are thankful to the numerous colleagues who read and criticized all or portions of our book during its long gestation, and who encouraged us to keep going. A partial list includes Drew Allen, Karen Bjorndal (and her students), Enrique Caviedes-Vidal, Denise Dearing, Jamie Gillooly, Chris Guglielmo, Yue-wern Huang, Rick Lindroth, Todd McWhorter, Ken Nagy, Bob Sterner, Sarah Pabian, Berry Pinshow, Warren Porter, Michele Skopec, Chris Tracy, Itzick Vatnick, and Joe Williams. Joel Kingsolver, Theunis Piersma, and Blair Wolf reviewed the entire book and made many useful suggestions. Of course, all the errors in the book, and especially our feeble humor attempts, are our responsibility. We are also thankful to our graduate students and postdoctoral associates for keeping our laboratories running when we periodically neglected them when immersed in writing. Annie Hartmann Bakken and Bruce Darken helped us draft many of the book's figures.

Finally, we would like to thank our wonderful wives (Corliss and Martha) and children (Ariela, Aurora, Cormac, and Talia) for sharing our enthusiasm for knowledge, and for kvetching only rarely about losing us to our scientific obsessions.

Overview

CHAPTER ONE

Basic Concepts: Budgets, Allometry, Temperature, and the Imprint of History

SEVERAL COMMON THEMES permeate the field of physiological ecology. Because little can be done in our discipline without touching on them, and because these themes serve as organizing principles later in the book, we devote this chapter to them. It can be read as a guide to the conceptual tools and questions that make animal physiological ecology a coherent field. These conceptual tools and questions also link physiological ecology with ecology at large and make the research that we do with organisms relevant for those who study populations, ecosystems, and the biosphere. Although physiological ecologists quantify the magnitude of a plethora of seemingly disparate traits, we believe that most of these traits can be characterized generally as pools, rates, fluxes, and efficiencies. We begin this chapter by describing how pools, rates, fluxes, and efficiencies can be integrated within the framework of biological budgets. The remaining portion of the chapter is devoted to characterizing how physiological ecologists investigate some of the most important factors that shape physiological traits (or biological budget elements, if you wish) and their variation within and among species. We use the metabolic rate to exemplify the effect of body size and temperature on a trait, and as a trait that is shaped by factors at a variety of time scales. We conclude the chapter by discussing why the evolutionary history of both species and traits needs to be taken into account in comparative studies.

1.1 The Input/Output Budget: A Key Conceptual Framework

Figure 1.1. A simple input/output budget for energy in a termite. The primary input is chemical potential energy in the food (I). The primary outputs include both chemical potential energy in excretory wastes (E) and new tissue produced (P), and heat evolved during respiration (R). Note that, of the total energy ingested only a fraction is assimilated, and of the fraction assimilated only a fraction is used in catabolic processes (R) or anabolic processes (P). Energy or material budgets can be used to estimate the ecological impacts of animals, such as this 3.5 mg termite. Based on these numbers, one estimates that in a savanna ecosystem in Western Africa with 4000 termites per square meter, termites breakdown 77 g wood per meter per year or at least 15% of the woody litter, and produce 48 mmol methane gas per year per square meter. Methane, a product of microbial fermentation in the termite, is one of the principal greenhouse gases, and termites worldwide may account for up to 15% of total production. (Based on data in Wood and Sands 1978 and Bignell et al. 1997.)

A powerful method for organizing a physiological ecology study is to construct a budget. A budget is a detailed accounting of the input and output of energy, or the mass of some substance through an individual, a population, and even an ecosystem (figure 1.1). A budget is not simply an accounting device but a tool that can be used to predict the magnitude of both inputs and outputs. An animal's requirements for energy, nutrients, and water can be predicted from knowledge of minimum obligatory loss rates and from assimilation efficiencies. Perhaps more significantly, budgets can have significant value in ecology because they can allow us to identify limiting factors to the fluxes of energy and materials through organisms. Also, because one organism's waste is another organism's bounty (chapter 10), knowledge about fluxes allows us to find out not only how much of a given resource animals need, but also how much they generate for other creatures. Budgets are important because they specify some of the key physiological questions that are important for ecologists: What are the rates of input and output of energy or matter? What are the efficiencies with which animals assimilate nutrients and energy? What are the patterns of allocation into different parts and functions? And what are the physiological mechanisms that regulate or limit the supply of or demand for energy and materials?

A fundamental reason for the importance of budgets for biologists is that the conservation law for mass and energy is one of the few hard laws that we have available (Kooijman 2000). Equally compelling is the practical utility of

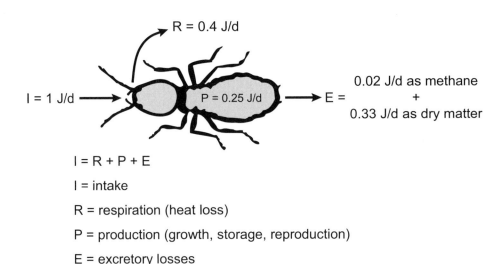

R = 0.4 J/d

I = 1 J/d →   P = 0.25 J/d → E =

0.02 J/d as methane
+
0.33 J/d as dry matter

I = R + P + E

I = intake

R = respiration (heat loss)

P = production (growth, storage, reproduction)

E = excretory losses

the approach. Budgets allow us to construct models that link mass and energy flow in individuals to population, community, and life history phenomena. Although we will emphasize individual organisms in this chapter, the same principles apply to collections of them. For example, budgets allow quantification of the flow of toxicants through ecosystems (Cairns and Niederlehner 1996), and plant physiological ecologists use community level budgets to estimate the impact of plant communities and animal communities on the composition of the atmosphere (Griffiths 1997). Budgets are one of the tools that allow physiological ecologists to scale up their measurements from individuals to the biosphere (Van Gordingen et al. 1997).

1.1.1 Pools, Fluxes, and Residence Times

The input/output format of a budget is straightforward and should be intuitive to anyone who handles a bank account. The size of the pool is the capital and the inputs and outputs are profits and expenditures, respectively. An expression for the flow of chemical potential energy through an invertebrate, for example, might include an energy input rate for feeding and energy output rates for digestive wastes, heat, and new tissue (e.g., progeny, figure 1.1). The chemical potential energy in food, digestive wastes, and new tissue is measured by bomb calorimetry, whereas the heat produced is typically measured by respirometry, which is also called indirect calorimetry (box 1.1). Although we will use the term "metabolic rate" as shorthand for the rate at which an animal catabolizes organic materials (usually aerobically) to generate energy, this term is strictly incorrect. Catabolism (breakdown) and anabolism (synthesis) collectively constitute metabolism. Perhaps we should use the term "respiration rate" instead. However, the term metabolic rate is so firmly entrenched that attempting to dislodge it probably confuses more than it enlightens.

The first law of thermodynamics, the conservation law for energy, requires that the rate of energy input equal the sum of the rates of outputs. A budget, or a balance equation, is an arithmetic expression of this principle (figure 1.1). Knowledge about the flow of materials through a system is necessary in the construction of a budget, but it is not sufficient. One must also know the sizes of the pools of energy and materials in an organism.

The change in the size of a pool of energy or material in an animal is an immediate consequence of the balance between inputs and outputs:

$$\frac{d\,(\text{size of the pool})}{dt} = \text{input rate} - \text{output rate}. \tag{1.1}$$

Box 1.1 Respirometry: The Measure of Heat Production Based on Oxygen Consumption and/or Carbon Dioxide Production

The reason oxygen or carbon dioxide can be used to measure heat production is because we understand the stoichiometry of catabolic reactions and therefore we can relate gas exchange directly to heat production. To illustrate this, consider the stoichiometry of the oxidation of 1 mole of palmitic acid—a fatty acid common in vertebrates:

$$CH_3(CH_2)_{14}COOH + 23\ O_2 \rightarrow 16\ CO_2 + 16\ H_2O$$
reaction's heat of combustion = 10,040 kJ
reaction consumes O_2 and produces CO_2
(23 moles O_2) × (22.4 liters O_2/mole at 760 torr and 0°C) = 515 liters O_2 STP
(16 moles CO_2) × (22.4 liters/mole) = 358 liters CO_2 STP
Therefore, 10,040/515 = 19.5 kJ/liter O_2
10,040/358 = 28.0 kJ/liter CO_2

Consult table 1.1 for average values of heat produced per gram of substrate and unit gas exchanged for each substrate

TABLE 1.1
Average values of heat produced per gram of substrate and per unit gas
exchanged for each substrate

Substrate	kJ/g substrate	kJ/liter O_2	kJ/liter CO_2	$RQ = \dfrac{CO_2\ formed}{O_2\ consumed}$
Carbohydrate	16.9	20.9	20.9	1.00
Fat	39.5	19.7	27.7	0.71
Protein (urea as end product)	23.6 (18.0)[a]	18.8[b]	23.1[b]	0.81[b]
Protein (uric acid as end product)	23.6 (17.76)[a]	18.4[b]	24.8[b]	0.74[b]

The last column in the table gives the respiratory quotient (RQ) which is the ratio of CO_2 produced to O_2 consumed. This can usually tell the investigator what fuel is being oxidized.

[a] The energy value per g for the average protein measured in a bomb calorimeter is about 23.6 kJ/g (Robbins 1993). However, protein's energy value to the animal depends on the excretion product of the animal because urea and uric acid have different combustible energy values which are subtracted from the combustion value of protein (see chapter 7 also).

[b] The energy value per unit gas for protein depends on the excretion product of the animal because urea and uric acid have different combustible energy values which are subtracted from the combustion value of protein.

continued

continued

Notice that the range in kJ/liter O_2 is very small; a value of 19.7 kJ/liter is customarily used and this results in an error of only 7% in the estimation of heat production. In contrast, kJ/liter CO_2 varies quite a bit with metabolic substrate. Therefore, the calculation of heat production from CO_2 production requires knowledge of the substrates being catabolized.

How is gas exchange actually measured? Aquatic animals are kept in closed containers of water, and the concentration of dissolved O_2 is periodically measured. Terrestrial animals are kept in closed containers and the concentration of O_2 and/or CO_2 in the air is measured, or air is pumped through the container and its flow rate and the change in O_2 and/or CO_2 between inflow and outflow are measured. It is even possible now to measure CO_2 and hence heat produced by free living animals by measuring the turnover of isotopes of oxygen and hydrogen. This method, called the doubly labeled water method, is described in chapter 13.

For example, an animal's body water is the result of the dynamic equilibrium between the rate at which water is gained by ingestion (in food and by drinking) and metabolic production and the rate at which it is lost by evaporation and in saliva, urine, and feces (chapter 12). For nongrowing animals, one can often make the assumption that the size of a given pool is in steady state. This assumption, together with the assumption that the "contents" of the pool are well mixed, allows establishing some useful equalities. At steady state the flow rate into the pool equals the flow rate out of the pool. You will find that some scientists call these rates the "fluxes" in and out of the pool. These fluxes equal the size of the pool divided by the average residence time of the material in the pool:

$$\text{input rate} = \text{output rate} = \frac{\text{size of the pool}}{\text{average residence time}} \tag{1.2}$$

Note that the reciprocal of residence time is the fraction of materials in the pool that turns over per unit time. This fractional turnover rate (or rate constant) plays a very important role in physiological ecology.

The ingredients of equations 1.1 and 1.2 summarize much of what physiological ecologists do. As we will see, a physiological ecologist interested in measuring the field metabolic rate of an animal needs to measure the animal's body water and the fractional turnover of two components in this pool: oxygen and hydrogen (see chapter 13). An ecotoxicologist interested in determining the output rate of a toxicant into an animal may need to measure the amount of toxicant in an animal (the pool) and its residence time (see chapter 9). Note that ecologists of a variety of persuasions also tend to think in terms of pools, rates, and fluxes. For example, a population ecologist can visualize a population as consisting of a pool of individuals. The birth, death, immigration, and emigration rates are the determinants of the input and output fluxes into and from the population. In a similar fashion, ecosystem ecologists spend a lot of time measuring the pools, fluxes, and residence times of energy and materials in ecosystems. If you are an ecologist, or would like to become one, then it is a good idea to become thoroughly familiar with the basic mathematics of pools, fluxes, and residence times. In box 1.2 we provide a basic introduction. Harte's (1985) wonderful little book *Consider a spherical cow* gives a more detailed account.

Box 1.2 Pools, Fluxes, and Residence Times

Imagine for a moment that your favorite organism is a well-mixed container full of a material that you are interested in (carbon, nitrogen, calcium, etc.). To keep things simple, let us assume that the amount (A) of this material in the organism is neither growing nor shrinking ($dA/dt = 0$). Let us denote by ρ_{in} and ρ_{out} the fractional rates at which the material enters and leaves the pool. At steady state, $\rho_{in} = \rho_{out} = \rho$. We have used the word "fractional" when referring to ρ_{in} and ρ_{out} because these numbers represent the fraction of A that gets in and out of the organism per unit time. Therefore, the units of ρ are time^{-1}. The fluxes of the material in and out of the organism are F_{in} and F_{out} and equal $F_{in} = \rho_{in} \cdot A$ and $F_{out} = \rho_{out} \cdot A$, respectively. The average time that it takes to fill up the pool of material in the organism (τ) must satisfy the following equation: $F \cdot \tau = \rho \cdot A \cdot \tau = A$. Therefore, $\tau = \rho^{-1}$. If the material within the organism represents a well-mixed pool, then we can say a good deal more about τ.

In a well-mixed pool of material, the time that each molecule stays in the organism is distributed as a negative exponential. That is, the probability density function of the age (x) of the molecules found in the pool at any given time is given by

$$f(x) = \rho e^{-\rho t} \tag{1.2.1}$$

Equation 1.2.1 implies that the expectation for the time that a molecule stays in the organism (or pool) equals $\tau = 1/\rho$, which is why the reciprocal of the fractional input rate is called the "average residence time". The variance in residence time (or age) of the molecules in the pool is given by $\sigma^2 = 1/\rho^2$. You will often find that people use the "half-life" ($\tau_{1/2} = \ln(2)/\rho = 0.69/\rho$)

continued

continued

instead of the average residence time to characterize how long molecules stay in the pool. The half-life estimates the median of the distribution of residence times, or ages, of molecules in the pool. Thus, 50% of all molecules stay in the pool for less than $\tau_{1/2}$ and 50% of the molecules stay longer. Because the negative exponential density function is asymmetrical and skewed to the left $\tau_{1/2}$ is smaller than τ. The theory that we have outlined here applies to well-mixed pools in which all the material is in a single compartment. These systems are referred to as having "one-compartment, first-order kinetics." Many physiologically and ecologically important systems satisfy these assumptions, or are close enough to satisfying them for the theory to apply reasonably well. We will apply this theory in chapter 4, when we talk about guts as chemical reactors, in chapter 8 when we discuss how the isotopic signature of a diet is incorporated into an animal's tissues, and in chapters 9, 12, and 13, when we describe how we measure toxicant and water turnover and the doubly labeled water method to estimate metabolic expenditure. In chapter 4 (box 4.2) we describe how you can use appropriate markers to estimate ρ. If probability density distributions intimidate you (or if you have no idea what they are), we strongly recommend that you read the first four chapters of the very nice introduction to probability theory for biologists by Denny and Gaines (2000).

Pool sizes and turnover or residence times vary within and among species with body size in a somewhat predictable pattern. Allometry is the term that refers to these patterns of change in some parameter with change in body size. In a subsequent section of this chapter we will examine allometry in some detail.

1.1.2 Efficiency

A useful metric that allows interpretation of the elements of a budget is efficiency. Efficiency is a ratio that describes the magnitude of one of the outputs relative to an input. All the inputs and outputs within a budget or balance equation must have the same units, and therefore efficiencies are either unitless proportions, or expressed as percentages if multiplied by 100. Throughout this book we will describe several efficiency indices that we believe are useful. Here we emphasize digestive efficiency only because it illustrates how the budget for a whole organism can be profitably viewed as made up of the combination of budgets of several subsystems. We will reexamine digestive efficiency in more detail in chapter 3.

The input/output format can be applied at different levels of a system or organism. Within the main system in figure 1.1, for example, one can specify several subsystems, each of which can have its associated subbudget. These subsystems allow the tracking of the flow of energy and materials through an animal, and hence allow the dissection of patterns of resource allocation (figure 1.1). For example, in a slightly more detailed view of the fate of ingested energy, we can define first a subsystem that represents the gastrointestinal tract (GIT). The input to it is food intake in kJ/d. Outputs include energy losses in feces plus an output representing the energy in kJ/d absorbed and retained in the animal, which is called digestible energy. With this framework we can define the apparent digestive efficiency of one of the components of an animal's diet as the ratio between output rate and input rate in the GIT. We will define digestive efficiency more rigorously and discuss why we call this digestibility "apparent" in chapter 3. Briefly, digestive efficiency is interesting to ecologists because a low efficiency translates into more time and effort spent collecting food to meet a fixed required net input: The animal retains a small fraction of the nutrients that it ingests. Furthermore, if food intake rate is fixed, for example if the gut's volume imposes limitations on the maximal amount that can be processed, then low digestive efficiency translates into less matter and energy available for maintenance, growth, and reproduction.

1.2 The Importance of Size: Scaling of Physiological and Ecological Traits

There are at least ten million kinds of organisms on earth and most of these have not yet been named, let alone studied. It is impractical to study the biology of all these kinds of animals; a selective sample may (and perhaps must) suffice. Are there any features of organisms that allow us to generalize from a few to the many? Although the diversity of life varies along a very large number of axes, body size is particularly important. From bacteria ($\approx 10^{-13}$ g) to whales (10^8 g), organisms vary in body weight by more than 21 orders of magnitude. This astounding variation led Brown and colleagues to assert that "biological diversity is largely a matter of size" (Brown et al. 2000). Body size influences virtually all aspects of an organism's structure and function. To a large extent, body size also shapes an organism's role in biological communities and ecosystems. The importance of body size for biology led Bartholomew (1981) to exclaim "the most important attribute of an animal, both physiologically and ecologically, is its size." This section links body size with the budget framework that we outlined at the beginning of the chapter. We describe how many traits (including pool sizes and turnover or residence times) vary with body size in a predictable fashion both within and among species.

The predictability of the relationship between body size and an organism's traits is fundamentally useful because it allows us to summarize and compare data. It also permits us to make educated guesses about an organism's biology simply from its size. The term that refers to these patterns of change in some parameter with change in body size is allometry. In this section we use the metabolic rate as an example of one of the many important biological rates determined "allometrically" by body size. Researchers have made thousands of measurements of metabolic rates of animals under standardized conditions. For endotherms such as mammals and birds, for whom the measurement is often called the basal metabolic rate (BMR), those conditions include a resting, fasted state, and an air temperature in the so-called thermal neutral zone where metabolism is not increased for body temperature regulation (figure 1.2 and box 1.3 explain the terms used by physiologists to characterize the thermal biology of organisms). For ectotherms such as reptiles, amphibians, fish, and arthropods, the measurement is typically called the standard metabolic rate (SMR); the conditions also include a resting, fasted state, but the temperature of the measurement is specified (e.g., SMR at 30°C). Figure 1.2 should convince you why it is fundamentally important to report the temperature at which the metabolic rate was measured. Using mammals for our illustration, and ground squirrels particularly, the BMR of a marmot (1.546 L O_2/h, or 30.5 kJ/h) is greater than that of a least chipmunk (0.073 L O_2/h or 1.43 kJ/h) by 21 times, which is not as large a factor as their absolute difference in size (respectively, 4.3 kg versus 46 g, or 93 times).

If we plot whole-organism metabolic rate of mammals as a function of body mass on arithmetic axes we find that this relationship rises in a decelerating fashion. Actually, we might not see much of a relationship in certain data sets if there are many data in the small-size categories and few in the large-size category. Sometimes, physiologists "standardize" metabolic rate by body mass and call this measurement the mass-specific metabolic rate. If you plot mass-specific metabolic rate against body mass you will find a decreasing relationship. We will see why in a moment. When you plot whole-animal metabolism against body mass in a double-logarithmic plot the increasing trend is obvious and the relationship appears linear (figure 1.3). The relationship between metabolic rate (B) and body mass (m_b) appears to be well described by a power function of the form

$$B = a(m_b)^b \qquad\qquad (1.3)$$

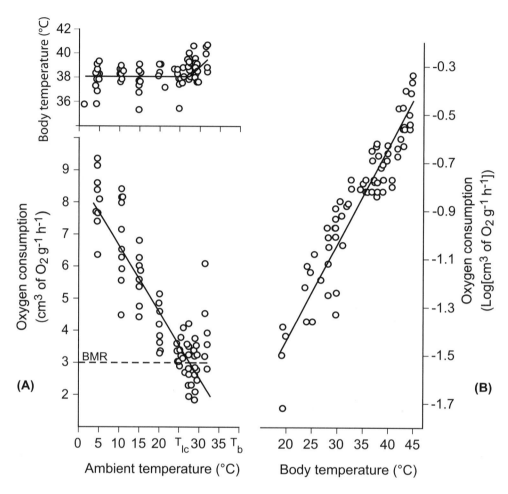

Figure 1.2. Endothermic homeotherms, like the heather vole (*Phenacomys intermedius*) maintain a constant body temperature by matching heat losses with metabolic heat production. The Scholander-Irving model shown in panel A describes the relationship between the metabolic energy production needed by an animal to maintain a constant core body temperature (T_b) and ambient temperature (T_a). Metabolic rate is measured by $\dot{V}O_2$, the rate of oxygen consumption. The slope in panel A, which has units of cm³ g⁻¹ h⁻¹ °C⁻¹, is sometimes called "thermal conductance," and characterizes feature(s) of heat exchange between the animal and its simple environment in the metabolic chamber. In a subsequent chapter we will add biophysical complexity to this idea (box. 13.4 in chapter 13). Note that the metabolic rate decreases linearly until $T_a = T_{lc}$ (lower critical temperature) and then remains relatively constant in the so-called thermal neutral zone (which our students often call the thermonuclear zone). The metabolic rate at the thermal neutral zone is called the "basal metabolic rate." (B) Resting metabolic rate increases roughly exponentially with body temperature. Because many ectothermic poikilotherms tend to have body temperatures dictated by ambient temperatures, metabolic rate often increases with ambient temperature. The data in panel B are for the desert iguana *Dipsosaurus dorsalis.* (Modified from McNab 2002.)

Box 1.3 Ectotherms, Endotherms, Homeotherms, and Poikilotherms

In the old days, animals were classified as warm blooded (birds and mammals) and cold blooded (all other animals). These terms are as well established as they are obsolete. They hide an astounding diversity in thermal biology. The thermal biology of animals can be classified by placing animals along two axes depending on the consistency in their body temperature, and on whether most of the energy used in the maintenance of their body temperature is derived from metabolism or from external heat sources. If an animal maintains a relatively constant body temperature, we call it a homeotherm (from the Greek adjective *homos* = same, uniform). If, in contrast, an animal has variable body temperature, we call it a poikilotherm (from the Greek adjective *poikilo* = varied). If an animal uses mostly energy from its own metabolism to keep its temperature high and above that of the ambient, the animal is an endotherm (from *endo* = within). If it uses primarily external heat sources, the animal is an ectotherm (*ecto* = outside).

Although it is often assumed that all homeotherms are endotherms, and that all poikilotherms are ectotherms, this is not necessarily true. An internal parasite of a bird maintains a very constant body temperature and hence it is a homeotherm. Yet, because its body temperature is maintained by the heat produced by its host, the parasite is an ectotherm. Conversely, many insects maintain a constantly high, and very tightly regulated, body temperature when they are active. These animals, however, do not regulate their body temperature when they are inactive (Heinrich 1993). At the scale of 24 hours, they are poikilotherms, but at the scale of the few hours during the active period of their daily cycle they regulate their body temperature by using metabolic heat production, and hence they are endotherms. Choosing the appropriate terminology is a matter of convenience, but terms must always be defined. Although the 2 × 2 possible combinations of the terms described here are often sufficient, in some cases it may be very difficult to find a good term to define the thermal biology of an animal.

where a and b are empirically derived parameters. Equation 1.3 is often referred to as the "allometric equation." Because metabolic rate is proportional to $(m_b)^b$, the mass-specific metabolic rate changes as $(m_b)^b/(m_b) = m_b^{b-1}$. The reason that mass-specific metabolic rate decreases with increasing body mass is that $b < 1$.

Both the pools of many materials in an animal's body and the rates of many physiological processes vary (or "scale") allometrically with body mass. Because linear relations are much easier to manipulate than power ones, allometric equations are frequently converted to their logarithmic form (see box 1.4):

$$\log B = \log a + b \log (m_b). \tag{1.4}$$

The benefits of presenting allometric data using log-log plots include a more even spread of data across the axes and the ability to plot a wide range of x and y values (or, respectively, m_b and B values in our case) in a relatively small space (figure 1.3). In spite of their usefulness, log-log transformations must be used judiciously. An important warning is that fairly large absolute differences can appear small on log-log scales. In addition, it is tempting to use log-log transformations to estimate the parameters a and b of the allometric equation from a data set. Although using normal least-squares regression on log-transformed data to estimate a and b is often done, it can sometimes yield results with strong statistical biases. Hence, fitting the parameters of power functions is *sometimes*

Box 1.4 Laws for Logarithms

Manipulating allometric laws requires remembering the four basic rules for logarithms and the seven algebraic laws for operating with expressions that contain powers and roots. If you ever feel self-conscious about having to refer to this box, your feelings may be assuaged by the following anecdote. When John T. Bonner, one of the most important biologists of the 20th century was compiling the data for his classic book *Size and cycle* (1965), he discovered that he could not remember the laws for logarithms. In a quandary, he crossed the hall to the office of Robert Macarthur, one of the founders of theoretical ecology. Macarthur devoted a few minutes to writing roughly the same rules that you see in this box. His handwritten page remained in Bonner's desk for many years.

Logarithm rules	**Rules for powers and roots**
$\log(ab) = \log(a) + \log(b)$	$x^a x^b = x^{(a+b)}$
$\log(1/a) = -\log(a)$	$x^{-a} = 1/x^a$
$\log(a/b) = \log(a) - \log(b)$	$x^a/x^b = x^{(a-b)}$
$\log(a^n) = n\log(a)$	$(x^a)^b = x^{ab}$
	$x^{1/a} = \sqrt[a]{x}$
	$x^{a/b} = \sqrt[b]{x^a}$
	$x^0 = 1$

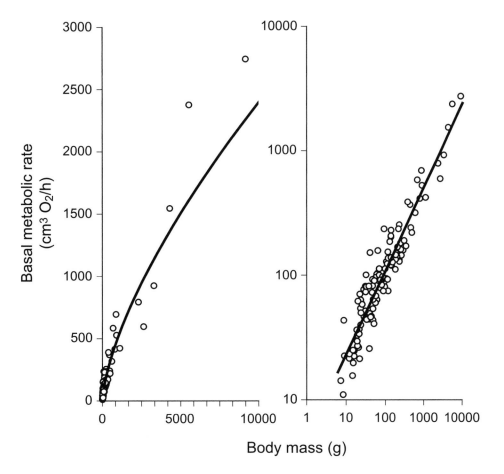

Figure 1.3. Basal metabolic rate is one of many variables that changes across body size in a nonlinear fashion but is linear when plotted on a log-log plot. Notice in the left-hand figure how the basal metabolic rate (BMR) increases in a sublinear fashion with increasing body size among 122 species of rodents (Hayssen and Lacy 1985). The same data appear linear in the right-hand log-log plot.

better done using nonlinear fitting procedures (see Motulsky and Ransnas 1987 for a friendly introduction).

Size has a pervasive role in determining the quantitative details of the distribution of materials within an organism, support against the force of gravity, and thermodynamics, and so size correlates with rates and capacities. Life history parameters such as developmental rate, reproductive rate, and lifespan are dependent in part on metabolic rate and tissue growth rate, each of which is size dependent, and so life history parameters also correlate to some extent with body size. Not only do variables such as metabolic rate, walking speed, and survival time during fasting scale with body mass, but also lifespan, age at first reproduction, gestation time, and birth rate.

The debate about the reasons for scaling laws started when Julian Huxley first formalized them in 1932 and has gone on for decades, but we can still apply them usefully even if we cannot explain them with complete assurance.

Allometric relationships have practical and theoretical importance for at least four reasons:

(1) Mountains of physiological, morphological, and ecological data can be summarized in them (e.g., Calder 1984; Peters 1983; Robbins 1993), thereby avoiding a sense of confusion from the quantity and complexity of raw data. (2) We can use the allometric relationships in a predictive manner to make allometrically educated guesses. For example, suppose one wants to model the energetic needs of a population of endangered rodents for whom metabolism has not been measured. Why not use the value predicted by an allometric model as the first approximation? Because it is unethical to assess the safety and toxicity of new therapeutic drugs for the first time on humans, allometric scaling has an interesting application in pharmacology (Mahmood 1999). Pharmacologists rely on measurements on small laboratory animals such as mice, rats, rabbits, dogs, and monkeys to predict the properties of therapeutic agents in humans. (3) We can use allometric relationships to derive new relationships, and sometimes they help us to formulate theoretical expectations. For example, knowing that the locomotory energetic cost to move 1 km scaled as $(m_b)^{3/4}$ allowed Tucker (1971) to predict that maximum migration distance of birds would increase as $(m_b)^{1/4}$. He arrived at this guess by reasoning that birds of all sizes could allocate a similar proportion of their body mass to fat (e.g., up to 50%), and that migration distance would be proportional to the ratio of energy stores (the kJ of energy in fat should be proportional to $(m_b)^{1.0}$) to the cost of transport (kJ/km, which should be proportional to $(m_b)^{3/4}$). To perform allometric manipulations such as these (migration distance is proportional to $(m_b)^{1.0}/(m_b)^{3/4} = (m_b)^{1/4}$) it is convenient to be familiar with the basic operations involving powers and logarithms (box 1.4). Juggling allometric equations is a sport favored by many theoretically inclined biologists. It is not only harmless fun, it can lead to remarkably interesting insights. Calder (1984) and Schmidt-Nielsen (1984), two distinguished comparative physiologists, have written useful manuals on this topic. (4) Finally, the fact that so many physiological and life history variables can be described by allometric equations tells us that size alone accounts for a large proportion of the variance among species. Thus, one can argue that almost any comparative analysis of physiological or life history data should begin with a consideration of size. Allometric laws not only force us to consider body mass in comparative analyses, they allow us to compare among organisms of different sizes.

Because many biological rates scale as power functions of m_b (i.e., rate = $a(m_b)^b$), the confounding effect of mass can be removed by expressing rates relative to $(m_b)^b$. A comparison of the BMRs of the chipmunk and marmot described in our previous example using metabolic rates corrected by $(m_b)^{0.75}$ reveals differences of only about 40% (respectively, 10.2 versus 14.4 kJ d^{-1} kg$^{-3/4}$). In a subsequent section, we will justify why the b value for the metabolic rate is often assumed to be 0.75. Using the ratio of a trait to $(m_b)^b$ for comparative purposes is simple and appealing. However, it can be fraught with statistical problems (see Packard and Boardman 1999). Box 1.5 outlines one of the statistically correct ways in which body size can be accounted for in comparative analyses. Calder (1984) pointed out that "adaptation" can be conceived as "adaptive deviation" from the basic size-dependent allometric pattern. In a subsequent section (1.4 "Using Historical Data in Comparative Studies") we will describe how body size and phylogenetic data can be combined to test Calder's size-dependent definition of adaptation.

Box 1.5 How Can We Eliminate the Effect of Body Size in a Comparative Analysis?

One possible way is to use the residuals of the allometric relationship in question. Panel A of figure 1.4 shows the relationship between lipid content (y) and body length (x) in rainbow trout (*Oncorhynchus mykiss*) that have been fasted for up to 150 days. Note that this is a case of intraspecific allometry; most of the other examples in this chapter are interspecific allometric relationships. Clearly, lipid content increases as an allometric power function with length (indeed, lipid content is roughly proportional to (length)^3). If we plot lipid content against time, we find no relationship (B). The effect of length overwhelms the effect of fasting time on lipid content. However if we plot the *residuals* of the expected relationship between length and lipid content (i.e., $y_{observed}(x_i) - y_{expected}(x_i)$) against days fasting, we find that lipid content decreases linearly with time (C). Residuals ($y_{observed}(x_i) - y_{expected}(x_i)$) are depicted as the dashed distance between a point and the curve in (A). Data are from Darin Simpkins (unpublished).

Because allometric relationships are often well described by power functions, researchers frequently work with log-transformed data. Hence residuals acquire the form

$$\log(y_{observed}) - \log(y_{expected}) = \log(y_{observed}/y_{expected}).$$

Because log-transforming data can lead to statistical biases, we often prefer to use the residuals of allometric curves fitted using nonlinear methods

continued

continued

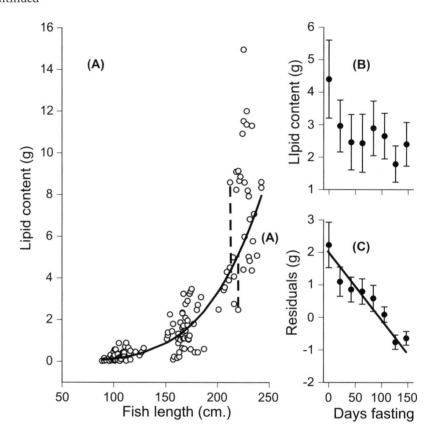

Figure 1.4. The relationship between body length and lipid content in trout that fasted for variable amounts of time is well described by a power function. Although lipid content is correlated with fasting time, the residuals of the relationship between lipid content and length clearly decrease with fasting time. To obtain a pattern, we must correct for the variation in the data set produced by differences in the body length of the fish.

(Motulsky and Rasnas 1987). However, not all researchers are as log-phobic as we are and there are situations in which it is more appropriate to log-transform data prior to analyses. Logarithmic-transformations can be absolutely essential when the variation along the allometric line increases with body size (as it often does).

1.2.1 Ontogenetic, Intraspecific, and Interspecific Allometry

As animals grow, their body mass changes. It is often useful to plot the change in a trait as a function of body mass throughout an animal's development. Huxley (1932) constructed a myriad of these ontogenetic allometric relationships. Ontogenetic allometry remains an essential tool in the analytical arsenal of developmental biologists (Stern and Emlen 1999). In a similar fashion, it is often useful to analyze the allometric relationships among the traits of fully grown adults within a species. We can call this form of allometry intraspecific allometry (it is also called static allometry). The boundaries between ontogenetic and intraspecific allometry can become blurred in animals that continue growing throughout their lives ("indeterminate growers"; see chapter 13). Therefore, on occasion you will encounter allometric analyses that combine animals not only of different sizes, but also of different developmental stages. Finally, many allometric analyses use the average values for a species as data points. In ontogenetic and intraspecific allometric analyses data points represent individuals. In interspecific allometries (also called evolutionary allometries), data points represent species averages (Cheverud 1982). On occasion, you will find that an allometric plot is a mélange of ontogenetic and intra- and interspecific allometries.

Interspecific allometries are very useful, but one must deal with them carefully for a variety of reasons. The primary one is that species are not statistically independent data points. They are related by descent. We will discuss in some detail how to deal with this important detail in a subsequent section (1.4 "Using Historical Data in Comparative Studies"). Also, interspecific allometries do not often satisfy the assumption that the independent variable (which more often than not is body mass) is measured without error. Hence, using standard least-squares procedures to analyze them can be inappropriate and one must rely on other statistical estimation procedures (Sokal and Rohlf 1995). Chapter 6 in Harvey and Pagel's (1991) book on the comparative method in evolutionary biology gives more detailed descriptions of these approaches. With surprising frequency authors mix ontogenetic, intraspecific, and interspecific data in a single allometric analysis (we avoid giving examples to evade the wrath of colleagues). This practice is statistically incorrect and hides interesting biological information (Cheverud 1982). The final reason why interspecific allometries must be treated with caution is that they assume that we can characterize the magnitude of a trait by a single number. Many traits, however, are phenotypically plastic. Phenotypic plasticity and its sometimes confusing terminology is the theme that we deal with next.

1.2.2 Phenotypic Plasticity: Physiological Variation at a Variety of Time Scales

Comparative physiologists have spent an inordinate amount of time describing the allometric relationship between the basal metabolic rate (BMR) of birds and mammals and body mass (McNab 2002). The BMR was not chosen because it is particularly relevant ecologically—it is not. Recall that it is measured on fasted animals at rest confined within dark boxes, which are not conditions that most animals experience naturally. Many factors (temperature, radiation, wind speed, activity, etc.) influence metabolic rate. The BMR provides a measurement done under controlled and standardized conditions that allows comparison without these confounding factors. The wide use of the BMR by comparative physiologists recognizes that the expression of physiological traits changes within an animal. But the BMR itself is variable; it changes when measured at different seasons in a single animal (reviewed by McNab 2002). The phenotypic variability that can be expressed by a single genotype is called phenotypic plasticity (Pigliucci 2001). Understanding how animals work and how their function influences their role in ecological systems demands that we recognize and investigate the phenotypic plasticity of physiological traits.

Comparative and ecological physiologists have investigated the phenotypic plasticity of animals for years. We will give many examples of ecologically relevant physiological phenotypic plasticity throughout this book. One of the confusing results of an otherwise wonderfully rich literature is a proliferation of terms. In this book we have adopted the terminology proposed by Piersma and Drent (2003; see table 1.2) to describe the different forms that phenotypic plasticity can take. We caution that this nomenclature has not been adopted widely, and that some biologists dislike it. We find it useful as an attempt to bring a semblance of order to a sometimes befuddling terminological tangle. According

TABLE 1.2
Phenotypic plasticity

Plasticity category	Phenotypic change is reversible	Variability in phenotype occurs in a single individual	Phenotypic change is seasonally cyclic
Developmental plasticity	No	No	No
Phenotypic flexibility	Yes	Yes	No
Life-cycle staging	Yes	Yes	Yes

Note: Often a single genotype can produce a variety of phenotypes if exposed to different environmental conditions (temperature, diet, presence of predators, . . ., etc.). The phenotypic variability that can be expressed by a single genotype is called phenotypic plasticity. Piersma and Drent (2003) defined three mutually exclusive categories of phenotypic plasticity. These categories depend on whether the phenotypic change is reversible, whether it occurs in a single individual, and whether it is seasonal or cyclic. We include polyphenism as a subcategory of developmental plasticity.

to Piersma and Drent (2003), phenotypic plasticity can be divided into 3 more or less mutually exclusive categories: developmental plasticity, phenotypic flexibility, and life-stage cycling.

Developmental plasticity is defined as the ability of a single genome to produce two or more alternative morphologies in response to an environmental cue such as temperature, photoperiod, or nutrition. The term developmental plasticity contains "polyphenism" as a subcategory. The term polyphenism refers to the ability of many arthropods to produce a sequence of generations with discrete phenotypes to accommodate sometimes seasonal environmental changes (Shapiro 1976). A nice example of developmental plasticity is the phenomenon of predator-induced morphological changes in which prey change their morphology in response to predation risk (Tollrian and Harvell 1999). For example, the presence of fish predators induces the development of defensive spines in a variety of aquatic arthropods and crustaceans (e.g., Harvell 1990). Figure 1.5 illustrates the use of allometry to document predator-induced morphological changes.

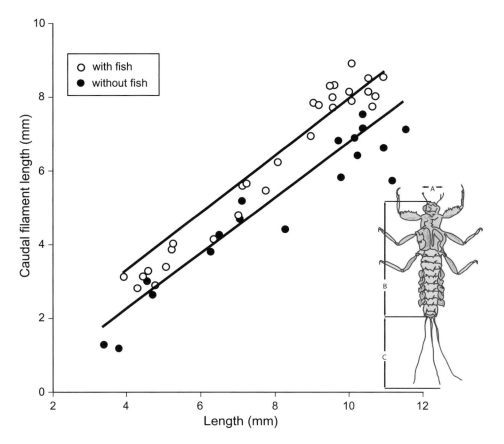

Figure 1.5. Dahl and Peckarsky (2002) discovered that when the larvae of the mayfly *Drunella coloradensis* grow exposed to predatory brook trout (*Salvelinus frontalis*), they develop longer caudal filaments (C in the image of the mayfly). Because caudal length increases with body length, the comparison between caudal filament lengths in streams with and without trout must be done allometrically. Long caudal filaments can be induced experimentally by dripping water from a container containing live trout into the water in which *D. coloradensis* are reared. Longer caudal filaments significantly reduce predation by brook trout on the mayfly larvae. (After Dahl and Peckarsky 2002.)

Developmental plasticity is irreversible. When environmental conditions change rapidly and over time scales that are shorter than a lifetime, animals can show reversible transformations in physiology and morphology. This phenotypic flexibility is widespread and ecologically very important. We will describe examples of it in many chapters of this book. In addition to changes in function that occur in direct response to a change in an environmental condition, many animals live in distinctly seasonal environments. In these, different activities tend to be predictably separated in time within individuals. Long-lived individuals must adjust their morphology, physiology, and behavior to fit the conditions of these seasonal changes. Piersma and Drent (2003) refer to these changes as life-cycle stages. They use the annual life cycle of the arctic rock ptarmigan *Lagopus mutus* as an example of such changes. These birds change plumage seasonally from white (which is cryptic in snow) to green and brown (which hides them in the brief summer of the tundra).

Although it is often assumed that phenotypic plasticity confers a selective advantage to individuals that have it, this assumption is better construed as a hypothesis that must be tested (Schmitt et al. 1999). It is called the adaptive plasticity hypothesis. Indeed, the intraindividual trait variation generated by different environmental conditions is ideally suited to test the criterion of "goodness of design" for phenotypic adaptation, which was proposed by the evolutionary biologist George C. Williams (1966). Intragenomic phenotypic variation is especially useful for testing the performance of alternative phenotypes when interspecific comparisons cannot be made.

Although we will not often use the terms acclimatization and acclimation in this book, you should be aware of them. Both terms are used frequently in the thermal biology literature to characterize the phenotypically flexible responses of animals to changes in their thermal environment. The term acclimatization refers to phenotypic responses in response to changes in the natural, and hence complex, thermal environment. The word acclimation refers to phenotypic changes in response to controlled changes in one or several thermal variables in the laboratory. You will also find the word adaptation used to refer to reversible phenotypic variation. In future chapters we will give examples of this misuse of the term adaptation, which are unfortunately in relatively common usage in some fields. In our view, the use of the term adaptation must be restricted to the Darwinian concept of adaptation as "an anatomical, physiological, or behavioral trait that contributes to an individual's ability to survive and reproduce ("fitness") in competition *and cooperation* with conspecifics in the environment in which it evolved" (Williams 1966) (emphasis added). Rose and Lauder's (1996) book gives more nuanced definitions of Darwinian adaptation and details about how we go about testing the hypothesis that a trait is adaptive.

1.2.3 Why Is b ≈ n/4? The Enigma of Quarter Power Scaling

There is little disagreement about the generality and value of allometric equations as phenomenological descriptions of a variety of biological relationships (but see Kooijman 2000 for a lone dissenting view). However, the mechanisms that lead to these relationships and the factors that determine the values of their parameters are unclear and the subject of some controversy (and sometimes of acrimonious criticism; Kozlowski and Konarzewski 2004). A perplexing pattern is the seeming ubiquity of simple multiples of 1/4 in the estimated value of the exponent b of allometric equations. A few examples are heartbeat frequency ($b \approx -1/4$), lifespan ($b \approx 1/4$), and the radii of both mammalian aortas and tree trunks ($b \approx 3/8$; West et al. 2000). Perhaps the best-known allometric patterns in physiology are the relationships between metabolic rate and mass. In 1932, Max Kleiber proposed that in mammals metabolic rate scaled with body mass to the 0.75 (i.e., 3/4) power (Kleiber 1932). Kleiber's pattern has come to be used as a standard in studies of mammalian metabolism, and researchers frequently describe their results in relationship to expectations based on Kleiber's original data set. You will still encounter in the literature reference to values that are x% higher or lower than Kleiber (McNab 2002).

Kleiber's observation should bewilder you. At first glance, one may assume that metabolic rate should increase in proportion (i.e., "isometrically" using allometric terminology) to body mass. After all, mammalian cells are built more or less from the same substances and the number of cells should increase isometrically with body mass. Jonathan Swift (1735) made this assumption when he estimated the amount of food that Lilliputians fed Gulliver. The Lilliputians found that Gulliver's height

> exceeded theirs in proportion to twelve to one . . . they concluded from the similarity of their bodies that Gulliver's must contain at least 1728 of theirs [12^3 assuming that Gulliver was geometrically similar to his treacherous hosts] and consequently would require as much food as necessary to support that number of Lilliputians.

There is a mistake in this argument. Animals do not fuel their metabolism by eating in proportion to their body mass. Had the Lilliputians fed Gulliver as much as Swift describes, he would have become enormously fat. Metabolic rate does not scale in proportion to body mass. The approach of Rubner, a German physiologist, was closer to reality. In the late 19th century, he proposed that animals feed to replace the heat that is dissipated through their body surface, which scales with $(m_b)^{2/3}$. The notion that metabolic rate scales with (mass)$^{2/3}$ is called "Rubner's surface law." Many, perhaps most, organisms, break Rubner's law. Their metabolism scales with mass to a value usually higher than 0.66.

Why should *b* equal 0.75 rather than 0.66? The 3/4 exponent seems to describe adequately the relationship between metabolic rate and body mass in animals ranging from protozoans to whales (figure 1.6). The value of *a*, the intercept of the log-log allometric relationship, varies significantly with the thermal biology of the beast—as we will discuss in a following section (1.3 "The Importance of Temperature"), and with other factors, but *b* remains remarkably constant and close to 0.75.

A veritable flock of theories have been proposed to explain why *b* should equal 3/4. Brown et al. (2000) review these theories in a historical context. We, very briefly, summarize the latest one, which can be called the fractal theory of quarter-power allometric laws (West et al. 1997). We must mention that West et al.'s theory has received significant challenges (see, e.g., Dodds et al. 2001; Darveau et al. 2002). West and his collaborators have addressed these criticisms in what has become a scientific ping-pong of ideas (Savage et al. 2004; Brown et al. 2005). Although we cannot guarantee that this theory will be accepted in the future, we present its foundations for three reasons: (1) We find its arguments compelling, (2) it has generated a significant amount of exciting novel research

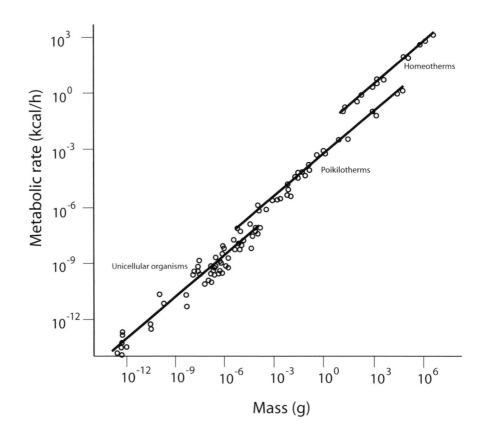

Figure 1.6. Hemmingsen (1960) compiled metabolic rate data for organisms ranging in size over 21 orders of magnitude. He standardized all the poikilothermic ectotherms to a temperature of 20°C and the endothermic homeotherms to one of 39°C. The relationship between metabolic rate and body mass is well described by a power function with exponent b = 3/4. The intercept of the log-log form of this relationship (log(a)), however, varies among groups in a seemingly predictable fashion. Homeotherms seem to have higher metabolic rates than poikilotherms, and poikilotherms appear to have higher rates than unicellular organisms.

that integrates physiology with ecology (Niklas and Enquist 2001; Gillooly et al. 2001; Carbone and Gittleman 2002), and (3) we will use some of its results in subsequent chapters. We again emphasize that West et al.'s hypothesis is one of many and that other competing hypotheses exist. Kozlowski and Weiner (1997) and Kooijman (2001) describe alternative mechanisms that can lead to the allometric scaling of metabolic rate and body mass. After consulting the literature and reflecting on what you have read, you may not find West et al.'s (1997) arguments convincing, and you may join the ranks of the 3/4-power-law skeptics. In that case, do not throw the allometry baby away with the 3/4-power-law water. Instead of using 3/4 (or multiples of 1/4) in your equations, use an empirically derived value for b. This value will almost certainly be between 0.6 and 1. Indeed, you may want to substitute in your mind an empirical value for 3/4 in many of our subsequent equations. Often, the conclusions of this substitution will not greatly change the conclusions of an analysis.

The derivation of West et al.'s model requires superficial acquaintance with fractal mathematics. Because in our experience this is not part of most biologists' backgrounds (although maybe it ought to be), we skip it. We simply list the model's assumptions and give a qualitative description. Interested readers may find West et al.'s (1999) general derivation of quarter power laws significantly more accessible than the original model published in 1997 (we do). West et al. (1997) base their fractal explanation of quarter power laws on the observation that biological structures and functions are determined by the rates at which resources (oxygen, nutrients, and water) can be delivered to them. Their main hypothesis is that to supply their metabolizing units (cells in organisms, respiratory molecules in mitochondria and cells) with resources, organisms use fractal-like, volume-filling, hierarchical transport systems (figure 1.7). West et al. (1997, 2000) make two additional assumptions: (1) The final branch of the transport network at the site where nutrients are exchanged is size invariant (i.e., the capillaries of elephants and those of shrews have the same radius), and (2) organisms have evolved so that the energy required to transport material through the network is minimized. These more or less reasonable premises lead to the 3/4 exponent in the relationship between metabolic rate and body mass and to its many 1/4-power-law corollaries.

One way to visualize what West and his collaborators suggest is the following: Organisms must exchange materials across surfaces. These surfaces must reach into all the corners of an organism's volume, and must have a system of delivery of materials that is efficient. To achieve these dual purposes, the circulatory system divides in a fractal-like fashion again and again so that its surfaces reach all the nooks and crannies of an animal's body. The circulatory system has a surface that "wants" to become a volume and that achieves volume-like characteristics by virtue

Figure 1.7. Many biological branching distribution networks show fractal-like structures. Common examples are vertebrate circulatory and respiratory systems (A) and plant vessel-bundle vascular systems (B). Diagram (C) shows the topological representation of these systems used by West et al. (1997) in their derivation of fractal theories for scaling laws. The parameter k specifies the order of the branching level. These levels in a vertebrate circulatory system would begin in the aorta ($k = 0$) and end in capillaries ($k = N$). The parameters used in West et al.'s (1997) model are summarized for a tube at level k in diagram (d) (l_k is the length of the tube, r_k is its radius, ΔP_k is the drop in pressure along its length, and U_k is the velocity of the material moved along the tube).

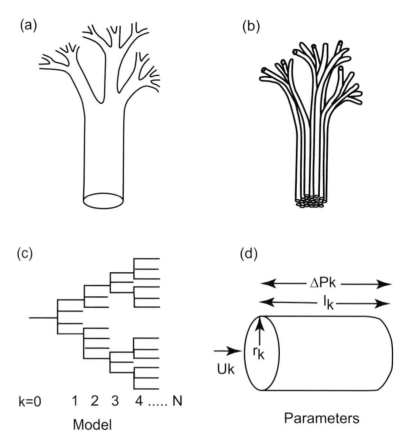

of its fractal-like structure (we apologize for the teleological language!). You can apply the same logic to a variety of "volume-filling" surfaces including the respiratory tree in mammals and the hydraulic system that conducts water from roots to leaves in plants.

Why is the quarter power law important? We believe that it is important because it helps us to account for a very large range of life's variation. It suggests that similar, maybe universal, design principles apply to organisms and that these principles allow us to understand one of life's important axes. West et al. (2002) provide a remarkable example showing that the fractal explanation may be extended to organelles and even to the molecules of the respiratory complex inside mitochondria (figure 1.8). The explanatory value of allometric quarter power laws, however, should not blind us to their limitations. Allometric laws work very well if one views the world through log-log glasses, but these glasses provide a peculiar perspective. The beautifully tight relationships in allometric data summaries plotted on log-log axes hide enormous amounts of variation. If you were to amplify an allometric plot, focusing only on a narrow range of body masses, and retransform the data to arithmetic axes, the relationship would not

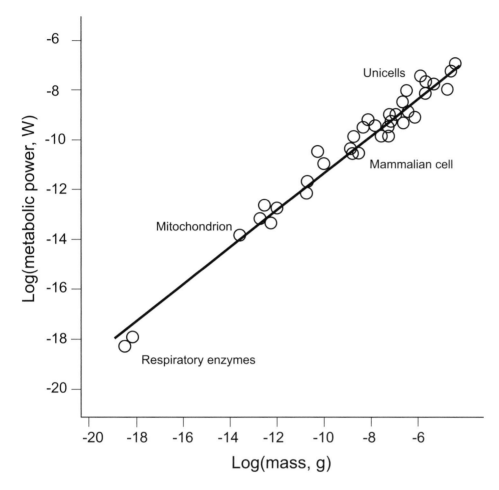

Figure 1.8. The metabolic power of isolated mammalian cells, mitochondria, respiratory complex, and cytochrome oxidase increases as a power function of mass with exponent equal to 3/4. The plot shows unicellular organisms for comparison. All data were adjusted to mammalian temperature using equation 1.12. (Data from West et al. 2002.)

look nearly as tight. For some body ranges there would still be severalfold variation remaining to be explained and predicted. Factors other than body mass still account for much of the functional variation that is of interest to physiological ecologists. One of these factors is, of course, temperature, but there are others. We will discuss the importance of temperature in a subsequent section, but before doing so, we illustrate the use of allometric laws in an area of ecology that has many of its foundations in physiological ecology: macroecology.

1.2.4 The Energetic Equivalence Principle: A Scaling Law for Macroecology

Macroecologists look for statistical patterns in the abundance, distribution, diversity, and biomass of individual organisms and species. They aim to understand why these patterns emerge, and to discover the processes that govern them

(Brown 1999). It is reasonable to conjecture that many of the patterns of abundance, distribution, and diversity of organisms over space and time can be attributed to resources (energy, nutrients, and water) and to the physiological mechanisms that shape how animals use them. Macroecology's wide-angle lens and physiological ecology's zoom are complementary. The processes studied by physiological ecologists should often reveal the mechanisms that underlie macroecological patterns. In this section we describe the energetic equivalence rule, a macroecological pattern that stems from physiological principles.

The energetic equivalence rule states that the total rate of energy use of a population per unit area (B_p) is invariant with respect to body size. In simpler terms, the energy used by all the herbivorous voles living in an island should not be very different from that of all the herbivorous deer that coexist with them. This surprising rule is a direct consequence of the dependence of the metabolic rate of an individual (B) on its body mass (m_b). The amount of energy used by a local population is its population density multiplied by the metabolic requirements of its individual members. Maximum population density for a given species (N_{max}) should be approximately equal to

$$N_{max} = \left(\frac{j}{B}\right) = \left(\frac{j}{a\,(m_b)^{3/4}}\right) \tag{1.5}$$

where j represents the rate at which resources become available to the population, B is the metabolic rate of each individual ($B = a(m_b)^{3/4}$), and a is a constant. This simple model predicts that population density should decrease with increasing species body mass to the $-3/4$ power. There is significant evidence for this pattern in creatures ranging from marine phytoplankton and terrestrial plants (Belgrano et al. 2002) to a variety of consumers (Damuth 1987). The energy used by a population (B_p) per unit time equals its density times the energy use of each of its members:

$$B_p \approx N_{max} B \approx \left(\frac{j}{a(m_b)^{3/4}}\right) a(m_b)^{3/4} = j. \tag{1.6}$$

Therefore, the amount of energy used by a population is independent of body mass but dependent on j, the ecosystem's productivity.

Although there is significant support for the energetic equivalence rule, there are also exceptions. Carbone and Gittelman (2002) used a remarkably complete data set to examine the dependence of mammalian carnivore densities on body mass and productivity. They found that, as expected, population density decreased as a power function of body mass (figure 1.9). However, they found that the exponent of this function was not –0.75; it was roughly –1. Why? A possible explanation may be that productivity j decreases with increasing body mass of the prey, which in turn is often (but not always) correlated with the mass of the carnivores (Carbone et al. 1999). The argument is as follows: Peters (1983) documented that productivity decreases as $(m_b)^{-1/4}$. Then, assume that the mass of the prey (m_{bp}) increases isometrically with the mass of the carnivore (m_{bc}), i.e., $m_{bp} \propto m_{bc}$, and hence $j = a'\, m_{bc}^{-1/4}$. Therefore,

$$N_{max} \approx \left(\frac{j}{B_c}\right) \approx \left(\frac{a'(m_{bc})^{-1/4}}{a\,(m_{bc})^{3/4}}\right) = \frac{a'}{a}(m_{bc})^{-1} \tag{1.7}$$

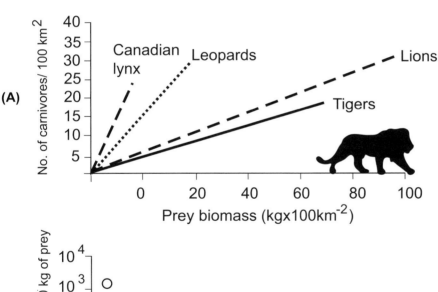

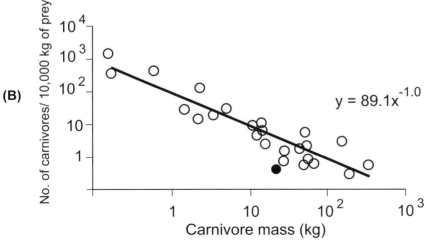

Figure 1.9. (A) The population density (numbers/km²) of mammalian carnivores increases as a function of prey biomass (in 10,000 kg/km²). Different lines represent different species. The lines are regressions through the origin. Because carnivore population density depends strongly on prey biomass, panel B standardizes density by prey biomass and plots it against carnivore mass. The relationship between standardized density and carnivore mass is well described by a power function with exponent $\approx$ –1. The filled circle in panel B is the Eurasian lynx (*Lynx lynx*). This population is subject to poaching, which may explain its relatively low density. (Modified from Carbone and Gittleman 2002.)

Unlike what seems to be the case in many other taxa (see Damuth 1987), the energy fluxes of mammalian carnivore populations seem to decrease with increasing body mass:

$$B_p \approx N_{max} B \approx \left(\frac{a'}{a} (m_{bc})^{-1} \right) a \, (m_{bc})^{3/4} = a'(m_{bc})^{-1/4}. \qquad (1.8)$$

1.3 The Importance of Temperature

The effect of temperature on biochemical reactions has been known for over a century and is summarized in the Arrhenius equation:

$$K = Ae^{-E_i/kT}, \qquad (1.9)$$

in which K is the rate constant of a reaction, A is a constant relating to molecular collision frequency, E_i is the activation energy, k is Boltzmann's constant, and T is absolute temperature (in °K). Because enzymatic processes mediate most biological processes, the importance of temperature as a determinant of biological rates should not be surprising. Within the range of temperatures at which most biological processes take place (0 to about 40°C) the Arrhenius equation behaves very roughly as an exponential. Thus, biologists often assume an exponential relationship and use the ratio of two biological rates 10°C apart ($Q_{10} = (K_{T+10})/K_T$) as an index of the thermal sensitivity of a process. Q_{10} is some times called the temperature coefficient and is related to the Arrhenius equation by

$$Q_{10} = \frac{K_{T+10}}{K_T} = e^{E_i/k(10/T((T+10)))} \qquad (1.10)$$

Although Q_{10} is often assumed to be constant and approximately equal to 2, equation 1.10 emphasizes that it is not. It varies with T among other factors. Like most physiologists, we have often used Q_{10} in back-of-the envelope calculations. However, we recognize that its use can be problematic.

The importance of body temperature for biological processes is well illustrated by an allometric application. Gillooly and colleagues (2001) used the following reasoning to incorporate temperature into the allometric equation relating metabolic rate (B) and body mass (m_b): Metabolic rate is the consequence of many biological reactions (B_i). Using the simplest possible assumption, we can postulate that the mass-specific metabolic rate is the sum of all the energy-consuming reactions taking place in the organism:

$$B = \sum_i B_i.$$ (1.11)

Each B_i depends on three major variables:

$B_i \propto$ (density of reactants)(fluxes of reactants)(kinetic energy of the system).

The first term of this product is mass independent, whereas the second scales with $(m_b)^{3/4}$. The third term is governed by the Arrhenius relationship. Hence equation 1.11 can be rewritten as

$$B(T, M) \propto (m_b)^{3/4} e^{-E_i/kT}$$ (1.12)

where E_i is the average activation energy for the reactions that govern metabolism. To compare among organisms of varying body masses, Gillooly et al. (2001) standardized equation 1.12 to $(m_b)^{3/4}$ as

$$\frac{B(T)}{(m_b)^{3/4}} = B_0 e^{-E_i/kT}.$$ (1.13)

Equation 1.13 makes two predictions, The first one is that plotting the logarithm of the mass-corrected metabolic rate ($\ln(B(T)/m_b^{3/4})$) against $1/T$ should yield a straight line with slope equal to $-E_i/k$:

$$\ln\left(\frac{B(T)}{(m_b)^{3/4}}\right) = \ln(B_0) - \left(\frac{E_i}{k}\right)\frac{1}{T}.$$ (1.14)

The graph of the logarithm of a rate constant against $1/T$ is called an Arrhenius plot. Arrhenius plots are used frequently in physical chemistry to estimate the values of energies of activation. Therefore, this first prediction states that the response of whole organisms to temperature should follow the same principles as enzymatic reactions. Gillooly and colleagues made a second, more precise prediction: Because E_i has an average value of about 0.65 eV ($\approx 1.12 \times 10^{-19}$ J) in biochemical reactions, and the Boltzmann factor k is 8.62×10^{-5} eV ($^\circ$K)$^{-1}$, they predicted that the slope of the plot of $\ln(B(T)/(m_b)^{3/4})$ against $1000/T$ (i.e., $E_i/k1000$) should equal approximately -7.5 $^\circ$K. Figure 1.10 shows only one example of the many plots made by Gillooly et al. (2001) for a variety of organisms. All these plots were linear, which we do not find terribly surprising. However, their slopes ranged from -5.02 to -9.15 $^\circ$K, which is a relatively narrow range and includes the value that Gillooly et al. (2001) had predicted.

Figure 1.10. An Arrhenius plot demonstrates that metabolic power (in watts = Joules/second) standardized by (body mass)$^{3/4}$ depends on temperature in the same fashion as biochemical reactions. The data set includes only birds and mammals, but data from unicellular organisms, plants, fish, amphibians, and reptiles show similar relationships. The relationship between metabolism and temperature indicates that a significant fraction of the reduction in metabolic rate experienced by animals in torpor or hibernation is simply the result of lowered body temperature.

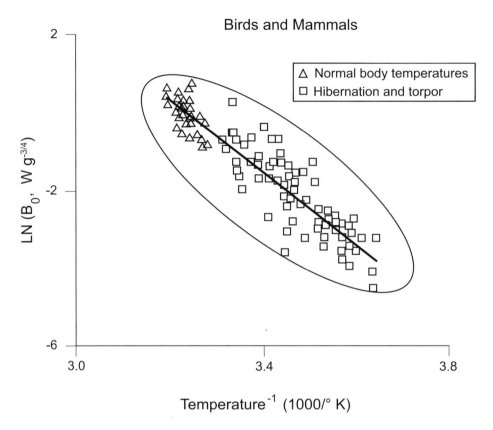

Figure 1.6 depicts the relationship between metabolic rate and body mass for animals ranging from unicells to whales. When Hemmingsen constructed this plot in 1960, he standardized the resting metabolic rates of all the poikilotherms to 20°C and those of endotherms to 39°C. Sensibly, Gillooly and collaborators (2001) standardized all metabolic rates to a common temperature (20°C). This standardization reduced the variation in metabolic rate enormously (figure 1.11). Temperature-standardized metabolic rates for unicells, invertebrates, and plants fell along a common line. The metabolic rates of fishes, amphibians, and reptiles were only slightly higher, and those for birds and mammals were still higher. In contrast with Hemmingsen's (1960) figure, which showed no overlap among groups, figure 1.11 shows a lot of overlap. Hemmingsen (1960) calculated a 225-fold range in mass-standardized metabolic rates. When temperature differences are accounted for, the range is reduced to a 20-fold range. The extraordinarily simple model summarized by equation 1.13 suggests that, as a first approximation, the metabolic rate can be estimated as the product of an allometric (quarter power) function of body mass and the Arrhenius relationship. A vast amount of variation in one of life's central traits seems to be accounted for by body mass and temperature.

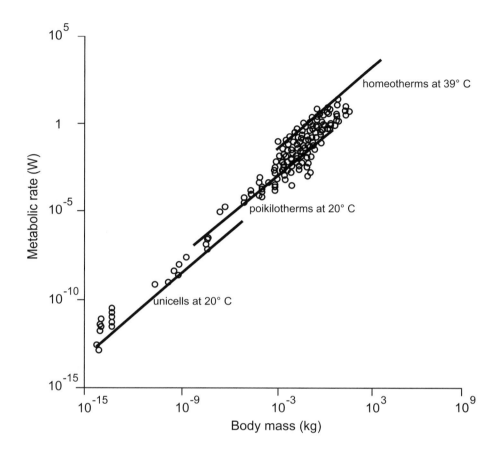

Figure 1.11. When all measurements are standardized to 20°C, differences in metabolic rate are reduced. The regression lines are the same lines shown in figure 1.5 and derived by Hemmingsen (1960). Birds and mammals have temperature-corrected metabolic rates that are higher than those of reptiles and amphibians, but the overlap in values between the two groups is extensive. The difference between unicellular organisms and ectothermic poikilotherms is very small. A significant fraction—albeit not all—of the variation in mass-corrected metabolic rate is explained by variation in body temperature.

1.3.1 Gillooly's Equation and the Metabolic Theory of Ecology

Equation 1.12 is very important because it summarizes the combined effect of body size and temperature on metabolic rate. We call it "Gillooly's equation" to recognize James Gillooly's insight about the multiplicative effect of allometry and temperature on metabolism. Gillooly's equation summarizes a fundamental property of organisms. It tells us that the rate at which energy flows through an organism depends on how big and how hot the organism is. Recall from box 1.2 that we can estimate the fractional rate of energy input into the energy pool of an organism as the ratio of the energy flux and the energy pool contained in the organism, which is proportional to body mass. Thus, the fractional rate of metabolism (or "mass-specific metabolism", $b = B/m_b$) can be expressed as

$$b \propto (m_b)^{-1/4} e^{-E_i/kT}.$$ (1.15)

Again, recall from box 1.2 that the residence time is the reciprocal of a fractional rate. Thus, the residence, or turnover, time of metabolic substrates (t_b) should be proportional to the reciprocal of equation 1.15:

$$t_b \propto (m_b)^{1/4} e^{E_i / kT}.$$

(1.16)

These equations summarize relationships that have been studied for a very long time. We know that large organisms require more resources, but use them (or "flux" them) on a mass-specific basis more slowly than smaller organisms, and that both resource requirements and flux rates are higher at higher body temperatures (Brown et al. 2004). These observations are not front-page news to most physiological ecologists. However, Gillooly's equation combines the effect of size and temperature in a single simple mathematical expression, which is useful. This expression allows us to compare across organisms that differ in size and temperature using an equation that is grounded on first principles of chemistry and physics.

Arguably, the rate at which organisms use energy (i.e., the metabolic rate) is at the heart of the processes in which they participate. Thus, Brown et al. (2004) have extended the metabolic framework embodied in equation 1.12 to document remarkable patterns in populations, communities, and even ecosystems. Let us look at one example at two of these levels: populations and interacting populations.

(1) Maximal population growth depends predictably on body mass and temperature. Although populations vary in number in complicated ways, one of the indisputable rules of population biology is that populations at low density and with large amounts of resources grow exponentially, with growth dictated by maximal intrinsic growth rates (r_{max}). Brown et al. (2004) found that, as predicted by equation 1.14, r_{max} scales with body mass to the $-1/4$ power, and depends on temperature with an energy of activation E_i equal to 0.68 eV. The populations of smaller and hotter organisms have the potential of growing faster than those of larger and cooler ones.

(2) The characteristics of interspecific interactions depend on temperature. Brown et al. (2004) compiled all the studies in which components of competitive or predator-prey interactions have been measured at several temperatures. They found that the rate of attack by predators and parasites, the feeding rate of herbivores, and the time that it takes for a species to exclude another one competitively all depend on temperature. Furthermore, the "energy of activation" of the temperature dependency of these components ranged from 0.56 to 0.81 eV. Brown et al. (2004) give more examples of the plethora of potential ecological applications of Gillooly's equation.

Equation 1.12 and the formulas that spin from it are the foundation of what Brown and his colleagues have called the "metabolic theory of ecology" or MTE. The MTE is a mechanistic synthetic framework that (1) characterizes the effects of body mass and temperature on metabolism, and (2) describes how metabolism dictates the effects of individuals on the pools, fluxes, and turnover of energy and materials in populations, communities, and ecosystems (Brown et al. 2004). Although we are enthusiastic about the MTE, we hasten to add two caveats. First, the MTE is a log-log theory that leaves a lot of residual variation unexplained. Most ecologists and physiologists are interested in understanding the factors that explain this residual variation. Second, the MTE concerns variation in pools, rates/fluxes, and times. We have argued in this chapter that these are important unifying themes in biology, but they are not the only themes. The MTE is not a theory of "everything" (Brown et al. 2004).

1.4 Using Historical Data in Comparative Studies

Previous sections emphasized that almost any prediction or comparison of physiological or life history data should begin with a consideration of the size and body temperature of the animal. There is another major factor that must be taken in consideration when comparing and even predicting traits among animals: we must have some information about the evolutionary history of the species under consideration. The use of phylogenetic information in comparative physiological studies is one of the main thrusts of evolutionary physiology, a vibrant, rapidly growing field that touches many, maybe most, of the topics included in this book. Because we emphasize ecology, our treatment of the importance of phylogenetics and evolution in ecological physiology will be cursory. We will not ignore evolution, but we will emphasize ecological applications of physiology. Readers must consult Feder et al.'s (2000) excellent review as a guide to evolutionary physiology. Because we will rely on phylogenetic data at several points in this book, a brief introduction to the tools that we will use is merited.

To motivate the use of phylogenetics in physiological ecology we will remind you of Calder's hypothesis (1984). He suggested that we could perceive adaptation in a trait as a deviation from the basic size-dependent allometric pattern resulting from selection. Calder suggested that animals may be similar because they are of the same size, but differ because they live in contrasting environments, feed on different foods, and so on. We often hypothesize that animals are similar because similar selective pressures have made them converge, and different because dissimilar selective pressures have made them diverge. How do we go about testing this hypothesis? In the old days (some say "good old days"), we would have simply chosen two species (preferably related and of roughly the same size) but with different ecological habits. We would have made a directional prediction about differences in the magnitude of a physiological trait, and then examined in the laboratory or the field whether this prediction was correct. An enormous amount of extraordinarily important research was done in this way for many years.

But times change and so do research approaches. As pointed out by Garland and Adolph (1994) the "two-species comparison" approach is limiting and plagued with difficulties of all sorts. Briefly, there are many possible reasons why two species could show a statistically significant difference in a physiological trait: First, the process of speciation by itself may lead to differences; second, genetic drift and founder effects may also lead to nonadaptive divergence between species that experience little or no genetic exchange; and third, adaptive change will lead to divergence. Although it is only the third of these reasons that we are testing, the other two can lead to differences as well. For any two species, comparison of any physiological trait is likely to reveal a difference, given that the researcher has chosen a sufficient large sample of individuals. The solution, then, is to conduct a multispecies comparative study. If we fail to account for phylogenetic relationships, a multispecies comparison is flawed as well (figure 1.12). The problem with two-species comparisons and naive phylogeny-free multispecies comparisons is that they assume that species are statistically independent data points, when they are not (figure 1.12). Species are related by descent in a hierarchical fashion and hence, when one uses them as independent data points in statistical comparisons, one commits phylogenetic pseudoreplication (figure 1.11). The term pseudoreplication was coined by Hurlbert (1984) and refers to a statistical mistake that is committed all too often. This mistake consists in assuming that data are statistically independent when they are not. Hurlbert (1984) illustrates pseudoreplication with the following example: Imagine taking many subsamples within a lake and treating them as replicates from all lakes. A similar logic applies to comparative studies. The different species within a genus (or any monophyletic lineage) are not proper independent replicates.

Testing the hypothesis that there is an evolutionary correlation between a potential selective pressure (an environment or a diet) and a trait is tricky. It requires that we identify the number of evolutionary transitions in which a change in the selective pressure was accompanied by a change in the trait that we have chosen as a response variable. This comparison requires recognizing the phylogenetic relationships among the species in our comparison. Box 1.6 describes what we believe is the most commonly used method to test for evolutionary correlations: Felsenstein's phylogenetically independent contrasts (or PICs; Felsenstein 1985). This is one of a growing number of statistical methods used to

(A)

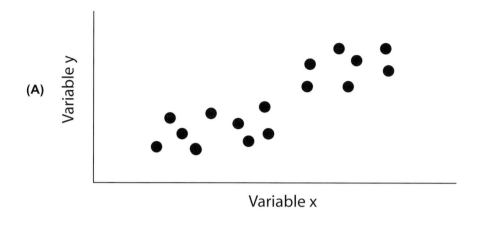

(B)

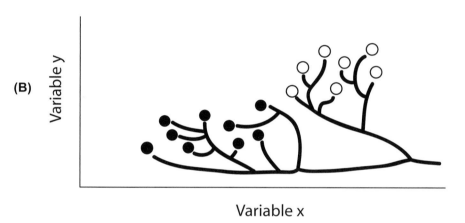

Figure 1.12. Imagine that you have hypothesized that when the diet (or any trait *x*) of an organism changes, then natural selection should favor a change in the expression of digestive enzymes (or any other relevant trait *y*). Thus, you predict that an interspecific comparison will reveal that *x* and *y* will be correlated across species. You assemble a sample of 15 species and plot *y* against *x*. When you perform a statistical test, you find that there is a highly significant correlation (panel A). You submit the result to a prestigious journal and one of the reviewers replies that your correlation is spurious, that the 15 species in your sample can be neatly divided into two monophyletic clades (i.e., two genera), and that hence your statistical test has inflated degrees of freedom. Your sample represents a single evolutionary divergence (panel B), and hence your sample size is reduced to $N = 2$. The reviewer is, of course, strictly right. Addressing her comment requires that you use the method described in box 1.6 (or one of its many alternatives; see Martins 2000). (Redrawn from Madison and Madison 1992.)

incorporate phylogenetic information in comparative studies. This book is not the place to review these methods. We suggest Martins' (2000) review and the references in it as a guide to the maze of phylogenetic comparative methods.

1.4.1 Limitations of Phylogenetic Analyses

Phylogenetic approaches to multispecies comparisons are not without drawbacks: They require physiological data on many species, and they require at least some phylogenetic information. Furthermore, the conclusions of phylogenetic comparative studies depend on the assumption that the phylogenetic tree is accurate and that the model of evolution assumed by the statistical method is correct. These are not minor considerations, and we have often worried that many of the methods available are too conservative (i.e., lack statistical power)

Box 1.6

Phylogenetically Independent Contrasts

Although the values of traits x and y in phylogenetically related species are not independent statistically, we can transform them into independent rates of evolutionary change between sister species or nodes. We will use the hypothetical phylogeny in panel A of figure 1.13 to describe the method developed by Felsenstein (1985) to estimate phylogenetically independent contrasts (PICs). We follow Felsenstein's (1985) explanation very closely. Step 1 is to choose two species at the tip that share a common ancestor (say 1 and 2), and to compute the contrast $X_1 - X_2$. Assuming that evolution proceeds by Brownian motion (in our view a fairly iffy assumption for traits of physiological and ecological significance), this contrast has an expected value of 0 and a variance proportional to $v_1 + v_2$ (v_1 and v_2 are branch lengths proportional to time since divergence). Step 2 is to remove these two tips, leaving behind ancestor 7, which now becomes a tip. In step 3, we assign this ancestor a value equal to the average of X_1 and X_2 weighted by the lengths of the branches, v_1 and v_2 (i.e., we place less value on tips with longer branches):

$$X_7 = \frac{[(1/v_1)X_1 + (1/v_2)X_2]}{(1/v_1 + 1/v_2)} \tag{1.6.1}$$

Finally, in step 4 we lengthen the branch below node 7 by increasing its length from v_7 to $v_7 + v_1 v_2/(v_1 + v_2)$. This lengthening occurs because X_7 is estimated with error. After doing this, we have constructed one contrast and reduced the number of tips by one. Then we continue to repeat these steps until we have only one tip left in the tree. If there were n species, this procedure would produce $n - 1$ contrasts (panel A of figure 1.13). To bring all the contrasts to a common variance, each contrast can be divided by the square root of its variance (i.e., contrast 1 can be divided by $(v_1 + v_2)^{1/2}$. Then you have to do it all over again for trait Y and correlate the contrasts in X with those in Y. For allometric relationships the value of a trait is log transformed before constructing contrasts. Because no evolutionary change in X should lead to no change in Y, the regression between contrasts is constructed through the origin.

The method described here is probably the most widely used of all comparative approaches. Therefore it is wise to lists some of its assumptions and the consequences that these can have. The most important assumption is that traits evolve by Brownian motion, that is, that traits (which are assumed to be continuous) diverge as a result of random wanderings through time. Panel B of figure 1.13 (after Harvey and Pagel 1991)

continued

continued

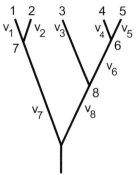

Contrast	Variance
$X_1 - X_2$	$v_1 + v_2$
$X_4 - X_5$	$v_4 + v_5$
$X_3 - X_6$	$v_3 + v'_6$
$X_7 - X_8$	$v'_7 + v'_8$

(A)

Where:

$$X_6 = \frac{v_4 X_5 + v_5 X_4}{v_4 + v_5}$$

$$X_7 = \frac{v_2 X_1 + v_1 X_2}{v_1 + v_2}$$

$$X_7 = \frac{v'_6 X_3 + v_3 X_6}{v_3 + v'_6}$$

$$v'_6 = v_6 + v_4 v_5 /(v_4 + v_5)$$

$$v'_7 = v_7 + v_1 v_2 /(v_1 + v_2)$$

$$v'_8 = v'_7 + v_3 v'_6 /(v_3 + v'_6)$$

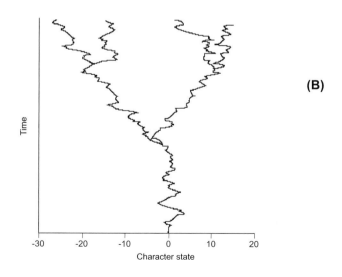

(B)

Figure 1.13. (panel A) Constructing phylogenetically 4 independent contrasts (PICs) in a phylogenetic tree with 5 species requires reconstructing 3 ancestors (nodes 6, 7, and 8). The table under the phylogeny demonstrates the calculations needed to estimate each contrast and its variance. (Panel B) Using PICs to establish an evolutionary correlation assumes that the traits under study evolve by Brownian motion. (Panel C) Using PICs to determine whether desert ringtails have lower basal metabolic rates (BMR) than expected from their body mass.

continued

continued

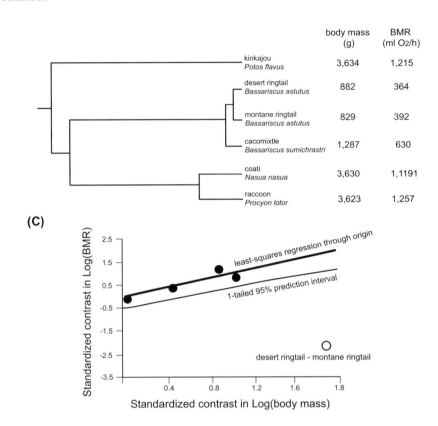

	body mass (g)	BMR (ml O2/h)
kinkajou *Potos flavus*	3,634	1,215
desert ringtail *Bassariscus astutus*	882	364
montane ringtail *Bassariscus astutus*	829	392
cacomixtle *Bassariscus sumichrastri*	1,287	630
coati *Nasua nasua*	3,630	1,1191
raccoon *Procyon lotor*	3,623	1,257

(C)

depicts the change in a trait in four lineages that follow random walks after splitting from a common ancestor at time 0. There is a lineage split at $n2$ and two more splits at $n4$ and $n5$. There are many reasons to be skeptical about using a random walk as a model for the evolution of traits. For example, there may be persistent selective pressures over time due to common selective regimes. This persistent selection may lead to trait conservatism and "clumping" in the values of the traits in some of the clades of the phylogeny. Cruz-Neto et al. (2001 and references there) discuss some of the problems that the potential collinearity between phylogeny and function has for a comparative analysis that relies on PICs. Briefly, under these conditions all the large evolutionary changes occur in deep nodes. This is problematic, because, as we have seen, contrasts that involved reconstructed ancestors tend to be small and hence close to the origin of a regression line. Collinearity between phylogeny and the traits reduces the power of comparative analyses. Garland et al. (1993) and Vanhoodydonck and van Damme (1999) discuss in some detail the issue of power in comparative analyses.

continued

continued

Panel C of figure 1.13 presents an example of the use of PICs developed by Garland and Adolph (1994) using data from Chevalier (1991). The question is whether desert ringtails (*Bassariscus astutus*) have evolved lower metabolic rate than would be expected for a procyonid mammal of their size. To answer this question, Garland and Adolph (1994) compiled a phylogenetic tree for a group of closely related procyonids. They then conducted all the steps described above to construct a regression line between the contrast in the logarithm of body mass and the contrast in the logarithm of basal metabolic rate. However, they left out the contrast between montane and desert ringtails. Because Chevalier had expected desert ringtails to have lower than expected metabolic rates, they constructed a one-tailed confidence interval for this regression line. In a final step, they plotted the contrast that includes the desert ringtail. This contrast was, as expected, lower than the 95% confidence interval. This result is the consequence of 3 possible processes: (1) the desert ringtail evolved low metabolic rates, (2) the montane ringtail evolved high ones, or (3) both groups diverged up and down from the average value of their common ancestor. Standardized phylogenetic contrasts are nondirectional and do not allow differentiating among these options. To answer which one of these options is the correct one, we require supplementary paleontological and/or biogeographical knowledge. Chevalier (1991) used this type of additional evidence to propose that desert ringtail lineage evolved lower metabolism. We encourage readers to use the method described in the first paragraph of this box to reconstruct the results of Garland and Adolph's (1994) analyses. If the idea of spending 30 minutes with a calculator is unappealing, we suggest using one of the many free computer programs available to conduct PICs analyses. We recommend PDAP (Phenotypic Diversity Analysis Programs, v. 5; Garland et al. (1993, 1999), CAIC (developed by Purvis and Rambaut 1995, http://www.bio.ic.ac. uk/evolve/software/caic), and COMPARE 4.4 (developed by Martins 2001, http://compare.bio.indiana.edu).

and assume an unrealistic model of evolution (see Schondube et al. 2001, for example). We fear that an undue emphasis on "phylogenetical correctness" will impede and stifle comparative studies. Many interesting groups lack adequate phylogenetic information and it would be folly to shun their study. It will be a sad day when all ecological and evolutionary physiologists limit their attention to a few well-studied model organisms and taxa. Our domain of study is life's functional diversity and hence we advocate an opportunistic approach: If a phylogenetic approach is possible, use it. If it is not because a phylogenetic tree is not available or because the phylogenetic structure of the group that you are interested in does not satisfy the assumptions of a comparative method, conduct the study anyway. Your study may inspire a systematist to study the phylogenetic relationships of your group or an evolutionary statistician to develop a method to test the hypotheses generated by your investigation. Phylogenetically informed comparative analyses are a central element of physiological ecology, but they should not be automatic or obligatory.

Having made a plea for phylogenetic tolerance, we emphasize the importance of considering all the phylogenetic data available when conducting a comparative study. Phylogenetic information does a lot more than allowing comparative physiologists the use of statistically rigorous methods. Perhaps more importantly, phylogenetics provides us with a map to the history of the traits that we study. David Winkler (2000) divides phylogenetically informed comparative approaches into two broad categories: the convergence approach (also called the "functional" approach), and the homology approach. The method of PICs is an example of the convergence approach that aims to look *across* phylogenetic lineages at repeated changes in traits and correlates these changes with changes in ecological circumstances or with changes in other traits. The homology approach explores unique events *within* lineages. The convergence approach strives to solve the problem of phylogenetic pseudoreplication. The homology approach recognizes that life's history has been punctuated by many important single events. Dealing with rare, or single, events precludes statistics and hence the homology approach has no pretense of statistical rigor. It simply attempts to map the evolutionary transitions that have occurred along the history of a lineage. Although the approach is descriptive, it causes us to focus carefully on the traits that we are interested in and to think about the kinds of forces that may have produced the observed changes in a lineage. We will provide many examples of the use of "homology thinking" throughout this book.

References

Alexander, R. M. 1998. Symmorphosis and safety factors. Pages 28–35 in E. W. Weibel, C. R. Taylor, and L. Bolis (eds.), *Principles of animal design.* Cambridge University Press, Cambridge.

Bartholomew, G. A. 1981. A matter of size: An examination of endothermy in insects and terrestrial vertebrates. Pages 45–78 in B. Heinrich, (ed.), *Insect thermoregulation.* Wiley, New York.

Belgrano, A., A. Allen, B. J. Enquist, and J. Gillooly. 2002. Allometric scaling of maximum population density: A common rule for marine phytoplankton and terrestrial plants. *Ecology Letters* 5:611–613.

Bignell, D. E., P. Eggleton, L., Nunes, and K. L. Thomas. 1997. Termites as mediators of carbon fluxes in tropical forests: Budgets for carbon dioxide and methane emissions.

Pages 109–134 in A. B. Watt, N. E. Stork, and M. D. Hunter (eds.), *Forests and insects.* Chapman and Hall, London.

Bonner, J. T. 1965. *Size and cycle: an essay on the structure of biology.* Princeton University Press, Princeton, N. J.

Brown, J. H. 1999. Macroecology: Progress and prospect. *Oikos* 87:3–14.

Brown, J. H., J. F. Gillooly, A. P. Allen, V. M. Savage, and G. B. West. 2004. Toward a metabolic theory of ecology. *Ecology* 85:1771–1789.

Brown, J. H., G. B. West, and B. J. Enquist. 2000. Scaling in biology: patterns, processes, causes, and consequences. Pages 1–24 in J. H. Brown and G. B. West (eds.), *Scaling in biology.* Oxford University Press, New York.

Brown, J. H., G. B. West, and B. J. Enquist. 2005. Yes, West, Brown, and Enquist's model of allometric scaling is both mathematically correct and biologically relevant. *Functional Ecology* 19:735–738.

Calder, W. A., III. 1984. *Size, function, and life history.* Harvard University Press, Cambridge, Mass.

Cairns, J., and B. R. Niederlehner. 1996. Developing a field of landscape ecotoxicology. *Ecological Applications* 15:608–617.

Carbone, C. and J. L. Gittleman. 2002. A common rule for the scaling of carnivore density. *Science* 295:2273–2276.

Carbone, C., G. M. Mace, S. C. Roberts, and D. W. Macdonald. 1999. Energetic constraints on the diet of terrestrial carnivores. *Nature* 402:286–288.

Chevalier, C. D. 1991. Aspects of thermoregulation and energetics in the procyonidae (Mammalia: Carnivora). Ph.D. thesis, University of California, Irvine.

Cheverud, J. M. 1982, Relationships among ontogenetic, static, and evolutionary allometry *American Journal of Physical Anthropology* 59:139–149.

Cruz-Neto, A. P., T. Garland, and A. S. Abe. 2001. Diet, phylogeny, and basal metabolic rate in phyllostomid bats. *Zoology* 104:49–58.

Dahl, J., and B. L. Peckarsky. 2002. Induced morphological defences in the wild: Predator effects on a mayfly, *Drunella coloradensis. Ecology* 83:1620–1634.

Damuth, J. 1987. Interspecific allometry of population density in mammals and other animals: the independence of body mass and population energy use. *Biological Journal of the Linnean Society* 31:193–246.

Darveau, C. A., R. K. Suarez, R. D. Andrews, and P. W. Hochachkas. 2002. Allometric cascade as a unifying principle of body mass effects on metabolism. *Nature* 417:166–170.

Denny, M., and S. Gaines. 2000. *Chance in biology: Using probability to explore nature.* Princeton University Press, Princeton, N.J.

Dodds, P. S., D. H. Rothman, and J. S. Weitz. 2001. Re-examination of the "3/4 law" of metabolism. *Journal of Theoretical Biology* 209:9–27.

Feder, M. E., A. F. Bennett, and R. B. Huey. 2000. Evolutionary physiology. *Annual Review of Ecology and Systematics* 31:315–341.

Felsenstein, J. 1985. Phylogenies and the comparative method. *American Naturalist* 125:1–25.

Garland, T., and S. C. Adolph. 1994. Why not to do two-species comparative studies: Limitations to inferring adaptation. *Physiological Zoology* 67:797–828.

Garland, T., Jr., A. W. Dickerman, C. M. Janis, and J. A. Jones. 1993. Phylogenetic analysis of covariance by computer simulation. *Systematic Biology* 42:265–292.

Garland, T., Jr., P. H. Harvey, and A. R. Ives. 1992. Procedures for the analysis of comparative data using phylogenetically independent contrasts. *Systematic Biology* 41:18–32.

Garland, T., Jr., P. E. Midford, and A. R. Ives. 1999. An introduction to phylogenetically based statistical methods, with a new method for confidence intervals on ancestral states. *American Zoologist* 39:374–388.

Gillooly, J. F., J. H. Brown, G. B. West, V. M. Savage, and E. Charnov. 2001. Effects of size and temperature on metabolic rate. *Science* 293:2248–2251.

Griffiths, H. 1997. *Stable isotopes integration of biological, ecological, and geochemical processes.* Bios Scientific, Oxford.

Harte, J. 1985. *Consider a spherical cow: A course in environmental problem solving.* William Kauffmann, Los Altos, Calif.

Harvell, C. D. 1990. The ecology and evolution of inducible defences. *Quarterly Review of Biology* 65:323–340.

Harvey, P. H., and M. D. Pagel. 1991. *The comparative method in evolutionary biology.* Oxford University Press, Oxford.

Hayssen, V., and R. C. Lacy. 1985. Basal metabolic rates in mammals: Taxonomic differences in the allometry of BMR and body mass. *Comparative Biochemistry and Physiology A* 81:741–754.

Heinrich, B. 1993. *The hot-blooded insects.* Harvard University Press, Cambridge, Mass.

Hemmingsen, A. M. 1960. Energy metabolism as related to body size and respiratory surfaces, and its evolution. *Reports of the Steno Memorial Hospital and Nordisk Insulin Laboratorium* 9:6–110.

Hurlbert, S. H. 1984. Pseudoreplication and the design of ecological field experiments. *Ecological Monographs* 54:187–211.

Huxley, J. 1932. *Problems of relative growth.* Methuen, London.

Kleiber, M. 1932. Body size and metabolism. *Hilgardia* 6:315–353.

Kooijman, S.A.L.M. 2000. *Dynamic energy and mass budgets in biological systems.* Cambridge University Press, Cambridge.

Kooijman, S.A.L.M. 2001. Quantitative aspects of metabolic organization: A discussion of concepts. *Philosophical Transaction of the Royal Society of London Series B* 356:331–349.

Kozlowski, J., and J. Weiner. 1997. Interspecific allometries are byproducts of body size optimization. *American Naturalist* 149:352–380.

Kozlowski, J., and M. Konarzewski. 2004. Is West, Brown, and Enquist's model of allometric scaling mathematically correct and biologically relevant? *Functional Ecology* 18:283–289.

Madison, W. P., and D. R. Madison. 1992. *MacClade: Analysis of phylogeny and character evolution.* Sinauer, New York.

Mahmood, I. 1999. Allometric issues in drug development. *Journal of Pharmaceutical Sciences* 88:1101–1106.

Martins, E. P. 2000. Adaptation and the comparative method. *Trends in Ecology and Evolution* 15:259–302.

Martins, E. P. 2001. COMPARE, version 4.4. Computer programs for the statistical analysis of comparative data. Distributed by the author via the WWW at http://compare.bio.indiana.edu/. Department of Biology, Indiana University, Bloomington IN.

McNab, B. K. 2002. *The physiological ecology of vertebrates: A view from energetics.* Comstock, Ithaca, N.Y.

Motulsky, H. J., and L. A. Rasnas. 1987. Fitting curves to data using non-linear regression: A practical and non-mathematical review. *FASEB Journal* 1:365–374.

Niklas, K. J., and B. J. Enquist. 2001. Invariant scaling relationships for intespecific plant biomass production rates and body size. *Proceedings of the National Academy of Sciences* 98:2922–2927.

Packard, G. C., and T. Boardman. 1999. The use of percentages and size-specific indices to normalize physiological data for variation in body size: Wasted time, wasted effort? *Comparative Biochemistry and Physiology A* 122:37–44.

Peters, R. H. 1983. *The ecological implications of body size.* Cambridge University Press, Cambridge.

Piersma, T., and J. Drent. 2003. Phenotypic flexibility and the evolution of organismal design. *Trends in Ecology and Evolution* 18:228–223.

Pigliucci, M. 2001. *Phenotypic plasticity: Beyond nature and nurture.* Johns Hopkins University Press, Baltimore.

Purvis, A., and A. Rambaut. 1995. Comparative analysis by independent contrast (CAIC): A Macintosh application for analyzing comparative data. *CABIOS* 11:247–251.

Robbins, C. T. 1993. *Wildlife feeding and nutrition.* Academic Press, New York.

Rose, M. R., and G. V. Lauder. 1996. *Adaptation.* Academic Press, New York.

Savage, V. M., J. F. Gillooly, W. H. Woodruff, G. B. West, A. P. Allen, B. J. Enquist, and J. H. Brown. 2004. The predominance of quarter-power scaling in biology. *Functional Ecology* 18:257–282.

Schmidt-Nielsen, K. 1984. *Scaling: Why is animal size so important?* Cambridge University Press, Cambridge.

Schmitt, J., S. A. Dudley, and M. Piglicci. 1999. Manipulative approaches to testing adaptive plasticity: Phytochrome-mediated shade-avoidance responses in plants. *American Naturalist* 154: S43–S54.

Schondube, J., C. Martínez del Rio, and L.G. Herrera. (2001). Diet and the evolution of digestion and renal function in phyllostomid bats. *Zoology* 104:59–74.

Shapiro, A. M. 1976. Seasonal polyphenism: *Evolutionary Biology* 9:259–333.

Sokal, R. R., and F. J. Rolf. 1995. *Biometry.* Freeman and Co., New York.

Stern, D.L.S., and D. J. Emlen 1999. The developmental basis for allometry in insects. *Development* 126:1091–1101.

Swift, J. 1735. *Travels into several remote nations of the world. In four parts. By Lemuel Gulliver, first a surgeon, and then a captain of several ships.* George Faulkner, Dublin.

Tollrian, R., and C. D. Harvell. 1999. *The ecology and evolution of inducible defenses.* Princeton University Press, Princeton, N.J.

Tucker, V. A. 1971. Flight energetics in birds. *American Zoologist* 11:115–124.

Van Gordingen, P. R., G. M. Foody, and P. J. Curran. 1997. Scaling up: From cell to landscape. Society for Experimental Biology Seminar Series 63, London.

Vanhoodydonck, B., and R. Van Damme. 1999. Evolutionary relationships between body shape and habitat use in lacertid lizards. *Evolutionary Ecology Research* 1:785–805.

West, G. B., J. H. Brown, and B. J. Enquist. 1997. A general model for the origin of allometric scaling laws in biology. *Science* 276:122–126.

West, G. B., J. H. Brown, and B. J. Enquist. 1999. The fourth dimension of life: Fractal geometry and allometric scaling of organisms. *Science* 284:1677–1679.

West, G. B., J. H. Brown, and B. J. Enquist. 2000. The origin of universal scaling laws in biology. Pages 87–112 in J. H. Brown and G. B. West (eds.), *Scaling in biology*. Oxford University Press, New York.

West, G. B., W. H. Woodruff, and J. H. Brown. 2002. Allometric scaling of metabolic rate from molecules and mitochondria to cells and mammals. *Proceedings of the National Academy of Sciences* 99:2473–2478.

Williams, G. C. 1966. *Adaptation and natural selection*. Princeton University Press, Princeton, N.J.

Winkler, D. W. 2000. The phylogenetic approach to avian life histories: an important complement to within population studies. *The Condor* 102:52–59.

Wood, T. G., and W. A. Sands. 1978. The role of termites in ecosystems. Pages 245–292 in M. V. Brian, (ed.), *Production ecology of ants and termites*. Cambridge University Press, Cambridge.

Chemical Ecology of Food

CHAPTER TWO

The Chemistry and Biology of Food

To EAT AND to be eaten are central themes in ecology. Food chemistry determines what animals eat and it often shapes the strategies that both animals and plants use to avoid being eaten. Thus, thinking about food's chemistry is something that ecologists do (or should do) quite often. This chapter is an introduction to a topic that we will refer to again and again throughout the book. Investigating the chemical composition of food can be done at a variety of levels of detail. The organization of this chapter reflects these levels. We begin by describing "proximate analysis" (PA), a commonly used (and misused) method that yields a very gross breakdown of food's constituents into "fractions" that are assumed to correspond to broad classes of chemical compounds. We describe the assumptions of proximate nutrient analysis and its many limitations. Then we use the fractions that PA yields to describe food's main chemical components (carbohydrates, proteins, lipids, minerals, and secondary metabolites; figure 2.1) and to comment on the biological roles that they play.

2.1 Getting Started; First Catch (Store and Prepare) the Hare

Suppose that you believe that it would be useful to know the chemical composition of the food eaten by the animal that you study. Maybe you suspect that variation in food chemistry explains food preferences, or that nutrient content correlates with some measure of performance (growth, reproduction, or survival). Unfortunately, analyzing a food's chemistry can rapidly become complicated and expensive. Thus, it is fairly unlikely that you will have the equipment necessary to do all the analysis that you want. Wisely, you decide to

Figure 2.1. The compounds present in food can be divided into broad categories. The dry matter in food can be divided into inorganic constituents (the total of which is recovered as ash in nutrient analyses) and organic constituents. The organic components of food can be divided into relatively indigestible fiber (please see section 2.3 for a more nuanced definition), macronutrients (carbohydrates, protein, and lipids), secondary compounds, and other organic substances (such as vitamins). The compounds that we have pooled into the "other" category may represent a small fraction of food's total mass, but can be crucially important for an animal's nutrition.

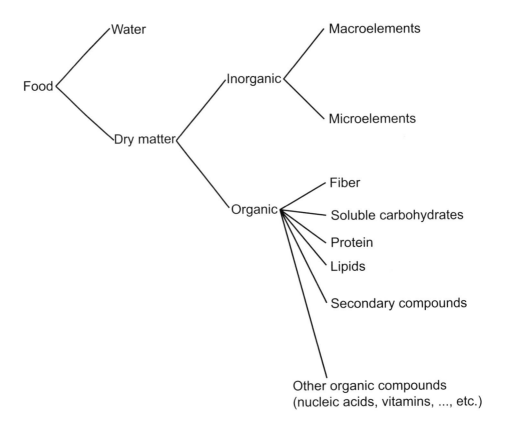

send the food out to a specialized laboratory that will do at least some of the analyses. Even before sending food out for analyses, you are confronted with a series of decisions: the first one is how to harvest, store, and prepare the food before it is analyzed or shipped to the laboratory that you have chosen. The procedures to properly harvest, store, and prepare samples vary from food type to food type, and can even depend on the type of nutrient or compound of interest. Box 2.1 outlines some of the general guidelines that must be followed before analyzing a food sample and table 2.1 shows a typical series of analyses that many food chemistry laboratories perform. The remainder of this chapter should give you a better idea about what analyses to perform on the food eaten by the animals that you study.

Deciding the detail that you need in a food analysis is not simple. The analysis should be dictated by the question that you are asking and by your knowledge about the digestive and metabolic capacity of your beast of choice. For example, if you work with animals with significant fermenting capacity, such as ungulates, analyzing the precise composition of carbohydrates in forage may not be particularly useful. Assessing the fraction of carbohydrates that are present as easily fermentable soluble carbohydrates and as refractory fiber may be sufficient.

Box 2.1.

Before you can do any chemical analysis on a material, you have to harvest it and store it. Ideally, the chemical composition of the food sample that you analyze should resemble that of ingested food. It is especially important that plant foods be collected at the times animals consume them, because plants typically undergo huge seasonal changes in composition as they progress through developmental stages towards senescence or dormancy. Figure 2.2 illustrates this for the N and water content of a number

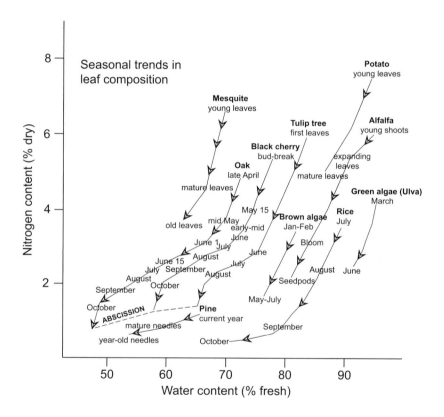

Figure 2.2. The nutritional composition of plant materials changes seasonally with the age of the tissue and the plant. The general trend is for plants to have drier tissues with less nitrogen as the growing season progresses. (Redrawn from Slansky and Scriber 1985.)

of plant species, and it is also well known that as plants senesce or prepare for dormancy their cell wall content typically rises as well.

When you collect a food sample often you will also have to dry it and homogenize it before analysis, and these procedures can modify its chemical composition. Thus, considering how samples are harvested, stored, and prepared before a chemical analysis is important. Although individual cases

continued

continued

may need special treatment, there are four generic rules that should always be followed: (1) Minimize the time between sampling and analysis. (2) Store samples at low temperatures in the absence of light and air. (3) Use vacuum freeze-drying. (4) Keep the grinding and homogenization procedure short (Quarmby and Allen 1989). The following paragraphs expand on these recommendations.

Harvest What Animals Eat

Animals can feed very selectively. Browsers and grazers may feed on leaves of a certain age. It is especially important that plant foods be collected at the times animals consume them, because plants typically undergo huge seasonal changes in composition as they progress through developmental stages toward senescence or dormancy. Figure 2.2 (from Slansky and Scriber 1985) illustrates the seasonal changes in the N and water content of a number of plant species. It is also well known that as plants senesce or prepare for dormancy their cell wall content typically rises as well. Carnivores may be particular about the tissue that they eat or about the life stage of the prey that they consume. Frugivores may choose fruit at certain degrees of ripeness, and so on. The samples that you analyze must represent what animals eat, not what is easy to collect. This recommendation may appear trite, but our experience indicates that it must be made. We have seen too many chemical analyses of unripe fruits, inedible twigs, and rotting flesh.

Storage

Samples must be processed as soon as possible after collection. Some samples are easy to process: Nectar, for example, can be spotted on filter paper and air dried. The soluble constituents are then eluted and analyzed (Kearns and Inouye 1993). Other types of sample may require a bit more work. A cooler that maintains samples at a temperature close to freezing reduces microbial action in the field. Ideally, samples should be placed in cold storage (-20 to $-10°C$) as soon as possible. Even in the cold, however, prolonged storage can lead to the denaturation of some proteins and to lipid autoxidation. To minimize changes during storage, we recommend storing dry samples in the cold and in the absence of light and air. A variety of chemical methods are available to minimize the activity of microbes (gamma and UV radiation, microbicides, gaseous sterilization; see Karel et al. 1985; Singh and Helman 1993).

Drying

Sometimes air-drying samples in the field is inevitable. Air-drying samples, however, can take a long time and because it is often done under completely uncontrolled conditions it is better to avoid it. Drying at high

continued

continued

temperatures (50, 60, 70, or 105°C) is also commonly done. Higher temperatures can result in breakdown and volatilization of some of a food's constituents. For leaves and stems, drying at high temperatures can increase the fiber content because carbohydrates and protein react nonenzymatically to form insoluble ligninlike complexes. Thus, if you must dry your samples at high temperatures, we recommend using a temperature below 50°C (40°C is the most often recommended temperature) and a forced-air oven to minimize drying time. Although high-temperature drying is acceptable for many analyses, we recommend freeze-drying food under reduced pressure whenever possible. Indeed, the vacuum freeze-drier is one of the most frequently used pieces of equipment in our laboratories. Banga and Singh (1994) provide detailed guidelines about how to dry food for preservation and analysis.

Homogenization

Samples must be ground and homogenized for analysis. A variety of grinding devises work well. We use a coffee grinder for large samples and mortar and pestle for smaller ones. For samples that must be made brittle to grind, or that must remain cold, we use a mortar and pestle designed to use with liquid nitrogen (do not try to place liquid nitrogen in a coffee grinder!). In addition to the relatively primitive methods that we have already described, there are a variety of commercial (and fairly expensive) ball mills available. Some materials are difficult to grind and homogenize (e.g., bird feathers and fish skins are horrible!) with even the most sophisticated grinding gadgets available. Although grinding materials seems straightforward enough, a few words of caution are merited. Overheating can take place, especially in ball mills in which the sample is contained within a grinding chamber. Although most food analyses require grinding dry materials, you may sometimes have to grind fresh samples. While grinding fresh samples, physical breakdown can bring an enzyme together with its substrate. Enzyme action can be prevented by grinding the sample at cold temperatures or by boiling the sample to inactivate enzymes. Finally, some analyses require sieving the dry samples through a series of screens to achieve particle size homogeneity.

In contrast, if your research involves animals that rely primarily on endogenous enzymes to digest food, such as fruit-eating birds, then knowing whether the soluble carbohydrates are present as starch, disaccharides, or glucose may be crucial. The importance of the interplay between food chemistry and an animal's physiological traits in determining digestion may force you to come

TABLE 2.1
Chemical analyses typically offered by a food analysis laboratories

List of Chemical Analyses

Proximate nutrient analysis	**Mineral analyses**
Moisture	Sodium
Crude protein (from Kjeldahl or combustion)	Potassium
Ash	Chloride
Ether extract	Calcium
Crude fiber	Phosphorus
Acid detergent fiber	Copper
Neutral detergent fiber	Zinc
	Iron
Special chemical analyses	Selenium
Energy (bomb calorimetry)	Cobalt and fluoride
Fatty acid profile	Molybdenum
Soluble carbohydrates	
Protein	

Note: We have eliminated prices and sample sizes required because they vary significantly among laboratories.

back to this chapter repeatedly as you read chapters 3, 4, and 6. In a similar fashion, the importance of some of the details of food chemistry described in this chapter may not be completely clear until you have finished reading those chapters. Attempting to reduce food's chemistry exclusively to chemistry is probably folly. Understanding the ecological importance of food's chemical composition requires considering both chemistry and physiology.

2.2 Proximate Nutrient Analysis

Most food analysis laboratories conduct proximate analysis (PA). This method was developed over 100 years ago to quantify the quality of forage for livestock. Although proximate nutrient analysis has many limitations, it is used widely and some of its components are informative if used judiciously. The system consists of the steps described in box 2.2 (Van Soest 1994). These steps yield several fractions (ash, crude protein, crude fat, crude fiber, and nitrogen-free extract (NFE)). Proximate analysis assumes that each of these fractions can be identified with a broad class of chemical compounds: ash represents minerals, crude protein represents protein, crude lipid represents fats, crude

Box 2.2.

Proximate analysis of food requires six steps.

(1) The sample is dried at 100°C to determine moisture content and dry matter. The dry matter is divided into three subsamples.

(2) Subsample 1 is ashed in a muffle furnace at 550°C. The fraction remaining after complete combustion is called "ash."

(3) The nitrogen content of subsample 2 is determined either by the Kjeldahl procedure or by combustion in an elemental analyzer. The estimates of total nitrogen by Kjeldahl and by combustion often differ slightly (Simmone et al. 1997). The Kjehldal procedure requires boiling the dry sample in a concentrated sulfuric acid solution with a catalyst until all organic matter is destroyed and all the nitrogen is transformed to ammonium. The solution is then cooled, sodium hydroxide is added, and the ammonia is volatilized and captured in a weak acid, where it is measured by titration, by colorimetry, or with an ammonia electrode (Herbst 1988). A measurement called "crude protein" is obtained by multiplying nitrogen content by 6.25. We will explain this mysterious (and often inaccurate) transformation from nitrogen to protein in the text.

(4) To estimate lipids, subsample 3 is extracted with a nonpolar solvent (typically diethyl ether) and filtered. The difference between the weight of the original sample and the filtered residue is called "crude fat."

(5) The fat-extracted residue is refluxed in 1.25% sulphuric acid followed by 1.25% sodium hydroxide. The insoluble residue is weighed and ashed. The ash-free dry weight of the residue is called "crude fiber."

(6) Finally, a fraction called nitrogen-free extract (NFE) is estimated as

$$NFE = dry\ matter - (ash + crude\ protein + crude\ lipid + crude\ fiber).$$

NFE is assumed to represent highly digestible, soluble carbohydrates.

fiber represents the least digestible fibrous structural materials, and NFE represents soluble carbohydrates. None of these assumptions is strictly true. Understanding why these assumptions are false can lead you to better understand the limitations of PA.

Because the preparation of crude fiber often leads to the solubilization and thus the loss of important fiber fractions, the largest error of the analysis is often in the estimation of fiber. Van Soest (1978) demonstrated that crude fiber correlates very poorly with other, more accurate, measurements of fiber content. Echoing the comments of many researchers in the past 20 years, we discourage the use of crude fiber in nutritional studies. In a following section we will define dietary fiber and describe better ways of estimating its content in food.

Converting from total nitrogen to crude protein can also lead to large errors. Part of the error is the result of multiplying the percentage of nitrogen by 6.25, a conversion factor derived from the assumption that animal foods

TABLE 2.2

Variation of nitrogen content in protein among foods, depending on their amino acid composition

Food Product	Protein-to-Nitrogen Ratio	Nucleic Acid Nitrogen (mg/gram of N)	Nonprotein Nitrogen (mg/gram of N)
Milk	6.02	4	27
Egg	5.73	1	4
Beef	5.72	4	15
Chicken	5.82	5	10
Fish	5.82	3	14
Wheat	5.75	21	12
Corn	5.72	21	17
Sorghum	5.93	21	8
Pea	5.40	15	35
Carrot	5.80	67	117
Beet	5.27	32	192
Potato	5.18	13	194
Lettuce	5.14	28	165
Cabbage	5.30	27	186
Tomato	6.26	14	174
Banana	5.32	70	125
Apple	5.72	96	57
Yeast	5.78	119	70
Mushroom	5.61	150	148

Note: The average conversion factor in this sample of foods is 5.68 ± 0.3. The nitrogen in nucleic acids varied from 1 to 15% and the nonprotein nitrogen ranged from 4 to 19.4%. Because conversion factors probably vary significantly between wild and cultivated species, and even among varieties of cultivated plants, readers must avoid using these conversion factors as representative of broad classes of food products (i.e., eggs, meat, fruit, etc.). Sosulski and Imafidon (1990) compiled the data tabulated here.

contain 16% protein (100/16 = 6.25). Table 2.2 shows examples of variation in the protein nitrogen content of foods consumed by *Homo sapiens*. Two observations spring from table 2.2: (1) Most of the conversion factors are lower than 6.25, and (2) the conversion factors vary by about 20% from 5.14 (in lettuce) to 6.26 (in tomatoes) with an average of 5.68. The assumption that all nitrogen in food is in proteins is also inaccurate. Food contains a variety of nonprotein nitrogenous compounds (including nucleic acids, amines, urea, ammonia, nitrate, and secondary compounds). Nucleic acids represent from 0.1% to 9.6% of total nitrogen (Imafidon and Sosulski 1990) and in some fruits up to 25% of all nitrogen can be in secondary compounds (Levey and Martínez del Rio 2001 and references there). The nitrogen content of biological materials is an important variable in itself. In our view, transforming nitrogen to crude protein adds no information to an analysis and may lead to misleading conclusions.

The estimation of lipid content by ether extraction is probably the least problematic of all the steps used in PA. This is, to a large extent, the consequence of the chemical definition of lipids: lipids are the materials that we can extract from food using a nonpolar solvent, like ether: "Lipids are defined as substances derived from living organisms that are insoluble in water but soluble in organic solvents such as chloroform, ether, or benzene" Kates (1972). Thus, perhaps the most important caveat about ether extractions pertains to their use in fecal samples. Ether does not recover soaps. This is problematic because soaps are the primary form in which undigested fatty acids are defecated. Thus, using ether extraction (or any other nonpolar solvent) can lead to overestimation of the efficiency with which lipids are assimilated. Finally, the estimation of NFE contains the combined errors of all other determinations. There is no good reason to use it.

At this point, you may wonder why we have spent so much time describing an obviously problem-ridden analytical approach. There are two reasons: The first one is that in spite of many years of criticism, PA is still widely used. The method is easy to use, relatively cheap, and many commercial laboratories perform it. If the manuscripts that reach our desk for review are a representative sample, ecologists will probably have to read, evaluate, and often recommend for rejection papers that rely on PA for years to come. The second reason is that knowing the moisture, dry matter, crude lipid, and nitrogen content of biological samples is valuable. As the next sections will demonstrate, several of the analyses that are performed in a PA are important and useful—if only as first steps in more detailed analyses. To escape the limitations of proximate analysis one must take a detailed look at the five classes of compounds that this method attempts to quantify: dietary fiber, carbohydrates, protein,

Figure 2.3. The composition of a plant cell according to detergent analysis and proximate analysis. In the Van Soest detergent analysis system the cell contents are soluble in neutral detergent and are called neutral detergent solubles (NDS), whereas the cell wall constituents are insoluble in neutral detergent and are called neutral detergent fiber (NDF). The cell wall constituents can be further divided according to whether they are soluble or insoluble in acid detergent, and are therefore called ADS and ADF, respectively. Notice that in the proximate analysis system crude fiber underestimates the amount of cell wall because it fails to measure many of the cell wall constituents. Pectin was originally considered part of the cell contents but according to Robbins (1993) is a major structural carbohydrate of the plant cell wall. Thus, if a plant sample has high pectin content, NDF may not be exactly synonymous with cell wall.

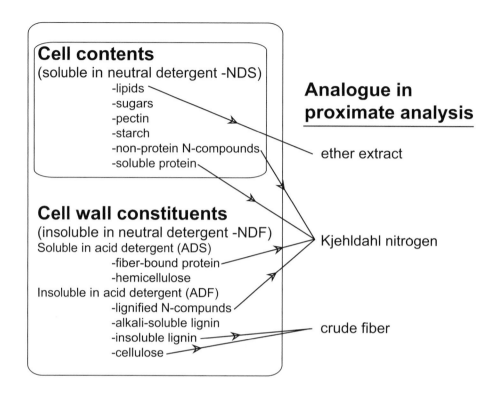

lipids, and minerals. Figure 2.3 illustrates the biological meaning of the results of proximate and detergent analysis when applied to a sample of plant material.

2.3 Dietary Fiber

The term dietary fiber is plagued by having two conflicting meanings. Some researchers adopt a strictly plant-anatomical definition of fiber: For them dietary fiber is all the material in food that is derived from plant cell walls (cellulose, hemicellulose, cutin-suberin, pectin, and lignin; box 2.3). Other researchers use fiber to mean the fraction of dry matter in food that is indigestible. For them, a definition of fiber must include some knowledge about the digestive capacities of animals. For example, some human nutritionists define fiber as "the group of substances that remain undigested until the ileum, but that are partly hydrolyzed by bacteria in the colon" (Trowell 1974). This inclusive definition would have to extend the chemical constituents of fiber to include chitin, bone, and some animal proteins that are difficult to digest by animals that ferment little and that have rapid passage rate (e.g., keratin in hair and feathers; Pritchard and Robbins 1990).

Box 2.3.

The term dietary fiber encompasses a heterogeneous group of compounds that make up plant cell walls. Plant cell walls are complex, and relatively poorly understood, structures (Carpita 1990). The ease with which herbivores assimilate plants depends not only on the chemical composition of cell walls, but also on the complicated physical and chemical interactions among their components. While reading about the properties of the most important chemical components of cell walls, you must keep in mind that these substances do not occur, and are not assimilated, in isolation in natural foods.

Lignin, cutin, and suberin are the main structural constituents of cell walls that prevent assimilation. Lignin is difficult to characterize chemically. The model of the structure of beech wood lignin shown in figure 2.4 is fairy hypothetical (Crawford 1981). Lignin is a high-molecular-weight, aromatic, nonsaccharide polymer that adds rigidity to cell walls. It is highly resistant to enzymatic and acid hydrolysis and even to breakdown by the gut symbionts of fermenting herbivores. The bonds that lignin can form with other cell wall components (such as cellulose and hemicellulose) can impede their fermentation.

Cutin-suberin, the main component of plant cuticles and bark, is a mix of aromatic and aliphatic polymers and waxes. The waxy fraction of cutin-suberin is assimilated only very slightly (Brown and Kolattukudy 1978). The cuticular surface of leaves offers a barrier to digestion and is the main refractory fraction of the nonlignified plant parts that animals eat (leaves, buds, and fruits). Like lignin, cutin-suberin blocks the breakdown of cellulose and hemicellulose.

Cellulose is the most abundant carbohydrate in terrestrial ecosystems. It amounts to 20–40% of the dry matter of all higher plants. Cellulose is a glucose polymer linked by β 1-4 bonds (B). It forms highly crystallized microfibrills among amorphous matrices of other cell wall components, which protect it from hydrolytic enzymes. Pure cellulose is fairly rare in nature (processed cotton fibers are an example). Most often, cellulose is combined to some degree with other cell wall components. Although it is often stated that multicellular animals do not secrete endogenous cellulases, this is not strictly true. The existence of endogenous cellulases from arthropods (insects), crustaceans (crayfish), and nematodes has been firmly established (Watanabe and Tokuda 2001). In spite of these reports, however, it is probably accurate to state that most cellulose is nutritionally unavailable to most animals that do not maintain fermentative symbioses with microbes and fungi (chapter 6). The nutritional availability of cellulose to fermenting herbivores depends strongly on the presence of lignin, cutin, and suberin. These substances reduce the availability of cellulose for fermentation (Van Soest 1994).

continued

continued

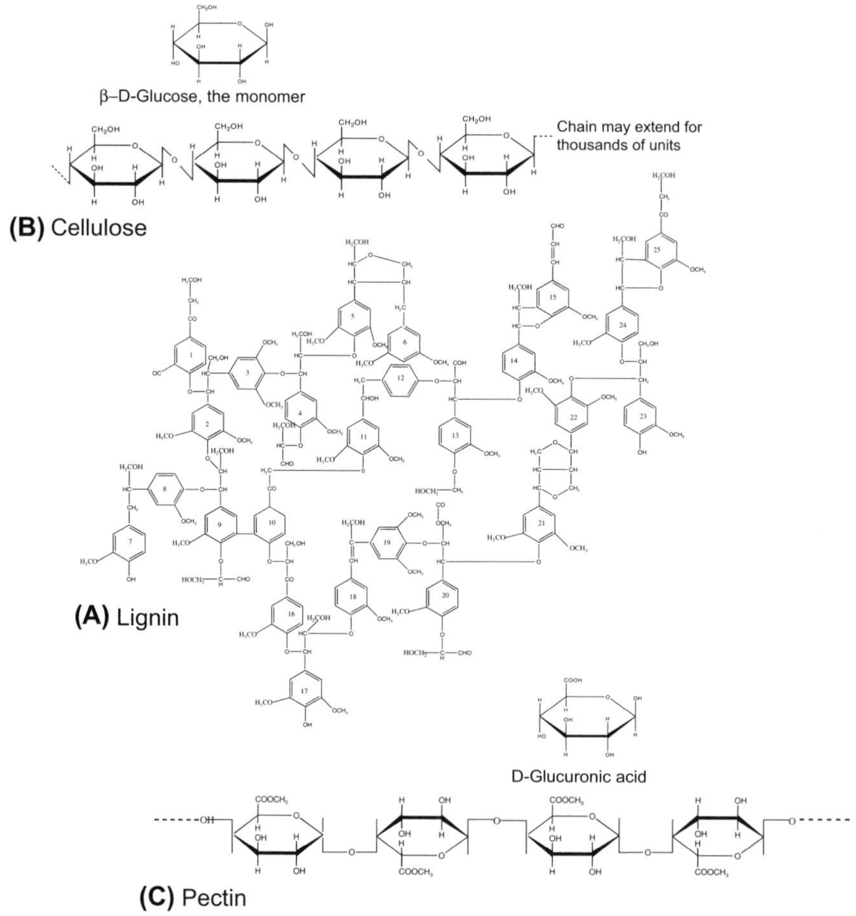

Figure 2.4. Chemical structures of three of the most abundant dietary fibers of plant origin: lignin, cellulose, and pectin.

β–D-Glucose, the monomer

Chain may extend for thousands of units

(B) Cellulose

(A) Lignin

D-Glucuronic acid

(C) Pectin

Hemicellulose is a complex and heterogeneous collection of polysaccharides that can be characterized as "xyloglucans." This means that they are a linear polymers of glucose linked by β 1-4 bonds, substituted regularly by xylose (a pentose or five-carbon sugar) and arabinose (another pentose). Hemicellulose is tightly linked with lignin, but is recovered by acid-detergent extraction from neutral detergent fiber. Because hemicellulose can be partially hydrolyzed by both acid and alkaline solutions, some of it may be digested in the acid environment of the stomach of some animals.

Pectin (C) is a water-soluble fiber that exists in the intercellular spaces (called the middle lamella) of plant tissues, where it probably helps to cement neighboring cells to each other. Pectin is a heterogeneous polysaccharide made of branched and unbranched chains. The main ingredient in pectin is galacturonic acid molecules joined by 1-4 bonds. The galacturonic units are

continued

continued

often estherified (an acid is combined with an alcohol) with methyl alcohol. In addition to galacturonic acid, pectins often have galactose and arabinose side chains. The diagram presented here (C) shows only the unbranched, linear polygalacturonic chains. Because pectins form gels with the addition of acid and sugar, they are used to make jam. The category "water-soluble" fiber also includes exudate and seaweed gums such as gum Arabic, agar, and carrageenan. These gums are heteropolysaccharides made of a variety of simple sugars including agarose, galactose, and mannose. They are readily fermentable, but can be refractory to assimilation by animals with little fermenting capacity. Figure 2.4 illustrates the chemical structure of lignin, cellulose, and pectin.

2.3.1 Detergent Analyses

As is usually the case with definitions, it is probably wise to use them operationally, taking care to articulate them as clearly as possible in each case. Here we will restrict our attention to the plant-anatomical definition of fiber. We will describe a procedure that defines fiber and its components in ways that are primarily useful to fermenting herbivores (chapter 6). This procedure can be called "detergent analysis." It was developed by Van Soest and classifies fiber into fractions that differ in their availability to fermenting microorganisms (box 2.4). Briefly, the method yields two fractions: neutral detergent fiber (NDF) and acid detergent fiber (ADF). NDF contains cellulose, hemicellulose, and lignin. ADF includes primarily cellulose, lignin, and, in the case of grasses, silica. The two fractions obtained in a detergent analysis can be extracted to obtain estimates of the amount of individual components (see Van Soest 1994 for a detailed description). Southgate (1976) proposes a very thorough alternative analytical approach to the determination of dietary fiber. We emphasize detergent analysis because it is faster and cheaper, and because more commercial laboratories perform it routinely.

Detergent analysis was developed to evaluate the quality of forage for fermenting herbivores. One of the cell wall components that detergent analyses fail to account for is pectin. Although pectin is nutritionally available for fermenting herbivores, many animals with poor fermenting ability do not assimilate it. Thus, a more inclusive analysis of fiber may be needed for nonfermenting animals. Asp et al. (1983) describe in detail a simple sequential analysis that allows determination of total fiber including soluble "fibers" such as pectin (see also Prosky et al. 1984).

Box 2.4. Van Soest's Detergent Analysis System

In the 1960s Peter Van Soest at Cornell University devised this method for dividing plant material into fractions containing either mainly cell contents (cytoplasm) or mainly cell wall consituents (Van Soest 1994). Detergents emulsify and bind the cell solubles, which are called neutral detergent solubles (NDS). The insoluble remnant is called a neutral detergent fiber (NDF; figure 2.3).

The basic steps are as follows (from (Robbins 1993)):

(1) Reflux a plant sample in neutral detergent. NDF = the insoluble remnant, and NDS = the mass of the original sample minus the mass of the insoluble remnant.

(2) Reflux the NDF in acid detergent. ADF = the acid insoluble remnant, and ADS = the mass of the original NDF sample mass minus the acid insoluble remnant.

The ADF remnant can be further partitioned:

(3) Treat the ADF with saturated potassium permanganate. Lignin = the mass of the original ADF minus the remnant.

(4) Treat the delignified remnant with 72% sulfuric acid. Cellulose = the mass of the original delignified remnant minus the residue. As an alternative, skip step 3 and treat the ADF with 72% sulfuric acid and the residue = lignin + cutin whereas the cellulose = mass of the original ADF minus the remnant.

There are some modifications in the method that depend on the characteristics of the plant sample (Robbins 1993). If the sample has high starch content, it may be pretreated with amylase before step 1. If the sample has a high tannin content, sodium sulfite is added to the neutral detergent solution in step 1, but is otherwise excluded. As originally devised, the method relied on reflux chemistry, but comparable detection and accuracy is achieved with the Ankom Technology method (http://www.ankom.com).

2.3.2 Refractory Materials

In the previous section, we defined fiber operationally to include only the components of plant cell walls. We did this because some of these components are the main determinants of the efficiency of plant assimilation by herbivores (chapter 6). Fiber content is one of the many criteria that we can use to assess the nutritional quality of food. For fermenting herbivores, high lignin content (partially estimated by ADF) is a good indicator of poor food quality. Digestive ecologists study a variety of animals, and dietary fiber represents only one of food's ingredients that some of these animals assimilate poorly.

The term "refractory materials" is often applied to food components that animals are unable to digest and hence are regurgitated or pass through the digestive tract intact. Whether a material is refractory to assimilation or not is the result of the interaction between the material's properties and the consumer's digestive traits. Thus, constructing a single definition of the term refractory is not possible. Digestive ecologists devote much of their time to finding out which food components are refractory to digestion. Having done this, they attempt to determine the physiological traits that allow animals to assimilate some compounds but not others, and to explore the ecological correlates of differences in digestive ability. In the following sections and in chapters 3, 4, and 6 we will provide many examples of this process.

2.4 Carbohydrates

2.4.1 Mono-, Di-, and Polysaccharides

Carbohydrates consist of carbon, oxygen, and hydrogen. The carbohydrates that are important in nutrition include monosaccharides, disaccharides, and polysaccharides. Monosaccharides are the simplest of sugars that cannot be broken down any further by hydrolysis. The most common in foods are the six carbon sugars (or hexoses) glucose, fructose, and galactose. You can see in figure 2.5(A) that all have a "D" prefix. This refers to the configuration of the hydroxyl group of the last carbon before the end (carbon 5 is a "chiral" carbon). If the sugar has a D configuration, the OH appears on the right. In the L configuration, the OH appears on the left. The monosaccharides glucose and fructose are found in nature as constituents of nectar and fruit pulp. Glucose is one of the major energy sources for animal metabolism, and galactose is found as part of the disaccharide lactose.

Disaccharides can be composed of two units of the same monosaccharide or of two different monosaccharides. Three of the major disaccharides in food are sucrose (table sugar), lactose (milk sugar), and trehalose (insect sugar; figure 2.5(B)). Because sucrose is used as the transport sugar by plants, it is very common in plant vegetative tissues (leaves, stems, and roots) and hence in a

Figure 2.5. (A) Glucose, fructose, and galactose are shown in both open chain and ring formulae. For glucose, we have also included two alternative anomeric forms. The position of the OH group in carbon 1 can be asymmetric (i.e., carbon 1 is anomeric). Hence, glucose can have two isomers: α-D-glucopyranose and β-D-glucopyranose. Monosaccharides can form either α or β bonds when they link with each other. As the differences in the biological properties of amylose and starch attest, this seemingly minor difference in structure has profound consequences. (B) The two sugars that make up a disaccharide can be different, as is the case with sucrose and galactose, or can be the same, as is the case with trehalose. The carbons involved in the linkage can differ as well. In trehalose, the two glucoses are linked by a 1-1 link; in sucrose, glucose and fructose are linked by a 1-2 links; and in galactose, glucose and are linked by a 1-4 link. Finally, the anomeric configuration of the hydroxyl group in each component monosaccharide can differ as well. (C) Amylose is a linear chain starch containing primarily α 1-4 glycosidic (glucose-glucose) bonds. The orientation of successive glucose residues favors a helicoidal structure. In amylopectin, the chain is branched every approximately 25 monomers by a α 1-6 linkage. Glycogen is even more branched. Amylose and amylopectin are the most

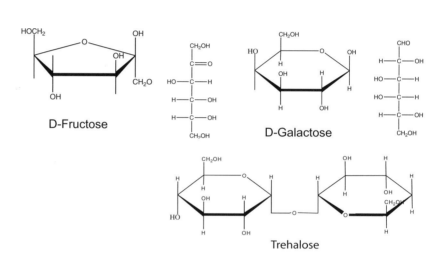

(A)

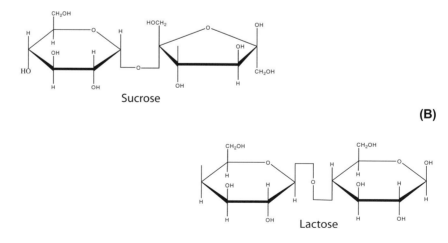

(B)

(figure continues)

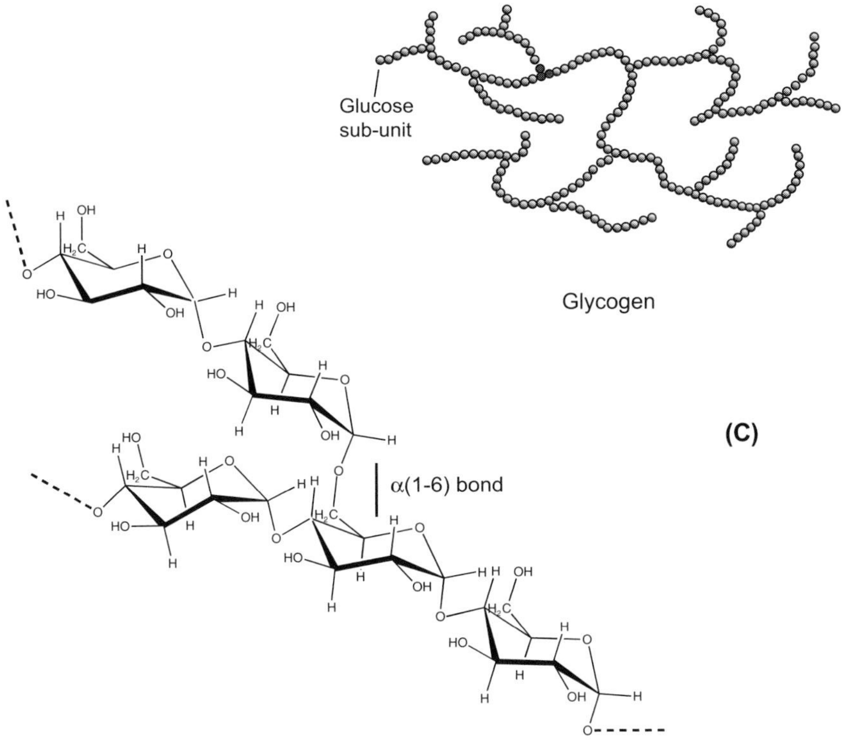

Glucose
sub-unit

Glycogen

(C)

α(1-6) bond

important storage carbohydrates in plants and are found in seeds, roots, and tubers. Glycogen is the primary storage carbohydrate in animals. In vertebrate tissues, glycogen is primarily stored in the liver, but small amounts are found in muscle. Sessile invertebrates (like clams, mussles, and oysters) can have very high glycogen contents.

variety of animal foods. It is also found in fruit pulp and as a constituent of floral nectar. Cane sugar is almost pure sucrose. Whenever a processed food's ingredient list refers to "sugar," it means sucrose. Trehalose is the major circulating sugar in insect hemolymph (Roser and Colaço 1993) and is present in large amounts in mushrooms (Kazuyuki et al. 1998). Lactose is made of glucose and galactose, and is a major sugar in the milk of most mammals. Curiously, the milk of some marsupials, including possums, and macropodid wallabies and kangaroos contains oligosaccharides of galactose and glucose (Messer et al. 1989). The term oligosaccharide includes the disaccharides and refers to carbohydrates with 2–10 monosaccharide residues.

Polysaccharides are polymers of simple monosaccharides. They are called homopolysaccharides or heterosaccharides depending on whether they contain only one or two or more, different monosaccharides in their long chains, respectively. The most important polysaccharides in food are the starches (figure 2.5(C)). Starches are storage carbohydrates: amylose and amylopectin are found in plants and glycogen is found in animals. Animals with well-developed fermentation systems assimilate starches rapidly and efficiently. In animals with low fermenting capacity, however, the efficiency with which starches are assimilated can vary significantly depending on the starch's properties and the animal's digestive abilities. Nutritionists recognize two types of naturally occurring hydrolysis-resistant starches (Snow and O'Dea 1981): (1) The starch in undisrupted cell walls of partially broken grains and seeds can be physically inaccessible to digestive enzymes. (2) in plants, starch is closely packed into granules arranged in a partially crystalline form. The crystalline pattern of starch in the granule, as determined by x-ray diffraction, seems to determine the susceptibility of starch to digestion (Englyst et al. 1992).

2.4.1 Chitin

Because chitin contains nitrogen, it is not strictly a polysaccharide. However, its structure is so similar to that of true polysaccharides (figure 2.6) that it is appropriate to discuss it with carbohydrates. Chitin is very widely distributed among all of life's kingdoms. It is found in the cell walls of many fungi, where it can sometimes play the role that cellulose plays in terrestrial plants. The most important role of chitin, however, is as the primary structural material in the exoskeleton (cuticle) of arthropods. Because many animals eat arthropods, chitin is an important food ingredient to consider. Chitin is abundant wherever arthropods are

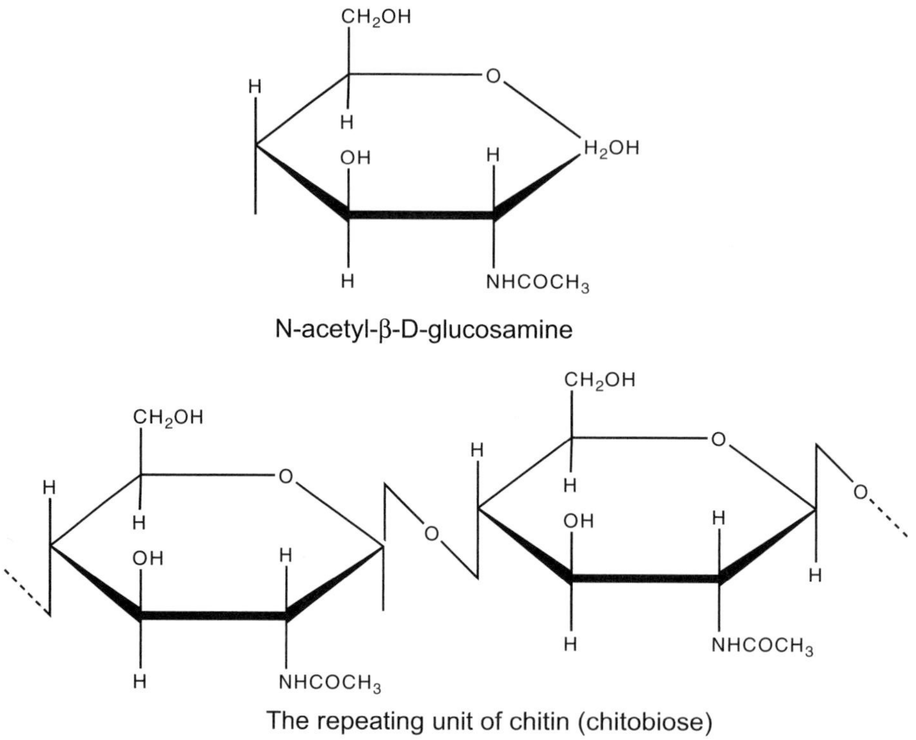

N-acetyl-β-D-glucosamine

The repeating unit of chitin (chitobiose)

Figure 2.6. Chitin is a homopolymer of N-acetyl-β-D-glucosamine, which is an amino sugar. Chitin has a structure that is very similar to that of cellulose, except that the hydroxyl on carbon 2 of each residue is replaced by an acetylated amino group. Like cellulose, chitin in food can be measured using a neutral detergent extraction (Weiser et al. 1997). However, it is probably wise to validate this technique with other more precise methods (Hachman and Goldberg 1981). For animals with calcified exoskeletons, like many crustaceans, the chitin extraction must be preceded by decalcification with NaOH.

abundant. In marine ecosystems, for example, estimates of krill biomass range from 80 to 500 million metric tons containing 1.6 to 10 million tons of chitin (Clarke 1980). In insects, chitin combines with proteins and pigments to constitute a cuticular exoskeleton that can vary enormously in strength, hardness, and resistance to digestion. In crustaceans, chitin often forms a matrix in which calcium carbonate mineralizes to provide added structural hardness.

As would be expected from a substance with a structure that resembles that of cellulose, chitin can be quite refractory to assimilation. Although many animals have endogenous enzymes that hydrolyze chitin (Jeauniaux and Cornellius 1978; Marsh et al. 2001), the efficiency with which animals assimilate this abundant substance varies significantly. On land, many animals have garnered the help of microorganisms to digest cellulose (chapter 8). In the ocean, a similar symbiotic relationship seems to have been established. Several whale species seem to rely on a symbiotic association with microorganisms to assimilate chitin (chapter 6; Olsen et al. 2001).

Herbst (1988) summarizes the large variety of methods available to analyze the "available" carbohydrates in food (available is often used as a synonym for "nonrefractory"). Briefly, for foods containing primarily mono- and disaccharides, like nectar and the juice of some fruits, field biologists use a hand-held refractometer to estimate the total concentration of simple sugars. Kearns and Inouye (1993) and White and Stiles (1985) describe the power and limitations of these handy instruments. A variety of simple spectrophotometric methods can be used to measure carbohydrate content. Most of these methods require measuring monosaccharides first, and then hydrolyzing the di- and polysaccharides with enzymes or acid (Herbst 1988, and references there). We often use high-pressure liquid chromatography (HPLC) to analyze simple sugars in biological samples. This method is of sufficient importance to physiological ecologists to merit a box describing it (box. 2.5).

Box 2.5. High-Performance Liquid Chromatography

HPLC is one mode of chromatography, which is probably the most widely used analytical technique. Chromatographic processes can be defined as separation techniques involving mass transfer between stationary and mobile phases. HPLC utilizes a liquid mobile phase to separate the components of a mixture. These components (or analytes) are first dissolved in

continued

continued

a solvent, and then forced to flow through a chromatographic column under high pressure. In the column, the mixture is separated by a variety of processes into its components. The amount of resolution is dependent upon the extent of interaction between the solute components and the stationary phase. The stationary phase is defined as the immobile packing material in the column. The interaction of the solute with mobile and stationary phases can be manipulated through different choices of solvents and stationary phases.

A more or less typical HPLC system has the following components: (1) one or several reservoirs containing the solvents, (2) one or more high-pressure pumps that deliver these solvents to a mixer, which is often the place where the sample is injected (we will presently describe why sometimes two reservoirs are needed), (3) a chromatography column often packed with microparticulate silica, (4) a recorder that measures the properties of the material flowing through it, (5) a recorder, and (6) a fraction collector (figure 2.7). The mixture of compounds in the sample is separated in the column. The most widely used columns separate materials as bonded phases or by exclusion. In a bonded phase the highly polar surface of silica is altered by the chemical attachment of the functional groups of the compounds in the mixture. The range of functional groups that can be bonded to silica is really wide. It is sometimes useful to have more than one reservoir of solvent

Figure 2.7. Diagram of a more or less typical high-pressure liquid chromatography system.

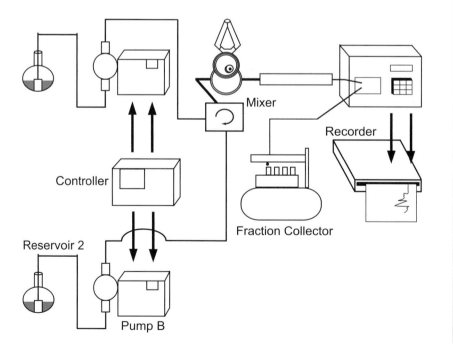

continued

continued

to be able to modulate the electrochemical characteristics of the mobile phase (pH, ionic strength, and so on) through time to achieve better separation. Exclusion chromatography separates molecules based on their effective size and shape in solution. Exclusion chromatography is also called gel permeation if used with organic solvents or gel filtration if used with aqueous solvents. Depending on the size and shape of the solute molecules they may or may not be able to enter into the pores of the stationary phase. Larger molecules are excluded from the narrower parts. In general, the smaller molecules are retained longer in the column. The function of detectors is to monitor the mobile phase emerging from the column. The output is an electrical signal that is proportional to some property of the mobile phase and/or the solutes. Some commonly used detectors sense the refractive index of the mobile phase and the absorption of UV/visible light of the compounds in the mobile phase. Other detectors measure fluorescence and conductivity. The output of the recorder is amplified and either displayed on a chart recorder, or, these days, stored on a computer's hard disk. Ideally the result is several peaks, each of which corresponds to an individual compound. These peaks can then be integrated to obtain the amounts and the components can be collected separately by a fraction collector. HPLC systems have a high degree of versatility not found in other chromatographic systems. HPLC can be used to separate a wide variety of chemical mixtures including carbohydrates, amino acids, lipids, and secondary metabolites. There are many guides to the use of HPLC. We recommend two: McMaster (1994) and Lindsay (1987).

2.5 Amino Acids and Proteins

Proteins are the primary constituents of animal tissues and occur at lower, albeit nutritionally important, levels in plant tissues. They are important functional and structural components of all organs, and are active as enzymes, hormones, and membrane ingredients. Proteins encompass a heterogeneous group of compounds with one unifying theme: proteins are composed of L α-amino acids (figure 2.8). Typically, proteins are chains of 20–25 types of amino acids linked by peptide bonds (figure 2.8). Amino acids can be classified depending on the polarity and charge of their side chain R. Figure 2.9 shows a commonly accepted taxonomy of amino acids.

Amino acids form peptides through an amide covalent bond between the α-amino and the α-carboxyl groups. Dipeptides are amino acid chains that

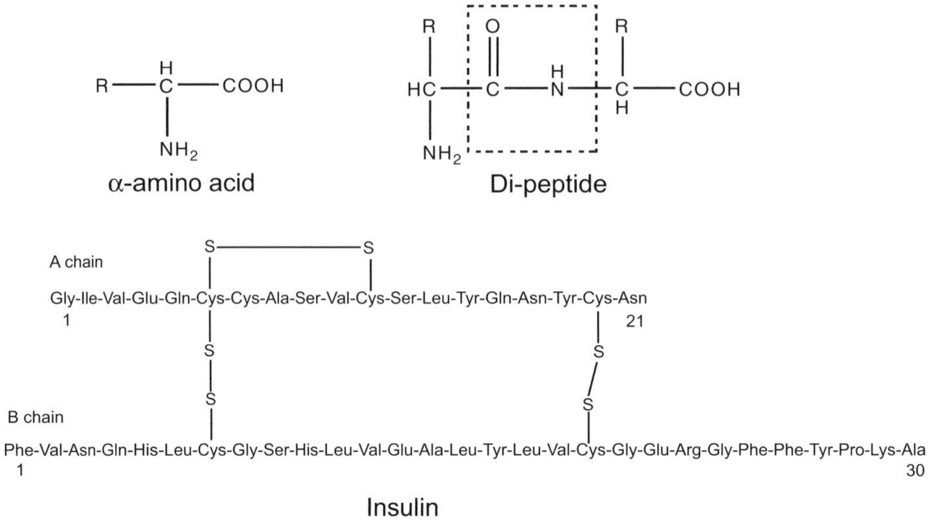

Figure 2.8. The classical α amino acids (listed in table 2.5) are those that are incorporated into proteins. An oligomer consisting of two amino acids is called a dipeptide. The amino acids are bound to each other by a peptide bond (enclosed by a box in the figure), which involves a keto group and an amino group. Long chains of amino acids are called polypetides or proteins. The figure shows the amino acid sequence of cow insulin. Insulin is a very small protein that consists of two polypeptide chains. The S-S bonds indicate disulfide bonds between cystein residues.

α-amino acid

Di-peptide

A chain

Gly-Ile-Val-Glu-Gln-Cys-Cys-Ala-Ser-Val-Cys-Ser-Leu-Tyr-Gln-Asn-Tyr-Cys-Asn
1   21

B chain

Phe-Val-Asn-Gln-His-Leu-Cys-Gly-Ser-His-Leu-Val-Glu-Ala-Leu-Tyr-Leu-Val-Cys-Gly-Glu-Arg-Gly-Phe-Phe-Tyr-Pro-Lys-Ala
1   30

Insulin

contain two amino acid residues, oligopeptides contain a few amino acids (like tetrapeptides), and very long chains are called polypeptides. Most peptides retain an unreacted amino group at one end (called the N-terminus) and an unreacted carboxy end at the other (called the C-terminus). The linear sequence of amino acids in a peptide is called its primary structure. Many proteins are made of two or more interacting polypeptide chains, and most have complex three-dimensional structures that determine their biological functions. Because food proteins are degraded during digestion, and transported into and through blood mainly as amino acids or di- and tripeptides (chapter 4), from a nutritional perspective the specific three-dimensional structure of a protein is not as important as its amino acid composition relative to the requirements of the consuming animal.

2.5.1 Dispensable and Indispensable Amino Acids

For over 60 years, amino acids have been conventionally divided into two categories: indispensable (or essential) and dispensable (or nonessential; table 2.3). The dispensable-indispensable dichotomy recognizes that some amino acids can be synthesized in adequate amounts by the animal and others must be ingested to meet the animal's requirements. Taking a peek at the original definition of indispensable amino acids is instructive because it allows a more nuanced classification of amino acids. In 1946, Borman and collaborators defined an indispensable amino acid as "one which cannot be synthesized by

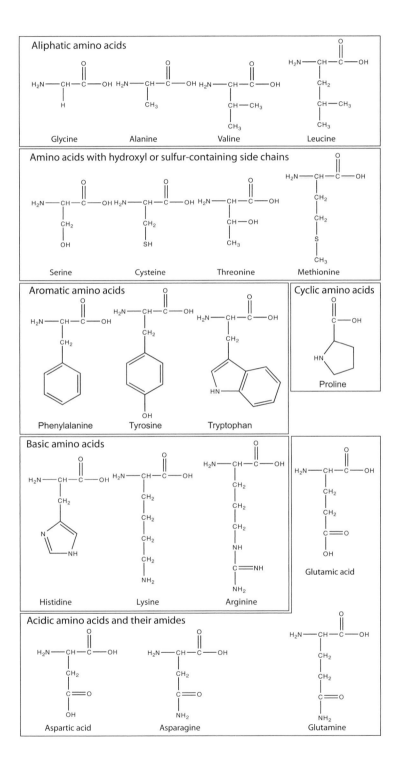

Figure 2.9. Amino acids can be classified along several axes depending on the characteristics of their side chains. These features include hydrophilic or hydrophobic character, and the presence of ionizable groups. We have chosen Matthews and van Holde's (1990). scheme to classify amino acids. This classification divides amino acids based on the properties of their side chains into six groups: amino acids with aliphatic side chains (Gly, Ala, Val, and Leu); amino acids with hydroxyl- or sulfur-containing side chains (Ser, Cys, Thr, and Meth); aromatic amino acids (Phe, Tyr, and Trp); a single cyclic amino acid (Pro); basic amino acids (His, Lys, and Arg); and acidic amino acids and their amides (App, Glu, Asn, and Gln). We will refer to this table several times in the book.

TABLE 2.3

Division of dietary amino acids into three groups: Nutritionally dispensable amino acids, amino acids that can become conditionally indispensable, and 3) amino acids that are indispensable under all conditions

Dispensable	Conditionally Indispensable	Indispensable	Structural feature
aspartic acid	cystine	histidine	imidazole ring
asparagine	tyrosine	isoleucine	branched aliphatic side chain
glutamic acid	taurine	leucine	branched aliphatic side chain
alanine	glycine	lycine	primary amine
serine	arginine	methionine	secondary thiol
	glutamine	phenylalanine	aromatic ring
	proline	threonine	secondary alcohol
		thryptophan	indole ring
		valine	branched aliphatic chain

Note: The table also lists the structural features that probably prevent some amino acids from being synthesized by animals (and thus render them indispensable). The structural feature corresponds only to the indispensable amino acids (after Reeds et al. 2000).

the organism out of materials *ordinarily available* to the cells *at a speed* commensurate with the demands for *normal growth*" (italics from Borman et al. 1946). The original definition seems to emphasize that the definition is relative and that there may be varying degrees of essentiality.

Several amino acids are indispensable because animals lack the enzymes that catalyze the synthesis of a specific, usually complex, structure in them (table 2.3). Plants and many bacteria have the ability to synthesize these "truly essential" amino acids from ammonia and nitrate, but insects and vertebrates lack it. In addition to these nutritionally indispensable amino acids, there is a heterogeneous group of amino acids for which, under some conditions and in some species, there are limits to the rate at which animals can synthesize them. When this limit is attained, the amino acid in question becomes an essential ingredient of the diet. These amino acids are called conditionally indispensable (table 2.3; Reeds 2000).

What are the conditions that may make synthesis rate limiting in conditionally indispensable amino acids? (1) The synthesis of some amino acids may require another amino acid either as a carbon donor, or as a donor of an accessory group such as sulfur. The synthesis of some amino acids is conditional on the availability of their precursors. (2) Some amino acids may be synthesized in only a limited number of tissues. Proline and arginine, for example, seem to be synthesized in the intestine, and primarily from dietary precursors. In rats with surgically reduced intestines, arginine becomes an essential nutrient (Reeds 2000). Some migratory birds appear to greatly reduce the size of their intestines during migration, and many fasting animals reduce their intestinal mass during prolonged fasting (chapter 4; Karasov and Pinshow 1998; Ferraris and Carey 2000). It is reasonable to ask if arginine becomes essential in recovering migrants and fasting animals. Arginine appears to be an indispensable amino acid for some strictly carnivorous mammals. Cats fed on arginine-free diets can die after just a few hours as a result of hyperammonemia (excessive ammonia in their blood). Figure 2.10 describes why arginine is an indispensable amino acid for strict carnivores, such as cats (MacDonald et al. 1984). To our knowledge, the essentiality of arginine has only been demonstrated in three species of strict carnivores: domestic cats, ferrets, and mink (Morris 1985). Strict carnivory has evolved repeatedly in mammals (for example in cetaceans and bats). It would be interesting to investigate whether the arginine deficiency syndrome described for cats has evolved in these groups as well.

The need for dietary arginine in carnivorous mammals illustrates what we believe is a relatively unrecognized pattern: The nine amino acids listed as indispensable in table 2.3 appear to be indispensable for many animals, perhaps for all metazoans. The list of amino acids that we have classified as conditionally indispensable includes those that can become indispensable depending on the physiological state of an individual or on the physiological characteristics of a species. Thus, we have a subset of dietary amino acids that are generally indispensable. The loss of the ability to synthesize them probably represents an ancient event in the history of animals. In contrast, we have other amino acids that have become indispensable for some species but not for others depending on the interaction between diet and physiology. They are conditionally indispensable in both a physiological and a phylogenetic sense. For example, the arginine requirements of carnivorous fish are more similar to those of omnivorous mammals than to those of strict carnivores (Kyu 1997). The reason, of course, is that with a few exceptions fish excrete nitrogen from catabolized protein primarily as ammonia rather than as urea. Consequently, they do not need arginine to fuel the urea cycle. In chapters 5 and 6 we will describe how some animals have circumvented the problem of acquiring indispensable amino acids even when feeding on diets deficient in them.

2.5.2 Assessing Protein Quality

The term "protein quality" refers to the ability of a dietary protein to provide the needs of an animal. Foods can vary in both protein quantity and quality.

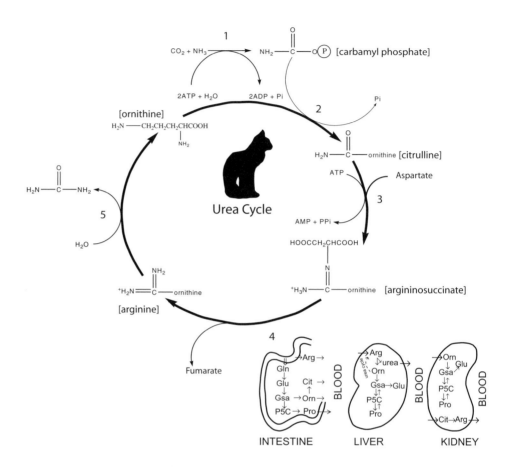

Figure 2.10. Why do cats need dietary arginine? In order to maintain constant blood glucose levels in the face of low carbohydrate intake, strict carnivores maintain extremely high constitutive levels of amino acid catabolic enzymes in their liver (Baker et al. 1991). The term "constitutive" means that the expression of these enzymes is not influenced by diet. These enzymes produce ammonia. Because ammonia is toxic and crosses cell membranes freely, mammals transform it into urea before excreting it. The transformation of ammonia into urea requires bonding ammonia with ornithine. As the diagram shows, arginine is a precursor of ornithine. In addition from its regeneration in the urea cycle, arginine (Arg) is synthesized in the kidney from citruline (Cit). Strict carnivores seem to miss some of the enzymes involved in the synthesis of citrulline from glutamine (Gln) and glutamic acid (Glu) in the small intestine. The missing enzymes are labeled with an asterisk in the diagram (after MacDonald and Rogers 1984; Pro is proline and P5C is pyrroline-5-carboxylate). In summary, arginine is an indispensable amino acid for strictly carnivorous mammals because these animals have exceedingly high protein requirements, accompanied by high levels of ammonia production resulting from high levels of amino acid catabolism, and reduced levels of enzymes that mediate the synthesis of arginine precursors.

Some foods may be rich sources of protein that has little value. Gelatin is a good example. Commercially available gelatin is almost pure protein. However, it lacks the indispensable amino acid tryptophan and hence has very low quality for humans. As the previous section illustrates, you should expect protein quality to be the result of the interaction between an animal's digestive and physiological traits (and those of its symbionts!) and the chemical composition of food. One of the indices of protein quality, the "chemical score," is based on a food's amino acid composition. Although this index was designed for humans, laboratory rodents, and poultry, it is probably useful to evaluate the food of animals with limited fermenting capacity and without nutritional symbioses. Of course, because our knowledge of the nutritional physiology of wild animals is so rudimentary, this statement should be considered more a hypothesis than an established fact.

Nutritionists who study humans and rats consider whole-egg protein the nutritionally complete "ideal" protein, and therefore it is used as a reference in the estimation of chemical scores. The protein to be assessed must be extracted and purified from food, hydrolyzed into its constituent amino acids, and then analyzed with an amino acid analyzer. The values for each of the indispensable amino acids for albumin and the test protein are then listed as shown in table 2.4 and compared. The amino acid that is at the lowest level in the test protein relative to egg protein is called the limiting amino acid. The percent content in diet protein of the limiting amino acid relative to that of egg protein is the chemical score. A chemical score is typically reexpressed as a "biological value," which is simply an index correlated with the chemical score that ranges from 0 to 1.0 (egg protein represents 1.0). Dietary protein often deviates from an ideal balance. Therefore, to ingest sufficient amounts of the most limiting amino acids, animals must ingest an excess amount of the nonlimiting ones, which are deaminated and catabolized. Therefore, the difference between the indispensable amino acid composition of the diet and that required by the animal dictates total protein needs. The larger the imbalance between the amino acids in diet and those required, i.e., the lower the chemical score or biological value, the higher will be the intake of protein needed to meet requirements.

2.6 Lipids

Fats get a bad rap from pop nutritionists who consider anything fatty superfluous and ugly (consider any issue of *Vogue* or *Cosmopolitan*). What is perhaps worse is that biologists and chemists have traditionally ignored lipids.

TABLE 2.4
Chemical score of the diet protein

Amino Acid	Oat Protein	Egg Protein	% Relative to Egg
	(grams of amino acid per 16 g of protein nitrogen)		
Lysine	**3.5**	**6.9**	**51**
Histidine	1.9	2.6	73
Threonine	4.7	4.9	96
Methionine	2.2	3.2	69
Cysteine	1.9	2.1	91
Valine	4.7	7.1	66
Leucine	5.8	8.8	66
Isoleucine	3.3	5.9	56
Phenylalanine	4.2	5.5	76
Tyrosine	2.6	3.8	68

Note: To make this calculation nutritionists compare the percentage of their indispensable amino acids with those of egg protein. The chemical score of the protein is the percentage value of the most limiting (least abundant) amino acid. In the case of oat (*Avena sativa*) protein, the most limiting amino acid (in bold) is lysine. Oat protein contains only 51% of the relative lysine content in egg albumin, and hence its chemical score is 51%.

Although we have understood the diversity of proteins and carbohydrates for over 100 years, and although fats are as central to life as are proteins and nucleic acids, it is only in the last 30 years that the intricacy and vital role of lipids have become clear. Lipids play a pivotal role in physiological ecology that demolishes the misperception of these substances as chemically boring stuff that organisms use only for fuel and to build waterproof barriers. We have devoted a significant amount of attention to lipids in this chapter because we believe that these ugly ducklings of biochemistry are rapidly transforming themselves into swans.

The term lipid refers to a tremendously heterogeneous group of organic compounds associated with living creatures that have the property of being insoluble in water, but soluble in nonpolar solvents such as hydrocarbons and alcohols. Lipids include fats and oils, waxes, phospholipids, sphingolipids, and sterols. All of these substances have important roles in food biochemistry and nutrition. Fats and oils are good energy storage molecules because they are energy rich, and because they are hydrophobic they can be stored "neat" (i.e., without large amounts of associated water). Lipids contain twice the amount of energy of carbohydrates and proteins (box 1.1) and can be stored with much less

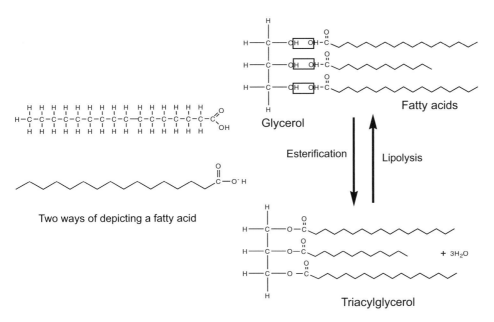

Two ways of depicting a fatty acid

Glycerol

Fatty acids

Esterification

Lipolysis

+ 3H₂O

Triacylglycerol

Figure 2.11. Fatty acids are perhaps the most important components of lipids. A fatty acid has a carboxylate head and a hydrocarbon chain. The hydrocarbon chain is often depicted as a zigzag line. A triacylglycerol is made of a glycerol backbone to which three fatty acids are attached by ester bonds. The physical characteristics of a triacyl-glycerol are determined by those three fatty acids.

water than, for example, glycogen, which is another energy storage molecule. With the exceptions of some sessile and parasitic species, animals store energy primarily as fat. The triacylglycerols that make up storage fats and oils are esters of three fatty acids and a glycerol molecule (figure 2.11). In the past, the term triglyceride was used as a synonym for triacylglycerol. Organic chemists discourage the use of this old term. Do not use it if you do not want to be considered a fat ignoramus. Other lipid components in food include phopsholipids, sphingolipids, waxes, and sterols (figure 2.12). The ingredients of phospholipids are glycerol, fatty acids, phosphoric acid, and, often, an amino alcohol.

Phospholipids share with some other lipids a form of molecular schizophrenia. They have a polar hydrophilic "head" and a nonpolar hydrophobic "tail" made of hydrocarbons. The technical term for substances with this particular characteristic is "amphipathic." From a biological perspective, the most important consequence of the phospholipids' molecular schizophrenia is that they spontaneously form micelles and membrane bilayers. Sphingomyelin is a more or less typical sphingolipid that is an important component of nervous tissue. In sphingolipids, there is a long-chain, nitrogen-containing alcohol instead of glycerol. Waxes are esters of long-chain (24–36 carbons long) fatty acids and monohydric alcohols. Both plants and animals use waxes to waterproof surfaces and to prevent evaporation (Hadley 1984; Gibbs 1998). Waxes are widespread and abundant as storage molecules in marine organisms, especially in deep-sea swimming forms like copepods, squids, chaetognaths, and, notably, the coelacanth

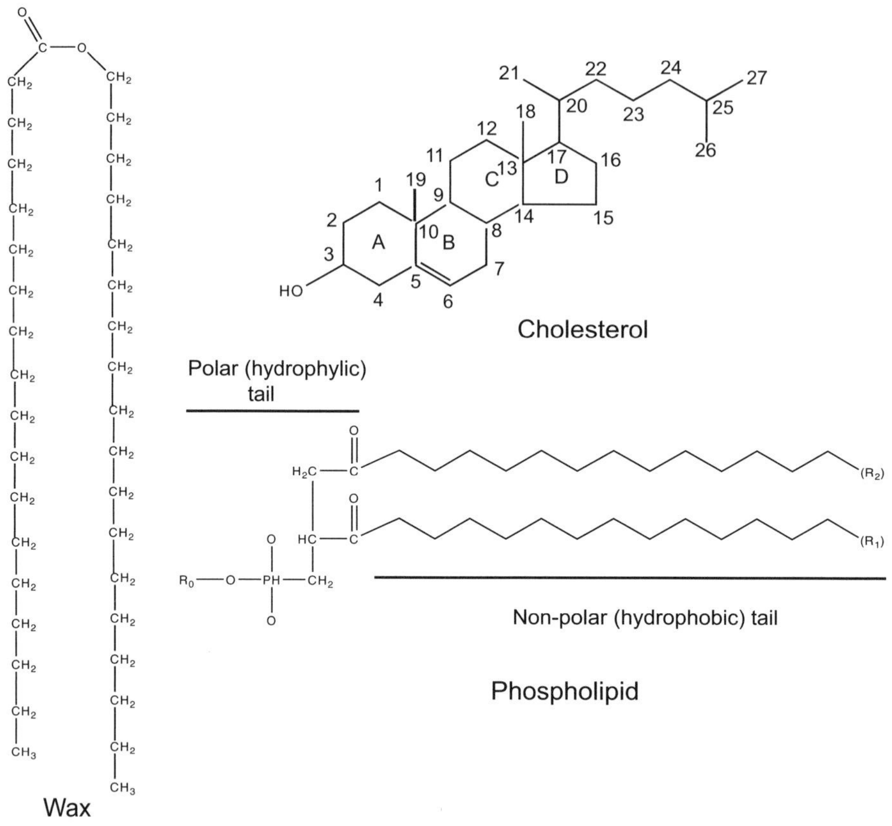

Figure 2.12. In natural waxes a fatty acid (in the diagram, oleic acid) is sterified to a long-chain alcohol. The small "head" group contributes little to hydrophilicity, as compared to the hydrophilic contribution of the two long hydrocarbon tails. In contrast, the phosphate-containing head of phospholipids (or more accurately glycerophospholipids) makes these compounds the quintessential "amphiphathic" compound, and hence an essential component of biological membranes. The structure of cholesterol bears little superficial resemblance to other lipids. The hydroxyl group at one end of the molecule makes cholesterol a weakly amphipathic molecule. The rest of the molecule is hydrophobic and readily soluble in the interior of cell membranes. The rigid and bulky structure of cholesterol provides rigidity to fluid membranes.

(Pond 1981). The jojoba (*Simmondsia chinensis*), a Sonoran desert shrub, is the only angiosperm known that stores energy in its seeds as waxes (Baker 1970). Sterols have profound physiological importance. Cholesterol is a constituent of cell membranes and of the bile salts which emulsify fats prior to digestion (chapter 4). Cholesterol is also the precursor of many extremely important steroid hormones. In this chapter we will emphasize the importance of fatty acids as determinant factors in the properties of biological membranes. Cholesterol is an essential component of biological membranes as well. It plays an important, albeit comparatively less studied, role in shaping their characteristics (Crockett 1998).

2.6.1 Fatty Acids

As we have seen, fatty acids are central constituents of a variety of lipids. The physical characteristics of these lipids are, to a large extent, a consequence of the

Fatty acid	Abbreviation	Melting point
Stearic acid	18:0	70° C
Oleic acid	$C_{18}1: n9$	16° C
Linoleic acid	$C_{18}2: n6$	-5° C

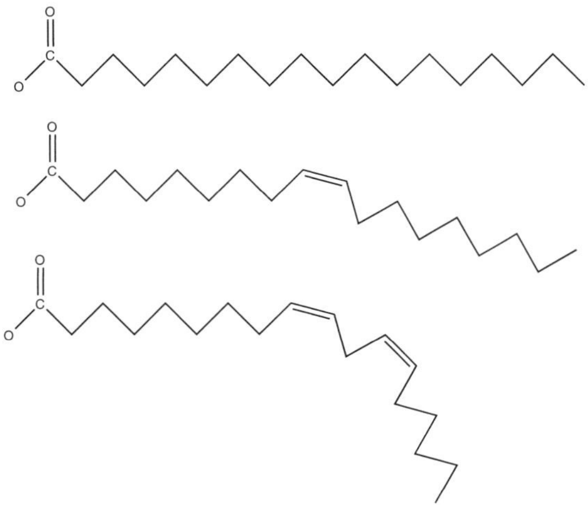

Figure 2.13. Introducing a double bond with cis geometric configuration, results in a change of approximately 30° from the linearity of a saturated chain. Because double bonds are non rotating, they restrict the movement of acyl chains. Theoretically, an acyl chain with six *cis* double bonds could be almost circular.

properties of fatty acids. Hence, to understand lipids, we must take a detailed look at fatty acids. Fatty acids have a hydrophilic carboxylate group that is attached to one end of a hydrocarbon chain of variable length. The length and the number of double bonds in the hydrocarbon chain determine the melting point of the fatty acid, and hence whether the triacylglycerol that they are part of will be a solid at room temperature (a "fat") or a liquid (an "oil"; figure 2.13). In general, the melting point increases with the number of carbons in a fatty acid and decreases with the number of double bonds (table 2.5). Fatty acids can be *saturated* if all the carbons of the tail are saturated with hydrogen atoms, or *unsaturated* if they contain one or more double bonds. Depending on the number of double bonds, fatty acids can be monounsaturated (MUFAs, one double bond), or polyunsaturated (PUFAs, two or more double bonds). In the aquaculture literature you will also find the term "highly unsaturated fatty acids" (HUFAs). The HUFAs are a subset of PUFAs that have more than two double bonds and 20 or more carbon atoms. Fatty acids play a very important role in ecological physiology, and therefore it is worth your while to become familiar with the confusing nomenclature used to name them. Box 2.6 describes the nomenclature of fatty acids.

If you compare the fatty acids in table 2.5 with your everyday experience with food lipids, an interesting correlation should be apparent. Lipids rich in unsaturated fatty acids like commercial vegetable oils are liquid at room temperature. In contrast, those with a higher content of saturated fatty acids like butter and lard tend to be solid, especially if the hydrocarbon chains are long. The reason is that long saturated chains can pack close together and form

TABLE 2.5
Nomenclature for fatty acids

Common Name	Systematic Name	Abbreviation	Structure	Melting Point
Saturated fatty acids				
Capric	n-Decanoic	10:0	$CH_3(CH_2)_8COOH$	31.6
Lauric	n-Dodecanoic	12:0	$CH_3(CH_2)_{10}COOH$	44.2
Myristic	n-Tetradecanoic	14:0	$CH_3(CH_2)_{12}COOH$	53.9
Palmitic	n-Hexadecanoic	16:0	$CH_3(CH_2)_{14}COOH$	63.1
Stearic	n-Octadecanoic	18:0	$CH_3(CH_2)_{16}COOH$	69.6
Arachidic	n-Eicosanoic	20:0	$CH_3(CH_2)_{18}COOH$	76.5
Behenic	n-Docosanoic	22:0	$CH_3(CH_2)_{20}COOH$	81.5
Lignoceric	n-Tetracosanoic	24:0	$CH_3(CH_2)_{22}COOH$	86.0
Cerotic	n-Hexacosanoic	26:0	$CH_3(CH_2)_{24}COOH$	88.5
Unsaturated fatty acids				
Palmitoleic	cis-9-Hexadecenoic	$16:1^{\Delta 9}$	$CH_3(CH_2)_5CH =$ $CH(CH_2)_7COOH$	0
Oleic	cis-9-Octadecenoic	$18:1^{\Delta 9}$	$CH_3(CH_2)_7CH =$ $CH(CH_2)_7COOH$	16
Linoleic	cis- cis-9-12-Octadecadienoic	$18:2^{\Delta 9,12}$	$CH_3(CH_2)_4CH =$ $CHCH_2CH =$ $CH(CH_2)_7COOH$	5
Linolenic	all cis-9-12-15-Octadecatrienoic	$18:3^{\Delta 9,12,15}$	$CH_3(CH_2)CH =$ $CHCH_2CH =$ $CHCH_2CH =$ $CH(CH_2)_7COOH$	−11
Arachidonic	all cis-5-8-11-14 Eicosatetraenoic	$20:4^{\Delta 5,8,11,14}$	$CH_3(CH_2)_4CH =$ $CHCH_2CH =$ $CHCH_2CH =$ $CHCH_2CH =$ $CH(CH_2)_3COOH$	−50

Note: Here we use the two most common abbreviations only. The melting point increases with the length of the hydrocarbon length but decreases with the number of double bonds.

regular, semicrystalline structures. The kink imposed by the *cis* double bonds in unsaturated fatty acids makes the packing more difficult. The difference in melting point between saturated and unsaturated fatty acids has important consequences for the biology of both plants and animals. We will devote several following sections to exploring the ecophysiological consequences of this difference for animals. Here we will briefly describe its consequences for the evolution of seed lipids in plants.

Box 2.6.

A fatty acid has a common name, a systematic name, and more than one abbreviation. For example, consider the following 18-carbon fatty acid:

$$CH_3-(CH_2)_7-CH=CH-(CH_2)_7-COOH$$

Because it is a major component of olive oil, its common (or "trivial") name is oleic acid. Its systematic chemical name is *cis*-octadecenoic acid. The name "octadecenoic" comes from the fact that fatty acids are regarded as derivatives of hydrocarbons with the same number of carbons. Saturated fatty acids are considered equivalent to alkanes (C_nH_{2n+2}) and unsaturated fatty acids are equivalent to alkenes (C_nH_{2n}). The corresponding alkene is octadecene ($C_{18}H_{36}$). The prefix *cis* refers to the relative position of the alkyl groups (*R*):

$$
\begin{array}{cccc}
H & H & R & H \\
| & | & | & | \\
C & = C & C & = C. \\
| & | & | & | \\
R & R & H & R \\
cis & & trans &
\end{array}
$$

(2.2)

Cis double bonds bend the chain of carbon atoms to form an angle of about 135°. Fatty acids with more than one *cis* bond can be bent into almost a U shape (figure 2.13). Most naturally occurring unsaturated acids have the *cis* orientation. In the anaerobic conditions of the rumen, microorganisms saturate unsaturated fatty acids and change a fraction of *cis* double bonds into the *trans* position. These modified fatty acids are absorbed and incorporated into the ruminant's tissue. In nature, *trans* fatty acids tend to be associated with microorganisms, and hence their presence in animal tissues is often an indicator of a nutritional fermentative symbiosis (Käkelä et al. 1996).

To provide a more convenient way of referring to fatty acids, and to confuse the unaware, several systems of abbreviations for fatty acids have been developed. For example, in one system, the abbreviation for oleic acid is $18:1^{\Delta 9c}$. This means that oleic acid has 18 carbons (18), 1 double bond (1), and this double bond is on carbon 9 and has *cis* orientation ($\Delta 9c$). The greek letters omega (ω) and delta (Δ) are sometimes used with special significance when naming fatty acids. Delta followed by a number, or numbers, is used to indicate the position of one or more double bonds counting from the *carboxyl carbon*. Omega is used to indicate how far a double bond is from the terminal *methyl carbon*. Table 2.5 lists some biologically important fatty

continued

continued

acids. Two of these are linoleic and linolenic acid. These fatty acids are considered indispensable (or essential). Linoleic acid has a double bond in the sixth carbon from the terminal methyl group, whereas linolenic acid has a double bond in the third carbon from the terminal methyl group. Hence, you will often find that these two fatty acids and their derivatives are referred as the ω–6 (or *n*6)and ω–3 (or *n*3) families of fatty acids. Using yet another abbreviation system, linoleic and linolenic acid are often abbreviated as C_{18}:2 *n*6 and C_{18}:3 *n*3, where 2 *n*6 and 3 *n*3 indicate that linoleic acid and linolenic acid have two and three double bonds, respectively. Counting from the terminal methyl carbon, the first double bond is in carbon 6 in linoleic acid and the second one is in carbon 3 in linolenic acid.

Linder (2000) hypothesized that ambient temperature during germination should be an important selective factor on the composition of seed storage lipids. Lower germination temperatures should favor unsaturated fatty acids with low melting point that can be more easily mobilized. Because they have to be enzymatically unsaturated from saturated precursors, on a per carbon basis, unsaturated fatty acids cost more to produce than saturated fatty acids. They also yield slightly less net energy when oxidized. Therefore, Linder (2000) predicted that, at higher ambient temperatures, seeds with higher proportions of saturated acids should be favored. A variety of lines of evidence support these predictions: The seeds of tropical species have significantly higher percentages of saturated fatty acids (about 40%) than those of temperate species (about 8%). Within a single clade, the sunflower genus *Helianthus*, Linder found a tight negative interspecific correlation between the proportion of saturated fatty acids and latitude (figure 2.14). It is significant that the plant species that humans use as sources of saturated fats, like coconuts (*Cocos nucifer*) and chocolate trees (*Theobroma cacao*), are tropical. It is perhaps also noteworthy, that the oldest oil-bearing crops, like sesame (*Sesamum indicum*), safflower (*Carthamus tinctorius*), and flax (*Linum usitatissimum*) grow best in the cooler regions of the world.

2.6.2 The Importance of Fatty Acid Structure in Animal Nutrition

Fatty acids have remarkably varied roles in animal physiology. As ingredients of triacylglycerides, fatty acids are used as sources of energy and as thermal

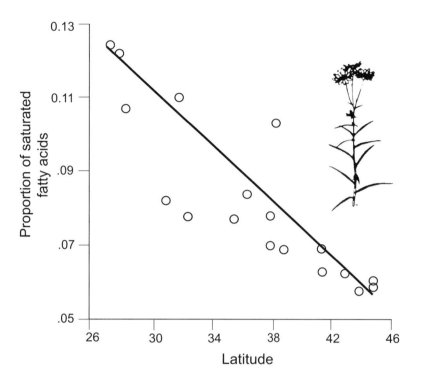

Figure 2.14. In North American sunflowers of the genus *Helianthus,* the proportion of unsaturated fatty acids in seed oils decreases with latitude—and presumably with germination temperature. This pattern should make you uncomfortable after having read in chapter 1 our stern warning about regressions that use species as data points and that do not take phylogeny into account. It should comfort you to know that Linder (2000) conducted a phylogenetically informed analysis and concluded that the pattern shown here was very robust.

insulation. As components of phospholipids and sphingolipids, they can determine the fluidity of cell membranes. Fatty acids are also important precursors of signaling molecules, like eicosanoids. As we will see, many of the biological roles that fatty acids play are the result of the way carbon atoms are joined together by either single or double bonds, and of the number of carbon atoms on the chain. In the following sections we will consider three interrelated topics: the variation in the composition of fatty acids among the different parts of the body within a single organism, the importance of fatty acids for thermal biology and hibernation, and the role that fatty acids in producers can have for ecosystem level processes. Before we can tackle these topics, we must briefly review how different types of fatty acids are produced and acquired. Although it is convenient to think of fatty acids individually, keep in mind that in the tissues of living organisms they are usually part of larger molecules such as tryacylglycerols and phospholipids.

2.6.3 The Biology of Desaturases

As we described in the previous section, fatty acids can be saturated or unsaturated. Unsaturated fatty acids are synthesized from saturated precursors by the action of enzymes called desaturases. Desaturases are named for the carbon after which they produce a double bond. For example, if a desaturase catalyzes the reaction that places a double bond between carbons 9 and 10 (counting from the carboxyl carbon), it is called a $\Delta 9$ desaturase. Desaturases are distributed unevenly among organisms (figure 2.15; Cook 1991). Most organisms have $\Delta 9$ desaturases, and the first double bond introduced into a hydrocarbon chain is generally in the $\Delta 9$ position. With a few exceptions, animals cannot introduce a double bond beyond the $\Delta 9$ position. Plants, on the other hand, can introduce a second and a third double bond between the existing $\Delta 9$ double bond and the terminal methyl groups. Diatoms and *Euglena* have the unusual ability to desaturate on either side of the $\Delta 9$ bond, and a few insect species have desaturases that place double bonds at the $\Delta 11$ and $\Delta 12$ positions. Insects use these desaturases to manufacture sexual pheromones (Weitian et al. 2002).

Recall that linoleic acid (C_{18}:2 *n6*) and linolenic acid (C_{18}:3 *n3*) are considered indispensable nutrients. They are indispensable because they have double bonds beyond carbon 9 (see table 2.5) and therefore, with few exceptions, animals lack the desaturases needed to produce them. The term indispensable (or essential) as applied to fatty acids is, to some extent, unclear. Recently, over 20 fatty acids in addition to linoleic and linoleic acids have been deemed essential (examples are arachidonic (C_{20}:4 *n6*) and docosahexaonic acid (C_{22}:6 *n3*; Cunnane 2000). These fatty acids can be synthesized by animals from linoleic and linolenic precursors. However, during certain life stages such as growth and reproduction, requirements exceed synthesis and animals must obtain at least a fraction of their total requirements from food. Because animals have the ability to synthesize these fatty acids, they cannot be considered strictly indispensable. In analogy with the classification that we introduced for amino acids, we can call these fatty acids conditionally indispensable (Cunnane 2000). Curiously, felids are deficient in $\Delta 6$ desaturases and therefore are unable to synthesize sufficient arachidonic acid to meet metabolic needs (MacDonald et al. 1984; Pawlosky et al. 1994; figure 2.16). Arachidonic acid is an indispensable nutrient for felids. As we will see in a subsequent section (2.6.6. "Unsaturated Fatty Acids in Aquatic Food Webs"), the inability of many animals to produce sufficient amounts of metabolically important polyunsaturated fatty acids to sustain maximal growth is a physiological detail that may have profound consequences for the structure and function of aquatic ecosystems.

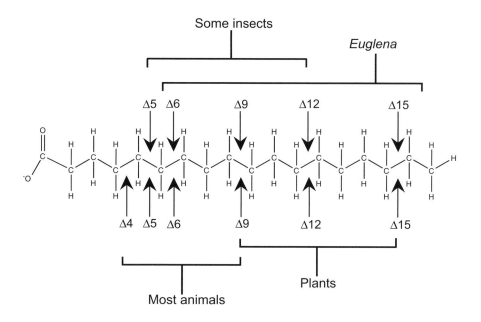

Figure 2.15. Unsaturated fatty acids are synthesized from saturated fatty acids by the action of desaturases. Specific enzymes are capable of creating a double bond in different positions of the acyl chain of a fatty acid. Different groups of organisms express different desaturases, and hence synthesize different unsaturated fatty acids.

2.6.4 Fatty Acids, Homeoviscous Adaptation, and Regional Heterothermy

The fatty acid composition of storage and membrane lipids seems to follow more or less the same rules as those for lipids in seeds. This observation should not come as a surprise. Because the mixtures of triacylglycerides in storage fats should remain fluid and metabolically usable in cooler tissues, cooler temperatures should favor increased levels of unsaturated fatty acids. Biological membranes exist in a "liquid-crystalline" state that is poised between extreme fluidity and low viscosity, on one hand, and a highly viscous, gel-like state, on the other. As you would expect, the fluidity of biological membranes is markedly influenced by temperature. To maintain adequate fluidity at low temperatures, you should predict an increase in the unsaturated fatty acid content in membrane phospholipids and sphingolipids. The idea that the fluidity of biological membranes is maintained relatively constant in the face of variation in temperature, and other influences such as pressure, by changes in fatty acid composition is called the homeoviscous adaptation hypothesis (Sinensky 1974).

Although there is still a significant amount of research to be done, the expectation that colder temperatures favor increased unsaturated fatty acid

Figure 2.16. Animals synthesize highly unsaturated fatty acids such as arachidonic acid, eicosapentaenoic acid, and docosahexaenoic acid by a process of sequential desaturations that add double bonds and elongations that add carbons to the acyl chain. The first step requires binding acetyl-coenzyme A to an indispensable fatty acid to produce an acyl-coenzyme A. Subsequent steps desaturate and elongate the fatty acid precursor. Although many animals have the enzymatic machinery to produce highly unsaturated fatty acids, requirements may exceed synthesis.

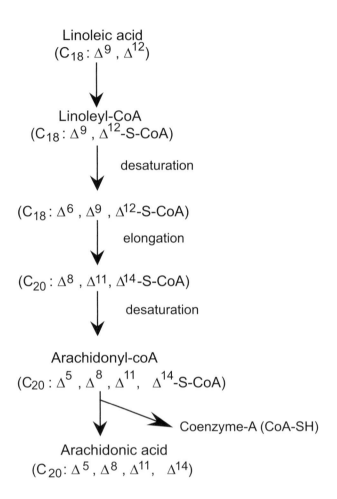

Linoleic acid
$(C_{18} : \Delta^9 , \Delta^{12})$

Linoleyl-CoA
$(C_{18} : \Delta^9 , \Delta^{12}\text{-S-CoA})$

desaturation

$(C_{18} : \Delta^6 , \Delta^9 , \Delta^{12}\text{-S-CoA})$

elongation

$(C_{20} : \Delta^8 , \Delta^{11}, \Delta^{14}\text{-S-CoA})$

desaturation

Arachidonyl-coA
$(C_{20} : \Delta^5 , \Delta^8 , \Delta^{11}, \Delta^{14}\text{-S-CoA})$

Coenzyme-A (CoA-SH)

Arachidonic acid
$(C_{20} : \Delta^5 , \Delta^8 , \Delta^{11}, \Delta^{14})$

content in both storage fat deposits and membranes seems to be correct. (e.g. Behan-Martin et al. 1993). The tissues of poikilothermic ectotherms seem to contain a higher proportion of unsaturated fatty acids than those of homeothermic endotherms. Furthermore, species from colder regions seem to contain more unsaturated fatty acids in their tissues than species that live in warmer climates (Pond 1998). The validity of the first part of these hypothetical patterns is supported by a cursory analysis of the lipids of the animals that humans eat. In general, the tissues of the poultry and ungulates that we eat have higher contents of saturated fatty acids than those of fish, crustaceans, and mollusks. Readers interested in the fatty acid composition of human foods can consult several chapters in Chow's (2000) comprehensive book. Although, the trends described above make intuitive sense, to our knowledge they have not been really tested rigorously. By rigorously, we mean with the use of a phylogenetically informed comparative approach. Linder's (2000) exemplary comparative study of lipids in seeds provides a model for the investigation of the fatty acid composition of animal lipids.

The comparative study of the fatty acid composition of animal tissues must take into account the potentially confounding effects of phenotypic plasticity. The fatty acid composition of both membrane and storage lipids is partially controlled by inheritance, but also by variation in availability in food, as well as by selective uptake, retention, and preferential oxidation of certain lipids in response to environmental variation. Both diet and ambient temperature influence fatty acid composition. This variation is well illustrated by domestic ungulates, and as we will see in a subsequent section, by hibernating mammals. In domestic cattle the ratio of polyunsaturated to saturated fatty acids (PUFA/SAT) is much higher in pasture-fed cattle (PUFA/SAT = 0.24) than in grain-fed cattle (PUFA/SAT = 0.16; Cordain et al. 2002). This difference is a direct consequence of differences in diet.

An additional complication is that fatty acids are not distributed evenly among the different parts of an endotherm's body. The skin surface and the appendages of an animal (ears, hooves, legs, and paws) are often cooler than the core body temperature. This phenomenon is called "regional heterothermy." The mixtures of fatty acids that are found in internal fat depots would solidify rapidly at the temperature of the limbs, which can be as low as 4°C in cold climates (Irving and Krog 1955). Pond and collaborators (1993) studied the triacylglycerol composition of Svalbard reindeer (*Rangifer tarandus platyrhynchus*) in the winter. This small, fat, stout, and shaggy subspecies of reindeer is well adapted to its extremely cold arctic habitat. Pond and colleagues found a greater proportion of unsaturated fatty acids in the adipose tissue near the skin than in the inner side of the subcutaneous fat deposits. They also found a similar gradient in triacylglycerol samples collected from bone marrow from the shoulder to the hooves (figure. 2.17). Käkelä et al. (1997) found similar distributions of fatty acids among the fat deposits of several species of subarctic aquatic and terrestrial species. The extremities contained higher levels of unsaturated fatty acids than the internal deposits. In seals, the triacylglycerols of the "pokilothermic" outer blubber had significantly more unsaturated fatty acids than the inner layers.

2.6.5 Fatty Acids and Thermal Biology

With few notable exceptions, aquatic ectotherms have body temperatures that vary with those of the water. Not surprisingly, there are many examples of modulation of fatty acid composition in response to changes in ambient temperature in these organisms. In general, as water temperature decreases, the proportion of unsaturated fatty acids in cell membrane lipids increases

Figure 2.17. The unsaturation index estimates the level of unsaturation in the fatty acids that make up a lipid mixture. It is calculated as the weighted average of the number of fatty acids with a given number of double bonds (unsaturation index =

$$\left\{ \frac{\%_1 + \%_2 + \%_3}{\%_0 + \%_1 + \%_2 + \%_3} \right\},$$

where $\%_i$ is the % of fatty acids that have i double bonds). You can interpret it as the average number of double bonds per fatty acid. The average unsaturation indices of triacylglycerols in Svalbard reindeer differed among different adipose depots. The fatty acids of the outer layers of subcutaneous deposits contained more double bonds than those of the internal layers. The fatty acids in the bone marrow of hind leg bones had more double bonds than those in subcutaneous deposits, and the degree of unsaturation increased from the femur to the tarsus and hooves. The most important unsaturated fatty acids in Svalvard reindeer fats were the monoenoic acids palmitoleic acid (C_{16}:2 n6) and oleic acid (C18:1). Linoleic acid (C_{18}:2 n6) was also present, albeit at fairly low levels (less than 1.5% of total fatty acids). (Constructed from data in table IV in Pond et al. 1993.)

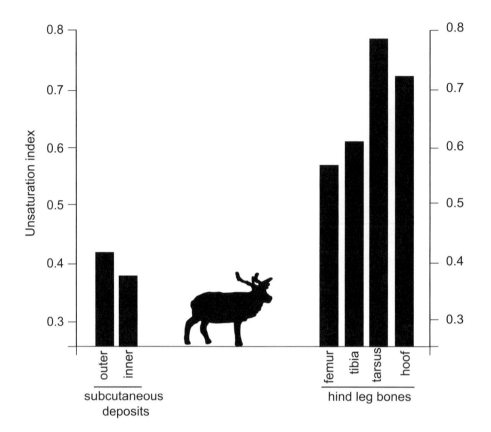

(reviewed by Hadley 1984; Gibbs 1998). Terrestrial ectotherms, such as many species of lizards, are capable of using behavior to regulate their body temperature. Many terrestrial ectotherms tend to select lower body temperatures during the winter. This selection of a lower body temperature is accompanied by changes in the fatty acid composition of tissues and membranes (Hazel and Williams 1990). Geiser and collaborators (1992) turned this observation on its head to ask the following question: Will animals forced to change the fatty acid composition of their tissues select different body temperatures? They fed shingle-backed lizards (*Tiliqua rugosa*) a diet containing either polyunsaturated (sunflower oil, 32% of all lipids are PUFAs) or saturated fatty acids (sheep fat, 8.6% PUFAs). The lizards chose very different body temperatures after being fed on either of these two diets for two weeks. The lizards fed on the PUFA-rich diet selected a body temperature that was about 3°C cooler than that chosen by the lizards fed on the saturated fatty acid diet (figure 2.18). Although the physiological regulatory mechanisms that mediate this shift in thermal preferences are obscure, this remarkable study has a clear implication: The composition of dietary lipids can influence an animal's thermal biology. This observation is well illustrated by hibernating mammals.

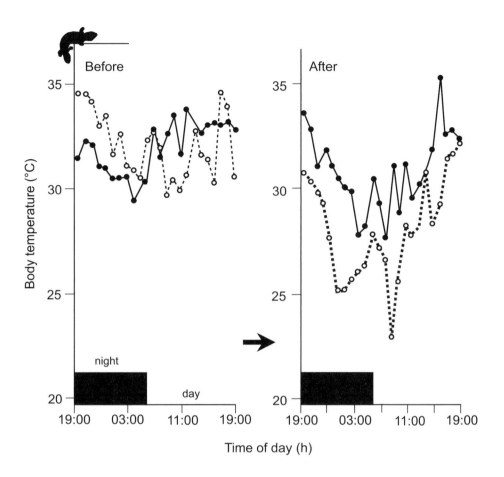

Figure 2.18. After two weeks of being fed a diet rich in either saturated (closed points) or unsaturated (open points) fatty acids, shingle-backed lizards chose significantly different temperatures. The lizards had thermocouples inserted in their cloacas and were placed in a 1.5 meter long gradient with temperatures varying continuously from 7 to 57°C. The black bar represents night. After 14 days on the saturated fatty acid diet, the lizards chose temperatures that were about 3°C warmer than those chosen by lizards fed on the diet rich in unsaturated fatty acids. (After Geiser et al. 1992.)

Many mammals, including echidnas, marsupials, and a large number of eutherian species, either enter daily torpor, or undergo seasonal hibernation (Lyman 1982). Throughout the summer and early fall, hibernating mammals become "hyperphagic" (increase food consumption), and convert the assimilated food into triacylglycerols. They often favor diets containing PUFAs and in consequence their fat storage deposits become rich in PUFAs. The fatty acid that occupies the middle position in these triacylglycerols is often a PUFA (Florant 1998). The inclusion of a PUFA in a central position of the triacylglycerols greatly lowers the melting point of the fat and allows it to be mobilized for energy, even when the animal is torpid and has a low body temperature.

A large number of laboratory studies, involving six rodent and two marsupial species, has shown that a diet containing relatively high levels of polyunsaturated fatty acids (primarily linoleic acid) is required for both hibernation and torpor. In a review of an abundant literature, Florant (1998) concluded that hibernators fed on diets containing high levels of PUFAs differ from those fed on low levels of PUFAs in three ways: (1) High PUFA animals are more likely to hibernate and to hibernate earlier in the season, (2) they also tend to have lower body temperatures and consequently lower metabolic rates while torpid, and (3) they tend to have longer torpor bouts (figure 2.19; Frank et al. 1998). Eating, and hence depositing a large number of PUFAs in fat deposits, is advantageous for a hibernator because it keeps its fat deposit fluid and hence mobilizable. However, PUFA-enriched fat deposits entail a cost. The double bonds in PUFAs react more readily with oxygen and oxygen free radicals than the single bonds in saturated fatty acids. Oxidized fatty acids cannot be catabolized to produce energy and are very toxic. They can cause hemolysis, muscular degeneration, and nervous system damage (Chow 2000). Thus, hibernators must eat sufficient PUFA-containing food, but not too much. Indeed, Frank and collaborators (1998) found that, when given a choice, hyperphagic golden-mantle ground squirrels (*Spermophilus lateralis*) ingested intermediate levels of PUFAs. When forced to ingest excessive amounts of PUFAs before hibernation, the squirrels' ability to hibernate was impaired (Frank and Storey 1995). The feeding mechanisms by which hibernators acquire fat deposits with optimal fatty acid mixtures are an interesting topic that has received comparatively little attention (Frank et al. 1998).

The process of oxidation of fatty acids, called autoxidation (or lipid peroxidation), is worth discussing briefly. Autoxidation is initiated by free radicals, such as the singlet oxygen ($^{•}O_2$), hydroxyl ($^{•}OH$), and hydroperxyl ($^{•}OOH$) radicals, found in most cells. If autoxidation is not stopped, it can lead to a chain reaction that involves adjacent lipids and even other molecules. Under normal circumstances the activity of antioxidant agents destroys or soaks up the free radicals that cause lipid peroxidation. A variety of enzymes detoxify lipid peroxides (Halliwell 1997). However, for unknown reasons, these agents and enzymes appear to be less active in torpid animals with low body temperatures. Consequently, lipid autoxidation increases during torpor (Buzadzic et al. 1990). It may be that there is a mismatch between the thermal sensitivity of autoxidation and that of the antioxidant and detoxification agents. Autoxidation may be less inhibited by low temperature than the effect of antioxidants, allowing lipid peroxidation to proceed uninhibited. Harlow and Frank (2001) found that the rate of lipid peroxidation was higher in white-tailed prairie dogs (*Cynomys leucurus*), which are spontaneous hibernators, than in black-tailed prairie dogs (*Cynomis ludovicianus*), which are facultative hibernators. Black-tailed prairie dogs show shorter and shallower torpor bouts than white-tailed prairie dogs (figure 2.19). It is tempting to speculate that the increased activity of antioxidant agents during the longer periods of arousal in black-tailed prairie dogs reduces the amounts of oxygen free radicals. The longer and more frequent torpor intervals in white-tailed prairie dogs may favor the accumulation of these substances.

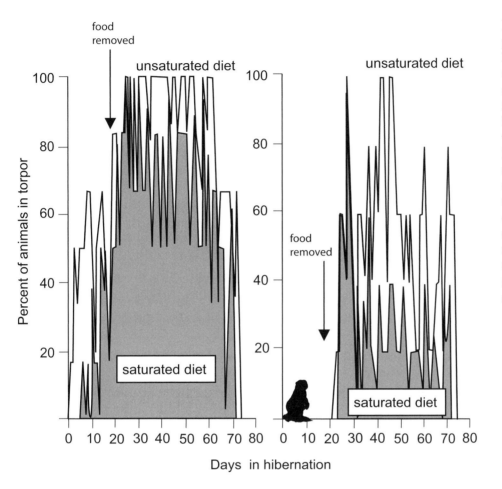

food
removed

unsaturated diet

Percent of animals in torpor

saturated diet

food
removed

unsaturated diet

saturated diet

Days in hibernation

Figure 2.19. White-tailed prairie dogs are obligate hibernators. In the winter, they enter torpor earlier and have longer and more frequent torpor bouts than black-tailed prairie dogs which are facultative hibernators. Both species show decreased frequency of torpor and shorter torpor bouts when fed on a diet rich in saturated fatty acids than when they are fed a diet containing sufficient polyunsaturated fatty acids. (After Harlow and Frank 2001.)

2.6.6 Unsaturated Fatty Acids in Aquatic Food Webs

In previous sections we have illustrated the interplay between thermal biology (including torpor and hibernation), food selection, and the fatty composition of diet. Clearly, the composition of dietary fatty acids can have interesting consequences for an individual animal's behavior. Here we will adopt a more expansive perspective and review evidence for a role of fatty acids in the regulation of the productivity and structure of pelagic ecosystems. The hypothesis that we will describe was first posed by Brett and Müller-Navarra (1997). In brief, these authors suggested that the production of highly unsaturated fatty acids by phytoplankton producers is one of the primary determinants of zooplankton productivity in pelagic ecosystems. This "bottom-up" ecosystem function hypothesis is founded on two observations: (1) diets with high content of highly unsaturated fatty acids (HUFAs) promote performance in zooplankton, and (2) ecosystems in which the phytoplankton includes primarily high HUFA-containing taxa tend to be more productive. A large body of research by aquaculturists supports the first observation: Highly unsaturated fatty acids (HUFAs), such as eicosapentaenoic acid (EPA, 20:5 *n*3) and docosahexaenoic acid (DHA, 22:6*n*3), are critical diet ingredients for high growth, survival, reproduction, and high food-to-tissue conversion rates in many marine and fresh water primary consumers (Müller-Navarra 1995). Adding semipure emulsions of HUFAs to algal monocultures improves their quality as food and increases the growth rate of the zooplankton feeding on them.

The second observation also has some empirical support. Phytoplankton fatty acid content and composition vary dramatically among taxonomic groups (figure 2.20). Phytoplankton with high nutritional quality for herbivorous zooplankton, such as diatoms and cryptophytes, tend to have high HUFA levels, whereas cyanobacteria, which are poor nutritionally, have low HUFA levels. Highly productive ecosystems, such as those supported by marine upwellings, tend to have HUFA-rich diatoms at their producer base. In contrast, unproductive eutrophic lake ecosystems seem to have phytoplankton dominated by cyanobacteria (Hecky 1984). Brett and Müller-Navarra (1997) point to the contrasting structures of the biomass pyramids of eutrophic lakes and of highly productive marine upwellings as a consequence of HUFA abundance in phytoplankton. In highly eutrophic lakes, zooplankton biomass averages 15% of phytoplankton biomass. In contrast, in ocean upwellings the zooplankton biomass is 2.7 times greater than the phytoplankton biomass (figure 2.21). Brett and Müller-Navarra (1997) argue that the diatoms of upwelling zones have high production-to-biomass ratios because high HUFA diatoms are converted to zooplankton very efficiently.

Understanding the processes that regulate the conversion of plant productivity into herbivore biomass is a central ecological question. It is critical, for example, in understanding the factors that control production in fisheries and lake eutrophication. Brett and Müller-Navarra's (1997) bold hypothesis places a single physiological process, the elongation and desaturation of polyunsaturated fatty acids to yield highly unsaturated fatty acids, at the heart of a fundamental ecological process. It is probably premature to accept the "HUFAs as limiting factors" hypothesis, without question. However, its inventive merit is unquestionable. It points to immediate applications and suggests novel research avenues. An example of an application is the possibility of using the content of HUFAs in phytoplankton as a quick and easy index of potential secondary production in pelagic food webs (Brett and Müller Navarra 1997). We are confident in predicting that future research into the role of HUFAs in aquatic ecosystems will contribute much to our understanding of how aquatic ecosystems

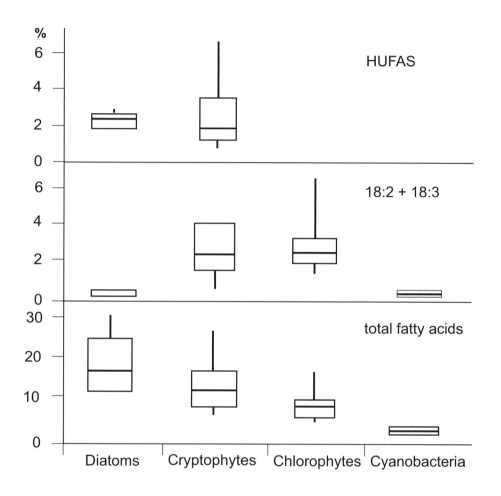

Figure 2.20. Although there is some variation within taxa, different groups of planktonic producers contain contrasting amounts of fatty acids, linoleic (18:2) and linolenic acid (18:3), and highly unsaturated fatty acids (HUFAs). Diatoms have both high proportions of HUFAs and high total fatty acid contents, but low levels of linoleic and linolenic acid. In contrast, cyanobacteria have low fatty acid contents, and low levels of all unsaturated fatty acids. Cryptophytes and chlorophytes have intermediate levels of total fatty acids, and relatively high levels of linoleic and linolenic acid, cryptophytes and chlorophytes differ in HUFA levels. (After Brett and Müller-Navarra 1997.)

function. As physiologists with an interest in ecology (or as ecologists with an interest in physiology?), we are fond of the hypothesis because it is an example of the pivotal role that biochemistry and nutritional biology can play in our attempts to explain broad-scale ecological processes.

2.7 Vitamins

Vitamins are relatively complex organic molecules that are present in minute amounts in food and that are indispensable for animals. They are indispensable because in the course of the animal's evolutionary history the ability to synthesize them has been lost. Vitamins consist of a heterogeneous group of compounds that are not as related to each other as are carbohydrates, proteins, and lipids. Their classification in a single group of nutrients depends more on

Figure 2.21. Aquatic food-webs with diatoms at their base, like the Peruvian upwelling, have inverted Eltonian biomass pyramids. In contrast aqatic foodwebs in eutrophic lakes, such as Clear Lake, California U.S.A., with cyanobacteria at their producer base, have right-side-up biomass pyramids. The boxes in this figure are proportional to the relative biomasses. Brett and Müller-Navarra (1997) hypothesized that the high HUFA content of diatoms permits more efficient conversion of phytoplankton into consumer biomass, and hence facilitates the transfer of materials across trophic levels.

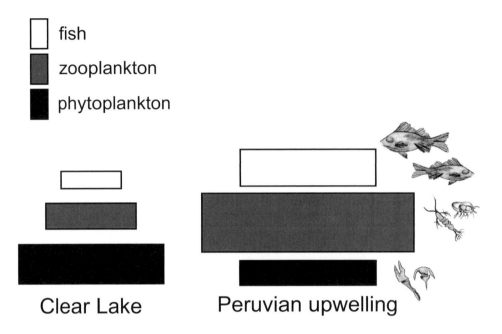

their function than on their chemical characteristics. They are differentiated from other essential trace elements (such as minerals) because they are organic substances. Although vitamins are required in very small amounts, their absence in an animal's diet will reduce its performance. Table 2.6 lists some of the most important vitamins, together with their function. An interesting feature of the vitamins listed in table 2.6 is the similarity between many (albeit not all) of the major vitamins and the major coenzymes. Coenzymes are organic molecules that bind to enzymes to help them carry out their specific reactions. You may recall from your elementary biology course that NAD^+ (nicotinamide–adenine dinucleotide), FAD^+, and coenzyme A are important coenzymes. In most cases, the vitamin is a component of the coenzyme molecule.

The vitamins listed in table 2.6 can be divided into water-soluble and fat-soluble vitamins. The water-soluble vitamins include vitamin C and vitamins in the B complex. The fat-soluble vitamins include vitamins A, D, E, and K. The distinction between water- and fat-soluble vitamins has physiological significance. Moderate overingestion of water-soluble vitamins is, in general, not harmful. The ingested excess is filtered by the kidney and excreted in urine. Linus Pauling (1970) recommended astoundingly high doses of vitamin C (more than 10 g per day) to prevent and treat diseases (in particular the common cold). Although the effectiveness of Pauling's prescription is controversial, many people heeded his advice, and so far a vitamin C hypervitaminosis epidemic has not happened among humans that follow nutrition fads. Fat-soluble vitamins

TABLE 2.6
Fat and water-soluble vitamins with synonyms and functions

Vitamin	Synonym	Function
Fat soluble		
Vitamin A	Retinol (A_1), dehydroretinol (A_2)	Vision (rhodopsin formation), maintenance of epithelia, reproduction, and bone development
Vitamin D	Ergocalciferol (D_2), cholecalciferol (D_3)	Stimulation of intestinal calcium and phosphorous transport, bone formation, renal calcium reabsorption
Vitamin E	Tocopherol	Biological antioxidant, blood clotting, protection against toxicants, disease resistance
Vitamin K	Phylloquinone (K_1), menaquinone (K_2), and menadione (K_3)	Blood coagulation (synthesis of prothrombin), synthesis and activation of a variety of proteins
Water soluble		
Thiamin	Vitamin B_1	Coenzyme in the Krebs cycle, involved in the synthesis of acetylcholine
Riboflavin	Vitamin B_2	Prosthetic group in flavoproteins (oxidases and dehydrogenases)
Niacin	Vitamin pp, vitamin B_3	Coenzyme (as NAD, nicotinamide) in many central oxidation-reduction systems (glycolysis, Krebs cycle, lipid and protein metabolism)
Vitamin B_6	Pyridoxol, pyridoxal, pyridoxamine	Precursor of important coenzymes that participate in amino acid metabolism
Panthotenic acid	Vitamin B_5	Important constituent of coenzyme A, one of the most important coenzymes in metabolism
Biotin	Vitamin H	Coenzyme of several carboxylating enzymes (it acts as a carboxyl carrier)
Vitamin B_{12}	Cobalamin	Coenzyme in purine and pyrimidine synthesis, transfer of methyl groups, synthesis of proteins from amino acids, and carbohydrate and lipid metabolism
Choline	Gossypine	Structural element of many phospholipids, synthesis of acetylcholine
Vitamin C	Ascorbic acid	Collagen synthesis, mixed function oxidation, may stimulate the immune system

Note: Often, vitamins have slightly different forms that differ in activity. Vitamin A has two forms (A_1 and A_2) and vitamin D has two forms (D_2 and D_3). These forms have different names listed in the table.

are a different story. They accumulate in membranes and fat deposits and when ingested in excess they can produce very toxic effects.

The seemingly arbitrary nomenclature of vitamins may perplex you. It is the result of many independent attempts at organizing a diverse group of compounds. Vitamins were originally isolated from certain foods and assigned letters of the alphabet. This system was modified when researchers discovered that a given physiological activity could be attributed to more than one substance. A system of suffixes was devised to solve this problem, which explains the existence of vitamins D_2 and D_3. Vitamins have also been named in reference to their function. Thus, vitamin K was named from the Danish word "koagulation" (coagulation), and vitamin H (biotin) was used because this substance protects the skin ("haut" in German). When the chemical structure of vitamins was determined, the letter designation was sometimes replaced by a chemical name. The result is a morass of names. Although vitamins are of profound nutritional importance, with few exceptions, their significance for ecological processes remains unexplored. Robbins (1993) reviewed the importance of vitamins for the furry and feathered. We recommend McDowell's (1989) book on the importance of vitamins for humans and domestic creatures as a readable comprehensive treatise.

Vitamins illustrate one of the themes in this chapter. Although vitamins are defined as "essential," their essentiality varies both within and among species. Most animals can synthesize choline, but in many cases not in sufficient amounts to satisfy an animal's needs. In vertebrates, choline requirements are higher in rapidly growing young than in adults, and for obscure reasons are higher in males than in females. Vitamin C can be synthesized by many species but not by others. Many invertebrates and most fish species are unable to synthesize it. Among terrestrial vertebrates amphibians and most reptiles can synthesize it. The phylogenetic distribution of the ability to synthesize vitamin C among birds and mammals is perplexing. Most mammals can synthesize it, but anthropoid primates, bats, guinea pigs, and maybe cetaceans cannot (Robbins 1993). Among birds the ability to synthesize vitamin C seems to have been lost repeatedly and in an unpredictable way (Martínez del Rio 1997). In mammals and birds, the loss of the capacity to synthesize vitamin C cannot be predicted from either phylogenetic affiliation or ecological factors such as diet. The need for vitamin C also varies among individuals. In willow ptarmigan (*Lagopus lagopus*), one of the few bird species in which vitamin C requirements during development have been studied systematically, the growing young need diets with about six times more vitamin C than adults (Hanssen et al. 1979). In summary, like amino acids and fatty acids, in some vitamins the definition of essentiality (or indispensability) may have to be qualified.

2.8 Minerals

Recall that one of the fractions of proximate nutrient analysis was ash. The total ash that remains after burning organic matter at high temperature represents its total mineral (or "inorganic") content. Ash contains a mixture of the solid, crystalline chemical elements that we call minerals. Some minerals are as central to life as the organic molecules that we have discussed in this chapter. Depending on the amounts that animals need in their diets, minerals are classified as major or macrominerals (or macroelements) and trace or microminerals (or microelements). Macrominerals are required in large amounts and microminerals are required in tiny ("trace") amounts (table 2.7). In chapter 11 we will describe in detail how mineral requirements are estimated and the ecological importance of several of the most important macronutrients. Here we outline very briefly the general function of minerals, we illustrate how some of them can have toxic effects if ingested in excess, and we exemplify (using selenium as a case study) how their abundance can vary geographically, and how this spatial variation can be so large that they can be limiting in some areas and toxic in others.

Unlike most of the nutrients discussed in the bulk of this chapter, living organisms cannot synthesize mineral elements. Minerals are essential without adjectives. Minerals participate in a variety of biological functions. They are central for:

TABLE 2.7
Classification of minerals as macro- or microminerals

Macrominerals		Trace or Microminerals	
Calcium (Ca)	Arsenic (As)	Iodine (I)	Nickel (Ni)
Phosphorus (P)	Boron (B)	Iron (Fe)	Selenium (Se)
Potasiums (K)	Chromium (Cr)	Lead (Pb)	Silicon (Si)
Sodium (Na)	Copper (Cu)	Manganese (1Mn)	Vanadium (V)
Sulfur (S)	Fluorine (F)	Molybdneum (Mo)	Zinc (Zn)
Chlorine (Cl)			

Note: As a general rule, to satisfy an animal's requirements the abundance of macrominerals in diet must be above 100 ppm (parts per million (μgrams per gram or mg/kg) , (whereas microminerals are required in concentrations lower than 100 ppm and sometimes in concentrations as low as parts per billion. As you might expect there is some overlap between these two classes of elements. Within the micronutrients, nutritionists recognize ultratrace minerals that are required in dietary concentrations lower than 1 ppm (and as low as 50 ng/g). The ultra-trace elements include As, Bo, Mo, Ni, and Va. We know exceedingly little about the ecological importance of these elements.

(1) information (electrical) transfer and storage (Na, K, Cl, Mg, Ca) and mechanical transmission (Mg, Ca);

(2) as elements of enzymes that catalyze acid-base and reduction-oxidation processes (S, Zn, Fe. Mg, C, Fe, Mn, Co, Mo, Se);

(3) as structural components of biomineralized skeletons (Si, B, P, S, Ca, Mg); and

(4) last, but not least, in the transmission and storage of chemical energy (P, S).

This book is not the place for an encyclopedic description of the biological roles of minerals. The magnificent chemistry textbooks by Williams and Frausto da Silva (1996) and Frausto da Silva and Williams (1994) describe the profound importance of these "inorganic" elements for life. McDowell (1992) has done us all a great service by compiling what we know about the importance of minerals for humans and their symbiotic domestic vertebrates in yet another readable book (recall from the previous section that he wrote a book on vitamins).

Minerals are most frequently measured by atomic absorption spectrophotometry (AAS). The sample is ashed and digested in acid. The sample is introduced into a flame where compounds become dissociated into unexcited, un-ionized individual atoms. Elements absorb radiation in specific narrow wavelength bands. When a beam of light at one of these wavelengths passes through the flame, the amount of light absorbed depends on the concentration of the element in question (Hughes et al. 1976). Atomic absorption spectrophotometry systems come in two guises depending on the atomization source used: a flame or a graphite furnace. The latter is orders of magnitude more sensitive and hence can be used in smaller samples. The former is cheaper to run. There are a variety of other spectroscopic methods to measure minerals (neutron activation analysis, plasma atomic emission spectroscopy, and proton-induced x-ray emission). Iyengar et al. (1997) describe in detail the criteria that you must consider when choosing the method and instrument to measure minerals.

2.8.1 From Essential to Toxic: The Effect Is in the Dose

As is the case with many compounds, the intake of an element determines its biological effect (figure 2.22). Because intake often depends on the concentration in the diet, mineral elements can have unfavorable effects if their concentrations are too high in an animal's food. The smallest amount of dietary minerals that will promote optimal responses is the minimum requirement. Below this amount performance increases with intake. For minerals, optimal dietary levels above this minimum are often fairly broad. Performance remains constant with variation in intake. The reason for these broad optima is the ability of animals to regulate homeostatically the efficiency with which some minerals are absorbed and excreted. In addition, animals can often store minerals and

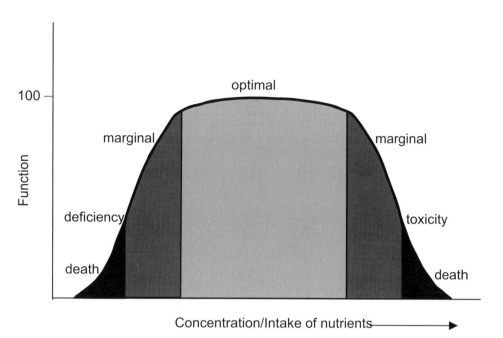

Figure 2.22. The relationship between an animal's performance (or function) and the intake or concentration of an essential nutrient is often depicted as a mesa with a broad plateau on top. This depiction applies well to many minerals and to many vitamins. The extent of the plateau estimates the tolerance of an animal for a nutrient and depends on the identity of the nutrient in question. For example, lipid-soluble vitamins have narrow plateaus, whereas water-soluble vitamins have very broad ones.

mobilize them in times of need. For example, vertebrates store calcium, phosphorus, and magnesium in bone and iodine in the thyroid gland. In figure 2.23 we use the regulation of iron levels in vertebrates to illustrate some of the mechanisms used by animals to achieve mineral homeostasis. Mineral homeostasis is a topic that we will return to in chapter 11. Nutritionists sometimes estimate "tolerance ratios" to estimate how much diets can vary in the content of a mineral and still satisfy an animal without having toxic effects. A tolerance ratio is the quotient between the maximum tolerable diet for a mineral in diet to its minimal requirement. These ratios vary over orders of magnitude depending on the mineral. Cobalt, for example, has a broad tolerance ratio of about 100 in cattle (minimal requirement and maximal tolerable concentrations in forage are ≈ 0.1 and 10 ppm, respectively). Fluorine, in contrast seems to have a very narrow tolerance ratio. In a large epidemiological study in China, Li et al. (2001) found that bone fractures in humans were lowest when fluoride in drinking water was ≈1.0 ppm. The incidence of fractures increased when fluoride levels were below 0.5 ppm or above 4 ppm, so the tolerance ratio was only 8.

2.8.2 Selenium, Custer's Defeat, and a Poisoned Reservoir

An essential nutrient can become a toxin at high intakes. If animals exceed their maximal tolerable intake of a mineral, then their performance drops (figure 2.22). Selenium is a good example of a mineral that is essential, but that at high

Figure 2.23. Iron homeostasis is regulated primarily by absorption and storage. Iron is absorbed in the brush border of small intestine's enterocytes by a divalent metal transporter (DMT1). Intestinal iron absorption is regulated in relation to circulating Fe and by hematocrit (the concentration of hemoglobin-containing red-blood cells). Anemic animals are significantly more efficient at absorbing iron than animals with normal iron levels. Within intestinal cells some of the Fe is converted to ferritin, but most is transported to tissues by the protein transferrin. Iron is stored primarily in the liver (and to a lesser extent in the spleen and the hematopoyetic tissue in bone marrow) bound to ferritins. Ferritins are iron storage proteins that form huge (8–12 nm in diameter!), hollow, spherical particles in which 2,000 to 4,500 iron atoms can be stored (Redrawn from Aisen et al. 2001).

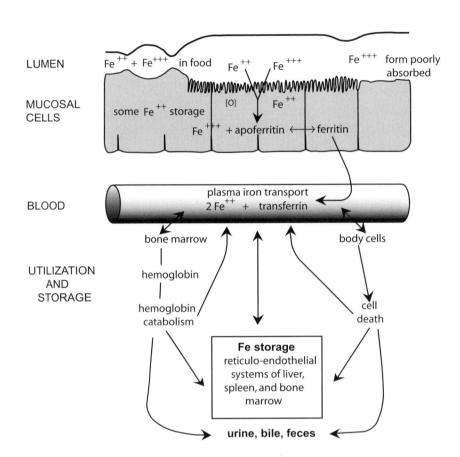

intakes can be toxic. Selenium is also interesting from an ecological point of view because its abundance in food plants varies geographically (van Ryssen 2001; figure 2.24). In some places animals can suffer from selenium deficiency whereas in others they suffer from ingesting it in excess. Selenium has very similar chemical properties to those of sulfur. It is biologically active when it replaces sulfur in the amino acids cystein and methionine, which become selenocysteine and selenomethyionine, respectively. The functions of selenium are closely linked to those of vitamin E. Both of these nutrients protect tissues from oxidative damage and both seem to be involved in the immune response. Selenium has an antioxidant function because selenocysteine is a component of the glutathione peroxidases, the enzymes that protect cells against oxidative damage (see 2.6.5. "Fatty Acids and Thermal Biology").

The bioavailability of selenium for plants depends on soil pH. When soils are acidic, selenium is in an insoluble form and the plants that grow in these soils have low selenium concentrations. When soils are alkaline and well aerated, selenium becomes available to plants as selenates (Flueck 1990). Most plants in most soils have low selenium levels, but a few species like the milkvetches and locoweeds

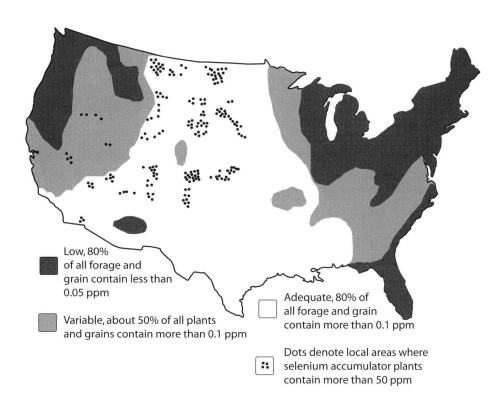

Figure 2.24. The concentration of selenium encountered by herbivores varies geographically. In some areas selenium concentration is sufficiently low to cause selenium deficiency. In other areas (mostly limited to arid environments), aggregations of selenium concentrating plants can make vegetation toxic. The map was redrawn after McDowell (1992).

Low, 80% of all forage and grain contain less than 0.05 ppm

Variable, about 50% of all plants and grains contain more than 0.1 ppm

Adequate, 80% of all forage and grain contain more than 0.1 ppm

Dots denote local areas where selenium accumulator plants contain more than 50 ppm

(genus *Astragalus*) accumulate selenium when selenium is bioavailable. These plants are normally unpalatable and herbivores eat them only when they have no other choice. Unfortunately, when selenium accumulators drop their leaves and these decompose, they make selenium available for absorption by palatable neighboring plants (Leininger et al. 1977; Lauchli 1993). Figure 2.24 shows the distribution of selenium levels in plants used by ungulates in the United States. Note that the plants in many large areas have low selenium levels whereas plants in many other (fortunately fairly localized) spots have exceedingly high levels.

Selenium deficiency is reported very often in herbivorous mammals, including cattle and wild ungulates (reviewed by Robbins 1993). The membranes of muscle cells often rupture in animals suffering from selenium deficiency. Because the damaged muscle turns white, the condition is called white muscle disease (its technical name is nutritional muscular dystrophy). Although selenium deficiency is common in herbivorous mammals, it is very rarely reported in other wild animals. To our knowledge, symptomatic selenium deficiency has been found in birds only once when it was reported in a captive flock of ducks (Dhillon and Winterfield 1983). Although wild herbivorous birds can have diets with very low selenium levels, these do not seem to be associated with negative effects (Fimreite et al. 1990). Why is selenium deficiency commonly found in mammalian herbivores but not in other species? There are two possible reasons: (1) Selenium deficiency is underdiagnosed in other creatures, or (2) it is much less common in other animals. It may be that selenium deficiency in ungulates is seasonally aggravated by the low vitamin E content of their herbivorous diets. Although fresh green foliage is rich in vitamin E, the content of dry foliage can be very low.

When it is ingested at high levels, selenium causes an abundance of symptoms (Spallholz and Hoffman 2002). Selenium produces toxic effects through (1) a mechanism that involves the formation of CH_3Se^- which either enters a redox cycle and generates superoxide and oxidative stress, or forms free radicals that bind to and inhibit important enzymes and proteins. (2) Excess selenium as selenocysteine leads to increased hydrogen selenide, an intermediate metabolite that causes liver damage. (3) The presence of excess selenium analogs of sulfur-containing enzymes and structural proteins plays a role in the malformations that are often seen in animals exposed chronically to selenium (see following section). When herbivores ingest large amounts of a selenium-accumulating plant (which can contain up to 9000 μg g^{-1} dry mass! [also called ppm]) they develop an acute condition that has the aptly descriptive name "blind staggers." Animals stumble around uncertainly until they die, which happens after only a few hours or days. This brings us to the apocryphal story of selenium's role in Custer's defeat (McDowell 1992). In the summer of 1876 General George A. Custer pushed his cavalry through a three-day forced march to Little Big Horn in Montana. The night before the battle, the cavalry's hungry and exhausted horses fed avidly on the selenium accumulators that grow in the alkali flats where Custer's soldiers camped. Because blind and stumbling horses do not make good war mounts, Custer lost the battle. Alternative (or maybe complementary) hypotheses fault Custer's arrogance and haste and invoke Sitting Bull and Crazy Horse's superior tactics. Although the selenium explanation for Custer's defeat is implausibly reductionistic, it makes a good story.

Animals can also develop chronic selenosis when they ingest selenium-containing vegetation at subacute levels for a long time. The clinical signs of chronic selenosis are loss of hair and elongated hooves in ungulates and deformed legs, toes, and beaks in birds. Selenium can pose a severe problem in arid wetlands that receive discharge from agricultural wastewater (Lemly et al. 1993). Evaporation in a closed drainage concentrates selenium and this high concentration of selenium is further magnified as selenium bioaccumulates up the wetland's food web (see chapter 9). The sad saga of the "poisoned" Kesterson Wildlife Refuge in the San Joaquin Valley of California provides a good example of a selenium-polluted arid wetland (Marshall 1985). This wildlife refuge was designed around a series of man-made reservoirs that were supposed to collect the water of the San Luis and Sacramento Rivers after it was used for irrigation. These reservoirs were designed to store the water temporarily before it was delivered to San Francisco

Bay. However, the drainage system from the reservoirs to the bay was never finished. The drainage that collected irrigation wastewater terminated in the shallow Kesterson Reservoir, which became an evaporation basin. Only ten years after the reservoir's construction, researchers found elevated selenium levels in every animal group that lived in or around Kesterson. A very large fraction of all the shorebirds and ducks that hatched in the reservoir had congenital malformations including missing eyes and feet, protruding brains, and very grossly deformed beaks, legs, and wings (Ohlendorf et al. 1990). Kesterson was designated a contaminated landfill and the total cost of mitigating its selenium contamination exceeded 60 million dollars (Lemly et al. 1993).

2.9 Secondary Metabolites

Plants synthesize an enormous variety of compounds that do not appear to play a major role in their primary nutritional or regulatory metabolism. A common term for these chemicals is secondary metabolites (SMs). Many invertebrates also synthesize such chemicals, or they may incorporate into their bodies the SMs from the plants or animals that they consume. The ability to sequester or synthesize what can be considered SMs extends to vertebrates such as fish, amphibians, and even birds (Dumbacher et al. 2000). Secondary metabolites play roles in communication, attraction, or defense against herbivores, predators, pathogens, and competitors (Harborne 1993). Secondary metabolites are so pervasive that it is almost a certainty that any thorough analysis of a plant food, and maybe even many animal foods, will identify some SMs. What is the nutritional and ecological significance of these compounds? Potentially, SMs can influence feeding behavior, interspecies interactions, and the distribution and abundance of animals.

An important part of figuring out the ecological consequences of SMs is to determine their identity and quantity in foods. Thousands have already been identified, and in our efforts to make some sense out of this almost bewildering diversity we try to categorize them according to their chemical structures, their chemical behavior, and/or the biosynthetic pathways by which they are produced. As an example, tannins, which have garnered a lot of attention from ecologists, are a heterogeneous group of plant phenolic compounds that are often placed into two different groups: condensed tannins and hydrolyzable tannins. Condensed tannins (also called proanthocyanidins) are large polymers consisting of a varying number of chains of flavan-3-ols. Hydrolyzable tannins are characterized as gallic or ellagic acid moieties esterified to a central carbohydrate molecule (most often glucose; figure 2.25). Tannins have also been characterized by their chemical behavior of binding and precipitating proteins,

Figure 2.25. Secondary
metabolites come in a vast
variety of chemical forms.
This table illustrates some
of the most common ones.

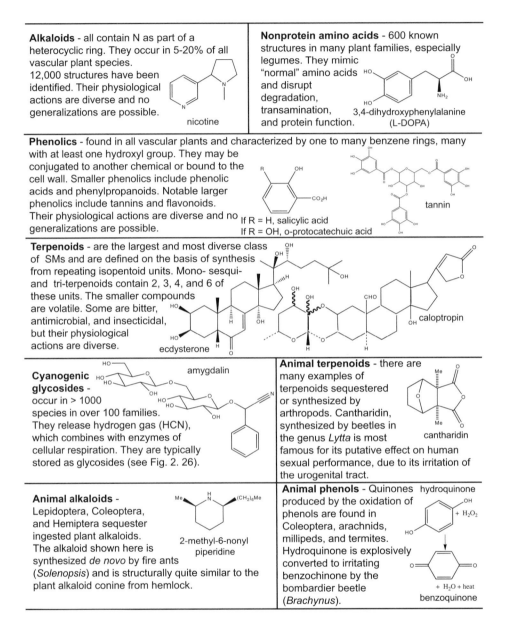

Alkaloids - all contain N as part of a heterocyclic ring. They occur in 5-20% of all vascular plant species. 12,000 structures have been identified. Their physiological actions are diverse and no generalizations are possible.
nicotine

Nonprotein amino acids - 600 known structures in many plant families, especially legumes. They mimic "normal" amino acids and disrupt degradation, transamination, and protein function.
3,4-dihydroxyphenylalanine (L-DOPA)

Phenolics - found in all vascular plants and characterized by one to many benzene rings, many with at least one hydroxyl group. They may be conjugated to another chemical or bound to the cell wall. Smaller phenolics include phenolic acids and phenylpropanoids. Notable larger phenolics include tannins and flavonoids. Their physiological actions are diverse and no generalizations are possible.
If R = H, salicylic acid
If R = OH, o-protocatechuic acid
tannin

Terpenoids - are the largest and most diverse class of SMs and are defined on the basis of synthesis from repeating isopentoid units. Mono- sesqui- and tri-terpenoids contain 2, 3, 4, and 6 of these units. The smaller compounds are volatile. Some are bitter, antimicrobial, and insecticidal, but their physiological actions are diverse.
ecdysterone
caloptropin

Cyanogenic glycosides - occur in > 1000 species in over 100 families. They release hydrogen gas (HCN), which combines with enzymes of cellular respiration. They are typically stored as glycosides (see Fig. 2. 26).
amygdalin

Animal terpenoids - there are many examples of terpenoids sequestered or synthesized by arthropods. Cantharidin, synthesized by beetles in the genus *Lytta* is most famous for its putative effect on human sexual performance, due to its irritation of the urogenital tract.
cantharidin

Animal alkaloids - Lepidoptera, Coleoptera, and Hemiptera sequester ingested plant alkaloids. The alkaloid shown here is synthesized *de novo* by fire ants (*Solenopsis*) and is structurally quite similar to the plant alkaloid conine from hemlock.
2-methyl-6-nonyl piperidine

Animal phenols - Quinones produced by the oxidation of phenols are found in Coleoptera, arachnids, millipeds, and termites. Hydroquinone is explosively converted to irritating benzochinone by the bombardier beetle (*Brachynus*).
hydroquinone
benzoquinone

and functional assays that measure protein-binding capacity of tannins have been developed (Robbins et al. 1987b). The organization of some of the SM classes in figure 2.25 emphasizes structural groupings. Notice how many SMs contain nitrogen in their structure (alkaloids, nonprotein amino acids, and cyanogenic glycosides), which should cause you to recall the fallacy of assuming that all N in a Kjeldahl analysis represents protein.

There are chemical assays that are fairly diagnostic for some of these classes of SMs. *Phytochemical methods* (Harborne 1980) is a classic general source of many such assays, some of which are simple enough to use under field conditions. Many books and journal articles are focused on specific groups of chemicals. When one relies on analyses of very broad classes of SMs, a big problem arises. Because broad classes of SMs include sometimes very heterogeneous groups of chemicals, it is difficult to interpret the nutritional and hence ecological significance of the measurements. The root of the problem is that the effects of specific compounds from within a class may range from nil to very toxic by any of an assortment of mechanisms. Let us use tannins again to illustrate this point. Condensed tannins apparently act in the gut to reduce the net availability of dietary protein in some but not all herbivores (Robbins et al. 1987a; 1991; Cork 1986) whereas some hydrolyzable tannins (or their moieties) are absorbed and act as metabolic toxins (Lowry et al. 1996). Gone are the days when ecologists could make broad assumptions such as that alkaloids are toxic or that phenolic compounds reduce nutrient digestibility! We simply cannot make accurate ecological predictions, such as food choice, based on data derived from very broad chemical assays. The best ecological studies involving SMs typically identify the specific compounds present in food and elucidate their effects on the animal experimentally (see chapter 9).

The identification process often begins with astute observation of animal behavior in the wild. For example, a biologist may notice that certain plants or plant parts are avoided by a herbivore, or perhaps predators avoid certain prey. The research goal then becomes to correlate the presence or absence of a specific chemical with animal performance, and later experimentally demonstrate it. Modern methods of extraction are based on the chemical's polarity (relative solubility in organic solvents) and solubility in water before and after modifications with salt and pH (Kaufman et al. 1999). Thus, an aqueous extract might be prepared by mixing frozen powder with hot water, followed by centrifugation and/or filtration, whereas an organic extract would be prepared by placing the powder in a Soxhlet extractor with organic solvent and refluxing for many hours (box 2.7). Sometimes these crude fractions are used in a first phase of animal testing to determine in which partition the most bioactive chemical(s) occur. Further analysis requires clean-up and purification followed by bioseparation of compounds within a fraction by chromatographic methods. If the position or retention time in chromatography matches that of known standards, then the identity of a compound might be tentatively made. But final analysis by mass spectrometry or nuclear magnetic resonance of a collected subsample, defined, for example, by its retention time in chromatography, is necessary in order to unambiguously identify the compound of interest. Finally, rather pure samples of

Box 2.7

The Soxhlet apparatus is used to extract lipids from a sample as well as to extract resins and lipophylic secondary metabolites. In essence, a Soxhlet is an efficient soaking device. The nonpolar solvent is heated in the boiler at the bottom, and the pure vapor rises up through the bypass tube and into the top part of the container. The vapor rises until it meets the cold condenser, and then the liquid drips back down into the thimble that holds the material from which you are extracting the SMs or lipids. This thimble is porous, so it not only holds the solid material, but also acts as a filter so that the material does not clog the siphon tube. When the level of the liquid in the Soxhlet container reaches the same level as the top of the siphon, then the liquid containing dissolved compounds is siphoned back into the boiler. Because the liquid in the thimble has to soak out of the porous sides, the level of liquid in the thimble does not fall as quickly as the level of the liquid surrounding it. The end result is that the material in the thimble is soaked repeatedly in pure solvent, and this is much more efficient than continuously soaking it in a solvent that steadily increases in concentration of the extracted compounds.

the compound can be used in experimental studies to test hypotheses about behavioral or physiological effects on animals. As you might imagine, research on ecological effects of SMs certainly benefits from collaboration of ecologists and chemists.

One interesting but complicating aspect of the scheme described above is that a SM may be stored in a plant or animal in what might be called an "inactive" form, perhaps in order to prevent autotoxicity. Examples include cyanogenic glucosides (figure 2.26), phenolic glucosides, and pyrimidine glucosides (Gopolan et al. 1992). Typically, toxic (or sometimes beneficial) effects are caused by the aglycone portion of the molecule following hydrolysis by β-glucosidases that are present in the plant and released upon maceration, in microbes inside the consumer's gut, or in the consumer's liver or intestine. Glucosides are not hydrolyzed by other types of gastrointestinal enzymes such as proteases or lipases or the stomach's acidity (Day et al. 2000). Because not all plants contain β-glucosidase, and many herbivores have primarily hindgut or nil intestinal microbial digestion, many herbivores

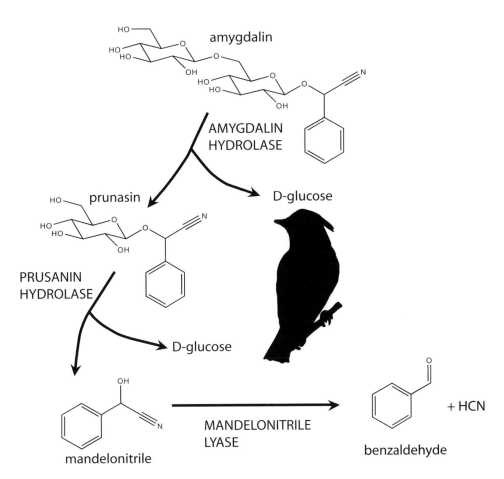

Figure 2.26. Cyanogenic glycosides are toxic to consumers through two possible mechanisms. In some plant tissues (i.e., the seeds of some plants in the Anacardiaceae), cyanogenic glycosides and the enzymes that are responsible for their hydrolysis are present in different cell and tissue compartments. Tissue damage by consumers leads to mixing of cyanogenic substrates (such as amygdalin and prunasin illustrated in the diagram) and to the release of cyanide (this phenomenon is sometimes called the "cyanide bomb"). Other plant tissues (i.e., the pulp of many fruits) contain only the cyanogenic glycoside. Ingestion of the unhydrolyzed glycoside can result in cyanide poisoning when these substances are hydrolyzed by the β-glucosydases of the consumer or of its gastrointestinal microbes. If the animals lack these enzymes, then the cyanogenic glycoside is absorbed and then excreted intact. Cedar waxwings can ingest five times the amount that kills a rat without any detectable effects. They simply excrete the compound intact (Struempf et al. 1999).

that lack intestinal β-glucosidase activity absorb amounts of glucosides intact in the intestine. Postabsorptive processing then determines whether the aglycone will be released and how toxic it will be (chapter 9). In a chemical analysis scheme, the glycoside may extract in the water fraction whereas the aglycone may actually be lipophilic. The higher water-solubility of the glycoside may permit a plant to store the compound in a vacuole with other water-soluble SMs, whereas lipophilic substances are typically sequestered in plant resin ducts, laticifers, on glandular hairs or trichomes, in thylakoid membranes, or on the cuticle (Wink 1999).

In a later chapter (chapter 9) we will review the effects of some SMs and how they are studied. For now, remember that SMs may increase the gross energy or total N content of a food, but their contribution is unlikely to be nutritive. If SMs have outright negative effects on consumers, it is a challenging research endeavor to define them and may require detailed research in diverse fields such as chemistry, physiology, and toxicology.

2.10 Words of Encouragement

Some ecologists are understandably discouraged when they discover that to answer many interesting ecological questions they have to spend a lot of time in the laboratory and master a variety of chemical techniques that rely on what can be cantankerous gadgets. After all, many of us became ecologists because it gave us an opportunity to spend time in the field. There are many solutions to this quandary. The first one (and the one that we have adopted) is to learn to enjoy the gadgets and the chemistry. It requires a peculiar mixture of tolerance to boredom with a childish compulsion to tinker with equipment. A second one (and one that we have also adopted) is to collaborate. It is close to impossible to master all the analytical techniques that ecologists can potentially need and it is difficult and inefficient to own (and maintain!) all the equipment needed to carry them out. Collaboration is a satisfactory, and much more efficient, way to go about getting the data. However, even when one does not conduct the necessary analysis in one's own laboratory, it is useful to understand the basic principles of the methods that our collaborators use. Understanding these can be intellectually challenging, but also it can be a lot of fun and it facilitates carrying out a successful collaboration. Third, you may find data in the literature that you can reasonably substitute for more accurate data obtained at the cost of new chemical analyses for your system. For example, there are several summaries of chemical composition of specific food types (e.g., arthropods (Bell 1990; Barker et al. 1998)). One of us (WHK) compiled a table for handy calculations in nutritional ecology (table 2.8).

TABLE 2.8
Average chemical characteristics of major food types

Food Type	Refractory Material[a]	N content (mg g^{-1} dry mass)	Energy Content (kJ g^{-1} dry mass)
Nectar (sucrose)	0	0	16.7
Vertebrate prey	0.04–0.17	122	23.1
Arthropods	0.01–0.5	86	24.5
Cultivated seeds		20	21.5
Wild seeds	0.18–0.53	10–28	21.5
Fruit, pulp, and skin	0.09–0.34	10	19.5
Fruit, pulp, skin, and seeds	0.4	10	21.6
Herbage (mainly leaves)	0.22–0.61	15	18.2

Note: These averages, or ranges, were determined from scores of studies of birds that eat a variety of food types (Karasov 1990), but can be reasonably applied for many other vertebrates.

[a] Refractory material is the proportion of the dry mass not easily digested by endogenous enyzmes of vertebrates, such as cell wall material of plants, cuticle of arthropods, hair, feathers, and bones of vertebrate prey.

References

Ainsen, P., C. Enns, and M. Wessling Resnick. 2001. Chemistry and biology of eukaryotic iron metabolism. *International Journal of Biochemistry and Cell Biology* 33:940–959.

Asp, N. G., C. G. Johansson, H. Hallmer, and M. Siljestrom. 1983. Rapid enzymatic assay of solube and insoluble dietary fiber. *Journal of Agricultural Food Chemistry* 31:476–482.

Baker, D. H., and G. L. Czarnecki-Maulden. 1991. Comparative nutrition of cats and dogs. *Annual Review of Nutrition* 11:239–263.

Baker, H. G. 1970. *Plants and civilization.* Wadworth, Belmont, Calif.

Banga, J. R., and R. P. Singh. 1994. Optimization of air drying of foods. *Journal of Food Engineering* 23:189–211.

Barker, D., M. P. Fitzpatrick, and E. S. Dierenfeld. 1998. Nutrient composition of selected whole invertebrates. *Zoo-Biology* 17(2):123–134.

Behan-Martin, M. K., G. R. Jones, K. Bowler, and A. R. Cossins. 1993. A near perfect temperature adaptation of bilayer order in vertebrate brain membranes. *Biochimica et Biophysica Acta* 1151:216–222.

Bell, G. P. 1990. Birds and mammals on the insect diet: A primer on diet composition analysis in relation to ecological energetics. *Studies in Avian Biology* 13:416–422.

Borman, A., T. R. Wood, H. C. Black, E. G. Anderson, M. J. Oesterling, M. Womack, and W. C. Rose. 1946. The role of arginine in growth with some observations on the effects of argininic acid. *Journal of Biological Chemistry* 166:585–594.

Brett, M. T., and D. C. Müller-Navarra. 1997. The role of highly unsaturated fatty acids in aquatic foodweb processes. *Freshwater Biology* 38:483–499.

Brown, A. J., and P. E. Koattukudy. 1978. Evidence of pancreatic lipase is responsible for the hydrolysis of cutin, a biopolyesther present in mammalian diets, and the role of bile salt and colipase in this hydrolysis. *Archives of Biochemistry and Biophysics* 190:17–26.

Buzadzic, B., M. Spasic, Z. S. Sazicic, R. Radojicic, V. M. Petrovic, and B. Halliwell. 1990. Antioxidant defences in the ground squirrel, *Ciellus citellus* 2. The effect of hibernation. *Journal of Free Radical Biology and Medicine* 9:407–413.

Carpita, N. C. 1990. The chemical structure of the cell walls of higher plants. Pages 15–30 in D. Kritchevsky, C. Bonfield, amd J. W. Anderson, eds., *Dietary fiber: Chemistry, physiology, and health effects.* Plenum Press, New York.

Chow, C. K. 2000. *Fatty acids in foods and their health implications.* Marcel Decker, New York.

Clarke, A. 1980. The biochemical composition of Krill, *Euphasia superba* Dana, from South Georgia. *Journal of Experimental Marine Biology and Ecology* 43:221–236.

Cook, H. W. 1991. Fatty acid desaturation and chain elongation in eucaryotes. Pages 141–169 in D. E. Vance and J. Vance, eds., *Biocehmistry of lipids, lipoproteins, and membranes.* Elsevier Science, New York.

Cordain, L., B. A. Watkins, G. L. Florant, M. Kehler, L. Rogers, and Y. Li. 2002. Fatty acid analysis of wild ruminant tissues: evolutionary implications for reducing diet-related chronic disease. *European Journal of Clinical Nutrition* 56:181–191.

Cork, S. J. 1986. Foliage of *Eucalyptus punctata* and the maintenance nitrogen require-ments of koalas, *Phascolarctos cinereus. Australian Journal of Zoology* 34:17–23.

Crockett, E. L. 199. Cholesterol function in plasma membranes from ectotherms: membrane specific roles in adaptation to temperature. *American Zoologist* 38:291–304.

Cunnane, S. C. 2000. The condtional nature of the dietary need for polyunsaturates: A proposal to reclasiffy "essential fatty acids" as "conditionally indispenable" or "conditionally dispensable" fatty acids. *British Journal of Nutrition* 84:803–812.

Day, A. J., F. J. Canada, J. C. Diaz, P. A. Kroon, R. Mclauchlan, C. B. Faulds, G. W. Plumb, R. A. Morgan, and G. Williamson. 2000. Dietary flavonoid and isoflavone glycosides are hydrolysed by the lactase site of lactase phlorizin hydrolase. *FEBS Letters* 468:166–170.

Dhillon, A. S., and R. W. Winterfield. 1983. Selenium-vitamin E deficiency in captive wild ducks. *Avian Disease* 27:257–263.

Dumbacher, J. P., P. F. Spande, and J. W. Daly. 2000. Batrachotoxin alkaloids from passerine birds: A second toxic bird genus (*Ifrita kowaldi*) from New Guinea. Pro-ceedings of the National Academy of Science. 97:12970–12975.

Englyst, H. N., S. M. Kingman, and and J. H. Cummings. 1992. Classification and measurement of nutritionally important starch fractions. *European Journal of Clinical Nutrition* 46 (suppl. 2):S33–S50.

Ferraris, R. P., and H. V. Carey. 2000. Intestinal transport during fasting and malnutri-tion. *Annual Review of Nutrition* 20:195–219.

Fimreite, N., E. K. Bart, and A. Munkeford. 1990. Cadmium and selenium levels in tetraonids from selected areas in Norway. *Fauna Norvegica Series C Cinclus* 13:79–84.

Florant, G. L. 1998. Lipid metabolism in hibernators: The importance of essential fatty acids. *American Zoologist* 38:331–340.

Flueck, W. T. 1990. Possible impact of emisions on trace mineral availability to free-rang-ing ruminants: Selenium as an example. *Zeitschrift fuer Jagdwissenschaft* 36:179–185.

Frank, C. L., E. S. Dierenfeld, and K. B. Storey. 1998. The relationship between lipid perox-idation, hibernation, and food selection in mammals. *American Zoologist* 38:341–349.

Frank, C. L., and K. B. Storey. 1995. The optimal depot fat composition for hibernation by golden-mantled ground squirrels (*Spermophilus lateralis*). *Journal of Comparative Physiology B* 164:536–542.

Frausto da Silva, J.J.R., and R.J.P. Williams. 1994. *The biological chemistry of the elements.* Oxford University Press, Oxford.

Geiser, F., B. T. Firth, and R. S. Seymour. 1992. Polyunsaturated dietary lipids lower the selected body temperature of a lizard. *Comparative Biochemistry and Physiology B* 162:1–4.

Gibbs, A. G. 1998. The role of lipid physical properties in lipid barriers. *American Zoologist* 38:268–279.

Gopolan, V., A. Pastuszyn, W. R. Galey, and R. H. Glew. 1992. Exolytic hydrolysis of toxic plant glucosides by guinea pig liver cytosolic beta glucosidase. *Journal of Biological Chemistry* 267:14027–14032.

Hachman, R. H., and M. Goldberg. 1981. A method for determination of microgram amounts of chitin in arthropod cuticles. *Analytical Biochemistry* 110:277–280.

Hadley, N. F. 1984. *The adaptive role of lipids in biological systems*. Wiley-Interscience, New York.

Halliwell, B. 1997. Antioxidants and human disease: A general introduction. *Nutrition Review* 55:S44–S52.

Hanssen, K., H. J. Grav, J. B. Steen, and H. Lysnes. 1979. Vitamin C deficiency in growing willow ptarmigan (*Lagopus lagopus lagopus*). *Journal of Nutrition* 109:2260–2278.

Harborne, J. B. 1980. *Phytochemical methods*. Chapman and Hall, London.

Harborne, J. B. 1993. *Introduction to ecological biochemistry*. Academic Press, New York.

Harlow, H. J., and C. L. Frank. 2001. The role of dietary fatty acids in the evolution of spontaneous and facultative hibernation in prairie dogs. *Journal of Comparative Physiology B* 171:77–84.

Hazel, J. R., and E. E. Williams. 1990. The role of alterations in membrane lipid compositionin enabling physiological adaptation of organisms to their physical environment. *Progress in Lipid Research* 29:167–227.

Hecky, R. E. 1984. African lakes and their trophic efficiencie: a temporal perspective. Pages 405–448 in D. G. Meyers and J. R. Strickler (eds.), *Trophic interactions within aquatic ecosystems*. Westview Press, Boulder, Colorado.

Herbst, L. H. 1988. Methods of nutritional ecology of plant-visiting bats. Pages 233–246 in T. H. Kunz, ed., *Ecological and behavioral methods for the study of bats*. Smithsonian Institution Press, Washington, D.C.

Hughes, M. J., M. R. Cowell, and P. T. Craddock, 1976. Atomic absorption techniques in Archaeology. *Archaeometry* 18:19–37.

Imafidon, G. I, and F. W. Sosulski. 1990. Nucleic acid nitrogen of animal and plant foods. *Journal of Agricultural and Food Chemistry* 38:118–120.

Irving, L., and J. Krog. 1955. Temperature of skin in the arctic as a regulator of heat. *Journal of Applied Physiology* 7:355–364.

Iyengar, G. V., K. S. Subramanian, and J.R.W. Woittiez. 1997. *Element analysis of biological samples: Principles and practice*. CRC Press, New York.

Jeauniaux, C., and C. Cornelius. 1978. Distribution and activity of chitinolytic enzymes in the digestive tract of birds and mammals. Pages 542–549 in R.A.A. Muzzarell and E. R. Parizier (eds.), *Proceedings of the first international conference on chitin/chitosa* (MIT Press, Cambridge, Mass.).

Käkelä, R., H. Hyväriven, and P. Vainiotalo. 1996. Unusual fatty acids in the depot fat of the Canadian beaver (*Castor canadensis*). *Comparative Biochemistry and Physiology B* 113:625–629.

Karasov, W. H. 1990. Digestion in birds: chemical and physiological determinants and ecological implications. *Studies in Avian Biology* 13:391–415.

Karasov, W. H., and B. Pinshow. 1998. Changes in lean mass and in organs of nutrient assimilation in a long-distance passerine migrant at a springtime stopover site. *Physiological Zoology* 71:435–448.

Karel, M., J. L. Wellbourn, and R. P. Singh. 1985. Innovative methods of food preservation and processing. *Food Technology* 39(6):17R–20R.

Kates, M. 1972. *Techniques in lipidology: isolation, analysis, and identification of lipids*. American Elsevier, New York.

Kaufman, P. B., L. J. Cseke, S. Warber, J. A. Duke, and H. L. Brielmann. 1999. *Natural products from plants*. CRC Press, Boca Raton, Fla.

Kazuyuki, O., I. Sawatani, C. Hiroto, F. Shigerahu, and K. Masashi. 1998. Trehalose content in foods. *Nippon Shokuhin Kagaku Kogaku Kaishi* 45:381–384.

Kearns, C. A., and D. W. Inouye. 1993. *Techniques for pollination ecologists*. University Press of Colorado, Niwot, Colo.

Kyu, K. 1997. Re-evaluation of protein and amino acid requirements of rainbow trout (Oncorhyncus mykiss). *Aquaculture* 151:3–7.

Lauchli, A. 1993. Selenium in plants: uptake, functions, and environmental toxicity. *Botanica Acta* 106:455–468.

Leininger, W. C., J. E. Taylor, and C. L. Wambolt. 1977. Poisonous range plants of Montana. Cooperative Extension Service, Montana State University Bulletin No. 348.

Li, Y., C. Liang, C. W. Slemenda, R. W. Ji, S. Sun, J. Cao, C. Emsley, F. Ma, Y. Wu, Ying-Po; Y. Zhang-Yan, S. Gao, W. Zhang, B. P. Katz, S. Niu, S. Cao, and C. C. Johnston. 2001. Effect of long-term exposure to fluoride in drinking water on risks of bone fractures. *Journal of Bone and Mineral Research* 16:932–939.

Linder, C. R. 2000. Adaptive evolution of seed oils in plants: accounting for the biogeographic distribution of saturated and unsaturated fatty acids in oils. *The American Naturalist* 156:442–458.

Lindsay, S. 1987. *High performance liquid chromatography*. John Wiley and Sons, London.

Lemly, A. D., S. E. Finger, and M. Nelson. 1993. Sources and impacts of irrigation drainwater contaminants in arid wetlands. *Environmental Toxicology and Chemistry* 12:2265–2279.

Levey, D. J., and C. Martínez del Rio. 2001. It takes guts (and more) to eat fruit: lessons from avian nutritional ecology. *Auk* 118:819–831.

Lowry, J. B., C. S. McSweeney, and B. Palmer. 1996. Changing perceptions of the effect of plant phenolics on nutrient supply in the ruminant. *Australian Journal of Agricultural Research* 47:829–842.

Lyman, C. P. 1982. Who is who among the hibernators. Pages 12–36 in C. P. Lyman, J. S. Willis, A. Malan, and L. S. Wangg (eds.), *Hibernation and torpor in mammals and birds*. Academic Press, New York.

MacDonald, M. L., Q. R. Rogers, and J. G. Morris. 1984. Nutrition of the domestic cat, a mammalian carnivore. *Annual Review of Nutrition* 4:521–562.

Marsh, R., C. Moe, R. N. Lomneth, J. D. Fawcett, and A. Place. 2001. Characterization of gastrointestinal chitinase in the lizard *Sceloporus undulatus garmani* (Reptilia: Phrynosomatidae). *Comparative Biochemistry and Physiology A* 128:675–682.

Marshall, E. 1985. Selenium poisons refuge and California politics. *Science* 229:144–146.

Martínez del Rio, C. 1997. Can passerines synthesize vitamin C? *Auk* 114:513–516.

Mathews, C. K., and K. E. van Holde. 1990. *Biochemistry*. Benjamin/Cummings, Redwood City, Calif.

McDowell, L. R. 1989. *Vitamins in animal nutrition*. Acedemic Press, New York.

McDowell, L. R. 1992. *Minerals in animal and human nutrition*. Academic Press, New York.

McMaster, M. C. 1994. *HPLC: A practical user's guide*. VCH, New York.

Messer, M., E. A. Crisp, and R. Czolij. 1989. Lactose digestion in suckling macropodids. Pages 217–221 in G. Grigg, P. Jarman, and I. Hume (eds.), *Kangaroos, wallabies, and rat-kangaroos*. Surrey-Beatty and Sons, Sydney, Australia.

Morris, J. G. 1985. Nutritional and metabolic responses to arginine deficiency in carnivores. *Journal of Nutrition* 115:524–531.

Müller-Navarra, D. C. 1995. Evidence that a highly unsaturated fatty acid limits *Daphnia* growth in nature. *Archiv für Hydrobiologie* 132:297–307.

Ohlendorf, H. M., R. L. Hothem, C. M. Bunck, and K. C. Marois. 1990. Bioaccumulation of selenium at Kesterson Reservoir, California. *Archives of Environmental and Contamination Toxicology* 19:845–853.

Olsen, M. A., A. S. Blix, T. H. A. Utsi, W. Sormo, and S. D. Mathiesen. 2001. Chitinolytic bacteria in the minke whale forestomach. *Canadian Journal of Microbiology* 46:85–94.

Pauling, L. 1970. *Vitamin C and the common cold*. Freeman, San Francisco.

Pawlosky, R., A. Barnes, and N. Salem. 1994. Essential fatty acid metabolism in the feline: Relationship between liver and brain production of long-chain polyunsaturated fatty acids. *Journal of Lipid Research* 35:2032–2040.

Pond, C. A. 1981. Storage, Pages 190–219 in C. R. Townsend, and P. Calow (eds.), *Physiological ecology: An evolutionary approach to resource use*. Blackwell Scientific Publications, Oxford.

Pond, C. A. 1998. *The fats of life*. Cambridge University Press, Cambridge.

Pond, C. A., C. A. Mattacks, R. H. Colby, and N.J.C. Tyler. 1993. The anatomy, chemical composition and maximum glycolytic capacity of adipose tissue in wild Svalbard reindeer (*Rangifer tarandus platyrhinchus*) in winter. *Journal of Zoology, London* 229:17–40.

Pritchard, G. T., and C. T. Robbins, 1990. Digestive and metabolic efficiencies of grizzly and black bears. *Canadian Journal of Zoology* 68:1645–1651.

Prosky, L., N. G. Asp, I. Furda, J. W. De Vries, T. F. Schweizer, and B. F. Harland. 1984. Determination of total fiber in foods, food products, and total diets: Interlaboratory study. *Journal of the Association of Analytical Chemists* 67:1044–1051.

Quarmby, C., and S. E. Allen. 1989. Organic constituents. Pages 160–200 in S. E. Allen ed., *Chemical analysis of ecological materials*. Blackwell Scientific Publications, Oxford.

Reeds, P. J. 2000. Dispensable and indispensable amino acids for humans. *Journal of Nutrition* 130:1835S–1840S.

Robbins, C. T. 1993. *Wildlife feeding and nutrition*. Academic Press, New York.

Robbins, C. T., A. E. Hagerman, P. J. Austin, C. McArthur, and T. A. Hanley. 1991. Variation in mammalian physiological responses to a condensed tannin and its ecological implications. *Journal of Mammalogy* 72:480–486.

Robbins, C. T., T. A. Hanley, A. E. Hagerman, O. Hjeljord, D. L. Baker, C. C. Schwartz, and W. W. Mautz. 1987a. Role of tannins in defending plants against ruminants: Reduction in protein availability. *Ecology* 68:98–107.

Robbins, C. T., S. Mole, A. E. Hagerman, O. Hjeljord, D. L. Baker, C. C. Schwartz, and W. W. Mautz. 1987b. Role of tannins in defending plants against ruminants: Reduction in dry matter digestion? *Ecology* 68:1606–1615.

Roser, B., and C. Colaço. 1993. A sweeter way to fresher food. *New Scientist* 138:25–28.

Shigenobu, S., H. Watanabe, H. Hattori, M. Sakaki, and H. Ishikawa. 2000. Genome sequence of the endocellular bacterial symbiont of aphids *Buchnera sp.* APS. *Nature* 407:81–86.

Simmone, A. H., E. H. Simmone, R. R. Eitenmiller, H. A. Mills, and C. P. Crissman. 1997. Could the Dumas method replace the Kjeldahl digestion for nitrogen and crude protein determination in foods? *Journal of the Science of Food and Agriculture* 73:39–45.

Sinensky, M. 1974. Homeoviscous adaptations — homeostatic process that regulates the viscosity of membrane lipids in *Escherichia coli*. *Proceedings of the National Academy of Sciences* 71:522–525.

Slansky, F., and J. M. Scriber. 1985. Food consumption and utilization. Pages 87–163 in G. A. Kerkut and L. I. Gilbert (eds.), *Comprehensive insect physiology, biochemistry and pharmacology*, vol. 4. Pergamon Press, Oxford.

Snow, P., and K. O'Dea. 1981. Factors affecting the rate of hydrolysis of starch in food. *American Journal of Clinical Nutrition* 34:2721–2727.

Sosulski, F. W., and G. I. Imafidon. 1990. Amino acid composition and nitrogen to protein conversion factors for animal and plant foods. *Journal of Agriculture and Food Chemistry* 38:1351–1356.

Southgate, D.A.T. (1976). *Determination of food carbohydrates*. Applied Science Publishers, London.

Spallholz, J. E., and D. J. Hoffman. 2002. Selenium toxicity: Cause and effects in aquatic birds. *Aquatic Toxicology* 57:27–32.

Streumpf, H. M., J. Schondube, and C. Martínez del Rio. 1999. Amygdalin, a cyanogenic glycoside in ripe fruit does not deter consumption by cedar waxwings. *Auk* 116:749–758.

Trowell, H. C. 1974. Definitions of fiber. *Lancet* 1:503.

van Ryssen, J. B. J. 2001. Geographical distribution of the selenium status of herbivores in South Africa. *South African Journal of Animal Science* 31:1–8.

Van Soest, P. 1978. Dietary fibers: Their definition and nutritional properties. *American Journal of Clinical Nutrition* 31 (Suppl.):S12–S20.

Van Soest, P. 1994. *Nutritional ecology of the ruminant*. Cornell University Press, Ithaca, N.Y.

Watanabe, H., and G. Tokuda. 2001. Animal cellulases. *Cellular and Molecular Life Sciences* 58:1167–1178.

Weiser, J. I., A. Porth, D. Mertens, and W. H. Karasov 1997. Digestion of chitin by Northern Bobwhites (*Colinus virginianus*) and American Robins (*Turdus migratorius*). *Condor* 99:553–556.

Weitian, L., H. Jiao, N. C. Murray, M. O'Connor, and W. L. Roelofs. 2002. Gene characterized for membrane desaturase that produces Δ11 isomers of mono and diunsaturated fatty acids. *Proceedings of the Natural Academy of Sciences* 99:620–624.

White, D. C., and E. W. Stiles. 1985. The use of refractometry to estimate nutrient rewards in vertebrate-dispersed fruits. *Ecology* 66:303–307.

Williams, R.J.P., and J.J.R. Frausto da Silva. 1996. *The natural selection of the chemical elements*. Oxford University Press, Oxford.

Wink, M. 1999. *Biochemistry of plant secondary metabolism*. Sheffield Academic Press, Sheffield, U.K.

Digestive Ecology

CHAPTER THREE

Food Intake and Utilization Efficiency

3.1 Overview of Section III: Why Study Digestion?

Digestion is important to ecologists for many reasons, but here are three big ones. First, digestive processes determine how efficiently animals extract energy and nutrients from their foods. A low digestive efficiency translates into more time and effort spent collecting food to meet a fixed required net input. In addition, a low digestive efficiency translates into less matter and energy available for maintenance, growth, and reproduction. Second, digestive efficiency determines the flux (i.e., amount per unit time) and makeup of materials transferred to other trophic levels in an ecosystem. Third, interactions of animals with plants, such as pollination and seed dispersal, sometimes hinge on particulars of the animals' digestive systems. You can think of more examples, but they all serve to emphasize that what animals eat and excrete shapes their role in ecological communities and determines their contribution to the flux of energy and materials in ecosystems (see also chapter 10 on ecological stoichiometry).

To begin our discussion about digestion we will link it to the previous chapter on food chemistry by showing that the efficiency of energy and nutrient extraction can be predicted from knowledge of food composition, particularly of the proportion that is refractory (i.e., relatively resistant to digestion; see chapter 2). However, it quickly becomes apparent that there is so much more than food composition to consider. If an animal's strategy is to maximize energy gain, is it best always to eat as much as the digestive system can possibly process, or do animals hold back and retain some digestive capacity as a buffer for emergencies? By the end of chapter 3, not only will you be acquainted with some useful tools for ecologists interested in feeding, but you will be introduced to these and other questions that in total make up a research area one might call "digestive ecology". We will continue describing the questions and approaches of digestive ecology in chapters 4–6.

To gain a satisfactory understanding of the ecological roles of digestion we must inevitably peer inside animals at specific digestive processes. To this end, we devote chapters 4 and 6 to two major types of digesters. In the first type, which we consider in chapter 4, animals digest relatively nonrefractory materials in foods using enzymes synthesized by themselves. In the second type, which we consider in chapter 6, animals digest relatively refractory materials by fermenting them with the aid of symbiotic microbes. Although there are empirical and theoretical reasons to establish this dichotomy between fermenters and nonfermenters, do remember that the separation is also heuristic and that often both systems operate concurrently in some animals. By the time that you have read all three chapters you will be familiar with what we call digestive ecology. You will have a greater appreciation for the important interplay between the digestive system and ecology that occurs over both ecological and evolutionary time.

3.2 Digestive Efficiency Is Inversely Related to "Fiber" Content

3.2.1 A Case Study: The Bear

Imagine performing balance trials with bears! This is something that wildlife ecologist Charlie Robbins has done with grizzlies. Digestive efficiencies are typically measured in balance trials in which rates of energy or material input and output of captive animals are carefully measured. Methods box 3.1 describes one such trial with bears (Pritchard and Robbins 1990). For our purposes, bears are a good example of species with minimal fermentative capacity. The data from the balance trial, when analyzed in the context of the scheme in figure 3.1, permit calculation of several types of apparent efficiencies such as the diet's dry matter digestibility (D_m^*), energy digestibility (D_e^*) or digestibility of some other particular component such as N digestibility (D_N^*). The basic calculation for any substance i uses the fluxes (or rates) of intake and elimination, which are either rates of food intake and feces production or the products of those dry mass fluxes and their respective concentrations of i per g dry matter:

$$D_i^* = \frac{\text{intake rate of } i - \text{fecal elimination rate of } i}{\text{intake rate of } i}. \tag{3.1}$$

Box 3.1. Measurement of Digestive and Metabolic Efficiency in Bears by the Balance Method

Black bears (*Ursus americanus*; mean body mass 48 kg) and grizzly bears (*Ursus arctos*; 105 kg) were placed into special cages designed so that their urine and feces could be collected separately, and they were fed various test foods (Pritchard and Robbins 1990). They were allowed a 3–5 day pretrial period during which other foods in their guts might be expelled (this takes less than 1 day in the bears) and during which their enzymatic processes might habituate to the new diet. In other animals, such habituation periods might be even longer if changes in gut structure are thought to occur or if processes are slower. After the period of habituation, the bears were offered as much food as they wanted for 7 days. Daily measures were made of food provided, food not consumed (called orts), feces produced, and urine produced. Subsamples of the food, and some or all of the feces and urine were dried to determine the ratio of dry mass to wet mass. The energy contents (kJ g^{-1} dry mass) of food, feces, and urine were measured in a bomb calorimeter, and the nitrogen contents were measured. Average values are shown in table 3.1 for two of the diets fed to both species, which performed very similarly. Pritchard and Robbins (1990) also collected urine and measured its energy content. For black bears eating blueberries, the flux of energy loss via urine was 11.63 kJ d^{-1} kg$^{-3/4}$, so applying the principle of equation 3.2,

$$MEC^* = ((57.4 \times 18.71) - (20.26 \times 19.24) - (11.63))/(57.4 \times 18.71) = 0.626.$$

Sometimes, the values of MEC^* are corrected to nitrogen balance because urinary energy is partly dependent on nitrogen (N) metabolism. For example, an animal in a feeding trial that is not eating enough (not in energy balance) will catabolize some body protein and some of its urinary energy will thus be derived from tissue catabolism. The correction is accomplished by adding 36.5 kJ per gram of N deficit to the numerator in equation 3.2 (Sibbald 1982). If the animal is storing N, this quantity is subtracted. The value 36.5 corresponds to the energy content of uric acid, but if urea is excreted then a lower value is used (31.2 kJ g^{-1} N deficit).

We do not calculate metabolizability of any food component other than energy because the calculation would have little meaning. To illustrate, suppose that the bears in our example were more or less in steady state for N (input = output). If equation 3.2 were applied to the data on N input and output, the calculated "metabolizability" of nitrogen would be 0, which is not a true reflection of the value of the food N. Also, the N content of

continued

continued

blueberries was so low that the bears' obligatory losses of N probably exceeded their intake. As we will discuss in a later chapter, if an animal's output rate exceeds its input rate then there will be a decrease in the N content of the body exactly equal to the difference between input and output. This N imbalance is sometimes called a negative N balance, or negative storage (which is not the most fortunate of terms). In this case equation 3.2 would yield a negative efficiency, which is again not a very useful calculation. Hence, equation 3.2 is not applied to any food component other than energy.

TABLE 3.1
Results from a balance trial with bears

	Trials with blueberries		Trials with trout	
	Black bear	Grizzly bear	Black bear	Grizzly bear
Food				
Intake (g d^{-1} $kg^{-3/4}$)	57.4	56.7	12.4	14.4
Energy (kJ g^{-1})	18.71	18.71	23.91	23.91
Nitrogen (mg g^{-1})	8.96	8.96	111.4	111.4
Feces				
Loss (g d^{-1} $kg^{-3/4}$)	20.26	20.53	1.60	1.48
Energy (kJ g^{-1})	19.24	19.28	14.27	12.90
Nitrogen (mg g^{-1})	21.04	20.07	50.93	52.40
Calculated apparent digestibility				
Dry matter D_m^*	0.647	0.638	0.871	0.898
Energy D_e^*	0.637	0.627	0.923	0.945
Nitrogen D_N^*	0.171	0.189	0.941	0.952

Data from Pritchard and Robbins (1990).

You may have noticed that several terms are used interchangeably for the utilization efficiencies from balance trials, e.g., digestive efficiency, apparent digestibility, digestibility coefficient. In the case of digestion, consider the first to be the conceptual ideal while the two latter terms apply to its empirical approximation. Because there is no uniformity in how these terms are used in the literature, it is important to always define them clearly in publications. Entomologists (Waldbauer 1964) and aquatic biologists (Penry 1998) use a somewhat parallel framework to study the efficiency with which terrestrial and aquatic invertebrates convert ingested food into new body substance. Readers should have no trouble interpreting the entomological terms using the framework described here.

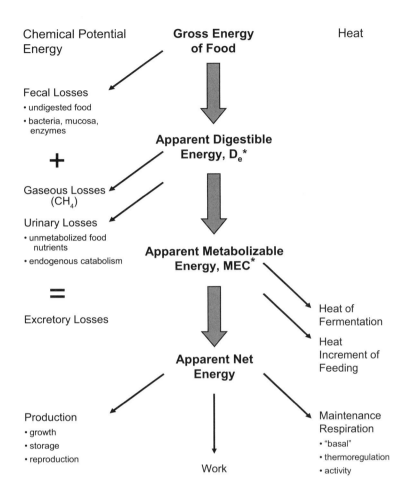

Figure 3.1. An input/output scheme of digestive and postdigestive processing of energy in a mammal. Outputs include chemical potential energy in several forms (aligned along the left side of the figure) and heat evolved from several processes (aligned along the right). Chemical potential energy outputs mainly from digestive processing include fecal and gaseous losses.

These digestibilities are called apparent, and therefore receive an asterisk (*), because the feces sometimes contain, in addition to the undigested food residue, so-called endogenous losses such as sloughed cells from the digestive tract, unabsorbed digestive secretions, or microbial material. Although the apparent digestibility of energy is usually a little lower than the true digestive efficiency for energy, its use in ecology is entirely appropriate because the endogenous losses represent real losses that must be replaced.

In the study on black bears the researchers fed nine different foods singly to the bears. They measured the fiber content of each food because they wanted to learn how to predict digestive efficiency from knowledge of food chemistry. As discussed in chapter 2, the term fiber is used in several ways by scientists and needs to be defined in each case. Here, we use an inclusive definition often used by human nutritionists that potentially includes plant cell wall materials, chitin, bone, and various other biochemicals that are difficult to digest by animals that ferment little. It was measured by an enzymatic technique and its fraction of dry matter is called total dietary fiber (TDF). It did indeed provide a useful measure for predicting the nutritional value of the foods, because the bears' digestive efficiency for dry matter or energy was highly inversely correlated with this chemical measure of the fraction of the food that is refractory to digestion (figure 3.2).

Figure 3.2. The digestive efficiency of black bears declines with increasing fiber content of their food. Nine foods, whose total dietary fiber was characterized by an enzymatic technique, were fed singly to bears and the bears' feces and urine were collected. Their dry matter digestibility (D_m^*) is indicated with filled circles and the least squares linear regression is shown by the solid line. Other efficiencies that were determined in the balance trials, digestible energy (D_e^*; unfilled triangle, dashed line) and metabolizable energy (MEC^*; unfilled square, dotted line) showed the same pattern with dietary fiber. (Data from Pritchard and Robbins 1990.)

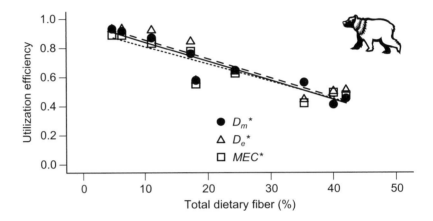

Cost-conscious ecologists sometimes substitute D_m^* as a reasonable estimate for D_e^*, which requires bomb calorimetry measurements, because the relationship between D_m^* and D_e^* is usually 1:1. But for high-fat foods D_e^* may exceed D_m^* by a few percent (Robbins 1993). Another utilization efficiency that can be calculated from the balance trial is called the apparent metabolizable energy coefficient (MEC^*), which includes the amounts lost in urine and gas (e.g., methane, figure. 3.1 and chapter 6):

$$MEC^* = \frac{\text{energy intake rate} - \text{fecal energy loss rate} - \text{urinary and gaseous energy loss rate}}{\text{energy intake rate}}.$$

(3.2)

MEC^* is "apparent" for the same reason that the digestibility coefficient is best considered "apparent"—it may reflect also endogenous losses. As will be discussed in chapters 7 and 9, several nutritional factors contribute to energy losses in the urine, and these losses must be accounted for in the energy budget. Also, methane produced in the digestive tract during fermentation cannot be oxidized by most animals and becomes an energy loss. Like D_e^*, the MEC^* is a proportion (or percentage if multiplied by 100) of the food energy, but because it includes additional losses in the urine and methane, metabolizability coefficients are necessarily lower than D_e^*, typically by 3–8 percentage points among nonfermenters, depending on diet (Robbins 1993). Because the differences are small, in relation to the very large differences in efficiency associated with diet differences, MEC^* was also highly inversely correlated with TDF (figure 3.2). This observation becomes very useful once we turn our attention to other animals besides mammals.

Many animals, including most nonmammalian vertebrates and invertebrates, produce excreta that are a combination of feces and urine. If urine and feces cannot be separated, then only energy metabolizability can be determined for these animals, not apparent digestibility. For the animals that do not produce feces separate from urine, many researchers use the MEC^*, sometimes also called an assimilation coefficient or efficiency, as an index for digestive efficiency. The logical rationale for this is that, typically, the majority of the chemical potential energy not retained in the body is lost in feces rather than in urine and methane. Analogously, researchers sometimes calculate an assimilable mass coefficient (AMC^*) using equation 3.1 but substituting all excreta for feces. Sometimes, the values of MEC^* or AMC^* are corrected to nitrogen balance because urinary energy is partly dependent on nitrogen (N) metabolism (box 3.1). Finally, for some animals it is impractical or impossible to collect quantitatively feces and urine, and some other less direct methods for measuring digestive efficiency are available (box 3.2).

Box 3.2. Inert Markers—An Alternative Method for Determining Utilization Efficiencies

It is possible to measure utilization efficiency by using a nondigestible reference marker in food. This is advantageous if all feces and urine cannot be collected or if total balance trials are too expensive in time, labor, and cost, or to satisfy a desire to measure the efficiency of a free-living animal eating a natural diet without the putative effects of captivity. Studier et al. (1988) used seeds of fruit as a natural marker to study digestion by cedar waxwings because the seeds pass through their gut uncrushed or digested. The birds often have the habit of quickly ingesting fruit at a fruit-bearing tree or shrub and retiring to a branch for a period of time, probably to take a "digestive pause," during which they defecate. By spreading plastic underneath roost (feeding) trees, Studier et al. collected excreta to compare with fruits gathered from branches where the birds were feeding. The ratio of seed

continued

continued

mass to total fruit mass was 1.256 g g^{-1} (dry mass basis) in fruit on the tree, but increased to 1.565 g g^{-1} as seeds became more concentrated in the excreta because pulp, but not seeds, was partially digested and absorbed during passage through the digestive tract. The dry matter utilization efficiency can be calculated from the relative concentrations of the inert marker (the seeds), and is equal to

$$D_m^* = 1 - \text{food marker concentration/excreta marker concentration} = 0.197.$$

The authors were even able to estimate the apparent assimilation of energy and glucose, by comparing their concentrations in the pulp of fruits (respectively, 17.8 kJ g^{-1} dry fruit and 229 mg g^{-1} dry fruit) with their concentrations in the dry excreta (respectively, 17.5 kJ g^{-1} and 20.3 mg g^{-1}). The utilization efficiency for energy (MEC^*), or D_i^* for a particular substance can be calculated if one knows those concentrations in food and excreta as well:

$$D_i^* = 1 - (((\text{mg marker g}^{-1} \text{ food}) \times (i \text{ g}^{-1} \text{ feces}))/((i \text{ g}^{-1} \text{ food})$$
$$\times (\text{mg marker g}^{-1} \text{ feces}))) \tag{3.2.1}$$

where i = substance in mg or energy in kJ. Therefore, the apparent assimilation was 0.21 for energy and 0.93 for glucose. At first glance, the value for energy seems remarkably low. For comparison, however, Holthuizen and Adkisson (1984) fed waxwings mixed fruits in a conventional balance trial and measured energy assimilation of only 0.37. Also for comparison, we measured glucose absorption in waxwings using radiolabeled D-glucose and a radiolabeled nonabsorbable marker, a large water-soluble compound called polyethylene glycol (Martinez del Rio et al. 1989). Using equation 3.2.1, but inserting the respective values for radioactivity per gram food and excreta, we found that 92% of the D-glucose was absorbed. This estimate is not confounded by endogenous losses, because there are none for radiolabeled glucose, but if some radiolabel was absorbed and then excreted that would create an underestimate. The take-home message from the comparisons is that you can measure digestive efficiency using inert markers.

There are other indigestible markers often mixed into the food fed to animals, such as chromic oxide (Muthukrishnan and Pandian 1987; Goodman-Lowe et al. 1999) and other trivalent metal oxides (Austreng et al. 2000), but sometimes there are natural components of food that might suffice. Many of the naturally occurring substances in foods that have been considered as nondigestible reference markers, such as lignin or various

continued

continued

measures of "fiber," are partially digested and/or difficult to determine chemically. Their use as reference markers thus introduces some error in the estimation of utilization efficiency. Long-chained n-alkanes, found in plant cuticular wax, have been used more successfully in the livestock industry (Mayes et al. 1986), and some elements, such as Mn, are so poorly absorbed that they have been validated and profitably used as reference markers in animals as diverse as marine mammals (Fadely et al. 1990), herbivorous monkeys (Nagy and Milton 1979), and omnivorous rodents (Kaufman et al. 1976).

3.2.2 Digestive Efficiencies of Wild Animals Eating Wild Foods

The patterns apparent in the bears' feeding trials are also broadly apparent across different animal taxa. To see this, we can compare efficiencies (D_m^*, D_e^*, or MEC^*, whichever is appropriate) of animals eating different diets. These have been determined in balance trials with hundreds of animals fed natural foods (we excluded formulated diets). Because researchers have rarely measured the proportion of the food mass that is refractory to digestion, our comparison is more qualitative. It simply ranks foods according to their relative "fiber" content. One of us (Karasov 1990) used data on neutral detergent fiber (NDF) contents to come up with such a ranking (figure 3.3A). Although within any single food category there can be tremendous variation, some generalities emerge. Animal foods tend to have the lowest amounts of refractory material (e.g., hair, feathers, bone, cuticle), seeds and fruits have intermediate levels (measured here as NDF), and herbage has the highest levels (especially mature leaves and structural parts). Detritus, which has not been analyzed in an analogous fashion, is included as a food type because ecologists have found that it may support over half the animal production in some ecosystems (Bowen 1987). Both aquatic (Bowen 1987) and terrestrial detritus have many organic compounds refractory to breakdown, such as lignin (up to 25% of mass in some cases; Melillo et al. 1982), and half or more of the mass of detritus can be minerals (ash; Bowen 1987; Jobling 1995). For detritus feeders we estimated the refractory material based on these data.

As we look across animal taxa (figure 3.3) we see that, although there are not data for every food type in each taxon, mean digestive efficiency is inversely related to the relative amount of refractory material in the food. The differences in energy value of foods become more marked when one considers that animal

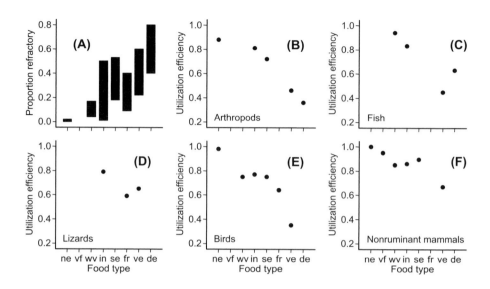

Figure 3.3. As a general rule in animals, digestive efficiency declines with increasing amount of refractory material in their food. (A; upper left) Food types can be ranked according to their relative content of refractory material, which, in this case, is based largely on neutral detergent fiber (NDF) contents (Karasov 1990). Ranges are given for the following food types: ne = nectar; vf = vertebrate flesh; wv = whole vertebrate; in = whole invertebrate; se = seeds; fr = fruit; ve = vegetation (grass, dicot leaves, twigs); de = detritus. (B; upper middle) Utilization efficiencies of immature arthropods (Slansky and Scriber 1985). (C; upper right) Utilization efficiences for fish species from (Brett and Groves 1979; Bowen et al. 1995). (D; lower left) Utilization efficiencies for lizards from (Zimmerman and Tracy 1989; Marken Lichtenbelt 1992). (E; lower middle) Utilization efficiencies for birds from (Karasov 1990); (F; lower right) Utilization efficiencies for nonruminant mammals from (Robbins 1993). The efficiencies plotted in figures B–E may be a mix of values of D_m^*, D_e^*, and MEC^*, but, as described in the text, these measures tend to be close to each other and highly correlated.

foods have higher gross energy content than most plant foods (table 2.8 in chapter 2) because they contain more fat and protein, which have higher gross energy content than carbohydrates (table 1.1 in chapter 1). The metabolizable energy per gram of food is the product of MEC^* and the gross energy content. On average, an herbivore must eat two to five times as much dry matter as an insectivore or carnivore to obtain the same amount of metabolizable energy (Karasov 1990; Jobling 1995). The very general summaries used to construct figure 3.3 are useful for highlighting broad generalizations, but researchers must consider carefully whether the averages meet their needs or whether more accurate data should be obtained for their particular species.

Comparison of patterns across the taxa also raises interesting questions for future research. For example, are birds a little less efficient digesting foods than nonruminant mammals (figure 3.3), and if so, what is the mechanistic explanation? The mean values for efficiency do not reflect the tremendous variation that exists among animals within any single taxon eating a specific diet type. For example, digestion of the disaccharide sucrose among bird species varies tremendously, from complete (0.98 ± 0.01 in rufous hummingbird, *Selasphorus rufous*; Karasov et al. 1986) to intermediate (0.61 ± 0.01 in cedar waxwings, *Bombycilla cedrorum*; Martínez del Rio et al. 1989) to nil (0 in American robins, *Turdus migratorius*; Karasov and Levey 1990). As another example, species of herbivorous rodents eating lucerne hay differ by a factor of more than two in digestive efficiency (range 0.18 to 0.39; Sakaguchi et al. 1987). These differences are due to differences in digestive system structure and function. Thus, although the composition of a food in many cases is more important than the species of consumer in determining digestive efficiency, the importance of structural and functional digestive characteristics of animals themselves cannot be denied. We will review in chapters 4 and 6 other examples of how animal anatomy and physiology, in addition to food chemistry, determine utilization efficiency. But before considering more closely digestive anatomy and physiology, there is another analytical framework for studying whole animal digestion for us to consider.

3.2.3 Utilization Plots: A Complementary Tool to Visualize Nutritional Efficiencies

Many nutritional questions of ecological importance can be addressed by using a framework called geometric analysis (Simpson and Raubenheimer 1995). Geometric analysis is powerful because it relies on graphical representations of many of the concepts that we have described. The geometric analysis of nutrition relies on two constructs, utilization plots and nutrient spaces, which we will describe in this and the next section. As mentioned before, a nutritional budget is a model that quantifies the relationship between the intake, utilization, and loss of nutrients. The budget for a given nutrient i can be described as

$$I_i = (A + L)_i \qquad (3.3)$$

where I is the amount of nutrient ingested per unit time, A is the amount retained in the body, and L is the amount lost per unit time ("dissociated" in the jargon of geometrical analysis). The terms in A and L can be expanded to identify the fate of the retained nutrient into somatic (s), reproductive (r), and storage tissues (st), for example,

$$A = A_s + A_r + A_{st}. \qquad (3.4)$$

L can be expanded into losses in feces (f), urine (u), and gases (e.g., methane) exhaled (e),

$$L = L_f + L_u + L_e. \tag{3.5}$$

Utilization plots examine nutritional efficiencies through the relationship between uptake in the x axis and the components of a budget in the y axis. To study digestion, the component of the budget plotted in the y axis is L_f. Let us see how the use of a utilization plot might alter our view of digestion by the bears.

Recall that the apparent digestibility of food nitrogen (D_N^*) was much lower for the blueberry diet (0.17) than for the trout diet (0.94) (methods box 3.1). In fact, the average D_N^* for all the plant foods (0.54 ± 0.06) was significantly lower than the average D_N^* for the animal foods (0.92 ± 0.07) in the feeding trials with the bears (Pritchard and Robbins 1990). Although you might conclude that plant protein is less digestible than animal protein, graphical analysis suggests otherwise. Figure 3.4A shows two hypothetical examples of possible relationships between I_i and L_i. The upper line shows the situation in which ingredient i is a substance that is defecated quantitatively (e.g., indigestible fiber). The slope of this line is 1, indicating that all the substance ingested is lost. The lower line has a much shallower slope of 0.25. In this case, a proportion p ($p = 1 - \text{slope} = 0.75$) of the substance ingested is retained (i.e., digested and absorbed). The intercept of this line is the minimal excretion of this substance, which can be considered the endogenous loss that was mentioned above. With this as background, now look at the right-hand figure 3.4B which includes plots of the data on N loss (L) versus N intake (I) from the feeding trials with both bear species fed plant- and animal-based diets. The slopes of L versus I are very shallow for both plant foods (0.047 ± 0.07) and animal foods (0.024 ± 0.023) and in fact are not significantly different from each other or zero. According to our interpretation, this indicates that essentially all the N in both types of food was 100% digested and absorbed (e.g., for animal foods $D_N^* = 1 - 0.024 = 0.976 \approx 1.0$). This is a conclusion completely different from our earlier conclusion that apparent digestibility of N was less than 1.0 and much lower in plant foods than animal foods. How can we explain this discrepancy? The explanation is that the calculation of apparent digestibility included the endogenous losses of N in addition to the unabsorbed N from food. Notice that the lines in figure 3.4B have positive intercepts and that the intercept for plant foods (0.37 ± 0.029) is significantly higher than that for animal foods (0.139 ± 0.031). Our interpretation of this is that the endogenous fecal N loss associated with consumption of plant foods is higher than for animal foods. Researchers typically express endogenous fecal N loss per unit food consumed, which for bears comes out to 8.4 mg N g^{-1} dry intake for plant foods and 5.8 mg N g^{-1} dry intake for animal foods. These

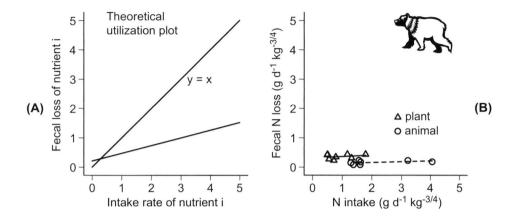

Figure 3.4. Digestive efficiency in feeding trials can be studied with utilization plots that regress fecal loss (the y axis) against intake (x axis). The plots in (A; left) show the theoretical pattern for two types of material in a food. An indigestible material, like lignin, will be defecated at the same rate as it is ingested, and so points will fall along the line $y = x$. A fairly digestible nutrient, designated by the lower line, will not be excreted to a great extent even as its intake increases. This particular line increases at a rate of 0.25 units for every 1.0 unit of intake, which corresponds to an efficiency of absorption of $1 - \text{slope} = 0.75$, or 75%. Notice that the line has a positive, small intercept. This represents an endogenous loss of the nutrient. The plots in (B; right) show data on nitrogen (N) intake and fecal loss from the feeding trials on bears (Pritchards and Robbins 1990). Their very shallow slopes indicate very high digestibilities of N, nearly 1.0, or 100%. Notice that they have small positive intercepts, indicating endogenous fecal losses of N, which differ according to diet. Higher endogenous N loss on plant diets probably relates to their higher fiber content.

values are very comparable to other data in mammals (Robbins 1993). Researchers think that the difference between the foods in endogenous losses is caused by the higher fiber content of plant foods possibly abrading more cells and microbes from the digestive tract.

Thus, our application of the utilization plot has shown that when there are endogenous losses of a substance, the calculated apparent digestibility is lower than the true digestibility. For energy or dry matter, apparent coefficients are generally only 0.01–0.03 below the "true" coefficients, unless animals are fed far below their maintenance food intake level, at which point the endogenous component of the feces becomes large relative to the undigested residue from the food. However, the calculation of apparent digestibility of particular nutrients can be very much confounded when the diets fed have low nutrient content (Robbins 1993). Herpetologists Beaupre and Dunham (1995) provide another example for lizards that compares conclusions based on utilization plots with those based on the approach using equation 3.1.

For the many animals that produce excreta that are a combination of feces and urine, the utilization plots necessarily plot L_{f+u} versus I. This adds another layer of complexity to their interpretation in regards to digestive efficiency. To discuss this further we refer you to box 3.3 and the many papers by Simpson and Raubenheimer. In later chapters we will use utilization plots again to evaluate the postabsorptive processing of nutrients.

Box 3.3. Utilization Plots

In utilization plots the components in the mass or energy budget of intake (I) and loss (L) or amount retained (A) are plotted against each other. Many animals, including most nonmammalian vertebrates and invertebrates, produce excreta that are a combination of feces and urine, which makes the interpretation of the utilization plot more complicated than the example discussed in figure 3.4. Figure 3.5 shows several hypothetical examples of possible relationships between I_i and L_i. The top panel shows the situation for a substance that is defecated and excreted quantitatively (e.g., indigestible fiber or a toxin absorbed but then all excreted). The slope of this line is 1, indicating that all the substance ingested is lost. The middle panel illustrates a substance that is required up to a certain level of ingestion m. After the level m is satisfied, the substance is excreted quantitatively. The I against L relationship for this substance is biphasic. The slope of the first part of the line has a slope that is lower than 1, meaning that a fraction p ($p = 1 - $ slope) of the substance ingested is retained. The intercept of this line is the minimal excretion of this substance, and the point at which the line changes slope is the maintenance level that permits a neutral balance. The bottom panel shows a substance that is stored or used for growth. Positive balance, allowing accumulation, is achieved for all levels of intake greater than m.

Utilization plots can also be used to investigate relationships among any of the components of nutrient budgets. Raubenheimer and Simpson (1994) illustrate this use with two nutritional indices commonly used with insects: efficiency of conversion to growth (ECI) and efficiency of conversion of digested food into growth (ECD) (Waldbauer 1964). Because growth can be defined simply as the ingested nutrient that is not dissociated, the plot of I against A (the amount retained) depicts the ECI graphically. The ECI is the slope from the origin into any point into the relationship between I and A. In a similar fashion the ECD can be obtained as the slope of the relationship between A and (I–L).

continued

continued

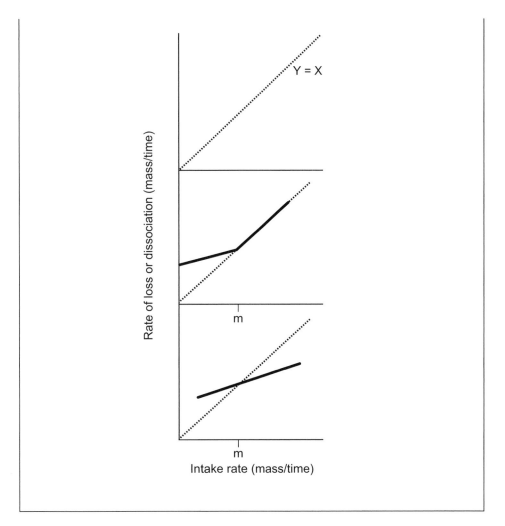

Figure 3.5. Utilization plots imbedded in box 3.3.

3.3 Both Digestion Rate and Digestive Efficiency Are Key Nutritional Variables

3.3.1 Trade-offs between Efficiency and Rate

Budgets are useful, but they can provide a static view of an organism. In particular, budgets, by themselves, fail to capture the dynamical interplay between rates and efficiencies. Consider the transfer of energy or material from one system (the gut) to another (the rest of the body) as defined in figure 3.1. Figure 3.6 shows a reasonable conjecture: the amount of nutrient transferred to the rest of the body

Figure 3.6. The trade-off between rate and efficiency is apparent when considering the transfer function for the release of the amount of material from one system (e.g., digestive system, called "S") to another system (e.g., the body, for postabsorptive processing, called "$S+1$") as a function of time. The dashed horizontal line on top represents the total amount available for processing. The greatest extent (efficiency) of transfer occurs at time $t2$. In that case, the average rate of transfer is reflected in the slope of line E. The slope of line P gives the maximum average rate of transfer, which is achieved by terminating processing earlier, at time $t1$, in which case the extent of transfer would be less. Maximum average rate is achieved at the cost of maximum efficiency. (Redrawn from Raubenheimer and Simpson 1998.)

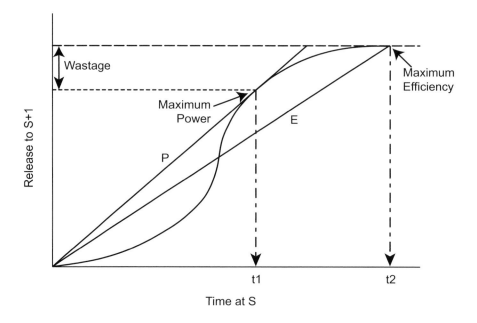

increases as a function of the time during which food is processed in the gut. As time increases, so does the amount of nutrient transferred, up to the total amount of nutrient available (shown as a dashed horizontal line in figure 3.6). The transfer function (Raubenheimer and Simpson 1998) is depicted as a solid sigmoid line. The approach is valid also for other shapes of transfer functions, but a sigmoid is useful to illustrate the point that we wish to emphasize, and this was the shape observed for carbohydrate and protein absorbed by locusts (Raubenheimer and Simpson 1998). The average transfer rate (in units of mass per unit time) is represented by the slope of a line from the origin to the transfer function. Notice that, for a sigmoidal curve, there is a maximal rate of transfer, which in this example occurs at time $t1$. Note that the maximal rate does not maximize the efficiency of transfer, which is maximized at time $t2$. If an animal is designed to maximize its rate of transfer, then this is achieved by terminating the process at $t1$, in which case some energy or material is wasted. Efficiency of transfer is below what would be possible had the process been extended to $t2$. If the animal is designed to maximize transfer efficiency, then this is achieved at the expense of the highest rate. As will be illustrated in the ensuing chapters, such trade-offs between rate and efficiency may apply to a variety of processes, such as the harvesting of food and the digestive and postabsorptive processing. While this example applies to a single nutritional event (e.g., a meal), the concept can be extended to longer periods of time and other kinds of situations (Raubenheimer and Simpson 1998). The rates and efficiencies determined in whole-animal balance trials can thus be viewed as a reflection of the animal's particular solution to the rate-efficiency trade-off inherent in a particular transfer function(s). Raubenheimer and Simpson (1998) point out that the basic concept of the rate-efficiency trade-off has been applied to digestion by several authors (Sibly 1981; Cochran 1987; Reynolds 1990) and has analogues in the many applications of the marginal value theorem (Charnov 1976) in foraging ecology.

One important implication of the rate-efficiency trade-off in digestion is that it leads us to anticipate that animals may differ in digestive efficiency, even when eating the same diet, according to the relative benefit(s) of maximizing either digestion rate or digestive efficiency or adopting some intermediate strategy. Although we often assume that there is a selective premium to high rate, i.e., achieving a given nutritional gain in a shorter time, efficiency might be favored for several reasons, including scarcity of a nutrient or high risk of feeding, so that it is most advantageous to satisfy requirements in the fewest number of meals (Karasov and Cork 1996; Raubenheimer and Simpson 1998; Hilton et al. 1998). As we mentioned in section 3.2.2, comparative feeding trials sometimes do illustrate that animals differ in digestive efficiency even when eating the same diet, presumably having responded over evolutionary time in different ways to these selective forces.

The rate-efficiency trade-off is apparent in comparisons among terrestrial and marine avian predators. Among species in both groups there are small but significant differences in digestive efficiency, even when feeding on the same diets (Hilton et al. 1999). In both groups, species that retain food in their gut for less time are less efficient digesters (figure 3.7). Other factors besides differences in retention time may explain the differences in efficiency, because lower efficiency was also associated with shorter intestine length (Hilton et al. 1999; 2000), and thus perhaps fewer intestine-bound enzymes and less surface area for digestion

Figure 3.7. In raptor species eating the same terrestrial prey (filled circles; from Hilton et al. 1999) and in seabird species eating the same types of fish (unfilled circles; from Hilton et al. 2000), the longer the food is retained in the gut (longer mean retention time) the more thoroughly it is digested. This trend suggests a trade-off between digestion rate (which is reciprocally related to retention time) and digestive efficiency.

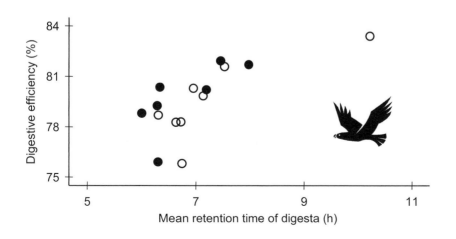

and absorption. Hilton and his colleagues note that the avian raptor species that have relatively less gut mass and shorter retention time tend to be species (e.g., falcons) that pursue active prey. They suggest that natural selection favored these traits because low gut mass and short retention time reduce the mass of tissue and digesta carried, and thus improve flight performance (acceleration, turning speed, agility, maximum velocity).

These examples highlight some major research questions in digestive ecology, such as "What anatomical and biochemical mechanisms lead to differences in efficiency?" and "Does the rate-efficiency trade-off apply within as well as between species?" These questions, in turn, lead us to another major research theme in this field, having to do with the flexibility of digestive structure and function.

3.3.2 Adjustable Rates and Their Limits—An Important Interpretive Paradigm

The transfer functions conceptualized in figure 3.6 are not necessarily fixed but there are likely limits to their adjustability. An illustration of this comes from a series of studies on the energetics of white-throated sparrows (*Zonotrichia albicollis*) in which one of us (Karasov) was involved. We took sparrows acclimated to room temperatures (21°C) and switched them to cold conditions (−20°C) either quickly or gradually over many days (figure 3.8A). In the cold, sparrows, like all homeothermic endotherms, must eat more to balance higher heat loss or they will catabolize their body tissues to supply the extra energy. Sparrows that were switched quickly increased their feeding and digestion rate only 45% and lost body mass, whereas sparrows that were acclimated slowly increased their rates by 83% and maintained body mass. Thus, the sparrows that were switched quickly experienced an energy deficit and should have been motivated to eat more, but did not. Figure 3.8C uses these experimental results to define two types of limits to energy flow.

First there was some immediate limit that was only 45% above what sparrows needed routinely for energy balance at 21°C. We can define the difference between that upper limit, which is the immediately available capacity to process energy (the upper line), and the routine energy demand (the lower line) as the sparrow's "immediate reserve capacity" (Diamond and Hammond 1992). Another name assigned to it is "safety margin" (Diamond 1991). A number of methods have been proposed to quantify immediate reserve capacity, but we think that the most reliable method is to quickly challenge animals to increase rate of digestion, through cold challenge, forced activity, or reduction in feeding time (Karasov and McWilliams 2005). This first limit is apparently

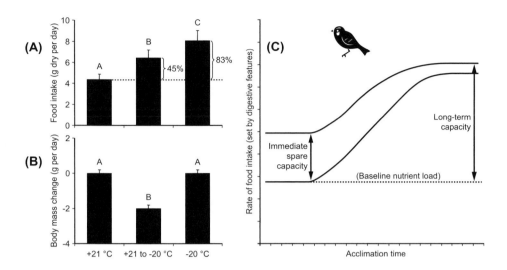

Figure 3.8. On the left top (panel A), white-throated sparrows (*Zonotrichia albicollis*) switched suddenly to cold temperature (+21 to −20°C) were able to increase their feeding rate only 45% rather than nearly doubling it, as did sparrows acclimated more slowly (−20°C). The rapidly switched birds lost mass (left, bottom; panel B). This result is interpreted in panel C. The baseline corresponds to the sparrow's routine energy demands (i.e., at +20°C). The solid upper line represents the physiological capacity for processing that nutrient load, set by feature(s) of digestion or postabsorptive processing. The *x* axis is time since shift to low temperature and hence increased energy demand. The sparrow can increase its food intake only within the limits set by the level of immediate reserve capacity, which decreases as the sparrow approaches its long-term capacity. When energy and nutrient demands increase, and if the animal has been given time to fully acclimate to these elevated energy demands, then phenotypic flexibility of the sparrow enables increased energy intake (shown as the increase of the solid lower line above the baseline nutrient load). (adapted from (Diamond, 1991; Diamond & Hammond, 1992).)

flexible, because sparrows that were switched more gradually were able to increase their feeding and metabolic capacity, by 83%. The capacity could be increased even more because Kontogiannis (1968) found that captive white-throated sparrows could tolerate temperatures down to −29°C. At this remarkably low temperature metabolizable energy intake was 126% higher than at +21°C. Because the sparrows could not tolerate lower temperatures without losing body mass, and could not increase their metabolizable energy intake even when challenged by forced activity (Kontogiannis 1968), their limit to energy flow after long-term acclimation was around 126% above "baseline". We can define the difference between the long-term capacity and the routine demand as the sparrow's "long-term reserve capacity".

Figure 3.8C leads to some interesting research questions. First, we might ask what factors dictate the limit, the maximum rate of energy flow through animals? In the wild, the possibilities include (1) the availability of food in the environment; (2) time available for foraging; (3) rates of food collection while foraging; (4) rates of digestion of food; (5) the ability to mobilize or deliver nutrients and oxygen to the tissues; and (6) the work capacity of various tissues. In a clever analogy, Weiner (1992) likened the partly hierarchical nature of these possibilities to fluid flow into and out of a barrel through funnels and spouts that might act as bottlenecks (figure 3.9). The factors listed fall partly or entirely within the domain of physiology, and three hypotheses have been proposed about apparent physiological limitations (Weiner 1992; Hammond and Diamond 1997). The central limitation hypothesis suggests that the bottleneck resides in physiological processes and systems, including the digestive system (represented by one of the funnels in the figure) and the liver, that are

Figure 3.9. January Weiner (1992) used the analogy of an animal's energy balance as the flow of a fluid (presumably good Polish beer) into a barrel. Input constraints like maximum rates of foraging, digestion, and absorption are arranged in series. In contrast, outputs like heat or tissue production and mechanical work are parallel and independently controlled. If the sum of outputs does not match the input, then the amount stored in the system goes up or down. Weiner left the "heat" spigot with a slow leak to represent basal metabolism (heat loss). (From Weiner 1992, with permission.)

involved in acquiring, processing, and distributing energy to energy-consuming organs such as muscle or mammary glands. As we will discuss in chapters 4 and 7, both the intestine and the liver become enlarged in cold-acclimated birds and mammals. The peripheral limitation hypothesis suggests that processes within the energy-consuming organs, such as thermoregulation, activity, and lactation in mammals, each have their own metabolic ceilings, represented by the spouts in the figure, and this determines the maximum energy flow. Finally, the idea of "symmorphosis" proposes that the capacities of several of these potentially limiting factors might be matched to each other and to natural loads (Taylor and Weibel 1981; Weibel 2000; box 3.4).

How much reserve digestive capacity should animals possess at any one time? Diamond has argued that the magnitude of reserve capacity is probably dictated by natural selection in relation to the costs of maintaining the digestive

---

Box 3.4. Symmorphosis

The symmorphosis principle applies an optimization principle from economics to the evolution of physiological design, and can be applied to many anatomical and physiological features. As applied to the energy budget, for example, it considers the maximum rates (capacities) of all the energy inputs and outputs. Symmorphosis assumes that natural selection will mold animal bodies according to three principles: (1) any given capacity will be matched by its natural load, (2) within a given metabolic pathway of sequential steps, the capacity of each step should be matched with those of others, and (3) capacities of different organ systems will be matched to each other (Diamond and Hammond 1992). Taylor and Weibel (1981) proposed the symmorphosis "hypothesis" (the reason for our quotation marks will be clear at the end of this section). They posited that symmorphosis is "the state of structural design commensurate to functional needs resulting from regulated morphogenesis, whereby the formation of structural elements is regulated to satisfy, but not to exceed, the requirements of the functional system."

In the title of one of his papers, Diamond (1991) translated Taylor and Weibel's rather formidable statement into plain English and applied it to the gut. Diamond entitled his paper "Evolutionary design of intestinal nutrient transport: Enough but not too much." Perhaps not surprisingly, the symmorphosis hypothesis has drawn a lot of criticism (Dudley and Gans 1991; Garland, and Huey 1987). Diamond and Hammond (1992)

continued

continued

summarized criticisms to symmorphosis and counterarguments in five major points, which we list here.

(1) It would be dangerous for capacities to be matched to peak loads. You do not design the strength of the cable that holds an elevator to hold just the maximal possible load. Physiological and biomechanical capacities should exceed peak loads by some reserve capacity.
(2) Reserve capacities should vary according to cost/benefit considerations. Cheap capacities that provide critical benefits should be present in larger excess.
(3) Reserve capacities that may seem excessive in healthy beasts may be essential in sick ones.
(4) Because many capacities serve multiple functions, a capacity may have to exceed its load by a lot in one pathway just to match the load by a narrow margin in another one.
(5) Capacities may be mismatched to current loads in order to arrive at an appropriate expected load at a later stage in development.

Clearly, the symmorphosis principle has too many potential caveats to constitute a bona fide, falsifiable hypothesis. Whenever we find an exception, we may invoke one of the five reasons listed above and, voila!, the symmorphosis hypothesis remains intact.

This does not mean that we consider the symmorphosis principle useless. Science is not constructed just of a collection of falsifiable propositions. The symmorphosis principle has spawned what we believe is enormously important research—some of which we will describe in subsequent chapters, and it has led to the creation of many significant testable hypotheses. We believe that attempting to predict when each of the five caveats listed above is likely to apply can lead to perfectly testable hypotheses that can be considered axillary to the symmorphosis principle. For example, Alexander developed a simple theoretical framework that allows making a priori predictions about the magnitude of reserve capacities (or "safety factors") depending on the design and maintenance cost of biological structures and on the predictability of failure of the materials of which these structures are made (Alexander 1998). Although Alexander's framework generates many falsifiable hypotheses, to our knowledge it has not been tested. The symmorphosis principle may not be a good Popperian hypothesis, but it serves as a handy framework. If the adjective "Popperian" is unfamiliar, we recommend the first two chapters of Hilborn and Mangel's book the *Ecological Detective* (Hilborn and Mangel 1997) as an introduction to the topics in contemporary philosophy of science that ecologists must be aware of.

reserve capacity and the risks of not having enough during a time of stress (Diamond 1993). The sparrows in figure 3.8 increased their digestive capacity by increasing their gut mass, as we will discuss in chapter 4, but what does this extra gut cost? No one is certain what it costs to maintain the gastrointestinal tract, but studies of mass-specific tissue-slice respiration (Martin and Fuhrman 1955) suggest that the intestine and liver are energy-intensive tissues, that 25% of the basal metabolic rate might be due to intestinal metabolism, and that within species the changes in the size of the intestine and liver are reflected in changes in basal metabolic rate (Konarzewski and Diamond 1995; Piersma 2002).

A third issue highlighted by figure 3.8B is the flexibility of whichever trait(s) are responsible for the limits to energy flow (i.e., digestion or heat production). The organ structures and their correlated biochemical and metabolic features are not necessarily fixed attributes of animals, even adults. As described in chapter 1, phenotypic plasticity is the term used to describe the capacity of single genotypes to produce a variety of phenotypes in response to changing environmental conditions, but it is usually used in reference to irreversible changes that occur during development (Piersma 2002; Piersma and Lindstrom 1997; Piersma and Drent 2003). Another term suggested for the situation of reversible changes in individual phenotypes is phenotypic flexibility (Piersma and Lindstrom 1997). Both types of phenotypic change are part of the life history of organisms, create ecological opportunities and constraints, and are possible subjects of evolutionary study (Piersma and Drent 2003). Thus, our consideration of flexible limits to energy flow has added several more to our growing list of major research questions in digestive ecology: Does digestion limit energy flow; If so, which step(s) of digestion (physical, chemical breakdown, absorption, etc.) are limiting?; Are limiting step(s) flexible?; If so, by what mechanisms, and what are the time courses for change? But before beginning to tackle those specifics we want to consider one more important feature of whole-animal feeding, which is the rate of food intake.

3.4 Daily Food Intake: Energy Maximization or Regulation?

Ecologists realize that sometimes there is a selective premium to maximizing rate (figure 3.6), but also that animals sometimes process energy at rates that are lower than maximal (figure 3.8C). Differences in these energy management strategies play out in all kinds of features of animal ecology that relate to feeding, such as growth and reproduction rates, and allocation of the time budget. How do we find out whether our study species are energy maximizers or regulators?

3.4.1 The Food Intake Pattern of Animals Can Reveal whether They Operate under Maximization and/or Homeostatic Principles

That energy and food intake are regulated variables is suggested by how animals vary them in response to changes in food quality. One common manipulation is to measure daily intake on a diet that contains some kind of nonenergetic, nonassimilable, material and then increase or decrease the energy density by, respectively, decreasing or increasing the material. These nondigestible materials are called "diluents". Examples of diluents are cellulose (which is refractory to digestion; chapter 2), kaolin (a clay), and agar. In the best-designed experiments these diluters are traded off against a specific nutrient in the diet, such as carbohydrate, because otherwise the concentration of all nutrients would be diluted. The common observation is that daily intake rate decreases with increasing digestible energy density, so the animal appears to defend a relatively constant energy absorption rate (figure 3.10). This inverse "intake-response relationship" (Castle and Wunder 1995) has been observed in a variety of species spanning major taxa and metabolic profiles (e.g., representatives of ectothermic slugs, insects, fish, and reptiles and endothermic mammals and birds) and feeding types (e.g., herbivores, carnivores, frugivores, and nectarivores). Another kind of experimental evidence with endothermic animals is that daily food and energy intake increase in a compensatory fashion for increases in thermoregulatory costs when temperature is lowered (figure 3.10B). Given this evidence for apparent compensatory feeding, it seems reasonable that this should be the initial assumption of ecologists in framing hypotheses, rather than the assumption of maximization of net energy gain.

Figure 3.10. Food intake-response relationships for (A) locusts (Dadd 1960) and (B) a vole (*Microtus ochrogaster*) (Castle and Wunder 1995). Notice that, for the locusts, as the diet is diluted with nondigestible fiber, total food intake ("total") increases in a compensatory fashion so that apparently digested dry matter intake ("digestible") is relatively constant. In the vole, total intake (lines with steepest negative slopes) and digestible intake (nearly horizontal lines) increase when energy expenditure increases due to thermoregulation at a lower air temperature. These intake responses are considered evidence for regulation of digestible energy intake.

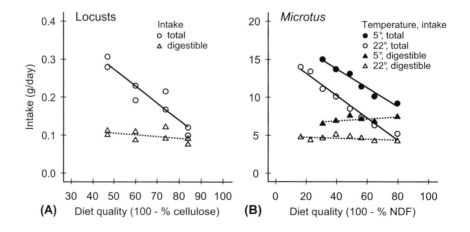

But further contemplation of the inverse intake-response relationship suggests that this interpretation might not be quite correct, and that there are circumstances under which the energy maximization hypothesis might be right, after all. First, what these relationships demonstrate most strongly is that these animals' feeding rates on foods with higher energy density are not dictated by a physical limit on the grams consumed per day, or else they would not so easily increase intake as the energy is diluted. The experiments do not rule out that these animals' feeding rates are dictated by some other limit on digestible energy processed per day, such as a limit imposed by rate of food breakdown, absorption, or postabsorptive processing. Unless those processes have some demonstrated reserve capacity, there is no way to know whether the animal is eating maximally to capacity or regulating intake, which implies feeding below capacity.

The higher digestible energy intake of the voles exposed to cold in figure 3.10B does not prove they had reserve capacity, because we know that during cold acclimation their digestive capacity was increased through enlargement of the gut (see chapter 4) and they may again be feeding maximally to capacity. Evidence that voles are indeed regulating their digestible energy intake comes from other kinds of studies that demonstrate that at 22°C they can be manipulated to consume and digest food faster than indicated in figure 3.10, although only by 9% (Zynel and Wunder 2002). To reiterate, the inverse "intake-response relationship" (Castle and Wunder 1995) that has been observed in a variety of species is consistent with the notion that animals regulate digestible energy intake, but it is not strong proof. What is more convincing is demonstrating that digestible energy intake is constant even when there is unused reserve digestive capacity.

A critical design feature of most studies measuring immediate reserve capacity is that they quickly challenge animals to increase rate of digestion, through cold challenge, forced activity, or reduction in feeding time (Karasov and McWilliams 2005). These kinds of studies have been done with only a handful of mammals and birds (Karasov and McWilliams 2005), and it is an intriguing question to consider how the test can be performed in ectotherms. Recall that in ectotherms a drop in temperature decreases metabolic rate. Can we devise experiments to increase an ectotherm's energy demands and then test whether it increases its digestion rate? Alternatively, several researchers have demonstrated that animals will increase digestion rate when their feeding time is restricted (Swennen et al. 1989; Iason et al. 1999), and this may be the better approach with ectotherms.

Although there is evidence for regulation rather than maximization of energy intake, there are cases in which the energy maximization hypothesis seems to apply. For example, in larvae of the phytophagous spruce budworm (*Choristoneura fumiferana*) eating synthetic diets diluted with cellulose, total intake did not significantly decrease with increasing digestible energy content of the diet, and so the budworms increased energy absorption rate rather than defending a constant rate (Trier and Mattson 2003), contrary to the patterns in figure 3.10A. This is the kind of relationship one might expect to see if energy maximization were the goal and there were also constraints on intake rate imposed, for example, by gut volume and turnover. Many ruminant herbivores appear to be energy regulators when eating diets with high digestible energy content (i.e., the inverse intake-response relationship shown in figure 3.10 applies), but on very low–quality vegetation their intake appears to be constrained and they behave as energy maximizers (i.e., there is a positive relationship between digestible energy intake and diet digestible energy content; Van Soest 1994; Illius and Gordon 1991).

The pattern for an energy maximizer can sometimes be disguised as the inverse intake-response relationship of an energy regulator. In an experimental study with hummingbirds, what at first glance appeared to reflect compensatory feeding was shown to be consistent with energy maximization constrained by digestive processing (figure 3.11). McWhorter and one of us (Martínez del Rio) randomly assigned eight broad-tailed hummingbirds (*Selasphorus platycercus*) to one of four sucrose concentrations within a natural range and switched them from 22°C (filled circles) to 10°C (unfilled circles) with only 1 day acclimation (McWhorter and Martínez del Rio 2000). The birds could greatly increase their volumetric intake as sucrose concentration declined (figure 3.11B), but two observations lead us to reject the hypothesis that the birds simply adjusted food intake to compensate for energy demands. First, notice that the actual amount of sugar ingested declined with dilution, i.e., with lower sucrose concentration (solid line, figure 3.11A). In other words, they compensated incompletely for dilution of their diet's energy content. Second, notice in figure 3.11A that they did not increase energy intake to match expenditure when they were exposed abruptly to low air temperature, where their heat loss would have increased. Indeed, they lost body mass. We speculated that there was a digestive constraint limiting the ability of the birds to digest sugar faster. Using a chemical reactor model of sucrose digestion to integrate our measures of sucrase enzyme activity with whole-animal measures of throughput time and extraction efficiency (see chapter 4), we predicted the maximum sucrose ingestion rate as a function of diet sucrose concentration (solid line, figure 3.11B). The prediction was very close to the observed values, supporting the notion of a digestive limitation. These results are thus contrary to the common notion that animals always engage in compensatory feeding.

Is it possible that animals regulate their food intake rates to satisfy requirements for specific nutrients in the food, rather than for energy? Probably not,

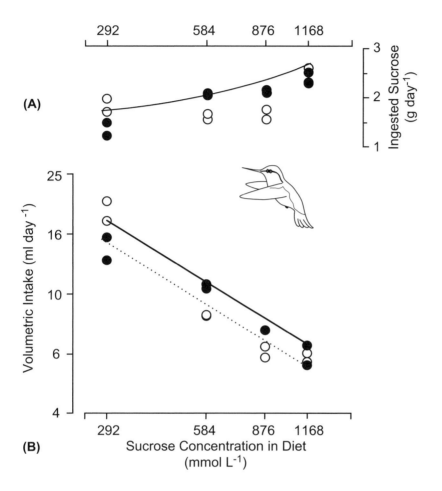

Figure 3.11. Volumetric food intake (lower panel B) and sugar ingestion (upper panel A) as a function of sugar concentration in broad-tailed hummingbirds. Filled circles represent data for birds at 22°C and unfilled circles represent birds at 10°C. Notice that, as sucrose was diluted with water, volumetric intake increased but not enough to keep sucrose intake constant. Furthermore, intake did not increase when temperature was reduced, even though normothermic birds would have had higher heat loss. Theses patterns and other evidence indicated that the birds' intake response was constrained by a digestive constraint (see text). The solid line represents the maximal intake rate predicted by a gut function model (see chapter 4). (Redrawn from McWhorter and Martínez del Rio 2000.)

because the mechanisms regulating energy intake are generally more influential than those regulating intake of specific nutrients. Voles, for example, do not overingest energy in order to compensate for either low protein (Trier 1996) or low minerals (Batzli 1986) in food. If they did overingest, they would have to decrease their energy digestibility or increase their heat production; otherwise they would gain too much mass. Yet there are examples of fish (Fu and Xie 2004) and insects (Trier and Mattson 2003) apparently "dumping" excess energy through heat production, and bears apparently depositing excess energy into fat when eating a low-protein diet (Felicetti et al. 2003). Phloem-sucking insects such as aphids and scale insects seem to ingest prodigious amounts of plant fluids with high sugar (mainly sucrose) and low amino acid content (White 1993). It has been claimed that aphids increase food intake on a N-poor diet in a compensatory fashion (Mittler 1987). One way of exploring whether animals eat for energy or for specific nutrients is through geometric analysis, a research framework for identifying "rules" that govern animal feeding (box 3.5). When this approach was applied to aphids, they showed marked ability to regulate sucrose (i.e., energy) intake by adjusting fluid intake in relation to sucrose content, whereas regulation for amino acid intake was not marked (Abisgold et al. 1994).

Box 3.5. Geometric Analysis

This is a research framework for identifying rules that govern animal feeding (Simpson and Raubenheimer 1995). As an illustration of the approach, consider how rules governing intake of energy and minerals were investigated in locusts (*L. migratoria*) (Raubenheimer and Simpson 1997). In the first experiment to determine whether fifth-stadium locusts regulated their food intake to meet some mineral need, they were fed synthetic foods containing a fixed concentration of protein and carbohydrate and various concentrations of salt mixture. The makeup of each food, labeled a–f in the left-hand panel of figure 3.12, is depicted as a line radiating from the origin whose slope is determined by the ratio of salt mixture to carbohydrate + protein. In five choice experiments the locusts were offered pairings of the foods, and the actual amounts of salt mixture and carbohydrate + protein they consumed as they ate combinations of the diets are depicted as points in the *x*–*y* plane ("nutrient space"). The pairings were a versus f, b versus f, c versus f, c versus e, d versus e. Empty circles depict the amounts consumed by the locusts, and their clustering implies that this is an "intake target." Notice that they mixed their diets in such a way that over two days they consumed about 5 mg salt mixture and 100 mg protein + carbohydrate.

In the second experiment to determine the relative strength of regulation of intake of mineral versus that for carbohydrate + protein, locusts were provided with foods one at a time that each contained a suboptimal salt level (Figure 3.12, right-hand panel). Once again, the four foods are depicted as lines from the origin and the amounts consumed are depicted as points along those lines. Notice that on each diet locusts ate enough to meet their intake target for carbohydrate + protein regardless of whether they met their intake target for minerals. As summarized by Simpson and Raubenheimer (1995) for this case, "the mechanisms regulating intake of non-mineral nutrients (mainly protein and carbohydrate) are so strong relative to those for mineral intake as to overwhelm the latter." This is one "rule of compromise," but others are indicated by other graphical patterns, as discussed by Raubenheimer and Simpson (1997).

continued

continued

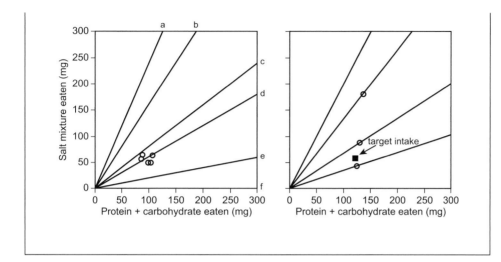

Figure 3.12. Geometric analysis of feeding choices in locust, embedded in box 3.5. Redrawn from Raubenheimer and Simpson (1997).

We do not claim that animals in every case regulate energy intake preferentially to intake of other substances, or that they always regulate rather than maximize their daily energy intake. But animals that fail to match energy intake to expenditure lose mass and probably suffer negative health effects more rapidly than when out of balance for a specific nutrient. We suspect that intake is determined by a hierarchy of factors and that energy is at the top of this hierarchy. This is why, in our subsequent modeling of nutrient and xenobiotic intake and flux, we will assume that animals consume just enough food to meet their daily or weekly energy needs. Our message is that, when the answer is important, there are straightforward and relatively simple tests to find it (box 3.5). In many instances, finding out will lead to better predictive models of animal feeding and ecological trophic relationships. Finding out leads also to further research. If an animal is found to maximize daily energy intake, then one next question to research is "What constrains it to that level rather than some higher level?"

3.4.2 Food Intake rate Can Be Estimated by Assuming that Animals Feed to Meet Their Energy Expenditure

Ecologists need to know animal feeding rates for all kinds of reasons, including determining nutrient requirements and fluxes, resource demands on the environment, and interspecies interactions like predation and competition. There are methods for directly measuring the feeding rate of animals, such as the use of isotope turnover as described later in methods box 12.1 (chapter 12), and researchers also sometimes backcalculate to the feeding rate that must equal observed levels of energy expenditure and production. But if all of the supporting measures are not available and an estimate might suffice, then the following steady state relationship can be used as an educated guess:

$$I = RC_e^{-1}[MEC]^{-1} \quad \text{or} \quad I = Rc_e^{-1}, \tag{3.6}$$

where I is daily intake in g dry matter d^{-1}, R is daily metabolic rate in kJ d^{-1}, C_e is the diet gross energy content (kJ g^{-1} dry matter), and MEC is the metabolizable energy coefficient of the diet (above). The product of the latter two variables is the metabolizable energy content of the diet, c_e, in kJ g^{-1} dry matter. All of the parameters may not be known for an animal of interest, but reasonable estimates are available. For most major taxa of vertebrates there are allometric equations to estimate daily field metabolic rate (FMR) as a function of body mass (Nagy et al. 1999). Preliminary estimates of diet gross energy and metabolizability can be made using information in these chapters on digestion if measures are not available for the animal of interest. Though the equation is for an animal in steady state, it can be adjusted upward if growth, storage, and reproduction are occurring, using procedures to be described in chapters 13 and 14. The relation in equation 3.6 is the basis for several predictive allometric equations for food intake of mammals, birds, and reptiles (Nagy 2001). Using equation 3.6 will not necessarily be accurate but may allow a preliminary estimate if more exact measures cannot be made or are too expensive.

3.4.3 Maximum Food Intake Is Sometimes Estimated as Some Multiple of Resting Metabolism

Depending on physiological and ecological circumstances, an animal's feeding rate may or may not be as high as is possible. The maximum energy intake is interesting because it could set limits on processes that determine distributional limits of species in cold climates (Root 1988) and it could constrain daily foraging efforts and growth and reproduction rates. How high can metabolizable energy intake go—that is, what is the asymptotic level in figure 3.8C?

It is interesting to place the maximum metabolizable energy intake (MEI_{max}) in relation to other levels of energy transformation in animals (Peterson et al. 1990; figure 3.13). The highest rates of energy transformation are fueled largely by anaerobic processes that cannot be sustained for more than seconds to minutes, probably because local substrates become exhausted and/or end products build up that are either toxic or inhibit the biochemical pathways. Sustained metabolism, measured as maximal oxygen consumption, can occur at fairly high rates for prolonged periods, as during long-distance running (10–15 times the BMR) or flight (up to 25 times the BMR). But no animals are known that maintain this power input rate over many days while keeping body mass

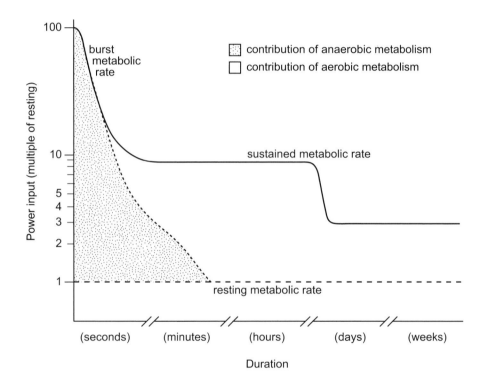

Figure 3.13. Burst metabolic rate is the maximum power input maintained by an animal. It is primarily anaerobic metabolism, and although it can reach as much as 100 times the resting power input, it can be sustained at that level for only a few seconds. Over longer intervals (seconds to minutes) moderate power input results from a mix of anaerobic and aerobic metabolism. Aerobic metabolism dominates when activity extends over hours, days, and weeks. Power input can be sustained at rates of up to 10–20 times resting for hours, but only 5–7 times resting over days or weeks. (Adapted from Peterson et al. 1990.)

constant (Hammond and Diamond 1997), reflecting that expenditure is not matched by metabolizable energy intake over the period.

There have been many measurements of feeding and digestion rates in mammals and birds highly motivated to feed, presumably at near-maximal levels. Typically they involve animals acclimated to very low temperatures, ideally at their limit of thermal tolerance, high levels of forced activity, or hyperphagic animals during lactation, while storing energy for migration and hibernation, or engaged in rapid growth. Some of these data have been summarized in allometric equations that express maximum metabolizable energy intake as a function of body mass (table 3.2). As a general rule, MEI_{max} scales

TABLE 3.2

Relationships between near-maximum metabolizable energy intake (MEI_{max}, kJ/d) and body mass (m_b, g)

Group	Energy demanding situation[a]	Allometric equation	Reference
Mammals and birds	C, L, G	$MEI_{max} = 11.84 m_b^{0.72}$	(Kirkwood 1983)
Mammals	L	$MEI_{max} = 18.49 m_b^{0.66}$	(Weiner 1992)
Birds	C	$MEI_{max} = 16.42 m_b^{0.66}$	(Karasov 1990)
Passerine birds	M	$MEI_{max} = 16.09 m_b^{0.70}$	(Lindstrom and Kvist 1995)
Shorebirds	M	$MEI_{max} = 11.7 m_b^{0.82}$	(Kvist 2001)[b]

[a] Energy demanding situations: C = cold acclimation; L = lactation; G = growth; M = migratory fattening.

[b] Equation calculated by us.

TABLE 3.3

Major research questions in digestive ecology

(1) What anatomical and biochemical mechanisms lead to differences in digestive efficiency and the net value of food?
(2) What is the relationship between the net value of food and food choice?
(3) Does digestion rate limit energy flow?
(4) If digestion rate limits energy flow, which step(s) (physical or chemical breakdown or absorption, etc.) are limiting?
(5) Do limiting step(s) of digestion exhibit reversible phenotypic flexibility?
(6) What are the mechanisms and time courses for reversible changes in digestive anatomy and biochemistry?
(7) What are the consequences of digestively determined food choice and rate of food processing for ecological processes such as population production, distributional limits, dietary niche?

with body mass in a fashion similar to other metabolic rates (i.e., with (mass)$^{2/3-3/4}$). Even in fish fed to satiation, i.e., to near maximal intake, intake scales with (mass)$^{0.6-0.8}$ (Jobling 1993). Because this allometric scaling is similar to the scaling for basal or resting metabolism (chapter 1), it is not surprising that when maximal intake rates of species are normalized to their respective basal metabolic rates the ratios cluster together. Typically, maximal rates of metabolizable energy intake are 4–7 times basal metabolic rate (Hammond and Diamond 1997). The highest value that we know of, measured in lactating mice exposed to low temperature, is 7.7 times the resting metabolism (Johnson and Speakman 2001). MEI_{max} can depend on the nature of the food. For example, shorebirds that must crush hard shellfish in their gizzards cannot sustain the very high rates they achieve when eating commercially prepared, soft trout food or mealworms (chapter 4). Also, Kvist and Lindstrom (2000) make an important caveat: the absolute amount of food digested per 24 h day is influenced not just by the hourly rate but also by the total hours available for feeding.

   The existence of apparent maximum sustained metabolic rates in animals raises many interesting questions in physiology, ecology, and evolution. We have combined some of those listed by Hammond and Diamond (1997) with questions that we have raised throughout this chapter (table 3.3). We have a good set of questions about digestive ecology, but to pursue many of them we need to peer inside animals at specific digestive processes, which is the subject of our next chapter.

References

Abisgold, J. D., S. J. Simpson, and A. E. Douglas. 1994. Nutrient regulation in the pea aphid *Acyrthosiphon pisum*: application of a novel geometric framework to sugar and amino acid consumption. *Physiological Entomology* 19:95–102.

Alexander, R. M. 1998. Symmorphosis and safety factors. Pages 28–35 in E. R. Weibel, C. R. Taylor, and L. Bolis (eds.), *Principles of animal design*. Cambridge University Press, Cambridge.

Austreng, E., T. Storebakken, M. S. Thomassen, S. Refstie, and Y. Thomassen. 2000. Evaluation of selected trivalent metal oxides as inert markers used to estimate apparent digestibility in salmonids. *Aquaculture* 188:65–78.

Batzli, G. O. 1986. Nutritional ecology of the California vole: Effects of food quality on reproduction. *Ecology* 67:406–412.

Beaupre, S. J., and A. E. Dunham. 1995. A comparison of ratio-based and covariance analyses of a nutritional data set. *Functional Ecology* 9:876–880.

Bowen, S. H. 1987. Composition and nutritional value of detritus. Pages 192–216 in D.J.W. Moriarty and R.S.V. Pullin (eds.), *Detritus and microbial ecology in aquaculture*. ICLARM Conference Proceedings 14. International Center for Living Aquatic Resources Management, Manila, Philippines.

Bowen, S. H., E. V. Lutz, and M. O. Ahlgren. 1995. Dietary protein and energy as determinants of food quality: Trophic strategies compared. *Ecology* 76:899–907.

Brett, J. R., and T.D.D. Groves. 1979. Physiological energetics. Pages 279–352 in W. S. Hoar, D. J. Randall, and J. R. Brett (eds.), *Fish physiology*. Academic Press, London.

Castle, K. T., and B. A. Wunder. 1995. Limits to food intake and fiber utilization in the prairie vole, *Microtus ochragaster*. *Journal of Comparative Physiology B* 164:609–617.

Charnov, E. L. 1976. Optimal foraging, the marginal value theorem. *Theoretical and Population Biology* 9:129–136.

Cochran, P. A. 1987. Optimal digestion in a batch-reactor gut: The analogy to partial prey consumption. *Oikos* 50:268–270.

Dadd, R. H. 1960. Observations on the palatability and utilisation of food by locusts, with particular reference to the interpretation of performances in growth trials using synthetic diets. *Entomologia Experimentalis et Applicata* 3:283–304.

Diamond, J. 1991. Evolutionary design of intestinal nutrient absorption: enough but not too much. *News in Physiological Science* 6:92–96.

Diamond, J. M. 1993. Evolutionary physiology. Pages 89–111 in C. A. R. Boyd and D. Noble (eds.), *The logic of life: The challenge of integrative physiology*. Oxford University Press, New York.

Diamond, J., and K. Hammond. 1992. The matches, achieved by natural selection, between biological capacities and their natural loads. *Experientia* 48:551–557.

Dudley, R., and C. Gans. 1991. A critique of symmorphosis and optimality models in physiology. *Physiological Zoology* 64:627–637.

Fadely, B. S., G. A. J. Worthy, and D. P. Costa. 1990. Assimilation efficiency of Northern fur seals determined using dietary manganese. *Journal of Wildlife Management* 54:246–251.

Felicetti, L. A., C. T. Robbins, and L. A. Shipley. 2003. Dietary protein content alters energy expenditure and composition of the mass gain in grizzly bears (*Ursus arctos horribilis*). *Physiological and Biochemical Zoology* 76:256–261.

Fu, S. J., and X. J. Xie. 2004. Nutritional homeostasis in carnivorous southern catfish (*Silurus meridionalis*): Is there a mechanism for increased energy expenditure during carbohydrate overfeeding? *Comparative Biochemistry and Physiology A* 139(3):359–363.

Garland, T., Jr., and R. B. Huey. 1987. Testing symmorphosis: does structure match functional requirements? *Evolution* 41:1404–1409.

Goodman-Lowe, G. D., J. R. Carpenter, and S. Atkinson. 1999. Assimilation efficiency of prey in the Hawaiian monk seal (*Monachus schauinsland*). *Canadian Journal of Zoology* 77:653–660.

Hammond, K., and J. Diamond. 1997. Maximal sustained energy budgets in humans and animals. *Nature* 386:457–462.

Hilborn, R., and M. Mangel. 1997. *The ecological detective*. Princeton University Press, Princeton, N.J.

Hilton, G. M., D. C. Houston, and R. W. Furness. 1998. Which components of diet quality affect retention time of digesta in seabirds? *Functional Ecology* 12:929–939.

Hilton, G. M., D. C. Houston, N.W.H. Barton, R. W. Furness, and G. D. Ruxton. 1999. Ecological constraints on digestive physiology in carnivorous and piscivorous birds. *Journal of Experimental Zoology* 283:365–376.

Hilton, G. M., R. W. Furness, and D. C. Houston. 2000. A comparative study of digestion in North Atlantic seabirds. *Journal of Avian Biology* 31:36–46.

Holthuizen, A.M.A., and C. S. Adkisson. 1984. Passage rate, energetics, and utilization efficiency of the Cedar Waxwing. *Wilson Bulletin* 96:680—685.

Iason, G. R., A. R. Mantecon, D. A. Sim, J. Gonzalez, E. Foreman, and F. E. Bermudez. 1999. Can grazing sheep compensate for a daily foraging time constraint? *Journal of Animal Ecology* 68:87–93.

Illius, A. W., and I. J. Gordon. 1991. Prediction of intake and digestion in ruminants by a model of rumen kinetics integrating animal size and plant characteristics. *Journal of Agricultural Science* 116:145–158.

Jobling, M. 1993. Bioenergetics: Feed intake and energy partitioning. Pages 1–44 in J. C. Rankin and F. B. Jensen (eds.), *Fish ecophysiology*. Chapman and Hall, London.

Jobling, M. 1995. *Environmental biology of fishes*. Chapman and Hall, London.

Johnson, M. S., and J. R. Speakman. 2001. Limits to sustained energy intake. V. Effect of cold-exposure during lactation in *Mus musculus*. *Journal of Experimental Biology* 204:1967–1977.

Karasov, W. H. 1990. Digestion in birds: Chemical and physiological determinants and ecological implications. *Studies in Avian Biology* 13:391–415.

Karasov, W. H., and S. J. Cork. 1996. Test of a reactor-based digestion optimization model for nectar-eating rainbow lorikeets. *Physiological Zoology* 69:117–138.

Karasov, W. H., and D. J. Levey. 1990. Digestive system trade-offs and adaptations of frugivorous passerine birds. *Physiological Zoology* 63:1248–1270.

Karasov, W. H., and S. R. McWilliams. 2005. Digestive constraint in mammalian and avian ecology. Pages 87–112 in J. M. Starck and T. Wang (eds.), *Physiological and ecological adaptations to feeding in vertebrates*. Science Publishers, Enfield, N.H.

Karasov, W. H., D. Phan, J. M. Diamond, and F. L. Carpenter. 1986. Food passage and intestinal nutrient absorption in hummingbirds. *Auk* 103:453–464.

Kaufman, D. W., M. J. O'Farrell, G. A. Kaufman, and S. E. Fuller. 1976. Digestibility and elemental assimilation in cotton rats. *Acta Theriologica*. 21:147–156.

Kirkwood, J. K. 1983. A limit to metabolizable energy intake in mammals and birds. *Comparative Biochemistry and Physiology* 75A:1–3.

Konarzewski, M., and J. Diamond. 1995. Evolution of basal metabolic rate and organ masses in laboratory mice. *Evolution* 49:1239–1248.

Kontogiannis, J. E. 1968. Effect of temperature and exercise on energy intake and body weight of the white-throated sparrow, *Lonotrichia albicollis*. *Physiological Zoology* 41:54–64.

Kvist, A. 2001. Fuel and fly: Adaptations to endurance exercise in migrating birds. Ph.D. dissertation, Lund, Sweden.

Kvist, A., and A. Lindstrom. 2000. Maximum daily energy intake: It takes time to lift the metabolic ceiling. *Physiological and Biochemical Zoology* 73:30–36.

Lindstrom, A., and A. Kvist. 1995. Maximum energy intake rate is proportional to basal metabolic rate in passerine birds. *Proceedings of the Royal Society (London) B* 261:337–343.

Marken Lichtenbelt, W.D.V. 1992. Digestion in an ectothermic herbivore, the green iguana (*Iguana iguana*): Effect of food composition and body temperature. *Physiological Zoology* 65:649–673.

Martin, A. W., and F. A. Fuhrman. 1955. The relationship between summated tissue respiration and metabolic rate in the mouse and dog. *Physiological Zoology* 28:18–34.

Martínez del Rio, C., W. H. Karasov, and D. J. Levey. 1989. Physiological basis and ecological consequences of sugar preferences in cedar waxwings. *Auk* 106:64–71.

Mayes, R. W., C. S. Lamb, and P. M. Colgrove. 1986. The use of dosed and herbage n-alkanes as markers for the determination of herbage intake. *Journal of Agricultural Science* 107:161–170.

McWhorter, T. J., and C. Martínez del Rio. 2000. Does gut function limit hummingbird food intake? *Physiological and Biochemical Zoology* 73:313–324.

Melillo, J. M., J. D. Aber, and J. F. Juratore. 1982. Nitrogen and lignin control of harwood leaf litter decomposition dynamics. *Ecology* 63:621–626.

Mittler, T. E. 1987. Applications of artificial feeding techniques for aphids. Pages 145–170 in A. K. Minks and P. Harrewijn (eds.), *Aphids: Their biology, natural enemies, and control*, vol. 2B. Elsevier, Amsterdam.

Muthukrishnan, J., and T. J. Pandian. 1987. Insecta. Pages 373–511 in T. J. Pandian and F. J. Vernberg (eds.), *Animal energetics*, vol.1. Academic Press, San Diego.

Nagy, K. A. 2001. Food requirements of wild animals: Predictive equations for free-living mammals, reptiles, and birds. *Nutrition Abstracts and Reviews B* 71:21R–31R.

Nagy, K. A., I. A. Girard, and T. K. Brown. 1999. Energetics of free-ranging mammals, reptiles, and birds. *Annual Review of Nutrition* 19:247–277.

Nagy, K. A., and K. Milton. 1979. Aspects of dietary quality, nutrient assimilation and water balance in wild howler monkeys (*Alouatta palliata*). *Oecologia* 39:249–258.

Penry, D. L. 1998. Applications of efficiency measurements in bioaccumulation studies: Definitions, clarifications, and a critique of methods. *Environmental Toxicology and Chemistry* 17:1633–1639.

Peterson, C. C., K. A. Nagy, and J. Diamond. 1990. Sustained metabolic scope. *Proceedings of The National Academy of Sciences* 87:2324–2328.

Piersma, T. 2002. Energetic bottlenecks and other design constraints in avian annual cycles. *Integrative and Comparative Biology* 42:51–67.

Piersma, T., and J. Drent. 2003. Phenotypic flexibility and the evolution of organismal design. *Trends in Ecology and Evolution* 18:228–233.

Piersma, T., and A. Lindstrom. 1997. Rapid reversible changes in organ size as a component of adaptive behavior. *Trends in Ecology and Evolution* 12:134–138.

Pritchard, G. T., and C. T. Robbins. 1990. Digestive and metabolic efficiencies of grizzly and black bears. *Canadian Journal of Zoology* 68:1645–1651.

Raubenheimer, D., and S. J. Simpson. 1994. The analysis of nutrient budgets. *Functional Ecology* 8:783–791.

Raubenheimer, D., and S. J. Simpson. 1997. Integrative models of nutrient balancing: Application to insects and vertebrates. *Nutrition Research Reviews* 10:151–179.

Raubenheimer, D., and S. J. Simpson. 1998. Nutrient transfer functions: The site of integration between feeding behaviour and nutritional physiology. *Chemoecology* 8:61–68.

Reynolds, S. E. 1990. Feeding in caterpillars: maximizing or optimizing food acquisition? Pages 106–118 in J. Mellinger (ed.), *Animal nutrition and transport processes. 1. Nutrition in wild and domestic animals,* vol. 5. Karger, Basel.

Robbins, C. T. 1993. *Wildlife feeding and nutrition.* Academic Press, San Diego.

Root, T. 1988. Energy constraints on avian distribution and abundances. *Ecology* 69:330–339.

Sakaguchi, E., H. Itoh, S. Uchida, and T. Horigome. 1987. Comparison of fiber digestion and digesta retention time between rabbits, guinea pigs, rats and hamsters. *British Journal of Nutrition* 58:149–158.

Sibbald, I. R. 1982. Measurement of bioavailable energy in poultry feeding stuffs: A review. *Canadian Journal of Animal Science* 62:983–1048.

Sibly, R. M. 1981. Strategies of digestion and defecation. Pages 109–139 in C. R. Townsend and P. Calow (eds.), *Physiological ecology.* Sinauer, Sunderland, Mass.

Simpson, S. J., and D. Raubenheimer. 1995. The geometric analysis of feeding and nutrition: a user's guide. *Journal of Insect Physiology* 41:545–553.

Slansky, F., and J. M. Scriber. 1985. Food consumption and utilization. Pages 87–163 in G. A. Kerkut and L. I. Gilbert (eds.), *Comprehensive insect physiology, biochemistry and pharmacology,* vol. 4. Pergamon Press, Oxford.

Studier, E. H., E. J. Szuch, T. M. Tompkins, and V. W. Cope. 1988. Nutritional budgets in free flying birds: cedar waxwings (*Bombycilla cedrorum*) feeding on Washington hawthorn fruit (*Crataegus phaenopyrum*). *Comparative Biochemistry and Physiology A* 89:471–474.

Swennen, C., M. F. Leopold, and L. L. M. de Bruijn. 1989. Time-stressed oystercatchers, *Haematopus ostralegus,* can increase their intake rate. *Animal Behaviour* 38:8–22.

Taylor, C. R., and E. R. Weibel. 1981. Design of the mammalian respiratory system. I. Problem and strategy. *Respiration. Physiology* 44:1–10.

Trier, T. M. 1996. Diet-induced thermogenesis in the Prairie Vole, *Microtus ochrogaster. Physiological Zoology* 69:1456–1468.

Trier, T. M., and W. J. Mattson. 2003. Diet-induced thermogenesis in insects: A developing concept in nutritional ecology. *Environmental Entomology* 32(1):1–8.

Van Soest, P. J. 1994. *Nutritional ecology of the ruminant.* Cornell University Press, Ithaca, N.Y.

Waldbauer, G. P. 1968. The comsumption and utilization of food by insects. *Advances in Insect Physiology* 5:229–288.

Weibel, E. R. 2000. *Symmorphosis, on form and function shaping life.* Harvard University Press, Cambridge, Mass.

Weiner, J. 1992. Physiological limits to sustainable energy budgets in birds and mammals: ecological implications. *Trends in Ecology and Evolution* 7:384–388.

White, T.C.R. 1993. *The inadequate environment.* Springer-Verlag, Berlin.

Zimmerman, L. C., and C. R. Tracy. 1989. Interactions between the environment and ectothermy and herbivory in reptiles. *Physiological Zoology* 62:374–409.

Zynel, C. Y., and B. A. Wunder. 2002. Limits to food intake by the Prairie Vole: Effects of time for digestion. *Functional Ecology* 16:58–66.

CHAPTER FOUR

Simple Guts: The Ecological Biochemistry and Physiology of Catalytic Digestion

THE DISCUSSION IN chapter 3 of whole-animal feeding rate and digestive efficiency raised a host of research questions and themes in digestive ecology (table 3.3). To pursue many of them we have to break down digestion into its components and dwell on specifics. We will provide some basic background material that can be augmented by consulting comparative physiology texts (Withers 1992; Randall et al. 1997; Stevens and Hume 1995) and reviews (Vonk and Western 1984; Karasov and Hume 1997; Wright and Ahearn 1997). As quickly as possible, however, we want to explore those questions and themes within ecological contexts like energetics and feeding choices of animals, and their roles in ecological communities. Along the way, we will see how some of the research questions and themes in digestive ecology (table 3.3) play out at each step of the digestive process.

4.1 Lots of Guts, But Only a Few Basic Types

4.1.1 Guts as Chemical Reactors

In vertebrates and most invertebrates the primary site of nutrient absorption is across the epithelium of a gastrointestinal tract. Most species in three phyla (Porifera [sponges], Coelenterata, Platyhelminthes) lack digestive tracts and rely on uptake of nutrients from water through the integument, as do some species in at least 11 other phyla (Wright and Ahearn 1997). Although you can see scores of anatomical gut designs in comparative physiology textbooks

(Withers 1992; Stevens and Hume 1995; Randall et al. 1997), Jumars and Penry pointed out that most guts are combinations of three basic functional types. These types can be considered analogous to three categories of ideal chemical reactors: batch reactors, plug-flow reactors, and continuously stirred tank reactors (Penry and Jumars 1987). We will describe the three types of gut/reactors briefly (figure 4.1)

(1) Batch reactors (BRs) process reactants (substrates) in discrete batches. Conversion efficiency can be high, but the digestive process is interrupted between batches and hence the overall rate of conversion of reactants into products can be low. Batch reactors occur in some coelenterates. The saccular organ, such as the stomach of a carnivore or the cecum of certain herbivores, may be analogous to a batch reactor if meals are processed separately in discrete intervals.

(2) In plug-flow reactors (PFRs) materials (digesta) flow in an orderly, unidirectional fashion through a tubular reaction vessel without axial mixing. The term "axial" in reactor engineering jargon refers to the longitudinal dimension of the reactor. If the digestive process follows saturable kinetics (discussed below) then the concentration of reactants, and therefore the rate of reaction, is higher near the reactor entrance and declines along the length of the vessel. This peculiar characteristic gives PFRs a very high conversion rate. Tubular organs such as the intestine are analogous to PFRs if they exhibit little mixing along their length.

Figure 4.1. Digestive systems show great anatomical variation but can be functionally classified according to three kinds of analogous chemical reactors. (From Hume 1989, with permission.) Beneath each reactor type is a biological analogue.

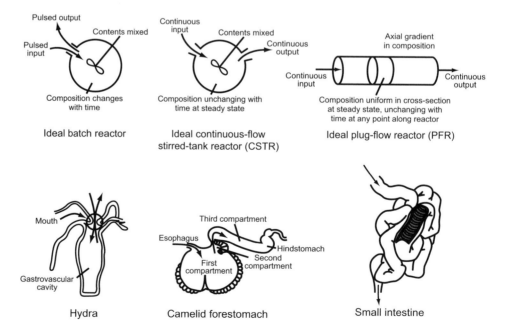

(3) Continuous stirred-tank reactors (CSTRs) are characterized by continuous digesta flow through a reaction vessel with complete mixing. Substrate concentration is diluted immediately upon entry into a CSTR by material recirculating in the reactor (residues from previous meals). This reduces the rate of reaction relative to the average reaction rate in a PFR with the same conversion efficiency. However, conversion efficiency of reactants into products in CSTRs can be high if digesta flow is low. The ruminant's rumen may be analogous to a CSTR.

By making some simplifying assumptions (box 4.1), the performance of guts/reactors can be evaluated in terms of a few characteristics such as conversion efficiency (the product formed and/or absorbed relative to the initial concentration of reactants), reaction rate, digesta retention time, volume of the reactor, and flow rate of digesta. To a first approximation we can estimate the conversion efficiency as

$$\text{conversion efficiency} \propto \frac{\text{reaction rate} \times \text{digesta retention time}}{\text{concentration of reactants} \times \text{reactor volume}}. \quad (4.1)$$

This equation can be used only as a first approximation because it assumes constancy in many parameters that in reality are relatively complicated functions of each other (see Penry and Jumars 1987 and Jumars and Martínez del Rio 1999 for examples of these functions). The equation is useful, however, because it identifies crucial qualitative relationships among digestive traits. All else being equal, conversion should be reciprocally related to initial concentration and gut volume, and positively related to both retention time and reaction rate. The flow rate of digesta through the gut/reactor, in relation to its size, determines the retention time:

$$\text{digesta retention time} \propto \frac{\text{reactor volume}}{\text{digesta flow rate}}. \quad (4.2)$$

Thus, conversion efficiency should be reciprocally related to flow rate (to convince yourself, substitute equation 4.2 in equation 4.1). Digesta retention time has a slightly different interpretation depending on the type of reactor. In BRs, digesta retention time is the time that it takes to process a batch. In PFRs, digesta retention time is the average time that it takes a particle to flow from the entrance to the exit of the vessel. In CSTRs digesta retention time is the average residence time of a particle in the reactor. Note that to use equation 4.1 in batch reactors you need to modify it. In batch reactors materials do not flow through the reactor; they are placed in and then taken out. Thus, in these reactors retention time has two components: the time during which the reactants (food) interact in the gut, plus the down time needed to take the materials in

Box 4.1. Guts as Chemical Reactors

Physiologists often use technological analogies. You may have read that the kidney is a filter, the heart is a pump, and the hypothalamus works like a thermostat. These technological analogies can be very helpful. We use something that we understand well, because after all, we built it, to understand something that we understand poorly. A fitting technological analogy for digestive physiology is the gut as a chemical reactor. Chemical reactors are vessels in which we place mixtures of reagents and from which we harvest useful products. Engineers have devised a large body of theory to describe and optimize the performance of chemical reactors. They have also developed a battery of empirical approaches to diagnose their function—and malfunction. The guts-as-reactors analogy was first proposed by biological oceanographers Penry and Jumars in 1987. They had the brilliant insight of applying the tools of chemical engineering to animal digestive systems. They argued that the question faced by digestive physiologists is to discover how various gut morphologies and digestive reactions may optimize an animal's digestive performance. This challenge is similar to that of the chemical engineer who has to evaluate the performance of reactors with different designs (gut functional morphologies) with the goal of optimizing yield or yield rate (energy or nutrient assimilation), and is given a series of chemical reactions (digestive hydrolyses and nutrient uptake) that take place within these reactors.

A sufficient description of the assimilation process requires two elements: the kinetics of enzymatic, microbial, and transport processes and a mass-balance equation that relates how these result in changes in nutrient and product composition in the gut. Penry and Jumars (1987) and many chemical engineering textbooks give derivations of mass-balance equations for the three ideal chemical reactor types. Here we simply present in graphical form the major differences in the concentration of reactants and products within these reactors (figure 4.2).

To characterize the flow of materials in a reactor (or within a gut), at a minimum we need to know how long individual particles or molecules stay in the vessel. This information can be determined by stimulus-response experiments in which the stimulus is a tracer input into the material entering the reactor and the response is the time that the particles of the tracer stay within it. Ideally, the tracer travels through the reactor at the same rate as the component of the feed whose pattern we are interested in tracking. In box 4.2 we will describe how we can estimate the average time that a particle stays within the gut or reactor (the so-called mean retention time MRT). Here we emphasize the form of the distribution of tracer concentration at the exit of the gut or reactor. Imagine that you introduce a pulse of tracer into a PFR. Because there is little mixing along the length of this reactor,

continued

continued

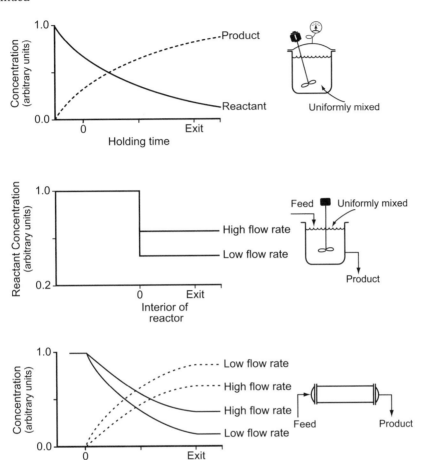

Figure 4.2. In batch reactors (top figure) reactants and products are mixed in a container and their concentration changes as reaction time proceeds. In continuous stirred tank reactors (middle figure), reactants and reagents flow in and out constantly, maintaining both the concentration and the rate of reaction constant. Reaction rate and product output rate are functions of flow rate through the reactor. Finally, in plug-flow reactors (bottom figure) materials flow continuously through a tube that has little axial mixing. Therefore at steady state there is a gradient in the concentration of reagents, products, and reaction rates along the length of the reactor. The steepness of this gradient (and thus the concentration of products at the exit of the reactor) depends on flow rate.

you eventually will find a pulse of tracer coming out of the distal end of the reactor. *MRT* is simply the time between the introduction and the exit of the pulse. Now imagine that you place a pulse of marker into a CSTR. The concentration of tracer in the output of the reactor will be high at first and will decline with time as the flow of materials into the reactor dilutes the tracer within the reactor. Indeed, in a well-mixed reactor the concentration of tracer exiting the CSTR will decline as a negative exponential with time (t):

$$C(t) = C(0)e^{-1/\tau} \qquad (4.1.1)$$

where $C(0)$ is the concentration of tracer at time 0 and τ is the *MRT* of the reactor. Chemical engineers call τ the throughput time. At steady state,

continued

continued

the throughput time is the ratio of the volume of the reactor and the flow rate in and out of it.

Many guts are more complicated than a PFR or a CSTR. One of the simplest gut configurations is a stomach (maybe analogous to a CSTR) followed by an intestine (a PFR). What is the output distribution of tracer in such a mixed reactor? As illustrated in figure 4.3, you should expect the

Figure 4.3. Output distribution of a marker in an ideal reactor consisting of a completely stirred tank reactor in series with a plug flow reactor.

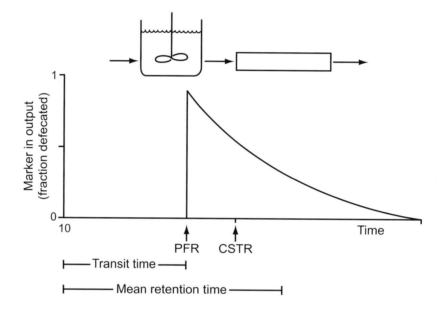

distribution to be a negative exponential that has been shifted to the right by the throughput time of the PFR. This figure is helpful because it illustrates two of the most widely used measurements that physiologists use to characterize residence time of digesta in the gut. Transit time refers to the time at which the tracer first appears in feces. In the case of the mixed reactor, *MRT* is the sum of the transit time (τ_{PFR}) plus the throughput time of the CSTR (i.e., $\tau_{tot} = \tau_{PFR} + \tau_{CSTR}$). For complex guts, the transit time and *MRT* are not equal (although many researchers wrongly assume that they are). It is tempting to use chemical reactor theory and simple input-output measurements of tracers to "diagnose" the geometry of digestive tracts (see Herrera and Martínez del Rio 1998). Unfortunately, guts are much more complicated than ideal chemical reactors and very often the distribution in time of defecated marker looks nothing like the nice distributions predicted by chemical reactor theory (Martínez del Rio et al. 1994). Retention time distributions must be interpreted cautiously.

continued

continued

In box 4.2 below we describe the most common method used to estimate *MRT* using tracers. Here we add a few methodological cautionary
notes on the measurement of both *MRT* and tracer retention time distributions. To be meaningful, these estimates must be done in animals that
are feeding normally. Often, for convenience, animals are fasted before
researchers give them the pulse of marker. What is probably worse, but
unfortunately not uncommon, is the practice of giving the marker to
fasted animals in a single meal and then fasting them until all marker is
defecated. Even a cursory consideration of the factors that affect *MRT*
should warn us about these practices. *MRT* is influenced by the volume
of the gut and by the rate of flow of digesta through it. Indeed, food
ingestion rate is one of the primary determinants of *MRT* (see Levey and
Martínez del Rio 1999). The biological meaning of "*MRT*" measured in
fasted animals is obscure.

and out of the reactor. As mentioned before, the existence of this down time
makes these reactors slow overall relative to reactors with continuous flow. The
physiological equivalent of digesta retention time in digestive systems is the
mean retention time (*MRT*) which is discussed in some detail in box 4.2.

Penry and Jumars (1987) used mass-balance equations to determine the ideal
gut-reactor configuration for two basic types of digestive reactions. In catalytic
(i.e., enzymatic) reactions, the reaction rate is a function of concentration according to the Michaelis-Menten equation (which we will introduce below). In
autocatalytic (e.g., microbial fermentation) reactions, the reaction rate is a complex function of substrate concentration and the concentration of the microbes. In
autocatalytic reactions the maximal rate of reaction occurs at an intermediate,
rather than at the highest, reactant concentration. Penry and Jumars (1987) concluded that for catalytic reactions a CSTR reactor requires either more time or a
larger volume to achieve a given conversion rate. Because PFRs maintain a gradient in reactant concentrations and thus of reaction rates from higher values near
the reactor entrance to lower values near the exit, they are a better design for
digestive processes that rely on catalytic enzymatic reactions. They suggested that
this is the reason why tubular guts predominate among complex, multicellular
animals. However, they also concluded that if, in addition to catalytic reactions,
fermentation autocatalytic reactions are important, then fermentation production
rate is maximized when a portion of the gut is a CSTR. These theoretical distinctions explain why we devote this chapter to digesters that rely largely on

Box 4.2. Measurement of Digesta Retention Time

In a critical review of a variety of methods, Warner (1981) concluded that the rate of passage of digesta through the gut is best described by the mean retention time. Typically, it is measured by feeding an indigestible substance, termed a marker, and measuring its appearance in excreta over time. Most markers are either solutes thought to stay in solution throughout the gut, particulate markers insoluble throughout the gut, or particle markers that become physically or chemically associated with food particles.

Roby et al. (1989) fed captive nestlings of Antarctic giant petrels (*Macronectes giganteus*) and Gentoo penguins (*Pygoscelis papua*) two radio-labeled markers simultaneously: a water-soluble marker called [^{14}C] polyethylene glycol (PEG) and a lipid soluble marker called [^{3}H]glycerol triether (GTE). They collected excreta at 4, 12, and 24 h postingestion and analyzed the markers by liquid scintillation. Most studies collect samples more frequently, but this small data set (table 4.1) is perfect for our demonstration.

TABLE 4.1
Passage of two inert markers through the digestive system of Antarctic giant petrels and Gentoo penguins

Time (h)	Proportion of dose recovered in excreta		Proportion of total amount of marker recovered in excreta	
	PEG	*GTE*	*PEG*	*GTE*
	Antarctic giant petrels			
4	0.005	0.004	0.006	0.009
12	0.653	0.237	0.832	0.500
24	0.125	0.223	0.160	0.481
	Gentoo penguins			
4	0.139	0.084	0.139	0.10
12	0.741	0.435	0.833	0.484
24	0.009	0.355	0.010	0.406

Note: PEG = polyethylene glycol, GTE = glycerol triether.

Some patterns are apparent even before making any calculations. First, the markers do not behave identically because the majority of the dose was eliminated by 12 h for PEG but not for GTE, especially in the petrel (compare panels A and C in figure 4.4). Second, marker elimination seemed slower in the petrel. Roby et al. (1989) and Duke et al. (1989) have described how the proventriculus can act as a separatory compartment and

continued

continued

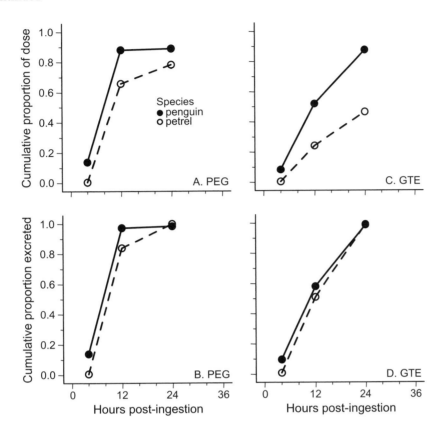

Figure 4.4. Pattern of excretion by penguins and petrels of the inert markers polyethylene glycol, PEG (panels A and B) and glycerol triether GTE (panels C and D).

prolong the retention of lipophilic material, especially in the petrel. Because of the prolonged retention of the lipid-phase marker, the recovery even by 24 h was incomplete in the petrel, whereas most of the aqueous-phase marker was recovered.

The calculation of mean retention time (MRT) permits a more quantitative comparison of the differences between the markers and the species. The MRT is calculated from the amount of marker excreted at the ith defecation (m_i) at time t_i after dosing:

$$MRT = \sum_{i=1}^{n} m_i t_i \Big/ \sum_{i=1}^{n} m_i. \qquad (4.2.1)$$

The calculated MRT will differ depending on what value(s) are used for m_i. A common method is to express all the marker data as a proportion (or percentage) of the total amount of marker recovered in excreta (as in panels B and D in figure 4.4). Notice how in this case this presentation of the data

continued

continued

> obscures the apparent differences between the species. Comparisons of
> *MRT* work out best if nearly all the marker is recovered. Warner (1981)
> described an extrapolation procedure for cases like that of the petrel where
> marker elimination was incomplete. Because most PEG was recovered in
> both species in our example, we will compare those: We calculate the
> aqueous-phase *MRT* as 11 h in the penguin and 14 h in the petrel.

endogenous enzymes to digest relatively nonrefractory materials in foods,
whereas we devote chapter 6 to digesters that typically ferment relatively refractory materials with the aid of symbiotic microbes. This chapter is about PFR-like
guts whereas chapter 6 deals with guts that have a CSTR-like compartment.

As long as we have analogized guts to chemical reactors, let us extend our
thinking a bit to include how the basic model might dictate the variation of
features of guts among animals. As our starting point, we assume that the purpose of guts is to provide energy and nutrients to fuel metabolism. Because
metabolism is related to the 0.6 to 0.8 power of body mass (chapters 1 and 3),
we might expect that the amount of gut and the processing time will vary in
some coordinated fashion. Also, because there are some large differences in
metabolism between endotherms and ectotherms (chapter 1), we might expect
some correlated large differences in these digestive characteristics. As you will
see in the following simple analyses, both of these expectations are borne out.

4.1.2 Allometry of Gut Size

The surface area of the gut, where breakdown of substrates and absorption of
their monomers occurs, scales with body mass to roughly the 0.75 power. The
allometry of the gut's area has been investigated intensively in mammals
(Chivers and Hladik 1980; Chivers 1989; Snipes and Kriete 1991; Snipes
1997), to some extent in other vertebrates (Ricklefs 1996; Karasov and Hume
1997), and not at all in invertebrates as far as we know. All these studies were
based on what has been called the nominal surface area, the surface area of the
intestine as a smooth tube, and they show that intestinal surface area is related
to the 0.6–0.8 power of body mass. But it is well known that the gut surface area
of invertebrates (Roche and Wheeler 1997) and vertebrates (Frierson and Foltz

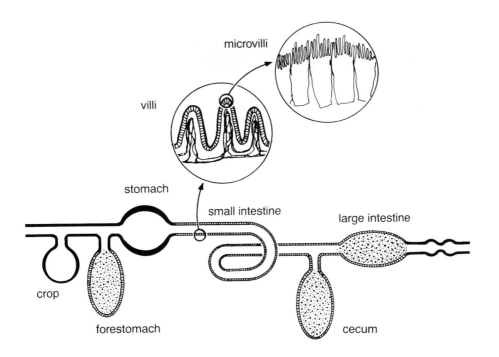

Figure 4.5. The primary nutrient digestive and absorptive region of the vertebrate gut is the small intestine, whose surface area is amplified by folds, villi, and microvilli. All vertebrates have a small intestine, but vary as to whether they possess other compartments such as crop, forestomach, stomach, cecum, and large intestine. As a general rule, catalytic enzymatic reactions occur in the small intestine, whereas fermentation can occur in the forestomach, cecum, and large intestine (shown with dotted areas). (From Weiner 1992, with permisson.)

1992; Karasov and Hume 1997; Makanya et al. 1997) is greatly elaborated by folds, finger- and leaflike extensions called villi, and even microvilli that increase membrane surface area of individual intestinal cells (figure 4.5). Methods to measure these components of intestinal area are described in a number of sources (Barry 1976; Karasov et al. 1985; Snipes 1994; Young-Owl 1994). When the villous and microvillous area of mammals is plotted as a function of body mass, the slopes of the relations do not differ significantly from each other or from that for the nominal area (figure 4.6A). The elevation of the lines of villous and microvillous area indicate that these structures on average increase potential absorptive area in mammals by 6.7 and 364 times, respectively, relative to nominal area.

The nominal surface areas of endothermic mammals and birds scale with mass similarly to those of ectothermic reptiles, amphibians, and fish (the pooled slope is 0.76), but the endotherms exceed the ectotherms in surface area by an average of 4.4 times (figure 4.6B). These observations are, of course, what we would expect from an organ that delivers nutrients to fuel metabolic rate (see chapter 1).

It is interesting that the intestinal surface areas of small birds tend to be lower than those of small mammals (figure 4.6C). Such a difference might be predicted based on the demands of minimization of the mass of gut contents and tissue in a flying organism. Indeed, our preliminary comparison of nonflying small mammals (< 500 g) with flying mammals (bats) and birds does indicate

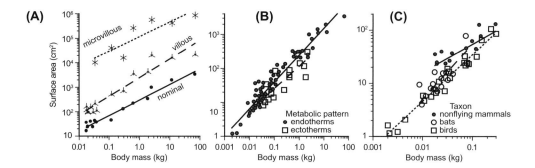

Figure 4.6. Relationships between intestinal surface area and body mass show the large enhancement due to villi and microvilli, and show that some groups seem to have greater surface area than other groups. In (A; left hand), measurements of all three types of surface area have been made in some mammals and they scale with body mass in a similar fashion (the pooled slope is 0.68). The calculated proportionality coefficients (intercept at unity, or where mass is 1 kg) are nominal area, 125 cm^2; villous area, 835 cm^2; microvillous area, 45,709 cm^2. In (B; middle), nominal surface areas of endothermic mammals and birds and ectothermic fish, amphibians, and reptiles scale with body mass in a similar fashion (pooled slope = 0.76), but the endotherms exceed the ectotherms in surface area by an average of 4.4 times. In (C; right hand), small flying endotherms (birds, bats) have less surface area than small nonflying mammals, although the comparison is confounded by different slopes. Data in (A) and (B) are from Karasov and Hume (1997), and the surface areas for fish and birds include caeca when present (see text). The data in (C) are from multiple sources: Martínez del Rio 1990a; Pappenheimer 1998; Barry 1976; Verzar and McDougall 1936; Makanya et al. 1997; Karasov and Levey 1990; Karasov 1988; Karasov et al. 1986; Dykstra and Karasov 1992; Karasov and Cork 1994; Afik et al. 1995; Karasov et al. 1996; Obst and Diamond 1989; Schondube et al. 2001; Snipes and Kriete 1991.

that fliers have less intestinal surface area (and presumably less intestine volume and gut contents) than nonfliers (figure 4.6C). We hasten to point out that the slopes and intercepts of such allometric analyses are sensitive to the particular data sets chosen, and our sets were not chosen with regard to controlling for defined subclass taxonomic groups, diet, etc. Many interesting analyses like these remain to be done in many animal groups in the best phylogenetically informed fashion.

4.1.3 Allometry of Digesta Retention Time

If guts are primarily tubes whose surface area is related to the 0.6–0.8 power of body mass, this relationship could dictate the allometry of other gut morphometrics like length and volume, because in a simple tube the three are related: $4\pi(\text{volume})(\text{length}) = (\text{area})^2$. We will leave this potentially fertile ground for the phylogenetically informed allometricians to plow. Volume is especially interesting to us because it is a variable in our simple gut-reactor models (equation 4.1). A number of analyses indicate that gut volume is related isometrically (i.e., to the 1.0 power) with body mass (Calder 1984). Consequently, according to equation 4.1, if digesta flow rates are related to metabolic need (i.e., digesta flow rate is proportional to $M^{3/4}$) it should follow that digesta retention time (MRT) should be related to $M^{1/4}$ (Calder 1984; Demment and Van Soest 1983, 1985).

This prediction turns out to be fairly well supported. First, the smooth muscle in the wall of the vertebrate alimentary tract produces an intrinsic cycle of electrical activity, called a basic electric rhythm, that can lead to a traveling wave of constriction which pushes the luminal contents distally. Within the individual, this basic determinant of digesta retention time can be modulated by a number of factors including the degree of stretch (how full the gut is), the nutrient makeup of the gut's contents (which are called chyme), locally produced peptide hormones, and nerves from the autonomic nervous system. But, looking across many species, Adolph (1949) found that the gut beat duration in mammals scaled with $M^{0.31}$, i.e., very close to the predicted 0.25 power scaling. Second, digesta retention time at the whole-animal level has been measured in many species by feeding animals indigestible markers and measuring marker excretion from the digestive tract as a function of time since feeding (box 4.2). Among birds and mammals that digest foods such as animal matter, fruits, and seeds using enzymatic processes, the mean retention time is related to the approximate 0.25 power of body mass, although the residuals of this allometric relationship can vary with diet (Robbins 1993).

The allometric summaries of gut size and processing time are interesting and provide a useful starting point for thinking about gut structure and function in relation to metabolic need. But do not let their dazzle blind you to the fact that these features are not necessarily static within an animal. In sections that follow we will emphasize how phenotypically flexible most features of the gut are, including size, processing time, and biochemistry. Using allometric summaries and the application of some principles from chemical reactor theory, we have simplified things quite a bit. From a first consideration of the biodiversity of gut anatomy and its seemingly bewildering complexity, we have refocused for now on a single basic tubular gut design modeled as a PFR whose size and characteristic time function (retention time) scale in relation to metabolic need. To finish up this section, we will provide a brief overview of the two major biochemical processes that occur in the tubular gut: chemical breakdown followed by absorption of breakdown products. Because both of these processes are saturable and can be described by enzymelike kinetics, we begin with a general discussion of how enzymes work. Our coverage is brief, and physiologically informed readers might choose to skip to the next section.

4.1.4 Overview of How Enzymes Work

Metabolism is a very complex business, and thousands of enzymes catalyze metabolic reactions. Depending on their function, enzymes can be divided into a variety of groups (oxidoreductases, transferases, hydrolases, lyases, isomerases, and ligases). Physiological ecologists often measure the activity of a variety of enzymes and hence some acquaintance with enzymes should be part of our toolbox. Many biochemistry textbooks present detailed descriptions of how enzymes work and we urge interested readers to consult one of them for further information. We are partial to the lucid chapter on enzyme function by Mathews and van Holde (1996). Although it would be inappropriate (and redundant) to add a detailed treatment of enzymes in this book, this section uses digestive enzymes as exemplars to outline just a few elementary, but fundamentally important, aspects of enzyme biochemistry that we will use in subsequent sections of the book.

Enzymes are organic catalysts that lower the energy barrier to a reaction, thereby making the reaction go faster. An enzyme, which is a protein, binds a molecule of substrate (or multiple substrates) into a region called the active site. The complex tertiary structure of enzymes makes possible the snug fit of a substrate into this molecular pocket. When substrates are bound within the active site both the enzyme and the substrate become deformed. The "stress" produced in the distorted enzyme pulls the substrate into a transition state, an intermediate state in which the molecule is strained or distorted. Molecules must pass through this transition state if the reaction is to occur. This model of enzyme function is called the induced fit hypothesis. Enzymes do more than just distort or position their substrates. Often, they also have specific amino acids or side chains poised in exactly the right place to help the catalytic process. Summarizing an immensely complex phenomenon, an enzyme (1) binds the substrate(s), (2) lowers the energy of the transition state, and (3) can directly promote the catalytic process. After this process is completed, the enzyme releases the products and returns to its original state. If you would like to see a lyrical description of the catalytic process (with music!), we recommend that you peek at the "Enzyme Warp" within Aimee Hartnell's delightfully quirky website (www.geocities.com/le_chatelier_uk/).

A minimal equation that describes a simple one-substrate reaction catalyzed by an enzyme is

$$E + S \underset{k_{-1}}{\overset{k_1}{\rightleftharpoons}} ES \overset{k_{cat}}{\longrightarrow} E + P, \qquad (4.3)$$

where E corresponds to the concentration of free enzyme, S to the substrate, and P to the product. This equation assumes that the reverse reaction between E and P is negligible. Therefore we can express the reaction rate (V) as a simple first–order reaction:

$$V = k_{cat}[ES] \qquad (4.4)$$

where k_{cat} is the dissociation rate constant. The problem that we have now is to figure out the form of the concentration of the enzyme substrate complex $[ES]$. Michaelis and Menten solved the problem for the first time at the beginning of the Twentieth century. They assumed that the dissociation rate as measured by k_{cat} was low compared to the rates of formation and dissociation of the enzyme substrate complex (k_1 and k_{-1}). Under these conditions $[ES]$ will be at steady state with $[E]$ and $[S]$ and hence

$$[ES]k_{-1} = k_1[E][S] \quad \text{and} \quad \Rightarrow \frac{[E][S]}{[ES]} = \frac{k_{-1}}{k_1} = K_m. \qquad (4.5)$$

$[E]$, the concentration of free enzyme, equals the total concentration of enzyme $[E]_T$ minus the concentration of enzyme that is bound to the substrate ($[E] = [E]_T - [ES]$). Therefore equation 4.5 transforms into

$$\frac{\big([E_T] - [ES]\big)[S]}{[ES]} = K_m \quad \text{and} \quad [ES] = \frac{[E_T][S]}{[S] + K_m} \qquad (4.6)$$

We can use equations 4.4 and 4.6 to obtain an expression for V:

$$V = k_{cat}[ES] = \frac{k_{cat}[E_T][S]}{[S] + K_m}. \qquad (4.7)$$

When the substrate concentration is very high, all available enzyme is complexed to the substrate and hence $[ES] = [E_T]$. Thus $k_{cat}[E_T]$ is the maximal possible rate of reaction when the substrate is "saturating." If we call $k_{cat}[E_T] = V_{max}$, we have the Michaelis-Menten equation in all its glory—and it is a glorious equation!

$$V = \frac{V_{max}[S]}{[S] + K_m}. \qquad (4.8)$$

Figure 4.7. Enzyme reaction rates vary with (A; left) substrate concentration, (B; middle) pH, and (C; right) temperature.

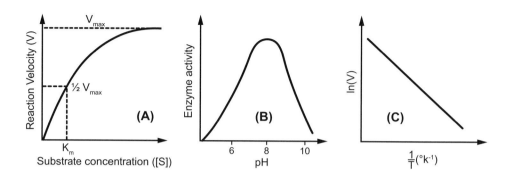

The shape of this relationship is shown in figure 4.7A. Haldane and Briggs used a different set of assumptions to construct the Michaelis-Menten equation. Although the final equation is the same, the interpretation of K_m is different. It is worth examining the interpretation of K_m in some detail.

K_m (called the "Michaelis constant") equals the substrate concentration at which the rate of reaction V is equal to $V_{max}/2$ (to convince yourself, substitute $V_{max}/2$ for V in equation 4.8 and solve for $[S]$). In Michaelis and Menten's derivation K_m has a straightforward interpretation. K_m equals the equilibrium constant k_{-1}/k_1. In Haldane and Brigg's derivation, which is perhaps more general, $K_m = (k_{-1}+k_{cat})/k_1$. The Michaelis constant is often associated with the strength of binding of substrate to enzyme. This is true in the Michaelis-Menten case in which k_{cat} is very small. Under these circumstances a large K_m means that k_{-1} is much larger than k_1 and the enzyme binds the substrate weakly. However, a very large value of k_{cat} can also lead to a large K_m, and hence the interpretation of K_m as a dissociation constant can be dodgy. The constant k_{cat} is a direct measure of the catalytic production of product (its units are $(time)^{-1}$). The larger k_{cat}, the faster the catalytic events in the enzyme must be. The reciprocal of k_{cat} can be interpreted as the time that it takes an enzyme to turn over one substrate molecule.

The meaning of K_m and k_{cat} becomes clearer if you consider what happens when $[S]$ is very small relative to K_m. Under these conditions most of the enzyme is free and hence $[E_T] \approx [E]$ and therefore

$$V \approx \frac{k_{cat}[E_T][S]}{K_m} = \frac{V_{max}}{k_m}[S]. \tag{4.9}$$

This equation tells you that, at low values of $[S]$, V increases linearly (rather than hyperbolically) with substrate concentration. The slope of this line equals V_{max}/K_m. More importantly, the ratio k_{cat}/K_m gives a direct measure of the performance of an enzyme when abundant enzyme sites are available. It measures an enzyme's efficiency and specificity.

At this point you may wonder why we have provided an almost interminable (although arguably perfunctory) analytical description of the mathematics used to describe how enzymes work. The primary reason is that an understanding of this process is absolutely central to the work of ecological physiologists. A secondary, albeit not unimportant, reason is that, as we will see in box 4.5, the Michaelis-Menten equation is a special case of formulas that describe many "saturable" processes that are prevalent throughout biology. For example, biologists use Michaelis-Menten-like equations to describe the transport of solutes across biological membranes, the relationship between predation rate and prey abundance (box 4.5 below), and the relationship between light intensity and photosynthesis rate in plants. In all these cases we have a collection of objects (enzymes, transport proteins, predators, and photosynthetic pigments) that "capture" and transform other objects (substrates, solutes, prey, and photons). Because processing the captured objects takes time, all these phenomena show saturation. When the density of "capturable" objects is very high, the rate of capture is limited by the capacity of the capturers to process their quarry.

A variety of factors influence the velocity of enzyme-catalyzed reactions (figure 4.7). It should not surprise you that pH and temperature are important determinants of V. Often an enzyme's rate of reaction shows a distinct optimal pH value. A pH stability study as shown in figure 4.7B is an essential component of an enzyme activity study. Recall from chapter 1 that many biological processes depend on temperature in a predictable fashion. The predictability of this temperature dependence is a consequence of the Arrhenius relationship between the rate constant of a chemical reaction, K, and temperature,

$$K = Ae^{-E_a/RT} \quad \text{and thus,} \quad \ln(k) = \ln(A) - \left(\frac{E_a}{R}\right)\frac{1}{T}, \tag{4.10}$$

where E_a (in J) and R (in J $°K^{-1}mol^{-1}$) are the activation energy and the gas constant, respectively. You may, or may not, recall this equation from chapter 1. For most enzyme-catalyzed reactions, V_{max} depends on several rate constants, each of which may be affected differently by temperature. Thus the calculated E_a from an Arrhenius plot will be an average value.

4.1.5 Overview of Chemical Breakdown

Most organic compounds ingested by animals require conversion into a limited number of smaller constituent compounds that can be absorbed. A general scheme of the enzymes of the vertebrate digestive system that accomplish the sequential breakdown of various nutrients is presented in figure 4.8. Enzyme

Figure 4.8. Principal nutrients in foods and the vertebrate endogenous digestive enzymes that hydrolyze them. (Modified from Stevens 1988.) Note that the digestion of complex carbohydrates and proteins requires two steps: luminal digestion by extracellular enzymes and membrane digestion by mucosal enzymes.

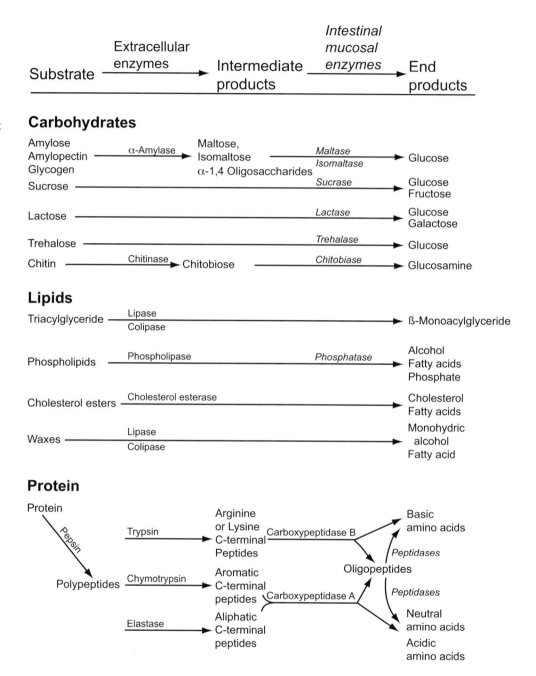

activities are measured by a variety of methods (box 4.3), and several reviews provide many details on the enzymes' structure, pH dependence, function, and distribution among animal taxa (Vonk and Western 1984; Stevens and Hume 1995; Karasov and Hume 1997). Chemical breakdown by many invertebrates follows a

Box 4.3. The Measurement of Enzyme Activity

The simplest method to study enzyme activity rates is under steady-state (or approximately steady-state) conditions. There are two ways of measuring the rate of activity of an enzyme. One can track the rate of appearance of product (or the disappearance of substrate) by sampling continuously. Alternatively, one can stop the reaction and measure the quantity of substrate consumed or product generated after a given incubation period. To give these descriptions a degree of reality, we detail how the activity of a more or less typical intestinal enzyme is measured.

Sucrase-isomaltase is a membrane-bound enzyme that breaks sucrose into its component monosaccharides glucose and fructose. Sucrase-isomaltase has a nice quirk; its hydrolytic action is strongly inhibited by Tris (tris-(hydroxymethyl) aminomethane), a common organic substance widely used to make buffers. The enzyme sample is incubated with the substrate at a given temperature and pH, and the reaction is stopped by adding a buffer containing Tris. This buffer also contains a mixture of reagents that transform glucose into a substance that absorbs light at a relatively narrow band of wavelengths. By measuring the concentration of this substance we can then determine how much glucose was produced. Of course, we must make sure that the rate of reaction is constant during the incubation time.

We can use sucrase–isomaltase to make a few comments about the intricacies and controversies of measuring enzyme activities. In the previous paragraph we used the following wording: "The enzyme sample is incubated. . . ." By "enzyme sample" we mean a variety of things. The purist among the biochemists will argue that the sample should be the pure enzyme in a suitable buffer. Indeed, many biochemists recite the following dictum as a mantra: "Thou shall not waste pure thoughts on an impure enzyme." We are not biochemists, we are ecological physiologists, and we all know that ecologists' heads are full of impure thoughts that a pure enzyme cannot answer. We are a lot more interested in finding out the capacity of an animal to assimilate or catabolize a nutrient in vivo than in knowing exactly the catalytic mechanism by which the enzyme breaks down a substrate. Purifying an enzyme can be a difficult, grueling process that leads to low yields. What biochemists mean by a low yield is that some of the enzyme that is in a tissue is not recovered in the process of purifying it. Some of it is lost between test tubes or stuck to a chromatography column. Clearly this is not a good thing if our goal is to know the capacity of the whole tissue to perform a certain function. Hence, physiological ecologists often work with "dirty" preparations such as homogenates of tissue (i.e., tissues that have been ground up in buffer) or with semipurified enzymes. The activity of membrane-bound digestive enzymes, for example, is often measured in isolated brush-border preparations (see box 4.4 below).

continued

continued

A biochemist might complain that by working with homogenates we ignore potentially important information. Consider the disaccharide maltose. It is the main product of the hydrolysis of starch by pancreatic amylase. It is further broken down into two glucose molecules by two enzymes: sucrase-isomaltase and maltase-glucoamylase (Alpers 1994). Although this fact is of unquestionable interest to us as physiologists, as ecological physiologists we are more interested in the functional capacity of the intestine to digest maltose than in the precise number of enzymes that perform this function. The hydrolytic activity that we measure in homogenates provides us with the answer. Some ecological questions may require cleaner preparations, and even purified enzymes, but many do not. The difference between biochemists and ecological physiologists in the units in which enzyme activities are reported is revealing: biochemists standardize enzyme activity (in μmoles of product produced per unit time) by the amount of enzyme protein (typically in mg). In contrast, ecological physiologists often standardize by the amount of wet (or dry) mass of the tissue. Using the mass of the tissue allows "scaling up" the measurements done in the test tube to the whole organ, and hence estimating its capacity to perform a certain function. Scaling up a given process from in vitro measurements to a whole organism involves a variety of assumptions, and requires that you consider a variety of questions such as: Is the K_m the same in vivo as in vitro? Are the conditions in the tissue (pH, substrate concentration, etc.) similar in vitro as they are in the living animal? Scaling up must be done judiciously.

Spectrophotometers

Spectrophotometric methods to measure the activity of enzymes are simple, cheap and accurate (see next section). They have the obvious requirement that either a substrate or a product absorb light in a spectral region where other substrates do not. Good examples are the myriad of reactions that consume or produce NADH. NADH absorbs strongly at 340 nm, but NAD^+ does not. Even if the reaction does not involve a light-absorbing substance, it is often possible to couple this reaction to another one that does.

A spectrophotometer is an instrument that measures the amount of light of a given wavelength that is absorbed or transmitted by a sample. The diagram in figure 4.9 shows its components, which include (1) a light source, (2) a collimator that focuses an intense beam of light, (3) a monochromator (usually a prism or grating) that divides the light beam into its component wavelengths, (4) a device to select the wavelength desired, (5) a compartment where the cuvette holding the sample is held, (6) a photoelectric detector, and (7) an electrical meter to record the output of the detector.

The fraction of incident light that is absorbed by a solution depends on the thickness of the sample, the concentration of the absorbing compound,

continued

continued

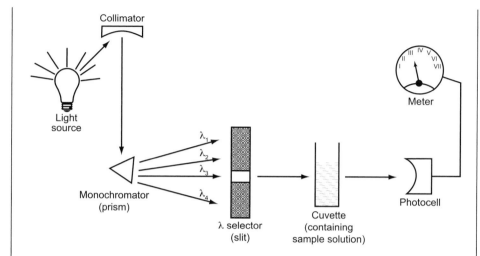

Figure 4.9. A diagram of the elements of a spectrophotometer.

and the nature of the compound. The ratio between the intensity of the transmitted light and the initial intensity of the light beam is called the transmittance. The logarithm of the inverse of transmittance is called the absorbance or optical density. For a given length (thickness of the cuvette, or path length, which usually equals 1 cm), the absorbance of a dissolved substance is a linear function of the substance's concentration:

$$\text{absorbance} = \log\left(\frac{1}{\text{transmittance}}\right)$$

$$= (\text{molar extinction coefficient}) \times (\text{path length}) \qquad (4.3.1)$$

$$\times (\text{concentration})$$

This relationship is called the Lambert-Beer law.

similar scheme based on the three categories of digestive enzymes: proteases, carbohydrases, and esterases (Withers 1992; Wright and Ahearn 1997). Within these categories there are many specific enzymes, but all are hydrolases (they break chemical bonds by the addition of water) and they typically have specificity for groups of substrates (e.g., certain proteins, carbohydrates, or lipids) rather than for specific substrates.

For example, the proteases hydrolyze peptide bonds between amino acids, but they can be divided into major groups, such as endopeptidases, which cleave bonds within a protein, and exopeptidases, which cleave the terminal amino acids with the free amino group (aminopeptidases), or the amino acid with the free carboxyl group (carboxypeptidases). Further specificity arises because some of the enzymes preferentially cleave bonds between particular combinations of amino acids. Some proteases are secreted in an inactive form called a "zymogen," which protects the digestive cell interiors from self-digestion. The zymogen is typically activated under acidic conditions (i.e., by H^+), or by other enzymes, or by the activated enzyme itself. This is another example of an autocatalytic reaction (see section 4.1.1).

Some digestive enzymes are secreted into the lumen of the gut and act there in what is called "luminal digestion." The proteases and amylase secreted by the pancreas are good examples of enzymes that act in the lumen. Others remain bound to the brush border of intestinal cells and effect what is called "membrane digestion." The enzymes that break down the disaccharides sucrose, maltose, and lactose are good examples of membrane-bound digestive enzymes. Not surprisingly, these enzymes are called disaccharidases. Often luminal and membrane digestion act in series. Starch, for example, is first broken down into oligosaccharides such as maltose by luminal amylases. These oligosaccharides are then broken down into glucose by membrane-bound enzymes.

Chemical breakdown of lipids differs from that of proteins and carbohydrates because lipids are not water soluble and they instead coalesce into large droplets in the water in the gut's lumen. Therefore, an important step in lipid digestion is emulsification, which reduces the size of the lipid droplets by mixing in amphipathic molecules (i.e., molecules that contain both polar groups soluble in water and nonpolar groups soluble in lipid and hence can act as detergents; see chapter 2). In vertebrates these molecules include bile salts, which are produced in the liver from cholesterol conjugated with amino acids (glycocholic and taurocholic acids). Although the bile acids help emulsification, they inhibit lipase activity by preventing binding of lipase to the substrate surface (Borgstrom and Patton 1991). The inhibition is counteracted by the presence of a wedgelike molecule called colipase, a small protein secreted by the pancreas of most vertebrates (Vonk and Western 1984). Colipase opens up spaces at the lipid globule's surface for lipase to act. Biliary secretions also include phospholipids and cholesterol. The net effect of emulsification is the dispersion of the lipid load into many very small droplets with high surface to volume ratio, called micelles, which can then be effectively attacked by esterases which hydrolyze lipids in solution, and by lipase enzymes which act only at the lipid-water interface. Lipases have higher affinity for lipids with unsaturated than those with saturated fatty acids, which helps explain why unsaturated fatty acids are sometimes more thoroughly assimilated than saturated fatty acids (Clifford et al. 1986; Place 1996), and why digestion is relatively slow and inefficient for wax esters, which generally contain long-chained saturated fatty acids (Jobling 1995). Micelles thus contain bile, many types of lipids, lipid breakdown products (e.g., mono- and diacylglycerols), and other liposoluble compounds such as lipid-soluble waxes, vitamins, and toxins. The micelles themselves are not absorbed, but approach close to the absorptive surface of the gut where their contents are also soluble in the lipid bilayer cell membrane, and thus their contents may be absorbed.

4.1.6 Overview of Nutrient Absorption: Modes and Sites

The products of chemical breakdown that are absorbed include monosaccharides and N-acetyl D-glucosamine (NAG) from carbohydrates including chitin, small peptides and amino acids from protein, and fatty acids and monoacyleeros from lipid (see figure 4.8). Also absorbed are short-chain fatty acids (SCFAs) from microbial fermentation and a variety of vitamins and minerals that are produced or released in the course of chemical digestion and fermentation. If foods contain natural secondary metabolites (SMs) and man-made toxicants then these xenobiotics may also be absorbed.

These substrates pass into circulatory fluids by a variety of pathways across the limiting barriers which are layers of cells of the gastrointestinal tract and, in many invertebrates, the epithelial cells of the integument (figure 4.10). At the cellular level, the limiting barriers for absorption are the membranes at the apical and basolateral surfaces and the junctions between cells. Although endocytic mechanisms like phago- and pinocytosis of food particles may be important in some animals, we will focus on absorption of dissolved compounds. Their flux (rate of transfer) involves either diffusion, solvent drag (bulk movement along with absorbed water), or a carrier-mediated mechanism. These mechanisms have been studied by various methods (box 4.4) in most groups of animals to varying degrees, and we will rely for "the basics" on two recent reviews of hundreds of studies (Karasov and Hume 1997; Wright and Ahearn 1997).

A carrier, or transporter, is a membrane protein that catalyzes the transmembrane movement of a substrate. Transporters are inherently selective, which means that only solutes with certain structural criteria gain access to that particular mediated pathway. There are two general categories of mediated transport: facilitated diffusion and active transport. In facilitated diffusion the transporter catalyzes the movement of the substrate, but only down its electrochemical gradient. In contrast, active transport can move an organic nutrient against its electrochemical gradient, although this "uphill" flux is achieved by coupling the nutrient flux to the flux of a solute such as Na, K, or H down its electrochemical

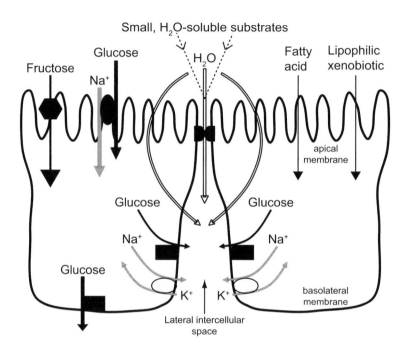

Figure 4.10. Absorption occurs mainly by either diffusion, solvent drag (bulk movement along with absorbed water), or a variety of carrier-mediated mechanisms. Lipophilic compounds can diffuse into and across the cell phospholipid bilayer membranes. Examples in this figure are fatty acids and some xenobiotics. A carrier, or transporter, is a membrane protein that catalyzes the transmembrane movement of a substrate. Examples in this figure are (i) the facilitative fructose transporter in the apical membrane (also called GLUT5, represented by a filled hexagon), (ii) the facilitative glucose transporter (called GLUT2, represented by a filled rectangle) in the basolateral membrane, and (iii) the active Na^+-glucose cotransporter (called SGLT1, represented by a filled spheroid) in the apical membrane which can move glucose against its concentration gradient. Na^+ ions, which move down their electrochemical gradient, are expelled from within the cell by Na^+-K^+ ATPase in the basolateral membrane (represented by an unfilled spheroid). It is thought that the extrusion of Na and glucose into the lateral intercellular space creates a region of high concentration that draws in water from the lumen through the tight junction between adjacent enterocytes. This flow of water can drag small water soluble chemicals, like glucose or amino acids, by a process called solvent drag. Water can also pass through pores through membranes. In this simple schematic, the exact stochiometries of solute fluxes are not depicted.

Box 4.4. The Measurement of Absorption

Our understanding of absorption depends on experiments applying a number of techniques. No single method does the trick, and the method of choice depends a lot on the question being asked. For example, what is the best way to estimate the capacity of a lab rat's intestine to absorb D-glucose? We illustrate two of these possible approaches in figure 4.11.

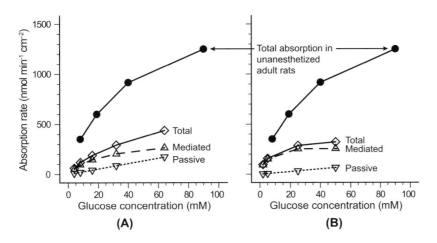

(A)   **(B)**

Figure 4.11. A (left-hand figure). Measurement of absorption in isolated intestine in vivo (Karasov and Debnam 1987). In an anesthetized rat a 25 cm length of intestine was cannulated at each end. Physiological saline containing different concentrations of glucose with and without phloridzin were infused, and the effluent was analyzed for glucose. Phloridzin is a plant secondary metabolite from apple (*Malus*) fruit, seed, and bark that is a known competitive inhibitor of the Na,D-glucose cotransporter (named SGLT1). Glucose absorption in the presence of phloridzin was relatively low, and was linear with glucose concentration (unfilled downward triangles, dashed line). The interpretation is that the phloridzin inhibits mediated glucose absorption, leaving only the passive, diffusive component. By subtracting out the apparent passive component from the total absorption (unfilled diamonds, solid line), one is left only with a saturable, mediated component (upward unfilled triangles, dashed line) which has a plateau value of 295 nmol min^{-1} cm^{-2} (this is the V_{max}) and a Michaelis constant (concentration yielding 1/2 of V_{max}) of 24 mM. This method seems physiologically realistic and offers some mechanistic detail. However, with this method we are not sure

(caption continues)

continued

continued

(*caption continued*)

whether the kinetics of glucose absorption, or the depressing effect of phloridzin, reflect processes at the apical or basolateral membrane of intestinal cells. Also, it is known that the anaesthesia can depress mediated absorption (Uhing and Kimura 1995). For comparison, look how much higher total absorption was in chronically perfused, unanesthetized adult rats studied by Ugolev et al. (1986) (interpreted by Pappenheimer 1998); filled circles, solid line. B (right-hand figure). Measurement of absorption in everted intestinal sleeves in vitro (Karasov and Debnam 1987). Rats were anesthetized and the small intestine was removed, rinsed with iced saline, and everted so that the villous surface was facing outwards. 1-cm-long sleeves of everted intestine were mounted on rods, and then these "dip sticks" were incubated in saline solution containing D-[^{14}C]glucose and L-[^{3}H]glucose for 2 min. The L-glucose is a steroiosomer that barely interacts with glucose transporters. Both isotopes are absorbed by the everted sleeves, but the uptake of the D-glucose was much greater than that of the L-glucose because the latter enters slowly by diffusion whereas the former enters by diffusion + active transport. The difference in their uptake is therefore due to active transport alone (upward unfilled triangles, dashed line), and this mediated uptake was 93% inhibited by phloridzin (not shown). Its V_{max} (276 nmol min^{-1} cm^{-2}) was similar to that in A, but the Michaelis constant (4 mM) was much lower than in the in vivo preparation. This is probably because at low glucose concentrations the boundary layer immediately adjacent to the membrane is a resistance to glucose uptake that is in series with the membrane itself, and the boundary layer is much thinner in the in vitro method than the in vivo method (Karasov and Diamond 1983a). The passive flux alone (unfilled downward triangles, dotted line) was measured by incubating the tissues with L-[^{3}H]glucose and an impermeant marker that corrected for radioactivity simply adhering to the tissue. With this method one knows that the uptake measured was across the apical membrane, and so the kinetics of glucose transport most likely correspond to transporter(s) there. With this method one can easily measure uptake all along the intestine length, and integrate the measures over the entire intestine for an estimate of the uptake capacity. But, in some species damage caused by everting the intestine can reduce the V_{max} (Starck et al. 2000), and the apparent passive flux was lower than in the in vivo method, partly because the in vitro method does a poor job of measuring the impact of absorption by solvent drag. Also, like the in vivo measure in anesthetized rats (A), the maximal mediated in vitro uptake in rats (here, or 431 nmol min^{-1} cm^{-2}; Toloza and Diamond 1992) and the total uptake were lower than absorption rates measured in chronically perfused, unanesthetized adult rats by Ugolev et al. (1986; filled circle, dotted line).

(*caption continues*)

continued

continued

(*caption continued*)

This latter measurement may include the effect of recruitment of additional glucose transporters (GLUT 2) that have lower affinity than the brush-border glucose transporter (SGLT1) (Kellett and Helliwell 2000), and it includes passive absorption which is typically neglected in calculations of absorption capacity but which becomes especially important as concentration increases (Pappenheimer and Reiss 1987; Pappenheimer 1993).

In addition, one can measure transport using intestinal brush-border membrane vesicles (BBMVs). We describe the method for D-glucose (Kessler and Toggenburger 1979). The mucosa from rat intestine are scraped and homogenized with a Mg-containing saline solution that causes the apical membrane of the enterocytes to precipitate. Following several high-speed centrifugation steps the membrane is isolated and it forms small vesicles, about 0.05–0.3 μm in diameter, which are suspended in solution. Very small amounts (10–30 μl) are mixed with solution containing tracer amount of D-[^{3}H]glucose, NaCl and other solutes, and unlabeled D-glucose and/or phloridzin. The vesicles take up D-glucose, but after just a few seconds of incubation the reaction is stopped and the vesicles are rinsed, filtered, and counted by liquid scintillation. Also, the protein content of the vesicle preparation is measured so that the glucose uptake can be normalized to a specific amount of membrane protein. Many kinetic features of glucose transport can be determined with this method (Hopfer 1987) without the confounding factor of high boundary layer resistance. Also, basolateral membrane vesicles can be isolated, making this one of the few techniques for studying glucose efflux from enterocytes by faciliative diffusion via the GLUT2 transporter. However, it is most difficult to extrapolate back from these vesicle measurements to describe the sugar absorption of the whole animal.

For a more complete discussion of these and other methods in vertebrates and invertebrates consult Kimmich (1981); Munck (1981); Wright and Ahearn (1997).

gradient. This is called secondary active transport, and it depends on the expenditure of metabolic energy to expel the ion from the cytoplasm by, for example, Na,K-ATPase or H-ATPase. We know from kinetic and molecular studies that there are scores of transporters including multiple transporters for specific monosaccharides (e.g., several glucose transporters, a fructose transporter), types of amino acids (e.g., neutral, basic, acidic) and peptides, vitamins, and minerals. Typically, a transporter shepherds a nutrient into the intestinal cell across the apical membrane and a different transporter shepherds it out the other side across the basolateral membrane (figure 4.10). Although it was once thought that fatty acids were absorbed only by diffusion, even fatty acid transporters have been identified. A pretty good general statement is that all the nutrients listed at the outset may be partly absorbed by a mediated process.

But only partly: lipophilic compounds can also diffuse into and across the cell phospholipid bilayer membranes. Recall that the by-products of lipid breakdown occur in the gut lumen in micelles. Micelles are not absorbed as intact structures, but release their products, such as fatty acids, which permeate according to their particular permeability coefficient and concentration. For example, long-chain fatty acids permeate across the lipid membrane at faster rates than those with shorter chains (due to higher lipid permeability), and protonated fatty acids permeate faster than nonprotonated fatty acids because the former are less polar. The aqueous layer overlaying the apical membrane, variously referred to as the unstirred layer, the mucus barrier, or the acidic microclimate, sometimes influences absorption rate. In vertebrates, intracellular events following uptake across the apical membrane include reesterification of absorbed lipolytic products followed by chylomicron formation and secretion into lymph.

What about passive absorption of water-soluble (hydrosoluble) nutrients? Because the transport of molecules across the phospholipid bilayer membrane is correlated with their lipid-water partition coefficient, this membrane generally limits the absorption of water soluble molecules larger than very small ions or glycerol, which may slip through membrane pores (Fu et al. 2000). Yet, there are reports in mammals and birds of considerable absorption of small- to medium-sized water-soluble compounds (reviewed by Chediack et al. 2003; McWhorter 2005). It has been presumed that these water-soluble molecules permeate across the epithelium primarily through the paracellular pathway, either by simple diffusion down their electrochemical gradient or by solvent drag, which means the molecules are dragged along in the volume of water that might be absorbed through this pathway (Pappenheimer and Reiss 1987; Madara and Pappenheimer 1987; Pappenheimer 1987; Powell 1987). Autoradiography (Ma et al. 1993) and confocal laser scanning microscopy (Hurni et al. 1993; Chang and Karasov 2004a) have visualized the appearance of water–soluble probes in the paracellular space. The major physical structure defining the permeability properties of the paracellular barrier is the tight junction (Anderson 2001). The barrier is created where protein particles (fibrils or "strands") in plasma membranes of adjacent cells meet in the paracellular space. Aqueous pores are thought to exist within the paired strands (Tsukita and Furuse 2000), and this is the putative path for water-soluble compounds. According to this model, absorption of water-soluble compounds will decline with increasing molecular weight (i.e., size) of the molecule, an observation that is consistent with movement through effective pores in epithelia ("sieving") (Chang et al. 1975; Friedman 1987). Research on this absorption pathway is not extensive but has been increasing over the past fifteen years, and it appears that its importance varies among species (figure 4.12). Some species may absorb most of the ingested D-glucose by this pathway (Chang and Karasov 2004b).

In summary, there are many pathways for absorption of nutrients and xenobiotics. Most nutrients have access to at least one transcellular-mediated pathway plus one of the nonmediated pathways; transcellular if lipophilic and paracellular if hydrophilic. Xenobiotics have access to the nonmediated pathways, depending on their lipid-water partition coefficient. The absorption rates for nutrients and xenobiotics depend on the luminal concentrations of the compounds, the kinetic characteristics of the mechanisms (box 4.4), and surface area. Because of surface elaboration by folds, villi, and microvilli (figure 4.5) surface area is greatest in the small intestine and most nutrient absorption typically occurs there. But studies have shown that other gut regions, like the pyloric caeca of some fish and the paired caeca of some birds, contribute large amounts of the nutrient absorption capacity of the entire tract (Karasov and Hume 1997). Consequently, the nominal surface areas plotted in figure 4.6 include those surface areas. In contrast, cecal and colon absorption of sugars and amino acids seem less important in mammals and so cecal and colon surface area were excluded when plotting mammals. Finally, it should be mentioned that absorption can be influenced by other materials in the diet. A noteworthy example of this is that

Figure 4.12. Absorption of small water-soluble compounds declines with increasing molecular weight (i.e., size) of the molecule, consistent with movement through effective pores in epithelia ("sieving"). In (A; left), calculated fractional absorptions by house sparrows are plotted according to increasing molecular weight (MW) of the probes (left to right; MW in Daltons; L-arabinose MW 150.1, L-rhamnose MW 164.2, perseitol MW 212.2, and lactulose MW 342.3). The probes were administered simultaneously to sparrows in three treatment groups designated as "D-mannitol" (no nutrient in the gut; gray bars), "D-glucose" (glucose solution in the gut; black bars), or "Food" (food in the gut; unfilled bars). The mechanism by which nutrients increase the absorption is uncertain. In (B; right), fractional absorption of water-soluble probes declines with increasing molecular size (probably due to "sieving"), but seems to vary among species. (Redrawn from Chediak et al. 2003.)

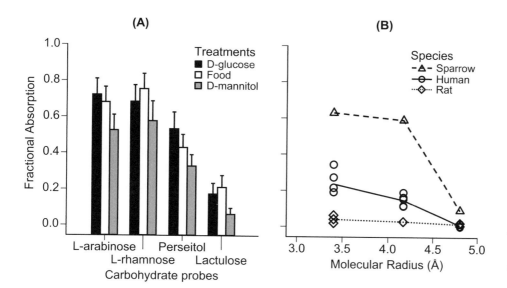

some herbivores and frugivores engage in soil consumption, called geophagy, and one function of this might be to bind toxic xenobiotics in ingested plant material (Diamond et al. 1999; Houston et al. 2001), and thus retard their absorption.

4.2 The Gut as a Bottleneck to Energy Flow

Extraction and metabolism of nutrients from food is a process that occurs sequentially in space and time; i.e., physical and biochemical breakdown of food begins in the mouth and/or proximal region of the gastrointestinal (GI) tract and precedes absorption, which occurs more distally, and overall those processes supply substrates to match metabolic needs. The sequential nature of the assimilation process invites investigation into the coordination between the various steps. In this section we will review three case studies that test whether breakdown and absorption capacities are well matched or whether one acts as a bottleneck on energy gain.

4.2.1 The Functional Response Raises the Question of Handling versus Digestion-Limited Consumption

There is a rich literature correlating the functional morphology of animal mouths with the physical properties of their food. For example, mammalian teeth, the size and shape of a bird's beak, and the structure of insects' mandibles typically reflect diet (Hiiemae 2000; Ricklefs and Travis 1980; Chapman 1995). But morphological features related to physical processing of the diet are often more than skin deep. Additional physical processing of foods occurs in the stomach of many animals, as in the grinding that can occur in the proventriculus (gizzard) of many invertebrates and some birds, which are lined with hard, abrasive surfaces and which sometimes contain sand or stones (gastroliths) to augment trituration (reduction to smaller particles). The diminution of food particle sizes in more distal chambers of the gastrointestinal system is a good measure of the extent of trituration that occurs (Chapman 1995). Smaller food particles with higher surface-to-mass ratios permit a faster rate of digestion. Thoughtful discussions of the hardness properties of foods in relation to structures for trituration can be found in Lucas (1994) and the final chapter of Chivers and Langer (1994).

One importance of all of this to ecologists is that the performance impact of these features for physically handling and dissembling foods might be incorporated into animals' functional responses which describe the relationship between consumption rate and food density. In one of the first and most often cited

Box 4.5. Functional Response Models

The relationship between how fast animals ingest food and food abundance is central to many questions in ecology. For example, ecologists are interested in knowing the mortality inflicted on prey by predators, or the rate at which a pollinator visits flowers. Biologists have long recognized that these rates depend on the density of prey. The relationship between intake rate and prey/resource density is called the functional response. In 1955 ichthyologist Victor Ivlev proposed the first functional response model to relate the number of prey eaten by fish in a day (y) and the density of prey (x):

$$y(x) = y_\infty (1 - e^{-\xi x}), \qquad (4.5.1)$$

where y_∞ is the maximal feeding rate and ξ is a constant that estimates "hunting success" (i.e., when ξ is high, y_∞ is reached at low x values). Ivlev assumed that y_∞ represented the maximal rate at which the animal could process food. A fish does not exceed y_∞ because it becomes satiated, presumably because it has a full gut. Thus, Ivlev's model recognizes physiological limitations imposed by digestion but ignores the limitations that handling prey before ingesting it may impose on feeding rate. Jeschke et al. (2002) review the many functional response models that include satiation.

The Canadian ecologist Buss Holling (1959) constructed a functional response model that assumes no digestive limitations and that emphasizes preingestional handling time. Holling's model assumes that a predator can feed for a time equal to T_t, and that it encounters prey at a rate proportional to prey density x. It also assumes that it takes the predator a time T_h to handle each prey. Therefore

$$Y(x) = ax(T_t - T_h y), \qquad (4.5.2)$$

where $Y(x)$ is the number of prey eaten, a is a constant that Holling called the instantaneous search rate, and $T_t - T_h y$ is the time that the predator has to search for prey. Equation 4.5.2 simplifies to what ecologists call Holling's disk equation:

$$y(x) = \frac{Y(x)}{T_t} = \frac{(ax)}{1 + aT_h x} = \frac{\left(\dfrac{1}{T_h}\right) x}{\left(\dfrac{1}{aT_h}\right) + x}. \qquad (4.5.3)$$

continued

continued

We rewrote the right-hand side of the disk equation in a peculiar, slightly more complicated-looking form. We did this, not to confuse you with algebra, but to emphasize the similarity of the disk equation with the Michaelis-Menten equation (equation 4.8). In the disk equation, as x becomes very large, the predation rate tends to $1/T_h$, which is analogous to $V_{\max}$ ($1/aT_h$ is analogous to K_m). The peculiar name of the disk equation stems from the experimental method that Holling used to develop and test it. Holling used a blindfolded laboratory technician as a predator of small sandpaper disks.

Holling recognized that the disk equation ignores potentially important physiological factors that limit intake. To address this deficiency, he developed a detailed and enormously complicated model (Holling 1965). Not very surprisingly, this model is less cited and less used than the disk equation. Ecologists use either Ivlev's or the disk equation to describe the results of their experiments. Which one of these equations is a better descriptor of data? If the intention is just to describe and summarize data, because the scatter in real data is so large and both equations have the same number of parameters, the choice depends on which of the two equations gives a better fit. Perhaps a more interesting and important question is which of the two equations is a better descriptor of what animals do. Are animals limited by digestive postingestional processes or by preingestional handling time? This question is important because it allows assessing the relative importance of digestive and behavioral mechanisms for an important ecological variable.

Jeschke et al. (2002) describe a fairly general model that allows answering this question. Because the derivation of Jeschke et al.'s (2002) model is lengthy, here we use a simpler model. Imagine an animal that eats in discrete bouts until it fills its gut to a level G. Then it waits a time T_d for the gut to process this batch, and then begins feeding again. In our simple model digestion and foraging are nonoverlapping processes. With some qualifications, the animal resembles the red knots that we described in the text. Our model is simplistic because, with the few exceptions of animals with guts that resemble batch reactors such as hydras and planarias, most animals, including red knots, can simultaneously forage and digest. We use it to illustrate the effects of adding digestive limitations to functional response models. You can consult Jenschke et al. (2002) for a more general model in which foraging and digestion are simultaneous processes. Assume that while the animal

continued

continued

is foraging its intake rate follows a functional response of the following form:

$$\text{intake rate} = \frac{ax}{1 + aT_b x}. \tag{4.5.4}$$

Therefore, it takes the animal a time T_f to fill its gut to a level G:

$$T_f = \frac{G(1 + aT_b x)}{ax}. \tag{4.5.5}$$

The animal's feeding rate equals the ratio of the amount of food processed (G) and the time that it takes to gather and digest it ($T_f + T_d$):

$$y(x) = \frac{G}{T_f + T_d} = \frac{G}{\dfrac{G(1 + aT_b x)}{ax} + T_d} = \frac{\left(\dfrac{G}{(GT_b + T_d)}\right)x}{\dfrac{G}{a\left(GT_b + T_d\right)} + x}. \tag{4.5.6}$$

Equation 4.5.6 looks and behaves like a disk equation, but it increases more slowly and reaches a lower asymptotic value. Although this is admittedly a very simple model, it tells us a few interesting things. The first one is that the use of functional responses depends on the time scale at which we measure intake rate. At the short scale of a foraging bout, handling time limits intake rate (maximal intake rate equals $1/T_b$). However, at the longer time scale of a foraging day, both digestion and handling time limit intake (maximal intake rate equals $G/(GT_b = T_d)$. At this longer time scale, the feeding rate declines with increasing T_d (food that is hard to digest lowers feeding rates).

Because in our simple model foraging and digestion are mutually exclusive processes, maximal intake rate is always limited by both of these processes. In contrast, in Jeschke et al.'s (2002) model, digestion and foraging occur simultaneously. In consequence, maximal intake rate is limited by either one of these processes. If handling time has a higher value than the time it takes to process a gut full of food, then maximal

continued

continued

intake rate equals the reciprocal of handling time (maximal intake rate = $1/T_h$). Conversely, if it takes longer to process a full gut than to handle a prey item, then maximal intake rate is given by the ratio of the gut's volume (G) and the time that it takes to digest it (maximal intake rate = G/T_d). Jeschke et al. (2002) defined animals as handling-time or digestion limited depending on whether animals handle prey more slowly than they can digest them ($1/T_h > G/T_d$) or digest prey more slowly than they can handle them ($G/T_d > 1/T_h$). It is worth emphasizing that the handling- or digestion-limited descriptions apply only to maximal rates of intake. If the density of food is low, the intake rate is limited by density.

How can we tell whether a consumer is limited by handling time or by digestion? If the animal is primarily limited by handling time, when prey are superabundant its short-term maximal intake rate and its long-term intake rate should be about the same. In a remarkably thorough review of a massive literature, Jeschke et al. (2002) found only a small handful of animals with handling time limitations. They predicted that handling-time limitations should be rare, and limited to animals with inordinately long handling times such as gastropods that drill through the shell of their prey. In contrast, they found that most published studies show evidence of digestion limitation. They concluded that most animals are digestion limited. Jeschke et al. (2002) make the point that in digestion-limited consumers foraging time should decrease with increasing food density, and that animals will often spend a major part of their time resting. Both of these predictions are consistent with many observations (reviewed by Jeschke et al. 2002).

We are taken by Jeschke et al.'s (2002) model because it nicely links digestive and preingestional limitations on intake rate and because it is consistent with chemical reactor theory. Recall that McWhorter and Martínez del Rio (2000) used a chemical reactor model to predict the maximal rate at which hummingbirds could ingest food. In this model, we used in vitro rates of sucrose digestion to predict the time (T_d) that it would take a hummingbird to digest a gut full of nectar (G). As suggested by both reactor theory and Jeschke et al. (2002), we estimated intake rate as G/T_d. We found that we could predict maximal intake rate from a combination of measurements that included the volume of the digestive sections of the gut and biochemical measurements of its hydrolytic capacity. We concluded that hydrolysis rates limited maximal ingestion rate. This example illustrates how a combination of guts-as-reactors estimates of maximal digestion rates and Jeschke et al.'s (2002) model can lead us to the integration of digestive physiology and functional responses.

examples of a functional response, Holling's disk equation (Holling 1959), consumption rate increases with prey density but reaches a plateau value beyond which consumption does not increase. Holling considered that the consumption rate was inversely related to handling time, which was the sum of attacking time and eating time. Although there have been scores of additional functional response models proposed and studied since Holling, only some have considered that maximal digestion rate, or some other physiological rate, could also dictate the plateau value (Jeschke et al. 2002). Box 4.5 outlines the equations used to quantify functional responses and outlines the need to identify the appropriate time scales at which either handling time or postingestional processing can limit feeding rate.

The easiest way to detect a digestion-limited animal is to measure food collection rate (or handling time) and digestion rate (or digestion time) directly and compare them, although there are some other methods as well (Jeschke et al. 2002). As an illustration, consider red knots (*Calidris canutus*) foraging on molluscs (clams in the genus *Macoma*) on mudflats along the Wadden Sea in northern Europe (figure 4.13). In these habitats, mollusc biomass ranges from 0.6 up to 10 g ash free dry mass (AFDM) per m^2, and the knots' instantaneous harvest rate increases with prey density but is also influenced by prey size and depth in the mud (Piersma et al. 1995). To assimilate molluscs, red knots crush

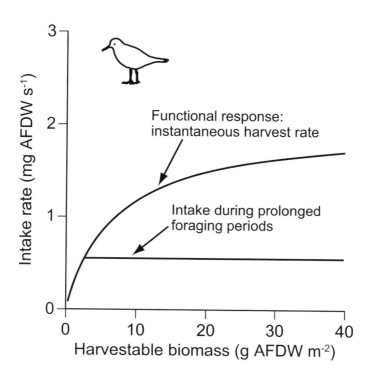

Figure 4.13. Hypothetical functional response of red knots (*Calidris canutus*) foraging on mollusks (*Macoma clams*) compared with intake rate observed during prolonged foraging periods. Though a functional response relationship like this has not been assembled (to our knowledge), observations reported by Zwarts and Piersma over the range of 2 to 40 g AFDM m^{-2} (Zwarts et al. 1992) enabled us to approximate it. The upper plateau value has been interpreted to reflect the limit set by handling, and the lower plateau has been interpreted to reflect the limit set by digestion.

the shell with their powerful muscular gizzards and excrete the fragments through the cloaca. Based on their rate of defecation, Zwarts and Blomert (1992) estimated that they could ingest and digest around 0.5 mg AFDM s^{-1}, though the intake they require to meet their energy needs during a normal day is about 0.3 mg AFDM s^{-1} (Piersma et al. 1995). Although knots can achieve instantaneous harvest rates as high as 2 mg AFDM s^{-1} (figure 4.13), field observations have shown that their sustained intake rate in the field is much lower, around 0.5 mg AFDM s^{-1} (Zwarts and Blomert 1992; van Gils et al. 2003). They do not sustain their high harvest rate for prolonged periods but instead they stop a lot and preen. Zwarts and Blomert (1992) interpreted stops as digestive pauses. Reviewing comparisons such as this, Jeschke et al. (2002) concluded that all the studies that have measured both digestion and handling time provide evidence for digestion limitations. We would add that it is more appropriate to conclude that they are physiology limited, unless there is a demonstration of digestive limitation per se. In our case study of red knots, how can we determine whether it is physical breakdown of food or some other step, or all steps together, that is limiting? Feeding limitation by the crushing step is suggested for red knots by the fact that they achieve lower rates of digestible energy intake (by less than half) when consuming whole bivalves than when consuming the flesh alone after it has been removed from shells (van Gils et al. 2003).

Red knots provide a fascinating example of phenotypic flexibility in internal structure related to mechanical processing of food. Red knots feed on soft-bodied prey (arthropods and spiders) in the summer and hard-shelled molluscs during migration and on their wintering grounds. The mass of the gizzard changes seasonally in red knots. It atrophies significantly after they arrive at their breeding grounds in the tundra when the birds shift their diets from shellfish to soft-bodied prey. After they leave the tundra, red knots resume their feeding on mollusks and their gizzards grow (Piersma et al. 1999a; Piersma et al. 1999b). Dekinga et al. (2001) have shown that the seasonal changes observed in the field can be replicated in the laboratory. When red knots were fed soft food, their gizzards lost mass (figure 4.14). When they were fed on molluscs, they increased their gizzard masses by 147% within six days. Red knots can feed faster on hard-to-break-down molluscs when their gizzards are enlarged, and this change is essential for them to meet their daily energy budget when foraging on molluscs (van Gils et al. 2003). These, and the other hallmark studies by Piersma and his colleagues on red knots (above), address for the mechanical breakdown step of digestion most of the major questions in digestive ecology that we listed in chapter 3 (2–6 in table 3.2).

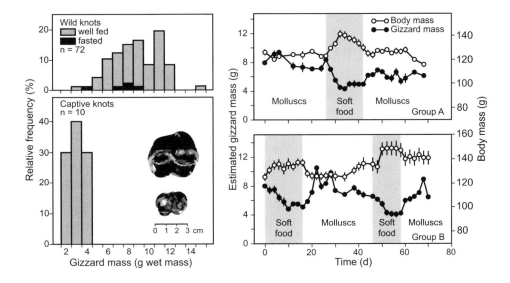

Figure 4.14. Red knots (*Calidris canutus*) provide a fascinating example of reversible phenotypic plasticity in mechanical food processing. These birds feed on soft-bodied prey (arthropods and spiders) in the summer at their breeding grounds, and on hard-shelled molluscs during migration and in their wintering grounds. The mass of the gizzard changes seasonally in red knots. It atrophies significantly after they arrive at their breeding grounds in the tundra when the birds shift their diets from shellfish to soft-bodied prey. After they leave the tundra, red knots resume feeding on molluscs and their gizzards grow (Piersma et al. 1999b). When red knots are kept in captivity and fed on soft trout pellets, their gizzards become much smaller than those of wild birds (two left-hand panels; redrawn from Piersma et al. 1993). These captive birds are reluctant to eat molluscs. When they are fed molluscs, their gizzards increase in mass (two groups shown in right-hand panels). Dekinga and collaborators used ultrasonography to measure gizzard sizes in red knots without harming the birds (redrawn from Dekinga et al. 2001; values are means ± standard error, 9–10 birds/group). Starck et al. (2001) describe the potential applications of ultrasonography technique in ecological physiology.

4.2.2 Testing for the Digestive Bottleneck in Lepidopteran Midguts

Digestion of leaves by herbivorous caterpillars involves primarily the assimilation of protein, carbohydrate, and fat of plant cell contents. Because these processes take place by catalytic rather than by fermentative reactions, Woods and Kingsolver (1999) modeled the overall process as a plug-flow chemical reactor. We will skip the mathematical details, including most of the assumptions, and rely on their graphical outputs to illustrate the model's major features. The rate of digesta flow in the gut is constant and, as digesta move along the midgut, successively larger amounts of food substrate (S) are converted to absorbable product (P) until no more substrate remains in the distal regions (figure 4.15A). The amount of absorbable product, which is zero at the mouth, initially rises but then falls in the distal region once substrate is exhausted. The scenario in figure 4.15A corresponds to a low feeding rate such that the capacities for breakdown and absorption are not overwhelmed by the rate of entry of substrate, and the overall efficiency is ≈100% (all nutrient is absorbed). Conceivably, if intake is intermediate all the substrate might be broken down but some absorbable product might escape unabsorbed (figure 4.15B), and if intake is very high then even substrate might escape undigested (figure 4.15C).

Because the operation of the hypothetical gut reactor clearly depended on level of intake, Woods and Kingsolver (1999) had to decide what level of intake

Figure 4.15. Chemical reactor model of a lepidopteran gut. Plots A (top left), B (middle left), and C (lower left) show the predicted axial concentrations of the protein substrate (labeled S) and its breakdown products (labeled P) at three levels of food intake (low, intermediate, high). The other plots show predicted axial concentrations for the intake rate that maximizes overall rate of absorption under four hypothetical scenarios: (D; top right) all steps are perfectly matched (no bottleneck) and small amounts of both substrate and absorbable product escape from the gut; (E; right, second from top) consumption is the limiting step, so the digestive capacity is underutilized and both substrate and absorbable product are completely extracted in proximal regions of the gut; (F; right, second from bottom) chemical digestion is the limiting step, so some substrate escapes the gut unaltered; and (G; right bottom) absorption is the limiting step, so absorbable product but not substrate escapes the gut unaltered. (Redrawn from Woods and Kingsolver 1999.)

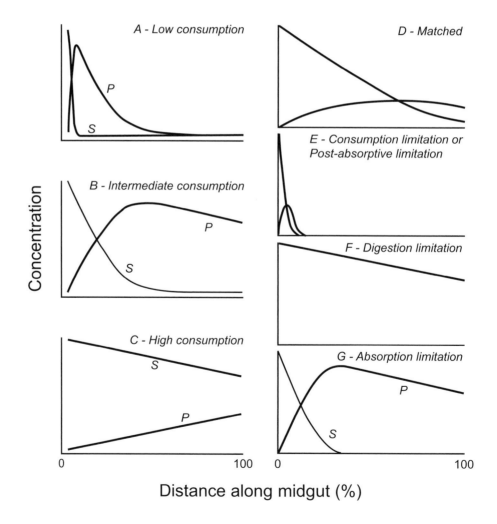

to impose. They did this for each run of the model by solving for the intake rate that gave the maximal overall rate of absorption from the gut. If you go back and look at our earlier discussion of transfer functions and the trade-off between rate and efficiency (section 3.3.1) you will see that this is synonymous with selecting the digestion time that maximizes rate rather than efficiency. Using this approach, Woods and Kingsolver predicted the concentration profile of protein (substrate) and absorbable product (peptides and amino acids) along the midgut under four hypothetical scenarios: (1) all steps are perfectly matched (no bottleneck) and small amounts of both substrate and absorbable product escape from the gut (figure 4.15D); (2) consumption is the limiting step, so the digestive capacity is underused and both substrate and absorbable product are completely extracted in proximal regions of the gut (figure 4.15E); (3) chemical digestion is the limiting step, so some substrate escapes the gut unaltered (figure 4.15F); and (4) absorption is the limiting step, so absorbable product but not substrate escapes the gut unaltered (figure 4.15G). For a variety of reasons they ruled out the second possibility. They made measurements of digesta flow and proteolytic activity in the midgut of the hawkmoth *Manduca sexta* that indicated that all protein would be hydrolyzed within the first few millimeters of the midgut. This seemed to rule out possibilities (1) and (3). Also, they reviewed a published study that provided evidence that no intact protein reached the distal gut of feeding caterpillars, whereas nonabsorbable products did (Shinbo et al. 1996). This supports the absorption limitation hypothesis (4). But also they reviewed one final study (Reynolds 1990) that suggested that postabsorptive processing rather than absorption was limiting. They discussed several other consequences of their model, including how it can help to predict feeding rate and growth rate under different conditions. The study by Woods and Kingsolver (1999) is a wonderful illustration of the utility of modeling in understanding a complex system composed of many features and in identifying how the features interact to determine the rate and efficiency of nutrient extraction from food, two parameters of the whole animal and hence of ecological importance. Notice that we have presented this model not because it provides a clear answer, but because it generates crisp questions and motivates experiments and observations to answer them.

4.2.3 Testing for the Digestive Bottleneck in Nectarivorous Hummingbirds

McWhorter and one of us tested whether food intake by migratory broad-tailed hummingbirds (*Selasphorus platycercus*) is limited by rates of hydrolysis (McWhorter and Martínez del Rio 2000). These birds, which need to maximize energy and hence fat gain prior to migration, digest mainly the sucrose in floral nectars and so sucrase activity in the intestine's brush border was measured in vitro. The in vitro measurement was made with homogenates of tissues collected along the length of the intestine under conditions that saturate the enzyme(s) so that the maximal reaction velocity (V_{max}) could be integrated along the length to yield a total hydrolytic capacity. This capacity was about 120% higher than the observed rates of sucrose intake and digestion, implying that the immediate spare capacity (sensu figure 3.8) was quite high. However, the common procedure of using the V_{max} over the entire intestine length is physiologically unrealistic because the sucrose concentration is progressively lowered as the digesta flows distally along the gut during digestion.

Using a more sophisticated model of the gut as a plug-flow chemical reactor (Jumars and Martinez del Rio 1999), and including estimates of K_m, we calculated a lower digestive capacity that was only 15–35% higher than observed rates of sucrose intake/digestion. This more realistic model accounted for the reduced rates of hydrolysis that occur when sucrose concentration is lowered as a result of digestion. The results of the model suggested that the hummingbirds had very little spare capacity to use if they had to suddenly increase their feeding rate. Support for this prediction came in trials in which we rapidly exposed the hummingbirds to low temperature (figure 3.11 in the last chapter). The cold-exposed birds did not (could not?) increase their intake but instead reduced their expenditure by utilizing torpor. In a similar kind of experiment rufous hummingbirds (*Selasphorus rufus*, 3.2 g) switched suddenly to low temperature did not (could not?) increase their intake sufficiently and lost body mass (Gass et al. 1999).

Our study underscores two important points. First, it provides another example that the bottleneck limiting energy flow can be a digestive process, in this case hydrolysis. Second, though most other studies estimating hydrolytic capacity (e.g., Hammond et al. 1994; Weiss et al. 1998; Martínez del Rio et al. 2001) have assumed constant saturating substrate concentrations, the newer, more physiologically realistic approach by McWhorter and Martínez del Rio (2000) showed that those studies overestimated the hydrolytic capacity.

4.3 The Gut in Energy Intake Maximizers

The previous sections provide evidence that the gut sometimes limits energy flow in endotherms and ectotherms eating a variety of diets. When animals face circumstances that favor very high rates of energy flow, how do they adjust their long-term limits upward?

4.3.1 Digestive Adjustments to Wintering
in Cool and Cold Environments

Let us skip back to the example in the last chapter of white-throated sparrows that lost body mass when they were shifted suddenly to a very low air temperature (figure 3.8). Do you suppose that the birds were limited by a peripheral limit, such as the ability to generate enough heat in their skeletal muscle to thermoregulate (figure 3.9), or by their ability to provide fuel to the heat generator (a central limit)? The peripheral limit can probably be ruled out because when they were switched to −20°C they only had to approximately double their heat production to match higher heat losses (Yarbrough 1971), and this is not particularly challenging even for a summer-acclimated north-temperate small passerine (e.g., Cooper and Swanson 1994; O'Connor 1995). Central limitation seems much more likely, and it could have been by the GI tract, considering the changes that occur in these organs during more gradual adjustment to cold (cold acclimation).

What changes would be consistent with GI tract limitation? Consideration of equations 4.1 and 4.2 suggests several responses to higher feeding rate (i.e., higher flow of digesta) at low temperature: (1) Higher digesta flow through a GI tract with no spare digestive capacity would cause shorter retention time and thus result in poorer nutrient extraction efficiency. (2) If the GI tract enlarges, the retention time might be unchanged, as would extraction efficiency. (3) If there were no change in gut size, increased biochemical reaction rates per unit gut might compensate for the reduction in retention time, leaving extraction efficiency unchanged.

Effective discrimination of these alternatives requires simultaneous measurement of all the variables, which we have done for the granivorous white-throated sparrows, frugivorous cedar waxwings (*Bombycilla cedrorum;* McWilliams et al. 1999), and insectivorous house wrens (*Troglodytes aedon;* Dykstra and Karasov 1992). The results uniformly match option (2). For example, when waxwings were switched from +21 to −20°C, daily food intake nearly tripled, mass of digestive organs increased 22–53%, and retention time and digestive efficiency declined only slightly. Intestinal tissue-specific rates of hydrolysis and nutrient absorption did not change significantly, but the total hydrolytic and absorptive capacity of the small intestine did increase because of the increase in intestinal mass. We have called this kind of response "nonspecific" because the increased intestine mass nonspecifically increased the capacity for hydrolysis and transport of all substrates in a more or less parallel fashion (Karasov and Diamond 1983b). Many other studies in birds (McWilliams and Karasov 2001) and mammals (Karasov and Hume 1997) acclimated to low temperature confirm that the most important adjustment to the higher feeding rate at low temperature is an increase in mass of the GI tract (and liver too), which has the dual effect of keeping retention time relatively constant in the face of higher digesta flow and increasing the intestine's biochemical capacity, all with the net effect of holding efficiency relatively constant.

4.3.2 The Gut in Reproducing Mammals and Birds

Lactation is the most energy-intensive period during mammalian reproduction because the female must process enough energy to meet her needs as well as those of her nursing young. Females that meet the high cost entirely by feeding, rather than by drawing down energy reserves, may have to double or triple their feeding rate compared with that when they are nonreproductive. In a fascinating series of studies Hammond and Diamond (Hammond and Diamond 1992; Hammond et al. 1994) manipulated the number of pups nursing on a female mouse to measure the mother's intake and digestive responses. The responses were analogous to those of birds and mammals during cold acclimation: the added cost of lactation led to enlargement of the GI tract (though see Johnson and Speakman 2001), to relatively little change in tissue-specific rates of hydrolysis or absorption, and to no decline in digestive efficiency even though feeding rate, and hence rate of digesta flow through the GI tract, increased markedly (figure 4.16). After lactation, the female's GI tract returns to the "normal" smaller nonreproductive size. Analogous up- and down-regulation of the GI tract mass may support increased feeding during egg laying in some birds (Karasov 1996). The adaptive interpretation of this reversible flexibility in gut size is, once again, that animals maintain an economical match between digestive capacity and dietary load ("enough but not too much"; section 3.3.2).

Figure 4.16. The higher food consumption of lactating mice is achieved mainly by increased gut size, which also increases biochemical capacity. (A; left hand figure) Food intake and small intestine mass progressively increased as more pups were added to the litters of reproducing mice, or when they were exposed to lower temperature (Redrawn from Hammond et al. 1994). (B; right hand figure) The greatly increased intestinal size of reproducing mice, relative to virgins, was associated with an increased summed capacity for sugar and amino acid absorption over the entire intestine, even if the uptake rate per unit mass was a little lower. (Redrawn from Hammond and Diamond 1992.)

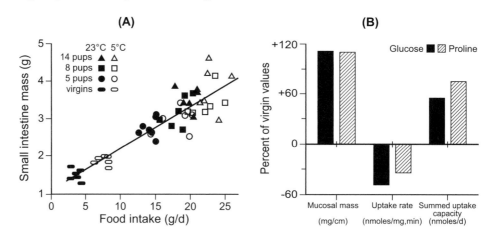

Whether the digestive capacity can be increased to match any demand put on it or whether the gut sometimes ultimately limits the energy budget is unknown for most animals. The issue has been thoroughly studied in laboratory mice (Hammond et al. 1994) and cotton rats (*Sigmodon hispidus*; Rogowitz 1998) challenged by cold acclimation, lactation, and by a combination of these factors (Johnson and Speakman 2001). With each increasing energetic challenge, mice increased their gastrointestinal mass and hydrolytic and absorptive capacity and, for the highest load of lactation in the cold, the energy budget limit was not set by the digestive system but more likely by the ability to produce milk (Hammond et al. 1996). Hammond et al. (1996) studied mice of a particular inbred strain (Swiss Webster) and other strains may show different patterns. Johnson and Speakman (2001) doubted that even milk production reached a limit in MF1 mice, at least during a female's first lactation. The authors speculated that females may have been restraining themselves during their first reproduction, perhaps to maximize lifetime reproduction.

It would be premature to generalize from these results with laboratory mice and conclude that digestive capacity can be increased to match any demand put on it, and will not be limiting in the ecological setting. As described above, there are interesting plausible examples of digestive bottlenecks involving animals eating foods quite different from the formulated laboratory chow fed to mice. How can we test whether the gut limits the energy budget for an animal in the field? The approach suggested by Jeschke et al. (2002) of comparing rates of food collection and handling against rates of digestion can be followed by integrating laboratory and field studies as exemplified by the studies with red knots (van Gils et al. 2003). There are other energy-intensive points in the life cycle, such as growth (Karasov and Wright 2002) and migration (see discussion below), during which digestion may prove to be the limiting factor in the energy budget.

4.3.3 The Gut in Growing and Reproducing Ectotherms

In ectotherms, body temperature (T_b) exerts a major influence on energy flow (chapter 1). Up to a point, ectotherms generally grow faster at higher temperatures (Atkinson 1994), but the physiological mechanisms responsible for the thermal effect on growth are generally unknown. In the wild, energy gain and growth result from a succession of behavioral and physiological processes that include food detection and collection, ingestion, digestion, absorption, and postabsorptive processing. Because not all the processes are equally sensitive to T_b, it is possible to isolate the specific process that is most limiting.

In the grasshopper *Melanoplus bivittatus*, between 15 and 35°C, metabolizable energy gain increased 4–5 times for each 10°C increase in temperature—this factor was called Q_{10} in chapter 1 (Harrison and Fewell 1995). For comparison, insect biologists Jon Harrison and Jennifer Fewell (1995) measured the temperature dependence of intake by grasshoppers with empty crops (the foregut chamber that holds food) over three intervals: (1) short feeding bouts, which reflect rates of chewing and ingesting; (2) 2 h intervals, which reflect rates of ingestion plus rates of crop and midgut filling; and (3) 8 h intervals, which reflect rates of ingestion, crop and midgut filling, plus rates of gut throughput. The temperature dependencies of intake over the first two intervals were low (Q_{10}'s ranged from 1.3 to 2.6) relative to that for metabolizable energy gain, whereas the dependencies for intake over 8 h and for feces production were similar (Q_{10}'s ranged from 3.3 to 5.2; figure 4.17). Therefore, Harrison and Fewell (1995) concluded that the high thermal dependency for grasshopper net energy gain and growth rate above T_b's of 15°C can be attributed primarily to strong thermal effects on some digestive feature(s) that influence gut throughput rather than effects of T_b's on ingestion rates, feeding behavior, or metabolism, which they also measured. At low T_b's (<15°C), intake is limited by rate of digestive processing as well. Harrison and Fewell (1995) thought that this may be a general phenomenon in heliothermic ("sun-loving") insects, which thermoregulate behaviorally to achieve high T_b's. Their analysis did not determine whether it was chemical breakdown, absorption, or gut motility that was most thermally limited, but another study comparing the thermal sensitivities

Figure 4.17. In feeding grasshoppers, the temperature dependence of each sequential step determining assimilation was inferred by the temperature dependence of dry mass flux over progressively longer intervals. Short feeding bouts reflect rates of chewing and ingesting, which are so high that they clearly are not the bottleneck to assimilation. Intervals of 2 h reflect rates of ingestion plus rates of crop and midgut filling. Eight-hour intervals of intake or feces production reflect rates of ingestion, crop, and midgut filling plus rates of gut throughput. The temperature dependence of these latter processes, which is reflected by the slope, matched that of growth rate. This is consistent with the idea that throughput limits growth rate, rather than ingestion, chewing, or midgut filling. Notice that temperature had little effect on efficiency of digestion (top plot). (Data from Harrison and Fewell 1995.)

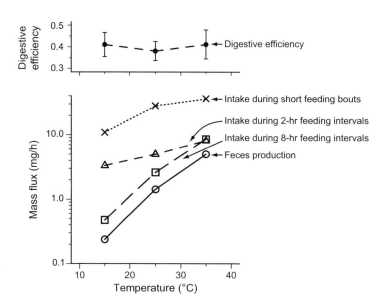

of some of these processes in tobacco hornworm (*Manduca sexta*) caterpillars concluded that it was most likely gut motility (Kingsolver and Woods 1997).

These two studies (Harrison and Fewell 1995; Kingsolver and Woods 1997) illustrate how the thermal sensitivity of whole-animal performance, such as rates of metabolizable energy gain and growth, emerge from the thermal sensitivity of the underlying component processes. The number of studies exploring the thermal sensitivities of specific digestive processes is rather limited in vertebrate ectotherms and endotherms (Karasov and Hume 1997) and in invertebrates (Kingsolver and Woods 1997), although there are many measures of how efficiency of digestion or assimilation varies with T_b. Efficiency shows little change with T_b in insects (Scriber and Slansky 1981) or in aquatic and terrestrial vertebrates (Karasov and Hume 1997). Is this invariance because rates of gut motility, breakdown, and absorption all have similar thermal dependencies, or because feedback mechanisms in the digestive tract ensure that the rate at which food enters the intestine from the stomach or crop and travels distally along the intestine does not exceed the rate at which it is broken down and absorbed? A related topic that has received little study has to do with possible compensatory adjustments in digestion (i.e., acclimation/acclimatization) to chronic changes in temperature (Karasov and Hume 1997).

It is unfortunate that there are so few studies exploring the thermal sensitivities of digestive processes, because the influence of T_b on those rates, and hence on the rate of energy assimilation, is probably a powerful determinant of many aspects of ectotherm ecology (Congdon 1989). T_b effects on processing rates have been invoked to explain behavior patterns in the field (e.g., Harrison and Fewell 1995; Zimmerman and Tracy 1989). Many species of reptiles, once they have fed, seek microclimates in which to bask and thereby increase their T_b (Waldschmidt et al. 1987; Cloudsley–Thompson 1991). In chapter 13 we will provide some examples of how temperature-dependent process limitation of energy flow is being incorporated into global models of ectotherm energy budgets as a function of altitude and latitude (Grant and Porter 1992) and into life history models (Dunham et al. 1989; Angilletta 2001).

4.4 Intermittent Feeders

4.4.1 Digestive Adjustments in Migrant Birds

It is well known that migrating birds typically carry a lot of fat to fuel their long flights. A major finding of nutritional ecology studies of migrating birds is that birds get fat mainly by hyperphagia, which means having a very high intake rate

(Blem 1980). Having learned that hyperphagia is typically supported by increases in size of the GI tract (sections 4.3.1 and 4.3.2), it would be logical to expect that migrants would have relatively large GI tracts. Logical perhaps, but wrong! One of us (Karasov) and our colleagues collected migrant blackcap warblers (*Sylvia atricapilla*) at a desert oasis stopover site in Israel (Karasov and Pinshow 1998), and migrant yellow-rumped warblers (*Dendroica coronata*) killed by colliding into a radiotower at night in central Wisconsin (with D. Afik). When we compared their intestinal masses with those of captives feeding ad libitum we found that the migrants had 25–40% lighter, not heavier, GI tracts (figure 4.18). We were actually clued into this by our colleague Theunis Piersma who was finding smaller GI tracts in migrant shorebirds relative to nonmigrants (Piersma 1998; Piersma et al. 1999b). Piersma and his colleagues thought that the shorebirds reduced digestive organs strategically, just prior to migratory departure, to accrue the benefits of reduced wing loading and reduced resting metabolic rate (Battley et al. 2000, 2001). These would lead to increased flight distance and a reduction in the number of stops enroute to refuel. But there is another possible explanation for a decreased gut size in migrants.

For most migratory songbirds, migration involves many flights interspersed with layovers at "stopover" sites where energy and nutrient reserves are rebuilt. Thus, birds during migration alternate between periods of high feeding rate at migratory stopover sites and periods without feeding as they travel between stopover sites. These intervals without food may be relatively short (e.g., <8 h) for birds like a yellow-rumped warbler migrating short distances at a given time, or they may last for days for birds like the blackcaps migrating over deserts or other large ecological barriers (e.g., oceans, mountains). Once at the stopover site food may be unavailable or difficult to find. The intervals without

Figure 4.18. Intestine mass was reduced in migrant songbirds relative to captives fed ad libitum. Captives that are fasted or have their food intake restricted to about one-half ad libitum also have reduced gut mass. (Data from Karasov and Pinshow 1998; Lee et al. 2002 and D. Afik and W. H. Karasov (unpublished).)

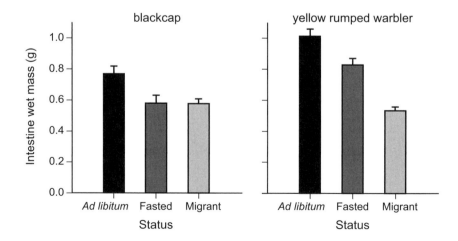

food, or with restricted feeding, can affect the GI tract. Laboratory studies with songbirds have shown that fasting or food restriction causes proportionally large declines in GI tract mass in comparison with other body components such as flight (pectoral) muscles and heart (Klaassen and Biebach 1994; Hume and Biebach 1996; Karasov and Pinshow 1998; Lee et al. 2002).

So, when a migrating bird lands at a stopover site to refuel, does it have the guts to do it? Ecological field studies of passerine birds have revealed that recovery of body condition after arrival at stopover sites is typically nil or slow for from 1 to 2 days, despite apparently abundant food resources, and then much more rapid (reviewed in (McWilliams and Karasov 2001, 2005)). Although ecological conditions like food availability influence rate of recovery, there may also be a physiological explanation, because birds exhibit the two-step recovery even when provided food ad libitum in the laboratory. In laboratory fasting studies, atrophy of organs important in food processing (e.g., intestine, liver) is associated with lower biochemical digestive capacity (Lee et al. 2002) and lower rates of feeding and digestion (Hume and Biebach 1996; Karasov and Pinshow 2000) which, during a migratory stopover, might limit the time course of mass gain. Such delays during stopover decrease an individual's overall migration speed and therefore its success. From theoretical models of optimal migration (Alerstam and Lindström 1990), it is apparent that the rate at which a species can add stores or restore a physiological deficit before the next flight is critical to migrating birds.

The gut-limitation hypothesis (see McWilliams and Karasov 2001 for alternative hypotheses) suggests that the initially slow rate of mass gain at stopover sites occurs because birds arrive at the site with reduced GI tract size and function, and rebuilding of the gut takes time and resources and restricts the supply of energy and nutrients from food. In the blackcaps, GI tract mass was restored to normal levels in birds that had previously been fasted or food restricted after just two days of ad libitum feeding (Karasov et al. 2004). An important mechanism was cell proliferation (increase in number of entero-cytes), which is also called hyperplasy (organ growth through increase in cell number). It can occur because intestinal mucosal cells have a short cell cycle time during which they are born in crypts at the base of the villi, migrate up the villi, and are eventually sloughed off into the lumen. The size of the mucosal tissue can be manipulated by increases in the rate of cell birth and/or declines in the rate of apoptosis (cell death). In small birds the intestinal turnover time (replacement time of the cells) is 2–3 days, though it may be longer in larger birds (Starck 1996).

A nagging question for passerine migrants is how well the digestive organ changes that occur in the fasting or food restriction protocols in the laboratory mimic the natural changes in digestive organ structure and function that apparently occur in migrants. Also, there may be differences in response between birds that were fasted compared with those food restricted (Karasov et al. 2004). For some shorebird migrants there are more data on changes in GI mass under natural conditions, and there is evidence that both red knots (Piersma et al. 1999b) and bar-tailed godwits (Piersma and Gill 1998; Landys et al. 2003) reduce their gut size *prior* to departure on long flights. There clearly is a need for more comprehensive study of GI tract structure and function in freshly captured migrants, experimentally fasted and food-restricted individuals, and normally feeding individuals.

4.4.2 Digestive Adjustments in Hibernating Mammals

Intermittent feeding also occurs naturally in the life cycle of some mammals. Examples include some ungulates during the rut (Miquelle 1990), polar bears in summer (Atkinson et al. 1996), and northern elephant seal pups that undergo a two to three month postweaning fast (Ortiz et al. 2001). The supreme examples of intermittent feeding occur in hibernators that spend portions of winter in torpor.

Some hibernating species, called "food storers," interrupt their periods of torpor with periods in which they consume food that they have stored up and whose energy is used to fuel the next period of torpor (Humphries et al. 2001). Other species, called "fat storers," have much longer periods of torpor and consume little or no food during periodic arousals because their seasonal torpor is fueled by stored energy (fat). The torpor seems to have similar effects on the GI tract in both food-storing hibernators and fat-storing species (Carey 2005), but the latter have been studied more. In both small thirteen-lined ground squirrels (*Spermophilus tridecemlineatus*) and large marmots (*Marmota marmota*) the intestinal mass is reduced 50–75% during hibernation, compared with peak values during the summer feeding season (Carey 2005).

The proximate cause of the reduction in intestine mass is both the absence of luminal nutrients (Carey and Cooke 1991) and possibly some feeding-associated neurohormonal cues that are known to influence gastrointestinal mucosal growth in all mammals (Johnson and McCormack 1994). The reduced intestinal mass is related to suppressed enterocyte proliferation rate and migration rates (Carey 2005). At a biochemical level, the intestinal tissue-specific activities of enzymes and nutrient transporters are similar to, or sometimes even a little higher than, those in nonhibernating animals, but the large reductions in intestinal mass cause a large reduction in intestinal digestive capacity (Carey 2005).

The reduction in intestine mass may be lessened in the food storers, but you can imagine that when they become normothermic and they need to eat but have reduced digestive capacity, they face the same feeding challenge as migrant birds just landing at a stopover site. In one food storer, the European hamster (*Cricetus cricetus*), small-intestine mass was reduced after 12 days of torpor compared with active animals before or after the hibernation season, but the intestine mass returned to near-normal levels two days after arousal when feeding had resumed (Galluser et al. 1988). This time course is similar to that for migrant food-restricted or fasted blackcaps to recover their normal intestinal mass, and the mechanism is similar—rapid hyperplasy. In laboratory rats, which have a one-day enterocyte turnover (Karasov and Diamond 1987), following a fast the villi return to their normal length within a day after feeding is initiated (Buts et al. 1990; Hodin et al. 1994).

4.4.3 The Gut in Intermittently Feeding Snakes

Among snake species one can define a continuum according to their frequency of feeding. At one end are species typified by the Burmese python (*Python molurus vivittatus*), which may swallow a large meal after a year-long fast, digest it in 10–14 days, and then remain digestively inactive for another long period of time (Secor and Diamond 1995; Starck and Beese 2001). At the other end are species typified by the garter snake (*Thamnophis sirtalis parietalis*), which searches actively and feeds frequently, even daily, although it maintains the ability to tolerate a long fast such as during its winter dormancy (Starck and Beese 2002). The intestines of these animals exhibit remarkable phenotypic flexibility in size and function that is perhaps quantitatively greatest in the infrequent feeders but qualitatively similar among them all, and that possibly reflects differences in modes of intestinal regulation than are observed in mammals and birds.

One signature observation of their digestive response is the up-and-down regulation of their intestinal mass that corresponds with their cycle of fasting-feeding-fasting (figure 4.19A). In the python, the intestine mass increases 2–3 times within 2 days of feeding (Secor and Diamond 1995; Starck and Beese 2001). Recall that birds and mammals can achieve this through a process of hyperplasy because they have a short cell cycle time of intestinal mucosal cells (above). But studies of intestinal cell cycle kinetics in snakes indicated that cell proliferation was not involved in the upregulation of gut mass (Starck and Beese 2001; Starck and Beese 2002), so how do they do it? Histological observations

Figure 4.19. Snakes that are intermittent feeders up- and down-regulate their intestinal mass, and consequently their biochemical digestive capacity, in correspondence with their cycle of fasting-feeding-fasting. (A; lower left plot) The flexible response of the thickness of the intestinal mucosa (mean for five snakes), as measured from ultrasonographs. Arrows indicate feeding events. (Replotted from Starck and Beese 2001.) (B; two right-hand plots) Amino acid absorption in the anterior small intestine is also up-regulated in the days immediately after feeding, and then regresses. Na-dependent uptake, which is active, was a relatively smaller component of uptake than Na-independent uptake, which represents passive absorption and possibly mediated absorption. Values shown are means ± standard error of the mean. (Replotted from Secor and Diamond 1995.)

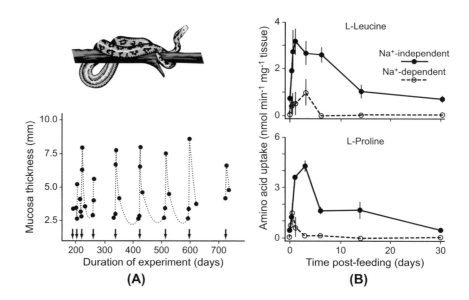

(A)   (B)

showed that intestinal cells of newly fed snakes have a greatly increased number of lipid droplets, and thus the increase in intestine mass was mainly due to hypertrophy (organ growth through increase in cell size). The functional significance of the lipid droplets is unknown, but it might relate to another signature observation of the digestive up-regulation. Within a few days of feeding in pythons, microvillous surface area increases (Starck and Beese 2001), and nutrient uptake rates per unit intestine mass increase up to 5 times (figure 4.19B; Secor and Diamond 1995). Could the lipid droplets represent a ready, local energy source for the rapid synthesis of new membrane, enzymes, and transporters (Starck and Beese 2002)?

When comparing snake species, the digestive responses to fasting seem increasingly dramatic according to the approximate duration in days between meals for adults in the wild, even after controlling for effects of phylogeny. Snake species that feed relatively frequently tend to exhibit smaller declines in intestinal structure and function between meals relative to infrequent feeders, and, consequently, when they are lucky enough to capture prey they initiate and complete its digestion relatively more rapidly (Secor and Diamond 2000). Probably what slows the infrequent feeders down somewhat is the extra time it takes to rebuild more of the gut. They may also expend more respiratory energy in rebuilding the gut (Secor and Diamond 2000), though some research indicates that this is a small cost (Overgaard et al. 2002). But, considering how infrequently they feed, and how expensive it is to maintain intestinal tissue, the energetic savings over the extended period of fasting presumably balance the cost of rebuilding the gut and the reduction in rate of energy gain once prey is captured.

4.5 The Gut in Diet Switchers

4.5.1 A Priori Expectations

Omnivory, the consumption of many rather than a single type of food, occurs over many time scales. You can probably think of animals that mix food types within a single foraging bout, between bouts within a day, between days within a season, and between seasons. But different food substrates are digested and absorbed along distinct biochemical pathways (figures 4.8 and 4.10) and may require different processing times. Can an animal maintain at all times a GI tract that is optimized for all the different substrates?

We were lucky enough to find an unambiguous answer ("no") to this question in our very first studies of digestive responses of birds switching diets. American Robins (*Turdus migratorius*) and European starlings (*Sturnis vulgaris*) switched from fruit to insect diets had reduced digestive efficiency the day after a diet switch and achieved diet-specific efficiency levels only after 2–3 days (Levey and Karasov 1989) (figure 4.20). This observation led to several years of research on which features of digestive physiology are flexibly matched to the prevailing diet, and whether the matches achieved made evolutionary and mechanistic sense.

One might predict a priori that for dietary components such as carbohydrates, protein, and lipids there will be a positive relationship between their level in the natural diet and the presence or amount of gut enzymes and transporters necessary for their breakdown and absorption. There are several arguments for expecting this relationship, which has been called the adaptive modulation hypothesis. One is an optimization argument based upon the assumption that

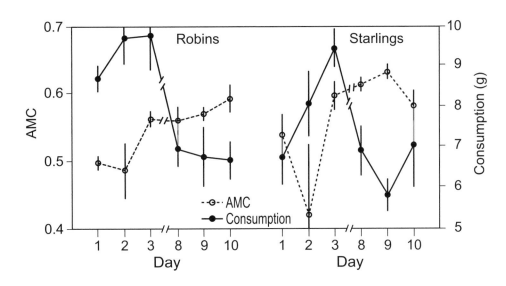

Figure 4.20. American robins (*Turdus migratorius*) and European starlings (*Sturnis vulgaris*) switched from fruit to insect diets had reduced digestive efficiency (AMC) the day after a diet switch and achieved diet-specific efficiency levels only after 2–3 days. Notice that on days that they were inefficient they compensated by eating more (higher consumption, in g). (Replotted from Levey and Karasov 1989.)

natural selection acts to maximize net energy gain (Karasov and Diamond 1988; Diamond and Hammond 1992). If a relationship between dietary level of an energy-containing substrate and its digestive enzymes did not exist, then valuable food energy might be wasted in excreta when feeding on diets with high substrate levels, and/or the metabolic expenses of synthesizing and maintaining the molecular machinery to break down the substrate would be wasted when feeding on diets with very low levels of substrate. Also, ingestion of large amounts of a substrate without the ability to digest and absorb it can lead to life-threatening diarrhea resulting from the solute's osmotic effect in the intestine. As an example, sea-lion pups developed severe fermentative diarrhea and died in five days when fed lactose and sucrose, substrates for which they have no hydrolytic enzymes (Sunshine and Kretchmer 1964). Finally, if enyzmes are crowded into the cell membrane and cytoplasm, then any surplus enzyme occupies space that could be devoted to something else (Diamond 1991). Not all animals should be expected to modulate enzyme and tranporter levels. Selection for phenotypic flexibility is low or nil in animals that do not switch diets or switch between diets that differ little in substrates, like carnivores (Buddington et al. 1991).

One might also predict a priori that, for foods composed primarily of carbohydrates versus proteins versus lipids, their respective retention times in the gut will increase in that rank order. Why is this? Recall that most of the nonrefractory carbohydrates (mono- and disaccharides, starch, glycogen) are broken down by two or fewer enzymes (e.g., amylase followed by maltase) and that their monomers are partly actively transported. Contrast this relatively simple digestion/absorption pathway with the more complex pathway for digestion of lipid, which involves formation of micelles and breakdown by lipase, followed by mainly passive absorption of monomers. Assuming that more processing steps result in a lower overall rate of release of absorbable monomers, and that passive absorption is slower than mediated absorption, a simple expectation is that lipid digestion/absorption will take longer than carbohydrate digestion/absorption, with protein digestion being intermediate, considering its longer series of enzymatic steps relative to carbohydrate digestion. When breakdown products accumulate in the GI tract lumen because of slow hydrolysis and absorption, they can inhibit stomach emptying and intestinal transit by negative feedback arising from intestinal receptors stimulated by those products of digestion. For example, high concentrations of amino acids and fats in the duodenum inhibit gastric motility in mammals (Malagelda and Azpiroz 1989) and poultry (Duke 1989) through hormonal (possibly CCK-8 and pancreatic polypeptide) and neural reflexes.

Let us consider how these expectations about flexibility in GI tract biochemistry and digesta retention match up with reality in several case studies with omnivorous animals.

4.5.2 Digestive Adjustments to Seasonal Diet Switching in Birds

Many of the New World warblers in the genus *Dendroica* are highly insectivorous during the summer breeding season but eat fruit and nectar in the fall and winter. Along with our colleague Doug Levey, we completed a series of studies to determine to what extent specific digestive features allow or constrain these shifts (Afik et al. 1997a,b; Afik and Karasov 1995; Afik et al. 1995; Ciminari et al. 2001; Levey et al. 1999). We fed yellow-rumped warblers (*D. coronata*) and pine warblers (*D. pinus*) for weeks one of three formulated diets composed primarily of fruit (banana), insects (mealworms, *Tenebrio molitor*), or seeds (sunflower). Many features of digestive anatomy, physiology, and biochemistry were influenced by diet. For example, seed eaters had significantly heavier gizzards whereas fruit eaters had significantly heavier small intestines. Here, we will compare some of the results on digestion of high fat foods with our a priori expectations.

Once the warblers were habituated to the diets, we compared their digestive efficiency on a dose of radiolabeled lipid ([^{14}C]glycerol trioleate) in each diet, which can be used as an index of their relative ability to assimilate lipid from each food. Both species were highly efficient in assimilating the triacylglycerol dose when it was mixed into the highest-fat seed diet, less so when it was in the insect diet with intermediate fat content, and least efficient when it was in the low-fat fruit diet (figure 4.21). Higher lipid digestive efficiency was partly explained by longer times for digestion, because birds habituated to the three diets had retention times that increased in the rank order originally predicted (figure 4.21). Also, the differences in lipid digestive efficiency were partly explained by differences in the birds' pancreatic lipase activity, which was significantly higher (by 60%) in birds fed the two higher-fat diets relative to that in birds fed the low-fat fruit diet (Levey et al. 1999). This pattern of higher lipase activity in birds on high-fat diets was also as predicted, and in section 4.5.3 below we briefly describe the mechanism by which animals modulate their digestive enzyme activity.

Figure 4.21. Warblers (*Dendroica pinus* and *D. coronata*) were highly efficient in assimilating a triacylglycerol dose ([^{14}C]glycerol trioleate) when it was mixed into the highest-fat seed diet, less so when it was in the insect diet with intermediate fat content, and least efficient when it was in the low-fat fruit diet. They achieved higher digestive efficiency by retaining digesta in their gut longer (longer MRT) and up-regulating pancreatic lipase activity (measured only in the pine warbler). Within each species' plot, bars with different lower case letters designate significantly different mean values. (Data from Levey et al. 1999 and unpublished; Afik et al. 1995.)

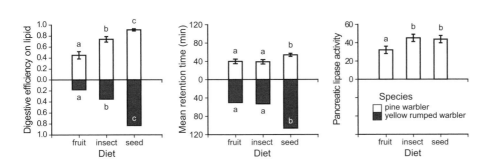

These experiments on warblers clearly showed how differences in digestive efficiency emerge from differences in the underlying component digestive processes, and that the same food substrate can have vastly different value to birds depending on their digestive characteristics. Furthermore, other experiments showed that when warblers are habituated to one food type they cannot immediately transition to the alternative digestive physiology that would permit them to use efficiently an alternate food type. For example, the very short mean retention times of warblers habituated to fruit diet were suitable ("enough, but not too much") for extracting simple-to-digest glucose and sucrose, which both species digested with high efficiency (70–90%) and which accounted for most energy in their fruit diet. But the very short mean retention time was not suitable for digesting lipid efficiently, although this was not a great handicap because lipid accounted for less than 10% of the energy in the fruit diet. When these fruit-adapted birds were suddenly switched to the high-fat seed diet and tested a few hours after they began eating it, they still had short mean retention times and low digestive efficiency (Afik and Karasov 1995). Thus, in the parlance we introduced earlier, their reserve digestive capacity for lipid was relatively small, although with adequate time for habituation their long-term capacity was increased to an adequate level. Their initial low capacity for lipid digestion could limit them ecologically from switching quickly from low-to high-fat diets, because ingestion of large amounts of a substrate without the ability to digest and absorb it can lead to life-threatening diarrhea resulting from the solute's osmotic effect in the intestine. The capacity to digest lipid was defined partly by the birds' lipolytic enzyme activity and by features of their gut motility that influence digesta retention time, and both were flexible. Analogously, the progressively improved capacity to digest high protein insects by American robins and European starlings switched from low-protein fruit diets (figure 4.20) was achieved by increases in intestinal aminopeptidase activity (Martínez del Rio et al. 1995) and/or digesta retention time (Levey and Karasov 1992).

4.5.3 Digestive Enzyme Activities are Reversibly Flexible in Polyphagous Insects and Vertebrate Omnivores

The ability of insects to modulate digestive enzymes rapidly and reversibly was shown in a study (Broadway and Duffey 1982) on two species that are polyphagous—this word means that they eat a range of host plants. Some of these insects were raised from neonate to fifth instar on a basal diet with 0, 0.6, 1.2, 2.4, or 4.8% (wet weight basis) casein protein added. In *Spodoptera exigua*, the activity of trypsin in the gut was positively correlated with dietary protein level (figure 4.22). In another, acute, experiment, larvae of both species were

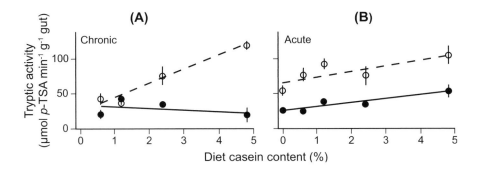

Figure 4.22. Dietary protein increased midgut tryptic activity in larval beet army worms (*Spodoptera exigua*; unfilled circles and dashed lines) and tomato fruitworms (*Heliothis zea*; filled circles and solid line). In the left-hand figure (A; chronic), larvae were provided diets with different casein amounts from hatching to fifth instar, but only the army worms exhibited the predicted increase in protease activity with increasing dietary protein. In the right-hand figure (B; acute), larvae were raised on 24% casein until the fifth instar and then switched for 24 h to diets with different protein contents. In this acute experiment, both species exhibited changes in tryptic activity correlated with the change in dietary protein. (Replotted from Broadway and Duffey 1982.)

raised from neonate to fifth instar on the basal diet with 2.4% casein, and then they were switched to the various test diets (0–4.8% added casein). After just 24 h of feeding on new diets, those that were switched to higher protein increased their tryptic activity whereas those switched to lower-protein diets decreased their tryptic activity (right panel in figure 4.22). In another insect example, Baker (1977) found that proteolytic activity in the beetle *Attagenus megatoma* was increased by dietary protein, whereas sucrase activity increased with dietary carbohydrates. Both patterns are as predicted by the adaptive modulation hypothesis.

Similar reversible changes in proteolytic activity correlated with dietary protein have been documented in more than a dozen fish, mammalian, and avian species, sometimes in less than 24 h after diet switch (Karasov and Hume 1997). There is even a report of digestive enzyme modulation in the marine worm *Nereis virens* switched between carnivorous and detritivorous diets (Bock 1999). Some animals, like *Heliothis zea* in figure 4.22, some other insects (Lemos et al. 1992), some mammals (Sabat et al. 1999), and many passerine birds (Caviedes-Vidal et al. 2000) do not modulate all their digestive enzymes in the predicted fashion. Could the differences among species in modulation patterns relate to differences in their life histories that differentially select for digestive flexibility? Many of the cases of animals matching their digestive enzyme activity to dietary substrate loads probably involve changes in levels of particular enzymes mediated by changes in transcription or translation of messenger RNA for those specific enzymes (Karasov and Hume 1997). This specific regulation can be contrasted with the earlier-discussed nonspecific regulation as when digestive capacity of all enzymes changes in the same direction due to changes in the mass of tissues that synthesize the enzymes (e.g., intestine or pancreas of vertebrates; midgut and hepatopancreas of many invertebrates).

Some plants produce secondary metabolites that inhibit protein digestion, and animals that eat those plants may respond in a compensatory fashion, providing another ecologically important example of digestive enzyme flexibility.

The best studied examples have to do with a class of secondary metabolites called protease or proteinase inhibitors (PIs) which bind to digestive proteins and reduce digestive efficiency and hence growth rate (Ryan 1979; Jongsma and Bolter 1997). Many insects, mammals, and birds respond by increasing secretion of proteolytic enzymes and, in the vertebrates, by increasing the size of the pancreas, which synthesizes many of the enzymes, often with the net effect of restoring digestive efficiency and growth rate. A competing hypothesis about the animals' response is that overproduction of digestive proteins is to the detriment of other essential proteins in the body, and that growth rate thus does not recover (Jongsma and Bolter 1997)). The most recent research in this area has revealed a coevolutionary war between plants and herbivores in which the plants produce a variety of PIs with specific action against different kinds of proteases and the animals produce digestive enzyme variants that are fairly insensitive to the PIs! The entire topic is not only interesting but also very important, because plant biologists are now experimentally manipulating in crop plants the genes that regulate PIs in order to enhance resistance to crop pests.

4.5.4 Absorption Activity Is Reversibly Flexible in Mammalian Omnivores

Nutrient transporters, like digestive enzymes, are a feature of digestive biochemistry that might be modulated according to ecological conditions. But we should not always expect positive relationships between nutrient transporter activity and dietary substrate levels. Opposing patterns of modulation of transport are expected for changes in dietary content of carbohydrate and protein, which provide metabolizable energy, versus water-soluble vitamins and essential minerals, which provide essential nutrients (Diamond and Karasov 1987). In the former case, transporters for monosaccharide, amino acids, or peptides worth calories should tend to be up-modulated by their substrates, or by dietary carbohydrate or protein, respectively. Transport of a water-soluble vitamin or mineral should be down-modulated by its substrate and up-modulated in deficiency, because such a transporter would be most needed at low dietary level and least needed at high level when requirements might even be met by passive diffusion down a concentration gradient. Another benefit of downward modulation of some minerals (e.g., iron) at high dietary level is avoidance of toxicity.

Ferraris and Diamond (1989) reviewed the contrasting patterns for effects of diet on intestinal apical uptake of sugars, dipeptide, and amino acids, minerals, and vitamins. The studies were performed mostly in laboratory rodents. The pattern of modulation varies greatly among solutes (figure 4.23). As predicted,

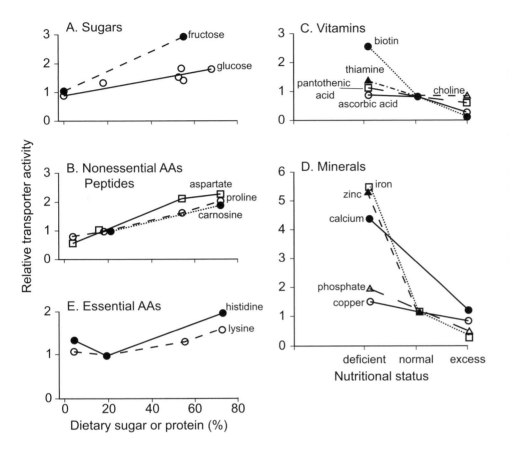

Figure 4.23. The modulation patterns for activities of intestinal nutrient transporters as a function of dietary levels or body stores of their substrates depend on the nutritional role of the substrate. In mouse intestine, as predicted for nutrients mainly worth calories, transport activity (the ordinate) for the monosaccharides glucose and fructose (shown in A), and transport activity for amino acids (aspartate, proline) or peptides (carnosine) (shown in B), are up-modulated by their substrates (the abscissa). In contrast, transport of required nutrients not worth calories, such as water soluble vitamins (C; biotin, choline, thiamine, ascorbate, and pantothenic acid for intestine of rat, guinea pig, and mouse respectively) and minerals (D; iron, zinc, calcium, phosphate, and copper transporters in rat intestine) is down-modulated by their respective substrate and up-modulated in deficiency. Modulation of transport of essential amino acids (E; lysine and histidine in mouse intestine), which are both required and worth calories, follows a composite pattern. (Replotted from Ferraris and Diamond 1989.)

dietary carbohydrate stimulates brush-border aldohexose uptake, and dietary protein or amino acids stimulate amino acid and dipeptide uptake. Also as predicted, intestinal absorption rates for minerals are modulated downward by high dietary levels and upward by low levels. Uptake of the essential amino acids histidine, lysine (figure 4.23), leucine, and methionine (Diamond and Karasov 1987) were maintained or even slightly enhanced in deficiency, and thus their modulation at low dietary levels was more similar to that of essential nutrients than to that of the nonessential amino acids aspartate and proline. Modulation of the latter amino acids can occur semi-independently of the essential neutral and basic amino acids because there are semi-private amino acid transporters for imino and acidic amino acids.

Modulation of vitamin transport is partly but not entirely consistent with the a priori predictions. Vitamin transporters tend to be down-modulated by their substrates and up-modulated in the absence of their substrates (figure 4.23), but not those for pantothenic acid (Stein and Diamond 1989), ascorbic acid (Karasov et al. 1991), or choline (Karasov 1992). Stein and Diamond (1989) suggested that modulation of a vitamin might not occur if a large proportion of the vitamin's absorption was passive and hence modulation of mediated absorption was not important. The uptake kinetics of pantothenic acid, ascorbic acid, and choline do indicate that mediated uptake makes a smaller contribution to total uptake than other vitamins whose transport is up-modulated in deficiency. As we will see below, this same logic can be used to explain why most birds do not modulate sugar absorption in the predicted pattern.

Molecular studies of the modulation of intestinal sugar absorption indicate that it is achieved largely by increasing and decreasing the apical membrane density of transporters for glucose (called SGLT1 or "sodium glucose transporter 1") and fructose (called GLUT5 or "glucose transporter 5" from its homology with glucose transporters) by altering transcription rates (Karasov and Hume 1997). Much of the response to a new dietary signal like "higher" or "lower" diet glucose content occurs in the newest cells that are born in the crypts and migrate up the villus length, finally to be shed into the lumen. In a laboratory rodent the cell population of the whole length of the villous is renewed in 1 to 2 days, which therefore determines the time for complete diet adjustment. But recent research has uncovered some response times to altered dietary glucose that can be much faster. Kellett and Helliwell (2000) described in rats the movement and insertion of additional glucose transporters (called GLUT 2) into the brush border stimulated by the presence of luminal sugar over the course of only a few hours.

Not all species modulate nutrient transporters as predicted. As we discussed for the digestive enzymes, we perhaps should not expect to observe much modulation in species that do not alter their diets, like obligate carnivores. Indeed, many carnivores that have been studied do not modulate sugar or amino acid absorption rates very much, whereas many of the omnivores that have been studied do modulate (Karasov and Hume 1997). But there are a number of exceptions (Buddington et al. 1991). Many omnivorous bird species that have been studied failed to exhibit modulation of mediated sugar absorption (Levey and Karasov 1992; Karasov et al. 1996; Caviedes-Vidal and Karasov 1996; Afik et al. 1997a) but do exhibit considerable passive absorption like that discussed in section 4.1.5 (see figure 4.12) (Caviedes-Vidal and Karasov 1996; Levey and Cipollini 1996; Afik et al. 1997b). Pappenheimer (1993) suggested that passive absorption may confer a selective advantage because it requires little energy and provides a mechanism by which rate of absorption is matched to rate of hydrolysis. A matching between the capacity for mediated absorption and dietary substrate level is not necessarily predicted if most absorption occurs by a passive pathway (Stein and Diamond 1989). As discussed above, vitamin transporters appear to be modulated only in the cases in which they make the dominant contribution to uptake.

Whilst passive absorption might be seen to have certain advantages, it has at least one potentially large disadvantage. A high intestinal permeability, permitting passive absorption, might be less selective than a carrier-mediated system and might permit xenobiotics to be absorbed from plant and animal material (Diamond 1991). This vulnerability to toxins could be an important ecological driving force, constraining exploratory behavior, limiting the breadth of the dietary niche, and selecting for compensatory behaviors such as searching for and ingesting specific substances that inhibit hydrophilic toxin absorption (Diamond et al. 1999). In this regard it is best to remember also that animals have an inherently high intestinal permeability to lipophilic xenobiotic compounds (see section 4.1.5 above) and have evolved a variety of mechanisms to biotransform and eliminate them, which we will review in chapter 9.

4.6 The Evolutionary Match between Digestion, Diets, and Animal Energetics

4.6.1 There Is an Evolutionary Match between Enzymes and Nutrient Transporters and the Dietary Loads of Their Corresponding Substrates

Many comparative studies with vertebrates have produced results that seem consistent with the a priori expectation (section 4.5.1) that for dietary components such as carbohydrates, protein, and lipids there will be a positive relationship between their level in the natural diet and the presence or amount of gut enzymes or nutrient transporters necessary for their breakdown (Vonk and Western 1984; Karasov and Hume 1997). Consider, as one example, amylase, which hydrolyzes the carbohydrate storage molecules of plants (amylose, amylopectin) and animals (glycogen) and which in vertebrates is synthesized mainly in the pancreas. At least six comparative studies in fish have found that amylase activity per mg pancreas was higher in herbivorous or omnivorous species that consume a lot of complex carbohydrates compared with carnivores that do not (Kapoor et al. 1975) (although see Chakrabarti et al. (1995) for an exception). In vertebrates the more carnivorous species have lower rates of intestinal glucose absorption (Karasov and Diamond 1988). This conclusion arises from comparisons among species of fish (Buddington et al. 1987) and birds (Karasov and Hume 1997) of ratios of glucose uptake to uptake rate of the amino acid proline when diet composition is controlled. So far so good, but the best comparative analysis should be guided by phylogenetic data (see chapter 1) and almost all comparative tests of this digestive adaptation hypothesis lack this feature.

In collaboration with Schondube and Herrera one of us (Martínez del Rio) studied the effect of evolutionary shifts in diet on digestive enzymes in a group of New World bats, the family Phyllostomidae (Schondube et al. 2001). Based on a phylogeny of this speciose group (49 genera and more than 140 species), it appears that from an original insectivorous ancestor several clades evolved in which species were either mainly insectivorous, nectarivorous, frugivorous, or sanguinivorous (blood eating) and carnivorous. By selecting representatives from each diet group we could test many a priori predictions about shifts in digestive enzymes with shifts in diet, based on the assumptions that our phylogeny was correct and that modern insectivores can be used to characterize the digestive traits of the putative insectivorous ancestor. We used the ^{15}N level of the bats' blood to characterize their diets, because bats feeding on animal tissues should (and did) have higher δ^{15}N values than those of bats feeding on plant products (see chapter 8). The enzymes we studied were those located in the lumen-facing (i.e., apical) brush-border membranes of intestinal cells.

Sucrase and trehalase hydrolyze the disaccharides sucrose (from plants) and trehalose (the principal blood sugar in insects). Maltase hydrolyzes to monosaccharides the oligosaccharides that are formed by amylase or that are directly ingested. Aminopeptidase N (also known as leucine aminopeptidase) hydrolyzes oligopeptides formed by proteases in the stomach and pancreas (figure 4.8). Amazingly, all 20 a priori predictions for these enzymes were borne out. For example, a shift from insectivory to sanguinivory and carnivory (i.e., reduction of insect trehalose in the diet) was accompanied by a ten- to fifteen fold decrease in trehalase activity (figure 4.24). A shift from insectivory to nectarivory or frugivory (addition of plant sugars to the diet) was accompanied by an increase in maltase and sucrase activity, a decrease in trehalase activity, and no change in aminopeptidase activity (because bats in all diet groups digest protein). The probability of such high concordance with predictions is so infinitesimally low that we concluded that evolutionary changes in diet in phyllostomid bats were indeed accompanied by adaptive shifts in digestive enzymes.

4.6.2 The Evolutionary Match between Digestive Traits and Animal Energetics

Fish, and the other ectothermic vertebrates, have daily energy requirements that are much lower than those of endothermic birds and mammals. Inspection of equations 4.1 and 4.2 suggests that the higher metabolic demands of endotherms might be met in either of two ways: (1) more rapid digesta flow through a

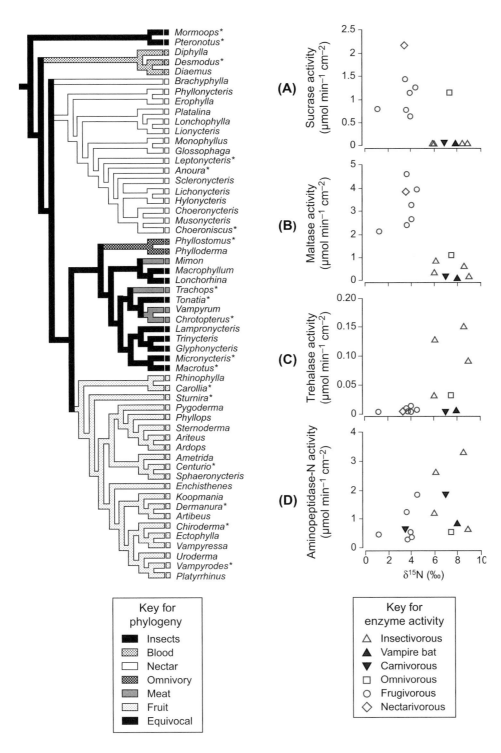

Figure 4.24. Within the New World bat family Phyllostomidae, the evolutionary shift from insectivory to nectarivory or frugivory was accompanied by an increase in sucrase (A; top right figure) and maltase (B; second from top) activity (which digest plant sugars in the diet), a decrease in trehalase (C; third from top) activity (digests insect sugar trehalose in the diet), and no change in aminopeptidase (D; bottom right) activity (because bats in all diet groups digest protein). In these plots, increasing animal matter in the bats' natural diet is indicated by increasing $\delta^{15}N$ in the bats' tissue (see text), and points are species means. The evidence that these correlations represent evolutionary transitions is based on the bats' diets mapped onto their hypothesized phylogeny, shown on the left. The genera marked with asterisk are included in the data set. Two of the bat genera (*Mormoops* and *Pteronotus*) are in a sister family, Mormoopidae. (Redrawn from Schondube et al. 2001.)

similar-sized GI tract with resultant shorter turnover time, possibly resulting in poorer nutrient extraction due to reduced exposure time of ingested particles to GI processes, or (2) more rapid digesta flow through a very large GI tract, in which case turnover time might be similar and the rate of energy absorption could increase as gut capacity increases. Comparative studies of mammals and reptiles indicate that elements of both ways evolved, because mammals tend to have GI tracts with more surface area and volume, but also exhibit much shorter turnover time and no decrease in extraction efficiency (Karasov and Diamond 1985). Are endotherms' higher ingestion rates met by a higher biochemical capacity of the intestine (the product of intestinal nominal surface area and hydrolysis and nutrient uptake per unit nominal area)?

This question has received limited attention for absorption and none for digestive enzymes, as far as we know. Within fish, reptiles, and mammals (Karasov 1987) there does not appear to be a dependence of uptake per cm^2 on body size for either proline or glucose when similar measurement techniques are used. Fish, but not reptiles, have lower tissue-specific uptake rates than mammals and birds (Karasov and Hume 1997), but much of this difference may be due to the lower measurement temperatures in the fish studies. The Q_{10} for mediated transport is 2–3 (Karasov 1988). The conclusion that follows from these observations is that absorptive capacity among vertebrates of different body size and taxa follows the patterns depicted in figure 4.6 for absorptive surface area. Absorptive capacity increases with body size to the approximate 0.75 power, and is higher in endotherms than ectotherms.

4.6.3 Sugar Preferences in Birds: The Behavioral and Ecological Consequences of Variation in Digestive Traits

The previous sections clearly showed how differences in digestive efficiency can emerge from differences in the underlying component digestive processes, and that the same food can have vastly different value to animals depending on their digestive characteristics. The activity of digestive enzymes and nutrient transporters in relation to retention time creates ecological opportunities and constraints. We have observed many examples of this in our studies of avian digestive physiology over the past decade, and we suspect that similar patterns would be revealed in studies of other taxa. The higher digestive efficiency for sucrose in hummingbirds relative to cedar waxwings can be traced to the former's higher brush-border sucrase-isomaltase activity (Martínez del Rio et al. 1989). The American robin's total inability to assimilate sucrose is due to its total lack of this enzyme (Karasov and Levey 1990). In our experiments, hummingbirds were more efficient even at absorbing monosaccharides like glucose and fructose than waxwings and robins. Although this difference has been questioned (Witmer 1999; Witmer and Van Soest 1998), it is somewhat expected considering that hummingbirds, on a mass-corrected basis, have relatively longer digesta retention times and higher glucose transport activity (Karasov 1990), which both act together to increase absorption efficiency (equation 4.1).

Perhaps not surprisingly, the variation that we have described in the ability to assimilate sugars has behavioral consequences. The preference of birds for different sugar mirrors the birds' digestive capacity to assimilate them (Martínez del Rio et al. 1992). American robins, and other closely related species such as European starlings and gray catbirds, all members of the large (≈600 species) and monophyletic sturnid-muscicapid lineage, lack intestinal sucrase activity (Martínez del Rio 1990a). These birds learn to reject sucrose in favor of hexoses because ingesting sucrose, which cannot be digested, can cause life-threatening diarrhea. Waxwings will ingest sucrose but prefer the hexoses because they digest these simpler sugars more efficiently (Martínez del Rio et al. 1989). In fact, among hexoses waxwings digest glucose slightly better than fructose and also prefer glucose over fructose (Martínez del Rio et al. 1989). Only the hummingbirds, with the highest reported sucrase activities among birds, prefer sucrose (Martínez del Rio 1990b). This example illustrates that the value of a food and its relative use depend on the consumer's digestive traits. It also exemplifies the potential consequences of digestive physiology for the study of animal-plant interactions.

Botanists Herbert and Irene Baker documented a curious dichotomy. They found that the nectar secreted by flowers pollinated by hummingbirds contained primarily sucrose, whereas nectar of flowers pollinated by passerines contained primarily glucose and fructose (reviewed by Martínez del Rio et al. 1992). Intriguingly, a similar pattern is found in fruit. The fruit pulp of bird-dispersed plants contains much less sucrose that that of fruits dispersed by mammals (Baker et al. 1998). One of us (Martínez del Rio) chose this pattern as a theme for a doctoral dissertation and decided to investigate the following working hypothesis: perhaps the sugar composition of floral nectars and fruit pulps was shaped by the sugar preferences of pollinators and seed dispersers. We have discovered that, indeed, birds differ in sugar preferences and we have demonstrated quite conclusively that the sugar preferences of animals are deeply influenced by digestive traits—as described in previous paragraphs. However, we and others have also discovered interesting nuances to what in retrospect seems a naive hypothesis. The hummingbird-passerine dichotomy in sugar composition is not as solid as once believed. In Africa, many flowers pollinated by sunbirds secrete sucrose (Nicolson and Van Wyk 1998). Furthermore, sunbirds, which are the African equivalent of hummingbirds, do not show a distinct preference for sucrose (Jackson et al. 1998). Recently we have found that sugar preferences may be concentration dependent. Schondube and Martínez del Rio (2003) discovered that the sugar preferences of a hummingbird (*Eugenes fulgens*) and a nectar-feeding passerine (*Diglossa baritula*) are distinctly concentration dependent. At low nectar concentrations the birds preferred hexoses, but at high concentrations they preferred sucrose. The physiological reasons for this concentration dependence in sugar preferences are unknown, but add a wrinkle to an already complicated story. Tantalizingly, Nicolson (2002) has documented a positive correlation between sucrose content and total sugar concentration in the nectar secreted by some bird-pollinated genera. Digestive traits can not only affect how animals behave, but also influence their interaction with their living and evolving resources, such as the plants that they pollinate and whose seeds they disperse.

4.6.4 Hydrolysis and Absorption Activity Vary during Ontogeny

Up to now we have considered only the traits of adult animals. Yet many animals have very different diets as adults than as larvae or juveniles. Perhaps the most immediately obvious example of an ontogenetic dietary shift is that between milk and the adult diet in young mammals. This example is unique in its details, as we will see in following paragraphs, but it is an example among many of predictable ontogenetic diet changes. Many birds that eat plant parts such as fruit, nectar, and seeds feed their growing young on high-protein insect diets. Among butterflies and moths, larvae are often follivorous whereas adults feed on nectar. Based on the plethora of examples of digestive responses to differences in diet that we have described, you should expect ontogenetic diet shifts to be often accompanied by changes in the expression of digestive traits. This hypothesis is to a large extent true, but there are sufficient twists and turns in the development of digestive function to merit the detailed description of a few examples. Two good examples of species that go through ontogenetic diet changes are a fish with the delightfully insulting name monkeyface prickleback (*Cebidichtys violaceus*) and the bullfrog (*Rana catesbiana*). Young pricklebacks are primarily carnivorous and, as they grow, they progressively adopt a vegetarian diet of algae. Bullfrogs show the opposite trend; tadpoles are primarily herbivorous and they metamorphose into carnivorous adults. Thus, in both of these species the proportion of carbohydrate relative to protein in the diet varies. It increases in pricklebacks and decreases in bullfrogs.

To examine the genetic differences in intestinal glucose and proline transport among fish species, Buddington et al. (1987) fed several species of fish on a standard diet. They found that the ratio of proline to glucose uptake in the small intestine decreased from carnivores to omnivores, and was lowest in herbivores. A surprising result of their study was that in one of their study species, this ratio varied with fish size (figure 4.25A). In monkeyface pricklebacks the ratio of proline to glucose uptake decreased with fish size. Because, as we mentioned before, pricklebacks are carnivorous as juveniles and become herbivorous as adults, the decrease in proline:glucose ratios illustrated in figure 4.25 makes functional sense. Buddington et al. (1987) fed the pricklebacks of all ages on exactly the same diet. Therefore their ontogenetic digestive changes cannot be attributed to adaptive modulation to the nutrient composition of a shifting diet. They are the result of a genetically programmed developmental program. As we will see in following paragraphs, this might be a common pattern in animals. It is not one found in bullfrogs, though.

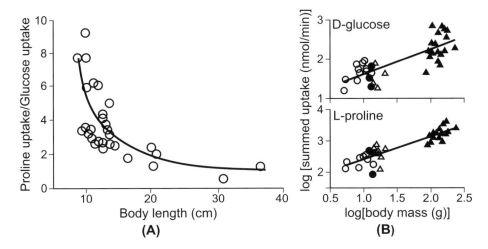

Figure 4.25. (A, left-hand figure) In monkeyface pricklebacks the ratio of proline to glucose intestinal uptake declines with fish size, even when fish are fed on exactly the same diet. In the wild these fishes change from being carnivorous as juveniles to herbivores as adults. Therefore, the changes in digestive function correspond to developmental changes in diet. (Replotted from Buddington et al. 1987.) In contrast, in bullfrogs (B, both right-hand figures) total intestinal proline uptake in tadpoles and adults are similar. Unfilled circles represent larvae, filled circles represent premetamorph tadpoles, and unfilled triangles represent larvae at early metamorphosis. Adults are represented as filled triangles. (Replotted from Toloza and Diamond 1990.) Note the contrasting ways of presenting similar data. We find the allometric approach used in (B) preferable over the deceptively simple ratios used in (A). Ratios not only have confounded statistical properties, they also hide information. From the data presented in panel A it is impossible to know if the proline:glucose ratio decreased, because glucose uptake increased, because proline uptake decreased, or because both changed. Although we are guilty of having used ratios in the past, we discourage their use.

Like most anurans, bullfrogs undergo a remarkable change in morphology and ecology during their metamorphosis from larvae to adult. Curiously, Toloza and Diamond (1990) found that, in spite of the change in diet from herbivory to carnivory that accompanies metamorphosis, the relative capacities of tadpoles and adults to transport glucose and proline in the intestine were equal (figure 4.25B). We find this result remarkable, and a tad perplexing, especially in view of the astounding differences in intestinal morphology between tadpoles and adults. Tadpoles have a long, highly coiled, thin-walled intestine with a very narrow bore. Adults, in contrast, have short, thick-walled intestines, with a wide bore. In contrast with Buddington et al.'s (1987) study in which fish of all ages were fed the same diet, Toloza and Diamond (1990) fed the tadpoles and the adults on different diets. They fed tadpoles on boiled lettuce and adults on mealworms. The absence of differences in nutrient transport between these two developmental stages occurred in spite of vast differences in the nutrient composition of their diets.

The transition in mammals from suckling to an adult diet has been very well studied from the perspective of digestion. The study of this transition in humans has revealed a remarkable tapestry that weaves genetics, digestive physiology, and cultural ecology. Lactation and its digestive implications merit a few paragraphs in a chapter on digestive ecology. Although a handful of bird species, including pigeons, some penguins, and flamingos, secrete a milklike substance (called crop milk) and feed it to their young (Fisher 1972), true lactation is a derived trait shared by, and found only in, mammals. Mammalian milk is a wonderfully complex secretion made of a variety of proteins, lipids, carbohydrates, and minerals (Blackburn 1993). In many species the carbohydrates in it are primarily in the form of lactose. Lactose is a disaccharide of galactose and glucose joined by a β-glycosidic bond. Lactose is uncommon in nature except in milk. The lactose content of milk varies a lot among mammal species. At the lowest end of the spectrum is the milk of sea lions, which contains no lactose. At the high end is human milk in which lactose represents about 50% of total solids (Jennes and Sloan 1970). Lactose is digested into glucose and galactose by the membrane-bound intestinal enzyme lactase-phlorizin-hydrolase (or lactase, for simplicity). Like milk, lactase is a unique mammalian invention.

Most adult mammals do not ingest milk and, as you would expect, express lactase not at all or at very reduced levels. Most mammals show a distinct transition in the expression of the digestive hydrolases of the intestinal microvilli during or immediately after weaning (figure 4.26). In omnivorous mammals, sucrase-isomaltase, which is expressed at low levels at birth, increases and remains high throughout life. Lactase activity, in contrast, is high at birth and declines sharply around weaning. Unlike many of the hydrolases that we have discussed in previous sections, lactase levels are not

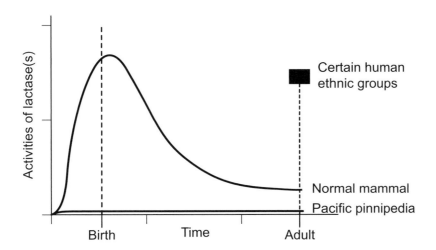

Figure 4.26. In most mammals the activity of the enzyme lactase is high at birth and declines as the animal ages. The ontogenetic pattern in this enzyme corresponds to the normal pattern in the intake of lactose-containing milk in young mammals. Because adding lactose to the animal's diet does not impede the decrease in intestinal lactase activity, we know that the development of intestinal lactase activity follows a fixed genetic program. If you were to measure the concentration of the messenger RNA that codes for lactase in intestinal cells, you would find that it closely parallels the pattern of lactase activity. Lactase activity is regulated at the transcriptional level. There are two exceptions to the widespread mammalian pattern of lactase expression: (1) some species never express high levels of lactase. These species are those whose milk lacks lactose such as some seals and sea lions. (2) A few human ethnic groups maintain high levels of lactose expression as adults.

regulated by the presence of its substrate in an animal's diet (Rings et al. 1994). If the young mammal is allowed to prolong suckling, or is fed on a lactose-containing diet after weaning, the onset of the decline in lactase is delayed, but only slightly. Thus, all evidence points to a genetically determined developmental program as the cause for the decline in intestinal lactase. The timing of this program can be modulated by corticosteroids, but these hormones can only hasten or delay the onset of the reduction in lactase activity (Henning 1981). Corticosteroids may be significant for mammals that are weaned prematurely. The corticosteroid surge that accompanies an untimely weaning may facilitate the early expression of the enzymes and allow the young mammal to eat an adult diet. In spite of the large "have milk" campaign in western societies, milk is not a suitable food for adult mammals, and it is not even suitable for the vast majority of adult humans. Undigested lactose passes through the small intestine into the large intestine where bacteria ferment it (see chapter 6). Fermentation leads to a sudden increase in the concentration of short-chain fatty acids and of gases (CO_2, H_2, and methane) in the large bowel's lumen and causes osmotic diarrhea and a bloated, gassy, sensation that most mammals find very uncomfortable.

There is a numerically minor, but nonetheless noteworthy, exception to the mammalian pattern that we have just described. Although most humans follow the ancestral mammalian pattern, a minority of all adult humans retain the high levels of lactase activity that are found in small children. The retention of high lactase levels is inherited as a single autosomal gene (called LAC*P for lactose persistence). The geographical distribution of this gene follows a distinctive pattern: its frequency is high in northwestern Europe, among nomadic pastoralists in the Middle East and Northern Africa, among the Khoi of South

Africa, and maybe among the aristocratic pastoral Tutsi of the East African interior (Sahi 1994). The gene is found at very low frequencies everywhere else except in North America, Australia, and New Zealand, where an individual's ability to digest lactose depends on his/her family's place of origin. In Australia, for example, white folks resemble Europeans in lactase persistence, whereas adult Aborigines are invariably lactose intolerant. Lactose tolerance appears either to have evolved independently several times or to have been acquired by migration in four distant populations.

It is perhaps not surprising that the frequency of lactase persistence is high in populations that depend on the milk of domesticated ungulates such as cattle, goats, and camels. Simoons (1969, 1970) proposed the "culture-evolution hypothesis" to explain this pattern. Simoons speculated that adults who retained the ability to use milk would gain a selective advantage in milking cultures. Because the growing number of milk-using adults would then be encouraged to devote more effort toward livestock, cultural practices and the evolution of a digestive trait would facilitate each other in a culture-evolution positive feedback. Flatz (1987) proposed a complementary hypothesis for northern Europe. Flatz noted that lactose facilitates calcium absorption in the intestine. The cold, cloudy winter climate of northern Europe dissuades most humans from exposing their skin to sunlight, and hence reduces the synthesis of vitamin D. Without milk and its effect on the bioavailability of calcium, individuals living in northern populations may have been vulnerable to rickets and osteomalacia. A mutant LAC*P allele would spread rapidly because not only would it allow adults to use a good source of calcium, but the lactose in milk would also facilitate its absorption.

Lactose intolerance is often perceived in western cultures as a disease. It is not. It is the ancestral mammalian condition, and the most common phenotype in humans. The western misconception of lactose intolerance as an ailment is widespread. It was recently highlighted by the media's response to the recent discovery of a genetic marker for lactose intolerance. Enatthah et al. (2002) found a DNA variant that associates completely with lactose intolerance in adults (or hypolactasia, a term that does sound like a disease). They found it in the members of several families of Finnish lactose-intolerant families and in all the individuals of several ethnic groups that are normally considered lactose intolerant. They concluded, reasonably, "that the variant occurs in distantly related populations indicates that it is very old." Because hypolactasia is the ancestral trait in humans, the variant must indeed be very, very old. The media hailed this discovery with revealingly inaccurate headlines. Our two favorites are the following: "Genetic Glitch May Cause Lactose Intolerance" and "UCLA and Finnish Scientist Identify Genetic Mutation that Causes Lactose

Intolerance." The "mutation" or "glitch" is the genetic condition found in tolerant individuals. In this chapter we have emphasized the interplay between digestive physiology and ecology. The distribution of lactose intolerance among ethnic groups emphasizes the need to construe ecology broadly when considering humans. Human ecology includes a dominantly large cultural component.

4.7 Summary: The Interplay between Digestive Physiology and Ecology

Many seemingly different kinds of guts in animals can be usefully modeled as just a few types of chemical reactors, and they share many of the same basic biochemical features of enzymatic breakdown and nutrient absorption. Physiological features, including the structure and function of the GI tract, are sometimes the most limiting steps in energy assimilation in ecological settings. They are perhaps most apparent to us when physiological bottlenecks enforce resting periods between feeding bouts, as occurs in herbivores (Kenward and Sibly 1977), in nectarivores (Diamond et al. 1986), and in predators of bivalves (Zwarts and Blomert 1992; Guillemette 1994), crabs (Zwarts and Dirksen 1990), and fish (Hilton et al. 2000). In order to mitigate the limitation, the guts of many animals exhibit considerable flexibility in the face of altered food composition and level of intake. In endotherms, the primary adjustment to increased feeding rate is an enlargement of organs important in nutrient assimilation, like stomach, intestine, pancreas, and liver. The net effect of a larger GI tract is to maintain retention time and hence digestive efficiency relatively constant with increased load. In ectotherms, maintenance of a high assimilation rate often requires behavioral thermoregulation at a relatively high body temperature, but GI tract adjustments to chronically altered temperature, as might occur during global warming, are practically unstudied. The response of endotherms and ectotherms to altered food composition involves a different response pattern than that to high food intake. Levels of specific enzymes, and sometimes transporters, are typically modulated in relation to dietary levels of their respective substrates. Additionally, increased concentrations of carbohydrates, protein, and fat tend to increase digesta retention time, at least in terrestrial vertebrates (Karasov and Hume 1997). Presumably, the net effect of increased reaction rates and increased contact time between digesta and the digestive processes is to maintain digestive efficiency when load is increased.

An animal's digestive flexibility is ecologically important for what it permits the animal to do and for how it constrains the animal's feeding ecology and behavior (examples are given in Bednekoff and Houston 1994; Hilton et al. 2000;

Whelan et al. 2000; Karasov 1996). Lack of or limited flexibility creates a static constraint, and explains why the same food can have vastly different value to different animal species depending on their digestive characteristics. When animals have digestive flexibility, the response time course may determine how quickly they can respond to changing food resources or environmental conditions. Whether a north-temperate mammal or bird lacking energy stores can survive a sudden cold snap could depend partly on how quickly it can increase the size of its GI tract to permit it to eat more. How much diet mixing occurs within or among feeding bouts, or how quickly diet switching occurs, could depend on the time course for induction of enzymes and transporters and/or the adjustment of digesta retention time. Although we have some knowledge of the mechanisms of digestive flexibility, we know relatively little about their time courses and how those may relate to these kinds of ecological scenarios.

References

Adolph, E. F. 1949. Quantitative relations in the physiological constitutions of mammals. *Science* 109:579–585.

Afik, D., E. Caviedes-Vidal, C. Martinez del Rio, and W. H. Karasov. 1995. Dietary modulation of intestinal hydrolytic enzymes in yellow-rumped warblers. *American Journal of Physiology* 269:R413–R420.

Afik, D., B. W. Darken, and W. H. Karasov. 1997a. Is diet-shifting facilitated by modulation of intestinal nutrient uptake? Test of an adaptational hypothesis in yellow-rumped warblers. *Physiological Zoology* 70:213–221.

Afik, D. and W. H. Karasov. 1995. The trade-offs between digestion rate and efficiency in warblers and their ecological implications. *Ecology* 76:2247–2257.

Afik, D., S. R. McWilliams, and W. H. Karasov. 1997b. A test for passive absorption of glucose in yellow-rumped warblers and its ecological significance. *Physiological Zoology* 70:370–377.

Alerstam, T., and A. Lindström. 1990. Optimal bird migration: the relative importance of time, energy, and safety. Pages 331–351 in E. Gwinner (ed.), *Bird migration: Physiology and ecophysiology*. Springer, Berlin.

Alpers, D. H. 1994. Digestion and absorption of carbohydrates and proteins. Pages 1723–1749 in L. R. Johnson (ed.), *Physiology of the gastrointestinal tract*. Raven Press, New York.

Anderson, J. M. 2001. Molecular structure of tight junctions and their role in epithelial transport. *News in Physiological Science* 16:126–130.

Angilletta, M. J. Jr. 2001. Thermal and physiological constraints on energy assimilation in a widespread lizard (*Sceloporus undulatus*). *Ecology* 82:3044–3056.

Atkinson, D. 1994. Temperature and organism size—a biological law for ectotherms? *Advances in Ecological Research* 25:1–58.

Atkinson, S. N., R. A. Nelson, and M. A. Ramsay. 1996. Changes in the body composition of fasting polar bears (*Ursus maritimus*): The effect of relative fatness on protein conservation. *Physiological Zoology* 69:304–316.

Baker, H. G., I. Baker, and S. A. Hodges. 1998. Sugar composition of nectars and fruits consumed by birds and bats in the tropics and subtropics. *Biotropica* 30:559–586.

Baker, J. E. 1977. Substrate specificity in the control of digestive enzymes in larvae of the black carpet beetle. *Journal of Insect Physiology* 23:749–754.

Barry, R. E. Jr. 1976. Mucosal surface areas and villous morphology of the small intestines of small mammals: functional interpretations. *Journal of Mammalogy* 57:273–289.

Battley, P. F., A. Dekinga, M. W. Dietz, T. Piersma, S. Tang, and K. Hulsman. 2001. Basal metabolic rate declines during long-distance migratory flight in Great Knots. *Condor* 103:838–845.

Battley, P. F., T. Piersma, M. W. Dietz, S. Tang, A. Dekinga, and K. Hulsman. 2000. Empirical evidence for differential organ reductions during trans-oceanic bird flight. *Proceedings of the Royal Society (London) B* 267:191–195.

Bednekoff, P. A., and A. I. Houston. 1994. Avian daily foraging patterns: Effects of digestive constraints and variability. *Evolutionary Ecology* 8:36–52.

Blackburn, D. G. 1993. Limits to lactation: lessons from comparative biology. *Journal of Dairy Science* 76:3195–3212.

Blem, C. R. 1980. The energetics of migration. Pages 175–224 in S. A. Gauthreaux (ed.), *Animal migration, orientation, and navigation.* Academic Press, New York.

Bock, M. J. 1999. Digestive plasticity of the marine benthic omnivore *Nereis virens. Journal of Experimental Marine Biology and Ecology* 240:77–92.

Borgstrom, B. and J. S. Patton. 1991. Luminal events in gastrointestinal lipid digestion. Pages 475–504 in M. Field and R. A. Frizzell (eds.), *Handbook of physiology, Section 6: The gastrointestinal system,* vol. IV. American Physiological Society, Bethesda, Md.

Broadway, R. M., and S. S. Duffey. 1982. The effect of dietary protein on the growth and digestive physiology of larval *Heliothis zea* and *Spodoptera exigua. Journal of Insect Physiology* 32:673–680.

Buddington, R. K., J. W. Chen, and J. M. Diamond. 1987. Genetic and phenotypic adaptation of intestinal nutrient transport to diet in fish. *Journal of Physiology* 393:261–281.

Buddington, R. K., J. W. Chen, and J. M. Diamond. 1991. Dietary regulation of intestinal brush-border sugar and amino acid transport in carnivores. *American Journal of Physiology* 261:R793–R801.

Buts, J. P., V. Vivjerman, C. Barudi, N. De Keyser, P. Maldague, and C. Dive. 1990. Refeeding after starvation in the rat: Comparative effects of lipids, proteins and carbohydrates on jejunal and ileal mucosal adaptation. *European Journal of Clinical Investigation* 20:441–452.

Calder, W. A. 1984. *Size, function, and life history.* Harvard University Press, Cambridge, Mass.

Carey, H. V. 2005. Gastrointestinal responses to fasting in mammals: lessons from hibernators. Pages 229–254 in J. M. Starck and T. Wang (eds.), *Physiological and ecological adaptations to feeding in vertebrates.* Science Publishers, Enfield, NH.

Carey, H. V., and H. J. Cooke. 1991. Effect of hibernation and jejunal bypass on mucosal structure and function. *Gastrointestinal and Liver Physiology* 261:G37–G44.

Caviedes-Vidal, E., D. Afik, C. Martinez del Rio, and W. H. Karasov. 2000. Dietary modulation of intestinal enzymes of the house sparrow (*Passer domesticus*): Testing an adaptive hypothesis. *Comparative Biochemistry and Physiology A* 125:11–24.

Caviedes-Vidal, E., and W. H. Karasov. 1996. Glucose and amino acid absorption in house sparrow intestine and its dietary modulation. *American Journal of Physiology* 271:R561–R568.

Chakrabarti, I., M. A. Ganim, K. K. Chaki, R. Sur, and K. K. Misra. 1995. Digestive enzymes in 11 freshwater teleost fish species in relation to food habit and niche segregation. *Comparative Biochemistry and Physiology* 112A:167–177.

Chang, M. H., and W. H. Karasov. 2004a. Absorption and paracellular visualization of fluorescein, a hydrosoluble probe, in intact house sparrows (*Passer domesticus*). *Zoology* 107:121–133.

Chang, M. H., and W. H. Karasov. 2004b. How the house sparrow, *Passer domesticus*, absorbs glucose. *Journal of Experimental Biology* 207:3109–3121.

Chang, R.L.S., C. R. Robertson, W. M. Deen, and B. M. Brenner. 1975. Permselectivity of the glomerular capillary wall to macromolecules. I. Theoretical considerations. *Biophysical Journal* 15:861–886.

Chapman, R. F. 1995. Mechanics of food handling by chewing insects. Pages 3–31 in R. F. Chapman and G. de Boer (eds.), *Regulatory mechanisms in insect feeding.* Chapman and Hall, New York.

Chediack, J. G., E. Caviedes-Vidal, V. Fasulo, L. J. Yamin, and W. H. Karasov. 2003. Intestinal passive absorption of water soluble compounds by sparrows: Effect of molecular size and luminal nutrients. *Journal of Comparative Physiology B* 173:187–197.

Chivers, D. J. 1989. Adaptations of digestive systems in non-ruminant herbivores. *Proceedings of the Nutrition Society.* 48:59–67.

Chivers, D. J., and C. M. Hladik. 1980. Morphology of the gastrointestinal tract in primates: Comparisons with other mammals in relation to diet. *Journal of Morphology* 166:337–386.

Chivers, D. J., and P. Langer. 1994. *The digestive system in mammals: Food, form, and function.* Cambridge University Press, Cambridge, U.K.

Ciminari, E., D. Afik, W. H. Karasov, and E. Caviedes-Vidal. 2001. Is diet-shifting facilitated by modulation of pancreatic enzymes? *Auk* 118:1101–1107.

Clifford, A. J., L. M. Smith, R. K. Creveling, C. L. Hamblin, and C. K. Clifford. 1986. Effects of dietary triglycerides on serum and liver lipids and sterol of rats. *Journal of Nutrition* 116:944–956.

Cloudsley-Thompson, J. L. 1991. *Ecophysiology of desert arthropods and reptiles.* Springer-Verlag, Berlin.

Congdon, J. D. 1989. Proximate and evolutionary constraints on energy relations of reptiles. *Physiological Zoology* 62:356–373.

Cooper, S. J., and D. L. Swanson. 1994. Seasonal acclimatization of thermoregulation in the black-capped chickadee. *Condor* 96:638–646.

Dekinga, A., M. W. Dietz, A. Koolhaas, and T. Piersma. 2001. Time course and reversibility of changes in the gizzards of red knots alternately eating hard and soft food. *Journal of Experimental Biology* 204:2167–2173.

Demment, M. W., and P. J. Van Soest. 1983. *Body size, digestive capacity, and feeding strategies of herbivores.* Winrock International, Morrilton, Arka.

Demment, M. W., and P. J. Van Soest. 1985. A nutritional explanation for body-size patterns of ruminant and nonruminant herbivores. *American Naturalist* 125:641–672.

Diamond, J. 1991. Evolutionary design of intestinal nutrient absorption: Enough but not too much. *News in Physiological Science* 6:92–96.

Diamond, J., K. D. Bishop, and J. D. Gilardi. 1999. Geophagy in New Guidea birds. *Ibis* 141:181–193.

Diamond, J., and K. Hammond. 1992. The matches, achieved by natural selection, between biological capacities and their natural loads. *Experientia* 48:551–557.

Diamond, J. M., and W. H. Karasov. 1987. Adaptive regulation of intestinal nutrient transporters. *Proceedings of The National Academy of Sciences* 84:2242–2245.

Diamond, J., W. H. Karasov, D. Phan, and F. L. Carpenter. 1986. Hummingbird digestive physiology, a determinant of foraging bout frequency. *Nature* 320:62–63.

Duke, G. E. 1989. Gastrointestinal motility and its regulation. *Poultry Science* 61:1245–1256.

Duke, G. E., A. R. Place, and B. Jones. 1989. Gastric emptying and gastrointestinal motility in Leach's storm-petrel chicks (*Oceanodroma leucorrhoa*). *Auk* 106:80–85.

Dunham, A. E., B. W. Grant, and K. L. Overall. 1989. Interfaces between biophysical and physiological ecology and the population ecology of terrestrial vertebrate ectotherms. *Physiological Zoology* 62:335–355.

Dykstra, C. R., and W. H. Karasov. 1992. Changes in gut structure and function of house wrens (*Troglodytes aedon*) in response to increased energy demands. *Physiological Zoology* 65:422–442.

Ennattah, N. S., T. Sahi, E. Savilahti, J. D. Terwilliger, L. Peltonen, and I. Järvela. 2002. Identification of a variant associated with adult-type hypolactasia. *Genetics–Nature* 30:233–237.

Ferraris, R. P., and J. M. Diamond. 1989. Specific regulation of intestinal nutrient transporters by their dietary substrates. *Annual Revision of Physiology* 51:125–141.

Fisher, H. 1972. The nutrition of birds. Pages 431–469 in D. S. Farner, J. R. King, and K. C. Parkes (eds.), *Avian biology*. Academic Press, London.

Flatz, G. 1987. Genetics of lactose digestion in humans. *Advances in Human Genetics* 16:1–77.

Friedman, M. H. 1987. *Principles and models of biological transport.* Springer-Verlag, Berlin.

Frierson, E. W., and J. A. Foltz. 1992. Comparison and estimation of absorptive intestinal surface areas in two species of cichlid fish. *Transactions of the American Fisheries Society* 121:517–523.

Fu, D., A. Libson, L.J.W. Miercke, C. Weitzman, P. Nollert, J. Krucinski, and R. M. Stroud. 2000. Structure of a glycerol-conducting channel and the basis for its selectivity. *Science* 20:481–486.

Galluser, M., F. Raul, and B. Canguilhem. 1988. Adaptation of intestinal enzymes to seasonal and dietary changes in a hibernator: the European hamster (*Cricetus cricetus*). *Journal of Comparative Physiology B* 158:143–149.

Gass, C. L., M. T. Romich, and R. K. Suarez. 1999. Energetics of hummingbird foraging at low ambient temperature. *Canadian Journal of Zoology* 77:314–320.

Grant, B. W., and W. P. Porter. 1992. Modeling global macroclimatic constraints on ectotherm energy budgets. *American Zoologist* 32:154–178.

Guillemette, M. 1994. Digestive-rate constraint in wintering common eiders (*Somateria mollissima*): implications for flying capabilities. *Auk* 111:900–909.

Hammond, K., K. C. K. Lloyd, and J. Diamond. 1996. Is mammary output capacity limiting to lactational performance in mice? *Journal of Experimental Biology* 199:337–349.

Hammond, K. A., and J. Diamond. 1992. An experimental test for a ceiling on sustained metabolic rate in lactating mice. *Physiological Zoology* 65:952–977.

Hammond, K. A., M. Konarzewski, R. M. Torres, and J. Diamond. 1994. Metabolic ceilings under a combination of peak energy demands. *Physiological Zoology* 67:1479–1506.

Harrison, J., and J. H. Fewell. 1995. Thermal effects on feeding behavior and net energy intake in a grasshopper experiencing large diurnal fluctuations in body temperature. *Physiological Zoology* 68:453–473.

Henning, S. J. 1981. Postnatal development: coordination of feeding, digestion, and metabolism. *American Journal of Physiology* 241:G199–G213.

Herrera, M.L.G., and C. Martínez del Rio. 1998. Pollen digestion by new world bats: Effects of processing time and feeding habits. *Ecology* 79:2828–2838.

Hiiemae, K. M. 2000. Feeding in mammals. Pages 411–448 in K. Schwenk (ed.), *Feeding. form, function and evolution in tetrapod vertebrates*. Academic Press, San Diego, Calif.

Hilton, G. M., G. D. Ruxton, R. W. Furness, and D. C. Houston. 2000. Optimal digestion strategies in seabirds: A modelling approach. *Evolutionary Ecology Research* 2:207–230.

Hodin, R. A., J. R. Graham, S. Meng, and M. P. Upton. 1994. Temporal pattern of rat small intestinal gene expression with refeeding. *American Journal of Physiology* 266:G83–G89.

Holling, C. S. 1959. The components of predation as revealed by a study of small mammal predation of the European pine sawfly. *Canadian Entomology* 91:293–320.

Holling, C. S. 1965. The functional response of predators to prey density and its role in mimicry and population regulation. *Memoirs of the Entomoligical Society of Canada* 45:1–60.

Hopfer, U. 1987. Membrane transport mechanisms for hexoses and amino acids in the small intestine. Pages 1499–1526 in L. R. Johnson (ed.), *Physiology of the gastrointestinal tract*. Raven Press, New York.

Houston, D. C., J. D. Gilardi, and A. J. Hall. 2001. Soil consumption by elephants might help to minimize the toxic effects of plant secondary compounds in forest browse. *Mammal Review* 31:249–254.

Hume, I. D. 1989. Optimal digestive strategies in mammalian herbivores. *Physiological Zoology* 62:1145–1163.

Hume, I. D., and H. Biebach. 1996. Digestive tract function in the long-distance migratory garden warbler, *Sylvia borin*. *Journal of Comparative Physiology B* 166:388–395.

Humphries, M. M., D. W. Thomas, and D. L. Kramer. 2001. Torpor and digestion in food-storing hibernators. *Physiological and Biochemical Zoology* 74:283–292.

Hurni, M. A., A.R.J. Noach, M.C.M. Blom-Roosemalen, A. De Boer, J. F. Nagelkerke, and D. D. Breimer. 1993. Permeability enhancement in Caco-2 cell monolayers by sodium salicylate and sodium taurodihydrofusidate: Assessment of effect-reversibility and imaging of transepithelial transport routes by confocal laser scanning microscopy. *Journal of Pharmacology and Experimental Techniques* 267:942–950.

Jackson, S., S. W. Nicolson, and C. N. Lotz. 1998. Sugar preferences and "side bias" in Cape Sugarbirds and lesser double-collared sunbirds. *Auk* 115:156–165.

Jennes, R., and R. E. Sloan. 1970. The composition of milks of various species: a review. *Dairy Science Abstracts* 32:599–612.

Jeschke, J. M., M. Kopp, and R. Tollrian. 2002. Predator functional responses: discriminating between handling and digesting prey. *Ecological Monographs* 72:95–112.

Jobling, M. 1995. *Environmental biology of fishes.* Chapman and Hall, London.

Johnson, L. R., and S. A. McCormack. 1994. Regulation of gastrointestinal mucosal growth. Pages 611–641 in L. R. Johnson (ed.), *Physiology of the gastrointestinal tract.* Raven Press, New York.

Johnson, M. S., and J. R. Speakman. 2001. Limits to sustained energy intake. V. Effect of cold-exposure during lactation in *Mus musculus*. *Journal of Experimental Biology* 204:1967–1977.

Jongsma, M. A., and C. Bolter. 1997. The adaptation of insects to plant protease inhibitors. *Journal of Insect Physiology* 43:885–895.

Jumars, P. A., and C. Martínez del Rio. 1999. The tau of continuous feeding on simple foods. *Physiological and Biochemical Zoology* 72:633–641.

Kapoor, B. G., H. Smit, and I. A. Verighina. 1975. The alimentary canal and digestion in teleosts. *Advances in Marine Biology* 13:109–239.

Karasov, W. H. 1987. Nutrient requirements and the design and function of guts in fish, reptiles, and mammals. Pages 181–191 in P. Dejours, L. Bolis, C. R. Taylor, and E. R. Weibel (eds.), *Comparative physiology: Life in water and on land.* Fidia Research Series vol. IX. Liviana Press, Padova, Italy.

Karasov, W. H. 1988. Nutrient transport across vertebrate intestine. Pages 131–172 in R. Gilles (ed.), *Advances in comparative and environmental physiology.* Springer-Verlag, Berlin.

Karasov, W. H. 1990. Digestion in birds: chemical and physiological determinants and ecological implications. *Studies in Avian Biology* 13:391–415.

Karasov, W. H. 1992. Tests of the adaptive modulation hypothesis for dietary control of intestinal nutrient transport. *American Journal of Physiology* 263:R496–R502.

Karasov, W. H. 1996. Digestive plasticity in avian energetics and feeding ecology. Pages 61–84 in C. Carey (ed.), *Avian energetics and nutritional ecology*. Chapman and Hall, New York.

Karasov, W. H., D. Afik, and B. W. Darken. 1996. Do northern bobwhite quail modulate intestinal nutrient absorption in response to dietary change? A test of an adaptational hypothesis. *Comparative Biochemistry and Physiology A* 113:233–238.

Karasov, W. H., and S. J. Cork. 1994. Glucose absorption by a nectarivorous bird: the passive pathway is paramount. *American Journal of Physiology* 267:G18–G26.

Karasov, W. H., B. W. Darken, and M. C. Bottum. 1991. Dietary regulation of intestinal ascorbate uptake in guinea pigs. *American Journal of Physiology* 260:G108–G118.

Karasov, W. H., and E. S. Debnam. 1987. Rapid adaptation of intestinal glucose transport: a brush-border or basolateral phenomenon? *American Journal of Physiology* 253:G54–G61.

Karasov, W. H., and J. Diamond. 1985. Digestive adaptations for fueling the cost of endothermy. *Science* 228:202–204.

Karasov, W. H., and J. M. Diamond. 1983a. A simple method for measuring solute uptake in vitro. *Journal of Comparative Physiology B* 152:105–116.

Karasov, W. H., and J. M. Diamond. 1983b. Adaptive regulation of sugar and amino acid transport by vertebrate intestine. *American Journal of Physiology* 245:G443–G462.

Karasov, W. H., and J. M. Diamond. 1987. Adaptation of intestinal nutrient transport. Pages 1489–1497 in L. R. Johnson (ed.), *Physiology of the gastrointestinal tract*. Raven Press, New York.

Karasov, W. H., and J. M. Diamond. 1988. Interplay between physiology and ecology in digestion. *BioScience* 38:602–611.

Karasov, W. H., and I. D. Hume. 1997. Vertebrate gastrointestinal system. Pages 409–480 in W. Dantzler (ed.), *Handbook of comparative physiology*. American Physiological Society, Bethesda, Md.

Karasov, W. H., and D. J. Levey. 1990. Digestive system trade-offs and adaptations of frugivorous passerine birds. *Physiological Zoology* 63:1248–1270.

Karasov, W. H., D. Phan, J. M. Diamond, and F. L. Carpenter. 1986. Food passage and intestinal nutrient absorption in hummingbirds. *Auk* 103:453–464.

Karasov, W. H., and B. Pinshow. 1998. Changes in lean mass and in organs of nutrient assimilation in a long-distance passerine migrant at a springtime stopover site. *Physiological Zoology* 71:435–448.

Karasov, W. H., and B. Pinshow. 2000. Test for physiological limitation to nutrient assimilation in a long-distance passerine migrant at a springtime stopover site. *Physiological and Biochemical Zoology* 73:335–343.

Karasov, W. H., B. Pinshow, J. M. Starck, and D. Afik. 2004. Anatomical and histological changes in the alimentary tract of migrating blackcaps (*Sylvia atricapilla*): A comparison among fed, fasted, food-restricted and refed birds. *Physiological and Biochemical Zoology* 77:149–160.

Karasov, W. H., D. H. Solberg, and J. M. Diamond. 1985. What transport adaptations enable mammals to absorb sugars and amino acids faster than reptiles? *American Journal of Physiology* 249:G271–G283.

Karasov, W. H. and J. Wright. 2002. Nestling digestive physiology and begging. Pages 199–219 in J. Wright and M. L. Leonard (eds.), *The evolution of begging: Competition, cooperation, and communication.* Kluwer Academic Publishers, Dordrecht.

Kellett, G. L., and P. A. Helliwell. 2000. The diffusive component of intestinal glucose absorption is mediated by the glucose-induced recruitment of GLUT2 to the brush-border membrane. *Biochemical Journal* 350:155–162.

Kenward, R. E., and R. M. Sibly. 1977. A woodpigeon (*Columbia palumbus*) feeding preference explained by a digestive bottleneck. *Journal of Applied Ecology* 14:815–826.

Kessler, M., and G. Toggenburger. 1979. Nonelectrolyte transport in small intestinal membrane vesicles: the application of filtration for transport and binding studies. Pages 1–24 in E. Carafoli and G. Semenza (eds.), *Membrane biochemistry: A laboratory manual on transport and bioenergetics.* Springer-Verlag, Berlin.

Kimmich, G. A. 1981. Intestinal absorption of sugar. Pages 1035–1061 in L. R. Johnson (ed.), *Physiology of the gastrointestinal tract.* Raven Press, New York.

Kingsolver, J. G., and H. A. Woods. 1997. Thermal sensitivity of growth and feeding in *Manduca sexta* caterpillars. *Physiological Zoology* 70:631–638.

Klaassen, M., and H. Biebach. 1994. Energetic fattening and starvation in the long-distance migratory Garden Warbler, *Sylvia borin*, during the migratory phase. *Journal of Comparative Physiology* B 164:362–371.

Landys, C.M.M., T. Piersma, and J. Jukema. 2003. Strategic size changes of internal organs and muscle tissue in the Bar-tailed Godwit during fat storage on a spring stopover site. *Functional Ecology* 17:151–179.

Lee, K. A., W. H. Karasov, and E. Caviedes-Vidal. 2002. Digestive responses to restricted feeding in migratory Yellow-rumped warblers. *Physiological and Biochemical Zoology* 75:314–323.

Lemos, F.J.A., F. S. Zucoloto, and W. R. Terra. 1992. Enzymological and excretory adaptations of *Ceratitis capitata* (Diptera: Tephritidae) larvae to high protein and high salt diets. *Comparative Biochemistry and Physiology A* 102:775–779.

Levey, D. J., and M. L. Cipollini. 1996. Is most glucose absorbed passively in northern bobwhite? *Comparative Biochemistry and Physiology A* 113:225–231.

Levey, D. J., and W. H. Karasov. 1989. Digestive responses of temperate birds switched to fruit or insect diets. *Auk* 106:675–686.

Levey, D. J., and W. H. Karasov. 1992. Digestive modulation in a seasonal frugivore, the American robin (*Turdus migratorius*). *American Journal of Physiology* 262:G711–G718.

Levey, D. J., and C. Martínez del Rio. 1999. Test, rejection, and reformulation of a chemical reactor-based model of gut function in a fruit-eating bird. *Physiological and Biochemical Zoology* 72:369–383.

Levey, D. J., A. R. Place, P. J. Rey, and C. Martinez del Rio. 1999. An experimental test of dietary enzyme modulation in pine warblers *Dendroica pinus. Physiological and Biochemical Zoology* 72:576–587.

Lucas, P. W. 1994. Categorisation of food items relevant to oral processing. Pages 197–218 in D. J. Chivers and P. Langer (eds.), *The digestive system in mammals: Food, form, and function.* Cambridge University Press, Cambridge.

Ma, T. Y., D. Hollander, R. Riga, and D. Bhalla. 1993. Autoradiographic determination of permeation pathway of permeability probes across intestinal and tracheal epithelia. *Journal of Laboratory and Clinical Medicine* 122:590–600.

Madara, J. L., and J. R. Pappenheimer. 1987. Structural basis for physiological regulation of paracellular pathways in intestinal epithelia. *Journal of Membrane Biology* 100:149–164.

Makanya, A. N., J. N. Maina, T. M. Mayhew, S. A. Tsachanz, and P. H. Burri. 1997. A stereological comparison of villous and microvillous surfaces in small intestines of frugivorous and entomophagous bats: Species, inter-individual and craniocaudal differences. *Journal of Experimental Biology* 200:2415–2423.

Malagelda, J. R., and F. Azpiroz. 1989. Determinants of gastric emptying and transit in the small intestine. Pages 409–437 in S. G. Schultz (ed.), *The handbook of physiology, section 6,* vol. IV. American Physiological Society, Bethesda, Md.

Martínez del Rio, C. 1990a. Dietary, phylogenetic, and ecological correlates of intestinal sucrase and maltase activity in birds. *Physiological Zoology* 63:987–1011.

Martínez del Rio, C. 1990b. Sugar preferences in hummingbirds: The influence of subtle chemical differences on food choice. *Condor* 92:1022–1030.

Martínez del Rio, C., H. G. Baker, and I. Baker. 1992. Ecological and evolutionary implications of digestive processes: Bird preferences and the sugar constituents of floral nectar and fruit pulp. *Experientia* 48:544–551.

Martínez del Rio, C., K. E. Brugger, J. L. Rios, M. E. Vergara, and M. C. Witmer. 1995. An experimental and comparative study of dietary modulation of intestinal enzymes in European starlings (*Sturnus vulgaris*). *Physiological Zoology* 68:490–511.

Martínez del Rio, C., S. J. Cork, and W. H. Karasov. 1994. Modeling gut function: An introduction. Pages 25–53 in D. J. Chivers and P. Langer (eds.), *The digestive system in mammals: Food, form and function.* Cambridge University Press, Cambridge, U.K.

Martínez del Rio, C., W. H. Karasov, and D. J. Levey. 1989. Physiological basis and ecological consequences of sugar preferences in cedar waxwings. *Auk* 106:64–71.

Martínez del Rio, C., J. E. Schondube, T. J. McWhorter, and L. G. Herrera. 2001. Intake responses in nectar feeding birds: Digestive and metabolic causes, osmoregulatory consequences, and coevolutionary effects. *American Zoologist* 41:902–915.

Mathews, C. K., and K. E. van Holde. 1996. *Biochemistry.* Benjamin/Cummings, Menlo Park, Calif.

McWhorter, T. J. 2005. Paracellular intestinal absorption of carbohydrates in mammals and birds. Pages 113–140 in J. M. Starck and T. Wang, (eds.), *Physiological and ecological adaptations to feeding in vertebrates.* Science Publishers, Enfield, NH.

McWhorter, T. J., and C. Martínez del Rio. 2000. Does gut function limit hummingbird food intake? *Physiological and Biochemical Zoology* 73:313–324.

McWilliams, S. R., E. Caviedes-Vidal, and W. H. Karasov. 1999. Digestive adjustments in Cedar Waxwings to high feeding rate. *Journal of Experimental Zoology* 283:394–407.

McWilliams, S. R., and W. H. Karasov. 2001. Phenotypic flexibility in digestive system structure and function in migratory birds and its ecological significance. *Comparative Biochemistry and Physiology A* 128:579–593.

McWilliams, S. R., and W. H. Karasov. 2005. Migration takes guts: digestive physiology of migratory birds and its ecological significance. Pages 67–78 in P. Mara and R. Greenberg (eds.), *Birds of two worlds.* Smithsonian Institution Press, Washington, D.C.

Miquelle, D. G. 1990. Why don't male bull moose eat during the rut. *Behavioral Ecology and Sociobiology* 27:145–151.

Munck, B. G. 1981. Intestinal absorption of amino acids. Pages 1097–1122 in L. R. Johnson (ed.), *Physiology of the gastrointestinal tract.* Raven Press, New York.

Nicolson, S. W. 2002. Pollination by passerine birds: Why are the nectars so dilute. *Comparative Biochemistry and Physiology B* 131: 645–652.

Nicolson, S. W., and B. E. Van Wyk. 1998. Nectar sugars in Proteaceae: Patterns and processes. *Australian Journal of Botany* 46:489–504.

Obst, B. S., and J. M. Diamond. 1989. Interspecific variation in sugar and amino acid transport by the avian cecum. *Journal of Experimental Zoology Supplement* 3:117–126.

O'Connor, T. J. 1995. Metabolic characteristics and body composition in House Finches: Effects of seasonal acclimatization. *Journal of Comparative Physiology B* 165:298–305.

Ortiz, R. M., C. E. Wade, and C. L. Ortiz. 2001. Effects of prolonged fasting on plasma cortisol and TH in postweaned northern elephant seal pups. *American Journal of Physiology* 280:R790–R795.

Overgaard, J., J. B. Andersen, and T. Wang. 2002. The effects of fasting duraton on the metabolic response in Python: An evaluation of the energetic costs associated with gastrointestinal growth and upregulation. *Physiological and Biochemical Zoology* 75:360–368.

Pappenheimer, J. R. 1987. Physiological regulation of transepithelial impedance in the intestinal mucosa of rats and hamsters. *Journal of Membrane Biology* 100:137–148.

Pappenheimer, J. R. 1993. On the coupling of membrane digestion with intestinal absorption of sugars and amino acids. *American Journal of Physiology* 265:G409–G417.

Pappenheimer, J. R. 1998. Scaling of dimensions of small intestines in non-ruminant eutherian mammals and its significance for absorptive mechanisms. *Comparative Biochemistry and Physiology A* 121:45–58.

Pappenheimer, J. R. and K. Z. Reiss. 1987. Contribution of solvent drag through intercellular junctions to absorption of nutrients by the small intestine of the rat. *Journal of Membrane Biology* 100:123–136.

Penry, D. L., and P. A. Jumars. 1987. Modeling animal guts as chemical reactors. *American Naturalist* 129:69–96.

Piersma, T. 1998. Phenotypic flexibility during migration: optimization of organ size contingent on the risks and rewards of fueling and flight. *Journal of Avian Biology* 29:511–520.

Piersma, T., M. W. Dietz, A. Dekinga, S. Nebel, J. van Gils, P. F. Battley, and B. Spaans. 1999a. Reversible size-changes in stomachs of shorebirds: When, to what extent, and why? *Acta Ornithologica* 34:175–181.

Piersma, T., and R. E. J. Gill. 1998. Guts don't fly: Small digestive organs in obese bartailed godwits. *Auk* 115:196–203.

Piersma, T., G. A. Gudmundsson, and K. Lilliendahl. 1999b. Rapid changes in the size of different functional organ and muscle groups during refueling in a long-distance migrating shorebird. *Physiological and Biochemical Zoology* 72:405–415.

Piersma, T., A. Koolhaas, and A. Dekinga. 1993. Interactions between stomach structure and diet choice in shorebirds. *Auk* 110:552–564.

Piersma, T., J. van Gils, P. De Goeij, and J. Van Der Meer. 1995. Holling's functional response model as a tool to link the food-finding mechanism of a probing shorebird with its spatial distribution. *Journal of Animal Ecology* 64:493–504.

Place, A. R. 1996. Birds and lipids: Living off the fat of the earth. *Poultry and Avian Biology Reviews* 7:127–141.

Powell, D. W. 1987. Intestinal water and electrolyte transport. Pages 1267–1306 in L. R. Johnson, J. Christensen, M. J. Jackson, and J. H. Walsh (eds.), *Physiology of the gastrointestinal tract.* Raven Press, New York.

Randall, D., W. Burggren, and K. French. 1997. *Animal physiology: mechanisms and adaptations.* W. H. Freeman and Co., New York.

Reynolds, S. E. 1990. Feeding in caterpillars: maximizing or optimizing food acquisition? Pages 106–118 in J. Mellinger (ed.), *Animal nutrition and transport processes. 1. Nutrition in wild and domestic animals,* vol. 5. Karger, Basel.

Ricklefs, R. E., 1996. Morphometry of the digestive tracts of some passerine birds. *Condor* 98:279–292.

Ricklefs, R. E., and J. Travis. 1980. A morphological approach to the study of avian community organization. *Auk* 97:321–338.

Rings, E. H., E. H. van Beers, S. D. Krasinski, and H. A. Buller. 1994. Lactase—Origins, gene expression, localization, and funtion. *Nutrition Research* 14:775–797.

Robbins, C. T. 1993. *Wildlife feeding and nutrition.* Academic Press, San Diego, Calif.

Roby, D. D., K. L. Brink, and A. R. Place. 1989. Relative passage rates of lipid and aqueous digesta in the formation of stomach oils. *Auk* 106:303–313.

Roche, R. K., and D. E. Wheeler. 1997. Morphological specializations of the digestive tract of *Zacryptocerus rohweri (Hymenoptera: Formicidae). Journal of Morphology* 234:253–262.

Rogowitz, G. L. 1998. Limits to milk flow and energy allocation during lactation of the hispid cotton rat (*Sigmodon hispidus*). *Physiological Zoology* 71:312–320.

Ryan, C. A. 1979. Proteinase inhibitors. Pages 559-618 in G. A. Rosenthal and D. H. Janzen (eds.), *Herbivores: Their interaction with secondary plant metabolites.* Academic Press, New York.

Sabat, P., J. A. Lagos, and F. Bozinovic. 1999. Test of the adaptive modulation hypothesis in rodents: Dietary flexibility and enzyme plasticity. *Comparative Biochemistry and Physiology A* 123:83–87.

Sahi, T. 1994. Genetics and epidemiology of adult-type hypolactasia. *Scandinavian Journal of Gastroenterology* 202 (supplement):7–20.

Schondube, J. E., L. G. Herrera, and C. Martinez del Rio. 2001. Diet and evolution of digestion and renal function in phyllostomid bats. *Zoology* 104:59–73.

Schondube, J. E., and C. Martinez del Rio. 2003. Concentration-dependent sugar preferences in nectar-feeding birds: Mechanisms and consequences. *Functional Ecology* 17:445–453.

Scriber, J. M. and F. Slansky. 1981. The nutritional ecology of immature insects. *Annual Review of Entomology* 26:183–211.

Secor, S. M., and J. Diamond. 1995. Adaptive responses to feeding in Burmese pythons: Pay before pumping. *Journal of Experimental Biology* 198:1313–1325.

Secor, S. M., and J. M. Diamond. 2000. Evolution of regulatory responses to feeding in snakes. *Physiological and Biochemical Zoology* 73:123–141.

Shinbo, H., K. Konno, C. Hirayama, and K. Watanabe. 1996. Digestive sites of dietary proteins and absorptive sites of amino acids along the midgut of the silkworm, *Bombyx mori*. *Journal of Insect Physiology* 42:1129–1138.

Simoons, F. J. 1969. Primary adult lactose intolerance and the milking habit: A problem in biological and cultural interrelations. I. Review of the medical research. *American Journal of Digestive Diseases* 14:819–836.

Simoons, F. J. 1970. Primary adult lactose intolerance and the milking habit: A problem in biological and cultural interrelations. II. A culture historical hypothesis. *American Journal of Digestive Diseases* 15:695–710.

Snipes, R. L. 1994. Morphometric methods for determining surface enlargement at the microscopic level in the large intestine and their application. Pages 234–263 in D. J. Chivers and P. Langer (eds.), *The digestive system in mammals: Food, form, and function.* Cambridge University Press, Cambridge.

Snipes, R. L. 1997. Intestinal absorptive surface in mammals of different sizes. *Advances in Anatomy, Embryology and Cell Biology* 138:1–88.

Snipes, R. L., and A. Kriete. 1991. Quantitative investigation of the area and volume in different compartments of the intestine of 18 mammalian species. *Zeitschrift fün Saugetierkunde* 56:225–244.

Starck, J. M. 1996. Intestinal growth in the altricial European starling (*Sturnus vulgaris*) and the precocial Japanese quail (*Coturnix coturnix japonica*). A morphometric and cytokinetic study. *Acta Anatomies* 156:289–306.

Starck, J. M., and K. Beese. 2001. Structural flexibility of the intestine of Burmese python in response to feeding. *Journal of Experimental Biology* 204:325–335.

Starck, J. M. and K. Beese. 2002. Structural flexibility of the small intsetine and liver of garter snakes in response to feeding and fasting. *Journal of Experimental Biology* 205:1377–1388.

Starck, J. M., W. H. Karasov, and D. Afik. 2000. Intestinal nutrient uptake measurements and tissue damage. Validating the everted sleeves method. *Physiological and Biochemical Zoology* 73:454–460.

Stein, E. D., and J. M. Diamond. 1989. Do dietary levels of pantothenic acid regulate its intestinal uptake in mice? *Journal of Nutrition* 119:1973–1983.

Stevens, C. E. and I. D. Hume. 1995. *Comparative physiology of the vertebrate digestive system.* Cambridge University Press, Cambridge.

Sunshine, P., and N. Kretchmer. 1964. Intestinal disaccharidases: absence in two species of sea lions. *Science* 144:850–852.

Toloza, E. M., and J. Diamond. 1992. Ontogenetic development of nutrient transporters in rat intestine. *American Journal of Physiology* 263:G593–G604.

Toloza, E. M., and J. M. Diamond. 1990. Ontogenetic development of transporter regulation in bullfrog intestine. *American Journal of Physiology* 258:G770–G773.

Tsukita, S., and M. Furuse. 2000. Pores in the wall: Claudins constitute tight junction strands containing aqueous pores. *Journal of cell Biology* 149:13–16.

Ugolev, A. M., B. Z. Zaripov, N. N. Iezuitova, A. A. Gruzdkov, I. S. Rybin, M. I. Voloshenovich, A. A. Nikitina, M. Y. Punin, and N. T. Tokagaev. 1986. A revision of current data and views on membrane hydrolysis and transport in the mammalian small intestine based on a comparison of techniques of chronic and acute experiments: experimental re-investigation and critical review. *Comparative Biochemistry and Physiology A* 85:593–612.

Uhing, M. R., and R. E. Kimura. 1995. The effect of surgical bowel manipulation and anesthesia on intestinal glucose absorption in rats. *Journal of Clinical Investigation* 95:2790–2798.

van Gils, J. A., T. Piersma, A. Dekinga, and M. W. Dietz. 2003. Cost-benefit analysis of mollusc-eating in a shorebird II. Optimizing gizzard size in the face of seasonal demands. *Journal of Experimental Biology* 206:369–3380.

Verzar, F., and E. J. McDougall. 1936. *Absorption from the intestine.* Longmans and Green, London.

Vonk, H. J., and R. H. Western. 1984. *Comparative biochemistry and physiology of enzymatic digestion.* Academic Press, London.

Waldschmidt, S. R., S. M. Jones, and W. P. Porter. 1987. Reptilia. Pages 553–619 in T. J. Pandian and F. J. Vernberg (eds.), *Animal energetics volume 2. Bivalvia through Reptila.* Academic Press, San Diego, Calif.

Warner, A. C. I. 1981. Rate of passage of digesta through the gut of mammals and birds. *Naturally Abstract and Review* 51:789–820.

Weiner, J. 1992. Physiological limits to sustainable energy budgets in birds and mammals: ecological implications. *Trends In Ecology and Evolution* 7:384–388.

Weiss, S. L., E. A. Lee, and J. Diamond. 1998. Evolutionary matches of enzyme and transporter capacities to dietary substrate loads in the intestinal brush border. *Proceedings of The National Academy of Sciences* 95:2117–2121.

Whelan, C. J., J. S. Brown, K. A. Schmidt, B. B. Steele, and M. F. Willson. 2000. Linking consumer-resource theory and digestive physiology: Application to diet shifts. *Evolutionary Ecology Research* 2:911–934.

Withers, P. C. 1992. *Comparative animal physiology.* Saunders College Publishing, Fort Worth.

Witmer, M. C. 1999. Do avian frugivores absorb fruit sugars inefficiently? How dietary nutrient concentration can affect coefficients of digestive efficiency. *Journal of Avian Biology* 30:159–164.

Witmer, M. C., and P. J. Van Soest. 1998. Contrasting digestive strategies of fruit-eating birds. *Functional Ecology* 12:728–741.

Woods, H. A., and J. G. Kingsolver. 1999. Feeding rate and the structure of protein digestion and absorption in lepidopteran midguts. *Archives of Insect Biochemistry and Physiology* 42:74–87.

Wright, S. H., and G. A. Ahearn. 1997. Nutrient absorption in invertebrates. Pages 1137–1205 in W. Dantzler (ed.), *Handbook of physiology section 13: Comparative physiology*. Oxford University Press, New York.

Yarbrough, C. G. 1971. The influence of distribution and ecology on the thermoregulation of small birds. *Comparative Biochemistry and Physiology A* 39:235–266.

Young-Owl, M. 1994. A direct method for measurement of gross surface area of mammalian gastro-intestinal tracts. Pages 219–233 in D. J. Chivers and P. Langer (eds.), *The digestive system in mammals: Food, form, and function*. Cambridge University Press, Cambridge.

Zimmerman, L. C., and C. R. Tracy. 1989. Interactions between the environment and ectothermy and herbivory in reptiles. *Physiological Zoology* 62:374–409.

Zwarts, L., and A. M. Blomert. 1992. Why knot *Calidris canutus* take medium-sized *Macoma balthica* when six prey species are available. *Marine Ecology Progress Series* 83:113–128.

Zwarts, L., A. M. Blomert, and J. H. Wanink. 1992. Annual and seasonal variation in the food supply harvestable by knot *Caldris canutus* staging in the Wadden Sea in late summer. *Marine Ecology Progress Series* 83:129–139.

Zwarts, L., and S. Dirksen. 1990. Digestive bottleneck limits the increase in food intake of whimbrels preparing for spring migration from the Banc D'Arguin, Mauritania. *Ardea* 78:257–278.

The whole is something over and above its parts, and not just the sum of them all.

— Aristotle

═ CHAPTER FIVE ═

Photosynthetic Animals and Gas-Powered Mussels: The Physiological Ecology of Nutritional Symbioses

WE LIVE IN a symbiotic world. The two most fundamental ecological processes, primary production and aerobic respiration, are the result of metabolic pathways that plants and animals acquired by symbiosis. Therefore without symbiotic interactions, life in this blue planet would be completely different. There would be no algae, no corals, no trees and grasses, and without them there would not be herbivores. Symbioses have shaped life. These are bold claims and this chapter is devoted to justifying them, and also to justifying why symbiosis must be the subject of our attention as physiological ecologists. First, however, we must define what we mean by symbiosis. The term derives from the Greek term *symbio,* which means living together. Symbioses are biotic interactions in which the interacting species live together in prolonged physical association. The term "prolonged" is, of course, relative and can only be gauged in relation to the generation time of the interacting organisms.

Although many ecologists consider many biotic interactions in which physical associations are fleeting and superficial, such as pollination and seed dispersal, as symbioses, we will adopt a more restrictive meaning. The symbiotic interactions that we will focus on are both long term and intimate. They are so intimate that the symbiont lives within the host. In some symbioses, the microorganisms are housed within the host but outside of the host's cells. Good examples of extracellular symbioses are fermentative digestive symbioses in which the symbionts live within the hosts' guts. More cozy relationships are the endosymbioses found among insects and bacteria in which the symbionts (which are now called endosymbionts) live within the cell membrane of specialized host's cells.

We devote chapter 6 to the partnership animals have established with other organisms to digest plant refractory materials such as cellulose. These symbioses, which we can call fermentative digestive symbioses, have received the attention of ecological physiologists for many years. Because the ecological importance of fermentative symbiosis is immense, this attention is justified. However, fermentative digestive symbioses are just a subset of a broader set of phenomena that can be characterized as nutritional symbioses. These are the associations that are the theme of this chapter. Because the symbionts in these associations are microbes, physiological ecologists have ignored them. If you take a look within almost any comparative physiology or ecological physiology textbook, you will not find the word "symbiosis." Most nutritional symbioses are studied by microbiologists and by a few bold ecophysiologists, most of whom work literally under water (see subsequent sections in this chapter). We believe that nutritional symbioses are too important for ecological physiologists to ignore. Accordingly, rather than following a traditional path and restricting our attention to fermentative digestion, we placed this phenomenon in the larger context of nutritional symbiosis between animals and microbes. This chapter begins with a cursory review of perhaps the most important symbiotic event in the history of life: the endosymbioses that led to the origin of eukaryotes. We then sketch several spectacular examples of nutritional symbioses, emphasizing their physiological characteristics. The next chapter gives a detailed account of the ecological and evolutionary importance of fermentative digestive symbioses.

5.1 A Symbiotic World

5.1.1 Endosymbiosis and the Origin of Eukaryotes

Photosynthesis and respiration, the two most globally important biological processes, were invented by prokaryotes. Indeed, prokaryotes invented most, if not all, important metabolic pathways. Eukaryotes acquired photosynthesis and respiration through symbiosis. We now believe that the organelles in eukaryotic cells that conduct oxidative energy metabolism (mitochondria) and photosynthesis (plastids including chloroplasts) arose from prokaryotic organisms that established endosymbioses with the ancestor of eukaryotes. Although many aspects of this hypothesis are still unknown, the body of evidence that supports it is persuasive: Both mitochondria and plastids have a single circle of DNA, which presumably represents the remnant of a bacterial chromosome. Furthermore, both mitochondria and plastids have ribosomes that resemble those of prokaryotes and which are sensitive to antibacterial antibiotics. Streptomycin blocks protein synthesis in bacteria as well as in chloroplasts and mitochondria.

Molecular evidence suggests that the mitochondrial genome arose from that of a purple bacterium and the plastid genome arose from that of a cyanobacterium (figure 5.1). These endosymbiotic prokaryotes lost the ability for an independent life and became reduced to the state of serving particular metabolic functions (oxidative energy metabolism and photosynthesis) within their hosts. Certainly, when the ancestors of plastids and mitochondria were free living, they must have possessed all the genes necessary for an independent life. But today, their genomes are greatly reduced. During evolution, organelles appear to have relinquished many of their genes to their hosts and learned to import back the products of these genes (Martin and Herrmann 1998). Although endosymbiosis adequately accounts for the origin of mitochondria and plastids, it has been much more difficult to trace the origins of the eukaryotic nucleus and cytoskeleton (which is composed or microtubules and associated proteins).

Figure 5.1. Although it is tempting to depict the genealogical relationships of the major groups of organisms as a tree as in (A), a more accurate representation recognizes that the branch anastomoses that result from endosymbiotic events create a tangled web (B). The origin of eukaryotes is the result of one or maybe two (not shown) fusions of archaeal and bacterial lineages. Diagram B also shows the later acquisition of chloroplasts in plants. Diagram A is based on sequences of the small subunit ribosomal RNA. In B, bold lines represent gene genealogies and broken arrows are potential lateral gene transfers. (After Katz 1998.)

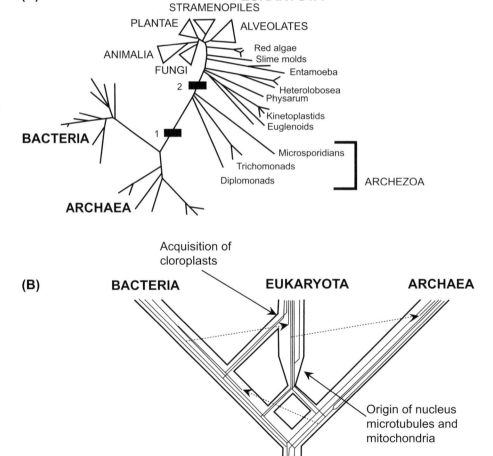

The notion that organelles were once free-living bacteria has a long and contentious history (Sapp 1994). Speculation about the endosymbiotic origin of plastids can be traced to the middle of the nineteenth century, but was fully developed by Mérejkovsky (1920), who claimed priority of the idea—apparently inappropriately (Sapp 1994). At about the same time that Mérejkovsky was busy staking intellectual priority, Portier (1918) not only proposed that the ancestors of mitochondria were bacterial symbionts, but also developed an all-embracing theory for the role of symbioses in biology. While many of the details of Portier's theory are wrong, it contains many elements that are markedly contemporary. He emphasized the autotroph/heterotroph dichotomy that ecologists have embraced wholeheartedly, and called for the development of a physiological and symbiotic microbiology to complement the emphasis on pathology that characterized the microbiology of his time. He would be pleased to see the emphasis that contemporary microbiology textbooks place on both microbial physiology and the role of microbes as symbionts.

The speculations of Mérejkovsky and Portier were not widely accepted by their contemporaries and languished for over 70 years. They would have probably languished for longer without the auspicious conjunction of two factors: a feisty biologist willing to champion the cause in the face of strong opposition, and the development of molecular methods to test them. We owe to Lynn Margulis the contemporary form of the endosymbiotic theory for the origin of eukaryotic cells. We also owe to her its wide dissemination, both within and outside academic science. Briefly, Margulis proposed the serial endosymbiotic theory (SET; Margulis 1993), which hypothesizes a series of symbiotic unions to account for the origin of eukaryotes. In the first one, the nucleus and cytoplasm arose from an archaebacterium (let us call this symbiont S1) that joined a spirochete-like eubacterium (S2). S1 contributed most of the protein-making machinery and S2 contributed the elements of the cytoskeleton. Both partners contributed the structures that led to the nucleus. This primitive eukaryote then acquired another endosymbiont, this time a purple bacterium capable of aerobic metabolism (S3). Photosynthesis was acquired in one eukaryotic lineage by endosymbiosis with a cyanobacterium (S4). The evidence for S1, S3, and S4 is compelling, but the evidence for S2, the putative spirochete ancestor of microtubules, remains tenuous. With characteristic chutzpah, Margulis has predicted that her spirochete hypothesis will be proved right within a decade (Margulis 1998). So far this has not happened, but she has been right many times before—against all odds.

SET is not the only hypothesis for the evolution of eukaryotes. Its main contestant, the hydrogen hypothesis, proposed by Martin and Müller (1998), eliminates S3. It poses that the ancestral merger of a eubacterium with broad metabolic capacity and an archaebatrium, led, in one step, to the ancestral eukaryote and to the evolution of mitochondria (Henze et al. 2001). Because the reputed eubacterium endosymbiont was a facultative anaerobe that had both the capacity to use aerobic metabolism to catabolize carbohydrates and, in the absence of oxygen, the capacity to catabolize the pyruvate resulting from glycolysis into CO_2, acetylcoenzyme-A, and H_2, this hypothesis is called the hydrogen hypothesis. The presumed host was an anaerobe that used H_2 and CO_2 to synthesize carbohydrates and to produce ATP through methanogenesis (see chapter 8). There are many appealing elements in the hydrogen hypothesis. For example, it explains the biochemical similarity of mitochondria and hydrogen-producing organelles, called hydrogenosomes, found in some anaerobic protists such as *Trichomonas*. Hydrogenosomes lack DNA but contain proteins with amino acid sequences that are found only in mitochondria and purple bacteria (Bui et al. 1996). It is premature to cast one's vote for either SET or the hydrogen hypothesis of eukaryotic origins. It is safe to predict, however, that the surviving hypothesis will have endosymbiosis as a central ingredient (Katz 1998).

There are at least two reasons why the endosymbiotic theory of eukaryote origins is of importance to ecological physiologists. The first one is that it reveals the chimerical nature of the organism that we study. By chimerical, we do not mean imaginary, which is the most widely used sense of the term. We mean made of the parts of several organisms like the chimera of Greek myth. The chimera was a fire-breathing monster that had the body of a goat, the head of a lion, and a serpent for a tail. The chimera was slain by the hero Bellerophon while he was riding another chimerical creature, the winged horse Pegasus. Animals are deeply chimerical. Recent molecular research has revealed that our genomes are a chimerical medley (Gupta and Golding 1996; Timmis et al. 2004). In addition, as we will describe in following sections, many, if not most, animals maintain symbioses that shape their ecological roles. The creatures involved in a close symbiosis act as a single ecological and evolutionary unit. They act as chimeras with unique morphological and physiological characteristics. This leads us to the second reason for paying attention to endosymbiotic theory.

This theory illustrates a fundamental feature of all symbioses: symbioses are sources of novel metabolic capacities (table 5.1). In a single symbiotic merger with a cyanobacterium, an ancient eukaryote acquired photosynthesis and spawned the radiation of plants. Establishing a symbiotic association allows organisms to acquire the genomes that code for whole metabolic pathways. Because these pathways have become central as mechanisms that mediate how animals acquire resources, the study of symbioses should be of pivotal interest to ecological physiologists. Chemist and occasional art historian Ugo Bardi tracked the history of the Greek chimera to the winged lion Anzu of the

TABLE 5.1
Novel metabolic and nutritional capacities acquired by animals in symbiotic relationships with microorganisms

Metabolic capacity acquired by symbiosis	Donor	Recipient
Photosynthesis	Alage and cyanobacteria	Several protists and invertebrates
Chemosynthesis	Bacteria	Several invertebrates
Nitrogen fixation	Bacteria	Termites
Essential amino acids and vitamins	Bacteria	Many animals, but especially insects on nutrient-deficient diets
Methanogenesis	Bacteria	Anaerobic protists
Methanotrophy	Bacteria	Mussels, sponges, and pogonophorans
Cellulose degradation	Bacteria and clams	Vertebrates, termites

Note: A detailed account of non-nutritional symbioses, such as bioluminescence, and of symbioses involving fungi and plants symbioses is given in Douglas (1994) from which this table was modified.

Sumerians. Anzu was the mount of Inanna, the Sumerian great goddess of fertility (http://www.unifi.it/surfchem/solid/bardi/chimera.htm). Like Anzu, the chimeras that result from biological symbioses are vehicles of fertile biological diversity.

5.2 A Diversity of Nutritional Symbioses

5.2.1 Photosynthetic Animals?

Heterotrophs are organisms that use organic materials as sources of energy and nutrients. We often think of animals as heterotrophic, which is strictly correct, but limiting. Many animals have established symbioses with autotrophic organisms and function ecologically as autotrophs. This leads to a peculiar and nicely interdisciplinary situation. Biologists who study autotrophic animals, such as corals, publish their research in both zoological and botanical journals. Their methodological tools are a medley of the tools used by plant and animal biologists. Recall that autotrophs come in two metabolic types: photoautotrophs, which obtain energy from light (figure 5.2) and chemoautotrophs, which oxidize inorganic compounds (often, but not exclusively, hydrogen sulfide) to obtain energy for synthesis of organic compounds. Microbiologists also call chemoautotrophs, chemolithotrophs or more simply lithotrophs. This section is devoted to the symbiosis of animals with photosynthetic organisms and the next one is about chemolithotrophic symbioses. Table 5.2 provides a taxonomy of the metabolic processes used by living organisms.

Animals have established symbioses with a diverse group of organisms with "oxygenic" (oxygen-generating) photosynthesis (figure 5.2). A few purple and green bacteria are anoxygenic phototrophs that use hydrogen as a source of electrons and CO_2 or organic compounds as sources of carbon. Because to our knowledge, these beasts have not established symbioses with animals, we will not consider them further. The photosynthetic organisms that have established symbioses with animals are generically, and incorrectly, called "algae." They are members of two groups of microorganisms: eukaryotic algae and prokaryotic cyanobacteria. These symbiotic organisms are diverse taxonomically and have evolved from free-living taxa many times. In the ocean, the most widespread (but by no means the only) symbiotic photoautrophs are dinoflagellates in the genus *Symbiodinium*, which are generically called zooxanthellae when they are members of symbioses. In freshwater systems, the chlorophyte *Chlorella* is the most common photosynthetic symbiont (Douglas 1994). Their heterotrophic partners include various protists (foramininferans, radiolarians, and ciliates), sponges, cnidaria (hydras, corals, and sea anemones), mollusks, and flatworms.

Given their ecological importance, it is perhaps not surprising that corals are the best-studied photosynthetic animals. Corals are a specious group of reef-forming cnidarians in the order Scleractinia. They live in shallow, warm, transparent, and well-illuminated oligotrophic waters. They cover some 2 million square kilometers of tropical oceans with an ecosystem of seemingly boundless diversity and spectacular beauty. The existence of coral reefs is the consequence of a symbiosis of the cnidarians with *Symbiodinium* dinoflagellates. The symbiotic association between these photosynthetic organisms and reef-forming cnidarians provides the driving energy for the high calcification rates typical of corals (Gattuso et al. 1999). Although corals are just one example of photosynthetic animals, they have been relatively well studied. Thus, many of the examples in this section will be about them.

5.2.2 Photosynthetic Symbioses: Protection and Nutrients

In most animal-photoautotroph associations, the photosynthetic partner is restricted to certain well-defined anatomical regions of the animal host. In cnidaria, such as corals, the zooxanthellae are almost always located within the cells of the inner (endodermal) layers and are enclosed by a membranous

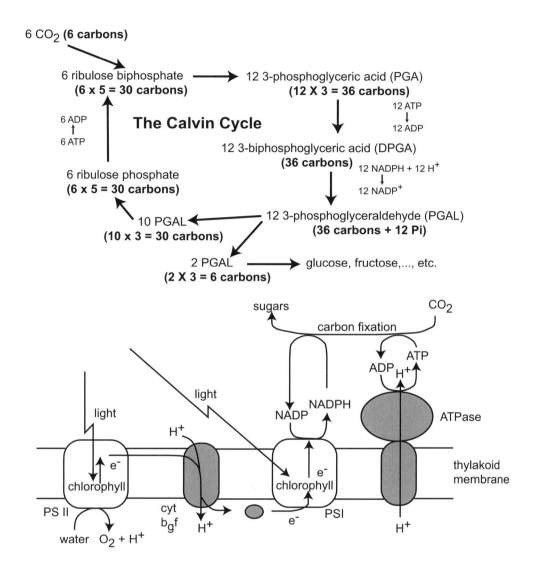

6 CO$_2$ **(6 carbons)**

6 ribulose biphosphate
(6 x 5 = 30 carbons) → 12 3-phosphoglyceric acid (PGA)
(12 X 3 = 36 carbons)

12 ATP
↓
12 ADP

6 ADP
↑
6 ATP

The Calvin Cycle

12 3-biphosphoglyceric acid (DPGA)
(36 carbons) 12 NADPH + 12 H$^+$
↓
12 NADP$^+$

6 ribulose phosphate
(6 x 5 = 30 carbons)

10 PGAL ← 12 3-phosphoglyceraldehyde (PGAL)
(10 x 3 = 30 carbons) **(36 carbons + 12 Pi)**

2 PGAL → glucose, fructose,..., etc.
(2 X 3 = 6 carbons)

Figure 5.2. Higher plants and algae, as well as cyanobacteria and their relatives, convert CO$_2$ (carbon dioxide) to organic material by reducing this gas to carbohydrates in a complex set of reactions. A simplistic equation summarizing these reactions is

$$CO_2 + 2H_2O + light \rightarrow O_2 + (CH_2O) + H_2O.$$

The parentheses indicate that CH$_2$O is not a molecule but a subunit of a sugar. Electrons for this reduction reaction ultimately come from water, which is then converted to oxygen and protons in a water-splitting reaction that turns engineers green with envy. The energy for this process is provided by light, which is absorbed by pigments (primarily chlorophylls and

(caption continues)

(caption continued)

carotenoids). The initial electron transfer (charge separation) reaction in the photosynthetic reaction center sets into motion a long series of redox (reduction-oxidation) reactions, passing the electron along a chain of cofactors and filling up the "electron hole" on the chlorophyll. In PS I, electrons are transferred eventually to NADP, turning it into NADPH which can be used for carbon fixation. The flow of electrons from water to NADP requires light and is coupled to the generation of a proton gradient across the photosynthetic membrane membrane. This proton gradient drives the synthesis of ATP. The ATP and NADPH that resulted from the light reactions are used for CO_2 fixation in a process that is independent of light. CO_2 fixation involves a number of reactions that are referred to collectively as the Calvin-Benson cycle:

$$6CO_2 + 12NADPH + 12H_2O + 18ATP \longrightarrow C_6H_{12}O_6 + 12NADP + 18ADP + Pi$$

The initial CO_2 fixation reaction involves the enzyme ribulose-1,5-bisphosphate carboxylase/oxygenase (RuBisCO), which can react with either oxygen (leading to a process named photorespiration and not resulting in carbon fixation) or with CO_2 (see chapter 8). We recommend Lambers et al.'s (1998) plant physiological ecology textbook for a more thorough introduction to photosynthesis than the one we can give here.

TABLE 5.2

Characterization of the metabolism of an organism by combining the terms that represent the source energy, carbon, and electrons with the term troph (from the Greek *trophos*, which means to eat)

Ingredient	Source	Term
Energy	Light	Photo
	Chemical	Chemo
Carbon	Inorganic	Auto
	Organic	Hetero
Electrons	Inorganic	Litho
	Organic	Organo

Note: As examples, a photosynthetic organism can be described as a photoautolithotroph, whereas an animal can be described as a chemoheteroorganotroph. The sulfur-oxidizing bacteria that form symbioses with animals at hydrothermal vents can be characterized as chemoautolithotrophs. Many organisms can use more than one source and hence are called mixotrophs (after van Dover 2000).

structure called the symbiosome (figure 5.3). In contrast, in the giant tridacnid clams of the Indo-Pacific reefs, they are extracellular and located in branching pouches of the gastrointestinal tract underneath the mantle tissue (figure 5.3). In all cases, the symbiont is located in thin sheets of tissue with the high surface-to-volume ratio needed to capture light. The structures used by hosts to house symbionts are often interpreted as evidence of the mutualistic nature of symbioses (mutualisms are ecological associations in which both interacting species derive benefits).

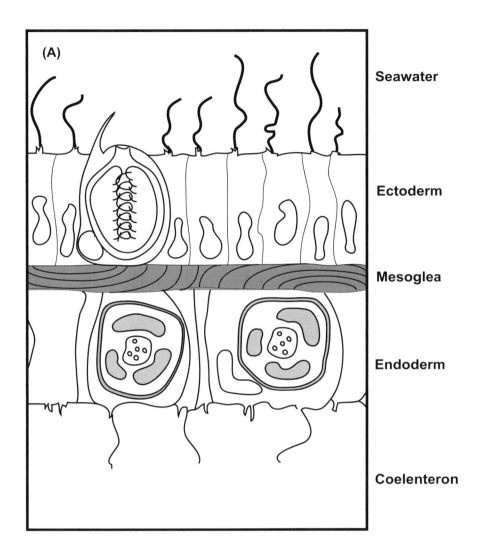

Figure 5.3. Animals that form symbiotic partnerships with photosynthetic organisms often have specialized structures to house them. The body of anthozoans is like a bag. A mouth surrounded by tentacles opens up the internal cavity, called coelenteron, to seawater. The walls of the bag are made of two, single-cell-thick, layers, the ectoderm and the endoderm, which are separated by a layer of connective tissue called the mesoglea. In most anthozoans, the zooxanthellae are housed in the endoderm surrounding the mouth. Each zooxanthellae is surrounded by a persymbiotic membrane. Panel A is after Allemand et al. (1998). In giant clams (B), the zooxanthellae are found within specialized tubules which originate from the stomach and which permeate the clam's mantle tissue. Panel B is after Legatt et al. (2002).

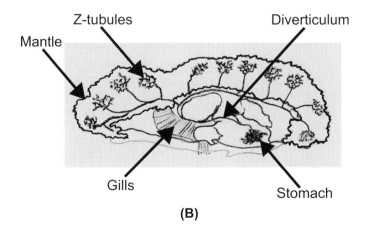

This may be too rosy an interpretation and one that can obscure the origins of these associations. Symbioses may arise from a variety of ecological interactions that include mutualism, but that also include parasitism and predation. Rather than emphasizing the mutualistic nature of nutritional symbioses, we should probably examine the factors and traits that permitted the long-term persistence and success of the association. As the evolution of specialized symbiont-housing structures attests, one of these factors is the protection of symbionts. Another nice example of a symbiont protection trait is the color of corals. The sometimes brightly colored pigments of corals dissipate the potentially damaging energy of light wavelengths of low photosynthetic activity and reflect potentially damaging UV-B and infrared light (Salih et al. 2000).

In nutritional symbioses, one of the primary factors that favors the success of a symbiosis is the reciprocal transfer of materials between host and symbiont. Photosynthetic symbionts potentially contribute to the nutrition of their hosts in two ways: (1) by providing organic products of photosynthesis, and (2) by assimilating the nitrogen wastes of the host and transforming them into useful compounds such as essential amino acids (see section 5.4). Many radiotracer studies (box 5.1) have revealed that the carbon synthesized by symbiotic algae is released into the host's tissues. In corals and giant clams, zooxantellae transfer from 78% to 96.6% of the carbon that they fix to their host (figure 5.4). When light levels are high, the energy contained in the organic compounds donated by the zooxanthellae to the host is sufficient to fuel all the corals' metabolic energy needs (figure 5.4). When the corals are shaded, however, the zooxanthellae contribute only $\approx$ 40% of the coral's needs (figure 5.4). In the nudibranch *Pteraolidia ianthina*, the transfer of photosynthates depends on both the population density of symbiotic zooxanthellae and light availability. In the winter, when irradiance is low and zooxanthellae are scanty, the symbionts translocate little organic carbon to their hosts and use their photosynthates to grow. In the summer, when symbiotic zooxanthellae populations are large and irradiance is high, their translocated photosynthates can satisfy all of the nudibranchs' energy needs (Hoegh-Guldberg et al. 1986). Organic carbon is translocated from photosynthetic symbionts to hosts, mostly, but not exclusively, in the form of a few low-molecular-weight compounds (Muscatine 1990). *Symbiodinium* zooxanthellae release glucose, organic acids (such as succinic and fumaric acid), and maybe amino acids or their precursors (Whitehead and Douglas 2003), whereas freshwater symbiotic *Chlorella* release sucrose and maltose (Douglas 1994).

Johnson et al. (1995) used compound-specific isotopic analyses (see chapter 8) to determine whether the fatty acids in the giant clam *Tridacna gigas* were synthesized by the clam or contributed by the zooxanthellae. This approach was possible because the fatty acids in the zooxanthellae are synthesized directly from acetate, rather than from acetyl-coenzyme A derived from pyruvate.

Box 5.1. Radiotracer Studies

These are commonly done to detect the fate of a compound. Here we describe their use to trace the fate of dissolved inorganic carbon in photosynthetic symbioses. Muscatine and Cernichiari (1969) presented the first direct evidence that zooxanthellae contributed carbon fixed by photosynthesis to a host. Briefly, these authors incubated a sea coral with radiolabeled sodium bicarbonate ($NaH^{14}CO_3$). They then separated the tissues of the animal from those of the zooxanthellae and measured radioactivity. They found significant radioactivity in the organic matter of the host's tissues. Trench (1971) used radioautography to identify the compounds transferred from the zooxanthellae to a sea anomone (*Anthopleura elegantissima*) and a zoanthid (*Palythoa townsleyi*). He separated the soluble fraction of tissues of hosts incubated with $NaH^{14}CO_3$ by chromatography and exposed the chromatograms to medical X-ray film. He found that glycerol was the major product transferred from the zooxanthellae to the hosts, but he also found labeled alanine, glucose, and a variety of other compounds. Although radiotracer studies allowed determining that photosynthates were transferred from zooxanthellae to the hosts, and gave a preliminary estimate of the magnitude of this transfer, these studies had a serious limitation: they underestimated transfer. The reason is that much of the carbon transferred to the animal host can be respired and either expelled or refixed by the zooxanthellae. The more involved methods used currently to estimate the flux of carbon fixed by zooxanthellae to hosts are described in detail by Muscatine (1990).

Whitehead and Douglas (2003) developed an ingenious method to determine the identity of the materials translocated from the symbionts to the host. The approach can be called the metabolite comparison approach. The approach requires a battery of experiments. In the first experiment, the chemical profile of substances found in an animal that has been supplied with ^{14}C-labeled bicarbonate is determined. Let us identify the radiolabeled substances found in the host's tissue as *A*, *B*, *C*, and *D*. In a subsequent series of experiments, the photosynthetic capacity of the zooxanthellae is blocked. (This can be done, for example, by exposing the animal to the commercial herbicide DCMU (dichlorophenyldimethyl urea)). Then the animal is exposed to a ^{14}C-labeled candidate mobile compound (let us call this candidate compound *S*). If *S* is a mobile compound, after incubation, the tissues of the animal will show some of the substances derived from it. If these substances correspond to all, or to a subset, of substances, *A*, *B*, *C*, and *D*, then there is reason to believe that substance *S* is a mobile compound that is translocated from the symbionts to the host. Using this approach, Whitehead and Douglas (2003) determined that glucose, succinate, and fumarate are important photosynthetic compounds transferred from *Symbiodinium* cells to the tissues of the anemone *Anemonia viridis*.

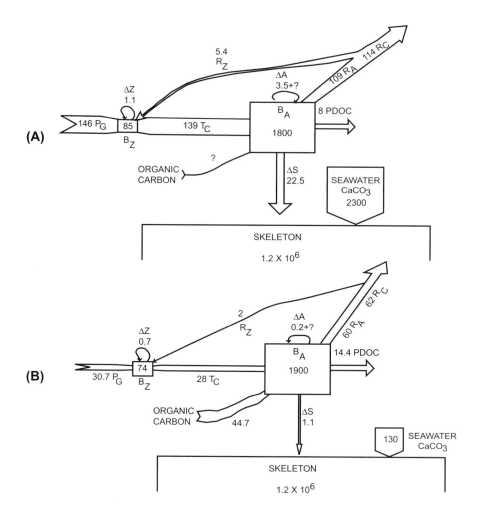

Figure 5.4. Falkowski and his collaborators (1984) constructed detailed carbon budgets for two colonies of the coral *Stylophora pistillata* growing either in the sun (A) or in the shade (B). The contribution of carbon fixed by zooxanthellae to the budget of the coral differed dramatically depending on light conditions. Both gross photosynthetic rate (P_G in μg of carbon per cm[2] per day) and the transfer of carbon from zooxanthellae to hosts (T_C) were higher in the light than in the dark. T_C was more than sufficient to satisfy the respiration (R_A) of a coral with a biomass (B_A) of ≈ 1800 μg of C per cm[2] in the light, but not in the shade. The respiration of zooxanthellae ($R_Z = R_C − R_A$) was also higher in the light than in the shade. This figure has two notable features: (1) the biomass of corals grew (ΔA) about an order of magnitude faster in the light than in the shade, and corals in the light deposited skeleton (ΔS) at 20 times the rate of those in the shade. However, corals in the shade excreted (or secreted) particulate or dissolved organic carbon at higher rates in the shade than in the light. More than 90% of the carbon fixed by zooxanthellae was transferred to the

(caption continues)

(caption continued)

host. This fraction is similar to that of chloroplasts within plants. However, the trophic level production efficiency, defined as the ratio of growth in the coral relative to gross photosynthesis of the zooxanthellae, was very low. A very large percentage of the carbon translocated was respired. Falkowski and collaborators (1984) argue that the low trophic production efficiencies of corals are the result of the high C:N ratios of the photosynthates produced by zooxanthellae. Zooxanthellae provide an energy source that must be supplemented with nitrogen-rich zooplankton or dissolved nitrogen.

Hence, these fatty acids are not depleted in ^{13}C ($\delta^{13}C$ ranged from −15.7 to −10.1). In contrast, the clam uses the "normal" lipid synthesis pathway, which involves the decarboxylation of pyruvate, which discriminates against ^{13}C (see section 8.2.2). Hence, the clam produces isotopically "light" fatty acids ($\delta^{13}C$ ranged from −14.2 to −25.2). As figure 5.5 indicates, the carbon isotopic composition of myristic (14:0) and palmitic acid (16:0) was indistinguishable between the host and the clams. All other fatty acids had more negative $\delta^{13}C$. Johnson et al. (1995) interpreted these results as evidence for direct translocation of myristic and palmitic acid from the zooxanthellae to the host. Curiously, the carbon isotopic composition of the essential unsaturated fatty acid γ-linolenic acid (18:3ω6) differed significantly between the clam and the zooxanthellae, suggesting that this nutrient was derived from heterotrophic filter feeding.

5.2.3 Light and CO_2 as Resources for Animals

Light and CO_2 are not often considered limiting resources for animals. Indeed, most animal physiology textbooks describe in detail how animals get rid of CO_2, not how they acquire it. Also, most animal physiology textbooks largely ignore light except as a source of radiant energy for thermoregulation. Because light and CO_2 are crucial resources in photosynthetic symbioses, we cannot afford to ignore them. Let us consider CO_2 first. Dissolved inorganic carbon (DIC) can be found in three forms: dissolved CO_2 ($CO_2 + H_2CO_3$), bicarbonate (HCO_3^-) and carbonate ions (CO_3^{--}). You may sometimes encounter the following nomenclature for DIC:

$$DIC = CO_2 + HCO_3^- + CO_3^{--} = \sum CO_2$$

For standard surface seawater conditions, most of the carbon ($\approx 90\%$) is as bicarbonate. CO_3^{--} and dissolved CO_2 contribute only 10% and less than 1%, respectively. Free-living marine microalgae inhabit an environment with a large

Figure 5.5. (A) The carbon isotopic composition of myristic (14:0) and palmitic acid (16:0) in the mantle of the giant clam *Tridacna gigas* is, within measurement error, equal to that of the same fatty acids in symbiotic zooxanthellae. In contrast, all other fatty acids have more negative $\delta^{13}C$ values. Clams probably obtain these two fatty acids from their symbionts. Closed squares are values for the zooxanthellae whereas open squares are values for the clam. The profiles of relative abundances of different fatty acids in zooxanthellae (closed bars) and clams (open bars) are shown in panel B. (After Johnson et al. 1995.)

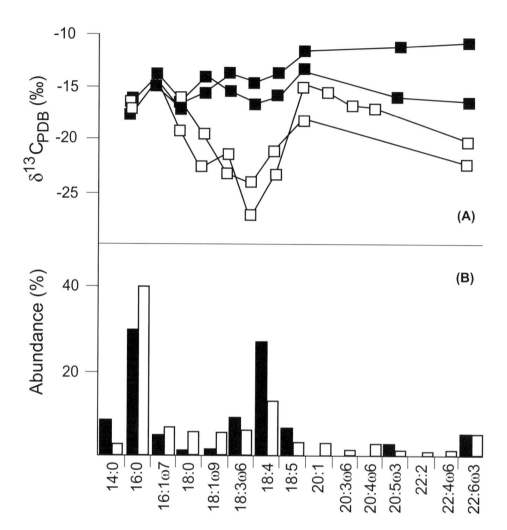

and relatively constant DIC pool. In contrast, algae that live in symbiotic associations do not have immediate access to the seawater DIC pool. Inorganic carbon (Ci) in all its guises must first pass through the host's tissues, and thus photosynthetic symbionts depend on their hosts to provide sufficient inorganic carbon.

The carbon concentration mechanisms (CCM) of both corals and giant clams appear to be different. Leggat et al. (2002) proposed the following model for giant clams: Ci moves from seawater to the zooxanthellae down a CO_2 gradient. This gradient is favored by the presence of carbonic anhydrase (CA) on the epithelial surface of both the mantle and the gills (figure 5.6A). At high HCO_3^- concentrations, carbonic anhydrase catalyzes the conversion of bicarbonate into CO_2:

$$HCO_3^- + H^+ \xrightleftharpoons{CA} CO_2 + H_2O.$$

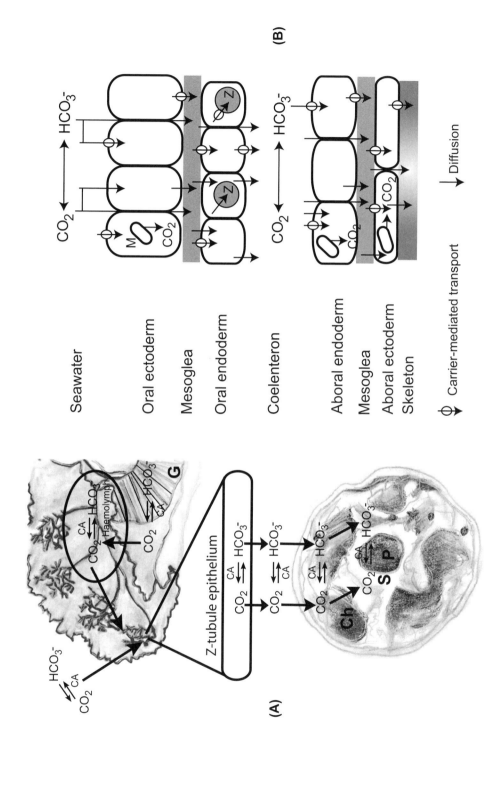

Figure 5.6. (A) Animals that house photosynthetic symbionts often have carbon concentration mechanisms to supply them with a source of dissolved inorganic carbon. Giant clams use carbonic anhydrase (CA) to hydrolyze abundant dissolved bicarbonate in seawater. The activity of CA produces a gradient that facilitates the flow of CO_2 into the zooxanthellae housed in z tubules. CA in the epithelium of the z tubules may also facilitate the entrance of CO_2 from hemolymph into the space occupied by zooxanthellae. (B) Scleratinian corals appear to concentrate HCO_3^- by transepithelial transport aided by protein carriers. These carriers may also supply the carbon needed to synthesize $CaCO_3$ that makes the coral's skeleton. (After Leggat et al. 2002; Allemand et al. 1998.)

CA creates a high concentration of CO_2 in the epithelial surfaces, which facilitates its diffusion into hemolymph. The delivery of CO_2 from the hemolymph into the z-tubules in which the zooxanthellae are housed is facilitated by the depletion of CO_2 in zooxanthellae that results from photosynthesis—and maybe by CA. Unlike giant clams that seem to use CO_2 preferentially as a substrate for photosynthesis, corals use bicarbonate as a preferred substrate. Figure 5.6B illustrates a simplified version of the CCM proposed by Allemand et al. (1998) for corals. These authors propose that HCO_3^- is transported across the ectoderm and into the endoderm by a combination of diffusion (which contributes only a small amount of the total DIC flux) and by several carrier-mediated processes. Although many of the molecular details of the processes proposed by Leggat et al. (2002) and Allemand et al. (1998) remain to be elucidated, their research has three clear findings: (1) the photosynthetic rate of zooxanthellae can be limited by the availability of Ci; (2) to increase Ci availability, hosts have carbon-concentrating mechanisms that differ between giant clams and corals; and (3) unlike most marine invertebrates which must dispose of CO_2, animals involved in photosynthetic symbioses must have mechanisms to acquire it.

Photosynthesis can be limited by either CO_2 or by light. Following Beer's law, light intensity (I_z) is attenuated exponentially with depth (z):

$$I_z = I_0 \, e^{-kz} \tag{5.1}$$

The attenuation coefficient k is a fudge parameter that combines the effect of two processes: the scattering and the absorption of light. The absorption properties of the ocean can be described by the joint absorption coefficients of many components including those of water, suspended particles, phytoplankton, detritus, and dissolved substances. Because each of these components has a distinct absorption spectrum, in addition to changes in the intensity of light, the spectral distribution of light changes with depth (figure 5.7). It is perhaps not surprising that coral reefs are only found in the euphotic zone, the upper, well-illuminated region of aquatic ecosystems (Falkwoski et al. 1990).

To understand the effects of light on the physiological ecology of photosynthetic symbioses, we must introduce one of the favorite tools of plant physiological ecologists: the light-response curve (box 5.2). The light-response curve is the decelerating relationship that relates the rate of photosynthesis measured as the rate at which either CO_2 is assimilated or O_2 is generated, and the light irradiance. Although the equations that describe the light-response curve will look unfamiliar, we hope that its general form will remind you of a functional response (chapter 4). As illustrated in box 5.2, plant physiological ecologists routinely use light-response curves to examine the integrated photosynthetic

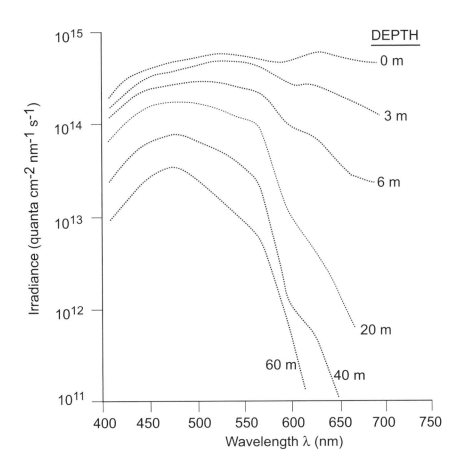

Figure 5.7. Light intensity declines exponentially with depth and its spectral distribution changes. This diagram is for a reef in Jamaica at noon on a sunny day. Note that it only includes the visible portion of the electromagnetic spectrum. Both vision and photosynthesis depend on this small portion of the spectrum. (After Dustan 1979.)

responses of plants to environments with contrasting light intensities. It is perhaps worth emphasizing that light-response tools are not ends in themselves. They provide an integrated, whole-leaf picture of the photosynthetic response. Further research must be done to elucidate the morphological and physiological mechanisms that shape differences in this response. These mechanisms include changes in the thickness and ultrastructure of leaves, and differences in the amounts and types of photosynthetic pigments (specifically chlorophyll *a* and *b*; Lambers et al. 1998).

Just as plant physiological ecologists are interested in finding out how the photosynthetic processes of plants respond to changes in the light environment (box 5.2), coral biologists are interested in finding out how corals respond to variation in light. Their research has unveiled remarkable similarities between plants and corals. Like plants, corals that grow under different light environments tend

Figure 5.8. (A) The rate of
assimilation of CO_2 by a
photosynthetic organism
increases in a decelerating
fashion with irradiance.
Panel A also shows the
geometrical meaning of the
parameters of equation 5.2.
(B) Plants grown at con-
trasting irradiances often
show differences in the
relationship between CO_2
assimilation and irradiance.

Box 5.2. The Light-Response Curve

The rate at which plants assimilate CO_2 (or generate O_2) as a result of photosynthesis increases asymptotically with increased irradiance (panel A of
figure 5.8, modified from Lambers et al. 1998). The relationship between
irradiance and photosynthetic rate is called the light-response curve. Irradiance is usually measured with instruments that are sensitive only to photons
with wavelengths between 400 and 700 nm. When this is the case, irradiance is called the photosynthetic photon flux (or photosynthetic photon flux
density) and is measured in μmol of light quanta per second (or sometimes
in μ Einsteins; an Einstein is 1 mole of quanta). Plant physiologists standardize photosynthetic rate per unit surface area of leaf. Coral biologists
sometimes standardize photosynthetic rate by coral area, but sometimes by
mg of coral biomass. Below the compensation point (I_C), irradiance is insufficient to compensate for CO_2 release (alternatively, the O_2 generated is

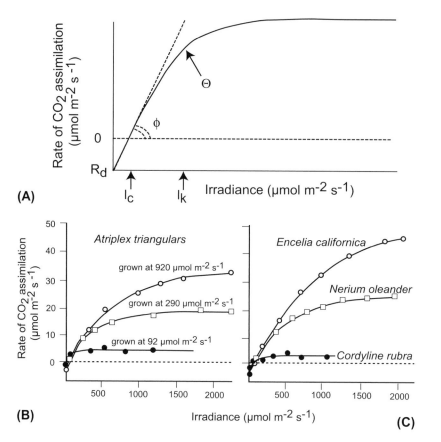

continued

continued

insufficient to compensate for the O_2 consumed). At low irradiances the rate of photosynthesis increases roughly linearly with irradiance. The initial slope of this line (φ) based on *absorbed* light describes the efficiency with which light is converted to fixed carbon and is called the quantum yield. The quantum yield estimates how much CO_2 is fixed per quantum of light. When the light response curve is based on *incident* light, this slope is called the apparent quantum yield. Plant physiologists and coral biologists use a variety of different equations to describe the relationship between rate of photosynthesis (P) and irradiance (I). Plant physiologists often use a non-rectangular hyperbola:

$$P(I) = \frac{\phi I + P_{max} - \sqrt{\{\phi I + P_{max}) - 4\Theta I P_{max}\}}}{2\Theta} - R \qquad (5.2.1)$$

where P_{max} is the maximal photosynthetic rate, R is the respiration rate, and Θ is a curvature factor that can vary from 0 to 1. Two other equations used by coral biologists (and sometimes by plant physiologists) are

$$P(I) = P_{max} \tanh\left(\frac{I}{I_k}\right) - R \qquad (5.2.2)$$

and

$$P(I) = P_{max} \tanh\left(\frac{\phi I}{P_{max}}\right) - R, \qquad (5.2.3)$$

where I_k is the irradiance at which a line of slope φ through the origin intersects P_{max}. Although these three equations give good fits and have interpretable parameters, they are phenomenological rather than constructed from first principles. It may not surprise you to learn that some rogue plant physiologists and coral biologists use the equivalent of Ivlev's equation to describe the relationship between photosynthetic rate and irradiance:

$$P = P_{max}\left(1 - \exp\left(\frac{I}{I_k}\right) - R\right). \qquad (5.2.4)$$

The two lower panels of figure 5.8 represent two situations commonly studied by plant physiological ecologists. Panel B shows light response curves for *Atriplex triangularis*, a plant from coastal salt marshes, grown

continued

continued

at three different light intensities. When data are standardized by leaf area, plants growing in the shade typically have lower P_{max} and lower R values, which lead to lower light-compensation points and higher rates of photosynthesis at low irradiances. When data are standardized by leaf mass, however, these differences often disappear. The reason is that plants grown in the shade often have thinner leaves with higher chlorophyll content than plants grown in full sun. The photosynthetic response of plants of a single species grown under different light conditions is paralleled by that of plants adapted (in an evolutionary sense) to grow either in the sun or in the shade. Panel C compares the light responses of species that occur naturally at high (*E. californica*), intermediate (*N. oleander*), low irradiance levels (*C. rubra*). Panels B and C were modified from Lambers et al. (1998).

to differ in morphology, concentration of photosynthetic pigments, and the overall response of the coral colony to light (reviewed by Barnes and Chalker 1990). The light-response curves of corals show the following changes with the light environment at which the corals grow: (1) the initial slope (ϕ) of the curve increases with decreased irradiance, (2) I_k increases with increasing irradiance (the photosynthetic rate of corals that grew in the shade saturates at lower irradiances), and (3) corals grown under low irradiances have lower compensation irradiances both because respiration is lower and because ϕ is higher (figure 5.9).

The responses of corals to variation in irradiance are lumped under the generic term "photoadaptation." This term encompasses time scales of minutes to hundreds of years, and phenomena that range from behavioral responses in the polyps to genetic differences in both the corals and their zooxanthellae (reviewed by Falkowski et al. 1990). We find the umbrella use of the term "photoadaptation" unfortunate. As Helmuth et al. (1997) have pointed out, it is more appropriate to distinguish between true photoadaptation and photoacclimatization. In this more nuanced terminology, photoadaptation would refer to a genetic change in the population structure of a coral or its symbionts resulting from the selective pressures exerted by different light environments. Photoacclimatization would refer to a physiologically plastic response of a coral or a clone of zooxanthellae inhabiting a coral to changing light conditions. Perhaps coral biologists should be wise to adopt the phenotypic plasticity terminology proposed in chapter 1. When dealing with corals it is easy to forget that each individual is a conglomerate of two, or as we will soon see, sometimes

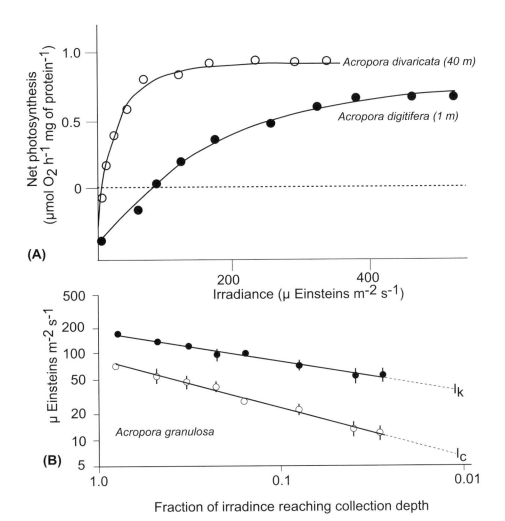

Figure. 5.9. The light-response curves of corals that live in shallow, well-illuminated waters (*Acropora digitifera*, A) and in deep water (*Acropora divaricata*) are different. The latter show reduced respiration rates, steeper curves, and lower compensation irradiances. The Australian coral *Acropora granulosa* grows at depths that range from 1 to 40 m. This coral shows, within a single species, the same trends found in interspecific comparisons. Both I_k and I_C decrease with depth in a remarkably predictable fashion. The *x* axis in panel B is the fraction of surface irradiance that reaches the depth at which corals were collected. (After Chalker et al. 1983.)

several, partners. The plastic response to light of the whole unit is the combined result of individual responses.

5.2.4 When It Is Dark, Eat! Phenotypically Flexible Heterotrophy in Corals

Although the responses of corals to light are similar to those of plants, there is one response that corals can adopt that most plants cannot. When light is limiting, corals can become more heterotrophic. Anthony and Fabricius (2000) noted that in many coastal areas inhabited by corals, there are periods during which water is turbid as a result of high concentrations of suspended particulate matter (SPM or "seston"). SPM is a catchall term for the stuff suspended in the water column, which includes zooplankton, phytoplankton, bacteria, dead algae, mucus, and feces. Although relative to zooplankton SPM is nutritionally poor, it is abundant and hence it is a potential source of nutrients. Because SPM attenuates light, coral photosynthesis is a negative function of its concentration. SPM is both a scourge and a potential resource. Anthony and Fabricius (2000) teased apart the effect of SPM on photosynthesis and heterotrophic feeding with a comprehensive experiment. They maintained colonies of two coral species (*Goniastra retiformis* and *Porites cylindrica*) in the light or in the shade and in water with four levels of SPM. In statistical parlance, they used a 2 × 4 completely randomized factorial design with two factors (light and SPM), two levels of light (low and high), and four levels of SPM.

Goniastra retiformis colonies responded to low light intensities by both typical photoacclimation, and by more than doubling the rate at which they ingested SPM (figure 5.10). In contrast, *P. cylindrica* did not photoacclimate significantly and did not increase its rate of SPM ingestion in the shade. Although *G. retiformis* experienced significant photosynthetic acclimation in the shade, it did not compensate for reduced light levels entirely, and respiration exceeded photosynthetic rate. However, at the high SPM treatment heterotrophic feeding compensated fully for the lower photosynthetic rate. Due to its phenotypic plasticity in photosynthesis and heterotrophy, *G. retiformis* gained tissue and skeletal mass at all treatments. *Porites cylindrica*, in contrast, fared well only in the treatments in which it was fully exposed to the sun.

The results of Anthony and Fabricius (2000) experiments illustrate two notable features of the light biology of corals: (1) Some corals have the ability to respond to changes in the light environment by adjusting photosynthesis and/or heterotrophic feeding, but some do not. Coral species differ in phenotypic plasticity. (2) Some corals are capable of fully compensating for losses in photosynthetic capacity due to low light levels by increasing their heterotrophic feeding capacity. Perhaps more importantly, the experimental results of Anthony and Fabricius (2000) pose novel questions and thus open interesting new avenues for research. What are the physiological and morphological traits that are up-regulated in the shade and that facilitate the increase in feeding capacity? Do polyps have the ability to up-regulate levels of digestive enzymes and transporters and/or the capacity of the coelenteron? Anthony and Fabricius (2000) documented dramatic changes in the functional response of *G. retiformis*

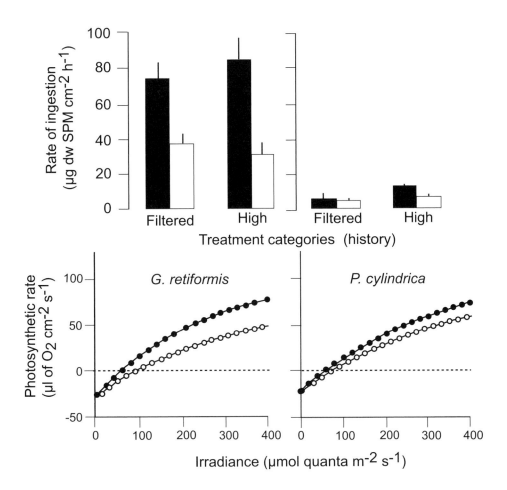

Figure 5.10. Not all corals respond to changes in irradiance in the same way. Anthony and Fabricius (2000) exposed coral colonies of two species (*Goniastra retiformis* and *Porites cylindrica*) to combinations of two light treatments (in the sun and in the shade) and several levels of suspended dissolved matter (SDM). *Goniastra retiformis* responded to reduced light exposure by both changes in its photosynthetic response and by changes in its ability to feed on SDM. After 2 months in the shade colonies had lower I_C (the irradiance at which photosynthesis rate equals respiration rate) and higher maximal photosynthetic rates than colonies in full sun. The light-response curves with closed dots are for shaded corals and those with open dots are for corals in the sun. Colonies in the shade also had higher SDM ingestion rates. Although corals in the shade (represented in the upper panel by the solid bars) had almost twice the ingestion rate of those in the sun (open bars), corals grown in the presence of SDM (labeled as "High" in the figure) had the same ingestion rate as those grown in filtered water. Increased feeding rate appeared to be a response to low light rather than to SDM availability. In contrast, neither light nor SDM levels had a significant effect on the light response curves and ingestion rates of *P. cylindrica*. (Figures are from data in Anthony and Fabricius 2000.)

under their different experimental treatments. Are these changes due to changes in "handling time" or in digestive ability? Does phenotypic plasticity in corals influence the range of habitats that different species can inhabit? Anthony and Fabricius (2000) speculate that species with low heterotrophic capacity and generally low trophic plasticity like *P. cylindrica*, may maintain growth only in places with consistently low turbidity. In contrast, more phenotypically plastic species like *G. retiformis* may prosper in both turbid environments and clear-water reefs.

5.2.5 Coral Bleaching: A Disease or an Adaptive Response

The stability of coral-zooxanthellae symbioses seems sometimes shaken by dramatic episodes of a condition called bleaching that sometimes occurs in very large areas and at epidemic scales (Glynn 1996). The symptoms of bleaching include a gradual loss of color that sometimes leaves corals bone white. This decoloration is the result of the loss of zooxanthellae. Coral bleaching has received significant attention, because it may be one of the many consequences of global climatic change (Dubinsky and Stambler 1996). Bleaching is experienced not only by corals, but also by other reef animals that have symbiotic relationships with zooxanthellae, such as soft corals, giant clams, and some sponges. Although the causes of bleaching are the subject of some debate, the factors most commonly associated with bleaching are elevated sea temperature and irradiance.

Rowan and his collaborators have clarified the possible causes of bleaching, and in the process have challenged a long-held view of coral biology (Rowan and Knowlton 1995; Rowan et al. 1997). Coral biologists usually assume a one-host, one-symbiont point of view in which host taxa alone are adequate units of study (Trench 1993). In previous sections, we adopted this view. We assumed that host-symbiont collectives can respond, as one, to changes in the environment by a variety of cooperative physiological acclimatization mechanisms. Rowan and his colleagues demonstrated that, at least for some corals, this view is wrong. Some corals maintain symbiotic relationships with diverse assemblages of symbionts. Using molecular markers of genetic differences (restriction fragment length polymorphisms in small ribosomal subunit RNA genes), they found that some corals host more than one symbiont taxa. The ecologically dominant Caribbean corals *Monstraea annularis* and *M. flaveolata*, each host up to four distantly related taxa of *Symbiodinium* zooxanthellae. Tersely, Rowan et al. (1995) christened their taxa *Symbiodinium A, B, C, D*, and *E*. Even more surprising than finding more than one symbiont per host was Rowan et al.'s (1995) finding of distinctive ecological zonation in each of the *Symbiodinium* taxa (figure 5.11). Taxon *C* inhabits deeper

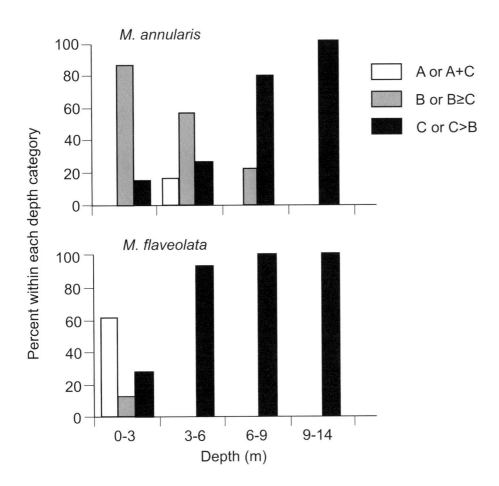

Figure 5.11. Three different zooxanthellae taxa, as defined by molecular markers, vary in relative abundance with depth in two Caribbean corals in the genus *Montastrea* (Rowan et al. 1997). Taxa *A* and *B* predominate at shallow depths and are replaced by *C* in deeper water. Many coral colonies had more than one symbiotic taxon.

waters than taxa *A* and *B*. Rowan et al. (1997) could even detect this zonation at very small spatial scales. Areas exposed to the sun had taxa *A* and *B*, whereas shaded zones only centimeters away had predominantly taxa *C*.

Rowan et al. (1997) witnessed a bleaching episode that revealed the potential importance of the polymorphic symbioses that they had documented. In 1995, temperatures rose above summer maxima and water was unusually clear in the Caribbean for several weeks, leading to unusually high irradiance. The corals bleached but the pattern of bleaching was not spatially homogeneous. Bleaching was rare or slight at both shallow (less than 2 m) and deep (more than 15 m) sites. Furthermore, colonies in shallow waters bleached preferentially in shaded places, whereas colonies in deep water bleached in unshaded places. Some colonies exhibited a ring of bleaching between the top and the side. What appears to have happened was that *Symbiodinium C* was expelled selectively at the limits of its distribution, which may correspond to the limits of its irradiance tolerance at high temperatures. In shallow waters, its abundance was reduced in low-irradiance areas. In deep waters it was reduced in areas of high irradiance.

Buddenmeier and Fautin (1993) proposed a hypothesis to explain bleaching and Rowan et al. (1997) contributed a potential mechanism. The "adaptive bleaching hypothesis" postulates that bleaching allows corals to replace their zooxanthellae with different ones that work better in the new conditions. Testing the adaptive bleaching hypothesis requires two ingredients: In response to changes in irradiance and temperature, (1) corals must get rid of the zooxanthellae taxa that perform poorly, and (2) they must eventually acquire symbionts that function better under the new situation (Baker 2003). Rowan et al. (1997) have provided evidence of selective loss of zooxanthellae taxa, and Rowan (2004) has demonstrated that temperature influences the photosynthetic capacity of different *Symbiodinium* taxa differently. *Symbiodinium D* seems to be heat tolerant and outperforms other taxa at high temperatures. Toller et al. (2001) experimentally reduced (defoliated?) the density of zooxanthellae in corals, and allowed these corals to be repopulated. The results were perplexing. When corals retained some zooxanthellae, they were repopulated by the same strain that they originally had, which is not surprising. However, when the corals were severely depleted, they were always repopulated by zooxanthellae that were atypical for their habitat. Toller et al. (2001) speculate that bleached corals acquire the appropriate set of symbionts only after the equivalent of an ecological succession.

Baker and collaborators (Baker et al. 2004) used a broad geographical scale approach to examine the adaptive bleaching hypothesis. They compared the *Symbiodinium* composition of coral reefs that regularly experience high temperatures. They also compared the composition of reefs before and several years after a bleaching episode. They found that the heat-tolerant *Symbiodinium D* was dominant at sites that experience high water temperatures frequently. They also found that *Symbiodinium D* became more abundant after bleaching episodes. They concluded that symbiont changes are a common feature of severe bleaching. They speculate that these shifts in the community of symbionts are "adaptive" in that they confer recovering reefs with some heat resistance that reduces the possibility of future bleaching.

Independently of whether Bauddenmeir and Fautin's (1993) hypothesis is correct, the findings described here are important because they reveal that there is neither a single taxon of symbiotic zooxanthellae, nor a unique obligate relationship between a host and a symbiont. At least some symbiotic corals contain communities of symbionts, whose taxa composition seems to be shaped by light and temperature. The recognition of diversity in the symbionts of corals complicates our interpretation of their "physiological" responses, but provides a potential mechanism to decipher a curious paradox. Coral reefs seem to be vulnerable to short-term environmental change, yet they have persisted over hundreds of millions of years through rapid and sometimes extreme environmental change. Maybe the possibility of creating symbiotic combinations gives corals not only ecological plasticity, but the evolutionary plasticity to weather global climatic change (see Baker 2003; Baker et al. 2004).

5.3 Hot Vents and Cold Seeps: Chemolithotrophs of the Deep Sea

In the previous section we explored symbiotic relationships that happen close to the ocean's surface. It is time now to go deep to explore the symbioses that sustain prolific biological communities in hydrothermal vents and cold seeps, far down in the deep sea, beyond the reach of sunlight. These communities are of profound interest to ecologists because their trophic structure appears to be based primarily—although probably not exclusively—on a peculiar primary production pathway. They are based on chemoautotrophy rather than on photosynthesis (or photoautotrophy).

Most biologists almost automatically assume that photosynthetic creatures form the productive base of ecosystems. This notion is largely true on land and in the photic, well-illuminated, zone of aquatic ecosystems. Although for a long time it was believed that it also applied to the vast deep sea, the notion was shaken in the 1980s. In 1977, scientists on board the deep sea submersible *Alvin* discovered hot springs at a depth of 2.5 km, on the Galapagos Rift (spreading ridge) off the coast of Ecuador. Because geologists had predicted that hot springs (geothermal vents) should be found at the active spreading centers along the mid-oceanic ridges, this discovery was not really a surprise. What was a surprise, however, was the discovery of almost incredibly abundant, and decidedly unusual, sea life. *Alvin*'s crew found giant tubeworms, huge clams, and ghostlike crabs prospering in densely packed communities around these hot springs.

The existence of productive communities in the Galapagos Rift was contrary to the then prevailing view of the deep ocean as a dark, featureless, and unproductive expanse where the microscopic bodies of dead planktonic creatures accumulate in kilometer-deep layers of muck. This picture of the ocean is not completely wrong. It is just incomplete. Although the deep sea is far from a biological desert, for the most part it is not a productive place. The many animals that live in it—and many do—rely on the energy-poor remains of organic materials that rain slowly from above. *Alvin*'s dive into the Galapagos Rift revealed that the deep ocean also contains spectacular islands of productivity. Since 1977 many more thriving hydrothermal vent communities have been discovered. In addition, similarly productive communities associated with "cold seeps" that release hydrocarbons (mostly methane) and/or hydrogen sulfide, have also been found (van Dover 2000). How can these thriving communities exist in the almost complete absence of light?

The dense populations of creatures that were discovered during the *Alvin* trips to the Galapagos Rift were first described as suspension feeders. Lonsdale (1977) hypothesized that these abundant organisms fed either on materials from the photic zone concentrated by currents around vents, or on suspended chemoautotrophic bacteria that rely on dissolved sulfide as a source of electrons (box 5.3). These bacteria are indeed abundant at vents, and are probably important producers. However, the nutritional ecology of the beasts that populate hydrothermal vents proved to be far stranger than anybody ever expected. At first, the chemoautotrophic free-living bacteria hypothesis was supported. The carbon isotopic composition of the mussel *Bathymodiolus thermophylus* ($\delta^{13}C \approx$ $-33‰$) was very different from that of marine seston (which ranges in $\delta^{13}C$ from $-21‰$ to $-15‰$; Rau and Hedges 1979), and similar to that of the few values of chemoautotrophic bacteria available at the time. Thus, at first, it appeared that free-living chemiosynthetic bacteria were at the producer base of hydrothermal vent ecosystems. Then, the first reports of the morphology of the giant worms and clams that inhabit the vents emerged.

5.3.1 Stable Isotopes and the Evidence for Symbioses in Riftia

Among the most bizarre creatures brought to light by the deep-sea expeditions to the Galapagos Rift were giant worms. These worms were exceptionally large. They were up to 2 m long and up to 4 cm thick. Their anterior ends, sporting blood-red tentacular plumes, protruded spectacularly from a maze of tubes. Their size and plume gave them their Latin name *Riftia pachyptila,* which can be loosely translated as fat (*pachy*) feathered (*ptila*) creature of the rift (*Riftia*). *Riftia pachyptila* is a close relative of pogonophorans, which are small benthic worms phylogenetically related to annelids. Because pogonophorans have no mouths, guts, or anuses, when they were discovered biologists assumed that they absorbed nutrients directly from the water. Like their pogonophoran relatives, *Riftia* lacks a digestive tract (figure 5.12), but a combination of morphological, biochemical, and stable isotopic evidence contradicted the suggestion that *Riftia* relies on the epithelial absorption of dissolved organic substances for sustenance. The evidence pointed in a different, and very surprising, direction. The giant

Box 5.3.

To place the sulfide-based deep sea symbioses in context, it is useful to consider where this sulfide comes from (throughout the book we use the term sulfide to refer to S^{2-} or its related compounds H_2S and HS^-; total sulfide is often denoted by ΣH_2S). Sulfide has two possible origins. It can be biogenic and produced by the anaerobic catabolism of organic substances and the reduction of sulfate by bacteria (Boetius et al. 2000), or it can be inorganic. Under the anaerobic conditions of sediments, the sulfate (SO_4^{2-}) contained in seawater is converted to sulfide:

$$SO_4^{2-} + 2[CH_2O] \longrightarrow S^{2-} + 2CO_2 + H_2O.$$

Boetius et al. (2000) provide novel evidence that suggests an alternative to this process. This alternative is important from a geochemical perspective because it may explain why little of the methane contained in vast stores under the ocean floor escapes into seawater. Methane is a potent greenhouse gas. Were it to escape, it would cause massive global warming (DeLong 2000). Briefly, Boetius et al. (2000) propose that methane-oxidizing archaea interact with sulfate-reducing bacteria. The archaea carry on the following reaction:

$$CH_4 + 2H_2O \longrightarrow CO_2 + H_2.$$

This reaction is thermodynamically favorable only if H_2 is removed from the system very rapidly. This is possible because H_2 produced by the anaerobic methane eaters becomes the energy substrate for the hydrogen-oxidizing sulfate reducers:

$$H^+ + 4H_2 + SO_4^{2-} \longrightarrow HS^- + 4H_2O$$

The sum of the cooperative processes of this seemingly tightly coupled bacterial consortium is

$$CH_4 + SO_4^{2-} \longrightarrow HCO_3^- + HS^- + H_2O.$$

Whereas a significant fraction of the sulfide in methane seeps is probably of biological origin, in hydrothermal vents most of the sulfide is probably of inorganic origin. In hydrothermal vents, seawater leaks through fractures in

continued

continued

the seafloor and as it travels down, it is heated to very high temperatures (higher than 400°C) by molten rocks. The hot fluid streams out of vent openings laden with dissolved metals, including hydrogen sulfide (H_2S). The sulfide in vent fluid is derived from seawater sulfate (SO_4^-) that is reduced at high temperature and from reduced sulfur leached from rocks (Jannasch and Mottl 1985).

In the presence of oxygen, some bacteria can use sulfide (or, again, its related compounds H_2S and HS^-) as a source of reducing power that can then be used to synthesize organic compounds. These bacteria are called thiotrophic. A one-step equation for this process is as follows:

$$S^{2-} + CO_2 + O_2 + H_2O \longrightarrow SO_4^{2-} + [CH_2O].$$

This reaction is simplistic because it represents autotrophy (the synthesis of organic compounds represented by $[CH_2O]$ from CO_2) as a single step. The actual biochemical transformations are more complicated. In these, the chemical energy in sulfide is used to generate reducing power through production of NADPH, which is then coupled to the Calvin-Benson cycle (see figure 5.2). For a more thorough introduction to the almost limitless intricacy of bacterial metabolism we recommend Paustian's (2003) microbiology web textbook (http://www.bact.wisc.edu/microtextbook/).

worms, and, as it turned out, their smaller pogonophoran relatives, rely on a nutritional symbiotic association with chemolithotrophic bacteria that use dissolved sulfide as an energy source.

Detailed microscopic examination revealed inclusions of elemental sulfur in a highly vascularized organ called the trophosome that fills the cavity of the trunk of *Riftia* and that accounts for $\approx 15\%$ of its total body mass (figures 5.12 and 5.14 below). In addition to having sulfur inclusions, the trophosome is packed with bacterial cells. The enzymatic pathways of these bacteria were consistent with the two steps of bacterial sulfur metabolism outlined in box 5.3. The final evidence for a nutritional symbiosis between *Riftia* and chemoautotrophic bacteria came from stable isotopes. Biological oceanographer Greg Rau (1981) found that the carbon isotopic composition of the tissues of worms ($\delta^{13}C \approx -11‰$) was equal to that of the bacteria in the trophosome and different from that of phytoplankton, which as we mentioned in a previous paragraph ranges from $-25‰$ to $-15‰$.

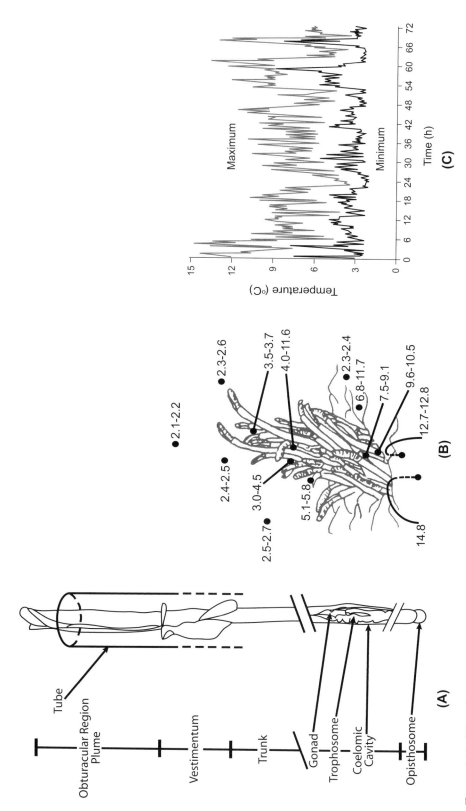

Figure 5.12. The tubeworm *Riftia pachyptila* is one of the strangest animals found at hydrothermal vents. They are large (1–2 m in length) animals without mouths or digestive systems (A). They have four body regions: (1) A blood-red, highly vascularized anterior tentacular plume (the "obturaculum") that allows the efficient exchange of dissolved molecules between blood and the environment; (2) a muscular collar ("vestimentum") that secretes the chitinous cylindrical tube in which the worms live; (3) the trunk, which contains the symbiont-containing trophosome lodged between two fluid-filled coelomic cavities; and (4) a short segmented ophistosome. When they are not disturbed, tubeworms expose the full length of the plume to the surrounding water. Tubeworms have a closed vascular system that circulates hemoglobin-rich blood from the trophosome to the plume. The tissue of the trophosome is richly vascularized and flecked with deposits of elemental sulfur. Tubeworms are found in areas where warm vent fluid mixes turbulently with cold seawater. Although water temperature is high at the base of the worm's tubes and low at the height of the plume (B), tubeworms live in a rapidly changing environment in which both temperature and the concentration of oxygen, sulfide, carbon dioxide, and bicarbonate change rapidly and unpredictably. Panel C shows changes in maximal and minimal temperature recorded by a probe placed on a clump of tubeworms by Johnson et al. (1988). The probe recorded water temperature every 5 seconds in 15-min-long intervals. The changes in temperature are good indicators of changes in dissolved chemicals. The concentration of sulfide and dissolved CO_2 are positively related with temperature, whereas the concentration of O_2 and bicarbonate are negatively related with temperature. (After *Van Dover* 2000.)

We hope that you will note a contradiction in our text. We argued that differences between the $\delta^{13}C$ of the creature and the $\delta^{13}C$ range of seston was evidence for reliance on chemolitothrophic bacteria as sources of organic substrates. However, in the case of mussels ($\delta^{13}C \approx -33‰$), $\delta^{13}C$ is more *negative* than in seston, whereas in the case of *Riftia* ($\delta13C \approx -11‰$), $\delta^{13}C$ is more *positive* than that of seston (if all these δ's are making your eyes glaze, you may want to read chapter 8). There is something askew in this line of reasoning! Marine biologists did not sweep this contradiction under the rug. Rather, they recognized that it reflected an isotopic composition dichotomy among the chemoautotrophic animals that live in hydrothermal vents. Because *Riftia* and a few other creatures that associate with thiotrophic symbionts have $\delta^{13}C$ values that range from $-9‰$ to $-15‰$, Childress and Fisher (1992) called them "the $-11‰$ group." In contrast, many mollusks from hydrothermal vents have $\delta^{13}C$ values that range from $-27‰$ to $-35‰$. Childress and Fisher (1992) call this group "the $-30‰$ group."

Robinson and Cavanaugh (1995) reviewed, and rejected, the many alternative hypotheses that have been posed to explain this dichotomy. They proposed that it may be due to the differential expression of two forms of rubisco (forms I and II). Rubisco is involved in the initial fixation of CO_2 in the Calvin-Benson cycle and the form in green plants (rubisco form I) discriminates against $^{13}CO_2$ (see section 8.4.1). Although both enzymes catalyze the same reaction and use the same substrates, they have different quaternary structures and apparently dissimilar kinetic isotope effects. Robinson and Cavanaugh (1995) found that rubisco I is expressed in the $-30‰$ symbioses whereas rubisco II is expressed in the $-11‰$ symbioses.

5.3.2 The Sulfide-Loving Bivalves

One of the most spectacular members of the $-30‰$ group is the clam *Calyptogena magnifica* in the family Vesicomyidae. These clams are huge. Their dinner-plate-sized shells enclose a red meaty body, large fleshy gills, a much reduced digestive system, and a hefty and well-vascularized foot. The gills account for almost 20% of the clam's tissues and house symbiotic sulfide-oxidizing bacteria. The bacteria are housed within vacuoles enclosed in gill cells called bacteriocytes. The absorptive apical surface of bacteriocytes is exposed to seawater and their basal membranes are immersed in blood-filled lacunae (figure 5.13). *Calyptogena magnifica* and *R. pachyptila* occupy different habitats within a vent. *Riftia pachyptila* is found in areas of strong vent water flow (figure 5.13), whereas *C. magnifica* occupy low-flow areas (figure 5.13). In contrast with those habitat specialists, the vent mussel *Bathymodiolus thermophylous* thrives in both the high-sulfide environment of the worms and in the low-sulfide environment of the clam. They are even found in regions where there seems to be no apparent flow of vent water. The traits of vent mussels seem to match their more generalized habitat use. Although they maintain bacteria in gill bacterocytes (figure 5.13), vent mussels have a functional filter-feeding apparatus, a full complement of digestive enzymes (Le Pennec et al. 1992), and the ability to take up dissolved amino acids from water. The contrasting traits of *C. magnifica* and *B. thermophylous* indicate that the clam is much more dependent on its symbionts and has more strict habitat requirements than the mussel.

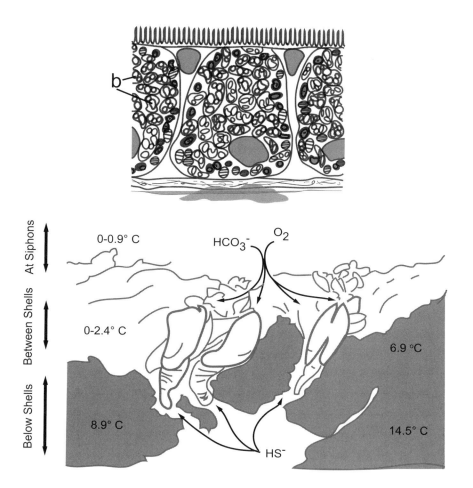

Figure 5.13. The vesicomyid clams *Calyptogena magnifica* live nestled snuggly within cracks of the newly formed basalt crust (lower panel). Their feet extend deeply into the crack and are bathed in warm, H_2S- and CO_2-rich vent fluid. The siphons of the clams, in contrast, are exposed to cold, O_2- and HCO_3^--rich seawater. This water is inhaled by the siphon and circulated through the gill. The bacteriocytes of the gill filaments house symbiotic thiobacteria (labeled by a b in upper panel). The apical surface of these bacteriocytes is exposed to mantle fluid whereas their basal ends are surrounded by blood-filled lacunae. (After Van Dover 2000.)

5.3.3 Living at the Edge

Like "photosynthetic" corals and giant clams, *Riftia* and *Calyptogena* must deliver CO_2 to supply the Calvin-Benson cycle of their symbionts (see Section 5.3.3). They also must take up O_2 not only to provide their own aerobic metabolism and that of their symbionts, but also to sustain the aerobic oxidation of sulfide (box 5.3). Animals that host thiotrophic ("sulfur-feeding") bacteria face another, significantly more unusual, requirement. They must take up dissolved sulfide from their environment and then they must transport it to their symbionts. Meeting this requirement is complicated because hydrogen sulfide is toxic and because its abundance in the environment is negatively correlated with that of oxygen, one of the essential substances needed to use sulfide as an energy source. When reduced sulfide compounds and oxygen co-occur, the sulfides are spontaneously oxidized with a relatively short half-life (McCollom and Shock 1997). Sulfide is toxic because it binds to the iron in the heme of cytochrome-*c* oxidase and blocks the last step in the respiratory electron transport system (Grieshaber and Völkel 1998). Animals that host thiotrophic symbionts must sequester and transport sulfide without poisoning their own aerobic respiratory system, and without allowing the sulfide to react with oxygen. How do the chemolithotrophic beasts of the deep juggle their conflicting demands for resources?

They do it by living at the edge armed with a battery of astounding physiological adaptations. Almost without exception, animals that host chemolithotrophic bacteria live in areas where well-oxygenated seawater meets relatively anoxic high-sulfide waters or sediments. The hot fluid that gushes out of vents contains high concentrations of dissolved sulfide and CO_2, but has low oxygen content. In contrast, the cold water surrounding the vent has low sulfide content, but is well oxygenated. *Riftia pachyptila* beds are found in areas of the vent where there is a strong flow of dilute, warm (15–20°C) vent fluid that mixes turbulently with cold (≈2°C) seawater. Although on average the long body of *Riftia* experiences a gradient in temperature and dissolved materials (figure 5.12), averages are poor descriptors of the turbulent milieu in which the worms live. Temperature and the concentration of dissolved substances change dramatically by the minute. Johnson et al. (1988) documented temperature fluctuations of 10°C (from 3°C to about 13°C) in the span of a few minutes (figure 5.12). *Calyptogena magnifica* clams reside in a more sedate environment. They live nestled in cracks in the basalt crust. Their siphons extend into the cold, O_2-rich seawater, but their foot is shoved deep into the crack where it bathes in warm sulfide-rich water (figure 5.13). Whereas *R. pachyptila* worms take advantage of temporal changes in sulfide, CO_2, and O_2, *C. magnifica* clams exploit the spatial separation of these resources in vent fluid and seawater (Cavanaugh 1994).

5.3.4 Carbon Uptake in Giant Worms and Magnificent Clams

Marine ecological physiologists are just beginning to piece together how *R. pachyptila* supplies its bacteria with inorganic carbon at the rates required to support its large size and very high growth rates. We described many of the elements of the puzzle in a previous section (section 5.3.3). Like animals that rely on photosynthetic symbionts, *Riftia* uses a carbon-concentrating mechanism (CCM) that depends on carbonic anhydrase (Kochevar and Childress 1996). As expected from the unusual environment in which *Riftia* lives, the details of its CCM differ from those of giant clams and corals. In relatively alkaline seawater (pH ≈ 8), the primary species of dissolved organic carbon is bicarbonate (HCO_3^-). Vent fluid is acidic (the pH can be lower than 6), and contains two to four times more dissolved inorganic carbon (DIC) than seawater (Goffredi et al. 1997a). Because the acidity of vent fluid favors CO_2 over HCO_3^- as the dominant inorganic carbon species, the plume of *Riftia* is often bathed in a fluid that facilitates the diffusion of CO_2 into the animal.

Although *Riftia* takes up CO_2 by diffusion (which is not concentrative), the concentration of DIC within the worm's vascular blood and coelomic fluids is often much higher than that of the environment. How can this be? Like other invertebrates that house DIC-consuming symbionts, *Riftia* concentrates carbon in its extracellular fluid with the aid of carbonic anhydrase (CA). Provided that the pH is relatively alkaline, CA transforms CO_2 into bicarbonate and maintains a CO_2 gradient that allows diffusive flow from the environment into the worm. This can be a problem because transforming CO_2 into bicarbonate generates carbonic acid and hence H^+. Goffredi and Childress (2001) and De Cian et al (2003) have provided the mechanism that keeps the pH of *Riftia*'s fluids relatively alkaline. They have found a protein that uses energy to pump the H^+ produced by CA's catalytic activity out of *Riftia*'s plume. Using immunolochemical localization, De Cian et al. (2003) have found that this protein (which in technical terminology is called a vacuolar-type H^+-ATPase) is coexpressed with carbonic anhydrase in the most apical epithelium of the plume's epithelia.

Riftia transports DIC as HCO_3^- in its blood, but its symbionts use CO_2 primarily. Both CA and vacuolar-type H^+-ATPase are expressed in the bacteriocytes of the trophosome and are probably responsible for the movement of DIC from blood into the symbionts. A missing piece of the puzzle is how the HCO_3^- generated in the apical cells of *Riftia*'s plume is translocated into the worm's circulating fluids (De Cian et al. 2003). The worms may have a HCO_3^- transporter in the basal membranes of these cells, but this transporter has not been characterized. Although the DIC within the fluids of *Calyptogena* clams is higher that that of the environment (Childress and Fisher 1992), the physiological details of the mechanism by which these clams concentrate carbon remain sketchy. Both *Calyptogena* and *Riftia* can feed their symbionts with recycled carbon. In *Riftia*, up to 50% of all carbon requirements can be met by carbon produced by the catabolism of the host and symbionts (Fisher et al. 1989). It is likely that the fraction of carbon recycled by the slower-growing *Calyptogena* is higher. Recall that the epithelium of the gill's filaments contains bacteriocytes and that the gill is exposed to seawater. CA expression is much higher in the gill than in the foot (Kochevar and Childress 1996), and hence it may be that the CCM of *Calyptogena* resembles that of photosynthetic giant clams. Recall that in these animals DIC moves from seawater to the zooxanthellae down a CO_2 gradient favored by the presence of carbonic anhydrase on the epithelial surface of gills (figure 5.6). It is unclear if (and how) *Calyptogena* clams take advantage of the abundant CO_2 in the warm vent fluid that bathes their feet.

5.3.5 Sulfide Uptake in Thiotrophic Animals

Animals that host thiotrophic symbionts must take up sulfide, and somehow sequester and transport it to their symbionts without poisoning their own aerobic respiratory system. The concentration of sulfide in the vents inhabited by clams is modest, at least by *Riftia* standards (see the following paragraphs). It ranges from 12 to 150 µmoles per liter. The sulfide concentration within the clam's vascular system is over an order of magnitude higher (from 100 to 2000 µmoles per liter; Scott and Fisher 1995). In *Calyptogena*, H_2S enters the foot's epithelia by diffusion and is promptly bound to a zinc-containing sulfide-binding lipoprotein (Zal et al. 2000). This sulfide-binding factor allows the clams to concentrate sulfide levels within their circulatory system, to transport it from the foot to the symbiont-bearing gills, and to keep it bound and hence in nontoxic form.

The transport of sulfide from blood to bacteriocytes remained a mystery until Pruski and Fiala-Médioni proposed a potential mechanism. Although this mechanism remains somewhat hypothetical, it is tantalizingly plausible. Pruski and Fiala-Médioni first found that animals that house sulfide-oxidizing bacteria had unusually high levels of a free sulfur amino acid called thiothaurine (NH_2-CH_2-CH_2-SO_2SH; Pruski et al. 2000). Thiotaurine represents 10–26% of the free amino acid pool in *C. magnifica* and up to two-thirds of the free amino acid pool in the trophosome of *R. pachyphyla*. Exposure to dissolved sulfide rapidly stimulates the synthesis of thiotaurine (Puski and Fiala-Médioni 2003). Thiotaurine seems a likely candidate for both a transmembrane carrier of reduced sulfur and an intracellular reduced-sulfur storage molecule. It crosses membranes readily, it can accumulate in the cell's cytoplasm without disturbing protein function, and it releases reduced sulfur readily (Pruski and Fiala Médioni 2003).

Although H_2S is the predominant species of sulfide in acidic vent fluid, *R. pachyptila* limit its diffusion into plume tissues. Instead, the worms seem to selectively take up HS^- (Goffredi et al. 1997b). Because H_2S is the species of sulfide that inhibits cytochrome-*c* oxidase function, by selectively excluding it *Riftia* may avoid some of its toxic effects. The mechanisms that allow HS^-, but not H_2S to diffuse into the plume are not known. Within the worm's blood, HS^- binds with high affinity to three different hemoglobins. Two of these are dissolved in vascular blood and the third is found in coelomic fluid (Zal et al. 1998). The hemoglobins of *Riftia* are extraordinary because they bind oxygen and sulfide simultaneously at two different binding sites (Zal et al. 1998). *Riftia*'s remarkable hemoglobins allow vent worms to concentrate sulfide from the environment and to transport it from the plume to the trophosome while maintaining low concentrations of free sulfide within its fluids. The same molecules also allow the worm to increase the oxygen carrying capacity of its fluids and hence facilitate its relatively high metabolism and growth rates.

The bacteria within the bacteriocytes of *Calyptogena* and inside of *Riftia's* trophosome use the energy from sulfide to generate reducing power to fuel the Calvin-Benson cycle. Table 5.3 lists the decrease in free energy of the sulfide-oxidation reactions that can potentially be performed by their symbiotic bacteria. Partial oxidation leads to the formation of elemental sulfur, but does not lead to as much energy production as the oxidation to sulfate. The sulfur inclusions often found in the tissues of sulfide-oxidizing animals may represent energy stores for periods of low sulfide or low oxygen availability (see the next paragraph and Arndt et al. 2001). The ultimate products of sulfide oxidation are sulfate ions and H^+. Because the affinity of *Riftia*'s hemoglobin for sulfide decreases rapidly with decreasing pH (Zal et al. 1998), the production of H^+ in bacteriocytes may facilitate the unloading of sulfide at the sites where it is needed.

So far, we have emphasized the role of sulfur-containing compounds as an energy source for creatures associated with sulfur-oxidizing bacteria. An intriguing set of observations points to a different role. During anoxia, several species of sulfide-oxidizing animals produce rather than take up large amounts of hydrogen sulfide. Arndt et al. (2001) speculate that the source of this sulfide is the elemental sulfate often deposited by these animals and that can serve as a terminal electron acceptor during anoxia. It remains unclear if the production of sulfide under anoxia is the result of the symbiont's anaerobic respiration, or of that of the host. Independently of who is the culprit, the reduction of sulfur

TABLE 5.3
Oxidation reactions of sulfur compounds and associated standard free energy changes

Reaction	Free energy change (kJ/mol)
$H_2S + (1/2)O_2 \rightarrow S^0 + H_2O$	−210
$HS^- + 2O_2 \rightarrow SO_4^2 + H^+$	−716
$S^0 + (1.5)O_2 + H_2O \rightarrow H_2SO4^-$	−496
$S_2O_3^- + 2O_2 + H_2O \rightarrow 2SO_4^2 + 2H^+$	−936

Note: The partial oxidation of sulfide into elemental sulfur is associated with the lowest change whereas the complete oxidation into sulfate ions is associated with the highest change (after Van Dover 2000).

may be important for the production of maintenance energy during periods of anoxia. *Riftia pachyptila* can survive for up to 60 h without oxygen and some sulfide-oxidizing clams (e.g., *Lucinoma aequizonata*) have an extraordinary capacity to tolerate environmental anoxia.

5.3.6 Farming and Milking: What Do the Hosts Get from Symbionts?

The mechanisms by which sulfide-oxidizing animals supply their symbionts with resources have received significant attention. Physiological ecologists have unraveled the exquisite details by which vestimentiferan worms and vesicomyid clams deliver dissolved inorganic carbon, oxygen, and sulfide to their symbiotic partners. What these partners give in return to the host has been much less studied. Symbionts can contribute to the nutrition of their hosts through two nonexclusive means. (1) They can translocate organic materials in a manner similar to that by which zooxanthellae translocate nutrients to their hosts; and (2) the host can digest symbionts (Fisher 1990). Streams et al. (1986) designate these two modes of nutrient transfer from symbionts to hosts with the colorful terms "milking" and "farming." Although there is evidence for both pathways, no study has yet estimated their relative importance in any particular symbiosis.

Evidence for symbiont milking comes from the translocation of radiolabeled organic materials from symbionts to hosts within minutes after the association has been exposed to $H^{14}CO_3^-$ (Fisher and Childress 1986; box 5.1). The presence of a variety of digestive proteases and glycosidases within the tissues of hosts

(Boetius and Felbeck 1995) suggests that bacteria may transfer not only small-molecular-mass organic molecules, but also more complex substances. Evidence for farming of bacterial symbionts comes from two sources: electron micrographs of host lysosomes in the process of digesting bacteria (Bosch and Grassé 1984; Fiala-Médoni et al. 1986), and high levels of expression of lysozyme (Boetius and Felbeck 1995; Fiala-Médioni et al. 1994). Lysozyme is a small enzyme that attacks the polysaccharides that make up the protective cell walls of bacteria and hence facilitates the assimilation of bacteria (lysozyme plays a prominent role in the next chapter). Per unit mass of tissue, the lysozyme activity in the gills of *C. magnifica* is almost 50 times higher than in those of clams without symbionts (Boetius and Felbeck 1995). The trophosome of *Riftia* also has very high lysozyme activity. Figure 5.14 shows a schematic view of the transfer and distribution of nutrients from symbionts to *R. pachyptila*.

5.3.7 Light in the Darkest Place: The Pervasive Global Influence of Photosynthesis

Jannasch and Motl (1985) wrote that perhaps the most important aspect of the discovery of chemiosynthetic communities in hydrothermal vents was "the

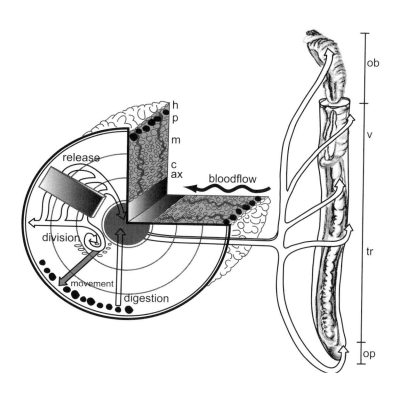

Figure. 5.14. *Riftia pachyptila* receives nutrients from the symbionts within its trophosome (tr) through two pathways: release of nutrients by the intact symbionts, and digestion of the symbionts. Digestion takes place in the peripheral layer of the tropohosome (labeled as p in the left panel), nutrient translocation takes place in the central (c) and medial (m) layers of the trophosome. Blood flow from the periphery to the axial blood vessel delivers assimilated nutrients to the rest of the worm's body. (After Bright et al. 2000.)

dependence of entire ecosystems on geothermal rather than solar energy." At the beginning of this section we echoed this statement and claimed that hydrothermal vents are unique ecosystems because their trophic structure is based primarily on chemoautotrophy. Our claim and Jannasch and Motl's (1985) are strictly incorrect. Although chemiosynthesis is at the trophic base of the fabulously weird communities found at hydrothermal vents, these communities also depend on photosynthesis. Sulfide in vent fluids is the primary power source for these communities, but sulfide without an electron acceptor yields no energy. The main oxidant is molecular oxygen dissolved in seawater and ultimately produced by photosynthesis. Vent ecosystems depend on solar energy. Without it, one half of the reactants of the oxidation-reduction, energy-producing reactions that support these ecosystems would be unavailable (Van Dover 2000). The influence of sunlight reaches even the darkest depths of the ocean.

5.3.8 The Isotopic Lightness of Some Deep-Sea Beings: Methanotrophic Symbioses at Cold Seeps

The discovery of flourishing ecosystems in the hydrothermal vents along the mid-ocean ridges has drawn much attention. This emphasis is justified. The discovery of hydrothermal vent ecosystems based on thiotrophy is one of the biological triumphs of the twentieth century. However, hydrothermal vents are not the only rich deep-sea ecosystems. The same technological advances that allow the exploration of the deep ocean and that led to the discovery of hydrothermal vent ecosystems in the late 1970s led, a few years later, to the finding of life thriving in other, colder deep-sea ecosystems. In 1984, the crews of submersibles exploring the bottom of the Gulf of Mexico discovered dense biological communities that were not associated with hydrothermal vents. Since 1984, many more of these ecosystems have been found all over the world wherever methane bubbles up to the ocean floor from undersea sediment layers.

The biological and geological processes that produce methane seeps are diverse. They include "cold seeps" where rainwater flows through the continental crust and emerges at the continent's underwater edge, hypersaline brine pools, and gas hydrate deposits (Van Dover 2000). Gas hydrates are solid icelike structures often associated with offshore oil deposits. They form under high pressure and low temperature and hold large amounts of methane. Some geologists estimate that the energy locked up in methane hydrate deposits is more than twice the global reserves of all conventional gas, oil, and coal deposits combined (Forrest 2003). However, no one has yet figured out how to pull out the gas inexpensively, which might be a blessing. Methane is a greenhouse gas and release of even a small percentage of the total deposits could cause serious mayhem in the Earth's atmosphere.

The methane in these seeps fuels rich biological communities. The communities found at methane seeps differ from communities at hydrothermal vents in many ways. The organisms at hydrothermal vent communities live fast, grow quickly, and often die young. Van Dover (2000) characterizes hydrothermal vents as high-disturbance environments. The animals that inhabit them are opportunists with a lifestyle that allows them to establish colonies and grow quickly to produce propagules. In contrast, the organisms that live in methane seeps can be extremely long-lived and the communities that they form part can be remarkably stable (Fisher et al. 1997). The vestimentiferan *Lamellibranchia lumeysi*, for example can live up to 250 years in communities that persist relatively unchanged for centuries (Cordes et al. 2003).

The dependence of these long-lived, slow-paced communities on methane can be assessed as a result of a quirk in methane's isotopic chemistry. Methane is isotopically light. It is strongly depleted in ^{13}C. How depleted in ^{13}C methane is depends on its origin. Biogenic methane is formed by the anaerobic degradation of organic materials by bacteria and its $\delta^{13}C$ ranges from $-110‰$ to $-50‰$. Thermogenic methane is produced by inorganic processes acting on hydrocarbon deposits at high temperatures and pressures deep within the earth's crust. Thermogenic methane is not quite as depleted in ^{13}C as biogenic methane, but still has a very negative composition relative to other carbon sources (its $\delta^{13}C$ ranges from $-50‰$ to $-16‰$).

Many bacteria use methane as a source of energy and carbon. In the presence of oxygen, these bacteria oxidize methane to formaldehyde, which is an intermediary for the generation of energy and for the synthesis of organic compounds. As most biologists have experienced when handling pickled specimens, formaldehyde is toxic because it inactivates proteins. Methanotrophic bacteria avoid pickling themselves by their own metabolic by-product by carrying out the methane-to-formaldehyde reaction in the periplasm, and then by further oxidizing formaldehyde into formate and CO_2 (figure 5.15). Formate is not toxic and can be translocated safely to the cytoplasm. It has been known for a long time that methane oxidizers are widespread and have profound ecological importance (Madigan et al. 1997). Methane-oxidizing bacteria are found wherever stable sources of methane are found and where there is some, but not too much, oxygen. They are particularly abundant in a narrow band of the thermocline, where the anoxic layer of a body of water meets the oxic zone. Until relatively recently, however, their symbiotic association with invertebrates were not known.

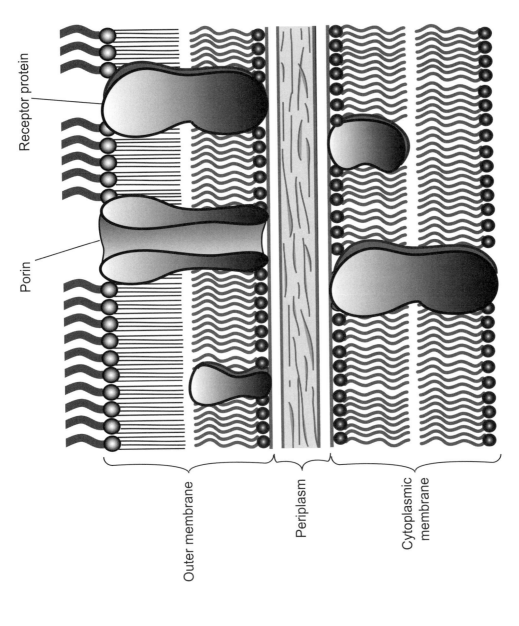

Receptor protein

Porin

Outer membrane

Periplasm

Cytoplasmic membrane

Figure 5.15. Methanotrophic bacteria oxidize methane to formaldehyde. To avoid the toxic effects of formaldehyde, they perform the methane to formaldehyde reaction in the periplasm, which is the space between the outer and the inner membrane in gram-negative bacteria. They then further oxidize the formaldehyde to formate and CO_2, which are nontoxic, in an extensive system of internal membranes. (After Salyers and Whitt 2001.)

(Figure Continues)

(Figure Continued)

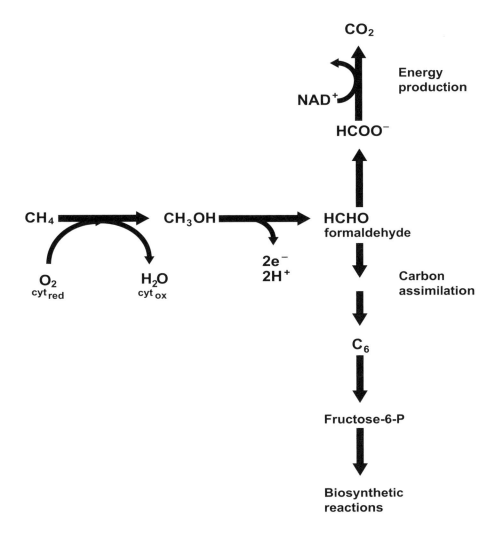

Shortly after the first abyssal methane seep ecosystem was discovered, Paull et al. (1985) used the submersible *Alvin* to pull out an undescribed mussel from the bottom of the ocean. The mussel was part of a dense bed at the base of the escarpment that drops 3300 m down from the continental shelf in western Florida. To their astonishment, the tissues of this mussel had the lightest carbon isotopic composition ever recorded in a living creature ($\delta^{13}C$ = −74.3‰). This observation strongly suggested that this organism was either feeding on methane-oxidizing bacteria, or it contained within its tissues a methane-oxidizing symbiont—a hypothesis favored by Paull and his collaborators.

Only a year after the appearance of these results, two groups published evidence of symbiosis between methanotrophic bacteria and mytillid mussels (Childress et al. 1986; Cavanaugh et al. 1987). The mussels consumed methane at relatively high rates and their gills contained bacteria with the stacked internal membranes typical of methanotrophs. These bacteria were present in high numbers and were housed in apical bacteriocytes. Extracts of the gills of the mussels contained the enzymes that are necessary to oxidize methane and to use formate to synthesize multicarbon compounds. Since these pioneering discoveries, methanotrophic symbioses have been also reported in a sponge (Vacelet et al. 1995) and a pogonophoran tubeworm (Schmaljohann and Flügel 1987).

The physiological ecology of methanotrophic symbioses has not been as intensely studied as that of sulfide-oxidizing symbiosis. Cary et al. (1988) found that a methanotroph symbiont-containing mussel was able to grow, albeit quite slowly, when methane was the sole source of carbon and energy. Because Cary and his collaborators fed their mussels only on methane with no additional sources of nitrogen or phosphorus, this result is notable. Maybe the mussel would have grown faster had it been fed on a more complete diet. It is hard to sustain growth solely with natural gas! Additional evidence for the use of symbiont materials by hosts comes from pulse-chase experiments in which $^{14}CH_4$ is bubbled into the environment of a methanotrophic mussel. After incubation, ^{14}C appears in the organic compounds of the host's tissues. However, the incorporation of radioactive carbon into the host's tissues takes a long time. Fisher and Childress (1986) found that transfer of radiolabeled carbon did not occur until at least 24 h after the end of incubation with ^{14}C-labeled methane. Streams et al. (1997) found the same pattern in another mussel using autoradiography.

Streams et al. (1997) suggest that the slow incorporation of ^{14}C into the host's tissues points to "farming" as the mode of nutrient transfer between symbionts and hosts, although direct translocation of nutrients cannot yet be discounted. Farming of symbionts has been convincingly demonstrated in a methanotrophic carnivorous sponge (Vacelet et al. 1995). This sponge *Cladorhiza* sp. was collected from a mud volcano at a depth of almost 5000 m in the Barbados Trench. The sponges formed large bushy aggregations around the periphery of the volcano's eye where methane concentration was the highest. Vacelet et al. (1995) found many partially digested bacteria in the phagocytic cells of the sponge, suggesting that the sponge farms more than milks it symbionts. The carnivorous sponges of the deep sea are usually small and rare. Vacelet et al. (1995) speculate that a symbiosis with methanotrophic bacteria allows the dense aggregations of large sponges found at the mud volcanoes.

5.3.9 Mixotrophic Animals

Not all of the "primary producing" animals that live in methane seeps possess the isotopic lightness of being that characterizes methanotrophy (Levin and Michener 2002). The isotopic signature that characterizes thiotrophic symbioses is often the most prevalent in these habitats (Van Dover 2000). However, the production of the hydrogen sulfide that feeds these thiotrophs is indirectly fueled by methane. Hydrogen sulfide seems to be generated by bacteria that use methane as a carbon substrate and that reduce the sulfate that percolates into methane-rich sediments from seawater (Masuzawa et al. 1992). Although there is variation from seep to seep, cold methane seeps can be characterized as having relatively high levels of both methane and hydrogen sulfide. Supporting two metabolically distinct types of symbionts seems like a good strategy for a generalist living in such a place. Indeed, several mussels in the genus *Bathymodiolus* and maybe two species of gastropods, reap the benefits of housing both thiotrophic and methanotrophic symbionts. A dual symbiont system can not only allow an animal to use two different resources, it can also give an animal the phenotypic flexibility to fit into a chemical environment that can change in both time and space. It may be that animals can change the relative abundance of each symbiont type in response to changes in the availability of sulfide and methane.

In support of this speculation, Trask and Van Dover (1999) found intriguing differences in two populations of *Bathymodiolus* clams living in two contiguous sites at the Mid-Atlantic Ridge hydrothermal vent. At one site, the mussels had higher relative abundances of methanotrophs and an isotopic composition that suggested higher reliance on methane. At the other site, the mussels had higher relative abundances of thiotrophic bacteria in their gills and their isotopic composition indicated that they relied on sulfide oxidation. Genetic analyses revealed that the two populations were part of a single deme, and hence it is unlikely that the results are the consequence of the misidentification of two separate species. Curiously, the warm vent fluid at the two sites did not differ in either sulfide or methane concentration. Thus, Trask and Van Dover's observations are suggestive but not conclusive evidence for "adaptive modulation of relative symbiont abundances" (see chapter 4). Whether animals with dual symbioses can modulate the relative numbers of each symbiont to fit the characteristics of their chemical environment remains a tantalizing hypothesis. Although the possession of dual symbionts seems like a win-win policy, the strategy seems remarkably rare. As far as we know, only a handful of species within one genus of mussel, and maybe two species of gastropods, have adopted it. Are there costs to either host or symbionts associated with having two coexisting symbionts with such contrasting metabolic capacities? The answer to these questions awaits the adventurous physiological ecologists willing to reach into the ocean's depths.

5.4 The Importance of Nitrogen in Nutritional Symbioses

By associating with bacteria, animals acquire the metabolic pathways that allow them to use light, sulfide, and methane as energy sources. Energy, however, is not the only important resource. One of the themes of this book is the need to adopt a multidimensional view of the nutritional resources required by animals. Here we describe one possible way to minimize the effects of nitrogen scarcity. Animals cope in a nitrogen-poor world with the aid of symbiotic microorganisms. Symbiotic microorganisms can help a host's nitrogen economy in several ways: (1) they can use the metabolic nitrogenous waste products of the host (ammonia, urea, and uric acid) to grow, and hence reduce competition with the host for nitrogen from the environment; (2) they can take these waste products and use them to manufacture compounds that the host needs but cannot manufacture itself, such as essential amino acids; and (3) they can access environmental sources of nitrogen that are inaccessible to the host (such as atmospheric N_2). We will provide examples of each of these potential advantages in the following paragraphs. First, however, we must review the terminology used by the physiological ecologists who study symbioses to characterize the effect of symbionts on the nitrogen economy of hosts.

5.4.1 The Terminology of Nitrogen Exchange in Symbiotic Interactions

The use of the host's nitrogenous wastes by symbionts is called nitrogen conservation (figure 5.16). In nitrogen conservation, the symbionts use the waste nitrogen of the host to grow, but do not give back nitrogenous molecules to the host. Nitrogen conservation has two potential advantages to the host: (1) it can reduce the costs of providing nitrogen to the symbionts, and (2) it promotes host growth because it increases the delivery of energy-containing material by a population of healthy symbionts. Nitrogen recycling refers to the situation in which the symbionts use the host's waste nitrogen to manufacture compounds that are then used by the host (figure 5.16). Nitrogen recycling is prevalent in hosts that farm and then digest their symbionts. Douglas (1994) has called this situation by the rather macabre name "nitrogen recycling by necrotrophy."

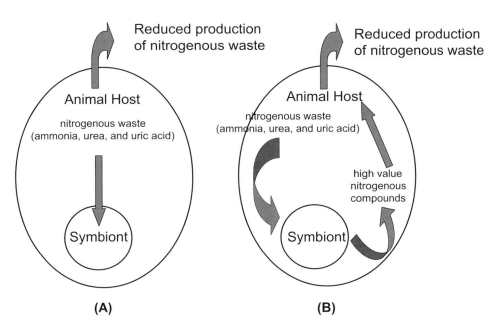

(A)   **(B)**

Figure 5.16. (A) The use of the host's nitrogenous wastes for the growth of symbionts is called nitrogen conservation. The net excretion of nitrogenous waste products can be reduced when the symbionts use these compounds as a nitrogen source, even when symbionts do not return nitrogenous compounds to the host. (B) Nitrogen recycling occurs when the symbionts use the host's nitrogenous waste products to manufacture and transfer compounds that are valuable to the host. (Modified from Douglas 1994.)

A necrotroph, literally, is a death eater (Rowlings 1997), which makes this terminology a bit incorrect. Hosts that farm probably start digesting symbionts while these are still alive. We will discuss necrotrophic recycling in significant detail when we deal with fermentative digestive symbioses. Nitrogen conservation and recycling have been almost unexplored in thiotrophic and methanotrophic symbioses (see Lee and Childress 1994 for an overview). If symbiont farming is prevalent in these symbioses (Streams et al. 1997), the question is moot. Death eating provides a relatively well-balanced diet. The following section focuses on photosynthetic symbioses.

5.4.2 Nitrogen Recycling in Photoautotrophic Symbioses: Is there Any?

Some marine biologists consider that both photosynthesis and nitrogen recycling underpin the tremendous ecological success of the symbiosis between *Symbiodinium* dinoflagellates and scleractinian corals (D'Elia and Wiebe 1990). Showing that an animal receives energy-containing organic compounds from its autotrophic host is relatively easy. All we need to do is provide the symbiotic association with a pulse of ^{14}C-labeled carbon sources (CO_2 or methane) and chase the fate of the labeled carbon. If ^{14}C-labeled organic compounds appear in the host's tissues, these (or their organic precursors) must come from the symbionts. Finding out whether a symbiotic association relies on nitrogen conservation and/or recycling, in contrast, is a lot more complicated. The biochemical basis of photosynthate release by zooxanthellae is more or less well established. The evidence for nitrogen conservation and recycling remains a tantalizing but still tenuously supported possibility.

Researchers first inferred nitrogen conservation in cnidarian-zooxanthellae symbioses when they observed increased ammonium excretion in sea corals and anemones incubated in the dark or from which zooxanthellae had been experimentally eliminated (reviewed by Wang and Douglas 1998). Reasonably, this observation was interpreted as evidence for zooxanthellae as a sink for the ammonium produced by the host. However, two alternative processes can contribute to this phenomenon. First, when the algae are not photosynthesizing actively, they do not translocate organic compounds to the host. Thus, the host may be forced to catabolize amino acids to fuel its metabolism, and excrete the resulting ammonium. Second, algae may stimulate the uptake of ammonium by the animals. Wang and Douglas (1998) tested these alternatives on anemones (*Aiptasia puchella*) kept in the dark or from which zooxanthellae were experimentally depleted. When they provided these animals with an organic carbon source, they partially overturned the increased ammonium excretion. They also found that providing an organic carbon source stimulated the expression of glutamine synthetase, an enzyme that catalyzes the assimilation of ammonium. Wang and Douglas' (1998) results suggest that alternative processes can contribute to explain an observation that was originally attributed solely to nitrogen conservation. This does not mean that nitrogen conservation does not occur or that it is unimportant in cnidarian-zooxanthellae symbioses. It does mean, however, that we must use more sophisticated methods to measure it.

Roberts et al. (1999a,b) gathered data that supports the idea of nitrogen conservation in the anemone *Anemonia viridis*. They found that zooxanthellae are the primary, if not the only site, at which ammonium is assimilated. They also found that the anemones grow when fasted, provided that they are provided with dissolved ammonium. Finally, when they supplied ^{15}N-labeled ammonium, ^{15}N appeared in the host's protein. Robert's et al.'s detailed studies highlight the difficulties of studying nitrogen recycling in some symbioses. In principle we can supply ^{15}N in labeled waste nitrogenous compounds, such as ammonium, and track its fate. However, because sometimes both the symbiont and the host have the ability of incorporating ammonium into amino acids and proteins, finding ^{15}N-labeled protein in the host is not sufficient evidence for recycling. Thus, Roberts et al.'s (1999b) finding that it is zooxanthellae, rather than hosts, that assimilate ammonium is crucial.

How can we know if the ammonium that we provide to the symbiosis is assimilated by the symbionts and recycled to the host? One possible way is to follow the incorporation of ^{15}N from labeled ammonium into essential amino acids, which by definition cannot be synthesized by the host. However, it is usually the carbon skeleton of an essential amino acid that cannot be synthesized, and an animal may have some capacity to produce essential amino acids by transaminating the carbon skeletons provided by the symbiont. A possible exception, are the essential amino acids lysine and threonine, for which there is no evidence that they can be produced by transamination in animals. Thus, showing that ^{15}N appears in essential amino acids is evidence that the host provides either essential amino acids or their precursors. In any case, the symbiont facilitates nitrogen recycling by the host. Although this compound-specific approach may hold the key to understand nitrogen recycling in many animal-microbe symbioses, it is difficult to implement. Measuring the ^{13}C/^{12}C ratio of specific compounds such as amino acids is now, if not routine, at least commonly done in a few laboratories (Hammer et al. 1998). Measuring ^{15}N/^{14}N ratios in amino acids is technically difficult. Roberts et al. (1999b) describe some of the difficulties one encounters when attempting to analyze the ^{15}N/^{14}N of individual amino acids. This is a case of a question for which we have the conceptual tools, but for which the technology is still inadequate. We must emphasize that analyzing the ^{15}N/^{14}N ratios of specific amino acids is possible, it is just so mired with difficulties that it is not a measurement that is conducted routinely.

Tracking the fate of ^{15}N from ammonium into the essential amino acids of hosts to test for nitrogen recycling relies on a crucial, albeit unsubstantiated, assumption. The assumption is that the nine essential amino acids that are essential for vertebrates and insects are also essential for the considerably less studied invertebrates that participate in nutritional symbioses (see chapter 2). Fitzgerald and Szmant (1997) incubated bleached scleractinian corals with ^{14}C-labeled glucose and, to their surprise, found that ^{14}C appeared in all the amino acids that are usually considered indispensable. Only threonine was not labeled in the corals' tissues. Fitzgerald and Szmant (1997) entitled their paper *"Biosynthesis of 'Essential' Amino Acids by Scleractinian Corals."* They used quotation marks to bracket the word "essential" to emphasize a disturbing possibility. Maybe the list of essential amino acids differs among metazoan taxa.

Wang and Douglas (1999) carefully repeated Fitzgerald and Szmant's (1999) protocol on the anemone *Aiptasia puchella*. Wang and Douglas's (1999) protocol differed slightly from that used by Fitzgerald and Szmant (1999) in that they supplied the anemone not only with labeled glucose, but also with aspartate and glutamate. Their results were very different. They found that radioactivity was recovered from all the amino acids of anemones with zooxanthellae. However, all but seven of the amino acids were found in the bleached anemone. They inferred that these seven amino acids (histidine, isoleucine, leucine, lysine, phenylalanine, tyrosine, and valine) were synthesized by the zooxanthellae, and translocated to the anemone (i.e., they were essential for the anemone). Why do the results of Wang and Douglas (1999) differ from those of Fitzgerald and Smantz (1997)? Maybe different species of cnidarians have distinct amino acid biosynthesis capabilities. Alternatively, cnidaria may be able to synthesize all essential amino acids, but use glucose, aspartate, or glutamate as precursors for the synthesis of different arrays of these compounds (Wang and Douglas 1999).

5.4.3 Essential Amino Acids and Insect-Bacteria Symbioses

In the previous two sections we emphasized symbiotic interactions in which the heterotrophic host established a partnership with an autotrophic microorganism. Many other animals have established what, for lack of a better term, can be called heterotrophic nutritional symbioses. In these, the symbiont supplies a compound that the diet provides in short supply or helps the host to assimilate a diet's component. Because dietary specialization can impose the risk of nutritional deficiencies, it should not be a surprise that many of these heterotrophic symbioses occur frequently in animals with specialized diets. Symbioses between insects and bacteria have evolved independently several times in connection with blood feeding (in flies, hemipterans, and lice) and with dependence on plant fluids (in ants and homopterans; Moran and Telang 1998). We will focus on the relatively well-studied association between phloem feeders and bacteria (Moran 2001).

Several groups of insects in the order homoptera (aphids, whiteflies, planthoppers, leafhoppers, cicadas, etc.) feed exclusively on the phloem sap of plants. Phloem is rich in sugar, and once tapped it is very abundant. However, although phloem contains free amino acids, its amino acid profile can be different from that required by insects (figure 5.17). To alleviate the nutritional deficiency of their specialized diet, phloem-feeding insects have adopted a common solution. They have established associations with bacterial endosymbionts that have the capacity to synthesize the required amino acids. These bacteria are hosted in specialized bacteriocytes (or mycetocytes) that together form an organ called a

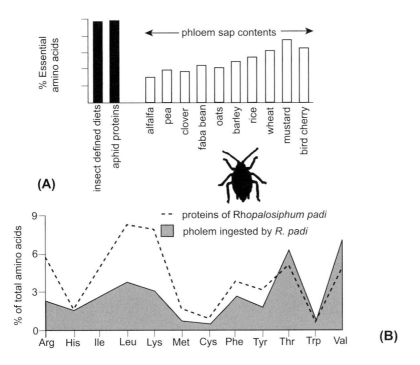

Figure 5.17. Phloem has low essential amino acid content relative to an optimal aphid diet and to aphid tissues (A). Phloem samples were sampled using aphid stylets. (B) Phloem has not only a low absolute amount of essential amino acids, but also many essential amino acids are at levels that are insufficient to satisfy the protein synthesis requirements of the aphid *Rhopalosiphum padi* (B; after Moran 2001). For example, the relative abundance of leucine and lysine in phloem is roughly three times lower than in proteins of the aphid.

bacteriosome. The bacteria are transmitted "vertically" by innoculation by the mother to the progeny before oviposition or at birth. Perhaps the most studied bacterial-insect nutritional endosymbiosis is that between aphids and the bacterium *Buchnera aphidicola*. There is now a remarkably complete body of evidence supporting the importance of *Buchnera* as a source of essential amino acids for aphids. The evidence consists of elegant physiological experiments and genomic analysis.

The physiological experiments compared the amino acid biosynthetic capacity of untreated aphids containing *Buchnera* (called symbiotic aphids) with those treated with antibiotics and hence without bacteria (called aposymbiotic aphids—the prefix apo means "away from" or "without" in Greek). Symbiotic, but not aposymbiotic, aphids incorporate radioactive sulphur (^{35}S) from inorganic sulfur into methionine; symbiotic, but not aposymbiotic aphyids, synthesize isoleucine, lysine, and threonine from ^{14}C-labeled glutamic acid and all essential amino acids from ^{14}C-labeled sucrose (reviewed by Douglas et al. 2001). The recently completed full genome sequence of *Buchnera aphidicola* confirms these results spectacularly. As expected from an endosymbiont that obtains a significant fraction of its needs from its host, *Buchnera* has a greatly reduced genome. However, it has retained the genes for the biosynthesis of all essential amino acids (Shigenobu et al. 2000; Moran and Bauman 2000). *Buchnera* appears to have evolved from a free-living bacterium (probably closely related to *Escherichia coli;* Moran 2001). Over 100 million years of an endosymbiotic association with aphids, *Buchnera* has lost many of its genes.

Within insects, all phloem feeders, all strict blood feeders, and many other groups with restricted diets are associated with endosymbionts. Recent molecular evidence suggests that nutritional endosymbioses date to the origin of the major insect taxa (Moran and Telang 1998). Some of the insect taxa that are members of nutritional symbioses are good examples of large adaptive radiations. Thus, the evolution of endosymbioses probably had a massive effect on insect diversity, and through it on the structure of terrestrial ecosystems.

5.4.4 Nitrogen, Nitrogen, Everywhere, Nor any Mole to Eat: Symbiotic Nitrogen Fixation in Animals

Dinitrogen (N_2) is the most abundant gas in the atmosphere, and yet nitrogen is a limiting nutrient in most ecosystems (see chapters 10 and 11). Although 86,250 tons of dinitrogen gas cover each hectare, only an elite assortment of bacteria and archaea possess the key enzyme complex nitrogenase needed to convert abundant dinitrogen into a form that can be used by eukaryotes and by other prokaryotes (Madigan et al. 1997). Wolfe (2001) has made the sobering observation that the entire supply of the Earth's nitrogenase would fill only one large bucket. Figure 5.18 shows the energetically expensive metabolic process used by microorganisms to convert N_2 into NH_3.

The magnitude of biological N_2 fixation in an ecosystem is estimated as the input of N that is needed to balance the budget after accounting for other N inputs and losses. Studies of the nitrogen budget of marine ecosystems, led to the prediction of undiscovered microbes involved in oceanic nitrogen fixation (Falkowski 1997). Since then, the nitrogen-fixing cyanobacteria that provide this fixation have been found (Zehr et al. 2001). In terrestrial ecosystems many free-living soil microbes fix atmospheric dinitrogen. In addition, many plant species have symbiotic relationships with nitrogen-fixing prokaryotes and contribute hundreds of kilograms per hectare per year of nitrogen to terrestrial ecosystems. Legumes and *Rhizobium* bacteria are the best studied, albeit by no means the only, examples of these associations. A reasonable upper boundary for the rate of natural nitrogen fixation on land is 140 million tons per year. The estimates for marine nitrogen fixation vary by an order of magnitude, from 30 to 300 million tons. Humans make a major contribution to the earth's

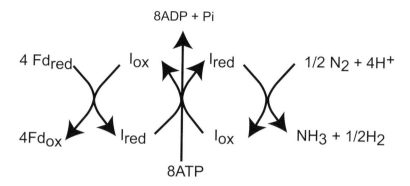

$$8ADP + Pi$$

$$4\ Fd_{red} \quad I_{ox} \quad I_{red} \quad 1/2\ N_2 + 4H^+$$

$$4Fd_{ox} \quad I_{red} \quad I_{ox} \quad NH_3 + 1/2H_2$$

$$8ATP$$

Figure 5.18. The reduction of nitrogen gas (N_2) to ammonia (NH_3) is catalyzed by the enzyme complex nitrogenase. Nitrogenase has two components, I and II, each of which is composed of two protein subunits. Nitrogen gas is reduced to ammonia by the reduced component (I_{red}) of nitrogenase which is oxidized in the process (I_{ox}). The I_{ox} component must be reduced to I_{red} in order for the process to continue. This reconditioning of I_{ox} is accomplished by nitrogenase component II. Ferredoxin regenerates I_{ox} into I_{red}. The regeneration of I requires large amounts of ATP. Fixing nitrogen is energetically very expensive. Twenty to 30 molecules of ATP are required for each molecule of N_2 fixed. Nitrogenase is not entirely specific for N_2, it also reduces cyanide (CN^-), acetylene (C_2H_2), and other triply bonded compounds. The reduction of acetylene by hydrogenase into ethylene (C_2H_4) is particularly useful. Ethylene and acetylene can be readily measured by gas chromatography and thus measuring the production rate of ethylene after supplying the symbiosis with acetylene is a nondestructive, simple, and relatively rapid way to measure the activity of nitrogen-fixing systems. (After Salyers and Whitt 2001.)

global nitrogen fixation. Fertilizer production, legume crops, and fossil fuel burning deposit approximately 140 million tons of new nitrogen per year. Hence over the last 100 years, humans have doubled the transfer of nitrogen from the atmosphere into the biological cycle of terrestrial ecosystems (Vitousek et al. 1997).

Invertebrate-bacteria symbioses are potentially important, albeit relatively unstudied, contributors to nitrogen fixation in terrestrial and aquatic ecosystems (Nardi et al. 2002). Many terrestrial arthropods including millipedes, beetles, and termites house nitrogen-fixing bacteria (Nardi et al. 2002). It is likely that many other arthropod taxa also host nitrogen-fixing bacteria, but the data are limited. In addition, nitrogen fixing has been reported in earthworms and shipworms (little marine boring clams that eat wood from the submerged wood of shipwrecks; Distel et al. 1991). As you would expect, nitrogen fixation is most prevalent in wood-eating invertebrates that must cope with astoundingly high C:N ratios (as high as 1000:1). A phylogenetic analysis documenting this putative association, however, is needed. The notion that some insects could fix nitrogen was, at first, regarded with healthy skepticism (Wigglesworth 1972). However, evidence accumulated, and now the existence of nitrogen-fixing is uncontroversial. It has been especially well documented and studied in termites, and we will consider it again in chapter 6.

The evidence for nitrogen-fixing symbioses in invertebrates is varied. It ranges from whole-animal unbalanced nitrogen budgets to molecular analyses of symbionts (reviewed by Nardi et al. 2002). The unbalanced-budget studies are instructive. Baker (1969) found that the amount of nitrogen accumulated by the wood-boring beetle *Anobium punctatum* was 2.5 times higher than the amount of nitrogen assimilated from food during development. Clearly, the beetles had to obtain the nitrogen from somewhere and the atmosphere appeared a likely source. In another series of studies, Becker (1942a,b) found that the growth rate of larvae of wood-boring beetles differed depending on whether the different species contained gut symbionts or not. If the beetles contained gut symbionts, their growth was independent of the nitrogen content of wood. If they did not, their growth rate was linearly related to the nitrogen content of wood. Again, it appeared that the beetles with symbionts had a nondietary source of nitrogen.

The evidence in support of symbiotic nitrogen fixation provided by these studies was strengthened when the acetylene reduction method was introduced (see figure 5.18). The acetylene reduction method has been used in nearly two dozen termite species. Breznack (2000) has emphasized the large intra- and interspecific variation in nitrogen fixation that these studies have measured. Some of this variation is undoubtedly biologically significant. For example, Curtis and Waller (1998) have shown that termite castes differ in nitrogenase activity and that the capacity to fix nitrogen varies seasonally (figure 5.19). Some variation, however, may be due to methodological quirks. The activity of nitrogenase is strictly and rapidly regulated by a variety of factors. It is blocked by oxygen and by fixed nitrogen. Hence assays for its activity in whole animals are sensitive to the physiological status of the animal and its symbionts. Many of the acethylene reduction measurements in the literature may underestimate nitrogen-fixation capacities (Breznac 2000).

More recently, nitrogen fixation has been traced to specific bacteria within the diverse communities that inhabit insect guts with the development of molecular and biochemical tools (in situ hybridization and immunolabeling described in chapter 6). These tools allow detection of nitrogenase (*nif*) genes and make it possible to determine the location and distribution of individual microbial cells and their nitrogenase transcripts and proteins (Nardi et al. 2002). These methods are useful, but they do not give estimates of fixation rates, which are what we need in order to find the magnitude of the contribution that these symbioses make to nitrogen fixation in the ecosystems in which they live.

Stable isotope analyses have revealed that some termites receive a significant fraction of their total nitrogen from nitrogen-fixing symbionts. Unlike many other transformations of nitrogen (see chapter 8), nitrogen fixation results in a relatively small fractionation (i.e., $\delta^{15}N_{atmosphere} - \delta^{15}N_{fixed} \approx 0$). In contrast, the soil and plants that do not fix nitrogen have $\delta^{15}N$ values that are more positive than $\delta^{15}N_{atmosphere}$. Tayasu et al. (1994) used mixing models (see chapter 8) to estimate the fraction of nitrogen obtained from the atmosphere in workers of the termite *Neothermes koshuensis*. They found that 30–60% was derived from nitrogen fixation. With the exception of this study, we know of no estimates of the contribution of nitrogen fixation to the nitrogen budget of hosts. Nitrogen

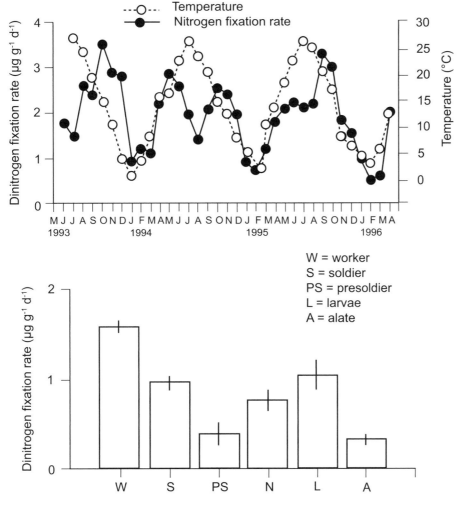

Figure 5.19. The capacity to fix nitrogen varies seasonally (upper panel) in colonies of *Reticulotermes* wood-eating termites. At any time, it also varies among termite castes (lower panel). It is highest among workers and lowest in larvae and presoldiers. Redrawn from Curtis and Waller (1998).

fixing bacteria are undoubtedly important for invertebrates that eat diets with high C:N ratios (such as wood), we just don't know how important.

It is perhaps not surprising that we also know little about how nitrogen fixed by bacteria is transferred to the host. Do termites farm or milk their symbionts? Does ammonia leak from the symbionts to the host or do hosts get more complex nitrogenous compounds from symbionts from either transfer or digestion? The hindgut is the putative site for the uptake of compounds produced by symbionts, but as pointed by Bignell (2000), there is virtually no published information on exchanges across the hindgut wall. Most arthropods, including termites, house their nitrogen-fixing symbionts in the anterior hindgut, which has low levels of hydrolytic enzymes. If one considers a termite alone, it is difficult to envision how termites can digest their symbionts.

The situation changes if one considers the intense sociality of these insects. In a tantalizing report, Fujita et al. (2001) found high lysozyme activity in the salivary glands and in the foregut of the wood-eating termite *Reticulitermes speratus*. This lyzosyme is probably secreted only in the salivary gland and is transferred to the foregut. Fujita et al. (2001) speculate that this lysozyme helps termites to digest bacteria transferred from termite to termite by trophallaxis. This term is derived from the Greek words *trophos* ("one who feeds") and *allax* ("crosswise"), and refers to the direct transfer of liquids, including suspended particles, such as bacteria, from one nest mate to another by either regurgitation ("stomodeal trophallaxis") or anal feeding ("proctodeal trophallaxis"). It is plausible that lysis of bacterial cell walls takes place in the foregut, hydrolysis of the bacterial cell contents occurs in the midgut where termites express high protease activities (Fujita 2001), and absorption of hydrolysates occurs in the hindgut. Usually the catalytic activity that allows protein digestion occurs in the fore- or midgut. Thus, animals that house symbiotic microbes in the hindgut face the problem of how to digest them (see chapter 6). Sociality, and the social exchanges of food that accompany it, may solve this problem. As we will describe in the next chapter, sociality and nutritional symbioses may be inextricably linked in termites.

How important is N_2 fixation by arthropod-symbiont associations for the nitrogen budget of terrestrial ecosystems? This question is difficult to answer for several reasons. The first one is that we do not even know how many arthropods house nitrogen-fixing symbionts. The other one is that there are few published accounts of the quantification of the rate of N_2 fixation by arthropods. To obtain an educated guess, Nardi et al. (2002) conducted informative back-of-the-envelope calculations. They estimated that at densities typical of a tropical rainforest (1 g/m^2), termites alone can account for from 2 to 15 kg of nitrogen fixed per hectare, which includes the average global N_2 fixation rate for terrestrial ecosystems (see also figure 5.20). These estimates, of course, must be taken with a grain of salt. However, they are suggestive of the global importance of insect gut microbes for the global nitrogen budget. Nitrogen-fixing symbioses may have another hidden, but perhaps as important, ecological role. Nitrogen is undoubtedly limiting for all terrestrial organisms that decompose wood and litter. Even small rates of N fixation can be critical to maintain or increase the rates of growth of such organisms and hence increase the rate of carbon cycling (figure 5.20). Although we have emphasized the importance of nitrogen fixation by animals either on a global basis, or per unit area in a given ecosystem, they may also be important because they create heterogeneity in the

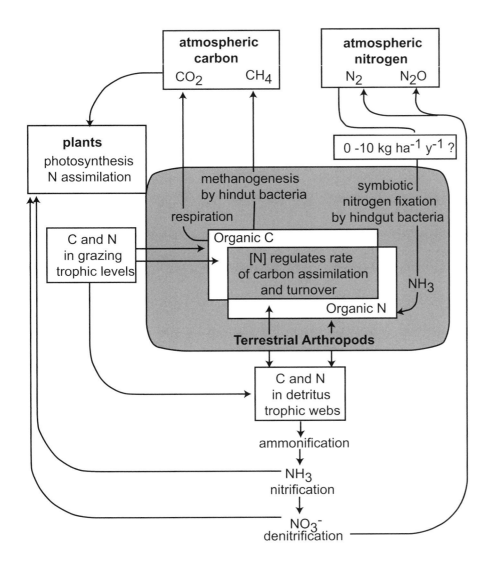

Figure 5.20. Approximately 68% of the nitrogen added to New Hampshire's Hubbard Brook Forest comes from nitrogen fixation ($\approx$ 14.2 kg/ha). However, light-demanding nitrogen-fixing plants exist in sufficient numbers in the forest's closed canopy. Free-living nitrogen-fixing heterotrophs in soil ecosystems can only account for a few kilograms of nitrogen per hectare per year. Nardi et al. (2002) assumed that only about 4 kg/ha of N_2 are fixed by free-living soil microorganisms. Thus, approximately 10 kg/ha remain unaccounted for and may be the contribution of symbiotic N_2 fixation by arthropods. The nitrogen fixed by arthropods is profoundly important because it is an important factor in the regulation of carbon assimilation and turnover in ecosystems. (After Nardi et al. 2002.)

spatial distribution of soil nitrogen. Termite mounds and termite-infested logs are probably hotspots of nitrogen availability (Curtis and Waller 1998).

References

Allemand, D., P. Furla, and S. Bénazet-Tambutté. 1998. Mechanisms of carbon acquisition for endosymbiont photosynthesis in Anthozoa. *Canadian Journal of Botany* 76:925–941.

Anthony, K.R.N., and K. Fabricius. 2000. Shifting roles of heterotrophy and autotrophy in coral energetics under varying turbidity. *Journal of Experimental Marine Biology and Ecology* 252:221–253.

Arndt, C., F. Gaill, and H. Felbeck. 2001. Anaerobic sulfur metabolism in thiotrophic symbioses. *Journal of Experimental Biology* 204:741–750.

Baker, A. C. 2003. Flexibility and specificity in coral-algal symbiosis: Diversity, ecology, and biogeography of Symbiodinium. *Annual Review of Ecology and Evolutionary Systems* 34:661–689.

Baker, A. C., C. J. Starger, T. R. McClanahan, and P. W. Glynn. 2004. Corals adaptive response to climate change. *Nature* 430:741.

Baker, J. M. 1969. Digestion of wood by *Anobium punctatum* DeGerr and comparison with other wood boring beetles. *Proceedings of the Royal Entomological Society of London C* 33:31–39.

Barnes, D. J., and B. E. Chalker. 1990. Calcification and photosynthesis in reef-building corals and algae. Pages 109–132 in Z. Dubinsky (ed.), *Ecosystems of the world 25: Coral reefs*. Elsevier, Amsterdam.

Becker, G. 1942a. Untersuchungen über die Ernährungphysiologie der Hausbokkäferlarven. *Seitschrift für Verleinchende Physiologie* 29:315–388.

Becker, G. 1942b. Öcologische und physiologische untersuchungen über die holzzerstörenden larven von *Anobium punctatum* DeGeer. *Seitschrift für Morphologie und Öcologie der Tiere* 39:98–52.

Bignell, D. E. 2000. Introduction to symbiosis. Pages 198–208 in T. Abe, D. E. Bignell, and M. Higashi (eds.), *Termites: Evolution, sociality, symbiosis, and ecology*. Kluwer Academic, Dordrecht, The Netherlands.

Boetius, A., and H. Felbeck. 1995. Digestive enzymes in marine invertebrates from hydrothermal vents and other reducing environments. *Marine Biology* 122:105–1113.

Boetius, A., K. Ravenschlag, C. Schubert, D. Rickert, F. Widdel, A. Gieseke, R. Amann, B. B. Jørgensen, U. Witte, and O. Pfannkuche 2000. A marine microbial consortium apparently mediating anaerobic oxidation of methane. *Nature* 407:623–626.

Bosch, C., and P. P. Grassé. 1984. Cycle partiel des bactéries chimioautotrophes symbiotiques et leurs rapports avec lés bactérioctes chez *Riftia pachyptila* Jones (Pogonophore Vestimentifére). II. L'evolution des bactéries symbiotiques et des bactériocytes. *Comptes Rendus Hebdomadaires des Sciences de l'Academic des Sciences* 200:413–419.

Breznack, J. A. 2000. Ecology of prokaryotic microbes in the guts of wood- and litter-feeding termites. Pages 209–301 in T. Abe, D. E. Bignell, and M. Higashi (eds.), *Termites: Evolution, sociality, symbiosis, and ecology*. Kluwer Academic, Dordrecht, The Netherlands.

Bright, M., H. Keckeis, and C. R. Fisher. 2000. An autoradiographic examination of carbon fixation, transfer and utilization in the *Riftia pachyptila* symbiosis. *Marine Biology* 136:621–632.

Buddenmeier, R., and D. G. Fautin. 1993. Coral bleaching as an adaptive mechanism. *Bioscience* 43:320–326.

Bui, E.T.N., P. J. Bradley, and P. J. Johnson. 1996. A common evolutionary origin for mitochondria and hydrogenosomes. *Proceedings of the National Academy of Science* 93:9651–9656.

Cary, S. C., C. R. Fisher, and H. Felbeck. 1988. Mussel growth supported by methane as sole carbon and energy source. *Science* 240:78–80.

Cavanaugh, C. M. 1994. Microbial symbiosis: patterns of diversity in the marine environment. *American Zoologist* 34:79–89.

Cavanaugh, C. M., P. R. Levering, J. S. Maki, R. Mitchell, and M. E. Lindstrom. 1987. Symbiosis of methylotrophic bacteria and deep-sea mussels. *Nature* 325:346–348.

Chalker, B. E., W. C. Dunlap, and J. K. Oliver. 1983. Bathymetric adaptations of reef-building corals at Davies Reef, Great Barrier Reef, Australia 2. *Journal of Experimental Marine Biology and Ecology* 73:37–56.

Childress, J. J., and C. R. Fisher. 1992. The biology of hydrothermal vent animals: Physiology, biochemistry, and autotrophic symbioses. *Oceanographic and Marine Biological Annual Review* 30:337–441.

Childress, J. J., C. R. Fisher, J. M. Brooks, M. C. Kennicut II, R. Bidigare, and A. E. Anderson. 1986. A methanotrophic marine molluscan (Bivalvia, Mytilidae) symbiosis: mussels fueld by gas. *Science* 233:1306–1308.

Cordes, E. E., D. C. Bergquist, K. Shea, and C. R. Fisher. 2003. Hydrogen sulphide demand of long-lived vestimentiferan tube worm aggregations modifies the chemical environment at deep-sea hydrocarbn seeps. *Ecology Letters* 6:212–219.

Curtis, A. D., and D. A. Waller. 1998. Seasonal patterns of nitrogen fixation in termites. *Functional Ecology* 12:803–805.

De Cian, M. C., C. Andersen, X. Bailly, and F. H. Lallier. 2003. Expression and localization of carbonic anhydrase and ATPases in the symbiotic tubeworm *Riftia pachyptilla*. *Journal of Experimental Biology* 206:399–409.

D'Elia, C. F., and W. J. Wiebe. 1990. Biochemical nutrient cycles in coral-reef ecosystems. Pages. 49–74 in Z. Dubinsky (ed.), *Ecosystems of the world 25: Coral reefs.* Elsevier, Amsterdam.

DeLong, E. F. 2000. Resolving a methane mystery. *Nature* 407:577–579.

Distel D. L., E. F. Delong, and J. B. Waterbury. 1991. Phylogenetic characterization and in-situ localization of the bacterial symbionts of shipworms (Terenidae, Bivalvia) by using 16S ribosomal RNA sequence analysis and olygodeoxynucleotide probe hybridization. *Applied and Environmental Microbiology* 57:2376–2382.

Douglas, A. E. 1994. *Symbiotic interactions.* Oxford University Press, Oxford.

Douglas, A. E., L. B. Minto, and T. L. Wilkinson. 2001. Quantifying nutrient production by the microbial symbionts in an aphid. *Journal of Experimental Biology* 204: 349–358.

Dubinsky, Z., and N. Stambler. 1996. Eutrophication, marine pollution and coral reefs. *Global Change Biology* 2:511–526.

Dustan, P. 1979. Distribution of zooxanthellae and photosynthetic chloroplast pigments of the reef forming coral Montanstrea annularis, Ellis and Solander, in relation to depth on a West Indian reef. *Bulletin of Marine Science* 29:79–95.

Falkowski, P. G. 1997. Evolution of the nitrogen cycle and its influence on the biological sequestratin of CO_2 in the ocean. *Nature* 387:272–275.

Falkowski, P., S. Dubinsky, L. Muscatine, and J. W. Porter. 1984. Light and the bioenergetics of a symbiotic coral. *Bioscience* 24:705–709.

Falkowski, P. G., P. L. Jokiel, and R. A. Kinzie III. 1990. Irradiance and corals. Pages. 89–108 in Z. Dubinsky (ed.), *Ecosystems of the world 25: Coral reefs.* Elsevier, Amsterdam.

Fiala-Médioni, A., C. Métivier, A. Herry, and M. Le Pennec. 1986. Ultrastructure of the gill of the hydrothermal vent mytilid *Bathymodiolus sp. Marine Biology* 92:65–72.

Fiala-Médioni, A., J. C. Michalski, J. Jollés, C. Alfonos, and J. Montreuil. 1994. Lysosomic an lisozyme activities in the gills of bivalves from deep-sea hydrothermal vents. *Comptes Rendus Hebdomadaires des Sciences de l' Academic des Sciences* 317:239–244.

Fisher, C. M. 1990. Chemoautotrophic and methanotrophic symbioses in marine invertebrates. *Reviews of Aquatic Sciences* 2:399–436.

Fisher, C. M., and J. J. Childress. 1986. Translocation of fixed carbon from symbiotic bacteria in the gutless bivalve *Solemya reidi. Marine Biology* 93:59–68.

Fisher, C. R., J. J. Childress, and E. Minnich. 1989. Autotrophic carbon fixation by the chemoautotrophic symbionts of *Riftia pachyptila. Biological Bulletin* 177:375–385.

Fisher, C. R., I. A. Urcuyo, I. A. Simkins, and E. Nix. 1997. Life in the slow lane: growth and longevity of cold-seep vestimentiferans. *Marine Ecology* 18:83–94.

Fitzgerald, L. M. and A. N. Szmant. 1995. Biosynthesis of "essential" amino acids by scleractinian corals. *Biochemical Journal* 322:213–221.

Forrest, M. 2003. The hunt for frozen gas fields. *Materials World* 11:22–24.

Fujita, A. I., I. Shimizu, and T. Abe. 2001. Distribution of lysozyme, protease, and amono acid concentration in the guts of a wood-feeding termite, *Retiuclitermes speratus* (Kolbe): Possible digestion of symbiont bacteria transferred by trophollaxis. *Physiological Entomology* 26:116–123.

Gattuso, P., D. Allemand, and M. Frankignouelle. 1999. Photosynthesis and calcification at cellular, organismal, and community levels in coral reefs: A review on interactions and control by carbonate chemistry. *American Zoologist* 39:160–183.

Glynn, P. W. 1996, Coral reef bleaching: facts, hypotheses and implications. *Global Change Biology* 2:495–509.

Goffredi, S. K., J. J. Childress, N. T. Desaulniers, R. W. Lee, F. H. Lallier, and D. Hamilton. 1997a. Inorganic carbon acquisition by the hydrothermal vent tubeworm *Riftia pachyptila* depends upon high external P_{CO_2} and upon proton-equivalent ion Transport by the worm. *Journal of Experimental Biology* 200:883–896.

Goffredi, S. K., J. J. Childress, N. T. Desaulniers, R. W. Lee, and F. H. Lallier. 1997b. Sulfide acquisition by the vent worm *Riftia pachyptila* appears to be via uptake of HS^- rather than H_2S. *Journal of Experimental Biology* 200:2609–2616.

Goffredi, S. K., and J. J. Childress. 2001. Activity and inhibitor sensitivity of ATPases in the hydrothermal vent tubeworm *Riftia pachyptila*: A comparative approach. *Marine Biology* 138:259–265.

Grieshaber, M. K., and S. Völke. 1998. Animal adaptations for tolerance and exploitation of poisonous sulfide. *Annual Review of Physiology* 60:33–53.

Gupta, R. S., and G. B. Golding. 1996. The origin of eukaryotic cells. *Trends in the Biochemical Sciences* 21:166–171.

Hammer, B. T., M. Fogel, and T. C. Hoering. 1998. Stable carbon isotope ratios of fatty acids in seagrass and redhead ducks. *Chemical Geology* 152:29–41.

Helmuth, B.S.T., B.E.H. Timmermann, and K. P. Sebens. 1997. Interplay of host morphology and symbiont microhabitat in coral aggregations. *Marine Biology* 130:1–10.

Henze, K., C. Schnarrenberger, and W. Martin. 2001. Endosymbiotic gene transfer germane to endosymbiosis: the origins of organelles and the origins of eukaryotes. Pages 342–352 in M. Syvanen and C. Kado (eds.). *Horizontal gene transfer*. Academic Press, London.

Hoegh-Guldberg O., R. Hinde, and L. Muscatine. 1986. Studies on a nudibranch that contains zooxanthellae. II. Contribution of zooxanthellae to animal respiration (CZAR) in *Pteraeolidia ianthina* with high and low densities of zooxanthellae. *Proceedings of the Royal Society (London) B* 228:511–521.

Jannasch, H. W., and M. J. Mottl. 1985. Geomicrobiology of deep-sea hydrothermal vents. *Science* 229:717–725.

Johnson, K. S., J. J. Childress, R. R. Hessler, C. M. Sakamoto-Arnold, and C. L. Beehler. 1988. Short-term temperature variability in the Rose Garden hydrothermal vent field. An unstable deep-sea environment. *Deep-Sea Research* 35:1711–1722.

Johnson, M., D. Yellowlees, and I. Gilmour. 1995. Carbon isotopic analysis of the free fatty acids in tridacnid-algal symbiosis: interpretation and implications for the symbiotic association. *Proceedings of the Royal Society (London)* B, 260:293–295.

Katz, L. A. 1998. Changing perspectives on the origin of eukaryotes. *Trends in Ecology and Evolution* 13:493–495.

Kochevar, R. E., and J. J. Childress. 1996. Carbonic anhydrase in deep-sea chemoautotrophic symbioses. *Marine Biology* 125:375–383.

Lambers, H., F. S. Chapin, and T. L. Pons. 1998. *Plant physiological ecology*. Springer-Verlag, New York.

Lee, R. W., and J. J. Childress. 1994. Assimilation of nitrogen by marine invertebrates and their chemoautotrophic and methanotrophic symbions. *Applied and Environmental Microbiology* 60:1852–1858.

Leggat, W., E. M. Marendy, B. Baillie, S. M. Whitney, M. Ludwig, M. R. Badger, and D. Yellowlees. 2002. Dinoflagellate symboses: strategies and adaptations for the acquisition and fixation of inorganic carbon. *Functional Plant Biology* 29:309–322.

Le Pennec, M., J. C. Martinez, A. Donval, A. Herry, and P. Beninger. 1992. Enzymologie du tractus digestif de la modiole hydrothermale *Bathymodiolus thermophylus* (Mollusque Bivalve). *Canadian Journal of Zoology* 70:2298–2302.

Levin, L. A., and R. H. Michener. 2002. Isotopic evidence for chemosynthesis-based nutrition of macrobenthos: the lightness of being at Pacific methane seeps. *Limnology and Oceanography* 47:1336–1345.

Lonsdale, P. 1977. Clustering of suspension-feeding macrobenthos near abyssal hydrothermal vents at oceanic spreading centers. *Deep Sea Research* 24:857–863.

Madigan, M. T., J. M. Martinko, and J. Parker. 1997. *Brock's, biology of microorganisms*. Prentice-Hall, Englewood Cliffs, N. I.

Margulis, L. 1993. *Symbiosis in cell evolution,* 2nd edition. Freeman, New York.

Margulis, L. 1998. *Symbiotic planet: A new look at evolution.* Basic Books, New York.

Martin, W., and R. G. Herrmann. 1998. Gene transfer from organelles to the nucleus: How much, what happens and why? *Plant Physiology* 118:9–15.

Martin, W., and M. Müller. 1998. The hydrogen hypothesis for the first eukaryote. *Nature* 392:37–41.

Masuzawa, T., N. Handa, H. Kitagawa, and M. Kusakabe. 1992. Sulfate reduction using methane sediments beneath a bathyal "cold seep" giant clam community off Hatsushima Island, Sagami Nay, Japan. *Earth and Planetary Science Letters* 110:39–50.

McCollom, T. M., and E. L. Shock. 1997. Geochemical constraints on chemolithoautotrophic metabolism by microorganisms in seafloor hydrothermal systems. *Geochimica et Cosmochimica Acta* 61:4375–4391.

Mérejkovsky, C. 1920. La plante considérée comme un complexe symbiotique. *Bulletin de la Societé Naturelle* 6:17–98.

Moran, N. A. 2001. The coevolution of bacterial endosymbionts and phloem-feeding insects. *Annals of the Missouri Botanical Garden* 88:35–44.

Moran, N. A., and P. Baumann. 2000. Bacterial endosymbionts in animals. *Current Opinions in Microbiology* 3:270–275.

Moran, N. A., and A. Telang. 1998. Bacteriocyte-associated symbionts of insects. *BioScience* 48:295–304.

Muscatine, L. 1990. The role of symbiotic algae in carbon and energy flux in reef corals. Pages 75–88 in Z. Dubinsky (ed.), *Ecosystems of the world 25: Coral reefs.* Elsevier, Amsterdam.

Muscatine, L., and E. Cernichiari. 1969. Assimilation of photosynthetic products of zooxanthellae to coral animal respiration. *Limnology and Oceanography* 26:601–611.

Nardi, J. B., R. I. Mackie, and J. O. Dawson. 2002. Could microbial symbionts of arthropod guts contribute significantly to nitrogen fixation in terrestrial ecosystems? *Journal of Insect Physiology* 48:751–763.

Paull, C. K., A.J.T. Tull, L. J. Toolin, and T. Linick. 1985. Stable isotope evidence for chemosynthesis in an abyssal seep community. *Nature* 317:709–711.

Portier, P. 1918. *Les symbiotes.* Masson, Paris.

Pruski, A. M., and A. Fiala-Médioni. 2000. Thiotaurine is a biomarker of sulfide-based symbiosis in deep-sea bivalve. *Limnology and Oceanography.* 45:1860–1865.

Pruski, A. M., and A. Fiala-Mèdioni. 2003. Stimulatory effect of sulphide on thiotaurine synthesis in threee hydrothermal-vent species from the East Pacific Rise. *Journal of Experimental* 206:2923–2930.

Rau, G. H. 1981. Hydrothermal vent clam and vent tubeworm $^{13}C/^{12}C$: Further evidence of a non-photosynthetic food source. *Science* 213:338–339.

Rau, G. H., and J. I. Hedges. 1979. Carbon-13 depletion in a hydrothermal vent mussel: Suggestion of a chemosynthetic food source. *Science* 203:648–649.

Roberts, J. M., P. S. Davies, and T. Preston. 1999a. Symbiotic anemones can grow when starved: nitrogen budget for *Anemonia viridis* in ammonium-supplemented seawater. *Marine Biology* 133:29–35.

Roberts, J. M., P. S. Davies, L. M. Fixter, and T. Preston. 1999b. Primary site and initial products of ammonium assimilation in the symbiotic sea anemone *Anemonia viridis*. *Marine Biology* 133:29–35.

Robinson, J. J., and C. M. Cavanaugh. 1995. Expression of form I and form II Rubisco in chemoautotrophic symbioses: Implications for the interpretation of stable isotope values. *Limnology and Oceanography* 40:1496–1502.

Rowan, B. 2004. Thermal adaptation in reef coral symbionts. *Nature* 430:742.

Rowan, R., and N. Knowlton. 1995. Intraspecific diversity and ecological zonation in coral-algal symbiosis. *Proceedings of the National Academy of Science* 92:2850–2853.

Rowan, R., N. Knowlton, A. Baker, and J. Jara. 1997. Landscape ecology of algal symbionts creates variation in episodes of coral bleaching. *Nature* 388:265–269.

Rowlings, J. K. 1997. *Harry Potter and the philosopher's stone*. Bloomsbury, London.

Salih, A., A. Larkum, G. Cox, M. Kühl, and O. Hoegh-Guldberg. 2000. Fluorescent pigments in corals are photoprotective. *Nature* 408:850–853.

Salyers, A. A., and D. Whitt. 2001. *Microbiology: Diversity, disease, and the environment*. Fitzgerald Science Press, Bethesda, Md.

Sapp, J. 1994. *Evolution by association: A history of symbiosis*. Oxford University Press, Oxford.

Schaljohann, R., and H. J. Flügel. 1987. Methane-oxidizing bacteria in pogonophora. *Sarsia* 72:91–98.

Scott, K. M., and C. R. Fisher. 1995. Physiological ecology of sulfide metabolism in hydrothermal vent and cold seep vesicomyid clams and vestimentiferan tubeworms. *American Zoologist* 35:102–111.

Shigenobu, S., H. Watanabe, H. Hattori, M. Sakaki, and H. Ishikawa. 2000. Genome sequence of the endocellular bacterial symbiont of aphids *Buchnera sp.* APS. *Nature* 407:81–86.

Streams, M. E., C. R. Fisher, and Fiala-Médioni. 1997. Methanotrophic symbiont location and fate of carbon incorporated from methane in a hydrocarbon seep mussel. *Marine Biology*. 129:465–476.

Tayasu, I., A. Sugimoto, E. Wada, and T. Abe. 1994. Xylophagous termites depending on atmospheric nitrogen. *Naturwissenschaften* 81:229–331.

Timmis, J. N., M. A. Ayliffe, C. Y. Huang, and W. Martin. 2004. Endosymbiotic gene transfer: Organelle genomes forge eukaryotic chromosomes. *Nature Review Genetics* 5:123–135.

Toller, W. W., R. Rowan, and N. Knowlton. 2001. Repopulation of zooxanthellae in the Caribbean corals *Monstraea ammularis* and *M. faveolata* follwing experimental and disease associated bleaching. *Biological Bulletin* 201:360–373.

Trask, J. L., and C. L. Van Dover. 1999. Site-specific and ontogenetic variations in nutrition of mussels (*Bathymodiolus sp.*) from the Lucky Strike hydrothermal vent field, Mid-Atlantic Ridge. *Limnology and Oceanography* 44:334–343.

Trench, R. K. 1971. The physiology and biochemistry of zooxanthellae symbiotic with marine coelenterates. I. The assimilation of photosynthetic products of zooxanthellae by two marine coelenterates. *Proceedings of the Royal Society (London) B* 177:225–235.

Trench, R. K. 1993. Microalgal-invertebrate symbioses: a review. *Endocytobiosis Cell Research* 9:135–175.

Van Dover, C. L. 2000. *The ecology of deep-sea hydrothermal vents.* Princeton University Press, Princeton, N.J.

Vacelet, J., N. Boury-Esnault, A. Fiala-Médioni, and C. R. Fisher. 1995. A methanotrophic carnivorous sponge *Nature* 377:296.

Vitousek, P. M., J. D. Aber, R. W. Hoarth, G. E. Likens, P. A. Matson, D. W. Schindler, W. H. Schlesinger, and D. G. Tilman. 1997. Human alteration of the global nitrogen cycle. *Ecological Applications* 7:737–750.

Wang, J. T., and A. E. Douglas. 1998. Nitrogen recycling or nitrogen conservation in an alga-invertebrate symbiosis. *Journal of Experimental Biology.* 201:2445–2453.

Wang, J. T., and A. E. Douglas. 1999. Essential amino and synthesis and nitrogen recycling in an alga-invertebrate symbiosis. *Marine Biology* 135:219–222.

Whitehead, L. F., and A. E. Douglas. 2003. Metabolite comparisons and the identity of nutrients translocated from symbiotic alage to an animal host. *Journal of Experimental Biology* 206:3149–3155.

Wigglesworth, V. B. 1972. *The principles of insect physiology.* Chapman and Hall, London.

Wolfe, D. S. 2001. *Tales from the underground: A natural history of subterranean life.* Perseus Books, New York.

Zal, F., E. Leize, F. H. Lallier, S. Hourdez, A. Toulmond, A.V. Dorsselaer, and J. J. Childress. 1998. S-sulfohemoglobin and disulfide exchange: The mechanisms of sulfide binding by *Riftia pachyptila* hemoglobins. *Proceedings of the National Academy of Science* 95:8997–9002.

Zal, F., E. Leize, D. R. Oroa, S. Hourdez, A. V. Dorsselaer, and J. J. Childress. 2000. Hemoglobin structure and biochemical characteristics of the sulphide-binding component from the deep-sea clam *Calyptogena magnifica. Cathiers de Biologie Marine* 41:413–423.

Zehr, J. P., J. B. Waterbury, P. J. Turner, J P. Montoya, E. Omoregle, G. F. Steward, A. Hansen, and D. M. Karl. 2001. Unicellular cyanobacteria fix N_2 in the subtropical North Pacific Ocean. *Nature* 412:635–638.

CHAPTER SIX

Digestive Symbioses: How Insect and Vertebrate

Herbivores Cope with Low-Quality Plant Foods

THE ORGANIC COMPOUNDS contained in plant cell walls are enormously abundant in terrestrial ecosystems and represent a vast potential source of energy for consumers. Just to give two examples, cell walls make up about 90% of the biomass of a tree and from 60 to 70% of the biomass of a grass stem. Unfortunately for terrestrial animals, cell walls are made primarily of cellulose, hemicellulose, and lignin, and these three substances can be very hard to digest. The chemical and physical structures of these compounds make their bonds relatively unreachable to the action of digestive hydrolytic enzymes. Consider cellulose, probably the most abundant organic compound on earth. As described in chapter 2, cellulose is a polymer of glucose molecules joined by β-1,4 linkages. These linkages are not only difficult to break by enzymes, they are also often protected from hydrolysis because they are embedded in highly crystallized microfibrils and surrounded by a more or less amorphous matrix of lignin and hemicellulose. The compounds that make up cell walls are so refractory to digestion that some authors consider them plant defense systems (Abe and Higashi 1991). And yet terrestrial herbivores consume an average of 18% and 55% of all primary production in terrestrial and aquatic ecosystems, respectively (Cyr and Pace 1993). How do animals cope with the refractory materials in plant cell walls?

A persistent and widespread dogma states that animals are unable to hydrolyze cellulose because they lack the required cellulases. Even when it is recognized that some animals can synthesize cellulases, it is often argued that these endogenous enzymes are ineffective (see Van Soest 1994). Thus, it is assumed that the only way that animals can assimilate cellulose is with the digestive aid of symbiotic microorganisms. This dogma is incorrect and needs to be revised (see Watanabe and Tokuda 2001 and following paragraphs), but there is a grain of truth in it. Cellulose is hard to hydrolyze and many, albeit by no means all, animals lack endogenous cellulases. Many animals assimilate cellulose because they have partnerships with microorganisms that are able to digest it. Although many animals assimilate cell wall compounds in partnership with microorganisms, the association between animals and the microorganisms that live within their digestive tracts is far richer than it is widely recognized. Gastrointestinal microorganisms can help hosts to assimilate cellulose, but often they do a lot more. This chapter is devoted to the physiological and ecological intricacies of the interaction between animals and the microorganisms that live in their guts. We divided the chapter into four sections. The first section describes the common metabolic pathways that microorganisms use to ferment cell wall compounds. After this general section, we focus on two fermentative digestive symbioses: We first examine the digestive symbioses of termites and other insects. Then we explore the role that the symbiosis with microbes plays in the digestive ecology of vertebrates. We conclude the chapter with a brief review of herbivory and detritivory in fishes.

Several themes recur throughout the chapter. Animals that feed on diets rich in refractory materials must have the guts to do it. Digestive symbioses provide an opportunity to explore how the ability to assimilate a given diet can be correlated with the morphology of the alimentary canal. The symbiotic microorganisms that inhabit animal guts modify the chemical composition of food. Consequently, the chemical composition of the materials that the host assimilates can be very different from that of the food that it ingests. This disparity often leads to peculiarities in the secondary metabolism of the host. Finally, animals that host fermentative symbionts often farm them. Thus, they must have a mechanism to digest bacterial refractory cell walls. Animals use a myriad of mechanisms to solve these problems. Some of these mechanisms are shared by several taxa; others are unique solutions adopted by only one or a few taxa. Because the many fermentative symbioses tend to be wonderfully idiosyncratic, we adopted a broadly taxonomic approach to take a look at them.

6.1 Fermentation of Cell Wall Materials

There is enormous variation in the details of how microbes degrade cell wall materials within the guts of animals. The gut of most animals that rely on microbes for digestion has a relatively large chamber that hosts a community of fermenting microorganisms. A large chamber facilitates the retention of plant material for a long time, which seems to be necessary because the hydrolysis of cellulose and other refractory cell wall materials can be a slow process

(Alexander 1991). In subsequent sections we will provide many examples of the morphology of animal fermentation vats. Often, but as we shall see not always, fermentation chambers are anoxic. Therefore, the microorganisms housed in these chambers obtain energy from the anaerobic fermentation of organic compounds, including cell wall materials (figure 6.1).

There are two problems that all organisms face when they catabolize energy-yielding organic compounds: (1) They must conserve some of the chemical energy released as ATP, and (2) they must dispose of electrons taken from electron donors. Fermentations are energy-yielding processes in which organic substrates are transformed in such a way that part of the substrate is oxidized at the expense of another part that is reduced (Fenchel and Finlay 1995). Unlike aerobic respiration, fermentations do not involve electron transport chains and do not depend on ultimate electron acceptors such as oxygen (Hochachka and Somero 2002). In fermentations, ATP is produced by substrate-level phosphorylation (figure 6.2). In this mechanism, high-energy phosphate bonds from organic intermediates of the fermentation pathway are transferred to ADP.

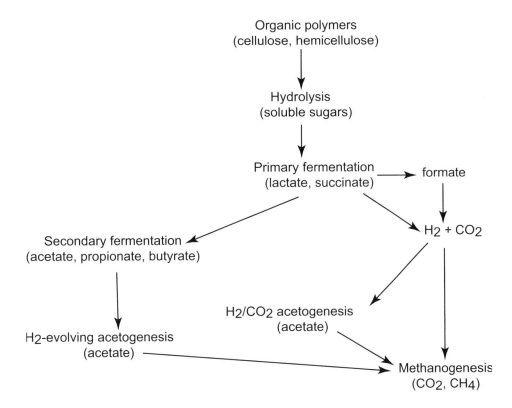

Figure 6.1. The pathways for the complete anaerobic degradations of organic materials involve the combined action of many organisms. The process is not completed in the gut of most organisms with digestive fermentation. In these organisms, the partially reduced products of primary and secondary fermentation and acetogenesis produced by microorganisms are absorbed by the host and catabolized. (Modified from Fenchel and Finlay 1995.)

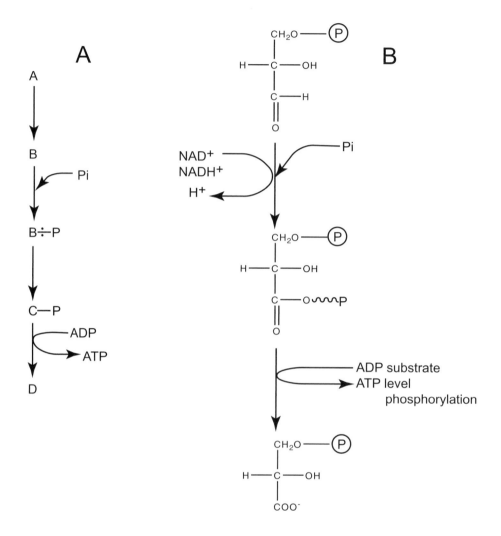

Figure 6.2. In fermentations, ATP is generated by substrate-level phosphorylation (A). In this process, phosphorylation is associated with redox transformations between organic compounds. For example, in the process of glycolysis (B), a six-carbon sugar is first phosphorylated in an ATP-consuming process and then split into two three-carbon compounds (glyceraldehyde–3-phosphate shown as an example in this figure). The first oxidation involves the removal of hydrogen from one of these glyceraldehyde–3-phosphates and the simultaneous reduction of NAD$^+$. The free energy released from this oxidation is conserved with the incorporation of phosphate into a high-energy bond. Now the three-carbon compound has two phosphate groups, one of which has a high-energy linkage. This high-energy phosphate is transferred by the substrate to ADP to make ATP. The remaining phosphate group is later converted to the high-energy linkage in phosphoenolpyruvate, which is then cleaved by the enzyme pyruvate kinase to form a second ATP, again by substrate level phosphorylation. (Modified from Fenchel and Finlay 1995.)

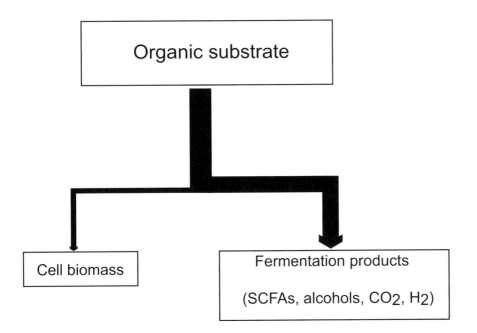

Figure 6.3. In typical microbial fermentations most of the carbon is released as a partially reduced end product of energy metabolism. These end products are absorbed and then oxidized by the host. Only a small amount is used for biosynthesis by the microorganisms.

The second problem is solved by the production and excretion of fermentation products from the original substrate (figures 6.1 and 6.2). In most fermentations, the fermenting organism excretes most of the original substrate's carbon as a partially reduced end product and uses only the remaining small fraction for biosynthesis (figure 6.3). These partially reduced end products are the energy-containing currency that fermenting microorganisms pay to the animals that host them. Fermenting microorganisms produce short-chain fatty acids (SCFAs) as waste products. These SCFAs diffuse across the gut wall into the animal tissues and are used as substrates for aerobic respiration (figure 6.1). The microbes that live within animal guts receive protection and a milieu that is relatively well buffered from the vagaries of the external environment. In return, the animal host receives microbial wastes. Recall from chapter 5 that we called the transfer of useful nutrients from the symbionts to the host "milking." Many fermentative digestive symbioses involve the milking of gut symbionts. Because the host often also digests a fraction of the symbiotic microorganisms that live in its guts, fermentative digestive symbioses can also involve "farming."

6.1.1 Guts as Ecosystems

Animals get a lot more than energy-containing SCFAs from microbes, but for now we will concentrate on these energy-yielding substances. Figure 6.1 illustrates the main pathways for the complete anaerobic degradation of organic polymers in anaerobic systems. It includes both the pathways found in animal guts and those found in freshwater and seawater anoxic environments. Although gut ecosystems share some traits with other anoxic environments, they have significant distinctive characteristics. All anaerobic communities are fueled by the degradation of organic polymers. The starting point in this process is the hydrolysis of complex polymers into smaller soluble products. The organisms responsible for the hydrolysis of these complex substances vary among anaerobic communities. In subsequent sections we will describe these micro-organisms in more detail, but now it is sufficient to mention that a variety of cellulolytic bacteria, protists, and fungi (not to mention a variety of animals) are capable of hydrolyzing complex cell wall materials into smaller soluble compounds. Let us examine how cellulose is hydrolyzed in a bit more detail.

The ease with which cellulose is degraded depends on its crystallinity and on its association with other polymers such as lignin. Therefore, the degradation of cellulose often depends on processes that disrupt the physical structure of cell walls. Grinding by vertebrate teeth and gizzards and by insect mandibles and the extraordinary alkalinity of the midgut of some insects achieve this disruption. Cellulose is hydrolyzed by the concerted action of three types of cellulases: endocellulases, exocellulases, and β-glucosidases. As their name indicates, endo-cellulases act on internal bonds whereas exocellulases act on external bonds. The action of exocellulases yields a 2-glucose disaccharide called cellobiose that is the substrate of the β-glucosidases. The soluble compounds that result from the hydrolysis of cellulose and other cell wall materials are then fermented.

The next fermentative steps in the anaerobic degradation of cell walls differ between animal guts and other anaerobic microbial communities. In gut microbial ecosystems, the principal end products are SCFAs that are absorbed into the bloodstream of the host. In other anaerobic systems, these SCFAs are degraded further into methane, CO_2, and water. Figure 6.1 shows that acetate, propionate, and butyrate are some of the SCFA fermentation end products of microbial fermentation. This figure glosses over an important detail, which is that the fermentation sequence from glucose to SCFAs often involves sequential fermentation by several microbial taxa. We can call the bacteria that initiate the fermentation process "primary fermenters." These primary fermenters produce compounds such as succinate and lactate that can be further fermented by other bacterial taxa, which we can call "secondary fermenters," into SCFAs such as propionate, acetate, and butyrate. These are the SCFAs that are absorbed by the host. The distinction between primary and secondary fermenters emphasizes the observation that the metabolic pathways that generate the compounds that the animal host absorbs are the result of a complex ecological food web whose members exchange substrates.

6.1.2 Discarding Gas: Methane Production in Sewers and Guts

The microorganisms that live within animal guts use only a small amount of carbon to produce methane. Although the production of "enteric" methane is not insignificant, it pales in contrast with that from other systems, such as the anaerobic sediments of wetlands or the sewer systems of water treatment plants. In these systems, microorganisms convert a very large fraction (90% or more) of the total inputs of organic carbon into methane and CO_2. The conversion of organic materials into methane illustrates a significant aspect of anaerobic fermentative systems. These systems include mutualistic associations of microbes that use their combined metabolic paths to degrade compounds. Although each member of the association cannot degrade the compound by itself, the collective action of all members can (Madigan et al. 2000). Microbiologists call such associations of mutualistic dining partners "syntrophs". The word syntroph is derived from the Greeks words *syn* (together) and *trophos* (feeder). Let us briefly consider one of these associations in sediments and then we will consider its counterpart in gut microbial ecosystems.

In sediments, primary fermenters convert carbohydrates to SCFAs and secondary fermenters convert these SCFAs into acetate. Methane producing ("methanogenic") organisms then convert acetate to methane, CO_2, and water. The reactions used by secondary fermenters are often thermodynamically unfavorable under standard conditions. For example, the transformation from butyrate to acetate,

$$\text{butyrate}^- + 2H_2O \rightarrow 2\text{acetate}^- + H^+ + 2H_2,$$

has a ΔG^0 of about $+46.2$ kJ. This reaction is made possible only by the action of methanogenic archaea that remove H_2 in the following reaction:

$$CO_2 + 4H_2 \rightarrow CH_4 + 2H_2O,$$

which has an advantageous ΔG^0 ($\Delta G^0 = -131$ kJ). Because methanogenic archaea remove the H_2 produced in the first reaction immediately, the oxidation of butyrate has a ΔG^0 of about -16. The energetic pickings of the secondary fermenters are lean, but the overall balance of the reaction is favorable. In sewer sludge and sediments, methane is also produced by methanogenic archaea that use acetate as a substrate in the reaction:

$$CH_3COO^- + H_2O \rightarrow CH_4 + HCO_3.$$

Although these acetotrophic ("acetate-feeding") microbes are not taxonomically diverse, they may be ecologically very important. In anaerobic sewer sludges a very large fraction of the total methane produced seems to originate from acetate rather than from H_2 and CO_2 (Ahring and Westermann 1985).

Although methane is not the primary end product in the microbial ecosystems that inhabit animal guts, the removal of hydrogen by methanogens and acetate-producing bacteria is also very important. It is important because it makes possible the regeneration of NAD^+ from NADH. Recall from figure 6.2 that fermentative processes lead to the reduction of NAD^+ into NADH. Accumulation of NADH causes a problem for anaerobes because, if they produce too much of it, it prevents further oxidation of substrate due to lack of a pool of NAD^+ to accept electrons. NADH has to be reoxidized to NAD^+ for the fermentative reactions to continue. If the concentration of H_2 is low, however, then the oxidation of NADH is favored.

In gut ecosystems, H_2 is removed by both methanogenic archaea and by eubacteria that perform the following reactions:

$$(\text{methanogenesis}) \quad CO_2 + 4H_2 \ \rightarrow \ CH_4 + 2H_2O,$$
$$(\text{acetogenesis}) \quad 4H_2 + H^+ + 2HCO_3^- \ \rightarrow \ CH_3COO^- + 4H_2O.$$

Although methanogenic archaea that convert acetate to methane play an important role in sediments, they do not play an important role in gut ecosystems. If they did, the host would suffer a very serious loss of energy. All the acetate that can be absorbed by the host would be transformed into gas and lost to the atmosphere. Van Soest (1994) speculates that methanogens cannot persist in large numbers in gut ecosystems because their generation time is too long relative to the time it takes the contents of the gut to turnover. If Van Soest is right, methanogens are washed out of the fermentation vat before they have time to reproduce.

6.2 Microbial Fermentation in Insect Guts

6.2.1 Cell Wall and Cytoplasm Consumers

Many plant-eating insects that feed on leaves, green shoots, and roots do not degrade cell wall materials to a significant extent, even when these represent a very large fraction of their intake. They skim the cream of their plant diets and assimilate only the cell's contents. Either they ingest enormous amounts of plant material, and defecate the cell wall components more or less intact, or they are very choosy and feed on plant parts that have low cell wall contents. Abe and Higashi (1991) call these animals cytoplasm consumers, and contrast them with cell wall consumers. Among insects, the cytoplasm consumers include most species that eat live plant tissues, whereas the cell wall consumers include most species that feed on dead plant tissues. Insects that chew on foliage, munch on grain and seeds, feed on nectar and pollen, or suck plant juices (phloem and xylem) can be considered cytoplasm consumers, whereas insects that feed on wood (dead and alive) can be considered cell wall consumers. This distinction is, of course, not hard and fast. There are some, perhaps many, species that straddle the two categories. For example, locusts (*Schistocerca gregaria*) are leaf chewers and yet are able to assimilate approximately 60% of the neutral detergent fiber present in leaves (Cazemier et al. 1997a). Recall from chapter 2 that neutral detergent fiber contains cellulose, hemicellulose, and lignin. Thus locusts are able to assimilate a significant fraction of the cell wall content of leaves in spite of having very low numbers of microorganisms within their guts (Cazemier et al. 1997b).

6.2.2 The Cell Wall Consumers and How They Digest Cellulose

The ability to assimilate the refractory compounds in cell walls has evolved independently numerous times among insects. Cellulose digestion has been reported in over 80 species of insects in 20 families belonging to eight orders (table 6.1). Cellulose digestion is common among, but not restricted to, insects that feed on wood. Firebrats and silverfish (order Thysanura) have notoriously catholic taste. They eat cornflakes, glue, wallpaper paste, photographs, and clothes, among many other things, and yet they are proficient cellulose digestors (Treves and Martin 1994). Cellulose digestion appears to be rare among detritus feeders, and very rare in insects that feed on foliage (Martin 1991). To our knowledge, no one has attempted a rigorous analysis that accounts for phylogenetic relationships to test the notion that the ability to digest cellulose, in all its guises, is correlated with feeding habits among insects. Therefore the assessment of the potential correlation between xylophagy (eating wood) and cellulose digestion must be considered plausible but tentative.

Insects use four different mechanisms to digest cellulose (Martin 1991). (1) They rely on an association with symbiotic protozoans that reside in their guts; (2) they exploit the cellulolytic capacity of gut symbiotic bacteria; (3) they can borrow cellulases from ingested fungi; and (4) they can secrete a complete cellulase system. The lower termites (see figure 6.4) and the wood roach (Cryptocercidae) maintain close associations with unique genera of flagellates that are effective

TABLE 6.1
Variety of mechanisms used to assimilate cellulose in many insect taxa

Order	Family	Feeding habit	Mode of cellulose digestion
Thysanura	Lepismatidae	O	EC
Orthoptera	Gryllidae	O	B
	Blattidae	O	EC, B
	Blaberidae	X	P
	Phasmidaea	L	EC
Isoptera	Mastotermitidae	F	FC, EC
	Kalotermitidae		P, EC
	Rhinotermitidae		P, EC
	Termitidae		EC, FE
Plecoptera	Pteronarcyidae	D	FE
Coleoptera	Scarabeidae	X	B, ES
	Bostrichidae	G	B
	Buprestidae	X	FE
	Anobiidae	X	FE
	Coccinelidae	L	EC
	Cerambycidae	X	FE
Trichoptera	Limnephilidae	D	FE
Diptera	Tipulidae	R	B
Hymenoptera	Siricidae		FE

Note: The acronyms used in this table to denote the mechanisms by which insects digest cellulose are P = symbiotic protists, B = symbiotic bacteria, FC = fungal cellulases, and EC = endogenous cellulases. Although it is widely believed that the ability to assimilate cellulose is found predominantly among xylophagous insects, species with other feeding habits also often show this capacity. The acronyms used to denote feeding habits are D = detritivore, F = fungivore, G = granivore, O = omnivore, R = root feeder, X = wood eater. These habits do not refer to the all species in each family, but to those species that have been examined for the digestion of cellulose. We suspect that the ability to digest cellulose will be found to be much more widespread among insects than previously believed. This table was modified from Martin (1991) with a few recent additions (Pellens et al. 2002; Cazemier et al. 1997 a,b; Vazques-Arista et al. 1997).

cellulose digestors (Inoue et al. 2000, and following sections). Cellulolytic bacteria, most often located in the hindgut, help cockroaches (*Periplaneta americana*, Blattidae) and rhinoceros beetles (*Oryctes nasicornis*, Scarabeidae) to digest cellulose. It is likely that bacteria also participate in the assimilation of celluloses in crane flies (Tipulidae). The role of bacterial digestion of cellulose in termites is controversial. Breznak (2000) suggests that cellulolytic bacteria play either no role or a very minor one, whereas Wenzel et al. (2002) hypothesize a large one, at least in higher termites that lack cellulolytic protozoans.

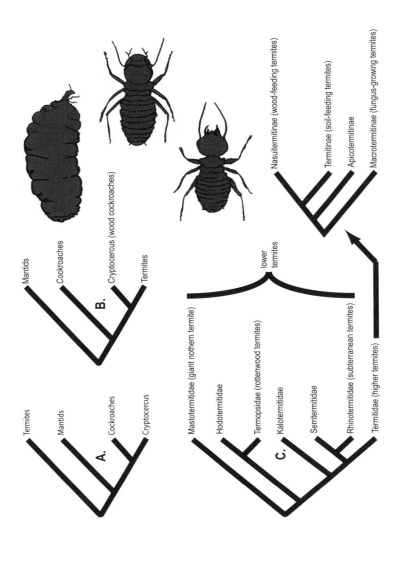

Figure 6.4. Although it is widely accepted that the termites, mantids, and cockroaches are related and form a monophyletic group (the "dictyopteran" insects), the relationships between them remain unclear. One hypothesis is that termites are the sister group of the mantid-cockroach clade (tree B). A more likely hypothesis is that cockroaches and termites are sister clades. The position of wood cockroaches in this scheme is also unclear. In tree B we have assumed that they are the sister clade of termites, a notion supported by many termite systematists, but this may not be the case. Tree C illustrates a hypothesis for the phylogenetic relationships of the main groups of termites. One of the notable differences between the lower and the higher termites is that the lower termites are associated with symbiotic intestinal flagellates whereas the higher termites are not. The drawings illustrate a termite queen (top), a worker (middle), and a soldier (bottom).

The third mechanism, the use of borrowed fungal enzymes, is also controversial. It was first proposed by Martin and Martin (1978) to explain cellulose digestion in the fungus-growing termite (*Macrotermes natalensis*) and then reported in the larvae of a siricid woodwasp (*Sirex cyaneus*), and in the larvae of several species of cerambycid long-horned beetles (Cerambycidae, Kukor et al. 1988). Sinsabaugh et al. (1985) proposed that this mechanism also accounts for the ability of some of the detritivorous nymphs of stoneflies and caddisflies to assimilate cellulose. Although the notion that wood-eating insects acquire the enzymes of their food to digest cellulose is intuitively appealing, the evidence in its favor remains relatively scanty and the hypothesis remains somewhat controversial. David Bignell and his collaborators have suggested that *Macrotermes* uses endogenous cellulases to break up cellulose (Bignell et al. 1994). They documented a fully competent cellulase complex in these termites and used measurements of hydrolysis rates to show that the activity of this complex can account for the complete degradation of cellulose. They argued that, at least in some termites (e.g., *Macrotermes*), the acquired enzyme hypothesis is superfluous. The termites can do it by themselves. The capacity to hydrolyze cellulose with endogenous cellulases, however, does not exclude a role for other mechanisms. We suspect that more than one mechanism is at work in many, and perhaps most, of the species of insects that can digest cellulose.

6.2.3 Termites

Termites are a magnificently diverse group. They are also the best-studied group of cell wall consumers among insects. They belong to the order Isoptera, which has over 2200 fossil and living species. Because termites are common household pests that often damage wood structures, they are frequently thought of as exclusively xylophagous. Although it is true that many termite species feed on wood in various stages of decay, not all do. Some species eat the dung of vertebrate herbivores and others eat humus-rich soil. Termites are divided into six families (figure 6.4), which in turn are split into two major groups: the so-called lower termites with six families (figure 6.4) and the higher termites. Although the higher termites are placed within a single family, they comprise roughly three-fourths of all species. An important difference between the lower and the higher termites is that the former are associated with symbiotic intestinal flagellates whereas the latter are not. Along with several hymenopteran taxa (ants, wasps, and social bees) all termites are eusocial. Eusocial insects share three traits: (1) mothers, along with individuals that may or may not be directly related, cooperate in the care of young, (2) individuals are divided into castes with reproductive division of labor, and (3) generations overlap, which allows for the older generations of offspring to help related, younger generations (Wilson 1971).

Figure 6.4 shows hypotheses for both the phylogenetic relationship of termites with their close relatives, the mantids and the cockroaches, and the relationships among termite families and subfamilies. The close relationship between termites, mantids, and cockroaches is well accepted and these three groups are often classified in the order Dictyoptera. Although the notion that these three groups of insects are monophyletic is not controversial, the phylogenetic hypotheses shown in figure 6.4 are still contentious (Eggleton 2001). We have included potentially conflicting phylogenetic information because the details of the digestive symbioses vary among termite groups. Also, digestive symbioses have been at the heart of fascinating questions about the evolution of termites for a long time, and these questions are best appreciated when framed in a phylogenetic context.

6.2.4 An Overview of Digestive Processes in Termites

Figure 6.5 is a cartoon of the fate of nutrients along the gut of a termite from either the lower or higher termite group. This figure illustrates the observation that termites, like many animals, have two coexisting digestive systems: one, the fore- and midgut, that deals with the soluble and relatively easy to digest cytoplasmic components of the diet, and another, the hindgut, that deals with the insoluble and more refractory cell wall components. The salivary glands and midgut epithelium of termites secrete enzymes that allow them to assimilate soluble protein, starch, and even microbes (see chapter 4). In addition to the normal catalytic digestive system that termites share with other insects, termites also have a symbiont-aided system that allows them to assimilate lignocellulose. In termites, the digestion of lignocellulose takes place with functional contributions of both the termite and its symbionts. Termites contribute to the degradation of lignocellulose in two ways: They use their mandibles and proventriculi to grind food into microscopic-sized particles, and they secrete endogenous cellulases. Milling food into smaller particles with larger surface-to-volume ratios facilitates their engulfing by gut protists and speeds up the action of the termite's endogenous enzymes. Termites secrete cellulases in both the salivary glands and the epithelium of the midgut. In lower termites the salivary glands seem to be more important than the midgut as sites of cellulase secretion but the opposite seems to be the case in the higher termites (Slaytor 2000).

In most insects, the hindgut is relatively short and serves primarily to recover water and electrolytes before feces are voided. The hindgut of termites

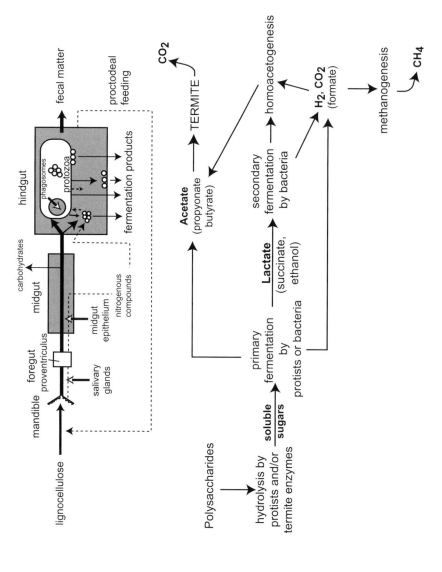

Figure 6.5. The digestive tract of a termite can be divided into the regions shown schematically in the upper panel. The thinner lines represent the materials that have been solubilized and that are absorbed by the termite. The hollow arrows represent sites where cellulolytic endogenous enzymes are secreted. The lower panel represents the metabolic pathways involved in the metabolic degradation of polysaccharides in the hindgut of a termite. (These two diagrams were modified from Brune 2003.)

and other xylophagous insects is very different. It is characterized by one or more greatly enlarged sections, which are colorfully called "paunches" (figure 6.5). These paunches increase the residence time of digesta and prolong the digestive and metabolic activities of the diverse microbiota that they lodge. The microbiota that inhabits termite guts is very diverse and microbiologists are still in the process of characterizing it. For example, the hindgut of *Reticulitermes flavipes* harbors six species of flagellates and from 20 to 30 distinct bacterial morphotypes (Brune and Friedrich 2000). However, because the majority of all prokaryotes in the termite hindgut resist cultivation, these numbers represent minimal estimates. In spite of years of work, it remains difficult to characterize all members of the termites' microbiota and to assign specific roles to all of its members. In subsequent sections we will describe the role of some of the best-studied taxa. Here it must suffice to mention cellulolytic protists, a large diversity of fermenting eubacteria, methanogenic archaea, and homoacetogenic eubacteria.

6.2.5 Symbiosis and the Evolution of Termites

In lower termites, the solid particles of food that enter the hindgut are engulfed by intestinal protists that break them up and ferment the resulting sugars. The protists are oxygen sensitive (exposure to oxygen kills them), make up a very large fraction of the volume of the hindgut, and are essential for the digestion of lignocellulose. The protists that live within the hindgut of lower termites are in unique genera in the orders Hypermatigida, Trichomonida, and Oxymonadida. All these protists are anaerobic and have no mitochondria in their cells. In the case of Hypermatigida and Trichomonida, the protist cells contain anaerobic energy-producing organelles called hydrogenosomes (chapter 5). If the termite's gut is defaunated, the termites die (Inoue et al. 2000). Because the defaunated termites can survive if they are fed on starch rather than cellulose, one of the important roles of protists must be to aid in the digestion of cellulose (Lebrun et al. 1990). This does not mean that termites are completely unable to digest cellulose or that cellulose digestion is the only role of the symbiotic protists.

A persistent dogma, and one that is found in almost every biology and biochemistry textbook, is that termites lack endogenous cellulases. This dogma is false. Without apparent exception, termites produce endogenous cellulases (Slaytor 2000). Not only have endogenous cellulases been purified and characterized, the cDNAs of many endogenous cellulases have also been cloned (Watanabe and Tokuda 2001). If termites can secrete endogenous cellulases, then why are the lower termites associated with protists and why do they die when these are removed from their guts? Protists probably contribute to the termite's capacity to hydrolyze cellulose in three ways: (1) by processing the lignocellulose in wood physically, (2) by expressing cellulases that have complementary catalytic functions to those of the termite's cellulase (remember that not all cellulases are equal), and (3) by supplementing the termite's total hydrolytic capacity. Lower termites may be some of the best examples of insects in which more than one cellulolytic mechanism contributes to the processing of cell wall materials (Nakashima et al. 2002). It is worth emphasizing again that, with very few exceptions, the higher termites do not host cellulolytic protists (Breznak 2000). Thus, cellulose hydrolysis in this large and very successful group is dependent on the action of endogenous cellulases, and perhaps also on the aid of cellulolytic bacteria—but the contribution of the latter has not been conclusively demonstrated.

The so-called lower termites are not the only insects that house cellulolytic protists. The wood cockroaches in the genus *Crytocercus* also house flagellates closely related to those found in termites (Grandscolas and Deleporte 1996). Although wood cockroaches cannot be considered social insects, they are gregarious and have relatively sophisticated brood care behavior. They live in small family groups in rotting moist logs. The family groups include one or both parents and their offspring (Nalepa 1984). Like lower termites, wood cockroaches indulge in proctodeal trophallaxis (see section 5.5.4). Because the adult symbiotic protists are very sensitive to exposure to oxygen, they must be transferred across individuals, presumably by proctodeal feeding. The need for transgenerational "protist transfaunation" may have favored the evolution of subsociality in *Cryptocercus* and of eusociality in termites (Thorne 1997, and figure 6.8 in this chapter).

Because wood cockroaches are subsocial, eat wood, and have protist symbionts that must be transferred from individual to individual, protozoologist L. R. Cleveland inferred that they are good candidates for what the ancestor of the termites might have looked like (Cleveland et al. 1934). In more contemporary terms, Cleveland proposed that termites and *Cryptocercus* were sister taxa and that the presence of gut protists was a shared derived trait (a "synapomorphy" in the arcane jargon of phylogenetic systematists) of the termite/*Cryptocercus* clade. The *Cryptocercus*/termite sister clade hypothesis is seemingly supported by the taxonomic similarity of the protists in the two insect taxa. However, Thorne (1997) observed that when species of termites live together with wood cockroaches, they often fight with each other and feed on the bodies of fallen enemies. Thus, the gut flagellates of termites and *Cryptocercus* may be similar as a result of either common ancestry or of one or many necrophagic transfaunation events.

Cleveland's hypothesis has generated lots of attention, but it has not been settled. The pendulum for a sister relationship between termites and *Cryptocercus* has swung back and forth (see Eggleton 2001 for a summary of the controversy). Either the presence of protists in *Cryptocercus* and termites will prove to be the result of a shared common ancestor (A in figure 6.4) or this trait will prove to be a nice example of evolutionary convergence (B and C in figure 6.4). Ultimately, this controversy can be decided only with a well-resolved phylogeny of the termites and the cockroaches. As the controversy simmered, Pellens et al. (2002) discovered another subsocial cockroach (*Parasberia boleriana*) in the Atlantic forests of Brazil. They found that this cockroach was xylophagous and had high numbers of symbiotic gut flagellates. Because this creature is unrelated to either termites or *Cryptocercus*, Pellens et al. (2002) stated with a great deal of certainty that the presence of protist symbionts in its gut is the result of an independent evolutionary event. The association between symbiotic protists and insects has thus evolved convergently more than once among insects.

6.2.6 Of Endocytobiotic Symbionts and Russian Stacking Dolls

Looking at a symbiosis can sometimes be like opening a Russian stacking doll. When you open the first doll, neatly stacked within it is another doll. If you were to dissect the hindgut of *Zootermopsis*, the pine woods termite, you would find that it contains vast numbers of a wood- chip-engulfing symbiont called *Trichomitopsis*, along with a number of other species such as *Streblomastix* and *Trichomonas*, a genus of protist commonly found in warm, moist interiors of virtually all species of animals. Now imagine that you focus on these symbionts. Within the cytoplasm of *Trichomitopsis* you would find endocytobiotic methanogenic archaea (Lee et al. 1987), and lining the exterior of *Streblomastix* you would find many ectobiotic rod-shaped bacteria undulating their flagella (Dyer and Khalsa 1993). The words "endocytobiotic" and "ectobiotic" mean living within the cytoplasm or on the external surface of the protists, respectively. In lower termites one or more kinds of endocytobiotic and ectobiotic prokaryotes are often intimately associated with each species of the flagellated protists.

Although the association between protists and bacteria has intrigued biologists for many years, we still have an incomplete understanding of the role that the prokaryotes play in the biology of their protist hosts. It is reasonable to assume that methanogens associated with anaerobic flagellates benefit by living in close proximity to a source of H_2 and CO_2, but the benefits for the protist are not obvious (Breznak 2000). Not all of the prokaryotes associated with gut protists are methanogenic. Some of the protist-bacteria associations can probably be characterized as motility symbioses. The coordinated undulations of thousands of spirochetes or flagellated rod-shaped bacteria propel the protist and presumably allow it to occupy the most favorable microhabitats within the termite's gut. Dryer and Khalsa (1993) observed that, when the ectobiotic bacteria were removed with antibiotics from the surface of *Streblomastix stix*, the protist was unable to orient in an acetate gradient. They hypothesized that the bacteria serve the dual role of propellers and sensory organs. The presence of sensory/motility symbioses within termite guts hints at the need for the flagellate symbionts to detect and then track changes in resources within the termite's gut. This observation suggests that the gut of termites is not a homogeneous well-mixed pot, but a heterogenous habitat with substantial variation in resources and environmental conditions. The next section examines this possibility.

6.2.7 Termite Guts: The Consequences of being Small

The fermentation chambers of large animals such as whales, kangaroos, and voles are kept anoxic by the respiratory activity of both facultatively and strictly aerobic members of the gut's microbiota. If the volume of the gut is large, the gut's surface to volume ratio is low and therefore the oxygen influx per unit volume is small (figure 6.6). If oxygen penetrates only a fraction of a millimeter

Figure 6.6. Because termite guts are tiny, they have huge surface-to-volume ratios. Consequently, the diffusive influx of O_2 to the termite gut is considerable. In contrast, the diffusive influx into the rumen of a cow is negligible. (After Brune 1998.)

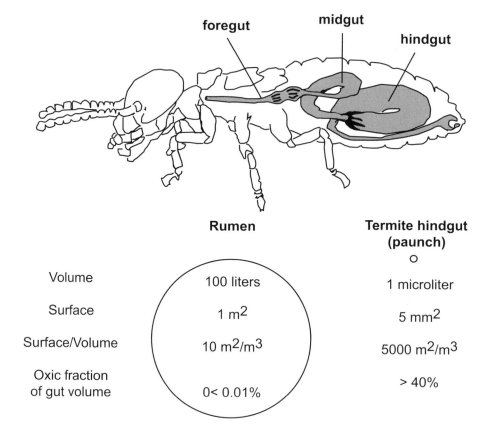

	Rumen	Termite hindgut (paunch)
Volume	100 liters	1 microliter
Surface	1 m²	5 mm²
Surface/Volume	10 m²/m³	5000 m²/m³
Oxic fraction of gut volume	0< 0.01%	> 40%

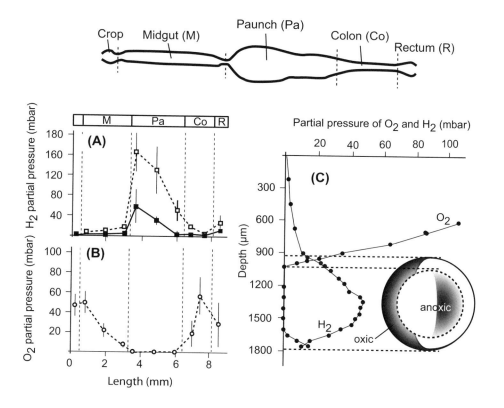

Figure 6.7. A schematic diagram of the gut of *Reticulitermes flavipes* shown unraveled (in the living termite the gut is coiled). Panel A shows the partial pressure of H$_2$ measured under air (closed squares) and under N$_2$ (open squares), and panel B shows the partial pressure of O$_2$ measured under air within the paunch of *R. flavipes*. Panel C illustrates the dramatic gradient in the partial pressure of oxygen and H$_2$ within the paunch of this termite. The dashed lines divide the gut into an oxic area close to the gut's wall and an anoxic area at the center of the gut's lumen. (After Ebert and Brune 1997.)

into the gut's interior, only a very small portion of the gut's volume is affected. Unlike the relatively large guts of vertebrates, termite guts are very small. Their volume is roughly 10^{-8} times smaller than that of a cow's rumen. Their small size greatly increases their surface-to-volume ratio and with it, the potential influence of oxygen. Microbiologist Andreas Brune (1998) estimated that the oxygen influx per unit volume into a termite's gut is 500 times higher than that into a cow's rumen.

Using microelectrodes, Brune (1998) demonstrated that oxygen penetrates a termite's gut to a depth of 50 to 200 μm. Although a fraction of a millimeter may seem a small distance, it is a long way deep into the tiny gut of a termite (figure 6.7). Consequently, the tubular sections of the termite's gut and even quite a large fraction of the periphery of the hindgut paunch are oxygenated (figure 6.7). Brune's measurements suggest that as much as 60% of the gut's volume of termites is oxygenated to some degree. In support of the idea that oxygen plays an important physiological role in termite gut ecosystems, Tholen et al. (1997) found that a very large fraction of the prokaryotes within the termite's gut were microbes that can tolerate oxygen. They found oxygen-tolerant lactic acid bacteria (58%), facultatively anaerobic enterobacteria (20%), and strictly anaerobic bacteria (20%).

The aerobic and aerotolerant microbes act as oxygen sinks and contribute to the existence of a strictly anaerobic core within the termite's gut (figure 6.7).

The presence of oxic environments within the termites' guts has consequences for the metabolic activities of the microbes that live within them. Tholen and her collaborators studied the metabolism of one of the most abundant microbes within the gut of the termite *Reticulotermes flavipes*, a strain of the lactic acid bacterium *Enterococcus* (Tholen et al. 1997). Like most lactic acid bacteria, this microbe could survive in the presence of oxygen. More interestingly, it changed its metabolism in the presence of oxygen. Under anaerobic conditions its final fermentation product was lactate, but in the presence of oxygen the bacteria generated acetate in a process that consumed oxygen and that was strongly dependent on hydrogen. Microbiologists often state that lactic acid bacteria seem to lack porphyrins and cytochromes and do not carry out electron transport phosphorylaton (Madigan et al. 2000). Yet, it appears that the *Enterococcus* that lives within the gut of *R. flavipes* can use oxygen as an electron acceptor. Like systems that depend on electron-transport respiration, the reduction of oxygen by *Enterococcus* is sensitive to cyanide, and seems to be widespread among termite symbiotic bacteria (Boga and Brune 2003). Although the biochemical basis of hydrogen-dependent, oxygen reduction in acetogens is still unclear, recent evidence suggests that lactic acid bacteria may express cytochromes and may have a functional electron transfer chain (Duwat et al. 2001). The guts of termites are still full of microbiological mysteries and surprises.

Figure 6.7 shows that the gut of *R. flavipes* has not only an oxygen gradient, but also a gradient in the concentration of hydrogen which, as we have discussed, is a key metabolic intermediate in fermentative symbioses. Brune and his collaborators have postulated that this gradient may provide suitable conditions for the coexistence of methanogenic archaea and homoacetogenic eubacteria. The gut epithelium of *R. flavipes* appears to be densely populated by a biofilm of methanogens, whereas the interior is populated by homoacetogens. They proposed that homoacetogens consume most of the H_2 within the gut's lumen. The residual hydrogen is scavenged by the biofilm of methanogens before it escapes into the atmosphere. They suggest that in *R. flavipes* methanogens and acetogens coexist in spite of competing for the same resource because they use two distinct microhabitats within the termite's hindgut.

The microheterogeneity of the termite's gut is undoubtedly responsible for the large diversity of microbial taxa that coexist within it. Berchtold et al. (1999) used fluoresecent in situ hybridization (FISH, box 6.1) to explore the distribution of microorganisms within the hindgut of *Macrotermes darwniniensis*. They found a remarkable amount of spatial segregation among the gut's

Box 6.1.

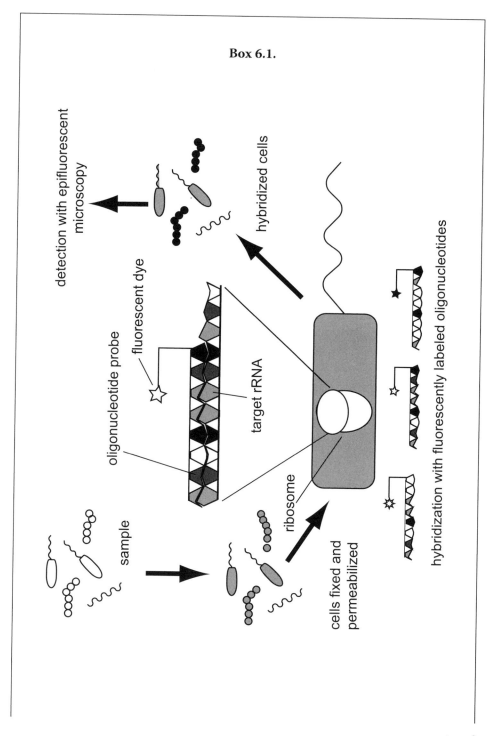

Figure 6.8. Steps required in the identification of bacteria using fluorescent in situ hybridization (FISH).

continued

continued

Fluorescent in situ hybridization (FISH) is one of the most widely used methods to identify and quantify microorganisms. FISH relies on fluorescently labeled oligonucleotide probes (figure 6.8). These probes are either a DNA or an RNA oligonucleotide probe that is complementary to a sequence in a target gene. For bacteria the target is often the 16S rRNA molecule. The 16S rRNA molecule is a major component of the small ribosomal subunit. It forms a part of the ribosomal structure that is the site of protein biosynthesis resulting in the translation of messenger RNA. Microbiologists have generated group-specific 16S rRNA probes against the major phylogenetic groups of bacteria (Amann et al. 1995; Amann and Ludwig 2000). A typical FISH protocol involves several steps: The first involves fixing the material containing the bacteria cells (e.g., a termite's paunch), embedding it in a suitable histological substrate, and making the bacterial membranes permeable to the probes. Membranes are often made permeable by repeatedly freezing and thawing the embedded sample. Then the sample is sectioned, hybridized to the probes, and washed to remove unbound probe. Finally, the sections are viewed under an epifluorescent microscope. The method allows not only quantifying the numbers of bacteria of each type, but also finding out where they are located.

microbes. Some taxa colonized preferentially the thick wall of the hindgut's posterior region, whereas other specialized in the lumen. The microorganisms found in the anterior part of the paunch were associated with flagellates. These represented 95% of the area that could be colonized by microorganisms. The protists are both symbionts of the termite and habitat for bacteria. For a long time the termite gut has been considered a homogeneous well-mixed anaerobic fermenting vat, similar to a beer-brewing kettle. This notion is rapidly being replaced by that of the termite gut as a spatially complex ecosystem characterized by sharp chemical gradients and by significant spatial microhabitat structure.

6.2.8 Symbionts and the Nitrogen Economy of Termites

The food of termites not only is hard to digest, it is also poor in nitrogen. Termites have traits that ameliorate the low nitrogen content of their diet. The Malpighian tubules of termites, like those of most insects, are slender tubular excretory structures that void their contents between the midgut and the hindgut (figure 6.5). Thus, the waste products of the metabolism of protein and nucleic acids, uric acid and urea, are mixed with digesta. After they enter the hindgut they are "mineralized" by microbes. The word "mineralization" is used in geochemistry and ecosystem ecology to denote the transformation of organic carbon into CO_2 and organic nitrogen into ammonia. The ammonia is assimilated by microbes and used to make more microbial protein. It is not known whether the intestinal protists of lower termites assimilate ammonia and use it to manufacture proteins or if they eat other microbes to obtain nitrogen. In either case, it is safe to assume that the symbiotic association between termites and their gut microbes facilitates nitrogen conservation. Do termite/symbiont associations also exhibit nitrogen recycling? (see chapter 5 for definitions of nitrogen conservation and nitrogen recycling).

Because termites secrete lysozyme in the salivary glands, they cannot digest the bacterial cell walls of their own nitrogen-rich microbes that live at the hind —inaccessible—end of the gut. Thus they must use other means to access this source of high-quality protein. As described in chapter 5, in the lower termites one of these means is by trophallaxis. Termites, like other social insects, deliver food to other termites from both the salivary glands and the crop ("stomodeal trophallaxis") and from the hindgut ("proctodeal trophallaxis"). Stomodeal food is delivered primarily to the royal pair and the larvae. Proctodeal food is distributed in a fashion that seems to follow the communist formula that goods must be distributed to each according to need. Machida et al. (2001) found that older instars of *Hodotermis japonica* donate proctodeal fluids to rapidly growing younger instars. They also found that the presence of nitrogen-deprived individuals increased the rate of proctodeal trophallaxis and thus homogenized the distribution of nitrogen among colony members. In lower termites proctodeal trophallaxis seems to contribute to the redistribution of recycled nitrogen within a colony. Proctodeal "food" is very different in composition from feces. In lower termites it contains large numbers of flagellates and bacteria and tends to have a high water content. Feces are dry and lack flagellates. The proctodeal food of lower termites is analogous to the highly nutritious cecal pellets ("cecotrophs") defecated by several species of herbivorous mammals with hindgut fermentation (see following sections).

Although the immature stages of higher termites are completely dependent on stomodeal feeding, adults do not indulge in anal feeding. Because higher termites lack symbiotic intestinal flagellates, it is tempting to infer causation from the association between the loss of protists and the loss of proctodeal trophallaxis (Wilson 1971). This inference is derived from the notion that the most important role of anal feeding is the transfer of symbiotic flagellates (see next section). This is probably not the case. In many lower termites, the gizzard of the young host is too small to allow the passage of flagellates. It takes several molting cycles before the gizzard is large enough to permit safe passage of the largest of these symbionts (Nalepa et al. 2001). Until then, the abundant flagellates in the proctodeal fluid are just good food. We find the loss of both flagellates and proctodeal trophallaxis in higher termites a mystery. Its consequences for nitrogen recycling are, to our knowledge, unknown. Under some circumstances, higher termites eat their own feces and this behavior may be associated with nitrogen recycling. However, the frequency of coprophagy and its effectiveness relative to proctodeal trophallaxis for nitrogen recycling have not been determined.

6.2.9. Detritivory, Coprophagy, and the Evolution of Digestive Mutualisms in the Dyctiopteran Insects

Proctodeal trophallaxis has two functions: (1) At the colony level, it has the consequence of increasing nitrogen recycling and facilitating the use of the nutritionally rich substances synthesized by symbionts. In brief, it allows termites to farm their symbionts. (2) Proctodeal trophollaxis also makes possible the transfer of symbionts among individuals. Termites share with most insects the trait of shedding the chitinous lining of the gut after each molt. Within one or two days after the molt, they eliminate this lining and with it the vital flagellates. To elicit the extrusion of a protist-rich proctodeal droplet, newly defaunated termites must caress the last abdominal segment of another individual with their antennae (Wilson 1971). In lower termites, proctodeal trophallaxis allows the propagation of the inoculum for the fermentative hindgut ecosystem among colony members. In chapter 5 we mentioned that many insects transmit their endosymbiotic bacteria vertically by innoculation by the mother to the progeny. In termites, inoculation of gut symbionts is primarily horizontal.

Entomologist Christine Nalepa proposed a compelling scenario for the evolution of proctodeal trophallaxis. This hypothesis is outlined in some detail in figure 6.9. We find her hypothesis persuasive because it integrates social biology with the intricacies of nutritional symbioses. In short, she argued that the evolution of proctodeal trophallaxis in termites is the culmination of a sequence that involved accidental coprophagy in a detritivore followed by regular intraspecific coprophagy in a subsocial ancestor, and ending in proctodeal trophallaxis. According to this hypothesis, the ancestor of termites was a detritivore that fed on rotting plant material ("detritus"). The term "rotting" has a negative connotation for humans. It denotes spoiled goods. Because we like beer, red wine, and smelly cheeses, we probably must qualify

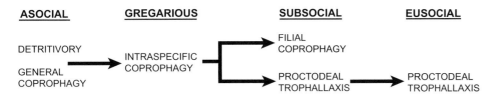

Figure 6.9. Nalepa et al. (2001) proposed the hypothesis that proctodeal trophallaxis in lower termites evolved from the internalization of the microbiota that colonizes rotting detritus and feces. The first step in their evolutionary sequence is from a general detritivore and coprophage to a gregarious creature that relies on intraspecific coprophagy for both nutrition and the transmission of gut microbes. Gregarious behavior favors the transmission of microbes and is exhibited by many unrelated terrestrial detritivores such as millipedes, collembolans, and cockroaches. In sub-social insects the transmission of gut mutualists can take place by either filial coprophagy, when young ingest their parents' feces, or by the more sophisticated proctodeal trophallaxis. Nalepa and her collaborators (2001) hypothesize that proctodeal trophallaxis in termites evolved because the protists that they share with *Cryptocercus* wood cockroaches must be transmitted in the adult (trophic) stage and not as cysts. Their encystement is triggered by the molting cycle of the host, and cysts are only found in the feces of termites in the nymph stage. Thus, the flagellates of termites and wood cockroaches cannot be transmitted from parents to offspring. However, it is possible that this quirk in the developmental physiology of these flagellates evolved after their association with termites and wood cockroaches. This hypothesis can only be tested by mapping the changes in developmental physiology of flagellates unto a well-supported phylogeny. The possibility that proctodeal trophallaxis may have evolved from filial coprophagy cannot be discounted yet.

the term. There are good rotting processes (those that lead to wine and Roquefort) and bad ones (those that lead to lutefisk). The action of microbes greatly improves the nutritional quality of detritus. The action of microbes reduces the C:N ratio of detritus by releasing CO_2 to the atmosphere and by incorporating mineralized nitrogen into the protein of their bodies. It also often initiates the degradation of lignocellulose. In the parlance of ecologists, microbes can transform "recalcitrant" carbon into "labile" carbon and inorganic mineralized nitrogen into organic protein. For detritivores, the expression "microbial conditioning" is much more accurate than the word "rotting" (Swift et al. 1979).

Feces and rotting organic matter are not all that dissimilar. Indeed, ecologists who work with detritivores do not distinguish between fecal pellets and detritus. Fecal pellets can be considered a form of detritus that is easier to colonize by bacteria because it is often constituted by finely ground material with large surface-to-volume ratios. Detritivores consume both microbially conditioned detritus and microbially conditioned feces. Because the microbiota that colonize detritus and feces can be considered equivalent to that inhabiting the guts of animals, Swift et al. (1979) called it the "external rumen." This term has, in our view, unfortunate bovinocentric connotations. Why not call it the external paunch? The biotic composition of the aerobic "external paunch" may be very different from that of its anaerobic internal couterpart. But the term external rumen appeals to the imagination and is widely used. In a nutshell, Nalepa and her collaborators (2001) hypothesized that the sophisticated microbiota of termites is the highly evolved outcome of the internalization of the inhabitants of the external rumen (figure 6.9).

6.2.10 Insect Farmers and Their External Paunches

We have used the word "farming" as a useful analogy to refer to hosts that cultivate and then digest some of their symbionts. The farming hosts that we have dealt with cultivate their symbionts inside of their own bodies. This section will consider the farming of symbionts in a much more literal sense. We will discuss three groups of insects that sow, weed, and harvest their food. These animals use plant cell wall materials as substrates for fungi. The ability to cultivate fungi for food evolved independently in termites, ants, and beetles. Fungus farming evolved once among termites (Aanen et al. 2002), once among ants (Mueller et al. 1998), and at least six times among beetles (Farrell et al. 2001). The analogies between these independently evolved insect farmers and humans are notable. Like humans, many of the insect farmers cooperate in huge agricultural enterprises that depend on complex division of labor. Insects, like humans, often tend monocultures that require weeding and the application of pesticides (Mueller and Gerardo 2002). A major difference between insect and human farming is age. Whereas humans developed agriculture only relatively recently, at the end of the Pleistocene 5 to 10 thousand years ago, fungus farming by ants and termites originated in the early tertiary, about 50 million years ago (Mueller et al. 1998). The agricultural insects not only play very important roles in the ecosystems in which they live, their activities are also very often in conflict with those of humans. The fungus-growing termites, ants, and beetles, are major agricultural, forestry, and household pests. Although the three groups of insect agriculturalists share some common characteristics, the fine details of the natural history of their interaction with fungi differ significantly. Our description of the biology of these remarkable interactions is by necessity cursory and glosses over a rich diversity. We refer readers to the reviews by Mueller and Gerardo (2002), Mueller (2002), Rouland-Lefevre (2000), and Farrell et al. (2001) for more detailed descriptions.

6.2.11 Termite Fungiculture

About 330 of the roughly 2200 species of termites cultivate fungi. The fungus-growing termites form a monophyletic group, the subfamily Macrotermitinae, within the higher termites (figure 6.4). Curiously, all the fungus-growing termites are found in the Old World whereas all the fungus-growing ants live in the New World. The fungus-eating termites cultivate, and depend obligately, on basidiomycete fungi in the genus *Termitomyces*. These fungi are grown in subterranean combs that the termites build within nest mounds. The fungi are inoculated and cultured on specialized termite feces called mylospheres. Mylospheres are sometimes called "primary" feces and are produced by termites from ingested material that is rapidly passed through the gut. They are different from "final" feces, which contain only the residues of more completely digested food. Because termites mix the spores of consumed fungus with ingested plant forage, and many of the spores survive transit through the gut, the addition of each mylosphere to the comb represents the sowing of a new fungal harvest (Mueller and Gerardo 2002).

Not all fungus combs are built with the same materials. Some termite species use dead or green leaves and others collect wood or grass stalks. A fungus comb is a conglomerate of substrate, mycelia, and round white structures called spherules. Combs are brain- or spongelike structures convoluted by ridges and tunnels. The termites use these passageways to access all depths of the structure. Although termites consume all parts of the comb, including the substrate, the mycelium of the fungi, and the spherules, they seem to be partial to the older, better-composted, comb material (Rouland-Lefevre, 2000). In termites, the comb acts almost literally like an external paunch in which fungi condition the plant material and detritus that the termites eat.

Fungus growing in termites is sometimes accompanied by astounding collective architectural feats. The huge mounds of fungus-growing termites are conspicuous landscape features of the African Savanna and the Australian outback. A mature *Macrotermes michaelseni* nest in Namibia can contain millions of individuals and the mound above it can rise up to thirty feet above the ground (Turner 2001). The nest where the termites conduct their agribusiness are below the gigantic cone of soil and sand. The vast mound is needed to ventilate the nest, which houses a huge CO_2-producing fungal enterprise. Turner (2001) has gathered compelling evidence that suggests that the termites' mound functions somewhat like a gigantic, but rather stiff, lung. The varying Savanna winds provide the forces that drive a form of tidal ventilation. We recommend Turner's fine popular account of the biophysics of gas exchange in the nests of *Macrotermes* (Turner 2002). We have dwelt on Turner's analogy of the termites' mound as a lung for the *Macrotermes-Termitomyces* association because it provides a nice parallel to the analogy that guides this section. The termite mound is the breathing organ for the external paunch of fungus-growing termite colonies. These two analogies emphasize a view that Turner (2000) has championed: the physiology of organisms extends well beyond the organisms' bodies (Turner 2000).

6.2.12 Ant Farmers

The more than 200 species of fungus-growing ants in the tribe Attini are found exclusively in the New World. They are widely distributed from the pine barrens of New Jersey to the temperate deserts of Argentina (Hölldobler and Wilson 1990). They are also enormously successful ecologically. They are the dominant ants and perhaps the dominant herbivores of the vast tropical and subtropical regions of the Americas. The fungus-growing ants are divided into several groups with different characteristics. The lower attines have relatively small colonies, typically with fewer than 100 workers that tend walnut-sized fungus gardens (Mueller 2002). The substrate for their gardens is an assortment of vegetable matter that the ants collect on the forest floor and that includes plant detritus, fallen flowers, seeds, and arthropod feces. The higher attines, which include the well-known leaf-cutter ants in the genera *Atta* and *Acromyrmex*, use fresh leaf parts as substrates for their fungal gardens. Their colony sizes can be enormous. Colonies reach several hundred thousand workers in *Acromyrmex* and several million workers in *Atta* (Mueller et al. 1998). Their ecological impact is vast. Cherret (1986) estimates that *Atta* leaf-cutter ants cut from 12 to 17% of all leaf production in tropical forests. Not surprisingly, leaf-cutter ants can also have an extraordinary impact on agriculture. The total losses due to ants to tropical agriculture probably amount to billions of dollars (Hölldobler and Wilson 1990).

Leaf-cutting ants are meticulous gardeners. When they bring fresh leaves into the nest, they first lick them and cut them into tiny pieces. They then chew the edges of the leaf pieces and sometimes place a droplet of anal fluid on them. This fecal droplet contains a concentrated cocktail of fungal enzymes that were ingested by the ant and that passed unharmed/undegraded through the digestive system (Martin 1987). The fungal enzymes recycled by the ants include proteases, amylases, and a chitinase. The details of how these enzymes escape digestion are still mysterious. It is likely that ants have reduced expression of proteases in the gut (Febvay and Kermarrec 1986). The larvae of *Acromyrmex spinosus* express a chymotrypsinlike endopeptidase in their midgut. Interestingly, the workers that tend the fungal garden do not express endopeptidases (Febvay and Kermarrec 1986). The loss of the ability to hydrolyze peptides in workers may have evolved to prevent the hydrolysis of recycled fungal enzymes that then can be used to condition the substrate and accelerate fungal growth. After preparing each leaf fragment, the ants place several small tufts of fungal inocula on the leaf piece and plant it into a newly formed garden. When the fungi cultivated by the higher Attini mature, they produce conspicuous clusters of hyphal-tip swellings called staphylae. The tips of these swellings are called gongylidia. The queen and larvae feed primarily on gonglydia, whereas workers feed on plant sap, on proctodeal food secreted by larvae, and to a small extent on gonglydia (Mueller 2002). The chemical makeup or the relative contribution of these different sources to the workers' nutritional budgets is barely known. Gonglydia are a striking, and for the ants very convenient, evolutionary innovation of the cultivars of higher Attini. Workers pluck them like ripe fruit and consume them or transport them and feed them to larvae and the queen. Because the fungi cultivated by lower Attini do not produce gongylidia, lower attine workers bring larvae to the fungal garden to graze on fungal hyphae.

Data on the nutritional composition of gongylidia has been studied in only one variety of ant-cultivated fungus and only with "proximate" analytical methods (Quinlan and Cherret 1979). As described in chapter 2, these methods give only very sketchy information about the nutritional quality of any food. Quinlan and Cherret (1979) found that gonglydia were richer in "lipids" and poorer in "crude protein" than hyphae, yet both larvae and workers prefer to feed on gonglydia than on hyphae. Furthermore, larvae grew faster when fed on gonglydia than when fed on hyphae. We suspect that more refined analytical methods will reveal large differences between the nutritional composition of hyphae and gonglydia. Mueller (2001) compared the proximate nutrient composition of hyphae from several species of fungi cultivated by both ants and termites and their free-living relatives and found no notable differences. Again, we predict that more refined analyses will reveal major differences. In particular we predict that insect cultivated fungi will contain higher levels of *assimilable* (in contrast to "crude," see chapter 2) protein, lipid, and carbohydrate than noncultivated fungi. Determining the content of vitamins and minerals of these fungi is also likely to prove very informative. In the case of fungi cultivated by the higher Attini, it is necessary to compare the composition of hyphae and gonglydia. Measuring only the composition of hyphae or whole cultures is like attempting to determine the nutritional content of corn grains from that of corn stalks.

Leaf-cutting ants must be meticulous in their gardening because, if they are not, the garden can be rapidly taken over by pathogens. The ants must continually be on guard against infection by a specialized weedy fungal pathogen in the genus *Escovopsis* (Currie et al. 1999). Without the vigilant tending by ants, this pathogen can rapidly devastate a fungal garden (Currie et al. 2003). As a defense against *Escovopsis*, leaf-cutting ants have garnered the help of a filamentous actinomycete bacterium. This microbe is closely related to bacteria in the genus *Streptomyces*, which are the source of most of the antibiotics used by humans. The ants culture these bacteria in modified regions of their cuticle. The bacteria produce chemicals that have very potent inhibitory effects on the parasitic invader, but that do not inhibit and may even stimulate the growth of the cultivated fungus (Currie et al. 2003). The actinomycete bacteria are more abundant in workers that tend the garden than in those that forage outside of the colony. The abundance of bacteria in each worker seems to be up-regulated in colonies infected with *Escovopsis* (Currie et al. 2003). The mechanism of up-regulation and its prevalence among leaf-cutting ants are unknown.

6.2.13 Differences between Leaf-Cutting Ants and Fungus-Growing Termites

Although there are remarkable similarities between fungus-growing ants and fungus-growing termites, there are also major differences. Noting the disconnected geographical ranges of these two groups, Hölldobler and Wilson (1990) asked themselves whether this complementary global pattern is caused by preemption due to competitive exclusion or simply the accidental outcome of the rarity of the evolutionary origin of fungus gardening among insects. They concluded that the latter is likely to be the case. They predicted that a reciprocal transplant of termites into the New World and leaf-cutting ants into the Old would likely result in the coexistence of these two taxa. Curiously, they do not mention the significant ecological mayhem that this experiment would almost certainly bring about!

The coexistence of these two seemingly similar groups would be possible because leaf-cutting ants rely primarily on fresh plant material, whereas the macrotermitines use dead plant material (although there are some exception to this generalization). Also, the attines forage above ground, often on trees, whereas the fungus-growing termites are subterranean. These differences are to a large extent dictated by differences in the nutritional physiology of the ancestor of each group. The unknown ancestor of the leaf-cutting ants was an ant —that much we can tell. Ants are nectar feeders, predators, and scavengers. Their digestive system reflects a commitment to easily digestible foods: It is full of filters that prevent the passage of coarse cell wall materials and their midgut is delicate and secretes a limited spectrum of digestive enzymes. Their hindgut is reduced and does not harbor large numbers of symbionts. Ants, including leaf-cutting ones, are ill suited to be herbivores. If they can use leaves, it is because they have allied themselves with fungi. They are also very poorly suited to be detritivores. The fungus-growing termites, in contrast, evolved from a detritivorous or xylophagous ancestor. They have a powerful grinding gizzard, they retain the expression of cellulolytic enzymes in the midgut, and their large hindgut houses a rich bacterial microbiota (Anklin-Mühlemann et al. 1995). The fungus-termite association is remarkably efficient at decomposing cell wall materials, including lignin (Ohkuma 2003). The association assimilates from 70 to 100% of the cellulose and up to 90% of the hemicellulose in the substrate that termites bring to the nest. It is often assumed that the fungi cultivated by attine ants digest cell wall materials. Recent evidence casts doubt on this assumption. The fungus–leaf-cutting ant association seems to be relative ineffective at digesting lignocellulose (see section 6.2.15).

6.2.14 Ambrosia Beetles

Many weevil species in the subfamilies Scolytinae and Platypodinae carve intricate tunnel systems—often called galleries—inside of trees. A very large number of species of these beetles (over 3,400 species) cultivate fungi on the walls of their galleries. The fungus-growing beetles are called "ambrosia beetles." The name seems to have originated when early researchers found that these burrowing beetles did not feed on wood but on fungi that, like the ambrosia that fed the gods of Greek mythology, satisfied all the beetles' nutritional needs (Farrell et al. 2001). Whereas the agricultural association between ants and termites evolved only once in each group, it has evolved independently at least six times in ambrosia beetles (Farrell et al. 2001). In all groups of ambrosia beetles, reproductive adults plant their fungus when they form new colonies. The beetles either ingest either mycelium or spores before dispersing, or, more frequently, they collect spores in specialized invaginations of the cuticle called mycangia. Spores germinate within the galleries excavated by the beetles and hyphae penetrate the phloem and wood of the host tree. Within the galleries the fungi produce an "ambrosial layer" that consists of a layer of hyphae that bear an erect palisade that sometimes bears reproductive structures (conidia) or that can form a tangled maze of strings of beadlike cells (Beaver 1989). The beetles and their larvae browse on the ambrosial layer. The development of the ambrosial layer seems to depend at least in part on contact between beetle and fungus.

The beetles are very strictly dependent on fungi. Feeding on fungi seems to be essential for growth, reproductive maturation, and in some species, for the production of pheromones (Beaver 1989). The best examples of the nutritional dependence of these beetles on their cultivated fungi are the beetles in the genus *Xyleborus*. These beetles cultivate the fungus *Fusarium solani*. Cholesterol is an essential nutrient for most insects. *Xyleborus* beetles depend instead on ergosterol, the sterol produced by fungi. Ergosterol is absent in wood, but *F. solani* produces it in abundance. *F. solani* also provides the beetles with a variety of essential fatty acids and phospholipids (Kok et al. 1970).

6.2.15 *What Do Insects Get from Cultivated Fungi?*

Insect farmers derive a variety of benefits from the cultivation of fungi. The relative importance of these benefits, however, differs among farming systems. Insect farmers can derive four types of potential benefits from the cultivation of fungi: (1) Fungi secrete powerful enzymes that aid in the degradation of cell walls, and thus make available the energy contained in cell wall compounds. In the case of termites and ambrosia beetles, the fungal enzymes may act in concert with the insect's endogenous enzymes. (2) The fungus may improve the C:N ratio of food by the metabolism of polysaccharides and lignin. (3) The fungus may transform the refractory carbon of cell walls into assimilable "labile" carbon. The criterion of "assimilability" is a minimal one. As described in the previous section, cultivated fungi often use the carbon skeletons of their substrates to manufacture nutrients that are of vital importance to the insect farmer. (4) The fungus may detoxify the allelochemicals contained in the materials harvested by the insects as fungal substrates. Much remains to be learned about the physiological details of the interaction between insects and their fungal cultivars. The four potential benefits that we have identified here must be construed more as plausible hypotheses than established facts. The nutritional foundations of the interaction between insect farmers and their fungi remain largely unexplored and a fertile territory for physiological ecologists to wander on.

The relative importance of each of these potential benefits varies among insect-fungus interactions. For example, Abril and Bucher (2002) compared the lignin:cellulose ratios in the refuse of several species of leaf-cutting ants with those of termite and cattle feces (figure 6.10). They found that these ratios were much lower than those found in the "refuse" of efficient cellulose digestors and indistinguishable from those of unprocessed harvested leaves. They concluded that the fungi cultured by leaf-cutting ants do not digest cellulose significantly. In support of Abril and Bucher's (2002) suggestion, D'Ettorre et al. (2002) found relatively low activity of cellulases in the fungi cultivated by leaf-cutting ants. The fungi cultivated by ants are probably less important as digestors of cellulose and lignin than those cultivated by termites. If fungi are not important in the assimilation of cellulose, what is their function? Fungi may be very important detoxifying agents that allow ants to use a very large spectrum of plant species.

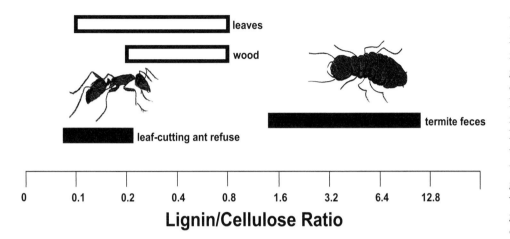

Lignin/Cellulose Ratio

Figure 6.10. Because lignin is refractory to digestion, the ratio of lignin to cellulose is a good indicator of the efficiency with which organisms hydrolyze cellulose. The bars represent the ranges of lignin/cellulose ratios measured in the feces of termites, the refuse material of the gardens of leaf-cutting ants, wood, and leaves —including grass blades. The feces of termites have high lignin/cellulose ratios (from Lee and Wood 1971). The lignin/cellulose ratios in the feces of domestic cattle are similar to those of termites (Abril and Bucher 2002, Van Soest 1994). In contrast, the lignin/cellulose ratios in the refuse that leaf-cutting ants carry from their gardens to their dump chambers is low (data from Abril and Bucher 2002 for two species of *Acromyrmex* and one species of *Atta*). Because this refuse is what remains after fungi have acted on harvested leaves, these low values indicate that the fungi are relatively inefficient at catabolizing cellulose. Data for leaves and wood are from Abril and Bucher (2002) and from table 11.9 in Van Soest (1994).

Insect ecologists have pointed out that a consequence of the tremendous chemical and physical diversity of plants is that most plant-eating insects have narrow diet tolerances (Strong et al. 1984). Leaf-cutter ants are an exception to this pattern. Cherret et al. (1989) report that *Atta* species living in diverse tropical forests cut the leaves of from 50 to almost 80% of all the species that live around their colonies. Leaf-cutting ants are not unselective. When offered choices among tree species, they exhibit distinct preferences (Cherret et al. 1989). However, they are significantly more polyphagous than other herbivorous insects that coexist with them. The ecological success of these ants is very likely the result of their ability to use a wide range of plant species. This ability is the result of their unholy alliance with fungi.

Plants must defend themselves against a variety of potential enemies that includes bacteria, fungi, insect larvae, and vertebrate grazers. To defend themselves, plants rely on a variety of stratagems that range from simple physical barriers to nasty chemicals and sophisticated hormone mimics. Two cooperating organisms with different morphologies and with very different physiologies have a good chance of breaking a larger array of plant defenses (Cherret et al. 1989). The mandibles of the insect can break down the barriers to hyphal penetration and the biochemical versatility of a fungus can breach the plant's chemical protective defenses (Dowd 1992). The insect and fungus working together can use a far wider range of species than either could alone (Cherret et al. 1989). Janzen (1975) has observed that polyphagy comparable to that of leaf-cutting ants is only found in vertebrates with foregut fermentation. The complex microbial fauna that lives in the stomach of these animals, like the external fungal rumen of attine ants, is also capable of detoxifying secondary compounds.

6.3 Terrestrial Vertebrates

Herbivory is common among vertebrates. In this section we focus on the nutritional ecology of terrestrial vertebrate herbivores, on their interaction with food plants, and on the gut symbionts that allow them to eat plants. Terrestrial vertebrate herbivores are quite different from the invertebrates that we have been considering so far. They are large, and some of them are endotherms and have very high metabolic rates. They are very diverse taxonomically and, like their invertebrate counterparts, they can play a very important role in the ecosystems in which they live. Traditionally, terrestrial vertebrate herbivores are divided into two major functional groups based on the position of the fermentation chamber along the alimentary canal: the hindgut and the foregut fermenters. In this section, we follow this functional taxonomy but add one more group to it. We first consider the cytoplasm consumers that rely little on fermentation, then we deal with the hindgut fermenters and we finish the section with the foregut ferementers. As we shall see, the position of the fermentation chamber has enormous physiological consequences.

6.3.1 Cytoplasm and Cell Wall Consumers

Like insects, vertebrate herbivores vary enormously in their reliance of the fermentation of cell wall components as a source of energy. Some vertebrate herbivores do not degrade cell wall materials to a significant extent. They ingest huge amounts of plant material and defecate the cell wall components more or less intact and/or they are very choosy and feed only on plant parts that have low cell wall contents. They are equivalent to the insects that we characterized as cytoplasm consumers in a previous section. The digestive system of cytoplasm consumers is very different from that of cell wall consumers in that it lacks a well-defined fermentation chamber (case 1 in figure 6.11). The absence of a well-defined fermentation chamber does not imply a complete absence of fermentative digestion, but it does probably mean that the capacity of these animals to degrade cell wall materials is very reduced.

Giant and red pandas (*Ailuropoda melanoleuca* and *Ailurus fulgens*) are good examples of cytoplasm-consuming herbivorous mammals. When feeding on bamboo shoots, pandas assimilate less than 10% of the cellulose and hemicellulose that they ingest (Dierenfeld et al. 1982; Wei et al. 2000). Among birds, the best examples of cytoplasm consumers are passerines in the genus *Phytotoma* ("plant cutters"). These odd little birds are among the smallest vertebrate terrestrial herbivores. They weigh only around 45 g and feed almost exclusively on young leaves (Bucher et al. 2003). Plant cutters have a sturdy serrated bill that they use to "masticate" leaves. Then they

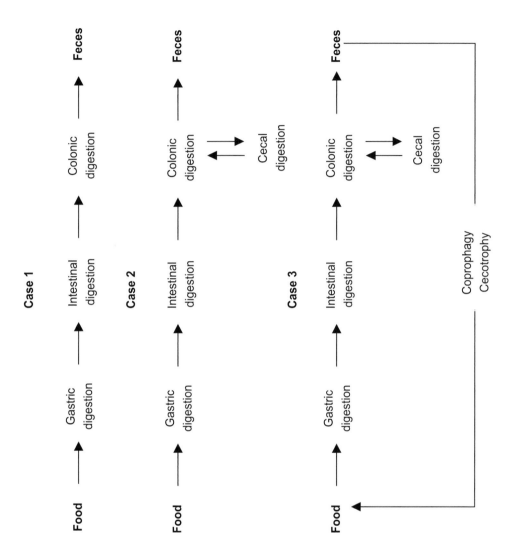

Figure 6.11. The gastrointestinal tracts of vertebrate herbivores with hindgut fermentation can be classified into three "designs" or cases. Case 1 is characteristic of cytoplasm consumers such as pandas and plantcutters. This design does not differ much from that of omnivores and carnivores. The digestive processes that dominate in this design are those that are mediated by endogenous enzymes (acid gastric digestion in the stomach and enzymatic digestion in the small intestine), but a small amount of fermentative digestion may take place in the colon. Case 2 characterizes animals with significant colonic and/or cecal digestion but that do not practice cecotrophy. The relative contribution of the colon and the cecae can vary significantly. Most large hingut fermenters, including horses, elephants, and probably dinosaurs can be characterized as case 2 hindgut fermenters. Case 3 is characteristic of the small vertebrate herbivores that practice coprophagy (including cecotrophy). In these animals the cecae typically contribute significantly to fermentative digestion. Good examples of this design are rabbits, hares, and many rodents. Note that in these 3 cases, the digestion of nutrients by endogenous enzymes precedes fermentative digestion by symbiotic microorganisms. (After Van Soest 1996.)

process the leaves in a short broad intestine that is characterized by unusually high rates of enzymatic hydrolysis (Meynard et al. 1999; figure. 6. 11). The maltase activity in the intestine of *P. rara* is about ten times higher than that of other passerines of equivalent mass. We know of no studies on the hydrolytic activities of panda's small intestines, but they may be unusually high as well. A fast passage rate demands a fast rate of biochemical processing. Because leaves have a relatively low content of soluble nutrients, cytoplasm consumers must ingest prodigious amounts of food. A 100 kg giant panda can eat between 12 and 34 kg of bamboo shoots per day (Dierenfeld et al. 1982). This amount is between two and six times the amount expected for a 100 kg mammal (Nagy 2001). López-Calleja and Bozinovic (1999) report the surprising finding that plant cutters can maintain body mass when feeding exclusively on lettuce, which has high water and fiber content and very poor nutritional quality. However, when feeding on lettuce leaf cutters must ingest 5–8 times the amount expected for their size (Lopez-Calleja and Bozinovic 1999).

6.3.2 Hindgut Fermenters

Pandas and plantcutters are exceptional herbivores in their lack of reliance on fermenting microorganisms. Most other herbivorous vertebrates rely on microbes to varying extents. The location of the primary site of the chamber that houses fermenting microbes differs among these herbivores. When the primary fermentative chamber is located in the first section of the stomach the animals are referred to as foregut fermenters (you will also find the terms pregastric and forestomach/foregut fermenters). Conversely, when the main fermentation chamber is located after the small intestine in the most posterior section of the gastrointestinal tract, the herbivores are called hindgut fermenters. In hindgut fermenters the fermentation chamber can be in the colon, the cecum, or in both the cecum and the colon (figures 6.12, 6.13, and 6.14). Therefore, the hindgut fermenters can be characterized as colonic fermenters, cecum fermenters, or ceco-colonic fermenters, respectively. Note that we have qualified the term "*fermentation chamber*" by the word "primary." The reason for this qualifier is that foregut fermenters almost always have significant fermenting microbial populations in the hindgut.

Hindgut fermentation is widespread among vertebrates. It is found in amphibians, reptiles, birds, and mammals (table 6.2). It may also have been prevalent among herbivorous dinosaurs (Farlow 1987). Although hindgut fermentation is found among some fishes, herbivory in these animals is sufficiently distinct to

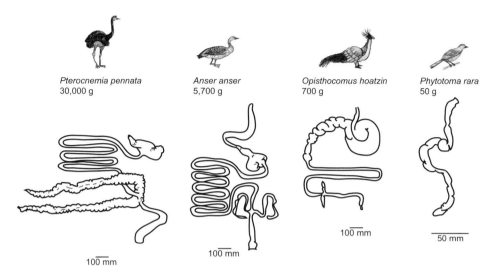

Pterocnemia pennata
30,000 g

Anser anser
5,700 g

Opisthocomus hoatzin
700 g

Phytotoma rara
50 g

Figure 6.12. The gut of herbivorous birds can have a variety of morphological arrangements. Plantcutters (*P. rara*) have very simple tubular guts, whereas rheas (*P. pennata*) and greylag geese (*A. anser*) have well-developed paired sacular cecae. The ostrich (not illustrated) is probably the only species with a sacular colon. Most herbivorous birds are hindgut fermenters with the notable example of the hoatzin (*O. hoatzin*), which is a foregut fermenter. (After López-Calleja and Bozinovic 2000.)

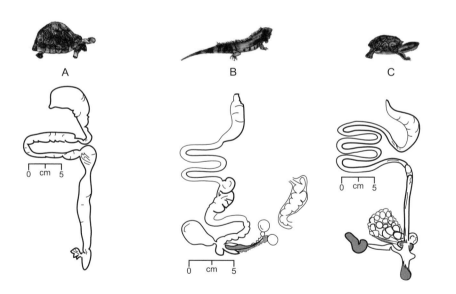

A   B   C

Figure 6.13. Most herbivorous reptiles are colonic fermenters. The colon may be saculated and simple (A, red-footed tortoise *Geochelone carbonaria*) or have a complex valve structure shown in a saggital section in panel B (green iguana, *Iguana iguana*). The freshwater turtle illustrated in panel C (red-eared slider, *Pseudemys scripta*) is unusual in that fermentation takes place both in the small intestine and in the colon, which is not sacular. The illustrations also show the juncture of the digestive, reproductive, and urinary tract at the cloaca (after Stevens and Hume 2004).

merit a section by itself. Note from table 6.2 and figures 6.13 and 6.14 that there is quite a bit of heterogeneity in the main site of fermentation in the hindgut. In some taxa the cecum is the primary site of fermentation, in others it is the colon. To our knowledge, a thorough phylogenetic analysis of the evolution of hindgut fermentation has not been conducted. However, it is safe to speculate that

Figure 6.14. Herbivorous marsupials can be colonic fermenters like wombats (A, common wombat *Vombatus ursinus*), cecal fermenters (B, greater glider *Petauroides volans*), ceco-colonic fermenters (C, koala *Phascolarctos cinereus*), and foregut fermenters (not shown). (Drawings after Stevens and Hume 2004.)

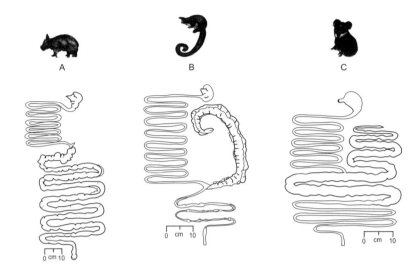

hindgut fermentation has evolved independently several times among vertebrates, presumably from ancestors with simple tubular guts. In hindgut fermenters the cytoplasmic contents and the cell wall constituents of food are digested sequentially. First, the cytoplasm contents are digested and absorbed by the endogenous enzymes of the herbivore, largely in the small intestine. Then, the cell wall materials are transferred to the hindgut where they can be fermented. This arrangement has advantages and disadvantages.

Having the fermentation vat after the stomach and the small intestine allows the herbivore to assimilate the soluble nutrients of food before microbes have a chance to ferment them. This is advantageous on two counts: First, the metabolism of microbes uses some of the energy in substrates. Thus, fermentation reduces the energy in food by between 10 and 20%. Hindgut fermenters assimilate soluble nutrients before the fermenting microbes get to them and consequently avoid paying the price of fermentation. Second, fermenting microbes modify some of the components that are abundant in plant diets and that are important for the nutrition of the herbivore. For example, gut microbes hydrogenate unsaturated fatty acids, including those that are essential to the host, such as linoleic and linolenic acids (see section 6.3.5). By placing the fermenting chamber in a posterior position, hindgut fermenters can assimilate these useful nutrients before microbes transform them.

Because the metabolic activities of microbes can destroy toxic substances and manufacture useful ones, the posterior position of the fermentation chamber poses some problems. Unlike some foregut fermenters, hindgut fermenters must contend with the toxins in food. They also must use a ploy to be able to

TABLE 6.2
Hindgut fermentation in terrestrial herbivores

Group —Family or order —Representative species	Reference	Notes
Frogs —Ranidae —*Rana catesbiana* (bullfrog)	Pryor (2003)	Foregut fermentation is found in the enlarged colon of many, perhaps all, herbivorous tadpoles. Nematodes may play an important role in the symbiosis.
Turtles —Cheloniidae —*Chelonia midas* (green sea turtle) —Testudinidae —*Geochelone denticulata* (yellow-foot tortoise) —Emydidae —*Pseudemis nelsoni* (Florida red-bellied cooter)	Bjorndal (1997)	With one exception all reptiles that have hindgut fermentation are colonic fermenters. Green sea turtles are the only herbivorous marine turtles. Some, but not all, herbivorous lizards have transverse valves in the proximal colon that slow the passage of digesta and increase the absorptive surface are of the fermentation chamber (figure 6.13).
Lizards —Iguanidae —*Iguana iguana* (green iguana) —Agamidae —*Uromastyx aegyptius* —Scincidae —*Egernia cunninhami*		Fresh water turtles such as *P. nelsoni* do not have an enlarged colon. They seem to carry on significant fermentation in the small intestine. Fermentation in the small intestine is a rarity among terrestrial vertebrates, but it occurs among herbivorous fishes (see *The herbivorous fishes*). Its existence is a bit perplexing. One of the advantages of having a fermenting hindgut is that the animal can still assimilate the soluble cytoplasmic contents of food. Having a fermenting small

(continued)

TABLE 6.2 (*continued*)

Group 　—*Family or order* 　　—*Representative species*	*Reference*	*Notes*
		intestine suggests that the host competes for these more easily assimilable, but also easily fermentable, substances with its symbionts.
Birds 　—Struthionidae 　　—*Struthio camelus* (ostrich) 　　—Rheidae 　　—*Pterochnemia pennata* (Darwin's rhea) 　　—Anatidae 　　—*Anser anser* (Greylag goose) 　　—Phasianidae 　　—*Bonasa umbellus* (ruffed grouse)	Vispo and Karasov (1997)	In herbivorous birds the paired cecae generally play a more important fermenting role than the colon and rectum. The colon and rectum are short and unsaculated. Exceptions are ostriches and screamers (*Chauna chavaria*, which are relatives of geese) which have a large sacculated colon in addition to the cecae. Herbivorous birds can be browsers such as ostriches and grouse or grazers, such as some species of geese.
Marsupials 　—Vombatidae 　　—*Vombatus ursinus* (common wombat) 　—Phascolarctidae 　　—*Phascolarctus cinereus* (koala) 　　　—Phalangeridae 　　—*Trichosurus vulpecula* (common brushtail possum) 　—Pseudocheiridae 　　—*Petauroides volans* (greater glider)	Hume (1999)	From a digestive perspective, the marsupial herbivores can be divided into the macropods which are foregut fermenters and the nonmacropods which are hindgut fermenters. Among the nonmacropods the wombats have a voluminous colon and a vestigial cecum, the greater glider has a greatly enlarged cecum and a small colon, and the koala has both an enlarged colon and an enlarged cecum (figure 6.14). With the exception of the wombats, which are grazers, the nonmacropod herbivorous marsupials are primarily arboreal folivores. Many of these folivores such

(*continued*)

TABLE 6.2 (*continued*)

Group —Family or order —Representative species	Reference	Notes
		as the koala and the greater glider, are *Eucalyptus* specialists. Eucalypt leaves contain high levels of lignin and phenolics and therefore are difficult to assimilate.

Placental mammals
—Rodentia
  —*Hydrochoerus hydrochaeris* (capybara)
—Lagomorpha
  —*Oryctolagus cuniculus* (rabbit)
—Primates
—*Alouatta palliata* (mantled howler monkey)
—Artyodactyla
  —*Sus scrofa* (wild boar)
—Perisodactyla
  —*Diceros bicornis* (black rhinoceros)
	Stevens and Hume (2004)	The full spectrum of hindgut fermentation styles can be found among placental mammals. In rodents and rabbits the volume of the cecum typically exceeds that of the colon. In ungulates, in contrast, the volume of the colon typically exceeds that of the cecum.

Note: This table provides only a few examples. The references cited give more comprehensive inventories. For amphibians, reptiles, birds, and marsupials we list by family the species that we have chosen to represent a particular type of hindgut fermenter. To avoid an extraordinarily long table, we listed the examples of placental mammal hindgut fermenters by order. To dodge the ire of either the proponents of traditional classifications or the advocates of phylogenetic taxonomies, we used common names for the higher taxa listed in this table (with luck, we may have managed to irk both traditionalists and phylogenetic revolutionaries). For a more detailed account of the gastrointestinal morphology of vertebrates we recommend Stevens and Hume's (2004) excellent book.

use the nutrients synthesized by their symbionts. Recall that termites, which can be considered the tiniest hindgut fermenters, face a similar problem. Termites secrete lysozyme in the salivary glands, but have a fermentation chamber in the hindgut. To take advantage of the nitrogen recycled by symbionts and of the nutritional goodies that they manufacture, termites indulge in proctodeal trophalaxis. Many hindgut fermenters use a similar strategy. They eat feces.

6.3.3 Ingesting the Dregs: Coprophagy, Cecotrophy, and Colonic Separation Mechanisms in Hindgut Fermenters

Coprophagy is prevalent among small herbivores and provides its practitioners with all the benefits that we enumerated for the invertebrates that farm microbes (see chapter 5). Microbe farming facilitates nitrogen recycling and makes possible the assimilation of vitamins and other nutrients synthesized by microbes. In hindgut fermenters, coprophagy is associated with colonic separation mechanisms (CSMs) that lead to the spatial segregation of digesta into two fractions: one that is highly nutritious and microbe rich, and another that is fiber rich and nutrient poor (Björnhag 1994). Thus, the herbivores can form two types of feces from these two fractions: a type designed for ingestion that includes the nutritional component of digesta and another that contains primarily undigested fiber and that is designed to be disposed of. Not all herbivores that have CSMs produce two types of feces, but many do.

CSMs have been reported in eight families of mammals and two families of birds (table 6.3) and come in two guises: particle-size dependent and mucus dependent (figures 6.15 and 6.16). In both cases, fluids, solubles, microbes, and small particles in suspension move back from the colon by antiperistalsis and accumulate in the cecae. The larger fibrous particles move in the opposite direction. Many of the animals that have CSMs are small and hence retain digesta in

TABLE 6.3
Colonic separation mechanisms

	Mucus-trap type	*Wash-back type*
Mammals		
Marsupials	Phalangeridae	
Rodents	Cricetidae	
	Muridae	
	Caviidae	
	Chinchilidae	
	Capromydiae	
	Hydrochaeridae?	
Rabbits and hares		Leporidae
Horses and donkeys		Equidae
Birds		Anatidae
		Phasianidae

Note: CSMs have been documented in eight families of mammals and two families of birds. CSMs are probably more widespread taxonomically than this table suggests (After Björnhag 1994).

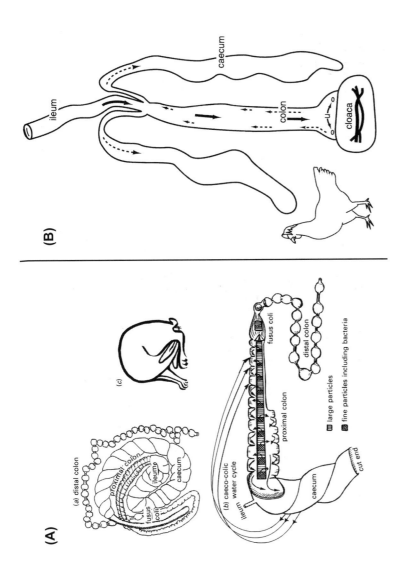

Figure 6.15. (A). In rabbits the contents of the ileum enter into the cecum. The cecal contents are then moved into the proximal colon in small batches. Intense muscular activity of the colonic wall then separates water, water-soluble molecules, and small particles from the larger and coarser fibrous particles. Water secretion by the colonic mucosa facilitates the separation. Antiperistaltic movements drive the fluid component of digesta towards the cecum. Simultaneously, the fibrous contents at the center of the proximal colon move towards the fusus coli where they are formed into fecal pellets. The colonic separation mechanism acts as long as material from the ileum enters the cecum. Cecotrophs are formed during the nonfeeding phase of the day cycle. When cecotrophs are formed no separation mechanism operates. The cecal contents pass through the colon, where a mucous envelope is deposited around each pellet. When the pellets reach the anus, they are eaten directly. (B). A similar separation mechanism takes place in herbivorous birds, except that the fibrous material does not enter the cecum, it remains in the center of the colon. In birds the ureters void the contents of the urinary tract (labeled as U in the illustration) in the cloaca. Urine can be refluxed back into the colon and cecae and provide a source of nitrogen for fermenting microbes. Australian digestive ecologists Steve Cork and Ian Hume call the CSMs of birds and lagomorphs the "wash back" type (Cork et al. 1999). (After Björnhag 1994.)

Figure 6.16. The colonic separation mechanisms in rodents are diverse and different from those found in rabbits. (A) Here we illustrate only the mechanism found in Scandinavian lemmings (*Lemmus lemmus*). Björnhag (1994) describes other CSMs present in rodents. In lemmings the proximal colon has two sections, the wide ampulla coli (AC) and the narrow spiral colon (SC). The first two "twists" of the spiral colon have a fold that divides the colon into two separate channels: a narrow one (b) and a broad one (a). The narrow channel contains almost exclusively bacteria and mucus. Björnhag (1994) hypothesizes that in the spiral section of the proximal colon digesta is mixed with a mucus secretion that originates in the mucosa. Bacteria (and maybe solubles and small particles) are differentially shunted by an unknown mechanism into the narrow channel and flow back into the caecum by antiperistalis. Undigested fibrous solids flow into the distal colon. (B) In support of Björhag's (1994) hypothesis, the concentration of nitrogen (in mg per gram of dry matter) decreases from the inner spiral of the proximal colon to the distal colon, suggesting that the protein-rich bacteria are recovered from digesta (after Björnhag 1994). Ian Hume and Steve Cork call the CSMs of rodents the "mucus trap" type.

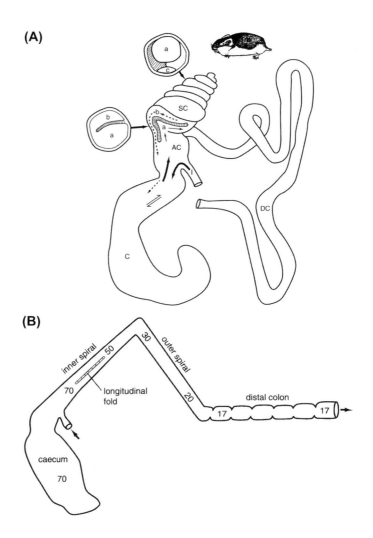

their gut for a small amount of time. CSMs allow these animals to recover the microbes that would otherwise be washed out in feces. CSMs permit the existence of large viable microbial populations in small herbivores in the face of relatively short digesta retention times.

A peculiar consequence of CSMs is that often the contents of the cecae are more nutritious than the food itself. Herbivores can take advantage of the nutritious contents of the cecae by forming special fecal pellets called cecotrophs (literally "food from the cecum") and then ingesting them. Cecotrophs differ greatly from the fibrous and nutritionally poor normal feces. Because cecotrophs tend to be softer than normal feces, they are sometimes called "soft feces." Cecotrophs contain lower fiber content and higher nutrient levels than normal "hard" feces (table 6.4). Lagomorphs produce true cecotrophs that are

TABLE 6.4
Comparative composition of kind feces, and cecotrophs

Component	Food	Feces	Cecotrophs
Water (%)	9.7	51.6	70.1
Protein (%)	11.0	7.0	26.2
Crude fiber (%)	19.7	29.6	17.8
Short-chain fatty acids (mmole/g)		45	180
Bacteria (10^{10} per g of dry matter)		31	142

Note: The cecotrophs produced by domestic rabbits contain more water, protein, SCFAs, and bacteria than feces. They also contain less fiber. Often, cecotrophs have a better nutritional composition than food (modified from Stevens and Hume 2004).

morphologically and nutritionally distinct. The cecotrophs of rabbits are enclosed within a mucous membrane, whereas those produced by hares are amorphous (Hirakawa 2001). Lagomorphs typically ingest cecotrophs without chewing them. In Leviticus 11:6, the bible provides a well-known alternative view: "And the hare, because he cheweth the cud, but divideth not the hoof, he is unclean to you." The bible is wrong in this biological opinion. Hares do not chew their cud and, when they ingest their own feces, they swallow them whole without chewing. Literalists intent on maintaining the bible's infallibility will argue that the we split hairs in differentiating between cecotrophy and rumination. However, our task as physiologists is to split hairs (and sometimes hares) to find out how animals work.

Not all coprophagus herbivores produce distinct cecotrophs that are swallowed whole. Kenagy and Holt (1980) reported that the feces eaten by kangaroo rats (*Dipodomis californicus*) and by voles (*Microtus californicus*) were visually similar and only slightly different in nutritional content from uneaten feces. They also observed that rodents carefully tasted and scrutinized feces before either discarding them or eating them, which suggests that rodents produce feces of varying quality and discriminate among them. Kenagy and Holt (1980) observed that all the species of rodents that they caught in coprophagous acts chewed the feces thoroughly before ingesting them. Not all rodents are feces chewers that produce morphologically indistinct feces. Capybaras (*Hydrocaeris hydrochearis*, Hydrochaeridae), the largest coprophagous vertebrates, behave like lagomorphs. They "lick [from their rear end] a *pasty material that differs from normal oval-shaped feces*" (Mendes et al. 2000).

The importance of cecotrophy for the nutrition of its practitioners has been studied in two ways: (1) by feeding animals diets deficient in an essential nutrient, and (2) by preventing the animals from ingesting their own feces. One can prevent animals from eating their own feces by placing them on a mesh floor that allows feces to fall through and by fitting a large rigid collar around their necks to thwart them from licking their anuses. The ruses that researchers use to prevent coprophagy often lead to deleterious effects even when the animals are fed on natural diets. The animals grow more slowly and show nutrient deficiencies. A plethora of studies demonstrate that cecotrophy can ameliorate the deleterious consequences of feeding laboratory rats on diets that lack a variety of vitamins (reviewed by Stevens and Hume 2004). Rats fed on diets deficient in biotin, riboflavin, pyridoxin, panthotenic acid, vitamin B_{12}, folic acid, or vitamin K experience no vitamin deficiencies if they are allowed to eat their feces, but experience serious symptoms of avitaminosis if they are not. Several studies on rats and rabbits have reported improved nitrogen balance and increased retention of a variety of nutrients in control animals than in animals that are prevented from eating their feces (Sakaguchi 2003). Chilcott and Hume (1985) reported similar results for common ringtail possums (*Pseudocheirus peregrinus*, Pseudocheiridae). The positive effects of coprophagy on nitrogen balance are, not surprisingly, more evident when animals are fed protein-poor diets (Cree et al. 1986). Interestingly, all these studies have consistently failed to document an effect on the efficiency with which animals assimilate fiber. This observation suggests that nitrogen recycling and the assimilation of nutrients manufactured by microbes, rather than increased fiber digestion, are the primary functions of coprophagy.

This section describes coprophagy in adults. The young of some species in which the adult is not normally coprophagic eat feces, often those produced by the parents. The young of horses, green iguanas (*Iguana iguana*), and koalas ingest parental feces. Foals ingest small amounts of maternal feces until they are about six moths old (Crowell-Davis and Houpt 1985). When young koalas first emerge from the pouch, they consume a special fecal treat that Australian biologists dotingly call "pap" (Osawa et al. 1993). Pap is moister and has a higher count of bacteria involved in the degradation of tannins than normal feces, which suggests that it is of cecal origin. Osawa et al (1993) speculate that the function of pap feeding in juvenile koalas is to inoculate them with the microbes that will allow them to feed on eucalypt leaves. In view of the possibility that juvenile coprophagy can aid in the establishment of the gut's microbiota, it is surprising that it has been not reported in more species.

Perhaps understandably, coprophagy is not a widely favored motif in literature (Pynchon 1973). A notable example is *Oryx and Crake*, Margaret Atwood's postapocalyptic novel (Atwood 2003). The novel takes place in a dystopic human world in which the living products of biotechnology have become feral. To replace a flawed humanity, a brilliant biotech scientist called Crake engineers a new species of hominid. With characteristic modesty, he christens his

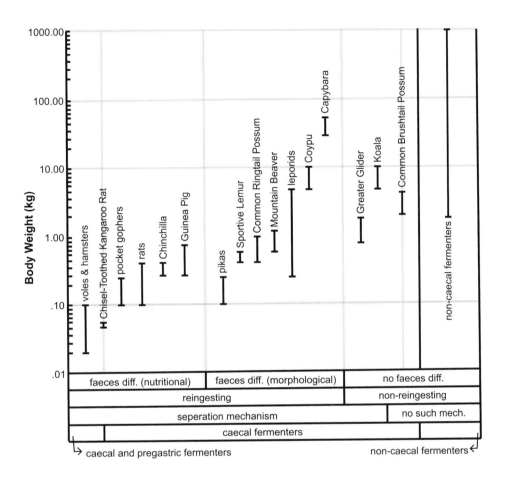

Figure 6.17. With a few exceptions (some rabbits and hares, coypus, and capybaras), most of the coprophagous mammals are small (less than 1 kg). This figure should include horses, which are not coprophagous but which have colonic separation mechanisms when feeding on natural grasses (Björnhag 1994). Both voles and hamsters have relatively complex mutichambered stomachs and they are sometimes considered "foregut fermenters" (see box 6.2). Foregut fermentation is probably not present in voles (B. Wunder, personal communication), but is more likely in hamsters. SCFAs have been reported in the stomach of Syrian hamsters (*Mesocrycetus auratus;* Dehorty 1997). Hamsters have voluminous cecae and hence if they have some foregut fermentation, its importance for the total energy budget is probably quite small. (Figure after Hirakawa 2002.)

creation "crakers." Crakers are strict vegetarians with a variety of unusual traits. They are distinctly generous and good natured. Their genitals turn blue when females enter estrous and insects do not bother them. Their lemony odor repels gnats and mosquitoes. As a gesture of affection, they exchange cecotrophs with each other. Crakers "look like retouched fashion photos, or ads for a high-priced workout program" (Atwood 2003). Atwood's crakers are illustrative because they are biologically implausible. The largest mammal that uses cecotrophy is the capybara, which weighs between 30 and 45 kg (figure 6.17). Contemporary humans range in mass from 45 to 85 kg (Ruff 2002). Furthermore, hindgut fermentation demands an ample gut that does not fit easily within the willowy body of a fashion model (figure 6.18). A cecotrophic hominid would have to be small and it would have the tubby figure of a gorilla or orangutan.

Figure 6.18. (A) When body mass is accounted for, the volume of the hindgut (colon and caecum, open circles) and of the total gut (open squares) of omnivorous humans (Hs, *Homo sapiens*) is significantly smaller than that of their closely related herbivorous relatives: chimpanzees (Pp, *Pan troglodytes*), gorillas (Gg, *Gorilla gorilla*), orangutans (Pp, *Pongo pygmaeus*), and three species of gibbons (*Hylobates pileatus* (Hp), *Hylobates lar* (Hl), and *Symphalangus syndactyllus* (Ss, from Van Soest 1994). (B) The phylogenetic hypothesis for the hominoid primates is very well supported by a variety of molecular data and was redrawn from Purvis (1995) and Ruvolo (1997). The illustrations of the gastrointestinal tract of an orangutan and a human were modified after Stevens and Hume (2004).

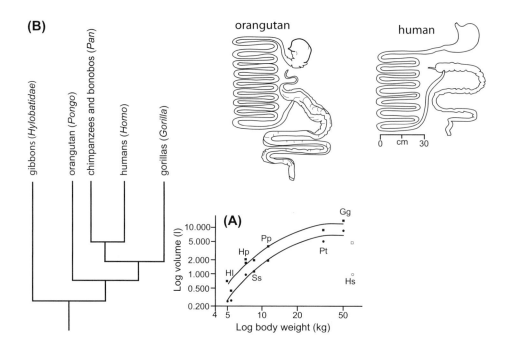

6.3.4 Nitrogen Recycling in Hindgut Fermenters

The association of herbivores with microbes facilitates both nitrogen conservation and nitrogen recycling. In mammals, urea is a terminal product of the catabolism of proteins. Mammals lack ureases and hence, once they synthesize urea, they cannot, by themselves, do much with it. What they can do is send it to the gut where microorganisms that do possess ureases can use it. Mammals send urea to the gut as an ingredient of saliva, bile, and other secretions. In the gut, bacteria hydrolyze it into ammonia and CO_2. The ammonia generated by microbes can follow two fates: it can be incorporated into bacterial biomass or it can diffuse into the host and be used to synthesize dispensable amino acids. In mammals with cecotrophy, the herbivore can ingest and assimilate the endogenously generated nitrogen that has been recycled into bacterial high-quality protein. A similar, but not identical, situation occurs in herbivorous birds such as geese, grouse, and chickens.

With some exceptions (see McWhorter et al. 2003), birds use uric acid as the terminal product of protein catabolism. The kidneys void uric acid as a component of urine into the bird's cloaca, where it mixes with the contents of the lower gastrointestinal tract. Uric acid is excreted in the form of tiny spheres (0.5–13 μm) made of alternate layers of urate (the salt form of uric acid) and mucopolysaccharides (Braun 1999). After reaching the cloaca, these urate spheres disintegrate.

Antiperistaltic movements reflux their constituents into the colon and then into the paired cecae (figure 6.12). The ammonia released by the action of microbial uricases in the colon and cecae can then either diffuse into the host or the microbes can use them to synthesize protein. In birds, colonic separation mechanisms facilitate nitrogen conservation, and perhaps recycling.

CSMs have been well documented in only two bird families, the ducks and geese (Anatidae) and the turkeys, grouse, and chickens (Phasianidae). However, the phenomenon is very likely widespread among herbivorous birds with well-developed cecae (see Clench 1999 for a detailed account of the diversity of avian cecae). Although the production of cecal feces is pervasive among birds and even though there are many anecdotes of coprophagy in chickens, quail, and ostriches, there are no well-documented published reports of regular coprophagy among herbivorous birds. Either birds are discrete when they eat their own feces, or the behavior is rare (Soave and Brand 1991). The paucity of observations of cecotrophy among herbivorous birds brings up the question of how these animals use the protein and nutrients synthesized by their symbionts. The same question might be asked for humans and other large mammalian hindgut fermenters that do not practice coprophagy. For example, what is the mechanism that permits humans to derive between 1 and 20% of their absorbed lysine from the microbes that live in their large intestines (Metges 2000)? This is a critical, unanswered question for all these organisms, and a situation in which research on one has the potential to increase knowledge for all.

Digesting microbes requires first breaking the bacterial cell walls and then hydrolyzing and absorbing the contents of the bacterial cell. Recall that bacterial cell walls are made primarily of peptidoglycan. This substance is hydrolyzed by the enzyme lysozyme. Most animals that assimilate the microbes that they farm use a compartment of the gut to culture the microbes and another one to digest them. In at least two mammalian lineages and one avian species, the latter can be a site of lysozyme secretion. Lysozyme's original function is to assist animals to fight harmful bacteria. Lysozyme is secreted in the small intestine, tears, and milk. Rabbits seem to be unusual in that they secrete lysozyme in the distal colon under a circadian schedule that follows tightly that of the production of cecotrophs (Cámara and Prieur 1984). Thus, the cecotrophs that reach the stomach contain large amounts of lysozyme and, presumably, of bacteria with partially hydrolyzed cell walls ready to be digested. A curious feature of the colonic rabbit lysozyme is that its pH optimum is very different from that of other lysozymes expressed in rabbits. It is acidic rather than neutral (Ito 1994). This observation suggests that in rabbits one of the lysozymes has been coopted from its original antibacterial role into the role of a digestive enzyme. This situation also occurred in some animals with foregut fermentation. Because the pH of the colon is typically neutral, it also suggests that the colonic lysozyme continues its digestive function in the acidic environment of the stomach.

In rats and humans lysozyme is secreted in the small intestine by specialized cells called Paneth cells located at the base of intestinal crypts (Porter et al. 2002). Paneth cells also secrete other antimicrobial peptides including α-defensins (also called cryptidins; Bevins et al. 1999). The chemical cocktail produced by Paneth cells is a powerful bactericide. Lysozyme hydrolyzes the bacterial cell walls and the defensins insert into membranes where they interact with one another to form pores that disrupt membrane function and lead to the death of the bacterial cell (Kourie and Shorthouse 2000). It is unknown if the colonic production of lysozyme (and maybe other antibacterial peptides) is a trait shared by all cecotrophic mammals.

The assimilation of bacterial protein by herbivorous birds is perplexing because birds do not seem to have spatial separation of culturing and digestion of microbes. Our colleagues Enrique Caviedes and Scott McWilliams have found significant activity of a membrane-bound peptidase (aminopeptidase-N) in the cecae of chickens, common quail, and Canada geese. Several studies have documented cecal carrier-mediated amino acid transport in birds (Moreto et al. 1991 and references therein). Thus, it is possible that there is some protein hydrolysis and amino acid uptake in the cecae of birds—and perhaps in the colon of noncecotrophic mammals. Moreover, even if birds can hydrolyze bacterial protein in the cecae, to do it they first must use lysozyme to crack bacterial cell walls. To our knowledge no one has yet measured the activity of lysozyme in the gastrointestinal tract of birds. Much remains to be learned about the mechanisms that vertebrate hindgut fermenters use to take advantage of the microbes that they farm.

6.3.5 Foregut Fermenters

Most vertebrate herbivores are hindgut fermenters. Exceptions are one bird species and several lineages of mammals that use foregut fermentation (box 6.2). The bird is the hoatzin (*Ophistocomus hoazin*; Grajal et al. 1989), which is a most unusual creature (figure 6.19). The essential features of foregut fermentation are outlined in figure 6.20. All foregut fermenters have a voluminous stomach that is partitioned into compartments with separate functions. The stomach often has one or more sacular compartments that act as fermentation vats. These compartments house symbiotic microbes and absorb the SCFAs that these microbes produce. The fermentation vat is often labeled a "forestomach" and its function is labeled as "pregastric fermentation" in figure 6.20. As we

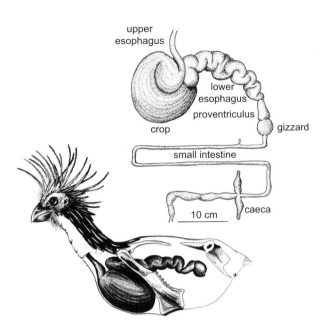

Figure 6.19. Most herbivorous birds carry their fermentation vat in the hindgut. Hoatzins (*Ophistocomus hoazin*) carry it in the foregut. The proventriculus (or glandular stomach) and the gizzard in hoatzins are small. In contrast, the crop is enormous. Its contents contain 75% of the total gut volume. The full crop and esophagous weigh about 10% of the total mass of the bird (Grajal 1995). Accomodating this huge fermentation vat requires significant anatomical modifications. Hoatzins possess strangely concave sterna that have an external callus at the lower end. This callus functions as a kneeling pad that supports the heavy crop while the hoatzin perches. Accommodating the crop leaves little room for the attachment of flight muscles in the reduced sternum. In consequence, hoatzins are weak flyers that often crash, rather than land, into perching branches. Because hoatzins smell faintly like cow manure, the Guyanese call them stinkbirds. The drawings in this figure are after Grajal and Strahl (1991).

shall see, many foregut fermenters retain large fibrous particles and allow small particles and fluids to move into the hindstomach. This process is labeled "digesta separation and bypass" in figure 6.20. The term "bypass" is used because these mechanisms can permit some component of food to evade metabolic transformations by the microbiota. The relative importance of the bypass mechanism differs greatly among species. Two conditions favor the maintenance of microbial populations in the forestomach. The forestomach is large and typically saculated which increases the retention of microbes. Its contents are buffered by the copious production of bicarbonate-rich saliva. Foregut fermenters have robust parotid salivary glands. Parotid saliva is not only rich in bicarbonate, it is also relatively high in sodium, calcium, and very importantly, in phosphate. Salivary phosphate is an important resource of phosphorus for the rapidly reproducing microbes of the gut.

The contents of the forestomach must be physically isolated from the acidic contents of the hindstomach, where gastric digestion takes place. This separation is accomplished by the presence of sphincters and/or tubular sections. A variety of digestive enzymes act on the slurry of digesta and bacteria that the herbivore transfers from the forestomach into the hindstomach (box 6.3). The epithelium of the hindstomach contains gastric pits lined with chief cells and parietal cells that secrete pepsinogen (the precursor of pepsin) and hydrochloric acid, respectively. The partially digested contents of the hindstomach flow into the small intestine, where digestion by pancreatic enzymes and the absorption

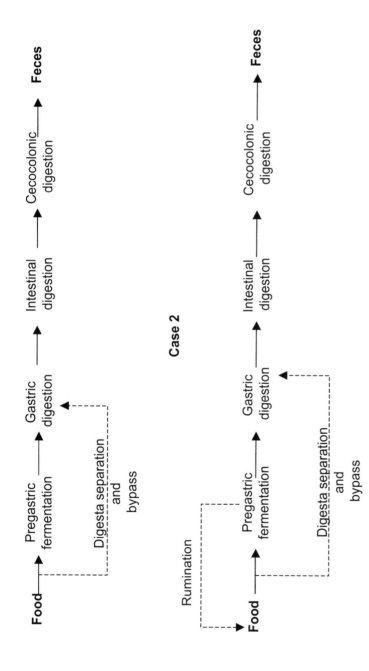

Figure 6.20. The design of the gastrointestinal tract of foregut fermenters can be divided into two models that differ only in the presence of rumination. Case 1 applies to all gut fermenters except true ruminants and camelids. In both models the chamber (or forestomach) that houses fermentative microorganisms and absorbs the SCFAs that they produce is located before the glandular stomach, where acid peptic digestion of proteins takes place. Many foregut fermenters secrete abundant lysozyme in the forestomach. This enzyme facilitates the digestion of bacterial cell walls. The nutrients that escape fermentation and those contained in the bodies of microbes are assimilated in the small intestine. Because bacteria contain high levels of RNA, the pancreas of foregut fermenters secretes high amounts of ribonuclease. Most, albeit not all, foregut fermenters have cecocolonic fermentation. (After Van Soest 1996.)

Box 6.2.

Foregut fermentation has evolved repeatedly among extant mammals (figure 6.21). It evolved once among marsupials in the kangaroos and their allies (family Macropodidae); once among the tree sloths (*Bradypus* and *Choloepus*, Bradypodidae); once among primates in several genera within the family colobinae (e.g., *Colobus, Presbytis, and Nasalis,* family Cercopithecidae); and one or several times within the large monophyetic group that includes the camels, pigs, whales, and ruminants (the "Cetartiodactyla," figure 6.22). Available phylogenetic information does not allow determining whether the lack of foregut fermentation in toothed whales and swine is the ancestral condition of their respective clades or the result of the loss of the trait (figure 6.22).

We mentioned foregut fermentation in baleen whales in chapter 2 when we described the chemical structure of chitin. We speculated that the function of foregut fermentation in whales was to digest chitin, which is relatively refractory. A detailed series of studies of Minke whales (*Balenoptera acurostrata*) by physiological cetologist Monica Olsen and

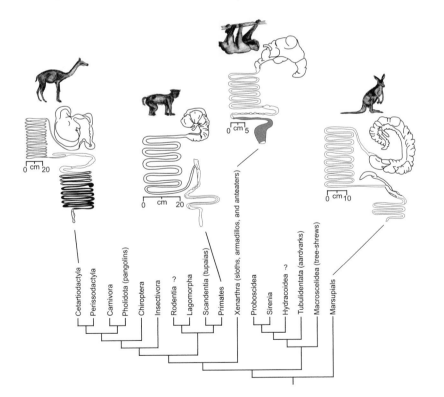

Figure 6.21. Foregut fermentation has evolved at least four times independently in mammals. We used question marks to emphasize the possibility of two more independent origins of foregut fermentation in rodents and hyraxes. The images of gastrointestinal tracts are after Stephens and Hume (1995).

continued

continued

Figure 6.22. Within the clade that includes the camels, pigs, whales, and ruminants, we find at least two cases of taxa without foregut fermentation (swine and toothed whales). The closed rectangles on top of each clade represent groups with (filled symbols) or without (unfilled symbols) foregut fermentation.

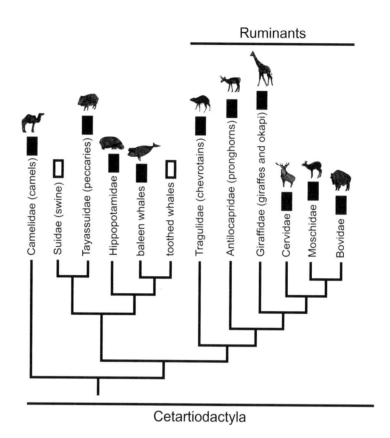

her colleagues supports this speculation (Mathiesen et al. 1995; Olsen et al. 2000). These researchers found large populations of indigenous anaerobic bacteria (including chitinolytic strains) in the forestomach of these whales. They also found relatively high concentrations of SCFAs and high fermentation rates. Because the forestomach of baleen whales is relatively small (less than 2% of total body mass compared with 9–17% in ruminants), Olsen et al. (2000) considered that the absorption of SCFAs contributed little to total energy metabolism. They argued that the foregut was important as a digestive organ where chitinolytic bacteria break the barrier to digestion represented by the exoskeleton of krill and other planktonic crustaceans. The function of chitinolytic bacteria is probably to "shell the shrimp."

Foregut fermentation has been reported in one rodent species, the Syrian hamster (*Mesocricetus uratus*, Cricetidae; Borer 1985), and in at least one species of hyrax, the rock hyrax (*Procavia capensis*, Procaviidae; Rübsamen et al. 1982). The consensus among researchers seems to be that foregut

continued

continued

fermentation plays a relatively minor role in these two groups (Dehorty 1997). The question marks in figure 6.19 indicate our hesitation to include these two taxa in our count of "proper" instances of the evolution of foregut fermentation. This does not imply that these creatures are uninteresting. It may well be that hyraxes and hamsters hold important clues to the incipient stages that animals must go through in order to become foregut fermenters.

The phylogenetic hypothesis for mammals depicted in figure 6.21 was derived from Murphy et al. (2001). That in figure 6.20 was modified from Montgelard et al. (1998) and Hassanin and Douzey (2003). The phylogenetic relationships within the cetariodactyls are still unresolved and seem to be difficult to elucidate. Thus, some aspects of the tree are well supported, other are not. Whereas the monophyly of the "pecora" or "true" ruminants is well supported (Hassanin and Dousey 2003), the sister group relationship between the pigs and peccaries and the hippos and whales is not (Geisler and Uhen 2003). Although the monophyly of the cetartiodactyls is widely accepted, some of the details of the relationships depicted in figure 6.22 must be considered preliminary.

of nutrients takes place (chapter 4). Many, albeit not all, foregut fermenters have well-developed secondary fermentation chambers in the hindgut. This fermentation chamber aids in the fermentation of refractory materials that escape fermentation in the foregut.

Figure 6.20 differentiates between foregut fermenters with and without rumination. True rumination, defined as the controlled and rhythmic regurgitation, remastication, and swallowing of digesta has evolved twice among extant mammals: once in the ancestor of camels, llamas, and their relatives, and once among once among the "true ruminants" (box 6.3). Rumination seems to allow herbivores to ingest in haste and masticate at leisure. If feeding must take place in an uncomfortable and/or dangerous place, rumination can potentially take place in a more favorable microclimate and in relative safety (Hume 1999). Nonruminants sometimes regurgitate food, chew it, and swallow it again. This behavior is called merycism to differentiate it from true rumination. Merycism is not associated with cyclic contractions of the forestomach and hence does not have the characteristic rhythmicity of rumination. Kangaroos and their relatives sometimes practice merycism. The frequency of merycism in koalas increases with tooth wear and hence may compensate for reduced molar effectiveness (Logan 2003). Merycism is sufficiently rare among humans to merit inclusion in Gould and Pyle's (1896) book on the anomalies and curiosities of medicine.

**Box 6.3. Molecular Tales of Two Enzymes: Lysozymes
and Ribonucleases in Foregut Fermenters**

A large fraction of the nitrogen and phosphorus assimilated by foregut fermenters comes from bacterial tissues that incorporated them from diet or recycled them from the host's metabolic by-products. Digesting bacteria efficiently demands the deployment of a battery of enzymatic tools that are not normally used in animals that do not rely on these microbes as food. Two enzymes that play important roles in the digestive physiology of foregut fermenters are lysoszyme and ribonuclease. We have encountered lysozyme before. It hydrolyzes the components of bacterial cell walls. Ribonucleases digest the abundant RNAs of bacteria. The molecular evolution of these two enzymes has received significant attention and provided two of the most spectacular examples of adaptive convergent evolution in biochemical function.

Ruminants, colobine monkeys, and hoatzins have evolved independently a lysozyme that functions as a digestive enzyme. This digestive lysozyme has many characteristics that distinguish it from the bacteriostatic lysozyme that is expressed in tears, milk, the Paneth cells of the small intestine, and the whites of bird eggs. The digestive lysozyme is expressed in the hindstomach, has an acidic pH optimum, and is relatively resistant to breakdown by pepsin (reviewed by Mackie 2002). Molecular evolutionary biologist Caro-Beth Stewart and her colleagues (Stewart et al. 1987) demonstrated that the colobine and ruminant lysozymes converge in the amino acid sequences that confer on these enzymes their unique pH optima and pepsin resistance (see also Swanson et al. 1991). The digestive lysozyme of hoatzins has a different genetic origin from that found in colobine monkeys and ruminants. The primate and ruminant digestive lysozyme evolved from a "conventional" lysozyme, whereas that in the hoatzin evolved from a calcium-binding one, like that found in the egg white of pigeons (Kornegay et al. 1994).

Most mammals and birds have a single gene copy that codes for lysozyme. Ruminants, in contrast, have many copies (Wen and Irwin 1999). In ruminants, large-scale production of digestive lysozyme entailed both gene duplication and changes in the molecular structure of the protein (Irwin et al. 1992). A common explanation for the origin of multiple gene copies is that these allow making more protein product. Indeed, lysozyme accounts for 10% of the total gastric mucosal protein and messenger RNA in ruminants (Irwin and Wilson 1989). The activity of lysozyme in the stomach of the foregut fermenters is over three orders of magnitude higher than that found in animals with no foregut fermentation (Dobson et al. 1984). Lest you consider the study of the vertebrate lysozymes a completed tale, we hasten to add that we still lack knowledge on the levels of expression

continued

continued

and molecular biology of lysozyme in kangaroos, peccaries, sloths, hippos, and baleen whales.

Rapidly growing bacteria contain high levels of RNA. Thus animals that "feed" on bacteria such as foregut fermenters, must have an efficient way to digest this substance, which a good source not only of nitrogen, but also of phosphorus. RNA is digested by ribonucleases secreted by the exocrine pancreas into the lumen of the small intestine. Available, albeit incomplete, evidence suggests that the ribonuclease content of the pancreas is higher in foregut fermenters and (perhaps not surprisingly) in some cecal fermenters that practice coprophagy than in omnivores and noncoprophagous herbivores (Barnard 1969; Beintema et al. 1973). Ruminants, camels, hippos, and langur have the highest contents of pancreatic ribonuclease of all mammals. Unfortunately, the species in which ribonuclease content has been measured represent a rag-tag phylogenetic sample. Therefore, the notion that foregut fermenters and coprophagic cecal fermenters have higher ribonuclease expression must be considered more a plausible hypothesis in need of testing than a well-established pattern. Measuring the levels of expression and characterizing the pancreatic ribonucleases of the hoatzin would be remarkably informative.

Although the putative association between digestive mode and expression of lysozymes remains to be established firmly, it is abundantly clear that ruminants and langur monkeys express extremely high levels of pancreatic ribonucleases. One must be a determined skeptic to doubt that these levels are an adaptive correlate of foregut fermentation. Ruminants, colobine monkeys, and camels seem to have converged in the levels of ribonucleases that they express (Beintema et al. 1973; Beintema 1990). In addition, the evolution of the digestive ribonuclease in these groups seems to have followed similar pathways. In both groups (and probably in camels as well), the gene for ribonuclease duplicated and one of the copies became specialized for the efficient digestion of bacterial RNA in the small intestine (Beintema 2002; Zhang et al. 2002).

6.3.6 Nitrogen Recycling in Foregut Fermenters

Fermenting microbes can manufacture good-quality protein from nonprotein sources such as ammonia, nitrates, and urea. This is useful to the herbivore, because it can conserve and recycle nitrogen and is of economic importance. Humans can include cheap nonprotein sources to partially replace expensive protein in the diet of domestic herbivores (Moore et al. 1999). We have already discussed how hindgut fermenters must often rely on coprophagy to be able to use the protein synthesized by their symbionts. Foregut fermenters do not need to eat their own feces to profit from the protein manufactured by their symbionts. Microbes cultured in the fermentative forestomach flow directly into the glandular hindstomach where their digestion begins.

With the exception of the hoatzin, most foregut fermenters are mammals whose primary nitrogenous product is urea. Urea is synthesized by the liver (and to a lesser extent by the kidneys) from ammonia arising from the deamination of amino acids and from nitrogen absorbed as ammonia in the forestomach. The absorption of ammonia from the rumen depends on the rate at which microbes produce it, which is determined by the availability of dietary nitrogen and carbohydrates. The liver extracts ammonia from the portal vein very efficiently and transforms it into urea (figure 6.23). In hindgut fermenters urea can be excreted into the forestomach directly from blood or by urea excretion into saliva (Kennedy and Milligan 1980). After urea reaches the rumen, it is quickly transformed into ammonia and enters the general pool of forestomach nitrogen from which microbial protein is synthesized. When dietary protein is in short supply, foregut fermenters are efficient conservers of nitrogen (figure 6.23).

6.3.7 Consequences of Foregut Fermentation: Bacterial Modification of Nutrients and Secondary Metabolism

In foregut fermenters, symbiotic microbes have a metabolic first pass at ingested substances. Thus, the materials that the host assimilates can be very different from those that it ingests. The degree of metabolic modification of nutrients by microbes depends, as we shall discuss in the following section, on how long digesta is retained in the forestomach and on the characteristics of the mechanisms of digesta separation and bypass. Here we will describe in very general terms the effects that microbes have on the metabolism of lipids and glucose, and on how animals cope with toxins. We will consider these topics in general and in mechanistic terms, and then we will reconsider them from a comparative perspective in the following section. Because mechanisms have been elucidated primarily on domestic animals, this section has a distinct bovinocentric slant.

6.3.8 Biohydrogenation of Unsaturated Fatty Acids

You may have heard that eating too much beef is bad for your health. The reason invoked to justify this warning is the high content of saturated fatty acids and *trans* unsaturated fatty acids in the tissues of domestic cattle (see chapter 2). This perception is partially correct. The red muscle of ruminants does contain

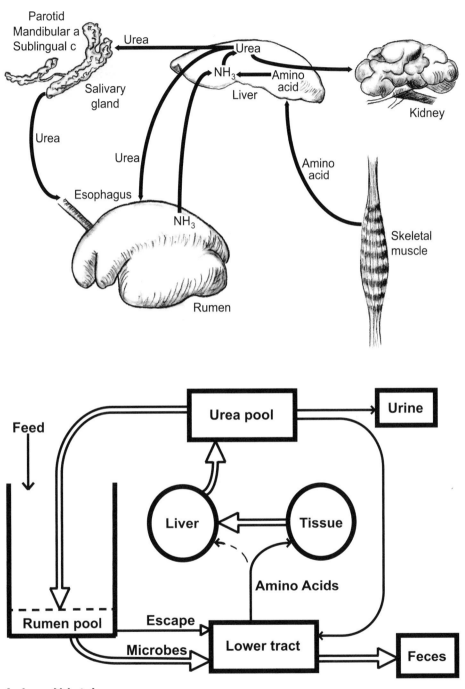

Figure 6.23. Foregut fermenters can recycle nitrogen with the aid of their fermentative symbionts. Urea produced in the liver from the catabolism of amino acids can either be excreted by the kidneys or transported into the rumen directly or via the salivary glands. The microbes of the rumen can then use urea to synthesize high-quality protein that can then be assimilated by the host. The efficiency with which nitrogen is recycled depends on the nitrogen content of food. (A) illustrates the flow of nitrogen in an animal feeding on a diet with low nitrogen. Under these conditions, a small fraction of the total urea pool is excreted by the kidneys and much is recycled. In contrast, when the animal feeds on a diet with high nitrogen content (B), a significant fraction of urea is excreted and only a small fraction is recycled. Note that the absolute amount of urea recycled and the size of the urea pool remain relatively constant (after Van Soest 1994).

(figure continues)

(figure continued)

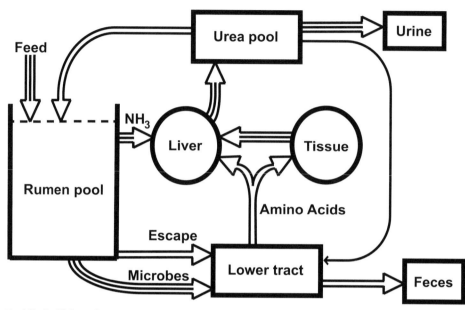

B. High N intake

relatively high levels of both saturated fatty acids and *trans* unsaturated fatty acids. Plant products and the flesh of nonfermenting or hindgut fermenting animals have significantly lower contents of these substances. Also, there is growing clinical and epidemiological evidence of an association between the ingestion of *trans* unsaturated fatty acids and factors that increase the risk of heart disease (Ascherio et al. 1999; Semma 2002). However, we must relieve cattle from part of the guilt. Beef is a relatively minor source of *trans* unsaturated fatty acids in western diets. Only from 1 to 7% of the total fatty acids in beef, butter, and milk are *trans*. In contrast, up to 35% of the total fatty acids in many highly processed foods (i.e., junk and fast food) are of the *trans* form (Pfalzgraf et al. 1994). In western diets, industrially processed foods are the main sources of the *trans* partially hydrogenated fatty acids that nutritionists warn us about.

In most diets the lipids ingested by herbivores are in the esterified form. Microbes in the fermenting forestomach first hydrolyze these lipids and then add hydrogen to the fatty acids with double bonds (figure 6.24). This process is called biohydrogenation to distinguish it from industrial hydrogenation, in which plant oils are exposed to hydrogen at high temperatures and pressures in the presence of a catalyst. The end product of both the biohydrogenation and industrial hydrogenation of linoleic and linolenic acid is stearic acid (figure 6.24). However, a variety of intermediate products are also formed, including conjugated linoleic acid and *trans* fatty acids. In animals these intermediate products are absorbed and deposited in tissues. Food technologists modify the conditions of the hydrogenation reaction to modulate the relative amount of intermediate products formed. Complete hydrogenation leads to a hard and brittle product whereas partial hydrogenation of varying degrees leads to institutional cooking oils, margarines, and shortenings rich in *trans* fatty acids.

Ruminants are like other mammals in that they are unable to synthesize essential fatty acids. Thus, they seem to have evolved mechanisms to conserve these nutrients and to use them in moderation. The essential fatty acid requirements of cows are an order of magnitude lower than those of nonruminants such as pigs and humans (Mattos and Palmquist 1977), but the mechanisms that mediate these lower requirements remain obscure. It is likely that ruminants use essential fatty acids sparingly and catabolyze them at lower rates than other fatty acids. Raclot and Groscolas (1995) describe the mechanisms that can explain the selective retention of essential fatty acids in mammals. To our knowledge, these mechanisms have not been elucidated in

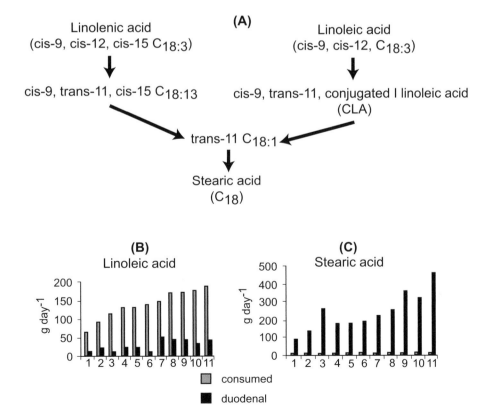

Figure 6.24. Unsaturated fatty acids, such as the essential linolenic and linoleic acids, are biohydrogenated by fermenting foregut microbes (A; after Khanal and Dhiman 2004). The end product of complete biohydrogenation is stearic acid, but incomplete biohydrogenation yields a variety of intermediary products that are absorbed by the host. Among them are trans fatty acids and conjugated linoleic acid (CLA). CLA has received a significant amount of attention because it appears to have all sorts of beneficial health effects. The ingestion of CLA has been associated with a reduction in the incidence of cancer, atherosclerosis, and diabetes (Kritchevsky 2000). Biohydrogenation greatly reduces the assimilation of unsaturated fatty acids (B) and increases the assimilation of stearic acid (C). Dairy cows ingest from 70 to 200 g per day of linoleic acid, but only from 10 to 50 g of this nutrient reach the duodenum (B). Even though cows consume only very small amounts of stearic acid, they assimilate as much as 500 g per day (C). Figures B and C were redrawn after Jenkins (2004) who compiled the results of 11 studies.

ruminants. In spite of the metabolic activity of microbes, and as a result of low requirements, sufficient essential unsaturated fatty acids escape hydrogenation to satisfy the animal's needs (Van Soest 1994).

6.3.9 Glucose Metabolism in Foregut Fermenters

Unsaturated fatty acids are not the only nutrients transformed by the microbes of the foregut. The dietary soluble carbohydrates (glucose and fructose, sucrose, and starch) that enter into the forestomach are rapidly fermented. Hence in many foregut fermenters little glucose reaches the small intestine. The relative physiological scarcity of glucose has led to significant digestive and metabolic differences between foregut fermenters and hindgut fermenters and nonfermenters. Foregut ferementers express comparatively low levels of disaccharidases, such as maltase and sucrase, and of glucose transporters in the small intestine (Kerry 1969; Wolffram et al. 1986). Foregut fermenters also tend to express very low levels of glucokinase (Ballard et al. 1969). Glucokinase catalyzes the phosphorylation of glucose in hepatocytes and hence mediates the entry of glucose into the liver. Adult foregut fermenters also tend to have low blood glucose levels relative to nonforegut fermenters (table 6.5).

In chapter 7 we will review the many striking metabolic changes that animals experience when they shift from the fed anabolic absorptive state to the fasted catabolic postabsorptive state. Briefly, absorptive animals use assimilated nutrients for synthesis whereas catabolic animals rely on stored substrates to fuel their metabolism. The liver in nonforegut fermenters shifts from using glucose to

TABLE 6.5
Difference in metabolic traits between foregut and nonforegut fermenters

	Foregut fermenters	*Non-foregut fermenters*
Blood glucose level	40–60 mg/dL. It falls when the animal fasts but is unchanged after feeding.	80–100 mg/dL. It remains high in fasting animals but increases after feeding.
Lipid synthesis	The main substrate is acetate The enzymes needed to synthesize fatty acids from glucose are not expressed.	Glucose is the main substrate
Liver glucokinase	Virtually absent	Present

Note: This table is based on differences between ruminants and non-foregut fermenters, but it probably applies to foregut fermenters in general (after Van Soest 1994).

synthesize glycogen and lipid in the absorptive state to the generation of glucose from the breakdown of glycogen and the transformation of amino acids in the postabsorptive state (gluconeogenesis). To maintain relatively constant blood glucose levels in the face of almost nil glucose assimilation, the liver of ruminants is in a perpetual state of gluconeogenesis. Ruminants rely primarily on propionate, the only glucogenic SCFA, as the substrate for gluconeogenesis (Van Soest 1994).

Young foregut fermenters undergo a physiological transformation when they shift from maternal milk to feeding independently. Their digestive system morphs from a preweaning one with limited fermenting capacity to a postweaning one with significant fermenting ability (Baldwin et al. 2004). This change is accompanied by other major physiological adjustments. In lambs, the expression of intestinal SGLT1 (the primary glucose transporter) peaks at two weeks of age and then drops to negligible levels by eight weeks (reviewed by Harmon and McLeod 2001). Curiously, unlike the fixed ontogenetic decline observed for lactase in most mammals (chapter 4), this decline in SGLT1 is reversible. When Shirazi-Beechey and her collaborators (1991) infused the small intestine of adult sheep with glucose they increased the expression of SGLT1 40 to 80 fold. The blood glucose levels of newborn calves and lambs are roughly those expected in nonforegut fermenters ($\approx$ 100 mg/dL, using the barbaric units used by clinicians to quantify the levels of glucose in blood). These levels decline by about 40% over the first three months of life to reach adult levels ($\approx$60 mg/dL; Nicolai and Stewart 1965). The changes in blood glucose levels are paralleled by changes in the expression of the liver enzymes that mediate glucose metabolism. From birth to weaning the liver of young ruminants transforms itself from a glycolytic and glycogenic organ to a glucogenic one (Baldwin et al. 2004). The enzymes that mediate the capture of glucose and its transformation into glycogen (e.g., hexokinase and glycogen synthase) decline in activity, whereas those that mediate gluconeogenesis increase in activity.

As you can imagine, these striking physiological changes are accompanied by equally striking morphological ones. The foregut of ruminants changes very rapidly from a non-fermenting one in newborns into a fermenting one in adults (figure 6.25). While young ruminants are feeding on milk, they rely on a small muscular structure called the esophageal (or ruminoreticular) groove to bypass milk directly from the esophagus into the hindstomach—which is called the abomasum in ruminants. As milk enters the esophagus the groove snaps shut and closes off the entrance to the rumen. Milk bypasses the rumen and goes straight into the abomasum. In some ruminants the importance of the reticular groove diminishes after the animal is weaned. In others, the reticular groove seems to retain its function as a bypass mechanism in the adult. Ørskov et al. (1970) found that sheep could be trained to allow fluids to bypass the

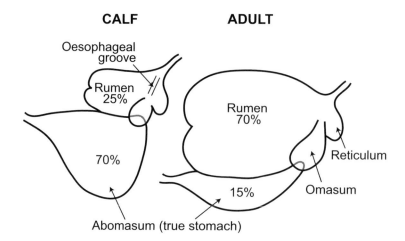

CALF   **ADULT**

Oesophageal
groove

Rumen
25%

Rumen
70%

70%

Reticulum

15%

Omasum

Abomasum (true stomach)

Figure 6.25. Ruminants are supposed to have four stomachs, They have only one which is divided into several compartments. The "true" ruminants have four: rumen, reticulum, omasum, and abomasum. The rumen and reticulum are considered a single organ called the reticulorumen. Most fermentative activity and absorption of fermentation products takes place in the reticulorumen. The omasum is characterized by a large number of leaflike folds which may absorb water and nutrients and prevent passage of large digesta particles. The role of the omasum is not entirely clear. It probably serves as a pump-filter combination that sorts out the liquid and fine digesta for passage into the abomasum. The reticuloomasal orifice connects the reticulum and the omasum. This orifice is adjacent to the posterior end of the esophageal groove. When the groove is closed, ingesta can bypass the reticulorumen and pass directly into the omasum. The abomasum secretes acid, pepsin, and lysozyme. Infant ruminants are functionally monogastric. The rumen is undeveloped and forms a small fraction of the total stomach volume. At birth, it represents from 24 to 35% of the total volume of the stomach (in sheep and deer it can be as low as 4%). The reticulorumen increases in size and relative importance as the animal grows. In adult ruminants, the reticulorumen represents 60–80% of the total stomach volume. (After Van Soest 1994.)

rumen and Hofmann (1984) hypothesized that the functional importance of the reticular groove differs between ruminants with different ecological habits—i.e., between grazers and browsers. In a subsequent section we will consider Hofmann's (1984) hypothesis in some detail.

6.3.10 Foregut Fermentation and Toxins

The microbes that inhabit the foregut modify not only nutrients but also toxins. In many instances the microbiota degrades the ingested toxins effectively (Foley et al. 1999); in others the opposite happens. Microbes can transform innocuous substances into toxins. McSweeney and Mackie (1997) provide a comprehensive account of how different toxins interact with the gut microbiota and with the host. Here we will provide a general description only. The primary pathways of microbial metabolism in the foregut are those that degrade and use polysaccharides, nitrogenated compounds, nucleic acids, and lipids. Some of the same microbial enzymes that participate in these processes have broad specificities and probably contribute to breaking up toxins. In other cases, microbes may have specific enzymes. As is the case with other nontoxic substances, the degradation of some substances can be the result of the concerted action of several microbes in a consortium. In other sections will describe the biochemical mechanisms that animals use to respond to the challenges posed by dietary toxins (chapter 9). One of the themes that we will dwell on is the inducible nature of these mechanisms (see chapter 9). In a similar fashion, the microbial community of the foregut is sometimes capable of changing in response to the presence of toxins in food.

Although there has been relatively little research on the responses of gut microbe communities to dietary toxins, the response of soil, sediments, and sludge microbial communities to xenobiotics has been relatively well studied. Because bacterial communities probably respond in the same way to similar stimuli, we can use the responses of bacterial communities in soils, sediments, and sludges to generate hypotheses about how gut bacterial communities should respond to dietary toxins. A word of caution on microbiological terminology is in order. When discussing consortia of microorganisms it is often tempting to assume that phylogenetically disparate collections of organisms respond "adaptively" to an environmental challenge. You will encounter the term "adaptation" in the microbial biodegradation literature to refer to the changes that microbial communities experience in response to exposure to toxins. Jan van der Meer and colleagues (van der Meer 1992) argue that this use of the term adaptation is confusing and we wholeheartedly agree. The view of microbial consortia as superorganisms that can respond collectively and adaptively to the environment sticks in our Darwinian gullet. However, the term "community adaptation" appears to be widely used by microbiologists and is unlikely to go away.

A variety of processes mediate the response of bacterial communities to xenobiotics and toxins: (1) some members of the microbial community can induce and/or up-regulate the activity of specific enzymes; (2) microbial populations able to metabolize the toxin can increase; (3) selection of mutants that have altered enzymatic properties, and/or exchange of mobile genetic elements among the community's members can take place (Top and Sprinkael 2003); (4) Finally, the microbial community may be invaded by organisms capable of degrading the toxin. Although there are many examples of these processes in the bioremediation literature (see Hinchlee et al. 1995), and although there are many studies that show that specific microorganisms detoxify specific plant secondary metabolites (Foley et al. 1999), we know of very few examples of "adaptation" of the guts's microbiota that lead to toxin tolerance in the host. The almost meteorically fast development of molecular methods to study microbial communities should now permit the investigation of a novel aspect of phenotypic plasticity: the plasticity conferred on a host by changes in the consortium of its evolving symbionts.

One of the most interesting examples of "adaptation" to a toxin by a herbivore involves the legume *Leucaena leucocephala*. The foliage of this fast-growing tree is rich in nitrogen and hence is used as a feed supplement in the tropics. The only problem is that the tissues of *L. leucocephala* contain mimosine, an amino acid which is transformed by many foregut microbes into a toxic compound with the formidable name 3-hydroxi-4-1(H)-pyrdon (or 3,4-DHP). In 1982 Australian nutritional ecologist Raymond Jones discovered that Australian goats that consumed leucena leaves developed toxicocity, but Hawaiian goats feeding on the same diets did not (Jones and Megarrity 1983). He used ruminal fluid from Hawaiian goats to inoculate Australian goats, and discovered that the newly colonized goats became tolerant to mimosine (Jones and Megarrity 1986). The case was closed when microbiologist Milton Allison and his colleagues (1992) isolated a bacterium that is capable of degrading 3,4-DHP from the rumen of mimosine-resistant goats. This bacterium was appropriately named *Synergistes jonesii*. Once established in a few individuals, *S. jonesii* can spread among the members of a herd by normal contact. Cultures of *S. jonesii* have been used to inoculate ruminants throughout the world. Weimer (1998) describes a few other examples of plant detoxification that result from the colonization of the foregut by a microbe with the right metabolic capacity.

The best example of the potentially negative effects of microbes is the transformation of nitrate into toxic nitrites. Some plants accumulate nitrates, especially when they are water stressed. Microbes transform dietary nitrate (NO_3) into nitrite (NO_2^-) and eventually into ammonia in the forestomach. However, if the concentration of nitrate in food is high, some of the nitrite is absorbed into the circulatory system. There it oxidizes oxyhemoglobin into a dark pigment called methemaglobin, which does not carry oxygen. Depending on the degree of oxidation, the color of blood ranges from gray-brown to black. Thus, animals suffering from nitrite poisoning can be diagnosed from the brown or black color of their mucosal membranes (Cawley et al. 1977). Because nonfermenters tend to be significantly more tolerant of dietary nitrate than foregut fermenters, this is an example of a case in which digestive symbionts decrease, rather than increase, the tolerance of animals for a toxicant.

6.3.11 The Many Ways of Being a Herbivore: Grazers and Browsers

Up to this point in the chapter we have adopted a very coarse functional classification of vertebrate herbivores. We have sorted them out by the morphology of their digestive tracts. Here we classify them by what they eat (table 6.6; figure 6.26). It is tempting to speculate that there is a good correlation between a classification based on gastrointestinal morphology and one based on diet. This speculation has generated a lot of useful research, which we will briefly review

TABLE 6.6
Herbirore classifications based on diet

Langer's herbivory rating	Preferred food	Hofmann's feeding type
1	Animal material, fruit, tubers; only occasionally buds and shoots	Omnivore
2	Occasional animal material, plant parts, fruit, tubers, seeds, buds, flowers, leaves, sap.	Concentrate selector
3	Plant parts, fruit, tubers, seeds, buds, flowers, leaves, sap; from 30 to 40% of diet is buds, blossoms, leaves, and young shoots.	Concentrate selector
4	More than 60% of the diet is leaves, shoots, blossoms, and other plant parts; the rest is fruits tubers, seeds, and buds.	Intermediate feeder
5	Predominantly leaves, buds, plant stems; considerable amounts of grass	Bulk and roughage feeder
6	Specialized obligate grazers	Bulk and roughage feeder

Note: Hofmann (1989) used a minimalist classification that divides herbivores into four categories: omnivores, concentrate selectors, intermediate feeders, and bulk and roughage feeders. Langer (1988) used a slightly more complex classification that establishes six ratings. Van Soest (1994) recognized the parallels between these two categories and created this table. Both Hofmann's and Langer's schemes suffer from the appearance of linearity. They assume that we can rank animals along a single axis. Hofmann's scheme emphasizes selectivity, whereas Langer's scheme conflates the ingestion of fibrous materials with the ingestion of grasses, which may not be accurate. Some browsers ingest woody materials that contain more refractory fibers than grasses. Figure 6.26 shows a two-dimensional classification of herbivores based both on the reliance on grass relative to browse and on selectivity.

Figure 6.26. Van Soest (1994) classified herbivores along two axes: the fraction of the diet comprised of grass relative to browse (defined as the foliage and twigs of bushes and herbaceous dicots) and selectivity. Some herbivores defy classification and happily move across this classificatory space seasonally and depending on environmental conditions. Because large animals may be clumsy at feeding on small things, the selectivity axis is strongly influenced by body size. In the right side panel we characterized six ungulates of the Rocky Mountains using data from Singer and Norland (1994). The arrows signify that some herbivores such as elk and deer are capable of changing the proportion of browses and grasses and of becoming more selective seasonally.

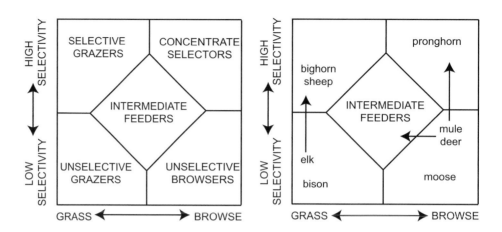

here. But first we must review the many ways in which ecologists have classified mammalian herbivores and the differences between grasses and browse.

Most of the ecological classifications of herbivores divide them into those that feed primarily on grasses (grazers) and those that feed on leaves and twigs of woody plant (browsers). Annual herbs and forbs are included in the category "browse," although they arguably are nutritionally quite different from the leaves of trees and shrubs. The grazer/browser dichotomy is coarse. To improve it, Hofmann (1989) divided ruminant herbivores into three categories based on the amount of grasses ingested: concentrate selectors ingest mostly browse (including annuals and forbs), grass and roughage eaters ingest mostly grass, and intermediate opportunistic feeders feed on both browse and grasses. Van Soest (1994) refined Hoffman's classification adding selectivity as another axis of classification (figure 6.26). The result is a system that differentiates among five types of vertebrate herbivores: selective and unselective grazers, unselective browsers and concentrate selectors, and intermediate feeders. It is noteworthy that many animals move across this two-dimensional space seasonally and depending on the availability of food plants. The dichotomy between grazers and browsers assumes that grasses and browse are distinct and well-defined categories. Although this assumption is crude, on average grasses and browse are different as food for herbivores (box 6.4).

6.3.12. Hofmann's Ruminant Diversification Hypothesis: Form, Function, and Diet in Ruminants

Most large-herbivore ecologists have embraced Reinhold Hofmann's classification of ruminants into three major types: concentrate selectors, grass and roughage

Box 6.4.

Grasses and browse are different types of food along a variety of axes. Grasses tend to have thicker cell walls with large amounts of cellulose of hemicellulose than browse. Because cellulose and hemicellulose must be fermented, grasses tend to be assimilated more slowly. Forbs and leaves tend to have thinner cell walls and higher cytoplasmic contents. While grasses have a propensity to have higher silica concentration, which wears teeth out and can reduce digestibility, browse is likely to have more lignin and secondary compounds such as tannins, terpenes, and alkaloids.

   Grasses and browse also differ in the design of the plant and in how individual plants are distributed in space. Grasses consist of leaves, stems, sheath, and fruit that sometimes cannot be easily differentiated by a foraging large herbivore (Shipley 1999). Browse, in contrast, can contain a more heterogeneous mix of buds, mature leaves, woody stems, and reproductive

TABLE 6.7
Differences between grasses (monocots) and browses
(herbaceous and woody dicots)

Characteristic	Grasses	Browses
Cell wall	Thick A greater proportion is cellulose and hemicellulose	Thin A greater proportion is lignin
Plant defense compounds	Silica	Phenolics and tannins. Terpenes Alkaloids Other toxins
Plant architecture	Fine-scaled heterogeneity in nutritional quality within a plant New growth added to the plant's base Low-growth form Three-dimensional volume	Coarse scale heterogeneity in nutritional quality within a plant New growth added at the tips Complex, diffuse, branching architecture
Spatial dispersion	Uniform	Dispersed

Note: Shipley 1999. There are many exceptions to the general differences outlined in this table. There are grasses that produce large amounts of secondary compounds (e.g., sorghum produces cyanide), and browses that grow in homogeneous stands (e.g., heather).

continued

continued

parts including flowers and fruit. Because grasses grow by adding new tillers to the plant's base, a sward of grass is a more or less homogeneous layer of vegetation. Browse grows by adding new plant tissue to branch or plant tips. This form of growth creates an irregular branched plant geometry that can sometimes permit selectivity, but can sometimes form an impenetrable maze, especially when the branches have spines and thorns (table 6.7). It is important to keep in mind that there are many exceptions to the differences between grasses and browses listed in table 6.7 and summarized here. Many ecological problems may require characterizing food plants using finer resolution. Van Soest (1994) summarized the classifications of forages used by range managers to characterize the plants eaten by livestock.

eaters, and intermediate feeders. The more nuanced classification proposed by Peter Van Soest (figure 6.26) has not been as widely accepted. In an important paper published in 1989, Hofmann proposed that his three-way classification of ruminants was associated with distinctive sets of morphophysiological traits. Briefly, Hofmann hypothesized that grazing ruminants depend more heavily on fermentation products than browsers. According to Hofmann this increased reliance on grass and roughage is associated with larger rumens, with structures of the rumen and omasum that retard the passage of food to the lower tract, and with the absence of a functional rumenoreticular grove in the adult (figure 6.25). As we discussed previously, this structure allows small particles and fluids to bypass the rumen to be digested in the abomasum and small intestine. He hypothesized that browsers not only have smaller and less complex rumens and larger functional reticular grooves, but also have larger parotid salivary glands that secrete copious saliva that helps to buffer the more rapidly fermentable soluble materials present in browse. Although Hofmann's "diversification of ruminants" hypothesis stimulated a lot of comparative research, the jury on its validity is still out (figure 6.27).

Gordon and Illius (1994, 1996) and Robbins et al. (1995) studied large comparative data sets on gut capacity and the retention time of fluids in the gut of ruminants and concluded that their data did not support Hofmann's hypothesis. The terse and definitive conclusion of Gordon and Illius' (1994) paper is worth quoting: "after controlling for body mass, there is little difference in digestive strategy between . . . ruminants with different morphological adaptations of the gut." More recently Javier Pérez-Barbería and his collaborators (Pérez-Barbería et al. 2001) used a phylogenetically explicit analysis to reach roughly the same conclusions, except that they concluded that the data set had a significant

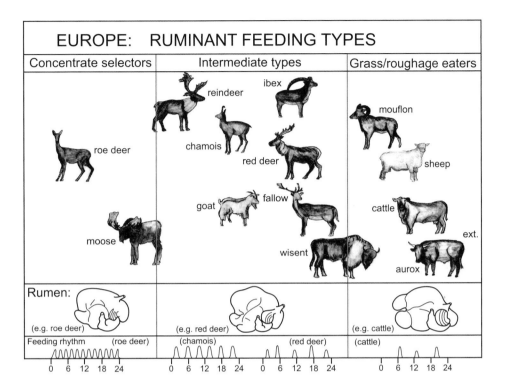

Figure 6.27. An ecological classification of European ruminants according to Hofmann (1989). Hofmann classified ruminants along a continuous feeding-type axis with concentrate selectors (browsers) such as roe deer at one extreme, and grass/roughage eaters (such as cattle) at the other. Hofmann predicted that species along this continuum would have differences in digestive morphology, physiology, and behavior.

phylogenetic signal. That is, closely related species tend to have both similar diets and similar gut morphologies. For example, topis (*Damalisus lunatus*), wildebeest (*Alcephalus spp*), and gnus (*Connochates taurinus*) are all grazers and are all members of a single monophyletic lineage. In a similar fashion, all the members of the genus *Tragelaphus* (which includes bongos, nyalas, kudus, bushbucks, and sitatungas) are browsers. As we mentioned in chapter 1, the power of phylogenetically explicit statistical analyses is low when the traits under consideration are clumped, rather than well dispersed within the phylogeny. Pérez-Barbería and his collaborators (2001) used a relatively small sample of 28 species of African mammals. Perhaps a larger sample of the over 190 worldwide ruminant species and the complete phylogeny for the group developed recently by Hassanin and Souzery (2003) would yield a different result.

In spite of the evidence against Hofmann's (1989) hypothesis, we concur with Ditchkoff's (2000) judgment that it is unwise to throw away it yet. Our reluctance to reject the diversification of ruminants hypothesis is based on two observations: (1) all the comparative studies that have aimed to test it have found patterns that agree with it; (2) recent functional evidence supports it. Here we elaborate briefly on each of these two points. Both Gordon and Illius (1994) and Robbins et al. (1995) observed patterns that agree with Hofmann's

Figure 6.28. A preliminary analysis of the size of salivary glands in several species of ruminants supported Hofmann's (1989) hypothesis. Browsers seem to have larger parotid glands than grazers and intermediate feeders seem to have glands of intermediate size. A notable exception is the greater kudu (*Tragelaphus sterpsiceros*). The graph relating parotid gland weight and body mass is after Robbins et al. (1995). The upper panel illustrates Hofmann's measurements of the percentage of body mass represented by salivary glands in nine selected species of ruminants.

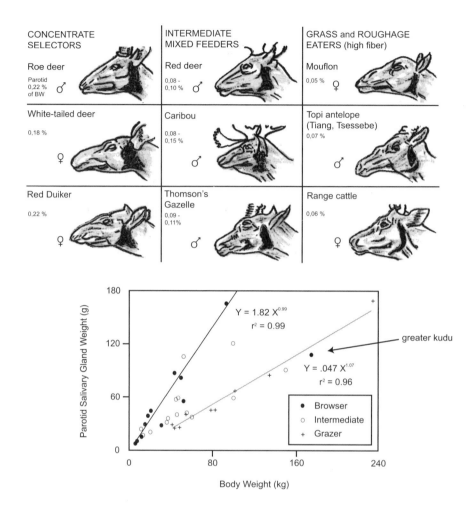

predictions. Gordon and Illius (1994) estimated that the supply of fermentation products in the rumen was insufficient to meet the energy requirements of browsers. This result implies that a deficit must be met by digestion in the intestine or by fermentation in the large intestine. For reasons described in the following sentences, we favor the hypothesis that soluble nutrients escape fermentation and are digested in the small intestine. Robbins et al. (1995) found that the salivary glands of browsers were approximately four times larger than those of grazers (figure 6.28). Robbins et al. (1995) measured the rate of saliva production in mule deer (*Odocoileus hemionus*), which are browsers, and domestic cows and sheep, which are grazers. They found no significant differences. However, they measured the slobbering rate of animals that were neither feeding nor ruminating. It is unknown if the basal rate of salivary production estimates the relative functional capacity of the salivary glands accurately. It is hard to believe that a four-fold difference in salivary gland mass does not translate into a difference in function.

A flurry of recent papers have also documented noteworthy differences between browsing and grazer ruminant species. Because these studies are based on a small and phylogenetically incomplete species sample, they do not constitute bona fide tests of the ruminant diversification hypothesis. Yet they are suggestive of functional differences between the three feeding classes of ruminants established by Hofmann. Rowell-Shäfer and her collaborators (2001) found that the activity of intestinal maltase (the enzyme that effects the membrane digestion of the disaccharide products of starch's digestion) and of the intestinal glucose transporter (SGLT1) was much higher in browsers such as roe deer (*Capreolus capreolus*) and moose (*Alces alces*) than in adult sheep and cows. They also found a much higher proportion of polyunsaturated fatty acids in the fat deposits of moose and roe deer than in those of fallow deer and sheep (Rowell-Schäfer et al. 2001). Clauss and his collaborators (2002) studied the distribution of fecal particles in 81 species of captive ruminants and found that browsers defecated a higher proportion of larger particles than intermediate feeders and grazers of the same size. They interpreted this result as evidence of a difference in the escape of larger particles from the reticulo-rumen of browsers. This result supports Hofmann's contention that browsers pass larger particles in their feces and are therefore unable to grind and thus assimilate the fiber contained in feed particles to the same degree as grazers. Unfortunately Clauss et al. (2002) relied on a comparative data set in which not all species were fed on the same diet. Because the size of fecal particles can vary within a species with diet, the results of this study can be attributed to differences in the diets that the animals were fed. In a thorough phylogenetic analysis of 24 species, Pérez-Barbería and his collaborators (2004) found a result that is consistent with the observation of Clauss et al. (2002). They found that ruminants with a high proportion of grass in their natural diets tended to have higher digestibility of neutral detergent fiber than those that feed on browse. Again, these results agree with Hofmann's hypothesis.

Hofmann's hypothesis has been fertile. It has generated a large amount of data and it has spawned many informative analyses. It is controversial and is likely to remain so until we have larger comparative data sets, and, perhaps more importantly, when we complement these with experiments. The availability of phylogenetic information for ruminants (e.g., Hassanin and Souzery 2003) and new phylogenetically explicit statistical comparative analyses should facilitate testing it.

6.4 Herbivory and Detritivory in Fish

So far we have emphasized herbivory by vertebrates on land. However, there are myriads of aquatic vertebrate herbivores in both freshwater and marine environments. Unfortunately, these interesting creatures have not been as thoroughly investigated. By necessity, this section is short and cursory. A following section deals with detritivorous fish that ingest slime and wood. The justification for the inclusion of this seemingly esoteric topic is that aquatic ecologists have realized that detritus is a fundamental resource in aquatic ecosystems, and we understand little about the mechanisms that animals use to assimilate it. Physiological ecologists can make a significant contribution to aquatic ecology by focusing on the nutritional ecology of detritivores.

6.4.1. Herbivorous Fishes and the Plants that They Eat

Several striking patterns emerge when one compares the mammalian herbivores that have such profound impact on terrestrial ecosystems with marine fish herbivores. The first one concerns diversity. Roughly 25% of the approximately 4,100 mammal species can be considered herbivorous. In contrast, only about 650 (or about 5%) of the 13,765 species of marine teleosts are considered herbivores (Choat and Clements 1998). The second one concerns size. There are many very large herbivorous mammals, and many mammalian faunas are dominated by large herbivores including a few megaherbivores that top the 1,000 kg scale. In contrast, most herbivorous fishes are relatively small (figure 6.29). Many, albeit not all, teleost fishes begin their herbivorous career at settlement and grow by over three orders of magnitude in mass until they reach adulthood. Thus the difference in mass between mammalian and piscine herbivores is accentuated when development is taken into account. Marine ecosystems can have huge schools of tiny herbivores, but there is no piscine equivalent to the herds of large herbivorous mammals that roam grassy terrestrial habitats.

In this chapter we have emphasized the difficulty that digesting cellulose poses to terrestrial herbivores. We have also emphasized that an association with microbial symbionts is one of the ways in which many terrestrial herbivores have overcome this difficulty. Marine ecosystems differ from terrestrial ones in cellulose abundance. Although cellulose is by no means absent in the cell walls of marine plants, it is not nearly as abundant in them as it is in terrestrial plants. This is not the best place to review in detail the diversity of polysaccharides found

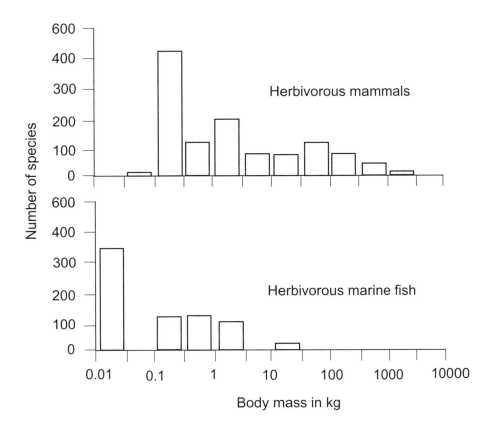

Figure 6.29. Compared with herbivorous mammals, herbivorous marine fish tend to be small. (After Choat and Clements 1998.)

in the cell walls of algae (we refer readers to Percival's 1979 review). It is sufficient to mention that fish have to contend with a diversity of potentially fermentable carbohydrates. The structural polysaccharides of algae include polymers not only of glucose, but also of manose (mannanes), xylose (xylanes), uronic acids (alginic acid), and galactose (agarose). Perhaps the best known of these polysaccharides is agarose, which is extracted from the red algae *Gelidium* and *Gracillaria* and used extensively in research laboratories to make electrophoresis gels. It remains unclear which of these must be hydrolyzed by the fishes' endogenous enzymes and which are broken by gut microbes.

Many algae, especially in the tropics, also use calcium carbonate ($CaCO_3$) as a "structural" element. We used quotation marks because $CaCO_3$ may have a dual function. It may indeed have a structural function, but it may also act as a deterrent against herbivory (Schupp and Paul 1994). Marine plants show enormous variation not only in the chemistry of the materials that make up their cell walls, but also in the production of allelochemicals, and in morphology. How do fish cope with the variety of food types represented by marine plants?

Ichthyologists Phillip Lobel (1981) and Michael Horn (1989) have proposed four mechanisms:

(1) The low pH in the stomach of fish can lyse, or at least perforate, cell walls and allow digestive enzymes to penetrate the nutrients in the cytoplasm (Zemke-White et al. 2000).

(2) Fish can use a gizzardlike stomach in which algal cells and inorganic materials are ground together to disrupt algal cells and release nutrients.

(3) Fish may have teeth (both mandibular or pharyngeal) that shred or grind algae before they enter the intestine (Wainwright et al. 2004).

(4) Fish may have symbiotic associations with microbes that assist in the digestion of algal cells and algal nutrients (Clements 1997).

These four mechanisms are not mutually exclusive. Figure 6.30 illustrates how these mechanisms can be combined in herbivorous fish species.

Because only a handful of researchers are devoted to exploring the functional biodiversity of herbivorous fishes, our knowledge of the relative importance of each of these mechanisms remains very limited. We know little about the phylogenetic distribution of the four mechanisms listed above, and less about whether these are associated with the use of particular algal resources. The exceptional work of Choat, Clements, and Mountfort has greatly increased our understanding of the functional biology of herbivorous fishes. These researchers have documented the microbial diversity in the guts of herbivorous fishes (including the largest prokaryote ever found; Angert et al. 1993). They also have measured the rate at which microbial symbionts produce short-chain fatty acids. They have discovered that they can divide "herbivorous" fishes into four broad categories: (1) Herbivores with a diet of macroscopic brown algae and with high concentrations of short-chain fatty acids (SCFAs) in their guts; (2) herbivores that feed on filamentous red algae and green algae and algae that form turfs and that have moderate SCFAs in their guts; (3) zooplankton feeders with moderate SCFA levels; and (4) species that feed on detritus and sediments with low levels of SCFAs. They found that fish in the first three categories feed slowly and retain food in the gut overnight. In contrast, the detritivores in category 4 have higher feeding rates (as would be expected from their low-quality diet) and start the day with an empty gut (Choat et al. 2004).

It is tempting to hypothesize (á la Reinhold Hofmann) that the morphophysiology of the four categories of herbivorous fishes identified by Choat et al. (2004) will correspond neatly with their food types. Perhaps disappointingly, Choat et al. (2004) have not found a clear correlation between gut morphology, feeding rate, concentrations of SCFAs, and diet. For example two of the herbivores that feed

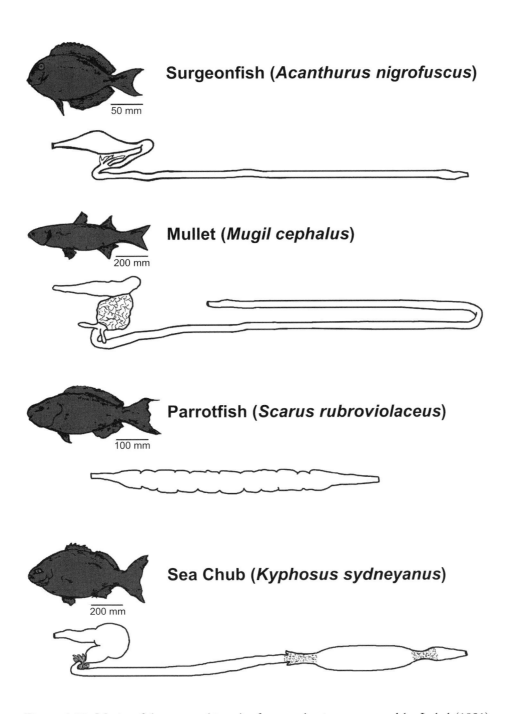

Figure 6.30. Marine fish can combine the four mechanisms proposed by Lobel (1981) and Horn (1989) to cope with algae. Both the surgeon fish (Acanthuridae) and the striped mullet (Mugilidae) have a well-developed muscular stomach that is best described as

(caption continues)

(caption continued)

gizzardlike (mechanism II), followed by pyloric cecae and a long narrow intestine. Pyloric cecae are blind tubes (hence the term cecae) connected to the anterior end of the intestine found in many fish species (including mullets and chubs). Pyloric cecae lack a fermentative microbiota and may secrete fluids that buffer the intestinal content (Clements 1997). Their function is a bit mysterious. The pH of the stomach of the surgeonfish is low enough to lyse algal cells (mechanism I). The pH of the mullet's stomach ranges from slightly acidic to slightly alkaline. The sea chub (Kyphosidae) has a distinct enlarged hindgut flanked by sphincters at the junctions with the midgut and the rectum. The hindgut of sea chubs contains a well-developed fermentative microbiota (mechanism IV). The parrotfish (Scaridae) scrapes reef surfaces and hence ingests not only algae but also a significant amount of inorganic material. Its digestive system consists of a powerful pharyngeal mill (Mechanism III) followed by a broad intestine. Parrotfish seems to have limited fermentative capacity. (After Stevens and Hume 1989.)

on macroscopic brown algae, *Naso unicornis* (Acanthuridae) and *Kyphosus vaigiensis* (Kyphosidae), have high SCFA concentrations in their guts. However, these two species have extremely different guts. The gut of *N. unicornis* looks like that of the surgeonfish *A. nigrofuscus* illustrated in figure 6.30, whereas the gut of *K. vaigiensis* looks like that of the sea chub *Kyphosus sydneyanus*, also illustrated in figure 6.30. Clements (1997) asserts that "there doesn't seem to be any direct relationship between intestinal anatomy and fermentation in fishes."

Kyphosus vaigiensis is exceptional in the degree of differentiation of the intestine. In most fishes the hindgut is difficult to distinguish from the midgut. *Kyphosus vaigiensis* shares with some species in its family (Kyphosidae, drummers) and with marine angelfishes (Pomacanthidae) the presence of a well-differentiated chamber at the posterior end of the gut (figure 6.30). This chamber is separated from the rest of the intestine by a sphincter and is often filled with a well-developed fermentative microbiota. However, many fishes that lack such a seemingly specialized "hindgut" also contain abundant symbionts that generate considerable amounts of SCFAs (Clements 1997). Choat et al. (2004) have pointed out that in the tropical perciform fishes that they studied, herbivory and fermentation are not associated with the gut structures that characterize herbivorous terrestrial vertebrates. Perplexingly, some herbivorous fishes with seemingly effective fermentation systems lack well-defined fermentation chambers.

The lack of a well-defined fermentative chamber does not seem to hinder fermentation in herbivorous fishes. Mountfort et al. (2002) have found that fermentation rates in three species of seaweed-eating New Zealand fishes (*Kyphosus sydneyanus*, Kyphosidae, *Odax pullus*, Odacidae, and *Aplodactylus arctidensis*, Aplodactylidae) are comparable to those found in the gut of herbivorous mammals, despite the much lower body temperatures of the fishes, and in spite of the fact that only one of these species has a well-defined fermentation chamber (Mountfort et al. 2002). Although these results suggest that symbiotic microbial processes in these fishes are much more important than previously thought, many questions remain unanswered. It remains unclear whether particular seaweed components are broken down by the fishes' endogenous enzymes or by those of their symbionts, and we still have only a very nebulous idea of the contribution that gut symbionts make to the energy and nitrogen economy of herbivorous fishes. The nutritional ecology of herbivorous fishes remains an open area of exploration for physiological ecologists.

6.4.2 Detritivores: The Slime and Wood-Eating Fish

The partnership between insects and microbes and fungi has led to several enormously successful insect radiations. We described in detail two of them in this chapter: the termites and the leaf-cutting ants. The termites can be considered an evolutionary radiation of terrestrial detritivores. There is no equivalent radiation of detritivores among terrestrial vertebrates. The only truly detritivorous vertebrates are found among fish and amphibian larvae. Detritus is defined as dead material produced by plants, algae, microbes, and animals. In aquatic systems detritus includes morphous (literally meaning "with form") particles that contain cell wall materials refractory to digestion, and amorphous particles. Amorphous particles contain trapped soluble organic matter and microorganisms embedded in a matrix of hydrated polysaccharides called exopolymers (Wotton 2004).

Bowen (1984) has shown that amorphous particles contain significantly lower contents of refractory materials and higher nitrogen content than morphous particles. He speculated that amorphous detritus is a readily assimilable source of energy and nutrients that sustains rich aquatic food chains. Indeed, detritivorous fish that feed on amorphous detritus can be both abundant and ecologically important. Detritivores comprise over 50% of the total fish biomass in tropical South American rivers (Bowen 1983). Although detritus-based food chains are undoubtedly important, we still know very little about the nutritional characteristics of amorphous detritus and about how it is assimilated. Detritivorous fish and aquatic invertebrates do not seem to rely on symbiotic microorganisms to assimilate amorphous detritus, but we don't know exactly how they assimilate it (see Bowen 1976 and 1983). The few aquatic vertebrates that eat morphous detritus do seem to depend on symbiotic microorganisms. Good examples are the wood-eating catfishes in the genera *Panaque* and *Cachliodon* (Loricariidae;

Nelson et al. 1999). These South American fishes feed on wood and seem to be able to grow on a wood-only diet (Nelson et al. 1999). Unfortunately, very little has been published about their digestive and nutritional physiology.

References

Aanen, D. K., P. Eggleton, C. Rouland Lefévre, T. Guldsberg-Frøslev, S. Rosendahl, and J. Boomsma. 2002. The evolution of fungus-growing termites and their mutualistic fungal symbionts. *Proceedings of the National Academy of Science* 99:14887–14892.

Abe, T., and M. Higashi. 1991. Cellulose centered perspective on community structure. *Oikos* 60:127–133.

Abril, A. B., and E. H. Bucher. 2002. Evidence that the fungus cultured by leaf-cutting ants does not metabolize cellulose. *Ecology Letters* 5:325–326.

Ahring, B. K., and P. Westermann. 1985. Methanogenesis from acetate: physiology of a thermophilic acetate-utilizing methanogenic bacterium. *FEMS Microbiology Letters* 25:47–52.

Alexander, R. McN. 1991. Optimization of gut structure and diet for higher vertebrate herbivores. *Philosophical Transactions of the Royal Society (London) B* 333:249–255.

Allison, M. J., W. R. Mayberry, C. S. McSweeney, and D. A. Stahl. 1992. *Synergistes jonesii*, gen. nov., sp. nov.: A rumen bacterium that degrades toxic pyridinediols. *Systematic Applied Microbiology* 15:522-529.

Amann, R. I., W. Ludwig, and K. H. Schleifer. 1995. Phylogenetic identification and in situ detection of individual microbial cells without cultivation. *Microbiological Reviews* 59:143–169.

Amann, R., and W. Ludwig. 2000. Ribosomal RNA-targeted nucleic acid probes for studies in microbial ecology. *FEMS Microbiology Reviews* 24:555–565.

Angert, E. R., K. D. Clements, and N. R. Pace. 1993. The largest bacterium. *Nature* 362:239–241.

Anklin-Mühlemann, R., D. E. Bignell, P. C. Veeivers, R. H. Leuthold, and M. Slaytor. 1995. Morphological and biochemical studies of the gut flora in the fungus-growing termite Macrotermes subhyalinus. *Journal of Insect Physiology.* 41: 929–940.

Ascherio, A., M. B. Katan, P. L. Zockk, M. J. Stampfer, and W. C. Willett. 1999. *Trans* fatty acids and coronary heart disease. *New England Journal of Medicine.* 340:1994–1996.

Atwood, M. 2003. *Oryx and Crake.* Anchor Books, New York.

Baldwin, R. L., K. R. McLeod, J. L. Klotz, and R. N. Heitmann. 2004. Rumen development, intestinal growth, and hepatic metabolism in the pre- and post-weaning ruminant. *Journal of Dairy Science.* 87 (E Supplement): E55–E65.

Ballard, F. J., R. W. Hanson, and D. S. Kronfeld. 1969. Gluconeoenesis and lipogenesis in tissues from ruminant and nonruminant animals. *Federation Proceedings* 28:218–231.

Barnard, E. A. 1969. Biological function of pancreatic ribonuclease. *Nature* 221: 340–344.

Beaver, R. A. 1989. Insect-fungus interactions in the bark- and ambrosia beetles. Pages 121–144 in N. Wilding, N. M. Collins, P. M. Hammond, and J. F. Webber (eds.), *Insect fungus interactions*. Academic Press, New York.

Beintema, J. J. 1990. The primary structure of langur (*Presbytis entellus*) pancreatic ribonucleases: adaptive features in digestive enzymes in mammals. *Molecular Biology and Evolution*. 7:470–477.

Beintema, J. J. 2002. Ribonucleases in ruminants. *Science* 297:1121–1122.

Beintema, J. J., A. J. Scheffer, H. van Dijk, G. W. Welling, and H. Zwiers. 1973. Pancreatic ribonuclease distribution and comparisons in mammals. *Nature—New Biology* 241:76.

Berchtold, M., A. Chatzinotas, W. Schöhuber, A. Brune, R. Amman, D. Hahn, and H. König. 1999. Differential enumeration and in situ localization of microorganisms in the hindgut of the lower termite *Mastotermes darwinensis* by hybridization with r-RNA-targeted probes. *Archives of Microbiology*. 172:407–416.

Bevins, C. L., E. Martin-Porter, and T. Ganz. 1999. Defensins and innate host defence of the gastrointestinal tract. *Gut* 45:911–915.

Bignell, D. E., M. Slaytor, P. C. Veivers, R. Muhlemann, and R. H. Leuthold. 1994. Functions of symbiotic gardens in higher termites of the genus *Macrotermes*: Evidence against the acquired enzyme hypothesis. *Acta Microbiological et Immunologica Hungarica* 41: 391–401.

Bjorndal, K. A. 1997. Fermentation in reptiles and amphibians. Pages 199–230 in R. I. Mackie and B. A. White, eds., *Gastrointestinal microbiology, vol. 1: Gastrointestinal ecosystems and fermentations*. Chapman and Hall, New York.

Björnhag, G. 1994. Adaptations in the large intestine allowing small animals to eat fibrous foods. Pages. 287–312 in D. J. Chivers and P. Langer, eds., *The digestive system in mammals*. Cambridge University Press, Cambridge, U.K.

Boga, H. I., and A. Brune. 2003. Hydrogen-dependent oxygen reduction by homoacetogenic bacteria isolated from termite guts. *Applied and Environmental Microbiology* 69: 779–786.

Borer, K. I. 1985. Regulation of energy balance in the golden hamster. Pages 363–408 in H. I. Siegel, ed., *The hamster—reproduction and behavior*, (New York). Plenum Press.

Bowen, S. H. 1976. Mechanism for digestion of detrital bacteria by the cichlid fish *Sarotherodon mossambicus* (Peters). *Nature* 260:137.

Bowen, S. H. 1983. Detritivory in neotropical fish communities. *Environmental Biology of Fishes* 9: 137–144.

Bowen, S. H. 1984. Evidence of a detritus food chain based on consumption of organic precipitates. *Bulletin of Marine Science* 35: 440–446.

Braun, E. J. 1999. Integration of renal and gastrointestinal function. *Journal of Experimental Zoology*. 283:495–499.

Brune, A. 1998. Termite guts: The world's smallest bioreactors. *Trends in Biotechnology* 16:16–21.

Brune, A. 2003. Symbionts aiding digestion. Pages 1102–1107 in V. H. Resh and R. T. Cardè, eds., *Encyclopedia of insects*. Academic Press, New York.

Brune, A., and M. Friedrich. 2000. Microecology of the termite gut: structure and function on a microscale. *Current Opinions in Microbiology.* 3:263–269.

Bucher, E. H., D. Tamburini, A. Abril, and P. Torres. 2003. Folivory in the white-tipped plantcutter *Phytotoma rutila*: seasonal variations in diet composition and quality. *Journal of Avian Biology* 34:211–216.

Cámara, V. M., and D. J. Prieur. 1984. Secretion of colonic isozyme of lysozyme in association with cecotrophy of rabbits. *American Journal of Physiology* 247:G19–G23.

Cawley, G. D., D. F. Collings, and D. A. Dyson. 1977. Nitrate poisoning. *Veterinary Record* 101:305–306.

Cazemier, A.E., J. M. Huub, H.J.M.O. den Camp, J. H. P. Hackstein, and G. D. Vogels. 1997a. Fibre digestion in arthropods. *Comparative Biochemistry and Physiology.* 118A:101–109.

Cazemier, A.E., J.H.P. Hackstein, H.J.M.O. den Camp, J. Rosenberg, and C. Van der Drift. 1997b. Bacteria in the intestinal tract of different species of arthropods. *Microbial Ecology* 33:189–197.

Cherret, J. M. 1986. History of the leaf-cutting ant problem. Pages 10–17 in C. S. Lofgren and R.K. Vander Meer, eds., *Fire ants and leaf-cutting ants: Biology and management.* Westview Press, Boulder, Colo.

Cherrett, J. M., R. J. Powell, and D. J. Stradling. 1989. The mutualism between leaf-cutting ants and their fungus. Pages 93–120 in N. Wilding, N. M. Collins, P. M. Hammond, and J. F. Webber (eds.), *Insect fungus interactions.* Academic Press, New York.

Chilcott, M. J., and I. D. Hume. 1985. Coprophagy and selective retention of fluid digesta: Their role in the nutrition of the common ringtail possum, *Pseudocheirus peregrinus. Australian Journal of Zoology.* 33:1–15.

Choat, J. H., and K. D. Clements. 1998. Vertebrate herbivores in marine and terrestrial environments: a nutritional ecology perspective. *Annual Review of Ecology and Systematics* 29:375–403.

Choat, J. H., W. D. Robbins, and K. D. Clements. 2004. The trophic status of herbivorous fishes on coral reefs. *Marine Biology* 145:445–454.

Clauss, M., M. Lechner-Doll, and W. J. Streich. 2002. Faecal particle distribution in captive wild ruminants: an approach to the browser/grazer dichotomy from the other end. *Oecologia* 131:343–349.

Clements, K. D., 1997. Fermentation and gastrointestinal microorganisms in fishes. Pages 156–198 in R. I. Mackie and B. A. White, eds., *Gastrointestinal microbiology, Vol. 1 Gastrointestinal ecosystems and fermentations.* Chapman and Hall, New York.

Clench, M. H. 1999. The avian cecum: Update and motility review. *Journal of Experimental Zoology.* 283:441–447.

Cleveland, L. R., S. R. Hall, E. P. Sanders, and J. Collier. 1934. The wood-feeding roach, *Cryptocercus*, its protozoa and the symbiosis between protozoa and roach. *Memoirs of the American Academy of Arts and Sciences.* 17:185–342.

Cork, S. J., I. D. Hume, and G. J. Faichney. 1999. Digestive strategies of nonruminant herbivores. Pages 210–260 in H. J. G. Jung and G. C. Fahey, eds., *Nutritional ecology of herbivores.* American Society of Animal Science, Savoy, Illi.

Cree, T. C., D. M. Wadley, and J. A. Malett. 1986. Effect of preventing coprophagy in the rat on neutral detergent fiber digestibility and apparent calcium absorption. *Journal of Nutrition*. 116:1204–1206.

Crowell-Davis, S. L., and K. A. Houpt. 1985. Coprophagy in foals: Effects of age and possible functions. *Equine Veterinary Journal*. J. 17:17–19.

Currie, C. C., U. G. Mueller, and D. Murdoch. 1999. The agricultural pathology of ant-fungus gardens. *Proceedings of the National Academy of Science*. 96:7998–8002.

Currie, C. R., A.N.M. Bot, and J. J. Boosma. 2003. Experimental evidence of a tripartite mutualism: bacteria protect ant fungus gardens from specialized parasites. *Oikos* 101:91–102.

Cyr, H., and M. L. Pace. 1993. Magnitude and patterns of herbivory in aquatic and terrestrial ecosystems. *Nature* 361:148–150.

D'Ettorre, P. D., P. Mora, V. Dibangou, and C. Rouland. 2002. The role of symbiotic fungus in the digestive metabolism of two species of fungus-growing ants. *Comparative Biochemistry and Physiology B*. 172:169–176.

Dierenfeld, E. S., H. F. Hintz, J. B. Robertson, P. J. Van Soests, and O. T. Oftedal. 1982. Utilization of bamboo by the giant panda. *Journal of Nutrition*. 112:636–641.

Ebert, A., and A, Brune. 1997. Hydrogen concentration profiles at the oxic-anoxic interface: a microsensor study of the hindgut of the wood-feeding lower termite *Reticulitermis flavipes* (Kollar). *Applied and Environmental Microbiology*. 63:4039–4046.

Dehorty, B. A. 1997. Foregut fermentation. Pages 39–83 in R. I. Mackie and B. A. White, eds., *Gastrointestinal microbiology, vol.1: Gastrointestinal ecosystems and fermentations*. Chapman and Hall, New York.

Ditchkoff, S. S. 2000. A decade since "diversification of ruminants": Has our knowledge improved? *Oecologia* 125:82–84.

Dobson, D. E., E. M. Prager, and A. C. Wilson. 1984. Stomach lysozymes of ruminants 1: Distribution and catalytic properties. *Journal of Biological Chemistry*. 259:11607–11616.

Dowd, P. F. 1992. Insect fungal symbionts: A promising source of detoxifying enzymes. *Journal of Industrial Microbiology* 9:149–161.

Dryer, B., and O. Khalsa. 1993. Surface bacteria of *Streblomastix strix* are sensory symbionts. *BioSystems* 31:169–180.

Duwat, P., S. Sourice, B. Cesselin, G. Lamberet, K. Vido, P. Gaudu, Y. Le Loir, F. Violet, P. Loubiére, and A. Gruss. 2001 Respiratory capacity of the fermenting bacterium *Lactococcus lactis* and its positive effects on growth and survival. *Journal of Bacteriology* 183:4509–4516.

Eggleton, P. 2001. Termites and trees: A review of recent advances in termite phylogenetics. *Insectes Sociaux* 48:187–193.

Farlow, J. O. 1987. Speculations about the diet and digestive physiology of herbivorous dinosaurs. *Paleobiology* 13:60–72.

Farrell, B. D., A. S. Sequeira, B. C. O'Meara, B. B. Normark, J. H. Chung, and B. H. Jordal. 2001. The evolution of agriculture in beetles (Curculionidae, Scolytinae, and Platypodinae). *Evolution* 55:2011–2027.

Febvay, G., and Kermarrec. 1986. Digestive physiology of leaf-cutting ants. Pages 274–288 in C. S. Lofgren and R. K. Vander Meer, eds., *Fire ants and leaf-cutting ants: Biology and management*. Westview Press, Boulder, Colo.

Fenchel, T., and B. J. Finlay. 1995. *Ecology and evolution in anoxic worlds*. Oxford University Press, Oxford.

Foley, W. J., G. R. Iason, and C. McArthur. 1999. Role of plant secondary metabolites in the nutritional ecology of mammalian herbivores: how far have we come in 25 years? Pages 130–209 in H.-J. G. Jung and G. C. Fahey (eds), *Nutritional ecology of herbivores*. American Society of Animal Science, Savoy, Il.

Geisler, J. H., and M. D. Uhen. 2003. Morphological support for a close relationship between hippos and whales. *Journal of Vertebrate Paleontology* 23:991–996.

Gordon, I. J., and A. W. Illius. 1994. The functional significance of the browser-grazer dichotomy in African ruminants. *Oecologia* 98:167–175.

Gordon, I. J., and A. W. Illius. 1996. The nutritional ecology of African ruminants: a reinterpretation. *Journal of Animal Ecology* 65:18–26.

Gould, G. M., and W. L. Pyle. 1896. *Anomalies and curiosities of medicine. Being an encyclopedic collection of rare and extraordinary cases, and of the most striking instance of abnormality in all branches of medicine and surgery*. W. B. Sanders, Philadelphia.

Grajal, A. 1995, Structure and function of the digestive tract of the Hoatzin (*Opisthocomus hoazin*): A folivorous bird with foregut fermentation. *Auk* 112:20–26.

Grajal, A., and S. D. Strahl. 1991, A bird with the guts to eat leaves. *Natural History* 100:48–54.

Grajal, A., S. D. Strahl, R. Parra, M.G. Dominguez, and A. Neher. 1989, Foregut fermentation in the Hoatzin, a neotropical leaf-eating bird. *Science* 245:1236.

Grandscolas, P., and P. Deleporte. 1996. The origin of protistan symbionts in termites and cockroaches: A phylogenetic perspective. *Cladistics* 12:93–96.

Harmon, D. L., and K. R. McLeod. 2001. Glucose uptake and regulation by intestinal tissues: implications and whole body energetics. *Journal of Animal Science* 79 (E Supplement): E59–E72.

Hassanin, A., and E. J. P. Souzery. 2003. Molecular and morphological phylogenies of Ruminantia and the alternative position of the Moschidae. *Systematic Biology* 52:206–226.

Hinchlee, R. E., F. J. Brockman, and C. M. Vogel. 1995. *Microbial processes for bioremediation*. Batelle Press, New York.

Hirakawa, H. 2001. Coprophagy in leporids and other mammalian herbivores. *Mammal Review* 31:61–80.

Hirakawa, H. 2002. Supplement: Coprophagy in leporids and other mammalian herbivores. *Mammal Review* 32:150–152.

Hochachka, P., and G. N. Somero. 2002. *Biochemical adaptation: Mechanisms and processes in physiological evolution*. Oxford University Press, Oxford.

Hofmann, R. R. 1984. Comparative anatomical studies imply adaptive vari-ations of ruminant digestive physiology. *Canadian Journal of Animal Science*. 64:203–205.

Hofmann, R. R. 1989. Evolutionary steps of ecophysiological adaptation and diversification of ruminants: A comparative view of their digestive system. *Oecologia* 78:443–457.

Hölldobler, B., and E. O. Wilson. 1990. *The ants*. Harvard University Press, Cambridge, Mass.

Horn, M. H. 1989. Biology of marine herbivorous fishes. *Oceanography and Marine Biology Annual Review*. 27:167–272.

Hume, I. D. 1999. *Marsupial nutrition*. Cambridge University Press, Cambridge.

Inoue, T., O. Kitade, T. Yoshimura, and I. Yamaoka. 2000. Symbiotic associations with protists. Pages 275–288 in T. Abe, D. E. Bignell, and M. Higashi (eds.), *Termites: Evolution, sociality, symbioses, ecology*. Kluwer Academic Publishers, Dordrecht.

Ito, Y. 1994. Colonic lysozymes of rabbit (Japanese white): Recent divergence and functional convergence. *Journal of Biochemistry—Tokyo*. 116:1346–1353.

Irwin, D. M., and A. C. Wilson. 1989. Multiple cDNA sequences and the evolution of bovine stomach lysozyme. *Journal of Biological Chemistry*. 264:11387–11393.

Irwin, D. M., E. M. Prager, and A. C. Wilson. 1992. Evolutionary genetics of ruminant lysozymes. *Animal Genetics* 23:193–202.

Janzen, D. H. 1975. *Ecology of plants in the tropics*. Edward Arnold, London.

Jenkins, T. 2004. Challenges of meeting cow demands for omega fatty acids. *Proceedings of the Florida Ruminant Nutrition Symposium* 52–66 (http://dairy.ifas.ufl.edu/flrns.html).

Jones, R. J., and R. G. Megarrity. 1983. Comparative toxicity responses of goats fed on *Leucaena leucocephala* in Australia and Hawaii (USA). *Australian Journal of Agricultural Research*. 34:781–790.

Jones, R. J., and R. G. Megarrity. 1986. Succesful transfer of dihydroxypyridine degrading bacteria from Hawaiian (USA) goats to Australian ruminants to overcome the toxicity of Leucaena. Australian Veterinary Journal 63:259–262.

Kenagy, G. J., and D. F. Hoyt. 1980. Reingestion of feces in rodents and its daily rhythmicity. *Oecologia* 44:403–409.

Kennedy, P. M., and L. P. Millign. 1980. The degradation and utilization of endogenous urea in the gastrointestinal tract of rumnants: A review. *Canadian Journal of Animal Science* 60:205–221.

Kerry, K. R. 1969. Intestinal disaccharidese activity in a monotreme and eight species of marsupials (with an added note on the disaccharidases of five species of seabirds). *Comparative Biochemistry and Physiology A* 52A:235–246.

Khanal, R. C., and T. R. Dhiman. 2004. Biosynthesis of conjugated linoleic acid: A review. *Pakistan Journal of Nutrition* 3:72–81.

Kok, L. T., D. M. Norris, and H. M. Chu. 1970. Sterol metabolism as a basis for mutualistic symbiosis. *Nature* 225:661–662.

Kornegay, J. R., J. W. Schilling, and A. C. Wilson. 1994. Molecular adaptation of a leaf-eating bird: stomach lysozyme of the hoatzin. *Molecular Biology and Evolution*. 11:921–926.

Kourie, J. I., and A. A. Shorthouse. 2000. Properties of cytotoxic peptide-formed ion channels. *American Journal of Physiology*. 278:C1063–C1087.

Kritchevsky, D. 2000. Antimutagenic and some other effects of conjugated linoleic acid. *British Journal of Nutrition* 83:459–465.

Kukor, J. J., D. P. Cowan, and M. M. Martin. 1988. The role of ingested fungal enzymes in cellulose digestion in the larvae of cerambycid beetles. *Physiological Zoology* 61:364–371.

Lebrun, D., C. Rouland, and C. Chararas. 1990. Influence of the defaunation on nutrition and survival of *Kalotermes falvicollis*. *Material und Organismen* 25:1–14.

Lee, K. E., and T. G. Wood. 1971. *Termites and soil*. Academic Press, London.

Lee, M. J., P. J. Schreurs, A. C. Messer, and S. H. Zinder. 1987. Association of methanogenic bacteria with flagellated protozoa from a termite hindgut. *Current Microbiology*. 6:337–342.

Lobel, P. S. 1981. Trophic biology of herbivorous reef fish: Alimentary pH and digestive capabilities. *Journal of Fish Biology* 19:365–397.

Logan, M. 2003. Effect of toothwear on the rumination-like behavior, or merycism, of free-ranging koalas (*Phascolarctos cinereus*). *Australian Journal of Zoology* 84:897–902.

Lopez-Calleja, M. V., and F. Bozinovic. 1999. Feeding behavior and assimilation efficiency of the rufous-tailed plantcutter: A small avian herbivore. *Condor* 101:705–710.

Lopez-Calleja, M. V., and F. Bozinovic. 2000. Energetics and nutritional ecology of small herbivorous birds. *Revista Chilena de Historia Natural* 73:411–420.

Machida, M., O. Kitade, T. Miura, and T. Matsumoto. 2001. Nitrogen recycling through proctodeal trophallaxis in the Japanes damp-wood termite *Hodotermis japonica* (Isoptera, Termopsidae). *Insectes Sociaux* 48:52–56.

Mackie, R. I. 2002. Mutualistic fermentative digestion in the gastrointestinal tract: Diversity and evolution. *Integrated and Comparative Biology* 42:319–326.

Madigan, M. T., J. M. Martinko, and J. Parker. 2000. *Brock's biology of microorganisms*. Prentice-Hall, New York.

Martin, M. M. 1987. *Invertebrate-microbial interactions*. Cornell University Press, Ithaca, N.Y.

Martin, M. M. 1991. The evolution of cellulose digestion in insects. *Philosophical Transactiong of the Royal Society (London) B* 333:281–286.

Martin, M. M., and J. S. Martin. 1978. Cellulose digestion in the midgut of the fungus-growing termite *Macrotermes natalensis*: The role of acquired digestive enzymes. *Science* 199:1453–1455.

Mathiesen, S. D., T. H. Aagnes, W. Sørmo, E. S. Nordøy, A. S. Blix, and M. A. Olsen. 1995. Digestive physiology of minke whales. Pages 351–359 in A. S. Blix, L Walløe, and ø. Ulltang (eds.), *Whales, seals, fish, and man*. Elsevier Science, Amsterdam.

Mattos, W., and D. L. Palmquist. 1977. Biohydrogenation and availability of linoleic acid in lactating cows. *Journal of Nutrition*. 107:1755–1761.

McSweeney, C. S., and R. I. Mackie. 1997. Gastrointestinal detoxification and digestive disorders in ruminant animals. Pages 583–634 in R. I. Mackie and B. A. White eds., *Gastrointestinal microbiology*, *Vol*.1:*Gastrointestinal ecosystems and fermentations*. Chapman and Hall, New York.

McWhorter, T., D. Powers, and C. Martínez del Rio. 2003. Ammonotely, nitrogen excretion and nitrogen requirements in three sympatric hummingbird species. *Physiological and Biochemical Zoology* 76:731–743.

Mendes, A., S. S. da Nogueira, A. Loborenti, and S.L.G. Nogueira Filho. 2000. A note on the cecotrophy behavior in capybara (*Hydrochaeris hydrochaeris*). *Applied Animal Behavioral Science* 66:161–167.

Metges, C. C. 2000. Contribution of microbial amino acids to amino acid homeostasis of the host. *Journal of Nutrition* 130:1857S–1864S.

Meynard, C., M. V. Lopez_Calleja, F. Bozinovic, and P. Sabat. 1999. Digestive enzymes of a small avian herbivore, the Rufous-tailed plant-cutter. *Condor* 101:904–907.

Montgelard, C., S. Ducrocq, and E. Douzery. 1998. What is a Suiforme (Artiodactyla)? Contributions of cranioskeletal and mitochondrial DNA data. *Molecular Phylogenetics and Evolution* 9:528–532.

Moore, J. E., M. H. Brant, W. E. Kunkle, and D. I. Hopkins. 1999. Effects of supplementation on voluntary forage intake, diet digestibility, and animal performance. *Journal of Animal Science.* 77 (Supplement 2J) 122–135.

Moreto, M., C. Amat, A. Puchal, R. K. Buddington, and J. M. Planas. 1991. Transport of L-proline and alpha-methyl-D-glucoside by chicken proximal cecum during development. *American Journal of Physiology* 260:G457–G463.

Mountfort, D. O., J. Campbell, and K. D. Clemets. 2002. Hindgut fermentation in three species of marine herbivorous fish. *Applied and Environmental Microbiology* 68:1374–1380.

Mueller, U. G. 2002. Ant versus Fungus versus mutualism: Ant-cultivar conflict and the deconstruction of the attine-fungus symbiosis. *American Naturalist* 160:S67–S96.

Mueller, U. G., and N. Gerardo. 2002. Fungus farming insects: Multiple origins and diverse evolutionary histories. *Proceeding of the National Academy of Science.* 99:15247–15249.

Mueller, U. G., S. A. Rehner, and T. R. Schultz. 1998. The evolution of agriculture in ants. *Science* 281:2034–2036.

Murphy, W. J., E. Eizirik, W. E. Johnson, Y. P. Zhang, O. A. Ryder, and S. J. O'Brien. 2001. Molecular phylogenetics and the origin of placental mammals. *Nature* 409:614–616.

Nakashima, K., H. Watanabe, H. Saitoh, G. Tozuda, and J. I. Azuma. 2002. Dual cellulose-digesting system of the wood-feeding termite *Coptotermes formosanus* Shiraki. *Insect Biochemistry and Molecular Biology* 32:777–784.

Nalepa, C.A. 1984. Colony composition, protozoan transfer and some life history characteristics of the wood roach *Cryptocercus punctulatus* Scudder (Dictyoptera: Cryptocercidae). *Behavioral Ecology and Sociobiology.* 14:273–279.

Nalepa, C. A., D. E. Bignell, and C. Bandi. 2001. Detritivory, coprophagy, and the evolution of digestive mutualisms in Dictyoptera. *Insectes Sociaux.* 48:194–201.

Nagy, K. A. 2001. Food requirements of wild animals: Predictive equations for free-living mammals, reptiles, and birds. *Nutrition Abstracts and Reviews Series B* 71:1R–12R.

Nelson, J. A., D. A. Wubah, M. E. Whitmer, E. A. Johnson, and D. J. Stewart. 1999. Wood-eating catfishes of the genus *Panaque*: gut microflora and cellulolytic enzyme activities. *Journal of Fish Biology* 54:1069–1082.

Nicolai, J. H., and W. E. Stewart. 1965. Relationship between forestomach and glycemia in ruminants. *Journal of Dairy Science* 48:56–62.

Ohkuma, M. 2003. Termite symbiotic systems: Efficient bio-recycling of lignocellulose. *Applied Microbiology and Biotechnology* 61:1–9.

Olsen, M. A., A. S. Blix, T.H.A. Utsi, W. Sormo, and S. D. Mathiesen. 2000. Chitinolytic bacteria in the minke whale forestomach. *Canadian Journal of Microbiology*. 46:85–94.

Ørskov, E. R., D. Benzie, and R.N.B. Kay. 1970. The effect of feeding procedure on closure of the esophageal groove in sheep. *British Journal of Nutrition* 24:785–795.

Osawa, R., W. H. Blanshard, and P. G. O'Callaghan. 1993. Microbiological studies of the microflora of the koala, *Phascolarctus cinereus*. II. Pap, a special maternal faeces consumed by juvenile koalas. *Australian Journal of Zoology* 41:611–620.

Pellens, R., P. Grandscolas, and I. Domingos da Silva-Neto. 2002. A new and independently evolved case of xylophagy and the presence of intestinal flagellates in the cockroach *Paraspheria boleiriana* (Dyctioptera, Blaberidaea, Zetoborinae) from the remnants of the Brazilian Atlantic forest. *Canadian Journal of Zoology*. 80:350–359.

Percival, E. 1979 The polysaccharides of green, red and brown seaweed: Their basic structures, biosynthesis and function. *British Physiology Journal* 14:103–117.

Pérez-Barbería, F. J., I. J. Gordon, and A. Illius. 2001. A phylogenetic analysis of stomach adaptation in digestive strategies in African ruminants. *Oecologia* 129:498–506.

Pérez-Barbería, F. J., D. A. Elston, I. J. Gordon, and A. W. Illius. 2004. The evolution of phylogenetic differences in the efficiency of digestion in ruminants. *Proceedings of the Royal Society (London) B*. 271:1081–1090.

Pfalzgraf, A., M. Timm, and H. Steinhart. 1994. Content of trans-fatty acids in food. *Zeitschrift für Ernährungswissenschaft* 33:24–43.

Porter, E. M., C. L. Bevins, D. Ghosh, and T. Ganz. 2002. The multifaceted Paneth cell. *Cellular and Molecular Life Sciences* 59:156–170.

Pryor, G. S. 2003. Roles of gastrointestinal symbionts in nutrition, digestion, and development of bullfrog tadpoles (*Rana catesbiana*). Ph.D. dissertation, University of Florida, Gainesville.

Purvis, A. 1995. A composite estimate of primate phylogeny. *Philosophical Transactions of the Royal Society (London) B* 348:405–421.

Pynchon, T. R., 1973. *Gravity's rainbow*, Viking Press, New York.

Quinlan, R. J., and J. M. Cherrett. 1979. The role of fungus in the diet of the leaf-cutting ant *Atta cephalotes* (L.). *Ecological Entomology* 4;151–160.

Raclot, T., and R. Groscolas. 1995. Selective mobilization of adipose tissue fatty acids during energy depletion in the rat. *Journal of Lipid Research* 36:2164–2173.

Robbins, C. T., D. E. Spalinger, and W. van Hoven. 1995. Adaptations of ruminants to browse and grass diets: Are anatomically-based browser-grazer interpretations valid? *Oecologia* 103:208–213.

Rouland-Lefèvre, C. 2000. Symbiosis with fungi. Pages 289–305 in T. Abe, D. E. Bignell, and M. Higashi (eds.), *Termites: Evolution, sociality, symbioses, ecology*. Kluwer Academic Publishers, Dordrecht.

Rowell-Schäfer, A., M. Lechner-Doll, R. R. Hofmann, W. J. Streich, B. Güven, and H.H.D. Meyer. 2001. Metabolic evidence of a "rumen bypass" or a "ruminal escape" of nutrients in roe deer (*Capreolus capreolus*). *Comparative Biochemistry and Physiology A* 28A:289–296.

Rübsamen, K., I. D. Hume, and W. v. Englehardt. 1982. Physiology of the rock hyrax. *Comparative Biochemistry Physiology* A 72A:271–277.

Ruff, C. 2002. Variation in human body size and shape. *Annual Reviews of Anthropology.* 31:211–232.

Ruvolo, M. 1997. Molecular phylogeny of the hominoids: inferences from multiple independent DNA sequence data sets. *Molecular Biology and Evolution* 14:248–265.

Sakaguchi, E. 2003. Digestive strategies of small hindut fermenters. *Animal Science Journal* 74:327–337.

Schupp, P. J. and V. J. Paul. 1994. Calcium carbonate and secondary metabolites in tropical seaweeds: variable effects on herbivorous fishes. *Ecology* 75:1172–1185.

Semma, M. (2002). Trans fatty acids: Properties, benefits, and risks. *Journal of Health Science* 48:7–13.

Singer, F., and J. E. Norland. 1994. Niche relationships within a guild of ungulate species in Yellowstone National Park, Wyoming, following release from artificial controls. *Canadian Journal of Zoology* 72:1383–1394.

Shipley, L. A. 1999. Grazers and browsers: How digestive morphology affects diet selection. Pages 20–27 in K. L. Launchbaugh, K. D. Saunders, and J. C. Mosley, eds., *Grazing behavior of livestock and wildlife*. Idaho Forest, Wildlife, and Range Experimental Station Bulletin No. 70. University of Idaho, Moscow, Idaho.

Shirazi-Beechey, S. P., B. A. Hirayama, Y. Wang, D. Scott, M. W. Smith, and E. M. Wright. 1991. Ontogenetic development of lamb intestinal sodium-glucose co-transporter is regulated by diet. *Journal of Physiology (London)*. 437:699–706.

Sinsabaugh, R. L., A. E. Linkins, and E. F. Benfield. 1985. Cellulose digestion and assimilation by three leaf-shredding aquatic insects. *Ecology* 66:1464–1471.

Slaytor, M. 2000. Energy metabolism in the termite and its gut microbiota. Pages 307–332 in T. Abe, D. E. Bignell, and M. Higashi, eds. *Termites: Evolution, sociality, symbioses, ecology*. Kluwer Academic Publishers, Dordrecht.

Soave, O., and C. D. Brand. 1991. Coprophagy in animals: A review. *Cornell Veterinarian*. 81:357–364.

Stevens, C. E., and I. D. Hume. 1996. Contributions of microbes in vertebrate gastrointestinal tract to productiona and conservation of nutrients. *Physiological Reviews* 78:393–427.

Stevens, C. E., and I. D. Hume. 2004. Comparative physiology of the vertebrate digestive system. Cambridge University Press, Cambridge.

Stewart, C., J. W. Schilling, and A. C. Wilson. 1987. Adaptive evolution in the forestomach lysozyme of foregut fermenters. *Nature* 330:401–404.

Strong, D. R., J. H. Lawton, and T.R.E. Southwood. 1984. *Insects on plants: community patterns and mechanisms*. Blackwell Scientific, New York.

Swanson, K. W., D. M. Irwin, and A. C. Wilson. 1991. Stomach lysozyme of the langur monkey: tests for convergence and positive selection. *Journal of Molecular Evolution*. 333:418–425.

Swift, M. J., O. W. Heal and J. M. Anderson. 1979. *Decomposition in terrestrial ecosystems.* University of California Press, Berkeley.

Tholen, A., Schink, B., and A. Brune. 199. The gut microflora of *Reticulitermes flavipes*, its relation to oxygen, and evidence for oxygen-dependent acetogenesis by the most abundant *Enterococcus* sp. *FEMS Microbiology and Ecology* 24:137–149.

Thorne, B. L. 1997. Evolution of eusociality in termites. *Annual Review of Ecology and Systematics* 28:27–54.

Top, E. M., and D. Springael. 2003. The role of mobile genetic elements in bacterial adaptation to xenobiotic compounds. *Current Opinions in Biotechnology* 14:262–269.

Treves, D. S., and M. Martin. 1994. Cellulose digestion in primitive hexapods: Effect of ingested antibiotics on gut microbial populations and gut cellulase levels in the firebrat, *Thermobia domestica* (Zygentoma, Lepismatidae). *Journal of Chemical Ecology* 20:2003–2020.

Turner, J. S. 2001. On the mound of *Macrotermes michaelseni* as an organ of respiratory gas exchange. *Physiological and Biochemical Zoology* 74:798–822.

Turner, J. S. 2002. A superorganism's fuzzy boundary. *Natural History* 111:62–67.

Van der Meer, J. R., W. M. De Vos, S. Harayama, and A. J. B. Zehnder. 1992. Molecular mechanisms of genetic adaptation to xenobiotic compunds. *Microbiological Reviews* 56:677–694.

Van Soest, P. J. 1994. *Nutritional ecology of the ruminant.* Cornell University Press, Ithaca, N.Y.

Van Soest, P. J. 1996. Allometry and ecology of feeding behavior and digestive capacity in herbivores: A review. *Zoo Biology* 15:455–479.

Vazques-Arista, M., R. H. Smith, V. Oldade-Portugal, R. E. Hinojosa, R. Hernandez, and A. B. Blanco. 1997. Cellulolytic bacteria in the digestive system of *Prostephanus truncates* (Coleoptera: Bostrichidae). *Journal of Economy Botany* 90:1371–1376.

Wainwright, P. C., D. R. Bellwood, M. W. Westneat, J. R. Grubich, and A. S. Hoey. 2004. A functional morphospace for the skull of labrid fishes: patterns of diversity in a complex biomechanical system. *Biological Journal of the Linnean Society* 82:1–25.

Watanabe, H., and G. Tokuda. 2001. Animal cellulases. *Cellular and Molecular Life Sciences* 58:1167–1176.

Wei, F., Z. Wang, Z. Feng, M. Li, and A. Zhou. 2000. Seasonal energy utilization in bamboo by the red panda (*Ailurus fulgens*). *Zoo Biology* 19:27–33.

Weimer, P. J. 1998. Manipulating ruminal fermentation: a microbial ecological perspective. *Journal of Animal Science* 76:3114–3122.

Wen, Y., and D. M. Irwin. 1999. Mosaic evolution of ruminant stomach lysozyme enzymes. *Molecular Phylogeny and Evolution* 13:474-482.

Wenzel, M., I. Schönig, M. Bertchtold, P. Kömpfer, and H. Känig. 2002. Aerobic and facultatively anaerobic cellulolytic bacteria from the gut of the termite *Zoostermopsis angusticollis*. *Journal of Applied Microbilogy.* 92:32–40.

Wilson, E. O. 1971. *The insect societies.* Harvard University Press, Cambridge, Mass.

Wolffram, S., E. Eggenberger, and E. Scharrer. 1986. Kinetics of D-glucose and L-leucine transport into sheep and pig intestinal brush-border membrane vesicles. *Comparative Biochemistry and Physiology A* 84:589–593.

Wotton, R. S. 2004. The ubiquity and many roles of exopolymers (EPS) in aquatic systems. Sci. Mar. 68 (supplement):13–21.

Zemke-White, W. L., K. D. Clements, and P. J. Harrios. 2000. Acid lysis by marine herbivorous fishes: Effects of acid pH on cell wall porosity. *Journal of Experimental Marine Biology and Ecology* 245:57–66.

Zhang, J., Y. Zhang, and H. F. Rosenberg. 2002. Adaptive evolution of a duplicated pancreatic ribonuclease gene in a leaf-eating monkey. *Nature Genetics* 30:411–415.

The Ecology of Postabsorptive Nutrient Processing

CHAPTER 7

Postabsorptive Processing of Nutrients

THE BIOCHEMICAL FATE of absorbed material from the digestive system determines its benefits and costs to the consumer. Some understanding of the biochemical details of metabolic processing of absorbed nutrients is critical to understand many features of nutritional ecology. The purpose of this chapter, and the next two, is to provide enough details of postabsorptive processing to explain further the costs and benefits of resources utilized by animals.

For most animals, carbohydrates, fats, and/or proteins are the major energy-yielding substrates. But the value of absorbed products of digestion of carbohydrates, proteins, and fats is not simply the chemical potential energy available in their bonds. Their gross energy values must be discounted for the energetic costs of postabsorptive processing. In some cases, particular organs are more or less incapable of utilizing the chemical potential energy in one of the substrates. For example, fatty acids (FAs) are not used by the brain. These compounds, or their components, also serve as the building blocks in the reconstruction of the animal body, which is continuously broken down and rebuilt in the dynamic process we call the catabolic-anabolic steady state. In some cases, particular FAs and amino acids are important building blocks that the animal synthesizes insufficiently and so they must be consumed in food (see chapter 2). Vitamins, which are also essential for normal metabolism but insufficiently synthesized, are not utilized for energy or synthesis of structural components but serve as cofactors or coenzymes in chemical reactions. Absorbed inorganic material has no energy value but yields some elements that are critical for animal function. As we will see in chapter 9, the elimination of some substances can also impart costs to the animal, and ingestion of too much of some elements can cause toxicity.

All these issues fall under the rubrics of nutritional biochemistry and toxicology. Although these are both large topics in their own right, the number of details that we need to understand is relatively small. Our goal is to illustrate how these details can have far-reaching ecological implications. If you are already familiar with the basics of nutritional biochemistry you can skip right to the implications, which begin in section 7.2.

7.1 Overview: The Postabsorptive Fate of Absorbed Materials

7.1.1 Metabolic Pools are Idealized Models Useful for Describing Postabsorptive Processing of Nutrients

We might imagine that, once a compound is absorbed, it mixes with some of the same material that is already in the body. The body's readily available quantity of that material, defined as a pool in chapter 1, also receives material from other pathways. For example, virtually all organic molecules in the body, with few exceptions (e.g., protein in hair, nails), are continuously broken down and contribute to the pool, and the pool may receive inputs due to conversions from one molecular type to another. Pools export material by several pathways, including incorporation of material into structures, conversion to other molecular types, catabolism to CO_2 and water (and heat), and loss or excretion from the body. The pool is mainly a conceptual device, but methods are available to measure some compartments in animals such as fat and lean mass (box 7.1 illustrates how to measure the lipid "pool" in an animal). Figure 7.2 summarizes major molecular interconversions of interest and the dynamic steady state of pools that relate to protein, carbohydrate, and fat metabolism. With this diagram as background, let us consider the fate of these major substrates.

7.1.2 Carbohydrates

Dietary sugars and starch are absorbed from the digestive tract mainly as monosaccharides, as described in chapter 4. A large fraction of monosachharides such as galactose and fructose are transformed by the liver and/or by the gastrointestinal tract to glucose, so we can conceptualize a glucose pool and consider its inputs and outputs. In insects and other arthropods it is necessary to consider also a trehalose pool (chapter 2).

(1) The energy in glucose can be stored in glycogen (e.g., in muscle or liver), or in fat (e.g., in adipose tissue or liver) via metabolic interconversions.

Box 7.1. Body Composition Analysis and Indices of Condition

Ecologists sometimes view animals' bodies as composed of a *structural part* plus *stores*; the latter are defined as nutrients accumulated in anticipation of periods of shortage (King and Murphy 1985). The term "reserves" is sometimes used synonymously with stores, but that term might be best used for body components necessary for normal life but used in emergencies (Van Der Meer and Piersma 1994). If ecologists measure stores, they can better predict resources available for major life cycle events such as reproduction, survival during resource limitation, migration, and hibernation. As a practical example, total lipid content may be a useful proxy for egg production in fish stock management (Marshall et al. 1999).

Although energy is stored as fat (section 7.1.4), stores are not equivalent to fat because some species store both fat and protein (Van Der Meer and Piersma 1994) and elements may be stored as well. But, in practice, ecologists typically make the following three simplifying assumptions: (1) stores = fat; (2) structural part of the body = lean mass; (3) lean mass = body mass - fat mass. This is essentially a two-compartment model of the body, and other models with three or more compartments subdivide the lean mass further (Speakman 2001). The measurement of fat by chemical extraction was described in chapter 2. Ecologists also determine fat and/or body composition by a variety of nondestructive methods. These can be evaluated according to their accuracy, given by

$$\text{mean relative error for } n \text{ comparisons} = 100 \left(\sum_{1}^{n} \left(\frac{|\text{predicted} - \text{observed}|}{\text{observed}} \right) \right) / n$$

precision, given by the mean intraindividual coefficient of variation CV = standard deviation/mean, or other metrics of model correspondence (Mayer and Butler 1993). An excellent, thorough treatment of methodological issues that influence the accuracy and precision of a variety of methods is available (Speakman 2001). Our comments about approximate accuracy and precision are for methods calibrated against direct measure by chemical extraction:

(1) *Palpation or visual determination.* In a few species it might be possible to assess adiposity by palpation, and a common method in birds is to visually score fat level in the abdominal region and in the intraclavicular depression using a nine-grade score (Kaiser 1993). A score of zero corresponds to no visible fat, and the highest score corresponds to fat visible everywhere (a veritable butterball!).

continued

continued

Figure 7.1. As shown for blackcaps (*Sylvia atricapilla*; an Old Word warbler), total body water measured by dilution of labeled water (deurterium, ^2H_2O) is strongly correlated with lean body mass (left-hand figure), and so fat mass can also be accurately estimated from the dilution space (right-hand figure). (Replotted from Karasov and Pinshow 1998.)

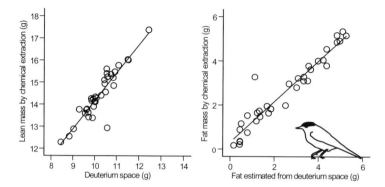

(2) *Labeled water dilution.* The water content of lean mass is 15–20 times higher than for fat. Thus, water content is highly correlated to lean mass, and water content can be nondestructively measured by injecting isotopically labeled water and taking a blood sample after the isotope has equilibrated with body water (as described in methods box 12.1 in chapter 12). The dilution of the labeled water can be calibrated to measure lean mass accurately (relative error = ± 5%). Investigators sometimes calculate fat mass as body mass – lean mass (figure 7.1), but the accuracy of the fat mass estimate is usually much poorer, usually ± 15–20%, because small errors in lean mass can cause proportionally larger errors in lipid mass when lean mass is much larger than lipid mass. The method could be influenced by hydration state, but important virtues are that it is relatively quick, it can be performed in the field, and it is relatively cheap.

(3) *Total body electrical conductivity (TOBEC®).* An oscillating magnetic field determines conductivity by detecting changes in a radiating coil's impedence. Whole-body conductance correlates mainly with lean mass, which contains conductive water and ions. Thus, TOBEC® devices can be calibrated to measure lean mass accurately (relative error = ± 5%; Scott et al. 2001) with a precision of about 10% (Karasov and Pinshow 1998), but precision is poorer for fat mass (usually > ± 20%) when it is calculated by difference, as explained for labeled water dilution. The method requires considerable care to control a number of factors that influence the measurement (especially posture and position of the animal, body temperature, hydration state), but a virtue is that repeated measures can be made even under field conditions. One of us (W.H.K.) has used it successfully; the other one (C.M.R.) has not been able to extract any meaningful measurements from the gadget.

(4) *Dual energy X-ray absorptiometry (DXA).* Different tissues have different absorptivities for X-rays, and DXA devices can be calibrated to measure fat mass, bone-free lean mass, total bone mineral, and bone mineral density (Nagy and Clair 2000; Korine et al. 2004). The precision

continued

continued

for all these measures is good (< 10%), and after appropriate calibration the relative error can be low (< 10%) for lean mass and for fat mass. Like TOBEC®, these devices permit repeated measures but they are more expensive.

(5) *Cyclopropane (or other liposoluble gas) dilution.* Cyclopropane gas is 34 times more soluble in lipid than nonlipid parts of the animal. Thus, when an animal is sealed in a cyclopropane-containing chamber the fraction absorbed is proportional to the animal's lipid mass. Relative accuracy is good, ±5–10% for fat (Henen 1991) and ± 1% for lean mass (Henen 2001), but a disadvantage is that the method is complicated and time consuming compared with other methods.

Glycogen can be broken up to contribute to the glucose pool. Ruminants, which absorb little dietary glucose because most is fermented, tend to have low levels of the enzymes that synthesize fatty acids from glucose. They conserve glucose to supply energy for the brain.

(2) Glucose may be catabolized aerobically for energy to CO_2 and water via the Krebs cycle in mitochondria, or anaerobically to lactic acid in the cytoplasm. In both aerobic and anaerobic metabolism, some of the energy released during catabolism is captured by the formation of a high-energy bond formed with the conversion of ADP + phosphate to ATP, though the majority is released as heat. Heat accounts for 59% of the energy released from complete oxidation of glucose to CO_2 and water, and 41% is stored as ATP. Under nonfasting conditions, glucose is the major and nearly only energy source for the nervous system.

(3) The carbon skeleton of glucose can be used to synthesize many amino acids or structural carbohydrates.

7.1.3 Proteins

Dietary proteins, absorbed as either amino acids or peptides (chapter 4), enter blood mainly as amino acids and so we can conceptualize an amino acid pool. Amino acids can be used as is or, if in excess, can be converted to other amino acids, glucose, or fatty acids. The first step in interconversion of amino acids is the removal of the α-amino group (mainly in the liver by degradative enzymes such as aminotransferases, dehydrogenase, dehydratase) which ultimately generates ammonium ions (NH_4^+). This represents a waste product whose excretion in urine, and its associated costs, will be considered subsequently.

Once stripped of their amino group, the carbon skeletons of some amino acids can reach the glucose pool or the fatty acid pool by metabolic interconversion and participate in metabolic processes already discussed for those compounds.

(1) Amino acids are used to synthesize other amino acids, protein (e.g., muscles, enzymes, hair) as well as other specialized derivatives (e.g., peptide hormones). Animal proteins are combinations of 20 α-amino acids, but only about half of these can be synthesized by animals (these are called dispensable or nonessential amino acids) and the other half must be consumed (these are called indispensable or essential; see chapter 2). Because of this fundamental biochemical difference, dispensable and indispensable amino acids can behave as if they are somewhat separate pools with separate dynamics (see the example of hawk moths in chapter 8). Some proteins incorporate metals such as Fe, Mg, and Mn, which also cannot be synthesized and must be ingested.

(2) The carbon skeletons of most amino acids can be entirely or partly converted to glucose (gluconeogenesis). These are called glucogenic amino acids. Hence, energy in their bonds can be stored in either glycogen or the glycerol of fat. In contrast, the carbon skeleton of the other group, called ketogenic amino acids (leucine and lysine in mammals), cannot be used in the net synthesis of glucose but can be converted to acetyl-CoA which can be used to synthesize FAs and ketone bodies (hence their name). Thus, energy in the bonds of ketogenic amino acids can be stored in fat.

(3) All amino acids can be completely oxidized via the Krebs cycle to CO_2 and water, with 25–35% of the released energy ultimately stored as ATP and 65–75% of total energy released as heat, depending on the specific amino acids involved. The glucogenic amino acids can be catabolized anaerobically.

7.1.4 Fats

We will conceptualize a pool composed of nonesterified or free fatty acids (NEFAs or FFAs) and glycerol, the monomeric units of triacylglycerols. Because most NEFAs and triacylglycerols have low water solubility, they occur in the blood associated with proteins (called apolipoproteins) in a complex called a lipoprotein. Thus, dietary fats enter the blood (or lymph) as triacylglycerols in fat droplets in chylomicrons (containing one type of lipoprotein), as triacylglycerols associated with proteins (other kinds of lipoproteins), and as NEFAs bound or unbound to plasma proteins. Herbivores with microbial fermentation chambers absorb the short-chain FAs released by the microbes

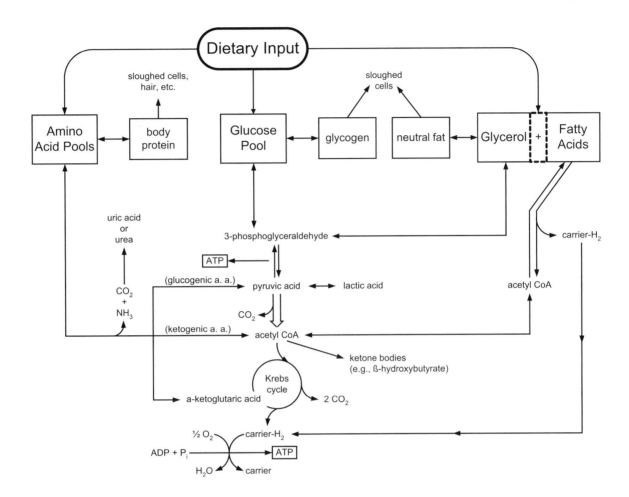

Figure 7.2. It is useful to think of the body as composed of pools that receive inputs from diet and other sources and that export material into structures or to biochemical pathways leading to catabolism, biotransformation, or excretion from the body. In many cases material can be transferred between pools involving carbohydrates, fats, and proteins. (Adapted from Vander 1975.)

(mainly acetate, butyrate, and propionate). Fat cells throughout the body release free glycerol and NEFAs, the latter of which travel, bound or unbound to protein, through blood to the liver which performs further repackaging as lipoproteins. In many animals release of NEFAs and glycerol occurs during fasting and exercise, and their levels in blood may rise. There are limited pathways for interconversion of FAs to molecular forms that can reach the carbohydrate or amino acid pools (figure 7.2).

(1) Except for the relatively small amount of energy in glycerol, most energy in fat cannot be converted to glucose and stored as glycogen. This is because the decarboxylation of pyruvate that leads to acetyl-CoA is irreversible (figure 7.2). Propionate is the only short-chain FA that is glucogenic.

(2) Glycerol and NEFAs may be catabolized. Most of the energy can be released by aerobic catabolism (about 60% as heat and 40% as ATP synthesis) and little by anaerobic catabolism. Acetyl-Coenzyme A (acetyl-CoA), derived from NEFAs, can be used to synthesize compounds called ketone bodies (e.g., acetoacetate, β-hydroxybutyrate) that can be an important source of energy, even for the brain, during fasting.

(3) FAs are used to synthesize other FAs, structures (e.g., membranes), and other molecules (e.g., steroids and acetylcholine), as described in chapter 2. Vertebrates can synthesize FAs containing no double bonds (unsaturated) or one double bond per molecule (monounsaturated). A small number of polyunsaturated FAs (those with two or more double bonds), including linoleic and α- linolenic acid, are required by the body but cannot be synthesized by most animals; they must be consumed and therefore are called essential FAs (chapter 2). Animals eating relatively more of these FAs in their food may store fat with relatively higher proportion of these FAs (Egeler et al. 2003).

(4) Fat is the primary storage form for energy in most animals, and several reasons have been identified why it has been so favored during natural selection. Because of its high energy density and because it is typically stored with very little water, the energy density per g wet adipose tissue (ca. 37.6 kJ g^{-1} wet tissue) is 7–10 times higher than either glycogen stores (3.5–4.4 kJ g^{-1} wet tissue), protein, or so-called lean mass (5.3 kJ g^{-1} wet tissue; Jenni and Jenni-Eiermann 1998). Because fuel stores must be carried by the animal, carrying stores as fat is cheapest energetically. The disadvantage of fat as a storage molecule, its low solubility, is accommodated through complex special mechanisms for its transport through the body.

(5) Let us consider in a little more detail the transport of FAs through circulation in two situations: when lipids are deposited in adipose tissue following intestinal absorption and when lipids are recruited from adipose tissue to supply energy for activity. The details shown in figure 7.3 correspond to well-studied mammals but may differ among species.

Moving from the upper left of figure 7.3 (the intestine) to the middle left (the capillary endothelium of peripheral tissues), you see that the specific apolipoproteins in chylomicrons produced in intestinal cells following lipid absorption interact in the capillary endothelium with lipoprotein lipases (LPL) that catalyze removal of dietary-derived triacylglycerols from the chylomicron.

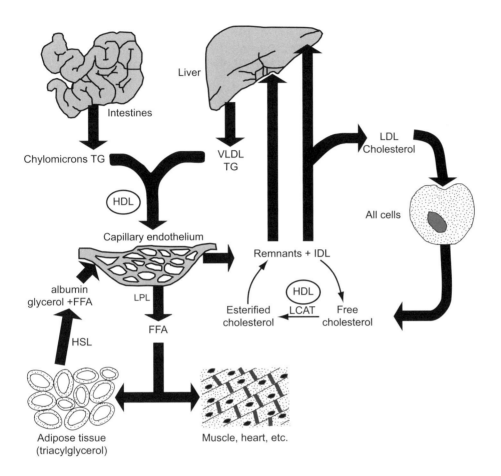

Figure 7.3. Schematic of transport of lipids between intestine, liver, adipose, and peripheral cells. As described in the text, chylomicrons transport triacylglycerols (TGs) to peripheral tissues or liver where they are taken up via action of the hormone lipoprotein lipase (LPL). The remnant chylomicrons themselves are absorbed by the liver. Triacylglycerols are released to the blood from adipose as free fatty acids (FFAs) through the action of hormone-sensitive lipase (HSL). They may be absorbed by tissues, but may be taken up by liver and reesterified to very low-density lipoproteins (VLDLs) and then released to the blood. Once VLDLs reach peripheral tissue, their triacylglycerols are removed via the action of lipoprotein lipase, and their remnants either become associated with cholesterol in substances called low density lipoproteins, (LDLs) or they are reabsorbed by the liver (called IDLs). Cholesterol itself is synthesized by the liver. (Redrawn and modified from Levy et al. 1998.)

NEFAs are absorbed into the tissue, and consequently, the chylomicron becomes progressively reduced in size until its remnant is absorbed and hydrolyzed by the liver.

When the energy in fat stores is needed, NEFAs are released by hormone-sensitive lipase (lower left corner of figure 7.3) and are transported away primarily bound to albumin. The amount bound depends on FA concentration, but one albumin molecule can bind at least eight or nine FAs. The FAs may then be taken up directly by cells around the body and used for energy. But also, the liver has a high capacity for NEFA uptake and reesterifies these to triacylglycerols which are subsequently delivered to blood as very low-density lipoproteins (VLDLs; shown as export from the liver in the middle of figure 7.3). VLDLs have specific apolipoproteins that interact with lipoprotein lipases in peripheral tissues that catalyze removal of triacylglycerols. When VLDLs become depleted of triacylglycerols their remnants (called IDLs) are taken up by the liver, or the remnants become other cholesterol-rich particles called low-density lipoproteins (LDLs; the cycle on the right-hand side of figure 7.3). The LDLs continue the delivery of FAs to tissues and they are also ultimately taken up and catabolized. The involvement of VLDLs in FA transport can greatly increase the overall flux of FAs because the proportion of FAs in VLDL is about 60–70% whereas the proportion of NEFAs in albumin is less than 10%. Furthermore, the increase in overall flux is achieved without greatly increasing the total protein content of the blood, which would increase blood viscosity and the circulatory work load. Small birds engaged in the energy-intense activity of flight appear to rely on VLDLs for FA transport to a greater extent than either larger birds or exercising mammals (Jenni-Eiermann and Jenni 1992).

7.1.5 Dietary Nutrient Requirements Arise Partly from the Limits to Molecular Interconvertibility and the Inefficiency of the Dynamic Catabolic-Anabolic Steady State

The limits of molecular interconvertibility define the requirements for organic nutrients. Amino acids and other materials are lost from the body in, for example, urine and sloughed epidermis (including hair and feathers) or cells of the gastrointestinal tract, and these losses must be replaced. The specific fatty acids, amino acids, and vitamins that are lost and which cannot be sufficiently synthesized must be replaced and thus become required nutrients that animals may seek out specifically in the course of diet selection. Analogously, the many inorganic nutrients required by animals cannot be synthesized at all, and their losses with excretion and sloughed material must be replaced.

The continuous breakdown and reassembly of the body's molecules is part of what renders biological systems dynamic and adaptable. For example, Schoenheimer, who championed the use of stable isotopes in metabolic studies (Schoenheimer 1942), showed that over the course of a few days many proteins in the body of rats were disassembled and reassembled using amino acids derived from the diet or endogenous sources. However, this continual turnover of molecules due to catabolism and anabolism is not 100% efficient and this influences the rate at which nutrients are required in the diet. In our example, most amino acids and metals in proteins that are released during protein turnover (degradation and resynthesis) are reused within the cell, but a small fraction may not be resynthesized into new protein. Some of those amino acids may be catabolized even if there is no shortage in the body of other energy substrates like glucose or fatty acids. The concomitant low-level loss of nitrogen excreted in urine is called endogenous urinary nitrogen (EUN). Another low-level loss of N, called metabolic fecal nitrogen (MFN), is the N associated with sloughed cells from the digestive tract, digestive secretions, or microbial material that is not absorbed prior to defecation. These obligatory losses through urine and feces (chapter 11, box 11.1) must be matched by nitrogen ingestion for the animal to be in nitrogen balance. An analogous low-level excretory "leak" of minerals must be matched by ingestion for mineral balance.

7.2 Controls over Postabsorptive Processing

7.2.1 Controls over the Postabsorptive Fate of Glucose, FAs, and Amino Acids

The biochemical controls for the flux of substrates into and out of storage or along different pathways are worth mentioning briefly because their features are sometimes used by ecologists as indices of metabolic status in vertebrates. (This discussion is limited to that taxonomic group.) The flux is catalyzed by enzymes whose levels change in response to dietary and hormonal signals. To illustrate, we will consider an example for each of the major substrates, followed by a short review of some of the most important hormones.

The movement of glucose into and out of storage as glycogen is catalyzed by two enzymes that act in opposition to each other and are especially influenced by the plasma ratio of glucagon to insulin (see table 7.1 and subsequent text for information about hormones). Availability of glucose for storage, as might occur after a meal, is signaled by high plasma glucose. The hyperglycemia directly stimulates insulin secretion by β-cells in the pancreatic islet of Langerhans, and the subsequent fall in plasma glucagon/insulin ratio stimulates cellular uptake of glucose by some tissues (e.g., muscle and liver) and its synthesis into glycogen by the enzyme glycogen synthase. An opposing enzyme, glycogen phosphorylase, breaks down glycogen to release glucose in order to counter low plasma glucose, as might occur during fasting or exercise. In this case, the hypoglycemia directly stimulates glucagon secretion by α-cells in the pancreatic islet of Langerhans, and the subsequent rise in the glucagon/insulin ratio simultaneously fosters glucose liberation and inhibits glucose sequestration. The binding of glucagon to its receptors on target cells initiates a cascade of cellular changes that include the activation (by phosphorylation) of glycogen phosphorylase and also the deactivation of glycogen synthase.

The movement of FAs into and out of storage as fat is also catalyzed by two enzymes that act in opposition to each other and are also especially

TABLE 7.1

Hormones that control deposition, mobilization, transformation, and utilization of major energy substrates

Hormone	Tissue of origin	Regulation: What stimulates secretion?	Primary action related to		
			Deposition/mobilization	Transformation	Utilization
Insulin	Pancreas (β-cells)	High plasma glucose and amino acid levels increase secretion	Increases glucose, amino acid, and fatty acid uptake by muscle and adipose Stimulates glycogenesis and lipogenesis	Inhibits gluconeogenesis	
Glucagon	Pancreas (α-cells)	Low plasma glucose increases secretion	Stimulates glycogenolysis, lipolysis Mobilization of amino acids	Stimulates gluconeogenesis from amino acids	
Glucocorticoids	Adrenal cortex	Physiological stress increases secretion	Stimulates mobilization of amino acids Glycogenesis in liver and muscle Lipolysis in adipose	Stimulates gluconeogenesis from amino acids Inhibits lipogenesis and protein synthesis	Inhibits peripheral uptake of glucose and amino acids and glucose oxidation
Norepinephrine and epinephrine	Adrenal medulla	Sympathetic stimulation	Stimulates glycogenolysis, lipolysis,		Stimulates glycolysis
Growth hormone (GH)	Adenohypophysis	GH-releasing hormone Low plasma glucose	Stimulates uptake of amino acids and glucose, protein synthesis Glycogenesis Lipolysis	Stimulates gluconeogenesis from glycerol	Stimulates fatty acid oxidation
Thyroxine	Thyroid	Thyroid-stimulating hormone	Promotes protein synthesis in presence of GH		Increases metabolism
Leptin	Adipocytes	Secretion proportional to amount of adipose tissue	Promotes catabolism and inhibits anabolism		Decreases feeding rate Sometimes increases metabolism

Source: Based on Randall et al. (1997) with additional information from Bentley (1998).

influenced by the plasma ratio of glucagon/insulin. The storage of fat in adipocytes is fostered by lipoprotein lipase (figure 7.3), which catalyzes the conversion of plasma triacylglycerols to NEFAs, which then enter an adipocyte and are packaged into triacylglycerols that are deposited in a liquid lipid droplet. The reverse breakdown of fat in the droplet and release of FFAs into plasma is catalyzed by the enzyme hormone-sensitive lipase (figure 7.3). The rising plasma glucagon/insulin ratio simultaneously fosters fatty acid liberation and inhibits fat sequestration, in part because the binding of glucagon stimulates the activation (phosphorylation) of hormone-sensitive lipase. The reverse process of fat sequestration is stimulated by a falling plasma glucagon/insulin ratio.

For amino acids there is not really any analogue to the energy-storage molecules glycogen or fat. Some animals may store protein seasonally in specific compartments such as skin and viscera, blood albumins, labile proteins in sarcoplasm between muscle fibrils, or even entire muscles that are not too critical for locomotion (Harlow 1995). On a more daily basis, when omnivores such as a rat or carp eat a high-protein diet, the excess N is not stored but is eliminated. This control is partly achieved through alterations in enzyme levels (Cowey and Sargent 1979). On a high-protein diet the activity of degradative enzymes (such as an aminotransferase) increases. When the diet is low in protein, those activities decrease. The activity of other enzymes associated with the disposal of nitrogen from deamination of amino acids (see section 7.1.3) also increases and decreases when the intake of dietary protein increases and decreases. Like glucose and FAs, the processing of amino acids is also notably influenced by the plasma ratio of glucagon/insulin. High plasma amino acid levels can stimulate secretion of insulin, which, in turn, stimulates uptake of amino acids by muscle and liver. Glucagon has the opposite effect. It stimulates the mobilization of amino acids, which can then enter plasma, though glucagon also stimulates gluconeogenesis from amino acids.

Considering the effects of glucagon and insulin on the uptake, release, and disposition of the major energy substrates, one can understand why insulin has been called the "hormone of plenty" (it promotes storage) whereas glucagon has been called the "hormone of retrieval." Besides insulin and glucagon, several other hormones control postabsorptive processing (table 7.1). Corticosteroids share many of the retrieval functions of glucagon, whereas catecholamines (norepinephrine and epinephrine) and growth hormone have more specialized functions and thyroxine has a general effect of increasing metabolism. Long-term energy homeostasis is achieved partly through the action of hormones that are secreted in proportion to the amount of adipose tissue in the body (adiposity), including leptin, which we discuss in the next section.

7.2.2 Control of Long-Term Energy Balance

The homeostatic controls for components of energy balance, such as food intake or the size of fat stores, are varied and complex. Decades of research, largely on laboratory rats but also on other vertebrates and invertebrates, have shown that regulation is achieved on several temporal scales and through a variety of mechanisms that are integrated and somewhat redundant (Havel 2001; Woods et al. 1998). In the case of food intake for example, gut distension will shorten the duration of any single feeding episode, as might also high circulating levels of the substances that were absorbed. In mammals and birds, plasma concentrations of glucose, amino acids, and gut peptides such as cholecystokinin (CCK) are components of the short-term system that modulates hunger and satiety, thus beginning and ending meals (Friedman and Halaas 1998; Denbow 1999; Woods et al. 1998). On a longer time scale there seems to be control of energy balance by feedback loop(s) that are usefully conceptualized using models from systems analysis (Cabanac 2001; figure 7.4). In the model of a generic regulated system there is some commodity that is regulated, a sensor that measures it directly or indirectly, and the measurement is compared with a target (a set point) by a controller which then modulates the activity of system features that can increase or decrease the amount of the commodity. As applied to energy balance, the commodity might be lipid stores, the controller is thought to be the ventromedial hypothalamus on the underside of the brain, and potential effectors are features such as food intake and digestive actions that determine the rate at which food energy is absorbed, and respiration (heat production) associated with resting and activity metabolism that modulate rate of energy loss.

It is most likely that the system monitors some variable(s) correlated with lipid mass rather than lipid mass per se, and the search for the variable(s) and their sensor(s) has been the subject of research for decades. Insulin and corticosteroids (acting via releasing factor; Cabanac 2001) are good candidates for the signal because their circulating concentrations are positively or inversely proportional, respectively, to total fat mass, and they can both influence food intake and body mass in a dose-dependent manner when administered into the central nervous system (Woods et al. 1998; Cabanac 2001). Corticosteroids also modulate energy expenditure. An even more recent candidate for the signal of body or fat mass status is the peptide hormone leptin.

There is much about leptin that makes it a compelling candidate for the signal. It is produced primarily by adipose tissue and its rate of secretion and its plasma concentration are correlated with total fat mass, making it a circulating biochemical signal of the size of body fat stores (Reidy and Weber 2000). Furthermore, experimental elevation of plasma leptin inhibits food intake, an effect

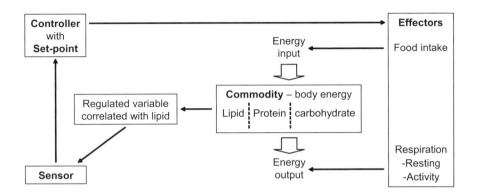

Figure 7.4. A simple model of regulation of body mass. The commodity that is regulated is probably total body lipid, probably indexed by a variable that is correlated with total lipid, such as a hormone like leptin in mammals and birds. The hormone is sensed by sensors (receptors in the brain or elsewhere), and the concentration is compared with a set point by a controller which then modulates the activity of effectors that can increase or decrease the amount of the commodity. As applied to energy balance in vertebrates, the controller is thought to be the ventro-medial hypothalamus on the underside of the brain, and potential effectors are features such as food intake and digestive actions that determine the rate food energy is absorbed, and respiration (heat production) associated with resting and activity metabolism that modulate rate of energy loss.

mediated through receptors in the hypothalamus, and stimulates respiration and catabolism of fat. The effects of leptin are mediated through effects on numerous other signaling pathways or physiological features, and a short list of some of them will impress you with the complexities. Leptin has been found to influence (1) neuropeptides and neurotransmitters expressed in the hypothalamic region, such as melanocyte-stimulating hormone (MSH) and neuropeptide Y (Friedman and Halaas 1998); (2) thyroid hormone, which stimulates respiration; (3) sympathetic nerves that influence production of uncoupling protein (UCP) which increases the proton leakiness of mitochondrial membranes; and (4) hormone-sensitive lipase which affects lipolysis (Lowell and Spiegelman 2000; Reidy and Weber 2000). There is much still to be learned about leptin, including its occurrence and/or role in other animals besides mammals and birds (Reidy and Weber 2000). Furthermore, every decade has brought the discovery of new putative signals in the regulation of food intake, energy balance, and body mass (e.g., ghrelin; Nakazato et al. 2001).

Animals appear to adjust their set point and thus their regulated fat mass and body mass using many types of information from their environment. Some animals are strategic about regulation of body mass and they integrate information about risk. Birds provide good examples of strategic regulation of body mass because excess body mass costs them extra energy in terms of higher flight costs and may also decrease flight speed and maneuverability and thus increase the risk of predation (Biebach 1996). Yet, if they simply minimize their body mass, then their energy reserves for unexpected food shortages may be inadequate. This dilemma, which you may recognize as another example of the issue of "enough but not too much," is a starvation-predation risk trade-off (Lima 1986). The starvation-predation risk hypothesis leads to the prediction that birds should regulate their body mass during a day and between seasons at levels directly related to the risk of energy shortages and inversely to predation risks (Pravosudov and Grubb 1997). Indeed, birds in captivity provided food at uncertain intervals regulated their body mass at a higher level than birds provided with unlimited food (Witter et al. 1995), though not in every case (e.g., Acquarone et al. 2002). In the wild, North American bird species that feed in wintertime on foods with unpredictable availability tend to maintain larger fat stores than do birds that feed on more reliable food sources (Rogers 1987). On a mechanistic level, we can easily imagine that birds' brains have a set point for leptin that corresponds to a body fat content that is optimal with regard to the costs and benefits of large fat stores. In this feedback scenario, circulating leptin levels, which signal the size of the fat stores, are compared with the set point and appropriate changes are made in food intake and energy expenditure to regulate the fat stores at the appropriate size (figure 7.4).

With this overview of postabsorptive nutrient processing and control as background, we will now consider two general classes of questions that emphasize the penetrance of these biochemical details into the ecological realm:

- First, what ecological costs are borne during the postabsorptive processing of nutrients?
- Second, can we use the biochemical details to infer the nutritional status of wild animals?

A third important issue is covered in chapter 11: how do these features determine the specific nutrients required by animals, and the rates at which they are required, all of which influence diet selection, mortality, and production?

7.3 Costs of Digestive and Postabsorptive Processing

The energy content of carbohydrate, fat, and protein absorbed across the gut wall is not all profit to the animal. Their gross energy values must be discounted for the energetic costs of postabsorptive processing, such as the urinary energy losses and heat increment of feeding that were introduced in figure 3.1 of chapter 3. In this section we consider these costs along with those that might be associated with digestion and absorption, because it is difficult to tease them all apart. One kind of cost relates specifically to feeding on protein, whereas another kind of cost is the heat increment following feeding on any type of food.

7.3.1 Energy Lost in Nitrogenous Excretory Products

As described above (section 7.1.4), the first step in interconversion of amino acids is the removal of the α-amino group, which ultimately generates ammonia.

In many aquatic organisms this toxic metabolite is disposed of cheaply by diffusion across skin and gills down a concentration gradient. In most terrestrial organisms, however, most of the ammonia is synthesized into either urea or uric acid, with an associated loss of energy (figure 2.10 in chapter 2). For this reason the energetic value of protein in nutrition is sometimes less than the gross energy content from combustion in the laboratory, which is usually given as 22.2 kJ g^{-1} protein. For animals in steady state, whose N intake equals its excretion, the relevant energy contents for protein are 18.0 kJ g^{-1} protein catabolized when urea is the metabolic end product and 17.79 kJ g^{-1} protein catabolized for uric acid. Hence, in many animals nearly 20% of the energy in amino acids and peptides absorbed is lost due to incomplete combustion. It also follows that animals that consume high-protein foods will excrete through their urine proportionally more of the energy that is absorbed across the intestine than do animals eating low-protein foods. In balance trials, like that described in box 3.1 in chapter 3, mammalian carnivores excreted in their urine 8% of digested energy, whereas mammalian fruit and seed eaters excreted only 2–3% of digested energy (Robbins 1993). (Note that urinary energy can also include the energy in metabolites of toxins, a topic considered in chapter 9.)

Ureotelic animals, those that excrete N mainly as urea, include mammals, adult amphibians, and elasmobranches (sharks and rays). Uricotelic animals (uric acid excreters) include birds, most reptiles, and insects. Ammonotelic animals, those that excrete N mainly as NH_4^+, include teleost fish, most amphibian larvae, cyclostomes, and aquatic invertebrates. The ammonotelic animals do not lose energy due to incomplete combustion of protein. The groupings are not strictly phylogenetic but depend on the ecological conditions of the places where the animals live. This is because disposal of ammonia and urea, which are soluble in water, requires more water than disposal of uric acid, which can be precipitated and excreted with minimal loss of water as a solvent. Aquatic tadpoles typically live under water and are ammonotelic, most semiterrestrial adult amphibians are ureotelic, and some xeric-adapted frogs are uricotelic. Needham argued that the low solubility of uric acid, and its tendency to precipitate, explains its occurrence in organisms that lay eggs on land (reptiles and birds), because the uric acid excreted by the embryos would precipitate and not build up to high concentration in the solution bathing the embryo (Needham 1938).

7.3.2 The Heat Increment of Feeding

Another type of postabsorptive energy loss occurs as heat production. The metabolic rate of a resting animal rises following ingestion of a meal and then over time falls back to the baseline fasting level of metabolism (figure 7.5). This raised metabolism following feeding, which has been assigned a number of names, including the "work of digestion," "specific dynamic effect (or action)," "heat increment of feeding," and "postprandial thermogenesis," occurs in vertebrates and invertebrates. The height of the peak metabolism relative to basal varies quite a bit depending on factors such as how low basal metabolism is (it is progressively lower the longer the fast and lower for ectotherms than endotherms) and how rapidly digestion and postabsorptive processing proceed. Most illuminating regarding apparent "costs" of food processing is the total area between the curves for postprandial and fasting metabolism. This area under the curve, which corresponds to energy, is generally proportional to the amount of metabolizable energy ingested in the meal (figure 7.5), although other factors such as meal size may also influence the proportion (Secor and Faulkner 2002). For Burmese pythons eating different sized meals this heat increment varies from 34 to 42% of the metabolizable energy (figure 7.5). These are probably the highest values observed among animals, because the python consumes proteinaceous prey at low frequency, and both factors influence the heat increment, as described below. Another method used to measure the heat increment of feeding (HIF) is to make a utilization plot (chapter 3) of resting metabolic rate (e.g., in units kJ/d as measured by respirometry) against metabolizable energy intake (from a balance trial, also in units of kJ/d), in which case the slope is a unitless proportion that is equal to the HIF (Sedinger et al. 1995).

The heat increment of feeding reflects numerous processes involved in the breakdown, absorption, and postabsorptive processing of food. The majority

Figure 7.5. The heat increment of feeding is directly related to meal size in animals, as illustrated in Burmese pythons (*Python molurus*). Oxygen consumption of pythons rose following ingestion of a meal and then over time fell back to the baseline fasting level of metabolism. The meal sizes, expressed as metabolizable energy, were 231 kJ (filled squares), 1,183 kJ (triangles), and 2,883 kJ (circles). The corresponding heat increments, 76, 424, and 1280 kJ, were calculated as the areas under each of the curves, minus the baseline oxygen consumption, multiplied by the respiratory equivalent of oxygen, 19.8 J^{-1}ml. (Replotted from Secor and Diamond 1997.)

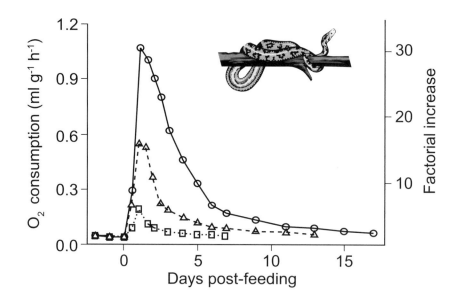

appears to relate to postabsorptive processing because most of the heat increment remains if digestion is entirely bypassed in vertebrates by infusing nutrients directly into the circulatory system. In particular, a major component of the heat increment relates to protein turnover, based on stochiometric calculations of the energy required in synthesis and degradation of proteins and correlations of the heat increment with N excretion. It has been estimated that the ingestion of 1 kJ of protein has an associated heat increment of 310 J in invertebrates and vertebrates (i.e., 31%). Heat increments, as a percent of gross energy, are generally considered to be smaller for carbohydrates (5% for simple sugars to 15% for starch) or lipid (10–15%) (Ricklefs 1974; Lucas 1996; Klasing 1998). This helps explain why the heat increment of feeding has been shown to correlate with either diet protein content or N excretion in mammals, fish, reptiles, and bivalves. Treatment of animals as diverse as copepods and fish with the protein synthesis inhibitor cycloheximide diminishes the heat increment of feeding.

The heat increment of feeding seems highest in some animals that feed infrequently and consume large meals that are separated by long periods of fasting. For example, in a phylogenetically informed comparison of snakes, those with longer intervals between feeding had higher heat increments of feeding (Secor and Diamond 2000). Python and boa snakes, which are infrequent feeders, exhibit large increases in masses of organs such as the small intestine and liver after feeding, and then once the meal is digested the organ sizes regress. Their heat increment of feeding, expressed as a percent of the metabolizable energy, is 1.8 times greater than that observed in other snake species that feed frequently and do not exhibit marked cycles in the masses of organs of nutrient assimilation. Hence, the difference may reflect the metabolic costs of building these organs.

The heat loss that accompanies the processing of a meal is not invariably wasted. In adult mammals and birds that are exposed to temperatures below their lower critical temperature, theoretically, it can be used to thermoregulate in endotherms. However, empirical evidence to support this "thermal substitution" effect is mixed (Rosen and Trites 2003). In ectotherms, the heat increment of feeding reduces the metabolizable energy that is available for maintenance and production by 10–30%.

7.3.3 Thermoregulatory Costs of Feeding

There are sometimes other ecological costs of food processing that animals must pay. When mammals and birds ingest cold or even frozen food they must generate additional heat to raise the temperature of the food bolus to body temperature. The quantity of heat required to warm up cold food is easily estimated as the product of the mass consumed, the temperature to which food must be increased (in °C), and the specific heat capacity of biological tissue, which varies depending on its composition and especially on its water content—a value of 3.35 J g^{-1} °C^{-1} is often a good approximation. Frozen food must first be thawed, which requires 334 J g^{-1} H$_2$O in the food. These costs can become significant in birds and mammals inhabiting cold environments. Wilson and Culik (1991) estimated that the cost of heating ingested food represents 13% of the daily energy requirement of Adelie penguins (*Pygoscelis adeliae*) and 5–13% of the daily energy budget of nonreproductive herbivorous meadow voles (*Microtus pennsylvanicus*), and even more if the food is frozen (Berteaux 2000).

Some ectotherms thermoregulate at higher body temperatures during digestion of food, and this raises their baseline energy expenditure. As was discussed in chapter 4, in most ectotherms the effect of raising the body temperature is mainly to speed up digestion and postabsorptive processing rather than markedly altering the efficiency of the processes. Both the metabolizable energy coefficient and the heat increment of feeding, expressed as a proportion of ingested energy, are typically independent of body temperature. Thus, when heat production (including the heat increment of feeding) and digestion rate have different temperature dependencies, one finds the body temperature that maximizes net energy gain by inspecting the temperature dependence of the difference between gain and expenditure (figure 7.6). An interesting question is whether selected basking temperatures of ectotherms correspond to these optima.

7.3.4 Liver Size and Maintenance Costs

The liver in vertebrates, or hepatopancreas in invertebrates, is one of the major organs responsible for the postabsorptive processing of nutrients. Its mass in vertebrates scales approximately as (body mass)$^{3/4}$ and is large relative to many other organs (Calder 1984), and its respiration rate per unit mass is also high relative to that of many other organs (Blaxter 1989). Thus, the liver accounts for a proportionally large amount of maintenance energy expenditure, and the possession of a larger than necessary liver is wasteful in energetic terms. One might predict that its size, like that of the gastrointestinal tract (GIT), will be modulated according to the maxim "enough but not too much" (Piersma 1998). This is indeed the case. There are numerous examples of up- and down-regulation of liver size in parallel with up- and down-regulation of the size of the gastrointestinal tract during hyperphagia. These include increases in rodents during

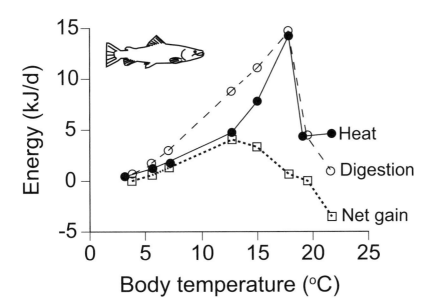

Figure 7.6. The net energy gain during digestion and postabsorptive processing is maximized at some temperature that is determined by the temperature dependence of rates of digestion and heat production. The net energy gain is calculated as the difference between rate of digestive energy gain and rate of heat production. The data shown are for brown trout (from Elliott 1976), whose body temperatures we assumed were equal to the water temperature in which they were held. Notice that the temperature that corresponds to maximum net energy gain is below the temperature of maximum gain from digestion.

cold acclimation (Toloza et al. 1991) and lactation (Hammond et al. 1994; Hammond and Kristan 2000), increases in migratory birds gaining body mass (Karasov and Pinshow 1998; Piersma et al. 1999; Karasov et al. 2004), and increases in snakes during feeding (Starck and Beese 2002). Unfortunately, for most of these examples we do not know whether the causal basis for change in liver mass is change in hepatocyte size or number, and whether changes in hepatocyte size cause much of a change in organ respiration. It is possible for hepatocyte size to increase due to incorporation of lipid (Starck and Beese 2002), which might have a relatively small effect on overall tissue respiration compared to changes that affect the number of mitochondria.

7.4 Feast and Famine: The Biochemistry of Natural Fasting and Starvation

The emphasis of this chapter, so far, has been on biochemical changes and associated metabolic costs incurred when animals consume, digest, and absorb nutrients. We want to consider also the changes that occur when animals do not feed. Many animals withstand long periods without food routinely during periods of their life cycle. Although examples of seasonal hibernation in vertebrates and invertebrates come first to mind, fasting even when normothermic (having "normal" rather than reduced body temperature) is a natural component of the life history of many species. The biochemical changes that occur as animals undergo progressively longer periods out of energy balance can be used to make inferences about their nutritional status.

7.4.1 Biochemistry of Starvation

Fasting and starvation have been studied in many species (Cherel and Le Maho 1998; Castellini and Rea 1992). Researchers divide fasting in vertebrates into three stages or phases (figure 7.7).

Stage/Phase 1. During the first hours to days of food deprivation, hepatic glycogen reserves are mobilized as the body defends circulating glucose levels, which are especially important for sustaining the central nervous system. In order to meet energy needs of other tissues the body mobilizes stored lipids and increases fat oxidation, but protein catabolism is reduced (figure 7.7). Thus, during stage 1, the corresponding changes in metabolite levels are that NEFAs in plasma begin to increase and plasma and urinary N (as urea or uric acid) begin to fall (figure 7.8).

Figure 7.7. During fasting animals increase their reliance on fat catabolism to meet energy needs, and reduce net protein catabolism to conserve protein. If fat becomes exhausted then protein catabolism rises. β-OHB represents the ketone body β-hydroxybutyrate. (Redrawn from Van Der Meer and Piersma 1994.)

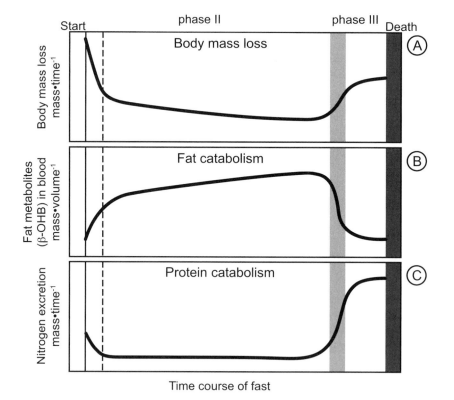

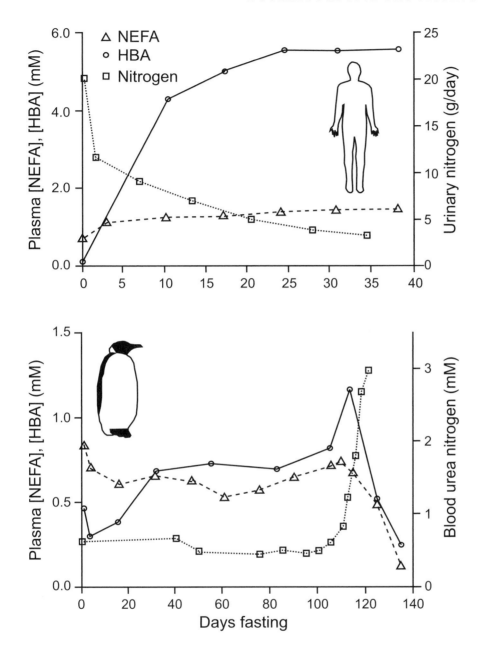

Figure 7.8. Patterns of change in plasma concentrations of β-Hydroxy-butyrate (HBA), nonesterified fatty acids (NEFA), and blood urea N (BUN) can be used to define stages of fasting and differences between nonfasting-adapted species (e.g., humans; top figure) and fasting-adapted species (e.g., emperor penguins, *Aptenodytes forsteri*; bottom figure). Notice that in penguins stage 2 is prolonged and that HBA does not rise to as high a level, which reduces acid-base complications. (Adapted from Castellini and Rea 1992.)

Stage/Phase 2. Once glycogen reserves are almost completely utilized, the nervous system's energy needs are met by the glucose formed by gluconeogenesis from glycerol and protein, and by ketone bodies, such as β-hydroxybutyrate, a product of lipid metabolism. A hallmark of stage 2 is rising levels of ketone bodies (figure 7.7), which can have the negative effect of altering acid-base balance. Protein catabolism remains reduced, but small amounts of protein are continually degraded to supply some glucose for the nervous system and to replace intermediates of the citric acid cycle that are constantly drained away and need to be replaced (this is called anaplerotic flux). Consequently, plasma and urinary N levels remain measurable but low (figure 7.8). The reduction in protein-containing "lean mass" contributes to lower basal energy expenditure because this tissue has a higher tissue-specific metabolic rate than adipose tissue.

Stage/Phase 3. Once lipid reserves are almost completely utilized, the nervous system's energy needs can be met only by glucose formed by gluconeogeneis from protein. Thus, a hallmark of stage 3 is falling ketone bodies and rapidly rising plasma and urinary N levels. Stage 3 can also occur when some lipid reserves remain but 30–50% of body protein has been catabolized. In this situation also, stage 3 is marked by declining ketone levels. The loss of so much lean mass in critical organs such as the heart and kidney makes it difficult or impossible to recover from stage 3.

This general scheme can vary among species as the length of the phases may differ. For example, small avian migrants catabolize protein relatively early when fasting (Karasov and Pinshow 1998).

7.4.2 Biochemistry of Fasting-Adapted Species

Penguins and snow geese go without food for days to weeks during egg incubation (Le Maho et al. 1991; Boismenu et al. 1992), and bears and phocid seals fast during lactation (Castellini and Rea 1992). These species prolong stage 2 with some modifications, and terminate their fast prior to stage 3. Stage 2 is prolonged by a substantial reduction in metabolic rate, which slows the rates of utilization of stored lipids and body protein. The fasting-adapted species exhibit increases in circulating ketone levels but these increases are relatively smaller than in starved animals, and thus acid-base complications are avoided. In fasting bears, which do not hibernate per se but remain responsive with relatively high body temperatures (30–35°C), the very low amounts of urea N produced by protein catabolism are apparently efficiently reincorporated into body protein, and it has been claimed, although never thoroughly tested, that this is achieved through a process of microbial production of ammonium from urea in the digestive tract, followed by reabsorption (Guppy 1986).

All the fasting-adapted species will exhibit stage 3 if forced to do so, but in the natural situation they terminate their fast prior to this point. Consider what happens to male king penguins (*Aptenodytes patagonicus*) during reproduction. They arrive fat at their breeding colony (mean mass 14.7 kg; Olsson 1998) and begin to court. After mating, the female lays a single egg. Courtship, mating, and egg laying last about 14 d. Then females leave to forage in the ocean while males take the first incubation shift, which lasts about 21 d. During these 35 (14 + 21) days of fasting on shore the males lose an average of 4 kg (29%) of their total body mass. If the female returns on time, the male's mass is still above the critical body mass (9.7–10 kg; Cherel et al. 1994) at which they switch from lipid to protein to fuel respiration. But what happens if she does not show up on time? Circulating ketone levels will abruptly rise and then fall, as occurs in stage 3 in starved animals (figure 7.8). Stage 3 will occur in penguins captured and restrained. If the males are not restrained by human investigators they may use changing ketone levels as an internal biochemical signal that they should abandon their eggs and return to the sea to feed (Castellini and Rea 1992). The mean mass of abandoning king penguin males (9.49 kg) is just below the critical body mass that represents the boundary between phases 2 and 3.

7.5 Biochemical Indices of Nutritional Status and Habitat Quality

A long-term goal of some ecologists has been to develop methods to "ask the animals themselves" about their nutritional status and how well their habitat supports their nutritional needs. One approach has been to develop morphological indices of condition, the crudest of which is a ratio of body mass to some measure of structural size such as length. For example, wildlife and fisheries biologists sometimes calculate a body condition index (BCI) as mass/a(length)^b (where $b \sim 3$) or plot the relationship between the residual of log(mass) against log(length). We presented an example of an alternative approach using residuals in box 1.4 in chapter 1. Severe problems have been identified with the simple ratios, and even the approach of using residuals is based on assumptions that are sometimes violated, and thus conclusions may be spurious (Green 2001). More informative indices measure specific components of the body such as fat content or lean mass content (box 7.1). When any of these measures are used to assess condition, it is important to recall that more is not necessarily better, and that stores may be strategically regulated according to their relative costs and benefits (see section 7.2.2).

   In this section we ask whether one might use the hormone or metabolite levels of wild animals to infer their nutritional status, and, if so, what are the best candidates? Among hormones the most obvious candidates are insulin and glucagon (see above). As might be expected, in controlled studies, insulin levels are lower and glucagon levels are higher in herbivorous and carnivorous mammals either fasted or on energy-poor diets (e.g., Delgiudice et al. 1987; Seal et al. 1978b). But because the hormone levels change in response to recent meals and in response to capture stress, it is not so practical to use these hormonal signals in wild-caught animals to assess their nutritional status.

Researchers have had more success focusing on the metabolites themselves. Many studies have sought to correlate indices of habitat condition with hematologic and blood chemistry values of wildlife captured in different habitats. It can become routine to collect blood from animals that are captured, and diagnostic laboratories are available to analyze dozens of blood parameters, but it can be difficult to make sense of it all without good baseline values and knowledge of what factors influence levels of the blood parameters. As illustrated in the following sections, progress has been made in research programs that couple controlled studies with field sampling, and that focus on biochemical features that have well-understood, direct, and fairly specific links to nutritional status that are empirically very evident.

7.5.1 Biochemical Signals of Nutritional Status in Ruminant Herbivores

In temperate, boreal, and arctic ecosystems there are pronounced seasonal changes in resource availability for wildlife. Winter is potentially a time of nutritional stress and is the season when most nutrition-related deaths occur, mainly due to chronic undernutrition rather than to starvation (Delgiudice and Seal 1988). Controlled studies with captive deer fed diets with low versus moderate levels of energy and protein indicate that during undernutrition, fat is mobilized for energy, as reflected by elevated levels of circulating NEFAs (Figure 7.9). Circulating blood urea levels (BUN = blood urea N) are not elevated in deer eating lower energy diets, and in fact in numerous studies there is a direct relationship between crude protein intake and BUN (Franzmann 1985) probably reflecting that dietary protein catabolism is the primary source of the blood urea (Delgiudice and Seal 1988). However, after fasting or prolonged undernutrition that exhausts lipid reserves, BUN levels will increase reflecting protein catabolism (Delgiudice and Seal 1988), akin to stage/phase 3 starvation (figure 7.8). Another metabolite that can signal stage/phase 3 is 3-methylhistidine, a methylated amino acid synthesized mainly from the catabolism of actin and myosin in skeletal muscle (Delgiudice et al. 1998). Elevated levels of 3-methylhistidine in blood or urine would reflect muscle catabolism, although this pattern would be confounded if the animal consumed a meal containing significant amounts of muscle. Also, molt in birds may increase excretion of 3-methylhistidine (Pearcy and Murphy

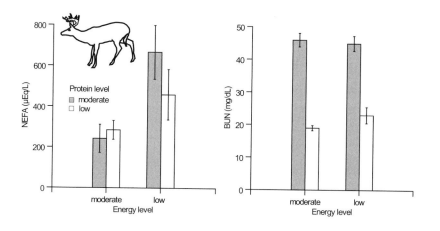

Figure 7.9. In white-tailed deer (*Odocoileus virginiana*) maintained on a low-energy diet, nonesterified fatty acid levels (NEFA) in blood are elevated, whereas blood urea N (BUN) is related to diet protein content. Therefore, NEFA may be more useful than BUN as an index to energy status. (Data from Seal et al. 1978b.)

1997). For these reasons, focusing on blood or urine urea or 3-methylhistidine levels to infer nutritional status is complicated. An elevated level can reflect either that the animal has a high intake of protein or is near starvation. A focus on plasma lipid metabolites would seem informative, with high levels of NEFA indicating chronic energy shortage and perhaps high ketone bodies in conjunction with high BUN or 3-methylhistidine reflecting severe undernutrition. In deer inhabiting wintering areas with older vegetation that is of lower nutritional quality, circulating NEFA levels were significantly higher (Seal et al. 1978a).

Most recently, researchers of ruminant herbivores have focused on the ratio of allantoin to creatine in serum or urine as a dietary index. Allantoin is the main purine derivative (along with small amounts of uric acid, xanthine, and hypoxanthine) arising from the catabolism of nucleic acid. Such compounds arise mainly from the digestion of rumen microbes (exogenous origin), and in both domestic and wild ungulates the plasma level or urinary excretion of allantoin increases mainly due to increasing ruminal microbial productivity which is positively influenced by dietary energy and protein levels (Vagnoni et al. 1996; Delgiudice et al. 2000). Higher urinary concentrations should be indicative of greater food intake and digestion, though the possibility of increased endogenous contributions due to accelerated tissue catabolism needs to be studied. The idea of sampling urine rather than blood is attractive because one avoids the time and expense of capturing animals, and urine that falls in snow is well preserved and collectable. However, when analyzing urine one needs to normalize the metabolite of interest to some other urinary component, because the simple concentration in urinary water may be diluted by snow or altered by the animal's hydration state. The method for achieving this normalization is described in box 7.2. Other recommendations for applying the method can be found in Ditchkoff and Servello (1999).

Box 7.2. Urinary Ratios as Nutritional Indices

Measuring metabolites in urine is a noninvasive technique for gaining information about nutritional status. Numerous investigations have explored the utility and biological meaning of levels of metabolites such as urea nitrogen, cortisol, allantoin, and glucuronides in urine collected from the bladder or from snow. Levels of the metabolites are often reported as ratios to urinary creatinine in an attempt to standardize for differences in urine concentration due to physiological factors such as state of hydration or due to dilution in snow (Delgiudice et al. 1988). What is creatinine and what makes it a good internal standard for materials in urine? Creatinine is the degradation product of creatine, a substance present in muscles and used, when phosphorylated as creatine-P, as a store of high-energy phosphate bonds. During burst activity, creatine-P donates its high-energy phosphate bond to replenish the ATPs lost in the contraction of the cell. During this process a small percentage of the creatine is spontaneously degraded into creatinine. Creatinine moves out of muscle cells and is excreted in urine as creatinine. In animals that do not eat a lot of muscle, the urinary excretion of creatinine can be used as an index of muscle mass. Creatinine is filtered by the kidney, it is only secreted at a low rate from the blood, and is not reabsorbed by renal tubules. Therefore, the ratio of creatinine to other metabolites should be similar in blood and urine, as long as the metabolites are not reabsorbed by the kidney. If some constant proportion of metabolites are reabsorbed by the kidney, ratios in urine and blood will at least be correlated.

(1) *Urea:creatinine ratio.* Blood and urine urea (or uric acid) levels can be elevated when an animal eating a high-protein food deaminates excess amino acids for oxidation (i.e., for energy), or when a starving animal in stage III fasting increases catabolism of its own tissues (see section 7.4.1). Thus, the ratio alone cannot be definitive, because a high urea:creatinine ratio could reflect either a high intake of dietary protein or near starvation. For accurate interpretation, other data on nutritional status are necessary. When food protein content is known to be low, and undernutrition due to low food availability is expected, relative urea:creatinine ration can be used to index the severity of energy restriction (Delgiudice et al. 1989).

(2) *3-methylhistidine:creatinine ratio.* 3-methylhistidine is an amino acid that occurs in skeletal muscle actin and myosin, and there is no exogenous source in herbivores (herbivores do not consume muscle!). Blood and urine 3-methylhistidine levels can be a reliable indicator of muscle degradation in many, but not all animals (see DelGiudice et al. 1998). Thus, if properly validated and calibrated, the relative 3-methylhistidine:creatinine ratio can be used to index of muscle degradation and thus of the severity of energy restriction.

continued

continued

(3) *Allantoin:creatinine ratio.* Blood and urine allantoin levels of ruminants rise with increasing ruminal microbial productivity which is
positively influenced by dietary energy and protein levels (Vagnoni et al.
1996; Delgiudice et al. 2000). Higher urinary concentrations should be
indicative of greater food intake and digestion, but as was the case for
urea, the possibility of increased endogenous contributions due to accelerated tissue catabolism needs to be considered.

(4) *Cortisol:creatinine ratio.* Cortisol, a glucocorticoid (see table 7.1), is
secreted according to a circadian rhythm, and almost any type of physical or neurogenic stress will also increase its secretion. Whether and
how it might be used as an index of nonspecific or specific nutritional
stress is an active area of research among biologists studying birds
(Kitaysky et al. 1999), mammals (Parker et al. 1993; Boonstra et al.
1998), reptiles (Dunlap 1995; Wilson and Wingfield 1994), and fish.
Because capture and handling generally increase plasma cortisol levels,
researchers often take serial blood samples and plot plasma cortisol
versus time since capture in order to define baseline level and/or rate of
increase and/or maximum level. Alternatively, the collection and measurement of cortisol or its metabolites in urine of undisturbed animals
may be the best way to measure baseline cortisol levels.

(5) *Glucuronide:creatinine ratio.* Glucuronic acid excretion reflects
intake of secondary metabolites (SMs) in food because conjugation with
glucuronic acid is a major pathway of excretion of SMs (see chapter 9).
A recent interesting proposal is to use the glucuronide:creatinine ratio as
an index to exposure to SMs in foods (Servello and Schneider 2002).

7.5.2 Biochemical Signals of Refueling Rates in Migratory Birds

The migration of birds can be divided into alternating phases of intense energy
use during flight and intense energy deposition during stopovers, though birds
do not necessarily feed at every stopover site. The overall rate of migration is
very much influenced by the fuel deposition rate during the stopover, which can
be limited by intrinsic physiological constraints (see chapter 4) and/or environmental conditions. Among the latter, differences in habitat quality (food
availability, competitors, predators) are undoubtedly important but they are very
difficult to measure. Could biochemical features of the migrants themselves be
used to assess the relative value of a habitat for supporting refueling? Currently,
this kind of knowledge is available by weighing many birds at a stopover site
throughout the day (Dunn 2000) or by reweighing the very small proportion of

Figure 7.10. Plasma triacyl-glycerol (left-hand figure) and β-Hydroxy-butyrate (β-HBA) levels (right-hand figure) were correlated with rates of feeding and mass gain in captive garden war-blers (*Sylvia borin*), which are long distance migrants. These data thus suggest that blood metabolite levels can be used as an index of rate of change in condition, and not only as a static measure. (Replotted from Jenni-Eiermann and Jenni 1994.)

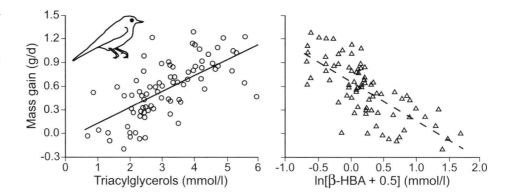

birds that are captured at a stopover site more than once. Perhaps biochemical analysis of blood sampled from the vast numbers of birds that are captured only once could be informative.

This possibility is increasing as we learn more about metabolic changes in migratory birds. Migrants at a stopover site that do not feed remain in a fasted catabolic state, whereas those that feed should be anabolic. Based on our previous discussion, the metabolic profiles of the two should differ. Not surprisingly, fasted migrant passerines exhibited levels of β-Hydroxy-butyrate (HBA) four times higher than those of fed birds, whereas plasma triacylglyc-erols were nearly two times higher in fed birds (Jenni and Jenni 1991). Indeed, studies with passerines (Jenni-Eiermann and Jenni 1994) and shore-birds (Williams et al. 1999) have shown that plasma triacylglycerol levels are positively related, and HBA levels negatively related, to rates of feeding and mass gain (figure 7.10; Guglielmo et al. 2002). Using such relationships, it is possible to capture a bird, sample and analyze its blood, and then calculate a fattening index that incorporates measures of both metabolites, with correc-tions for body mass and time of day that also influence the metabolite levels (Schaub and Jenni 2001). In the first field studies of this kind, the fattening indices of wild birds do seem to correspond to rates of mass gain and habitat quality (Schaub and Jenni 2001; Guglielmo et al. 2002; Guglielmo et al. 2005). Thus, the biochemical features of migrants themselves could be used to assess the relative value of a habitat for supporting refueling.

As promising as this approach may appear, its success depends critically on a sound understanding of the metabolic responses of the particular species, because species do respond differently. Examples of such interspecies differences abound: migrating Great White Pelicans (*Pelecanus onocrotalus*) exhibited higher plasma triacylglycerols when fasted than when fed, which is exactly the opposite of the pattern found in the small passerines that were studied (Shmueli et al. 2000). Flying pigeons relied less on VLDLs for lipid transport than small passerines (Schwilch et al. 1996), and short-distance migrants appeared to rely more on glucose and protein and less on lipid oxidation than long-distance migrants (Jenni and Jenni 1991). This cautionary note should be considered a call for careful calibration and validation of biochemical indices of condition.

References

Acquarone, C., M. Cucco, S. L. Cauli, and G. Malacarne. 2002. Effects of food abundance and predictability on body condition and health parameters: Experimental tests with the Hooded Crow. *Ibis* 144:155–163.

Bentley, P. J. 1998. *Comparative vertebrate endocrinology.* Cambridge University Press, Cambridge.

Berteaux, D. 2000. Energetic cost of heating ingested food in mammalian herbivores. *Journal of Mammalogy* 81:683–690.

Biebach, H. 1996. Energetics of winter and migratory fattening. Pages 280–323 in C. Carey (ed.), *Avian energetics and nutritional ecology.* Chapman & Hall, New York.

Blaxter, K. 1989. *Energy metabolism in animals and man.* Cambridge University Press, Cambridge.

Boismenu, C., G. Gauthier, and J. Larochelle. 1992. Physiology of prolonged fasting in Greater snow geese (*Chen caerulescens atlantica*). *Auk* 109:511–521.

Boonstra, R., D. Hik, G. R. Singleton, and A. Tinnikov. 1998. The impact of predator-induced stress on the snowshoe hare cycle. *Ecological Monographs* 68:371–394.

Cabanac, M. 2001. Regulation and the ponderostat. *International Journal of Obesity* 25 (supplement 5):S7–S12.

Calder, W. A. 1984. *Size, function, and life history.* Harvard University Press, Cambridge, Mass.

Castellini, M. A., and L. D. Rea. 1992. The biochemistry of natural fasting at its limits. *Experientia* 48:575–582.

Cherel, Y., J. Gilles, Y. Handrich, and Y. Le Maho. 1994. Nutrient reserve dynamics and energetics during long-term fasting in the King Penguin (*Aptenodytes patagonicus*). *Journal of Zoology* 234:1–12.

Cherel, Y., and Y. Le Maho. 1998. Physiology and biochemistry of long-term fasting in birds. *Canadian Journal of Zoology* 66:159–166.

Cowey, C. B., and J. R. Sargent. 1979. Nutrition. Pages 1–69 in W. S. Hoar, D. J. Randall, and J. R. Brett (eds.), *Fish physiology* vol. 8. Academic Press, New York.

Delgiudice, G. D., K. D. Kerr, L. D. Mech, M. R. Riggs, and U. S. Seal. 1998. Urinary 3-methylhistidine and progressive winter undernutrition in white-tailed deer. *Canadian Journal of Zoology* 76:2090–2095.

Delgiudice, G. D., K. D. Kerr, L. D. Mech, and U. S. Seal. 2000. Prolonged winter undernutrition and the interpretation of urinary allantoin:creatinine ratios in white-tailed deer. *Canadian Journal of Zoology* 78:2147–2155.

Delgiudice, G. D., L. D. Mech, and U. S. Seal. 1988. Chemical analyses of deer bladder urine and urine collected from snow. *Wildlife Society Bulletin* 16:324–326.

Delgiudice, G. D., L. D. Mech, and U. S. Seal. 1989. Physiological assessment of deer populations by analysis of urine in snow. *Journal of Wildlife Management* 53:284–291.

Delgiudice, G. D., and U. S. Seal. 1988. Classifying winter undernutrition in deer via serum and urinary urea nitrogen. *Wildlife Society Bulletin* 16:27–32.

Delgiudice, G. D., U. S. Seal, and L. D. Mech. 1987. Effects of feeding and fasting on wolf blood and urine characteristics. *Journal of Wildlife Management* 51:1–10.

Denbow, D. M. 1999. Food intake regulation in birds. *Journal of Experimental Zoology* 283:333–338.

Ditchkoff, S. S., and F. A. Servello. 1999. Sampling recommendations to assess nutritional restriction in deer. *Wildlife Society Bulletin* 27:1004–1009.

Dunlap, K. D. 1995. Hormonal and behavioral responses to food and water deprivation in a lizard (*Sceloporus occidentalis*): Implications for assessing stress in a natural population. *Journal of Herpetology* 29:345–351.

Dunn, E. H. 2000. Temporal and spatial patterns in daily mass gain of magnolia warblers during migratory stopover. *Auk* 117:12–21.

Egeler, O., D. Seaman, and T. D. Williams. 2003. Influence of diet on fatty acid composition of depot fat in western sandpipers (*Calidris mauri*). *Auk* 120:337–345.

Franzmann, A. W. 1985. Assessment of nutritional status. Pages 239–259 in R. J. Hudson and R. G. White (eds.), *Bioenergetics of wild herbivores.* CRC Press, Boca Raton, Fl.

Friedman, J. M., and J. L. Halaas. 1998. Leptin and the regulation of body weight in mammals. *Nature* 395:763–769.

Green, A. J. 2001. Mass/length residuals: measures of body condition or generators of spurious results? *Ecology* 82:1473–1483.

Guglielmo, C. G., D. J. Cerasale, and C. Eldermire. 2005. A field validation of plasma metabolite profiling to assess refueling performance of migratory birds. *Physiological and Biochemical Zoology* 78:116–125.

Guglielmo, C. G., P. D. O'Hara, and T. D. Williams. 2002. Extrinsic and intrinsic sources of variation in plasma lipid metabolites of free-living Western Sandpipers (*Calidris mauri*). *Auk* 119:437–445.

Guppy, M. 1986. The hibernating bear: Why is it so hot, and why does it cycle urea through the gut? *TIBS* 11:274–276.

Hammond, K. A., M. Konarzewski, R. M. Torres, and J. Diamond. 1994. Metabolic ceilings under a combination of peak energy demands. *Physiological Zoology* 67:1479–1506.

Hammond, K. A., and D. M. Kristan. 2000. Responses to lactation and cold exposure by deer mice (*Peromyscus maniculatus*). *Physiological and Biochemical Zoology* 73:547–556.

Harlow, H. J. 1995. Fasting biochemistry of representative spontaneous and facultative hibernators: the white-tailed prairie dog and the black-tailed prairie dog. *Physiological Zoology* 68:915–934.

Havel, P. J. 2001. Peripheral signals conveying metabolic information to the brain: Short-term and long-term regulation of food intake and energy homeostasis. *Experimental Biology and Medicine* 226:963–977.

Henen, B. T. 1991. Measuring the lipid content of live animals using cyclopropane gas. *American Journal of Physiology* 261:R752–R759.

Henen, B. T. 2001. Gas dilution methods: elimination and absorption of lipid-soluble gases. Pages 99–126 in J. R. Speakman (ed.), *Body composition analysis of animals.* Cambridge University Press, Cambridge.

Jenni-Eiermann, S., and L. Jenni 1992. High plasma triglyceride levels in small birds during migratory flight: A new pathway for fuel supply during endurance locomotion at very high mass specific metabolic rates? *Physiological Zoology* 65:112–123.

Jenni-Eiermann, S., and L. Jenni. 1994. Plasma metabolite levels predict individual body-mass changes in a small long-distance migrant, the garden warbler. *Auk* 111:888–899.

Jenni, E. S., and L. Jenni. 1991. Metabolic responses to flight and fasting in night-migrating passerines. *Journal of Comparative Physiology B* 161:465–474.

Jenni, L., and S. Jenni-Eiermann. 1998. Fuel supply and metabolic constraints in migrating birds. *Journal of Avian Biology* 29:521–528.

Kaiser, A. 1993. A new multi-category classification of subcutaneous fat deposits of songbirds. *Journal of Field Ornithology* 64:246–255.

Karasov, W. H., and B. Pinshow. 1998. Changes in lean mass and in organs of nutrient assimilation in a long-distance passerine migrant at a springtime stopover site. *Physiological Zoology* 71:435–448.

Karasov, W. H., B. Pinshow, J. M. Starck, and D. Afik. 2004. Anatomical and histological changes in the alimentary tract of migrating blackcaps (*Sylvia atricapilla*): A comparison among fed, fasted, food-restricted and refed birds. *Physiological and Biochemical Zoology* 77:149–160.

King, J. R., and M. E. Murphy. 1985. Periods of nutritional stress in the annual cycle of endotherms: Fact or fiction? *American Zoologist* 25:955–964.

Kitaysky, A. S., J. C. Wingfield, and J. F. Piatt. 1999. Dynamics of food availability, body condition and physiological stress response in breeding Black-legged Kittiwakes. *Functional Ecology* 13:577–584.

Klasing, K. 1998. *Comparative avian nutrition.* CABI Publishing, New York.

Korine, C., S. Daniel, I. G. van Tets, R. Yosef, and B. Pinshow. 2004. Measuring fat mass in small birds by dual-energy X-ray absorptiometry. *Physiological and Biochemical Zoology* 77:522–529.

Le Maho, Y., J. P. Robin, Y. Cherel, Y. Handrich, and R. Groscolas. 1991. Long-term fasting in penguins as a nutritional adaptation to breed or molt. *Acta XX Congressus Internationalis Ornithologici,* vol. IV, pp. 2177–2185, New Zealand Ornithological Trust Board, Wellington, N.Z.

Levy, M. N., B. M. Koeppen, B. A. Stanton, and R. M. Berne. 1998. *Physiology.* Mosby-Year Book, New York.

Lima, S. L. 1986. Predation risk and unpredictable feeding conditions: Determinants of body mass in birds. *Ecology* 67:377–385.

Lowell, B. B., and B. M. Spiegelman. 2000. Towards a molecular understanding of adaptive thermogenesis. *Nature* 404:652–660.

Lucas, A. 1996. *Bioenergetics of aquatic animals.* Taylor & Francis, London.

Marshall, C. T., N. A. Yaragina, Y. Lambert, and O. S. Kjesbu. 1999. Total lipid energy as a proxy for total egg production by fish stocks. *Nature* 402:288–290.

Mayer, D. G., and D. G. Butler. 1993. Statistical validation. *Ecological Modeling* 68:21–32.

Nagy, T. R., and A. L. Clair. 2000. Precision and accuracy of dual-energy X-ray absorptiometry for determining in vivo body composition of mice. *Obesity Research* 8:392–398.

Nakazato, M., N. Murakami, Y. Date, M. Kojima, H. Matsuo, K. Kangawa, and S. Matsukura. 2001. A role for ghrelin in the central regulation of feeding. *Nature* 409:194–198.

Needham, J. 1938. Contributions of chemical physiology to the problem of reversibility in evolution. *Biological Review* 13:225–251.

Olsson, O. 1998. Divorce in king penguins: Asynchrony, expensive fat storing and ideal free mate choice. *Oikos* 83:574–581.

Parker, K. L., G. D. Delgiudice, and M. P. Gillingham. 1993. Do urinary nitrogen and cortisol ratios of creatinine reflect body-fat reserves in black-tailed deer? *Canadian Journal of Zoology* 71:1841–1848.

Pearcy, S. D., and M. E. Murphy. 1997. 3-methylhistidine excretion as an index of muscle protein breakdown in birds in different states of malnutrition. *Comparative Biochemistry and Physiology* A 116:267–272.

Piersma, T. 1998. Phenotypic flexibility during migration: optimization of organ size contingent on the risks and rewards of fueling and flight. *Journal of Avian Biology* 29:511–520.

Piersma, T., G. A. Gudmundsson, and K. Lilliendahl. 1999. Rapid changes in the size of different functional organ and muscle groups during refueling in a long-distance migrating shorebird. *Physiological and Biochemical Zoology* 72:405–415.

Pravosudov, V. V., and T.C.J. Grubb. 1997. Energy management in passerine birds during the nonbreeding season. Pages 189–234 in V. J. Nolan, E. D. Ketterson, and C. F. Thompson (eds.), *Current ornithology,* vol. 14. Plenum Press, New York.

Randall, D., W. Burggren, and K. French. 1997. *Animal physiology: Mechanisms and adaptations.* W. H. Freeman and Co., New York.

Reidy, S. P., and J. M. Weber. 2000. Leptin: an essential regulator of lipid metabolism. *Comparative Biochemistry and Physiology A* 125:285–297.

Ricklefs, R. E. 1974. Energetics of reproduction in birds. Pages 152–292 in R. A. Paynter, Jr. (ed.), *Avian energetics.* Nutthal Ornithological Society, Cambridge, Mass.

Robbins, C. T. 1993. *Wildlife feeding and nutrition.* Academic Press, San Diego, Calif.

Rogers, C. M. 1987. Predation risk and fasting capacity: Do wintering birds maintain optimal body mass? *Ecology* 68:1051–1061.

Rosen, D.A.S., and A. W. Trites. 2003. No evidence for bioenergetic interaction between digestion and thermoregulation in Stellar sea lions *Eumetopias jubatus.* *Physiological and Biochemical Zoology* 76:899–906.

Schaub, M., and L. Jenni. 2001. Variation of fueling rates among sites, days and individuals in migrating passerine birds. *Functional Ecology* 15:584–594.

Schoenheimer, R. 1942. *The dynamic state of body constituents.* Harvard University Press, Cambridge, Mass.

Schwilch, R., L. Jenni, and S. Jenni-Eiermann. 1996. Metabolic responses of homing pigeons to flight and subsequent recovery. *Journal of Comparative Physiology B* 166:77–87.

Scott, I., C. Selman, P. I. Mitchell, and P. R. Evans. 2001. The use of total body electrical conductivity (TOBEC) to determine body composition in vertebrates. Pages 127–160 in J. R. Speakman (ed.), *Body composition analysis of animals.* Cambridge University Press, Cambridge.

Seal, U. S., M. E. Nelson, L. D. Mech, and R. L. Hoskinson. 1978a. Metabolic indicators of habitat differences in four Minnesota deer populations. *Journal of Wildlife Management* 42:746–754.

Seal, U. S., L. J. Verme, and J. J. Ozoga. 1978b. Dietary protein and energy effects on deer fawn metabolic patterns. *Journal of Wildlife Management* 42:776–790.

Secor, S. M., and J. Diamond. 1997. Effects of meal size on postprandial responses in juvenile Burmese pythons (*Python molurus*). *American Journal of Physiology* 272:R902–R912.

Secor, S. M., and J. M. Diamond. 2000. Evolution of regulatory responses to feeding in snakes. *Physiological and Biochemical Zoology* 73:123–141.

Secor, S. M., and A. C. Faulkner. 2002. Effects of meal size, meal type, body temperature, and body size on the specific dynamic action of the marine toad, *Bufo marinus.* *Physiological and Biochemical Zoology* 75:557–571.

Sedinger, J. S., R. G. White, and J. Hupp. 1995. Metabolizability and partitioning of energy and protein in green plants by yearling lesser snow geese. *Condor* 97:116–122.

Servello, F. A., and J. W. Schneider. 2002. Evaluation of urinary indices of nutritional status for white-tailed deer: Tests with winter browse diets. *Journal of Wildlife Management* 64:137–145.

Shmueli, M., I. Izhaki, O. Zinder, and Z. Arad. 2000. The physiological state of captive and migrating Great White Pelicans (*Pelecanus onocrotalus*) revealed by their blood chemistry. *Comparative Biochemistry and Physiology A* 125:25–32.

Speakman, J. R. 2001. *Body composition analysis of animals.* Cambridge University Press, Cambridge.

Starck, J. M., and K. Beese. 2002. Structural flexibility of the small intestine and liver of garter snakes in response to feeding and fasting. *Journal of Experimental Biology* 205:1377–1388.

Stevenson, R. D., C. R. Peterson, and J. S. Tsuji. 1985. The thermal dependence of locomotion, tongue flicking, digestion, and oxygen consumption in the wandering garter snake. *Physiological Zoology* 58:46–57.

Toloza, E. M., M. Lam, and J. Diamond. 1991. Nutrient extraction by cold-exposed mice: A test of digestive safety margins. *American Journal of Physiology* 261: G608–G620.

Vagnoni, D. B., R. A. Garrott, J. G. Cook, P. J. White, and M. K. Clayton. 1996. Urinary allantoin: Creatinine ratios as a dietary index for elk. *Journal of Wildlife Management* 60:728–734.

Vander, A. J. 1975. *Human Physiology.* McGraw-Hill, New York.

Van Der Meer, J. and T. Piersma. 1994. Physiologically inspired regression models for estimating and predicting nutrient stores and their composition in birds. *Physiological Zoology* 67:305–329.

Williams, T. D., C. G. Guglielmo, O. Egeler, and C. J. Martyniuk. 1999. Plasma lipid metabolites provide information on mass change over several days in captive Western Sandpipers. *Auk* 116:994–1000.

Wilson, B. S., and J. C. Wingfield. 1994. Seasonal and interpopulational variation in plasma levels of corticosterone in the side-blotched lizard (*Uta stansburiana*). *Physiological Zoology* 67:1025–1049.

Wilson, R. P., and B. M. Culik. 1991. The cost of a hot meal: Facultative specific dynamic action may ensure temperature homeostasis in post-ingestive endotherms. *Comparative Biochemistry and Physiology A* 100:151–154.

Witter, M. S., J. P. Swaddle, and I. C. Cuthill. 1995. Periodic food availability and strategic regulation of body mass in the European Starling, *Sturnus vulgaris*. *Functional Ecology* 9:568–574.

Woods, S. C., R. J. Seeley, D. Porte Jr., and M. W. Schwartz. 1998. Signals that regulate food intake and energy homeostasis. *Science* 280:1378–1383.

CHAPTER EIGHT

Isotopic Ecology

Finding out what animals eat often consumes a significant fraction of a field biologist's time. Many field biologists will tell you that the time is well spent. Although few will admit it, at least part of the reason why biologists indulge in diet studies is intimacy. Finding out what animals eat often places you in intimate contact with them—which is why many of us are biologists in the first place. However, the job of describing a species diet and its changes in space and time can be tedious, time consuming, and expensive. Spending countless hours tied to a microscope counting stomata and identifying seeds or insect parts from fecal or stomach content samples can be trying even for the most devoted naturalist. Furthermore, the methods that we use are restricted. Fecal samples and stomach contents provide only snapshots of what animals just ate and defecated, and with only a few exceptions we cannot use them on extinct animals.

The analysis of natural ratios of stable isotopes in animal tissues provides a convenient and useful method to analyze some aspects of animal diets that are intractable with traditional approaches. The method is based on a simple observation: animals often—albeit not always—incorporate the isotopic composition of what they eat in their tissues. By analyzing the isotopic composition of animal tissues and of their food we can track what animals are eating. Animal ecologists like to say that, isotopically, "animals are what they eat" (note the parallel of this metaphor in isotopic ecology with its use in ecological stoichiometry). Because the analytical methods to determine stable isotope ratios in biological samples are rapidly becoming cheaper and more accessible, the technique is becoming widespread among animal biologists. No method is without assumptions, and the use of stable isotopes in animal ecology is loaded with them (see Gannes et al. 1997). This chapter highlights the power of stable isotopes but also pinpoints the assumptions that must be recognized when using them in ecological research. Dietary reconstruction of extant and extinct animals is probably the most widely used application of isotopic ecology. However, the method can be used for a variety of other research purposes ranging from examining patterns of nutrient allocation to the determination of areas of origin for migratory animals. This chapter provides examples of these applications.

8.1 Basic Principles

8.1.1 Stable Isotope Measurements: A Primer

The relatively short-lived radiogenic isotopes of hydrogen (^{3}H), carbon (^{14}C), phosphorus (^{32}P), and sulfur (^{22}S) have been used extensively in animal physiological ecology. As we have seen, radio-labeled amino acids, lipids, and sugars are used to measure hydrolysis and absorption (chapter 4), as well as to measure the turnover of nutrient reserves (chapter 7). This chapter is not about these radiogenic elements. It is about the stable isotopes of the light elements hydrogen, carbon, and nitrogen (and to a lesser extent oxygen and strontium). These elements exhibit natural variation in the abundance of stable isotopes in the parts per thousand or even in the percent range (table 8.1). Following advances in the design of isotope-ratio-monitoring mass spectrometers that permit extremely precise analysis of isotope abundances (box 8.1), there has been a tremendous increase in the use of naturally occurring isotopic variation to trace biological, chemical, and physiological processes (see Lajtha and Michener 1998). Because the notation and basic principles of stable isotope chemistry were developed by geochemists, a good isotope geochemistry book is a useful addition to an isotopic ecologist's bookshelf (we recommend Hoefs' 1997 short text).

The isotopic composition of a sample is measured as the ratio of one isotope to another. In most cases the abundance of one isotope (generally the lightest) exceeds the abundance of the other by a large margin (table 8.1). Consequently, this ratio can be a very small number. To make measurements of the relative abundances of two isotopes intelligible, geochemists express the isotopic composition of most materials as the normalized ratio of the sample to a standard in parts per thousand (per mil, ‰):

TABLE 8.1

Average terrestrial abundances and worldwide standards of the stable isotopes of major elements of interest in ecological studies

Element	Low mass	% Abundance	High mass	% Abundance	Standard
Hydrogen	^{1}H	99.984	^{2}H (Deuterium)	0.016	SMOW Standard Mean Ocean Water
Carbon	^{12}C	98.89	^{13}C	1.11	PDB PeeDee Belemnite
Nitrogen	^{14}N	99.63	^{15}N	0.37	N$_2$ (atm.) Air nitrogen
Oxygen	^{16}O	99.759	^{18}O	0.204	SMOW
Sulfur[a]	^{32}S	95.00	^{34}S	8.22	CD Triolite from the Canyon Diablo meteorite
Strontium[a]	^{86}Sr	9.86	^{87}Sr	7.02	

[a] Both sulfur and strontium have more than two stable isotopes. The additional isotopes of sulfur are ^{33}S and ^{36}S. Those of strontium are ^{84}Sr and ^{88}Sr. We have listed in the table only those isotopes that are measured more frequently.

$$\delta X = \left(\frac{R_{sample} - R_{standard}}{R_{standard}} \right) \times 1000 \qquad (8.1)$$

where X is an element, and R_{sample} and $R_{standard}$ are the ratios of the heavy to the light isotopes for the sample and standard, respectively. Some of the standards chosen by geochemists seem capricious to biologists, but at this point we have no say in the matter. Atmospheric nitrogen, a marine belemnite from the Pee Dee Formation (VPDB) and ocean water (standard mean ocean water, SMOW) are used as standards for nitrogen, carbon, and hydrogen, respectively. Thus, you will find that isotope ratios are commonly expressed as ‰ AN, ‰ SMOW or ‰ VPDB (table 8.1). The words "depleted" and "enriched" are useful pieces of isotopic jargon. These terms often refer to the heavy, and often less abundant, isotope of a pair: depleted means a more negative δ value whereas enriched means a more positive δ value.

Box 8.1. Basic Principles of Stable Isotope Mass Spectrometry

A mass spectrometer separates charged atoms and molecules on the basis of their masses based on their motion in magnetic and electrical fields. Speakman (1997) uses a nice analogy to explain how a mass spectrometer works. Suppose that you are assigned the task of separating heavy from light golf balls without a balance. If you could stand at the end of a cliff and hit these balls with identical force and identical swing, the light balls would fall further than the heavy ones (figure 8.1). Of course, not even Tiger Woods can achieve this amazing consistency in swing and force—which precludes using cliffs and clubs to weigh golf balls, and makes golfing fun to some misguided souls. The theory behind the mass spectrometer is as simple as the effects of hitting balls (atoms or molecules) of different masses with exactly the same force into a constant gravitational field (in the mass spectrometer a magnetic field). The practice is a lot more complicated. Mass spectrometers are notoriously fickle machines. You should recognize all the elements of Speakman's analogy in the following depiction of a gas isotope ratio mass spectrometer (GIRMS). We borrowed the GIRMS sketch and description presented here from Boutton's (1991) lucid account. Speakman (1997) and Hoefs (1997) give slightly more technical descriptions of a GIRMS that complement Boutton's (1991).

A GIRMS can be divided into six parts: (1) an inlet that handles the sample; (2) an ion source where gases are ionized by a hot filament; (3) a curved flight tube located in a magnetic field that resolves ions of different masses; (4) a set of Faraday cup detectors and amplifiers for collecting and amplifying the resolved beams; (5) a pumping system that can maintain a good vacuum in the flight tube; and (6) a computer system that acquires data and controls the instruments of the GIRMS. The molecules analyzed in a GIRMS are analyzed as gases (e.g., CO_2, N_2, and H_2). To illustrate how a GIRMS works, we will follow a cohort of CO_2 molecules through the instrument's guts. The CO_2 sample flows from the inlet into the chamber containing the ion source. An electron beam generated by a heated filament ionizes the CO_2 molecules. Propelled by an accelerating potential, the positive ions fly down the analyzer tube into the curved sector located in a magnetic field. The trajectory of the ions passing through the magnetic field becomes a function of their mass and energy. Because they have less momentum, CO_2 ions with a mass of 44 ($^{12}C^{16}O^{16}O$) are deflected to a greater extent than ions of mass 45 ($^{13}C^{16}O^{16}O$, $^{12}C^{17}O^{16}O$) or mass 46 ($^{12}C^{18}O^{16}O$). Thus, the major species of CO_2 are resolved into three separate ion beams according to their mass. Each beam strikes separate Faraday cup collectors (see the diagram). When these ions strike the collectors

continued

continued

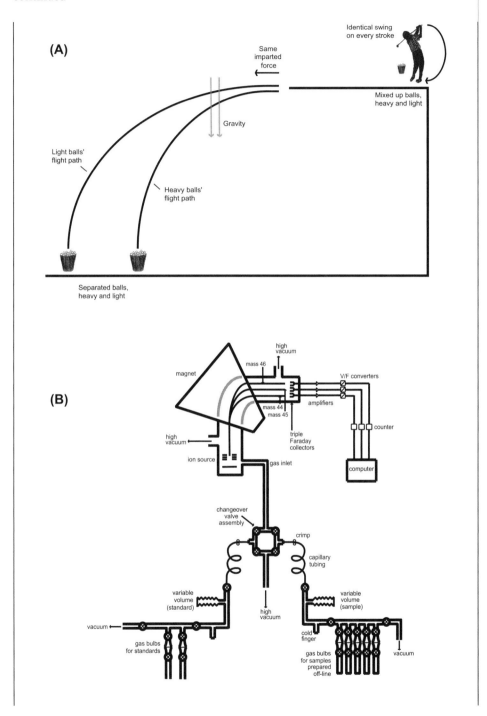

Figure 8.1. Golf balls hit consistently by a golfer at the end of a cliff can be separated as heavy and light, depending on the distance that they travel (A). A mass spectrometer separates heavy and light molecules based on the same principle. The elements of the diagram in B are described in text.

continued

continued

and are neutralized, they generate electrical currents, which are amplified and used to compute the stable isotope ratios. The $^{13}C/^{12}C$ ratio is calculated from a combination of the mass 45/44 and mass 46/44 ratios and corrected for the very minor ^{17}O contributions to the mass 45 signal.

The GIRMS system illustrated in this box has a dual inlet. Dual-inlet systems were developed to get very high precision. In a dual-inlet GIRMS, a sample gas and a reference gas are analyzed alternately many times. Because a single sample is analyzed several times, dual-inlet GIRMS systems require large samples. For organic materials, sample masses range from 5 to 20 mg. These samples have to be prepared off line. Typically, samples are thoroughly mixed with an oxidant (CuO), sealed under vacuum, and combusted at 900°C for 1–2 h. Temperature is then reduced to 650°C and the tubes are baked for two more hours. Finally, the tubes cool down overnight at room temperature and the combined resulting gases (CO_2, N_2, and H_2O) can be separated and purified to be introduced into the GIRMS. Labs that conduct isotopic analyses are invariably cluttered by purification gas lines. Sample preparation for a dual-inlet GIRMS is laborious and lengthy.

Continuous-flow systems are designed to analyze a large number of very small samples (from 0.1 to 5 mg). Sample preparation is minimal. A sample that has been dried and ground is loaded into a tin boat, folded, and placed in an autosampling carousel. When its turn arrives, the sample is flash-combusted at 1600–1800°C in an oxygen stream. The resulting gases are carried by a helium stream through purification columns to a gas chromatography column, which separates these gases. Reference gases must be analyzed periodically, to check for precision, but the precision gained by analyzing a reference and a sample alternately is lost. Continuous-flow systems are less precise, but they are widely used because many small samples can be analyzed rapidly with minimal previous preparation.

8.1.2 The Problems and Benefits of Fractionation

The natural variation in the relative abundance of stable isotopes in any substance is the consequence of tiny mass differences that cause the isotopes to behave differently in both physical processes and chemical reactions. In general, the lighter isotope (^1H, ^{12}C, or ^{14}N) tends to form weaker bonds and to react faster than the heavier isotope (^2H, ^{13}C, or ^{15}N). As a consequence, the abundance of stable isotopes of an element will vary among the interacting molecules that participate in chemical reactions and physical processes. The change in isotopic abundance among these molecules is called fractionation. Fractionation ($\alpha_{A\text{-}B}$) between the chemical species (geochemists call molecules or physical forms of a molecule, such as liquid water and vapor, "species") A and B is described in terms of the ratio in delta (δ) values between the species:

$$\alpha_{A-B} = \frac{R_A}{R_B} = \frac{(1000 + \delta_A)}{(1000 + \delta_B)}. \tag{8.2}$$

Values of α are usually very close to 1, so the difference between two δ values is often reported and denoted by the discrimination factor Δ_{A-B} ($\Delta_{A-B} = \delta_A - \delta_B$). If Δ_{A-B} is small (less than 10, or α less than 1.010), then the approximation $1000 - \ln(\alpha_{A-B}) \approx \Delta_{A-B}$ is accurate. A few researchers express fractionation in terms of isotopic enrichment (ε) where ε is defined as $\varepsilon_{A-B} = (\alpha_{A-B} - 1) \times 1000$. If Δ_{A-B} is small then ε_{A-B} and Δ_{A-B} are approximately equal.

There are two types of fractionation that are relevant to biologists. Equilibrium fractionation occurs among chemical molecules linked by equilibria as a result of bond strength differences between the isotopic species. For example, carbonate in bone is derived from blood bicarbonate (a fact that will be of significance for dietary reconstruction shortly). Carbon and oxygen isotopes are rapidly exchanged among blood bicarbonate, dissolved blood carbon dioxide, and body water by the following equilibria:

$$CO_2(aq) + H_2O \longleftrightarrow H_2CO_3 \longleftrightarrow H^+ + HCO_3^-.$$

The isotope equilibrium of bone carbonate is controlled by the composition of dissolved CO_2, which is produced by respiration, and fractionation associated with equilibrium exchanges of carbon. Suppose that one is attempting to estimate the isotopic composition of the diet of an extinct mammal from the carbon in the apatite of its teeth. At mammalian body temperatures, the fractionation (ε) from CO_2 to HCO_3 is about 8‰ (Mook 1986). Assuming that ε between dissolved bicarbonate and carbonate in apatite is 1‰ or 2‰ (the ε value for calcium carbonate), then apatite carbonate should have a $\delta^{13}C$ value approximately 9–10‰ greater than that of respired CO_2, which presumably reflects that of diet.

Kinetic fractionation effects occur because of differences in the rate of transport or rate of reaction of isotope species. For reactions catalyzed by enzymes, the magnitude of fractionation can be used to approximate the relative affinity of an enzyme for a compound with one isotope or another. An example of a kinetic fractionation is the reaction that catalyzes the oxidation of pyruvate into acetyl coenzyme-A (chapter 7). The enzyme responsible for this reaction (pyruvate dehydrogenase) oxidizes pyruvate with a ^{12}C-carbonyl group faster than pyruvate with a ^{13}C-carbonyl group. The result is the production of coenzyme-A that is depleted in ^{13}C relative to its pyruvate source (Hayes 1993). An important consequence of the higher affinity of pyruvate dehydrogenase for ^{12}C-carbonyl groups is that lipids synthesized from the carbon skeletons of carbohydrates tend to be depleted in ^{13}C relative to their carbohydrate source (DeNiro and Epstein 1977). To avoid the potential variation in carbon isotopic composition that results from variable lipid content in a tissue, carbon isotopic ratios are ideally measured in fat-free tissues.

Because the isotopic ratio of a consumer's tissues generally resembles that of its diet, one of the major uses of stable isotopes is dietary reconstruction. Because diet is typically inferred from the isotopic composition of a single tissue, the processes that can lead to differences in composition between tissues and diet must be understood. For example, the enzymatic fractionation resulting from the action of pyruvate dehydrogenase described above, produces lipids that tend to be depleted in ^{13}C relative to diet. Some of the differences between the isotopic composition of diet and a consumer's tissues are the result of fractionating processes. Others are the result of what has been called isotopic routing. Isotopic values of specific tissue components may not always follow the values of bulk diet. The reason is that the carbon skeletons, amino groups, and hydrogen atoms of different dietary components (protein, lipids, and carbohydrates) can be shunted to different tissue constituents. Because several processes can lead to differences in the isotopic composition of diet and animal's tissues, it is inappropriate to call these differences fractionation. Following Cerling and Harris (1999), we use the term discrimination factor ($\Delta_{\text{tissues-diet}} = \delta_{\text{tissues}} - \delta_{\text{diet}}$) for the difference between the isotopic composition of diet and that of the consumer's tissues.

8.2 Mixing Models

8.2.1 Simple Mixing Models

The use of the isotopic signal in an animal's tissue to determine the relative contribution of different food items to its diet often relies on an important assumption. We assume that, in isotopic terms, animals are what they eat, and thus the isotopic composition of their tissues equals the weighted average of the isotopic composition of the diet's constituents. For two diet constituents:

$$\delta X_{\text{tissues}} = p(\delta X_A + \Delta_A) + (1-p)(\delta X_B + \Delta_B) \qquad (8.3)$$

where p equals the fraction of diet A, δX_A and δX_B are the isotopic compositions of diet components A and B, and Δ_A and Δ_B are the discrimination factors for diets A and B. Provided that the isotopic compositions of two elements are used, equation 8.3 can be extended to estimate the fraction (p_i) of the diet comprised of three types of item (Phillips 2001; Ben David and Schell 2001). This requires solving the following system of linear equations in which A, B, and C are three different food types, $X1$ and $X2$ are two elements, and p_A, p_B, and p_C are the contributions of each food type to the animal's diet:

$$\delta X1_{\text{tissues}} = p_A\,\delta X1_A + p_B\delta X1_B + p_C\delta X1_C,$$
$$\delta X2_{\text{tissues}} = p_A\,\delta X2_A + p_B\delta X2_B + p_C\delta X2_C,$$
$$1 = p_A + p_B + p_C. \tag{8.4}$$

8.2.2 Mixing Isotopes with Stoichiometry

Phillips and Koch (2001) refined mixing models to incorporate differences in food stoichiometry (see chapter 10). They reasonably argued that the contribution of a given dietary item to an animal's carbon (or nitrogen) pool depends on how much carbon (or nitrogen) that item contains and on how efficiently each dietary item is assimilated. Let us denote by M the total assimilated biomass, p the fraction of total assimilation contributed by diet 1, $(1-p)$ the fraction of total assimilation contributed by diet 2, and $[C_1]$ and $[C_2]$ the concentrations of element X in diets 1 and 2, respectively. The relative contribution of diet 1 to the pool of element X in the consumers' tissues will be

$$p_1 = \frac{Mp[C_1]}{M(p[C_1]+(1-p)[C_2])} = \frac{p[C_1]}{p[C_1]+(1-p)[C_2]}, \tag{8.5}$$

and the isotopic composition of the pool of element X will be

$$\delta X = \left(\frac{1}{p[C_1]+(1-p)[C_2]}\right)(\delta X_1 p[C_1]+\delta X_2(1-p)[C_2]). \tag{8.6}$$

The concentration-dependent mixing model can be easily modified for more than one isotope and more than one diet. Again, $n-1$ isotopes can be used to differentiate among n diets. For more than one isotope, p_i is the fraction of total assimilated biomass (M) contributed by item i, and p_{X_i} represents the fraction of assimilated element X,

$$p_{X_i} = \frac{Mp_iX_i}{M\sum_{j=1}^{n} p_jX_j} = \frac{p_iX_i}{\sum_{j=1}^{n} p_jX_j}, \tag{8.7}$$

For three food sources, and two elements—for example, carbon and nitrogen with concentrations $[C_i]$ and $[N_i]$ (i = 1, 2, and 3) and isotopic compositions $\delta^{13}C_i$ and $\delta^{15}N_i$—we have that:

$$\delta^{13}C_{tissues} = \frac{1}{(p_1[C_1] + p_2[C_2] + p_3[C_3])}(p_1[C_1]\delta^{13}C_1$$
$$+ p_2[C_2]\delta^{13}C_2 + p_3[C_3]\delta^{13}C_3),$$
$$\delta^{15}N_{tissues} = \frac{1}{(p_1[N_1] + p_2[N_2] + p_3[N_3])}(p_1[N_1]\delta^{15}N_1$$
$$+ p_2[N_2]\delta^{15}N_2 + p_3[N_3]\delta^{15}N_3),$$
$$1 = p_1 + p_2 + p_3. \tag{8.8}$$

Of course, equation 8.8 reduces to equation 8.5 if all the diet components have the same elemental composition (i.e., $[C_1]$ = $[C_2]$ = $[C_3]$ and $[N_1]$ = $[N_2]$ = $[N_3]$). The error caused by neglecting concentration dependence increases as the differences in elemental composition among dietary components increase (Martínez del Rio and Wolf 2005). With a bit of algebra, equation 8.8 can be written in matrix form as a system of three linear equations in three unknowns:

$$\mathbf{AP} = \mathbf{B} \tag{8.9}$$

where

$$A = \begin{bmatrix} (\delta^{13}C_1 - \delta^{13}C_{tissue})[C_1] & (\delta^{13}C_2 - \delta^{13}C_{tissue})[C_2] & (\delta^{13}C_3 - \delta^{13}C_{tissue})[C_3] \\ (\delta^{15}N_1 - \delta^{15}N_{tissue})[N_1] & (\delta^{15}N_2 - \delta^{15}N_{tissue})[N_2] & (\delta^{15}N_3 - \delta^{15}N_{tissue})[N_3] \\ 1 & 1 & 1 \end{bmatrix}$$

and

$$P = \begin{bmatrix} p_1 \\ p_2 \\ p_3 \end{bmatrix}, \quad B = \begin{bmatrix} 0 \\ 0 \\ 1 \end{bmatrix}.$$

Phillips and Koch (2001) provide an algorithm to solve for the vector **P** in equation 8.9 (http://www.epa.gov/wed/pages/models.htm) and Martínez del Rio and Wolf (2005) describe in detail how to use and interpret the results of mixing models.

Incorporating a food's stoichiometry into mixing models requires a tad more work. It requires analyzing the food's stoichiometry and determining the efficiency with which different items are assimilated. However, concentration-dependent models are more realistic and may be particularly useful in omnivorous animals that eat mixed diets with contrasting C:N ratios.

8.2.3 Isotopic Routing

The problems of dietary reconstruction are especially acute when one attempts to do it for extinct animals. For example, anthropologists and paleontologists have traditionally used bone collagen, which is largely composed of protein, to analyze isotopic composition for dietary reconstruction. Collagen has two problems: (1) it contains 33% glycine, which is a relatively ^{13}C-enriched amino acid. Hence collagen tends to be ^{13}C enriched relative to other tissues. (2) Collagen is largely composed of protein and the composition of body protein in omnivores often reflects the isotopic composition of dietary protein (Ambrose and Norr 1993). Recognition of the principle that in omnivores the isotopic composition of tissue protein often reflects that of dietary protein, and not that of bulk diet, has led researchers to the analysis of the carbonates in bone apatite (Tieszen and Fagre 1993). These are synthesized from circulating bicarbonate derived from CO_2 and hence probably reflect the components of the diet that are catabolized for energy (Ambrose and Norr 1993). Omnivorous animals feeding on diets with low protein content often allocate dietary protein for tissue maintenance and repair, rather than catabolizing it for energy. Consequently, apatite carbonates probably underestimate the contribution of dietary protein. To get an accurate idea of an extinct animal's diet it is probably best to measure the isotopic composition of both collagen and bone apatite carbonates. Mixing models do not recognize the potential complications of routing. Martínez del Rio and Wolf (2005) outline how these models can be (1) used to diagnose routing, and (2) modified to incorporate knowledge about routing. In essence, one can use concentration dependent mixing models as "null" models that estimate the expected isotopic composition of a tissue given perfect mixing of the elements in the different diets, and compare the output of these models with observed values.

The problems that we have just described apply to a lesser degree to foregut fermenting herbivores and strict carnivores. In some—albeit not all (see chapter 6)—foregut fermenting animals rumen microbes mix the protein of all dietary components with the carbon skeletons of all dietary components. In strict carnivores protein is used for both energy and tissue synthesis and hence isotopic routing poses fewer problems. Isotopic ecologists like to say that animals are what they eat plus and minus fractionation. We hope to have demonstrated that this statement is overly optimistic. Stable isotopes are invaluable tools in animal physiological ecology but like all tools, they must be used properly. Using them properly means recognizing their limitations.

8.3 Isotopic Signatures

A wide variety of fractionating phenomena lead to distinctive isotopic ratios in the materials, biotic and abiotic, used by animals. We call these distinct ratios "isotopic signatures." Because animals incorporate these signatures in their tissues, we can use them to track resource use. The following sections review some of the signatures that have proven useful to animal ecologists. In addition to describing the processes that give rise to the distinctive isotopic ratios of some biological materials, we provide examples of how these have been used to investigate ecological phenomena. Our examples range from nutrient allocation to the determination of the origin of migratory animals.

8.3.1 Carbon Fractionation during Photosynthesis

Perhaps the most intensively studied fractionating process in biology is carbon fractionation during photosynthesis (Farquhar et al. 1989). We recommend that you read box 8.2 to make sense of the following sentences. Carbon isotopes can be used to differentiate among plants employing C_3, C_4, and CAM modes of photosynthesis (figure 8.3, box 8.2). Carbon fractionation in C_3 plants occurs at two steps in the photosynthesis process: diffusion/dissolution and carboxylation. CO_2 diffusion through the stomata and then dissolution into the mesophyll water produces a fractionation (against ^{13}C) of between 4‰ and 7‰. The second fractionating step, carboxylation by Rubisco, causes a much greater fractionation, once again, against ^{13}C ($\varepsilon_{CO_2\text{-}3C_{sugar}} = 19‰$). The result of these two fractionating processes is that C_3 plants have tissues that are greatly depleted in heavy carbon relative to the atmosphere. In C_3 plants $\delta^{13}C$ varies from −35‰ to −22‰, whereas the $\delta^{13}C$ of atmospheric CO_2 equals about −8‰. The extreme $\delta^{13}C$ values in this range are from plants growing under closed canopies and in water-stressed habitats, respectively. In closed canopies, the CO_2 available for photosynthesis is

Box 8.2.

During the light reactions of photosynthesis the energy of the sun is garnered to produce ATP and NADH. The energy contained in these molecules is then used to transform atmospheric CO_2 into organic molecules such as sugars:

$$CO_2 + 3ATP + 2NADPH + 2H^+ \rightarrow [CH_2O]n.$$

In all plants investigated so far, the assimilation of CO_2 ultimately proceeds through a cyclic process which commences with the carboxylation of the five-carbon compound ribulose 1,5-biphosphate (RuBP). The carboxylation of RuBP is catalyzed by a tremendously important enzyme with the formidable name ribulose-1,5-biphosphate carboxylase/oxigenase (Rubisco). RuBP is cyclically regenerated as hexose sugar is produced. The overall reaction can be formulated as

$$6RuBP + 6CO_2 + 18ATP + 12NADPH + 12H^+ \rightarrow hexose + 6RuBP$$
$$+18P_i + 12NADP^+$$

This equation skips an important step. Immediately after RuBP is carboxylated, it is split into two molecules of 3-phosphoglycerate (PGA). The cyclical process by which hexoses are produced and RuBP is regenerated is called the Calvin cycle. Because each PGA molecule contains three carbons, the Calvin cycle is also called the three-carbon (or C_3) pathway. In plants with C_3 photosynthesis, atmospheric CO_2 diffuses into chloroplasts and is directly incorporated into RuBP (figure 8.2).

Two ecologically important modes of photosynthesis (C_4 and CAM) rely on an additional intermediate step: atmospheric CO_2 is first fixed into organic acids. These two acid-mediated forms of photosynthesis are associated with characteristic arrangements of cells in photosynthetic organs. Let us begin by contrasting the leaf morphology of C_3 and C_4 plants. In C_3 plants, chloroplasts are distributed throughout the mesophyll tissue. In contrast, in C_4 plants chloroplasts are located in leaf cells in two concentric layers surrounding the vascular bundle. The inner-layer of bundle sheet cells in C4 plants contains numerous chloroplasts. The chloroplasts in the outer ring of mesophyll cells are not only less numerous, they are also smaller than those found in bundle cells. In the C_4 (four-carbon) mode, atmospheric CO_2 is used to carboxylate phosphophenylpyruvate (PEP, a three-carbon molecule) into oxaloacetate (a four-carbon molecule). The enzyme that catalyzes this step is called PEP carboxylase. The oxaloacetate

continued

continued

Figure 8.2. Schematic representation of the processes and morphology associated with C3 and C4 photosynthesis.

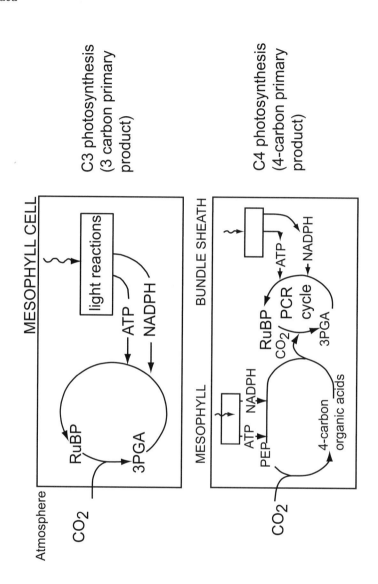

continued

continued

is then reduced into another four-carbon molecule (malate) or aminated to make aspartate (a four-carbon amino acid). The malate (or aspartate, depending on species) moves from mesophyll cells to bundle-sheath cells. Malate or aspartate is decarboxylated to yield pyruvate. The resulting CO_2 enters the Calvin cycle. In C_4 photosynthesis the fixation of atmospheric CO_2 is spatially separated from the Calvin cycle. C_4 photosynthesis has evolved repeatedly, presumably from C_3 photosynthesis. At least 19 families of flowering plants use C4 photosynthesis.

The plant photosynthetic mode seems to be related to nutritional content and accessibility for herbivores. Plants with C_4 photosynthesis are generally considered to be a poorer food for herbivores than plants with C_3 photosynthesis (Caswell et al. 1973). Three reasons are often postulated to explain this nutritional difference: (1) The total amount of protein is lower in C_4 than in C_3 species; (2) much of the protein in the leaves of C_4 plants is enclosed within the thick cell walls of bundle-sheath cells, making it relatively inaccessible; and (3) C_4 plants contain larger amounts of structural compounds such as lignin and silica that reduce digestibility. Because there are many exceptions to these three differences the C_3-C_4 nutritional dichotomy must be taken with a grain of salt (see Scheirs et al. 2001). However, it remains generally true that the photosynthetic modes are associated with differences in their quality as food for different herbivores.

enriched in ^{12}C as a result of respiration by plants and microbes. To understand why water-stressed plants tend to be less enriched in ^{12}C it is necessary to dwell a bit on the details of the factors that determine gas exchange in plants.

Farquhar et al. (1989) proposed the following simple equation to relate the isotopic composition of a plant with that of air:

$$\delta^{13}C_{plant} = \delta^{13}C_{air} - d_d - \frac{(d_c - d_d)P_i}{P_a} \tag{8.10}$$

where d_d is the fractionation occurring due to diffusion through the stomata and into leaf mesophyll water, d_c is the net discrimination caused by carboxylation, and P_a and P_i are the ambient and intracellular partial pressures of CO_2. In water-stressed plants stomata tend to close to prevent desiccation; hence the stomatal conductance is small in relation to the capacity for CO_2 fixation, P_i is small and $\delta^{13}C_{plant}$ tends toward $\delta^{13}C_{air}-d_d$. Conversely, if plants are well watered, stomatal conductance is high, P_i approaches P_a, and $\delta^{13}C_{plant}$ approaches $\delta^{13}C_{air}-d_c$. In C_3 plants there is a positive relationship between water stress and the $\delta^{13}C$ of fixed carbon.

Photosynthesis by C_4 and CAM pathways leads to lower carbon isotope fractionation than C_3 photosynthesis (figure 8.2). The reason for this lower fractionation is that the action of Rubisco occurs in relatively closed systems in which the ^{13}C-enriched CO_2 cannot be liberated to the atmosphere and hence has to be utilized. In both C_4 and CAM plants, CO_2 is first carboxylated into organic acids by the enzyme phosphophenylpyruvate (PEP) carboxylase. Fractionation in this step is slight (≈ 2‰). These organic acids are subsequently decarboxylated, and the resulting CO_2 is utilized in the photosynthetic carbon reduction (Calvin) cycle by Rubisco. In CAM and C_4 plants, the acquisition of atmospheric CO_2 is separated from the Calvin cycle. Because the fractionations that result from PEP carboxylation and by diffusion/dissolution are modest, the tissues of obligate CAM and C4 are isotopically more similar to the atmosphere (which has a $\delta^{13}C$ between -8‰ and -6‰) than those of C_3 plants (figure 8.3). Some plants use the CAM pathway facultatively. When water is scarce they close their stomata during the day and fix organic acids at night (their $\delta^{13}C$ is relatively positive). If water is available they rely on C_3 photosynthesis during the day (their $\delta^{13}C$ is more negative). Their carbon isotopic composition reflects the fraction of their carbon budget that is acquired by each photosynthetic mode.

Marine and freshwater plants show an enormous range of carbon isotopic values (Table 8.2). The reason is that in addition to variation in photosyntethic mode, other factors can influence fractionation. Diffusion of bicarbonate and CO_2 across a stagnant water boundary layer at the plant's surface may prevent discrimination against $^{13}CO_2$. If all CO_2 is used, there is no room for isotopic discrimination by Rubisco and hence plant tissues resemble the isotopic composition of available carbon. Finlay et al. (1999) have taken advantage of this boundary layer effect to show that the carbon isotopic composition of insect herbivores depends on the velocity of the stream in which they live. In fast-flowing streams, the boundary layer effect is smaller, and hence both the producers and the herbivores that feed on them show more negative $\delta^{13}C$ values (figure 8.4). The same processes that affect photosynthetic ^{13}C discrimination in streams take place at a much larger scale in the ocean. France (1995) found highly significant differences between the $\delta^{13}C$ of benthic and planktonic algae and Hobson et al. (1995) used these differences to distinguish between inshore and offshore feeders in seabirds. Although aquatic plants show significant variation, there are enough differences in carbon isotopic composition between terrestrial and marine producers to allow separating the relative contribution of terrestrial and marine sources to a consumer's diet. Because carbon in marine environments is ultimately derived from ^{13}C-enriched dissolved bicarbonate ($\delta^{13}C \approx 0$‰; Ambrose et al. 1997), marine plants (and hence also the consumers that eat them) tend to be enriched in ^{13}C relative to terrestrial plants (table 8.2).

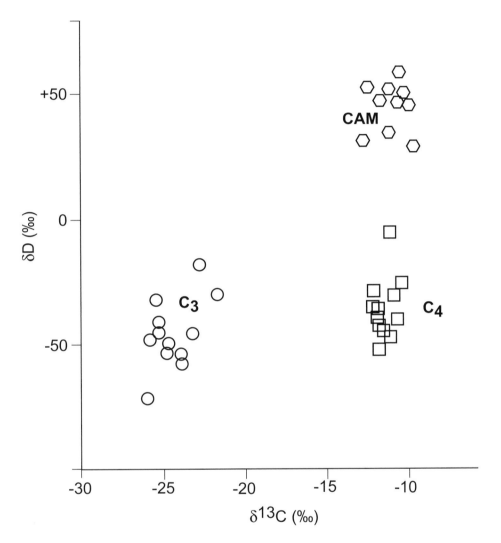

Figure 8.3. Plants with different photosynthetic pathways show contrasting carbon signatures. The graph shows the $\delta^{13}C_H$ and δD of cellulose (from Sternberg et al. 1984). Plants with C_3 photosynthesis show a more negative (^{13}C depleted) $\delta^{13}C$ than plants with C_4 and CAM photosynthesis. For reasons that remain to a large extent unknown, CAM plants show very positive (D enriched) δD.

8.3.2 Atmospheric Carbon Dioxide, C_4 Plants, and Herbivorous Mammals

Plants with the C_3 pathway account for the majority (85%) of terrestrial plants. About 10% of the earth's terrestrial flora uses CAM photosynthesis. C_4 plants dominate warm to hot open ecosystems, but on a floristic basis comprise the lowest percentage of species of the terrestrial flora (Ehleringer et al. 1991). The C_3 pathway is probably ancestral in plants and appears to have evolved under high atmospheric CO_2 concentrations. The C_4 and CAM pathways appear to have evolved independently several times in plants (Ehleringer et al. 1991). These pathways seem to have evolved more recently, and their abundance

TABLE 8.2

Variation of carbon isotopic composition of producers in coastal ecosystems

Source	Usual range in $\delta^{13}C$ ‰
Terrestrial C_3 plants	−23 to −30
Terrestrial C_4 plants	−10 to −14
River seston (particulate organic matter)	−25 to −27
Peat deposits	−12 to −28
C_3 marsh plants	−23 to −26
C_4 marsh grasses	−12 to −14
Seagrasses	−3 to −15
Macroalgae	−8 to −27
Benthic unicellular algae	−10 to −20
Temperate marine phytoplankton	−18 to −24
River-estuarine phytoplankton	−24 to −30
Autotrophic sulfur bacteria	−20 to −38
Methane-oxidizing bacteria	−62

Note: This table was compiled by Fry and Sherr (1989) from a variety of sources.

Figure 8.4. In three rivers, the isotopic composition of herbivorous invertebrates becomes more ^{13}C depleted as the current velocity increases. Because discrimination against ^{13}C during photosynthesis increases with CO_2 availability, the effect of current velocity on $\delta^{13}C$ can be attributed to the increased supply rate of CO_2 to the benthic algae that the herbivores feed on. (After Finlay et al. 1999.)

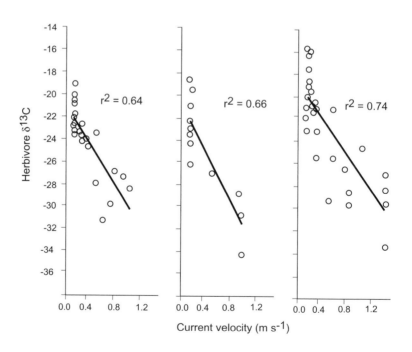

seems to have increased in response to lower atmospheric levels. The performance of C_4 plants relative to C_3 plants is very dependent on atmospheric CO_2 concentration (figure 8.5): low CO_2 conditions favor C_4 species and high levels of CO_2 favor C_3 species (Ehleringer et al. 1997).

Understanding the contrasting effect of CO_2 levels on the performance of different photosynthetic pathways requires defining two terms: quantum yield and photorespiration. Our definition of these terms will be sketchy and elementary (after all, our focus is animals!). Interested readers should consult a plant physiological ecology textbook (we recommend Nobel 2000). Quantum yield is a measure of photosynthetic efficiency and is defined as the ratio of CO_2 molecules gained to the number of photons absorbed. Photorespiration is a byproduct of the dual activity of Rubisco as a carboxylase and an oxygenase (see Ehleringer et al. 1997 for details). When CO_2/O_2 ratios in the chloroplast are low, O_2 is absorbed and CO_2 is given off. Under current atmospheric conditions, the concentration of CO_2 is 1000 times less than that of O_2, allowing a significant amount of photorespiration to occur and reducing the quantum yield and efficiency of C_3 photosynthesis. As described in box 8.2, C_4 plants fix CO_2 in organic acids in mesophyll cells and then transport those to bundle sheath cells where they are decarboxylated to release CO_2. The CO_2 concentration in bundle sheath cells is 10- to 20-fold higher than in mesophyll cells. This CO_2-concentrating mechanism improves the carboxylation efficiency of RUBISCO and abolishes photorespiration. Under low atmospheric concentrations plants with C_4 photosynthesis typically have higher photosynthetic rates than C_3 plants. However, at higher CO_2 concentrations C_4 photosynthesis becomes saturated whereas C_3 photosynthesis does not. Figure 8.5 shows the combinations of atmospheric and growing season temperatures that favor either C_3 or C_4 grasses.

Figure 8.5 suggests that the relative prevalence of C_3 and C_4 plants in terrestrial ecosystems should be dictated by the abundance of atmospheric CO_2. Over the last 100 million years atmospheric CO_2 seems to have decreased from

Figure 8.5. In C_4 plants atmospheric carbon is initially fixed inside leaf mesophyll cells. The resulting C_4 acid is decarboxylated inside the bundle cell, providing a source of CO_2 for rubisco and the C_3 photosynthetic cycle (upper panel). Because C_4 photosynthesis acts as CO_2 concentrating mechanism, C_4 grasses tend to be more efficient at fixing carbon at high temperatures and low atmospheric CO_2 levels. Thus, they are presumably competitively superior than C_3 grasses when it is hot and atmospheric CO_2 is low. (After Cerling et al. 1998.)

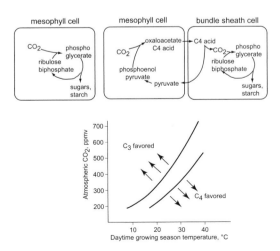

well above 1000 parts per million by Volume (ppmV) to the approximately 280 ppmV of the preindustrial revolution atmosphere (figure 8.6). In the pleistocene, CO_2 levels in interglacial periods were about 270 ppmV and dropped to about 180 ppmV during glacial maxima (Cerling et al. 1998). The last major change in atmospheric CO_2 has been due entirely to the human inordinate appetite for energy, with its accompanying burning of fossil fuels. CO_2 levels have increased from preindustrial levels ($\approx$280 ppmV) to 360 ppmV today. When coupled with the history of atmospheric CO_2 levels (figure 8.6), the model depicted in figure 8.5 has important implications for the photosynthetic composition of terrestrial ecosystems. The planet has been in a low-CO_2-level mode for some 7 million years, a condition rarely (if ever) attained in the early history of the atmosphere. A low level of CO_2 favors C_4 over C_3 plants. Consider the level of CO_2 in Holocene interglacial periods—about 280 ppmV. In the warmer parts of the planet, this level spells CO_2 starvation for C_3 plants and allows C_4 monocots to flourish (Ehleringer et al. 1997).

    Herbivorous mammals are good indicators of the presence of C_4 biomass. Their fossils are abundant and the enamel in their teeth preserves the isotopic signature of the plants that they eat. The isotopic composition of tooth enamel is enriched by about 14.3‰ over diet (Cerling et al. 1997) so mammals with average C_3 diets should have $\delta^{13}C$ values that range from about −21‰ to −8‰, depending on whether they feed on plants growing under a closed canopy or on water-stressed plants from open habitats. Consequently, unambiguous evidence

Figure 8.6. Evidence from a variety of sources (modeling, fossil plant stomatal densities, and soil carbonates) suggests a steady decline in atmospheric CO_2 concentration beginning about 100 million years ago. For about 7 million years, the earth has been in a "CO_2 starved" mode in which C_4 plants seem to be relatively favored (see figure 8.5). The middle panel shows a reconstruction of fluctuations in the concentration of atmospheric CO_2 over the last 420 thousand years from ice cores. Atmospheric CO_2 concentrations recorded at Mauna Loa, Hawaii, demonstrate that the atmosphere's CO_2 is rapidly increasing probably as a consequence of the burning of fossil fuels. (After Cerling et al. 1997.)

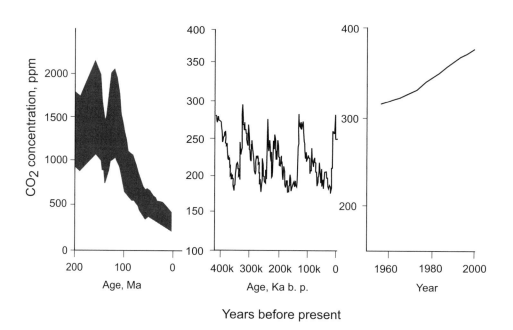

for the presence of C_4 plants in mammalian diets requires $\delta^{13}C$ values higher than $-8‰$ in modern mammals (figure 8.7). Because fossil-fuel burning by humans has changed the value of atmospheric $\delta^{13}C$ by about $1.5‰$, the cutoff diet for a pure C_3 diet in extinct animals should be slightly more positive. Cerling et al. (1997) suggest $-7‰$ as this cutoff point. As figure 8.7 indicates, the distribution of $\delta^{13}C$ of tooth enamel in mammalian herbivores that lived more than 8 million years ago (mya) shows a single distinctively C_3 peak. In contrast, modern herbivores show a bimodal distribution, which corresponds to a mammalian fauna that includes animals feeding on C_3 and C_4 plants.

This modern bimodal distribution in tooth enamel $\delta^{13}C$ values does not appear in mammalian fossil assemblages until the late Miocene and early

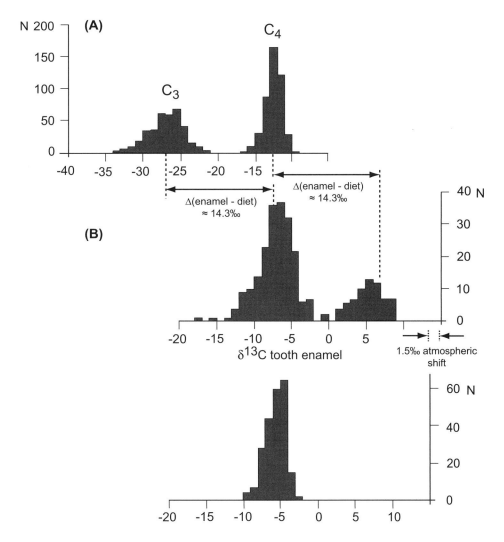

Figure 8.7. Modern C_3 plants exhibit more negative $\delta^{13}C$ values than C_4 plants (A). The tooth enamel of the teeth of modern mammalian herbivores shows a clear bimodal distribution (B). This distribution indicates that some species feed primarily on C_3 plants whereas others feed primarily on C_4 grasses. The distribution of $\delta^{13}C$ values in enamel is shifted to the right from that of plants because this tissue is enriched over diet by $14.3‰$. The bottom histogram shows the distribution of $\delta^{13}C$ values in the tooth enamel of mammalian fossils older than 7 million years ago. These animals were feeding primarily on C_3 plants. (After Ehleringer et al. 2001; Cerling et al. 1998.)

Pliocene. Between 8 and 6 mya, there was a worldwide change in the carbon isotopic composition of mammalian teeth (Cerling et al. 1997). Between 8 and 7 mya many woodland mammals were replaced by more open-habitat species with isotopically C_4 tooth enamel (figure 8.8). By 5 mya there were fewer C_3 eaters left, indicating strong C_4 grass dominance (figure 8.8). Pakistan, East Africa, North America, and South America showed a simultaneous increase in C_4 grazers between 8 and 6 mya. Western Europe is a notable exception in that herbivores retained a C_3 diet. Western Europe may be the exception that proves the rule. According to figure 8.5, C_3 plants are favored in spite of lowered CO_2 levels if temperature is relatively low, as it is in Western Europe. In some places the C_3 to C_4 shift in diet was accompanied by significant faunal changes. In East Africa, for example, early Miocene mammalian faunas included hyraxes, suids, rhinos, and proboscideans. The Pliocene witnessed the replacement of these animals by a mixture of grazers and browsers typical of a savanna mosaic. Grazing antelopes and hippos replaced chevrotains. Three-toed equids replaced rhinos and hyraxes, and high elephantoids with high-crowned teeth well suited to grind tough siliceous grasses replaced low-crowned teethed gomphoteres. The iso-

Figure 8.8. The $\delta^{13}C$ value of fossil tooth enamel of equids from Texas, Pakistan, and East Africa shows a dramatic change around 7 million years ago. This change indicates a widespread increase in the global biomass of C_4 plants. (Adapted from Ehleringer et al. 2002.)

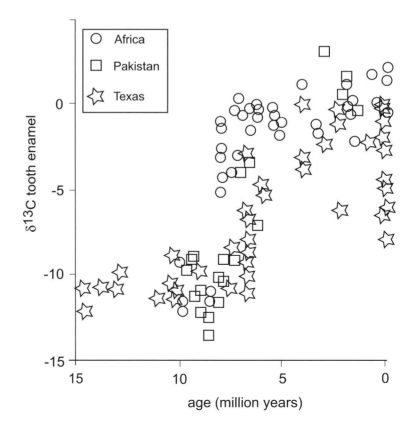

topic composition of the teeth of mammalian herbivores provides evidence of the vegetation changes brought by changes in atmospheric CO_2. Stable isotopes allow us to conduct physiological ecology studies in extinct animals!

8.4 The Dynamics of Isotopic Incorporation

8.4.1 Stable Isotopes and Reproductive Allocation of Nutrients

The allocation of resources to reproduction is a fundamental aspect of life history with profound ecological and evolutionary consequences. Unfortunately, mechanistic studies of nutrient allocation to reproduction have been hampered by the lack of quantitative methods. Radiotracers fed or injected into animals as a pulse have been used (e.g Boggs 1997). However, without knowledge of the specific activity of the nutrient pool and of its turnover dynamics, the data from radiotracer methods can be very difficult to interpret. Two ingenious studies illustrate how carbon stable isotopes can be used to study nutrient allocation. The first study investigated the allocation of nutrients to reproduction in an insect, the hawkmoth *Amphion floridensis* (O'Brien et al. 2000).

In hawkmoths and butterflies larval and adult diets are often nutritionally very different: larvae eat leaves and adults drink nectar. Nectar is a dilute sugar solution with only trace amounts of electrolytes and amino acids. O'Brien and her collaborators took advantage of this dietary difference to investigate the relative allocation of larval and adult resources to eggs. They fed larvae on the leaves of wild grapes (*Vitis*, $\delta^{13}C \approx -30‰$). After females emerged, they were fed on sucrose solutions that mimicked the composition of the floral nectar that adults feed on. The sucrose in these solutions was obtained from either sugar cane, a C_4 plant with $\delta^{13}C = -11.3‰$, or sugar beet, a C_3 plant with $\delta^{13}C = -28.8$. The isotopic composition of the first eggs produced by these females had the carbon isotopic composition of larval food (figure 8.9). As time went on, the isotopic composition of the eggs started to resemble that of the adult diet, suggesting that the carbon in nectar was contributing to egg manufacture. Indeed, after 6 days, the incorporation of carbon from nectar into eggs reached a plateau at slightly over 60% of the total carbon in the eggs (figure 8.9). O'Brien et al (2000) proposed a simple model to account for their results (figure 8.9). They hypothesized that in hawkmoths there are two carbon pools contributing to egg production (figure 8.9). One is a mixing pool that shows significant turnover and contributes with a fraction α to the carbon in eggs (figure 8.9). The other one is a nonmixing pool that maintains the larval carbon isotopic composition ($\delta^{13}C_0$)

Figure 8.9. Isotopes can be used to assess the incorporation of adult food into reproduction. The carbon isotopic composition of the eggs of the hawkmoth *Amphion floridensis* is very similar to that of the larval diet ($\delta^{13}C$ = −30‰) at the beginning of egg laying. However, as time goes by the $\delta^{13}C$ of eggs changes and begins to resemble the composition of the sugars in the nectar that adults feed on ($\delta^{13}C$ = −11.26‰ and $\delta^{13}C$ = −28.8, for cane sugar and beet sugar, respectively). Eggs never resemble the $\delta^{13}C$ of the sugars in the adult diet completely, suggesting that a fraction (about 40%) of the carbon in eggs must be acquired from larval stores. A simple model with two carbon pools can explain these results. One carbon pool (presumably representing dispensable amino acids) receives inputs from adult diet and hence shows turnover. The other pool (presumably representing indispensable amino acids) does not receive inputs from adult diet. The parameters α represent the fraction of carbon in eggs that comes from the mixing pool. The parameters β represent the fraction of carbon in the mixing pool that moths allocate to eggs rather than to other purposes. (After O'Brien et al. 2000.)

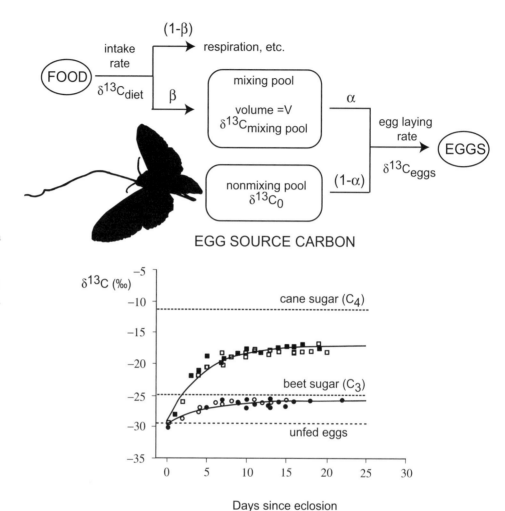

and contributes with a fraction $(1-\alpha)$ to the carbon in eggs. O'Brien et al. (2000) described their data with the following equation:

$$\delta^{13}C_{eggs} = \alpha\{\delta^{13}C_0 + [(\delta^{13}C_{diet} - \delta^{13}C_0)(1 - e^{-r(\text{time})})]\}$$
$$+ (1 - \alpha)(\delta^{13}C_0). \tag{8.11}$$

The first term in this equation shows the temporal change in the isotopic composition of the mixing pool. The parameter r in this term represents the fractional turnover of this pool. The second term represents the constant contribution of the nonmixing pool.

What is the possible physiological identity of the mixing and nonmixing pools? O'Brien et al. (2000) speculated that the nonmixing pool represents the essential (indispensable) amino acids that cannot be synthesized from amination of the carbon skeletons derived from nectar sugars (chapter 7). O'Brien and her collaborators (2002) used compound-specific isotopic analysis to test this hypotheses (see Macko 1998 for a brief introduction to the procedures involved in compound-specific isotopic analyses and Hammer et al. 1998 for an interesting application of their use). They analyzed the isotopic composition of individual amino acids in eggs and found that indispensable amino acids retained the larval carbon isotopic composition whereas dispensable amino acids acquired that of adult diet. O'Brien and her collaborators concluded that provisioning of essential amino acids by larvae is a critical determinant of egg production and longevity in hawkmoths. When the stores of essential amino acids are depleted egg-laying ceases and the productive life of the moth ends (O'Brien et al. 2002).

Boutton et al. (1988) used a similar method to study the transfer of dietary carbon to milk in cows. In mid lactation, cows were fed either alfalfa-barley ($\delta^{13}C$ = –25.0‰, diet A in figure 8.10) or corn-based diets ($\delta^{13}C$ = –11.5‰, diet C in figure 8.10). After a 7-week equilibration period on each of these diets, cows were switched to the other diet. The $\delta^{13}C$ of milk rapidly approached the isotopic value of dietary carbon in a pattern that is very similar to that of nectar incorporation in hawkmoths (box 8.3). Box 8.3 uses Boutton et al.'s (1988) data to illustrate the use of simple budget models to investigate carbon allocation into reproduction.

8.4.2 Isotopic Turnover and the Choice of Tissues in Diet Reconstruction Studies

The allocation of dietary carbon to eggs in hawkmoths and to milk in cows points to an important consideration: The dynamics of incorporation of an isotopic "signature" into a given pool of nutrients depend on the rate at which the materials in the pool turn over. If we view tissues as nutrient pools, this observation is of relevance in diet reconstruction. The isotopic composition of a tissue is the result of the integration of isotopic inputs over some time in the past. Thus, using tissues with different turnovers will give information of the past diet of an animal over different time intervals. The turnover rate of a tissue's constituents governs the time window of isotopic incorporation (Tieszen et al. 1983).

Some tissues, such as liver and plasma proteins, have high turnover rate, and their isotopic composition reflects integration of recent inputs. For example, in

Figure 8.10. A simple one-pool mass balance model represents the change in the carbon isotopic composition of cows' milk after they shift diets. Cow number 3 was first fed on diet C ($\delta^{13}C \approx$ −11.5‰) and then shifted to diet A ($\delta^{13}C \approx$ −26.0‰, Panel I). The same cow was fed on diet A for two months (Panel II) and then shifted to diet C. Dashed lines in Panels I and II represent values of diets and the solid curves are the first-order one-compartment model fitted to the data. (Modified from Boutton et al. 1988.)

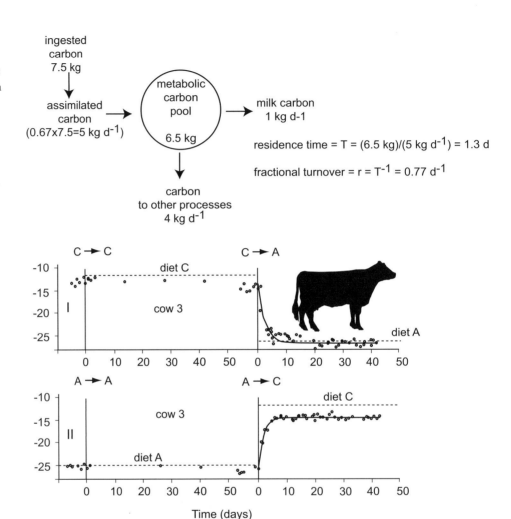

Japanese quail, carbon in liver and plasma proteins has a half-life of approximately 3 days (Hobson and Clark 1992). Thus, 90% of the carbon found in liver in these animals at any moment was incorporated approximately over the past 9 days. Other tissues have low turnover and their isotopic composition reflects integration of inputs over a longer time period. In young quail, carbon in collagen has a half-life of 173 days (Hobson and Clark 1992). Therefore 90% of the carbon in this tissue was incorporated over the past 575 days. The choice of tissue for a dietary reconstruction study depends on the question asked. Tissues with high nutrient turnover rates will track isotopic changes in diet closely, whereas tissues with low nutrient turnover rates will integrate signature from a large temporal window resulting in a smoother, less steep, curve (figure 8.11). If the question is to determine how animals track the availability of resources, then a tissue with high turnover rate (i.e., plasma proteins and liver) must be used. Conversely, if the ques-

Box 8.3.

The kinetics of isotopic incorporation of the carbon in diet into a tissue are determined by the size of the pool and by the flow of materials in and out of the pool. The diagram in figure 8.10 simplifies Boutton et al.'s (1988) data of incorporation of dietary carbon into milk. Because only 0.67 of the dietary carbon was assimilated, 5 kg of carbon are incorporated into the metabolic carbon pool. Therefore at steady state, the instantaneous fractional turnover rate of metabolic carbon pool (r) equals 0.77 day^{-1} (i.e., $5/6.5 = 0.77$). If one assumes that the isotopic composition of milk is equal to that in the metabolic carbon pool then the $\delta^{13}C$ of milk after a diet shift from diet A to diet C can be predicted by the equation

$$\delta^{13}C_{milk} = (\delta^{13}C_A + \Delta_A) + [(\delta^{13}C_C + \Delta_C) - (\delta^{13}C_A + \Delta_A)](1 - e^{-f(\text{time})},$$

The curves in figure 8.10 were fitted by this equation using $r = 0.77$ d^{-1}, and isotopic compositions and discrimination factors (Δ) from Boutton et al. (1988). This simple "one-compartment" model provides a good fit to the data. Note the resemblance of the equation above with the mixing term in the equation used by O'Brien et al. (2000).

tion is to determine the importance of different diet items in the diet of an animal over a long time period, then a tissue with low turnover must be used. We must emphasize here that turnover depends on both the characteristics of a tissue and on the species in question. Pearson et al. (2003) found that in yellow-rumped warblers (*Dendroica coronata*) the carbon in plasma proteins had a half-life of less than a day. These warblers are much smaller ($\approx$12 g) than the quail ($\approx$90 g) studied by Hobson et al. (1992). If you recall the arguments in chapter 1, you should not find this result surprising. Indeed, as you would expect, Carleton and Martínez del Rio (2005) found that the rate of incorporation of ^{13}C and ^{15}N into the blood of birds scaled with body mass to the −0.25 power.

8.5 Stable Isotopes and Migration

8.5.1 Deuterium, Rayleigh Distillations, and Migration

The study of animal movements is a fundamental aspect of ecology. Not only are seasonal movements a crucial aspect of many migratory animals' life histories,

Figure 8.11. Equation 8.11 can be used to investigate the carbon isotopic turnover of different tissues. We used it to explore the consequence of changing the isotopic composition of diet. The plot was modified from data presented in Hobson and Clark (1992). Japanese quail (*Coturnix japonica*) raised on a diet with $\delta^{13}C = -24‰$ were shifted to a diet with $\delta^{13}C = -19.5‰$. Panel A shows a trace of the change in diet's isotopic composition. In Panel B we used equation 8.11 to simulate the effect of changing the $\delta^{13}C$ of diet on the isotopic composition of three tissues with different turnovers: liver, muscle, and bone collagen. Liver has a high turnover ($r = 0.27$ d^{-1}) and its isotopic composition tracks diet changes. In contrast, tissue with low turnover like collagen ($r = 0.004\ d^{-1}$) reflect past diets rather than actual diet, even after 100 days after a diet change. For simplicity, we assumed that there was no isotopic discrimination between diet and tissues—which is not the case. In quail, collagen was enriched in ^{13}C by about 2.4‰ relative to diet (Hobson and Clark 1992).

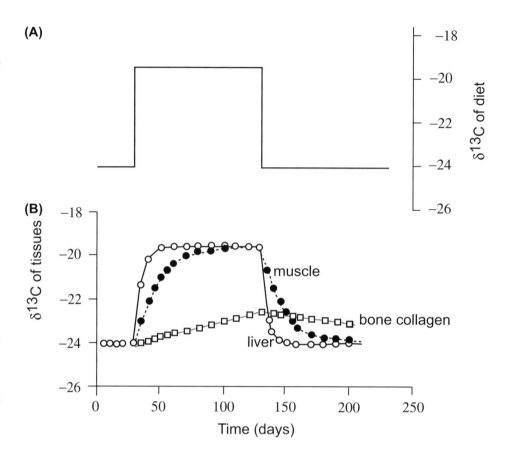

animal movements are critical tools for conservation and to understand an animal's role in an ecosystem. If we know the breeding, wintering, and stopover sites of migratory animals we can conserve these animals and their habitats more effectively. Many animals move daily among habitats and transfer nutrients among them. Others, such as migrant pollinators and seed dispersers, move long distances and perform their ecological functions at places thousands of kilometers apart, linking communities across continents. Although animal movements are central to ecology, our ability to follow individual animals is restricted. Marking, following, and recapturing appropriate numbers of individuals across vast geographical areas is difficult. Not surprisingly, data on animal movements are limited and the species sample for which we have adequate data is biased to large and economically important animals (Hobson 1999). Recently, the toolbox of methods available for study of animal movements has increased with the addition of molecular methods, with the use of smaller and smaller radio transmitters—recently satellite transmitters, and stable isotopes. Readers interested in an inclusive perspective of these methods can consult Webster et al. (2002). We focus on stable isotopes.

The isotopes of carbon, deuterium, and strontium have been used to track animal movements. Hobson's (1999) excellent article provides a comprehensive summary of the uses of all these elements. Deuterium is a useful element to study latitudinal migration because the deuterium composition of rainwater shows a broad continent-wide pattern in North America. Rainwater is relatively enriched in deuterium relative to hydrogen in the southeast and becomes more depleted in the northwest (figure 8.12). Because δD and $\delta^{18}O$ in meteoric water covary (figure 8.12), the same geographical pattern is found for ^{18}O. The reason for this geographical pattern involves several important issues in isotopic ecology, and hence is worthy of detailed examination.

During evaporation, "light" water molecules escape more readily from a body of water into the atmosphere than do molecules of "heavy" water. Although the degree of fractionation depends on temperature, at any place on earth atmospheric water vapor is always depleted (i.e., more negative δD and $\delta^{18}O$ values) relative to oceanic water. The reverse process occurs when water condenses in the atmosphere. Condensation is nearly an equilibrium process, favoring heavy water molecules in rain and snow. The first rain to fall from a new cloud formed over the ocean has δD and $\delta^{18}O$ values that are close to those of ocean water. The water vapor remaining in the atmosphere, however, is systematically depleted in deuterium and ^{18}O. Thus subsequent precipitation is derived from a vapor reservoir that has delta values even more negative than freshly evaporated seawater. The isotopic ratio (R) in the remaining vapor is given by

$$R = R_0 f^{(\beta-1)} \qquad (8.12)$$

where R_0 is the initial $D/^1H$ value in vapor, f is the fraction of vapor remaining, and β is the fractionation factor. Because $\beta > 1$, as the fraction of vapor remaining decreases, R becomes smaller and δD becomes more negative (i.e., vapor becomes more depleted in the heavy isotope, figure 8.13). This equation is known as the *Rayleigh distillation equation* and plays a very important role in stable isotope physical chemistry. The isotopic separation between rainwater and seawater becomes more pronounced as air masses move farther inland or are lifted to higher elevations (figure 8.12). In addition, because the fractionation factors for both oxygen and hydrogen isotopes become larger with decreasing temperature, the difference between seawater and precipitation increases towards the poles. These processes sometimes make it possible to recognize the isotopic signature of meteoric waters in a particular region.

Figure 8.12. Values in δD and δ^{18}O (in parentheses) in rainwater show a clear gradient in North America (map from Meehan et al. 2004). Analyzes of δD and δ^{18}O in water are commonly displayed on a two isotope plot. Meteoric waters (meaning rain and snow, in geochemist's jargon) on such a plot lie along a straight line given approximately by δD = 8 δ^{18}O + 10. Although oxygen and hydrogen isotopic values vary with temperature and the composition of clouds, they lie along this relatively tight relationship. This relationship is known as the "global meteoric water line." (After Richardson and McSween 1989.)

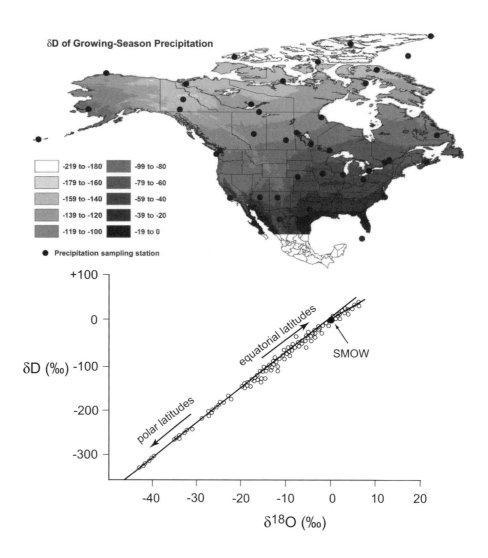

What does this have to do with animal migration? Many studies have established strong correlations between growing season average δD signatures in precipitation and those in plant biomass (Yapp and Epstein 1982). Animals eat this biomass and incorporate its isotopic signature into their tissues. Many bird species molt and then grow their wing feathers on or close to their breeding grounds. Because feathers are inert after synthesis, they retain the isotopic composition of the bird's diet during the period of growth. Hobson and Wassenaar (1997) and Chamberlain et al. (1997) found good relationships between the deuterium isotopic composition of feathers in insectivorous birds and the average δD of precipitation during the feathers' growing season (figure 8.14). These authors used δD in feathers to determine the origin of birds wintering at a single wintering

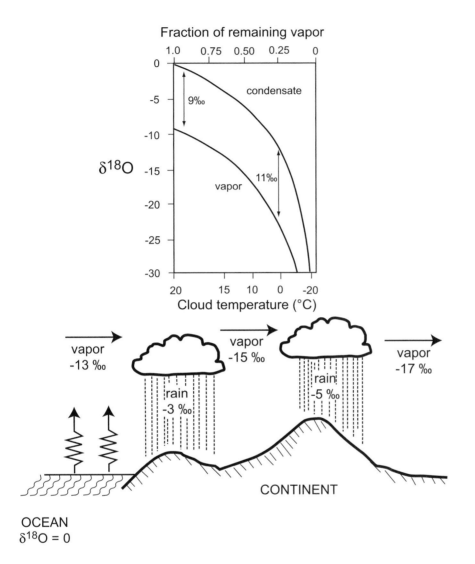

Figure 8.13. $\delta^{18}O$ in a cloud's vapor and condensate plotted as a function of the fraction of remaining vapor. The temperature of the cloud is shown in the lower axis. Note that fractionation increases with decreasing temperature. During removal of rain from a moist air mass, the residual vapor is continuously depleted in the heavy isotopes because the rain leaving the system is enriched in D and ^{18}O.

sites (figure 8.14). A similar approach has been used to determine the geographical origin of monarch butterflies wintering in Mexico (Wassenaar and Hobson 1998).

Unlike other elements considered in this chapter, hydrogen forms relatively weak bonds with nitrogen and oxygen and is able to exchange with ambient water vapor. For feather keratin, only 60% of the hydrogen forms nonexchangeable bonds. Hence 40% of all the hydrogen can be exchanged, and this uncontrolled exchange between the sample and ambient water vapor can lead to a change in the feather's signal after growth. Isotopic exchange does not pose a problem if all feathers used in a study are kept under the same conditions and a correlation is established between δD at the site of origin and that of feathers (Hobson 1999).

Figure 8.14. The composition of feathers (δD_f) increases linearly with the average composition of precipitation (δD_w, panel B). The hydrogen isotopic composition of the feathers of migrant passerines captured at a single site in Guatemala reveal a mixture of sites of origin (panel C). The numbered triangles in the map (A) represent sites at which birds feathers were collected. (After Hobson and Wassennar 1997.)

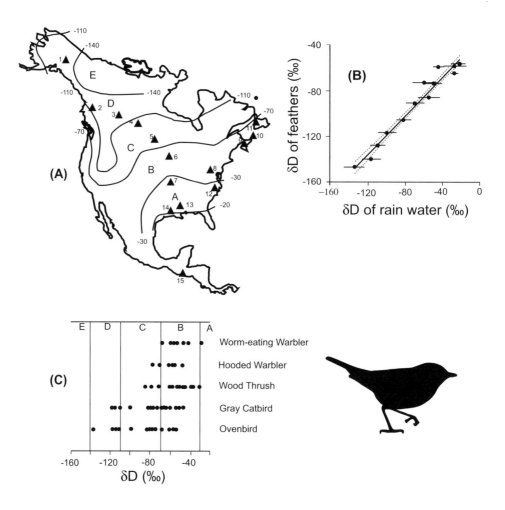

However, uncontrolled isotopic exchange limits comparing the values among studies. Wassenar and Hobson (2000a) proposed a simple method to standardize the treatment of complex organic samples for deuterium content.

8.5.2 Strontium and Animal Movements and Migration

Because the deuterium composition of rainwater forms a rather sloppy gradient across continents (figure 8.12), the deuterium content of feathers provides only broad estimates for the site of origin of migratory birds. In some continents, such as Africa and Europe, the gradients are complex or nonexistent and the method is difficult to apply. Finer geographical discrimination, or any discrimination at all, can be achieved with the use of other elements in conjunction with deuterium. Two isotopes that show significant promise for this purpose are carbon and strontium (Sr). Wassenaar and Hobson (2000b) used the deuterium and ^{13}C composition of feathers to delineate the geographical origins of red-winged blackbirds (*Agelaius phoeniceus*). They found that deuterium composition correlated with latitude, whereas ^{13}C correlated with the relative reliance of a population on different food types such as C_4 crops (sorghum and corn) and C_3 crops and wild plants and insects.

Plants absorb Sr from soils and, in turn, supply the Sr that is deposited in bone and, in mammals, in tooth minerals. Strontium has two stable isotopes: ^{87}Sr, which is produced by radioactive decay of ^{87}Rb, and ^{86}Sr. The usefulness of strontium's isotopes hinges on the existence of very variable $^{87}Sr/^{86}Sr$ ratios in soils. The magnitude of these ratios depends on the initial Rb/Str ratio and on the age of the underlying bedrock from which the soils are derived. Koch et al. (1995) used a combination of analyses of carbon and strontium isotopes in bones and teeth to estimate temporal changes in diet and location of elephants in Amboseli National Park in Kenya. Tooth dentin grows by accretion with very limited remodeling after synthesis. By analyzing a series of samples obtained by microdrilling from the edge of the tooth to the edge of the pulp cavity, which contains dentin deposited immediately before death, Koch et al. (1995) could assess seasonal changes in location and diet. They estimated that a 12–15 mm long core of dentin collected from molars represented 4 or 5 years of sequential tissue deposition. Each microsample integrated a record of about 3 months. The isotopic composition in structures that are deposited sequentially, such as teeth and mollusks shells, allows the amazing opportunity of glancing at the changes in resource use by a single animal over time (Goodwin et al. 2001). Although strontium shows much promise as a tracer of animal movements, it is found in bone and teeth. Unfortunately, collecting these tissues from living animals is often more invasive than plucking a feather, clipping a bit of hair, or even drawing a blood sample.

These examples provide a glimpse of the potential of the use of the isotopic composition of many elements to investigate animal movements. Other elements can be added to a stable isotopic analysis to provide further geographical resolution (lead and sulfur are potential candidates; Wassenar and Hobson 2000b). It is likely that there is vast, unknown and untapped, geographical variation in the isotopic composition of a variety of elements that makes its way up through local food webs. This variation may one day allow pinpointing the site of origin of many migratory animals from a single feather, hair tuft, or scale.

8.6 Nitrogen Isotopes

8.6.1 Nitrogen Isotopes I: Nutrient Transfers
across Ecosystem Boundaries

As is the case with other isotopes, the utility of nitrogen isotopes (^{15}N and ^{14}N) in animal ecology relies on their distribution in food. In carbon isotopes, variation in photosynthetic mode provides the source of isotopic variation among food types. In contrast, many processes generate variation in δ^{15}N. Because several of these fractionating processes are poorly understood, the use of nitrogen isotopes as tracers for ecological processes is often phenomenological. We rely on the existence of isotopic differences and patterns, but we often have only a foggy idea of the processes and mechanisms that create them. Nitrogen isotopes are useful, but using them requires special care and can be risky (Robinson 2001). In this section we outline some of the many uses of nitrogen isotopes in animal ecology. We emphasize lack of mechanistic knowledge in several of our examples because we believe that it is the task of animal physiological ecologists to provide this knowledge.

As usual throughout this chapter, we begin by looking at the base of food chains. Plants vary widely in δ^{15}N for at least four reasons: (1) soils vary widely in δ^{15}N, (2) deep-rooted plants tend to be enriched in ^{15}N relative to shallow-rooted plants, (3) nitrogen-fixing plants tend to be depleted in ^{15}N by about 2–4‰ relative to non-nitrogen-fixing plants, and (4) marine phytoplankton tends to be enriched by about 4‰ relative to terrestrial plants (Kelly 1998). The range in δ^{15}N generated by these four causes is fairly broad. It spans almost 26‰, and can potentially provide a variety of signatures that, in turn, can allow tracking the contribution of different plants into consumers. For example, Schoeninger et al. (1998) found that δ^{15}N decreased significantly with increasing percentage of time that lemurs (*Lepilemur leucopus*) fed on ^{15}N depleted nitrogen fixing legumes (figure 8.15). Unfortunately, sometimes the isotopic variation among individuals and species may be too large to allow identifying the relative contribution of various plant types to food chains on the basis of nitrogen isotope ratios alone. Thus, using nitrogen isotopes in conjunction with other isotopes can be informative—and is often safer.

Recall that the tissues of marine plants are enriched in ^{13}C relative to terrestrial plants. Because marine plants are also relatively enriched in ^{15}N, a combination of carbon and nitrogen isotopes can be used to estimate the contribution of terrestrial and marine sources to a food web. Anadromous salmonid fishes migrate from the ocean into their natal streams and rivers to spawn, and then die.

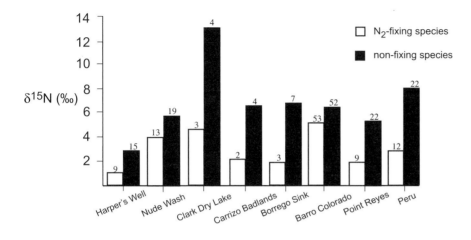

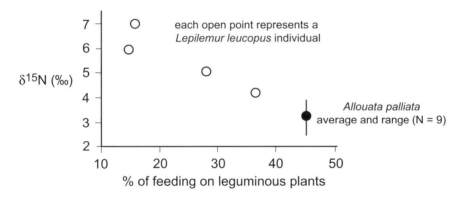

Figure 8.15. On average, the foliage of nitrogen fixing plants has a more ^{15}N-enriched isotopic composition than that of nonfixers. The $\delta^{15}N$ of the hair of lemurs (*Lepilemur lucopus*) in Madagascar decreased with the percentage of time that these animals fed on nitrogen fixing tamarind trees (*Tamarindus indica*). The data set also includes data for several howler monkeys from Barro Colorado island in Central America. (After Schoeninger et al. 1998.)

Salmonid carcasses represent a potentially large transfer of nutrients from the ocean into freshwater and terrestrial foodwebs. Because spawning salmon contain a higher proportion of the heavier carbon and nitrogen isotopes than most temperate terrestrial plants (figure 8.16), $\delta^{15}N$ and $\delta^{13}C$ provide natural markers to track the transfer of nutrients from salmon to freshwater and terrestrial food webs. Stable isotopes have revealed that spawning salmon have a dramatic fertilizing effect on both fresh water and terrestrial ecosystems: Salmon carcasses increase the productivity of stream invertebrate communities and thus the growth of freshwater fish (Bilby et al. 1996). On land, a number of mammalian carnivores rely on salmon seasonally (Ben-David et al. 1998; figure 8.16) and nitrogen from salmon flows into plants in riparian communities and fertilizes them.

Helfield and Naiman (2001) estimate that from 15 to 25% of the nitrogen budget of riparian forests adjacent to salmon-spawning streams comes from

Figure 8.16. δ¹³C and δ¹⁵N analyses of American marten blood (*Martes americana*) revealed seasonal variation in diet. In the summer most marten individuals eat primarily terrestrial animals (B). In the fall, marten show much higher interindividual variation in diet and a larger ingestion of food of marine origin (A; probably salmon carcasses; Ben-David et al. 1998).

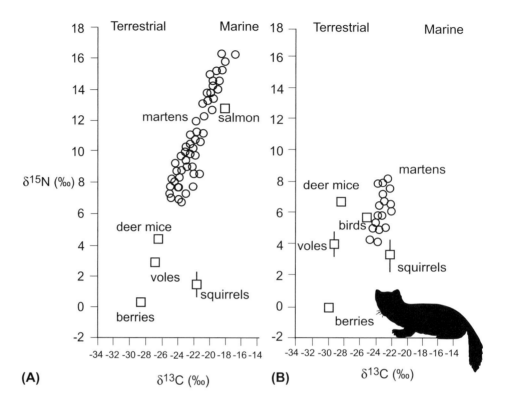

salmon carcasses. The effect of salmon carcasses extends into riparian terrestrial communities through three potential transfer pathways (Ben David et al. 1998): (1) During flash floods, fish carcasses are deposited on stream banks and, as they decompose, they can act as fertilizers of riparian vegetation. (2) Dissolved organic and inorganic nitrogen from the streams and rivers filter into the ground water (the hyporheic zone) and thus can be taken by riparian plants. (3) Terrestrial and semiaquatic predators such as brown and black bears, bald eagles, otters, mink, and marten carry carcasses into the riparian forests where they often cache them. In addition, these animals may deposit their urine and feces away from streams, extending the influence of marine fertilization of terrestrial systems (Hildebrand et al. 1999). Helfield and Naiman (20021) demonstrated the effect of salmon fertilization on riparian forests in southeast Alaska. They found that the C:N ratios were lower and the δ¹⁵N was higher in the foliage of Sitka spruce (*Picea sitchensis*) in watersheds with spawning salmon than in reference sites without salmon. They also found that basal growth in these trees was higher in areas with spawning salmon than in areas with no salmon (figure 8.17).

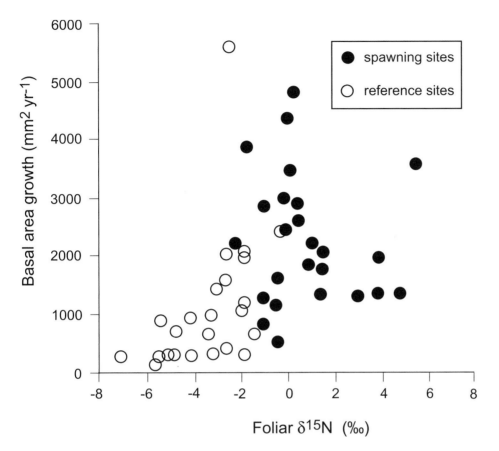

Figure 8.17. Spawning migrations of anadromous Pacific salmon (*Oncorhynchus* spp.) in the coastal watersheds of south-eastern Alaska bring nitrogen that can be used by riparian vegetation. The foliage of Sitka spruce (*Picea sitchensis*) at sites with spawning salmon runs had the characteristically enriched isotopic composition of a marine source. These sites also had higher growth rates. (From Helfield and Naiman 2001.)

8.6.2 Nitrogen Isotopes II: Trophic Relationships

Traditionally, the primary use of nitrogen isotope ratios in animal ecology has been the determination of trophic level. In a classic paper DeNiro and Epstein (1981) documented an average 3.4‰ enrichment in $\delta^{15}N$ value for whole-body samples over diet (we will call this enrichment $\Delta^{15}N_{tissues-diet}$). The 3.4‰ enrichment per trophic level has acquired a magical status (see Eggers and Jones 2000). As chapter 1 exemplifies, biologists are often seduced by the simplicity of phenomena that appear to be characterized by single numbers. Although there is significant evidence suggesting that there is a ^{15}N enrichment with trophic level, the magnitude of this enrichment is quite variable. $\Delta^{15}N_{tissues-diet}$ ranges from −1‰ to 6‰ (Peterson and Fry 1987; Post 2002). In a meta-analysis, Vanderklift (2002) found consistent differences in ^{15}N enrichments among taxa and tissues. A variety of factors, including growth rate and diet quality, can determine variation in ^{15}N enrichment and most of them remain unexplored (Martínez del Rio and Wolf (2005) review some of these factors). Robbins et al. (2005) demonstrated that $\Delta^{15}N_{tissues-diet}$ declined with the biological value of dietary protein across a variety of taxa and Trueman et al. (2005) found that $\Delta^{15}N_{tissues-diet}$ decreased linearly with growth in Atlantic salmon (this is an effect that we believe is very common). In spite of the wide use of the "3.4‰ $\delta^{15}N$ enrichment per trophic level" rule, we urge caution.

$\delta^{15}N$ enrichment varies within an individual among tissues (reviewed by Kelly 1998), among individuals depending on the C:N ratios of diet (Adams and Sterner 2000), and among species depending on diet type (vertebrate and invertebrate diets; Kelly 1998). An added complication is the potentially large variation in the nitrogen isotopic composition at the base of trophic levels. Vander Zanden and Rasmussen (1999) reported that primary consumers in lakes varied in $\delta^{15}N$ from −2‰ to 9‰. Most of this variation could be explained by the habitat of the consumers: For reasons that remain obscure, there were highly significant isotopic differences among primary consumers feeding in the littoral (mean $\delta^{15}N = 1.6$), pelagic (mean $\delta^{15}N = 3.1$), and pro-fundal (mean $\delta^{15}N = 5.2$) habitats. Thus without knowledge of the specific habitat(s) used by a consumer, trophic level cannot be deduced from $\delta^{15}N$ in these lacustrine food webs.

Why is there a ^{15}N enrichment? It appears that answering this question in a mechanistic fashion is key to understand, and then to interpret correctly, the ^{15}N enrichment associated with increased trophic level. Surprisingly, the physiological mechanisms that determine this enrichment are poorly under-stood. It is believed that during catabolism, amino acids with amine groups containing ^{15}N are disproportionately retained relative to those with amine groups containing ^{14}N (Macko and Estep 1984; Gaebler et al. 1966). The result is that excreted urinary nitrogen tends to be isotopically light (^{15}N depleted) relative to an animal's tissues (Steele and Daniel 1978; Minagawa and Wada 1984). Figure 8.18 illustrates this effect in a very simplified mass-balance model. This model suggests that ^{15}N enrichment will increase with the fraction of the total nitrogen budget of an animal that is excreted by a fractionating route like ammonia, urea, and uric acid (Martínez del Rio and Wolf (2005), analyzed a much more detailed version of this model). In the model, we used the notation $\Delta(p)$ to emphasize that it is possible that the discrimination factor from tissue to urinary nitrogen depends on the protein budget of an animal. The simple model depicted in figure 8.18 suggests that the construction of a detailed nitrogen isotopic budget (something that to our knowledge has not been done for an animal) would be very informative. Determining enrichment, and the isotopic composition of nitrogen inputs and outputs, while varying protein intake would provide the range of ^{15}N enrichments that an animal can experience.

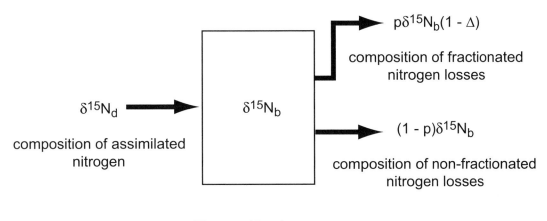

$$\delta^{15}N_b = \delta^{15}N_d/(1 - p\Delta)$$

Figure 8.18. In steady state, the $\delta^{15}N$ of assimilated nitrogen ($\delta^{15}N_d$) must equal the weighted average $\delta^{15}N$ of voided nitrogen. We assumed that a fraction p of all nitrogen was excreted in urine and a fraction $(1-p)$ through other means. If urinary nitrogen is isotopically lighter than body nitrogen ($\delta^{15}N_b$) by a discrimination factor $-\Delta$ ($\Delta > 0$), then body nitrogen will be heavier than assimilated diet nitrogen by a factor equal to $1/(1-p\Delta)$. Note that as p increases so does $\delta^{15}N_b$.

Ambrose and DeNiro (1987) reported significant ^{15}N enrichment in the bone collagen of drought-tolerant mammalian herbivores in East Africa. This pattern has received some support (reviewed by Kelly 1998). Within a site, drought-tolerant species seem to have more positive collagen $\delta^{15}N$ than animals that are obligate drinkers (Sealy et al. 1987). Among sites, Sealy et al. (1987) reported a negative correlation between the $\delta^{15}N$ of herbivores and annual precipitation. If this pattern is robust, it is potentially useful to paleontologists attempting to determine the ecology of extinct mammals. What is the mechanism that underlies the ^{15}N enrichment that accompanies drought tolerance? Because dehydrated animals produce more concentrated urine (chapter 8), Ambrose and DeNiro (1987) postulated that this enrichment was the result of the higher *concentration* of ^{15}N-depleted urea in the urine produced by drought tolerant animals. Figure 8.18 suggests, however, that enrichment depends on the fraction of nitrogen excreted as fractionated urea, but not on urea concentration—unless the renal mechanism of urea concentration discriminates against the excretion of ^{15}N (Ambrose 1991). Sealy et al. (1987) proposed that urea recycling in the gut (see chapter 6) can lead to ^{15}N enrichment and postulate that this process occurs with more frequency and intensity in animals under dry conditions (the mechanism by which urea recycling can lead to ^{15}N enrichment eludes us, but we mention this hypothesis for completeness).

The model depicted in figure 8.18 is for an animal in steady-state nitrogen balance. A slightly more complicated model can be constructed for animals that are in either positive or negative balance (Martínez del Rio and Wolf 2005). During fasting, nitrogen intake is nil, but fractionated nitrogen losses persist—albeit at low levels. Therefore, a simple, though important qualitative consequence resulting from the budget depicted in figure 8.18 is that the whole-body $\delta^{15}N$ of a fasting animal should become increasingly positive as its nitrogen reserves become increasingly depleted (Hobson et al. 1993). One more possible explanation for the $\delta^{15}N$ enrichment in drought-tolerant animals and animals living under dehydrating conditions is that they are more often in negative nitrogen balance than animals living under more mesic conditions. As far as we know, no one has used experiments to determine the importance of the three nonexclusive mechanisms proposed here to explain why mammals that face dehydrating conditions have higher $\delta^{15}N$ values.

8.7 Concluding Remarks and (Yet Again) a Call for Laboratory Experiments

Natural processes produce an enormous amount of variation in the isotopic composition of a variety of elements in a plethora of biological materials. This variation, and the fact that isotopic signatures are often incorporated more or less faithfully between resources and consumers, makes stable isotopes very useful. We hope that this chapter convinced you of the power of stable isotopes in animal physiological ecology. We also hope to have identified some of the assumptions that are invoked when stable isotopes are used in animal ecology. Often, we have no way of knowing if these assumptions are met. As the previous section attests, we still have enormous knowledge gaps in animal isotopic ecology. We know of the existence of apparently clear isotopic patterns, but we have not yet begun to explain the mechanisms that create them.

The use of nitrogen isotopes to determine trophic position illustrates this point poignantly. Why do animals show, on average, a 3.4‰ increase in $\delta^{15}N$ over their diet? We suspect that it is because animals excrete light nitrogen preferentially. This conjecture is probably correct, but it leaves room for significant variation. Indeed, the $\delta^{15}N$ enrichment of an organism over its diet varies over fivefold even within a single species (Adams and Sterner 2000). Why do we find this variation, and, more importantly, what does it mean? Does the $\delta^{15}N$

enrichment depend on an animal's nitrogen balance, and/or on whether it is growing, fasting, or in steady state? Can we provide some rules of thumb to associate trophic enrichment in $\delta^{15}N$ to animals at different life stages or to species with contrasting life histories? Without knowledge of mechanisms, we are stuck with a large amount of unexplained variation and forced to rely on an average magical number. We conducted a cursory survey of articles that included the words "stable isotopes," "nitrogen," and "trophic level" in their abstracts. In 88 articles published between 1993 and 2001, we found that only two used experiments to explore the factors that can influence the $\delta^{15}N$ enrichment across trophic levels (Hobson et al. 1993; Adams and Sterner 2000). Both papers demonstrated significant effects of physiological condition on $\delta^{15}N$. The vast majority of the remaining 86 papers assumed that a $\delta^{15}N$ increase of 3.4‰ implied an increase in trophic level.

Plant physiological ecologists have relied on the use of stable isotopes for a long time and with much success. To a large extent, the success of plant isotopic ecology hinges on a firm foundation of mechanistic studies that have ascertained in detail how the isotopes of different elements are fractionated along physiological pathways such as photosynthesis, nitrogen fixation, and evapotranspiration (Lajtha and Marshall 1994). Animal physiological ecologists have not yet developed this mechanistic background. Without it, the validity of many of the inference of isotopic studies is unclear. Calling for more mechanistic, laboratory studies, has become an almost a mandatory ritual in reviews of animal isotopic ecology (Gannes et al. 1997, 1998; Hobson 1999; Adams and Sterner 2000). Trite as it may be, we repeat this call here. Animal physiological ecologists have a wonderful opportunity to have a positive impact on science by providing the mechanistic foundations for isotopic ecology.

References

Adams, T. S., and R. W. Sterner. 2000. The effect of dietary nitrogen content on trophic level ^{15}N enrichment. *Limnology and Oceanography* 45:601–607.

Ambrose, S. H. 1991. Effects of diet, climate and physiology on nitrogen isotope abundances in terrestrial foodwebs. *Journal of Archaeological Science* 18:293–317.

Ambrose, S. H., B. M. Butler, D. B. Hanson, R. L. Hunter-Anderson, and H. H. Krueger. 1997. Stable isotopic analysis of human diet in the Marianas Archipelago, Western Pacific. *American Journal of Physical Anthropology* 104:343–361.

Ambrose, S. H., and M. J. DeNiro. 1987. Bone nitrogen isotope composition and climate. *Nature* 325:210.

Ambrose, S. H., and L. Norr. 1993. Carbon isotopic evidence for routing of dietary protein to bone collagen, and whole diet to bone apatite carbonate: Purified diet growth experiments. Pages 1–37 in J. Lambert and G. Groupe (eds.), *Molecular archaeology of prehistoric human bone.* Springer-Verlag, Berlin.

Ben-David M., T. A. Hanley, and D. M. Schell. 1998. Fertilization of terrestrial vegetation by spawning Pacific salmon: the role of flooding and predator activity. *Oikos* 83:47–55.

Ben-David, M., and D. M. Schell. 2001. Mixing models in analyses of diet using multiple stable isotopes: a response. *Oecologia* 127:180–188.

Bilby, R. E., B. R. Franzen, and P. A. Bisson. 1996. Incorporation of nitrogen and carbon from spawning salmon into the trophic system of small streams: Evidence from stable isotopes. *Canadian Journal of Fisheries and Aquatic Science* 53:164–173.

Boggs, C. L. 1997. Dynamics of reproductive allocation from juvenile and adult feeding: radiotracer studies. *Ecology* 78:181–191.

Boutton, T. W. 1991. Stable carbon isotope ratios of natural materials: I. Sample preparation and mass spectrometric analysis. Pages 155–171 in D. C. Coleman and B. Fry (eds.), *Carbon isotope techniques.* Academic Press, New York.

Boutton, T. W., H. F. Tyrrell, B. W. Patterson, G. A. Varga, and P. D. Klein. 1988. Carbon kinetics of milk formation in Holstein cows in late lactation. *Journal of Animal Science* 66:2636–2645.

Carleton, S. A., and C. Martínez del Rio. 2005. The effect of cold-induced increased metabolic rate on the rate of ^{13}C and ^{15}N incorporation in house sparrows (*Passer domesticus*). *Oecologia* 144:226–232.

Caswell, H., F. Reed, S. N. Stephenson, and P. A. Werener. 1973. Photosynthetic pathways and herbivory: a hypothesis. *American Naturalist* 107:465–480.

Cerling, T. E., J. R. Ehleringer, and J. M. Harris. 1998. Carbon dioxide starvation, the development of C4 ecosystems, and mammalian evolution. *Philosophical Transactions of the Royal Society* (*London*) *B* 353:159–171.

Cerling, T. E., J. M. Harris, B. J. MacFadden, M. G. Leakey, J. Quade, V. Eisenmann, and J. R. Ehleringer. 1997. Global vegetation change through the Miocene/Pliocene boundary. *Nature* 389:153–158.

Cerling, T. E., and J. M. Harris. 1999. Carbon isotope fractionation between diet and bioapatite in ungulate mammals and implications for ecological and paleontological studies. *Oecologia* 120:347–363.

Chamberlain, C. P., J. D. Blum, R. T. Holmes, X. Feng, T. W. Sherry, and G. R. Graves. 1997. The use of isotope tracers for identifying populations of migratory birds. *Oecologia* 109:132–141.

DeNiro, M. J., and S. Epstein. 1977. Mechanism of carbon fractionation associated with lipid synthesis. *Science* 197:261–263.

DeNiro, M. J., and S. Epstein. 1981. Influence of diet on the distribution of nitrogen isotopes in animals. *Geochimica et Cosmochimica Acta* 45:341–351.

Eggers, T., and T. H. Jones. 2000. You are what you eat . . . or are you? *Trends in Ecology and Evolution* 15:265–266.

Ehleringer, J. R., T. E. Cerling, and B. R. Cerling. 1997. C_4 photosynthesis, atmospheric CO^2 and climate. *Oecologia* 112:285–299.

Ehleringer, J. R., and T. E. Cerling. 2001. Origins and expansion of C4 photosynthesis. Pages 186–190 in H. A. Mooney and J. Canadell (eds.), *Encyclopedia of global environmental change*, vol. II. John Wiley and Sons, London.

Ehleringer, J. R., T. E. Cerling, and D. Dearing. 2002. Atmospheric CO_2 as a global change driver influencing plant-animal interactions. *Integrated and Comparative Physiology* 42:424–430.

Ehleringer, J. R., R. F. Sage, L. B. Flanagan, and R. W. Pearcy. 1991. Climate change and the evolution of C4 photosynthesis. *Trends in Ecology and Evolution* 6:95–99.

Farquhar, G. D., J. R. Ehleringer, and K. T. Hubick. 1989. Carbon isotope discrimination and photosynthesis. *Annual Review of Plant Physiology and Plant Molecular Biology* 40:503–537.

Finlay, J. C., M. E. Power, and G. Cabana. 1999. Effects of water velocity on algal carbon isotope ratios: implications for river food web studies. *Limnology and Oceanography* 44:1198–1203.

France, R. L. 1995. Differentiation between littoral and pelagic food webs in lakes using stable isotopes. *Limnology and Oceanography* 40:1310–1313.

Fry, B., and E. B Sherr. 1989. d^{13}C measurements as indicators of carbon flow in marine and freshwater ecosystems. Pages 196–229 in P. W. Rundel, J. R. Ehleringer, and K. A. Nagy (eds.), *Stable isotopes in ecological research*. Springer-Verlag, New York.

Gaebler, O. H., T. G. Vitti, and R. Vukmirovich. 1966. Isotope effects in metabolism of ^{14}N and ^{15}N from unlabeled dietary proteins. *Canadian Journal of Biochemistry* 44:1249–1257.

Gannes, L. Z., D. O' Brien, and C. Martínez del Rio. 1997. Stable isotopes in animal ecology: Assumptions, caveats, and a call for more laboratory experiments. *Ecology* 78:1271–1276.

Gannes, L. Z., C. Martínez del Rio, and P. Koch. 1998. Natural abundance variation in stable isotopes and their potential uses in animal physiological ecology. *Comparative Biochemistry and Physiology* 119A:725–737.

Goodwin, D. H., K. W. Flessa, B. R. Schone, and D. L. Dettmen 2001. Cross-calibration of daily growth increments, stable isotope variation, and temperature in the Gulf of California bivalve mollusk *Chione (Chionista) cortezi*: Implications for paleoenvironmental analysis. *Palaios* 16: 387–398.

Hammer, B. T., M. Fogel, and T. C. Hoering. 1998. Stable carbon isotope ratios of fatty acids in seagrass and redhead ducks. *Chemical Geology* 152:29–41.

Helfield, J. M., and R. J. Naiman. 2001. Fertilization of riparian vegetation by spawning salmon: Effects on tree growth and implications for system productivity. *Ecology* 82:2403–2409.

Hilderbrand, G. V., T. A. Hanley, C. T. Robbins, and C. C. Schwartz. 1999. Role of brown bears (*Ursus arctos*) in the flow of marine nitrogen into a terrestrial ecosystem. *Oecologia* 121:546–550.

Hobson, K. A. 1999. Tracing origins and migration of wildlife using stable isotopes: A review. *Oecologia* 120:314–326.

Hobson, K. A., R. T. Alisaukas, and R. G. Clark. 1993. Stable nitrogen isotope enrichment in avian tissues due to fasting and nutritional stress: Implications for isotopic analysis of diet. *Condor* 95:388–398.

Hobson, K. A , W. G. Ambrose, and P. E. Renaud. 1995. Sources of primary production, benthic-pelagic coupling, and trophic relationships within the Northeast Water Polynya: Insights from $\delta^{13}C$ and $\delta^{15}N$ analysis. *Marine Ecology Progress Series* 128:1–10.

Hobson, K. A., and R. G. Clark. 1992. Assesing avian diets using stable isotopes I: Turnover of ^{13}C in tissues. *Condor* 94:181–188.

Hobson, K. A., and L. I. Wassenar. 1997. Linking breeding and wintering grounds of neotropical migrant songbirds using stable hydrogen isotopic analysis of feathers. *Oecologia* 109:142–148.

Hoeffs, J. 1997. *Stable isotope geochemistry.* Springer-Verlag, New York.

Kelly, J. F. 1998. Stable isotopes of carbon and nitrogen in the study of avian and mammalian trophic ecology. *Canadian Journal of Zoology* 78:1–27.

Koch, P. L., J. Heisinger, C. Moss, R. W. Carlson, M. L. Fogel, and A. K. Behrensmeyer. 1995. Isotope tracking of change in diet and habitat use in African elephants. *Science* 267:1340–1343.

Lajtha, K., and J. D. Marshall. 1998. Sources of variation in the stable isotopic composition of plants. Pages 1–21 in K. Lajtha and R. H. Michener (eds.), *Stable isotopes in ecology and environmental science.* Blackwell Scientific Publications, Oxford.

Lajtha, K., and R. H. Michene. 1998. *Stable isotopes in ecology and environmental science.* Blackwell Scientific Publications, Oxford.

Macko, S. A. 1998. Compound-specific approaches using stable isotopes. Pages 241–247 in K. Lajtha and R. H. Michener (eds.), *Stable isotopes in ecology and environmental science.* Blackwell Scientific Publications, Oxford.

Macko, S. A., and M.L.F. Estep. 1988. Microbial alteration of stable nitrogen and carbon isotopic compositions of organic matter. Organic Geochemistry 6:787–790.

Martínez del Rio, C., and B. O. Wolf. 2005. Mass balance models for animal isotopic ecology: Linking diet's stoichiometry and physiological processes with broad scale ecological patterns. Pages 141–174 in M. A. Starck and T. Wang (eds.), *Physiological and ecological adaptations to feeding in vertebrates.* Science Publishers, Enfield.

Meehan, T. D., J. T. Giermakowski, and P. M . Cryan. 2004. GIS-based models of stable hydrogen isotope ratios in North American growing season precipitation for use in animal movement studies. *Isotopes in Environmental and Health Issues* 40: 291–300.

Minagawa, M., and E. Wada. 1984. Stepwise enrichment of ^{15}N along food chains: further evidence and the relation between $\delta^{15}N$ and animal age. *Geochimica et Cosmochimica Acta* 48:1135–1140.

Mook, W. G. 1986. ^{13}C in atmospheric CO_2. *Netherlands Journal of Sea Research* 20:211–223.

Nobel, P. S. 2000. *Physicochemical and environmental plant physiology.* Academic Press, New York.

O'Brien, D. M., D. P. Schrag, and C. Martínez del Rio. 2000. Allocation to reproduction in a hawkmoth: a quantitative analysis using stable isotopes. *Ecology* 81:2822–2831.

O'Brien, D. M., M. L. Fogel, and C. L. Boggs. 2002. Renewable and non-renewable resources: Amino acid turnover and allocation to reproduction in lepidoptera. *Proceedings of the National Academy of Science* 99:4413–4418.

Pearson, S. F., D. J. Levey, C. H. Greenberg, and C. Martínez del Rio. 2003. Effects of elemental composition on the incorporation of dietary nitrogen and carbon isotopic signatures in an omnivorous songbird. *Oecologia* 135:516–523.

Peterson, B. J., and B. Fry. 1987. Stable isotopes in ecosystem studies. *Annual Review Ecology and Systematics* 18:293–320.

Phillips, D. L. 2001. Mixing models in analyses of diet using multiple stable isotopes: A critique. *Oecologia* 127:166–170.

Phillips, D. L., and P. L. Koch. 2001. Incorporating concentration dependence in stable isotope mixing models. *Oecologia* 130:114–125.

Post, D. M. 2002. Using stable isotopes to estimate trophic position: Models, methods, and assumptions. *Ecology* 83:703–718.

Richardson, S. M., and H. Y. McSween. 1989. *Geochemistry: Pathways and processes.* Prentice-Hall, New York.

Robbins, C. T., L. A. Felicetti, and M. Sponheimer. 2005. The effect of dietary protein quality on nitrogen discrimination in mammals and birds. *Oecologia* 144:534–540.

Robinson, D. 2001. $\delta^{15}N$ as an integrator of the nitrogen cycle. *Trends in Ecology and Evolution* 16:153–162.

Scheirs, J., L. De Bruyn, and R. Verhagen. 2001. A test of the C3-C4 hypothesis with two grass miners. *Ecology* 82:410–421.

Schoeninger, M. J., U. T. Iwaniec, and L. T. Nash. 1998. Ecological attributes recorded in stable isotope ratios of arboreal prosimian hair. *Oecologia* 113:222–230.

Sealy, J. C., N. J. Van der Merwe, J. A. Lee-Thorp, and J. L. Lanham. 1987. Nitrogen isotope ecology in southern Africa: implications for environmental and dietary tracing. *Geochimica et Cosmochimica Acta* 51:2707–2717.

Speakman, J. R. 1997. *Doubly-labeled water: Theory and practice.* Chapman and Hall, London.

Steele, K. W., and R. M. Daniel. 1978. Fractionation of nitrogen isotopes by animals: A further complication to the use of variation in the natural abundance of ^{15}N for tracer studies. *Journal of Agricultural Science* 90:7–9.

Sternberg, L., M. J. DeNiro, and H. B. Johnson. 1984. Isotope ratios of cellulose from plants having different photosynthetic pathways. *Plant Physiology* 74:557–561.

Tieszen, L. L., T. W. Boutton, K. G. Tesdahl, and N. A. Slade. 1983. Fractionation and turnover of stable carbon isotopes in animal tissues: implications for $\delta^{13}C$ analysis of diet. *Oecologia* 57:32–37.

Tieszen, L. L., and T. Fagre. 1993. Effect of diet quality and composition on the isotopic composition of respiratory CO2, bone collagen, boapatite, and soft tissues. Pages

123–135 in J. Lambert and G. Groupe (eds.), *Molecular archaeology of prehistoric human bone.* Springer-Verlag, Berlin.

Trueman, C. N., R.A.R. McGill, and P. H. Guyard. 2005. The effect of growth rate on tissue-diet isotopic spacing in rapidly growing animals. An experimental study on Atlantic salmon (*Salmo salar*). *Rapid Communications in Mass Spectrometry* 19:3239–3247.

Vander Klift, M. A., and S. Pousard. 2003. Sources of variation in consumer-diet $\delta^{15}N$ enrichment: A meta-analysis. *Oecologia* 136:168–182.

Vander Zanden, J., and J. B. Rasmussen. 1999. Primary consumer $\delta^{13}C$ and $\delta^{15}N$ and the trophic position of aquatic consumers. *Ecology* 80:1395–1408.

Wassenar, L. I., and K. A. Hobson. 1998. Natal origins of migratory monarch butterflies at wintering colonies in Mexico: New isotopic evidence. *Proceedings of the National Academy of Science* 26:15436–15439.

Wassenar, L. I., and K. A. Hobson. 2000a. Improved method for determining the stable hydrogen isotopic composition (δD) of complex organic materials of organic interest. *Environmental Science and Technology* 34:2354–2360.

Wassenar, L. I., and K. A. Hobson. 2000b. Stable carbon and hydrogen ratios reveal breeding origins of red-winged blackbirds. *Ecological Applications* 10:911–916.

Webster, M. S., P. S. Marra, S. M. Haig, S. Besch, and R. T. Holmes. 2002. Links between worlds: unraveling migratory connectivity. Trends in Ecology and Evolution 17:76–83.

Yapp, C. J., and S. Epstein. 1982. A re-examination of cellulose carbon-bound hydrogen D measurements and some factors affecting plant-water D/H relationships. *Geochimica Cosmochimica Acta* 46:955–965.

CHAPTER NINE

How Animals Deal with Poisons and Pollutants

THIS CHAPTER CONTINUES our consideration of how the biochemical fate of absorbed material from the digestive system determines its benefits and costs to the consumer. We discussed earlier how secondary metabolites (SMs) and/or anthropogenic toxicants are nearly ubiquitous in animal foods (chapter 2), and we briefly reviewed the evidence that there are pathways for their absorption whether they are lipo- or hydrosoluble (chapter 4). Some of the mechanisms described in this chapter can reduce the absorption of SMs (Dearing et al. 2005). The foreign chemicals, or xenobiotics, that are absorbed may not yield any energy and they cannot be used as building blocks. In this chapter we discuss how ingestion of xenobiotics may drain energy and/or essential nutrients as the animal expels them from the body. Through a variety of mechanisms, including the induction of specific enzymes that biotransform xenobiotics, animals attempt to adjust elimination rates to match absorption rates. However, the energy and material lost in the course of xenobiotic elimination must be replaced, which increases resource demands. If elimination rate is less than absorption rate the xenobiotic level can build up in the body, which can lead to harmful effects, or toxicity.

Some ingested xenobiotics from plant, animal, or anthropogenic (manmade) sources may be toxic, but a simple definition of this word, or of "toxins" or "toxicants," does not come easily. As expressed five hundred years ago by Paracelsus (1493–1541): "All substances are poisons; there is none which is not a poison. The right dose differentiates between a poison and a remedy" (Deichmann et al. 1986). Most chemicals have the potential to produce injury if ingested in sufficient amount. In this chapter we will consider toxins to be naturally occurring, and toxicants manmade, chemicals that can cause harmful effects at relatively low doses. The physiological responses of animals to even low levels of toxicants sometimes tell us much about their exposure to contaminants in the ecological setting.

9.1 Overview: The Postabsorptive Fate of Absorbed Xenobiotics

9.1.1 Compartmental Models Are Useful for Describing Postabsorptive Processing of Xenobiotics

Just as we did for nutrients, we might imagine that a xenobiotic that is absorbed from the gastrointestinal tract mixes with some of the same material that is already in the body (figure 9.1). The body's pool also receives material from other absorptive pathways such as the dermis or respiratory organs (gills, lungs). Sometimes a central pool exchanges material with a peripheral pool (e.g., fat stores in the case of lipophilic compounds). Besides incorporation into such a peripheral reservoir, material may also leave the central pool by conversion to other molecular types (biotransformation) and by loss or excretion from the body. We will first use this simplified conceptualization of absorption, distribution, biotransformation (metabolism), and excretion to organize our discussion about processing of xenobiotics. Later, we will use such a compartmental model to develop a simple mathematical model that predicts steady-state levels of the xenobiotic in the body, which is an important step in predicting toxicity.

Figure 9.1. After absorption, a xenobiotic is distributed among one or more compartments in the animal, possibly biotransformed to a metabolite, and ultimately excreted as the parent compound and/or metabolite.

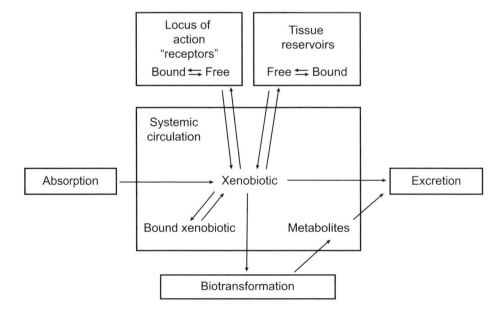

9.2 Distribution of Xenobiotics in the Body

9.2.1 Absorbed Xenobiotics Are not Necessarily Distributed Evenly throughout the Body

Besides distribution into body water, or one of its three compartments, plasma, interstitial tissue, or intracellular water, consider some important ways that absorbed chemicals might be differentially distributed:

- Several plasma proteins reversibly bind organic xenobiotics by molecular interactions such as ionic bonds, hydrogen bonds, hydrophobic interactions, and van der Waal's forces. The most important of such proteins in vertebrates is albumin because it is the most abundant protein in plasma. Ninety-nine percent of some organic xenobiotics, such as the insecticide dieldrin, may be reversibly bound to plasma proteins.
- Some elements, such as lead, may substitute for calcium and accumulate in the hydroxyapatite lattice matrix of bone by exchange-absorption reactions.
- Lipophilic compounds will partition into fattier tissues where concentrations can become much higher than in other tissues in the body. Typically, lipophilicity is measured as the n-octanol/water partition coefficient (K_{ow}, or its log value). This coefficient compares the concentration of the compound in either of two solvent phases in a separatory funnel, octanol and water. Octanol is used because its solvent properties are similar to those of biotic lipid.
- Metals sometimes accumulate to high levels in certain tissues because they bind to tissue components and not because they have high K_{ow}'s. For example, more than 90% of cadmium, copper, lead, and zinc is contained within the hepatopancreas of woodlice (an isopod), although the organ comprises less than 10% of the animal's body mass (Walker et al. 1996).

These instances of uneven distribution of xenobiotics can have ecological implications. Let us consider some implications of sequestration.

9.2.2 The Ecological Significance of Xenobiotic Sequestration

Sequestration of xenobiotics can confer protection from predators. This has been well studied in insects, in which scores of species have been shown to sequester and store at least ten classes of toxins from their plant foods and then benefit from increased avoidance by their predators (Harborne 1993). A classic example of this is the interaction of milkweeds (*Asclepias* spp.), monarch butterflies

(*Danaus plexippus*), and their avian predators (Brower et al. 1968). The plants produce several cardiac glycosides (a type of terpenoid compound) which deter feeding by many insect species. The monarch caterpillar, however, feeds on the plants, sequesters the compounds in its body, and is relatively insensitive to them due to a modified Na^+–K^+ ATPase (Brattsten 1979). Following metamorphosis, the adult monarch butterfly contains the cardiac glycosides in its abdomen and wings. Naive birds that attack the adult get a mouthful of the bitter-tasting chemical and/or sensations from its effects, which causes them to vomit, and they associate the coloration of the butterfly with the experience and avoid them as prey subsequently.

If an animal sequesters a toxicant in a nonsensitive compartment of the body it can tolerate more of the chemical, but this may also increase transfer of toxicant to higher trophic levels. The sequestration of metals by metallothioneins provides good examples. Metallothioneins are medium-molecular-weight (6000–10,000 daltons) single-polypeptide chains, rich in cysteine and poor in aromatic amino acids and histidine, that have high affinity for metals (e.g., Zn, Cd, Cu, Hg). They occur in the cytosol in many tissues, but especially in cytosol of the liver of vertebrates and midgut gland or hepatopancreas of invertebrates (Roesijadi 1992). Transgenic or metallothionein-null mice that lack this cytosolic protein are ten times more sensitive to cadmium toxicity than normal mice. In many invertebrates, heavy-metal tolerance correlates with levels of metallothionein. Consistent with the idea of resistance, in many vertebrates and invertebrates the levels of metallothionein and/or its mRNA increase several times following exposure to elevated levels of certain metals (figure 9.2). Because the protein leads to metal sequestering, high levels of metal may be passed on to the animals' predators. For example, grass shrimp (*Palaemonetes pugio*) feeding on Cd-enriched *Artemia salina* absorbed 30–40% of the Cd, mainly the cytosolic fraction. Due to induction of metallothionein in the shrimp, a sixfold increase in Cd content of their prey was magnified into a 33-fold increase in their own Cd content, which can then be passed on to their predators further along the trophic chain (Wallace et al. 2000). Furthermore, the shrimp exposed to metal have reduced ability to avoid predators, and so their increased predation risk acts synergistically with the magnification of Cd content to increase the trophic transfer of Cd.

The tendency of metallothionein to bind metals, and to be induced by certain metals, has led to the suggestion that its measurement in some species may be useful in the environmental monitoring of metals. But success in this approach requires thorough understanding of metallothionein dynamics, because in some species their synthesis depends also on temperature, other organic environmental contaminants, and stressors such as infection and starvation (Berger et al. 1995).

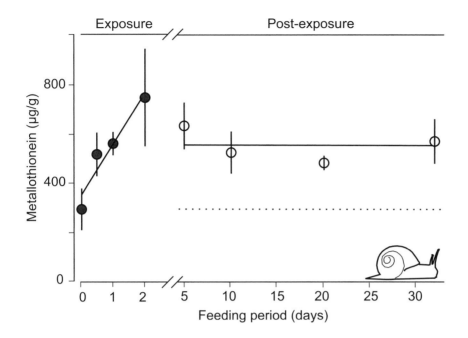

Figure 9.2. Metallothionein (MT) content in the midgut glands of Roman snails increases following feeding on heavy metals. During the first two days ("Exposure"; filled circles) the snails were fed agar with added cadmium (Cd) followed by a 30-d period ("Post-exposure", unfilled circles) during which they were fed uncontaminated lettuce. Even 30 post-Cd-feeding the MT content remained elevated above that of controls (the dotted line). (Redrawn from Berger et al. 1995.)

Sequestration of xenobiotics to increase tolerance, as in the case of metallothioneins, occurs in some but by no means the majority of animals. A far more general response is enzymatic biotransformation of absorbed xenobiotics.

9.3 Biotransformation of Absorbed Xenobiotics

9.3.1 Two Major Phases of Biotransformation Progressively Increase the Water Solubility of Xenobiotics

Compounds are more likely to be excreted if they are water soluble because they may be simply filtered by the kidneys and excreted dissolved in watery urine, or because they are more likely to interact with special membrane-bound transport proteins that specifically bind and remove compounds from blood or other watery fluids. In most organisms, lipophilic xenobiotics are often converted into more polar, hydrophilic compounds through either of two types of reactions called phase I and phase II reactions, or through both types acting in sequence (Walker and Ronis 1989; Livingstone 1991) (figure 9.3). Phase I reactions (also called functionalization) increase polarity and may prepare the compound for phase II, which involves conjugation to endogenous substrates. Presumably, many of the enzyme systems involved in these reactions were

Figure 9.3. General pathways of xenobiotic biotransformation increase the polarity and water solubility of xenobiotics, which speeds their excretion. An example for phenol is shown. Phase I pathways involving cytochrome P450 enzymes (P450s), microsomal epoxide hydrolase (mEH), or NAD(P)H quinine oxidoreductase (QOR) form more reactive metabolies. Phase II pathways involve conjugating these metabolites to endogenous substrates such as glucose through the action of UDP-glucuronosyltransferase enzymes (UGTs) or amino acids through the action of γ-glutamylcsyteine synthase (YGCS) and glutathione-S-transferase (GSTs). (Adapted from Lamb et al. 2001.)

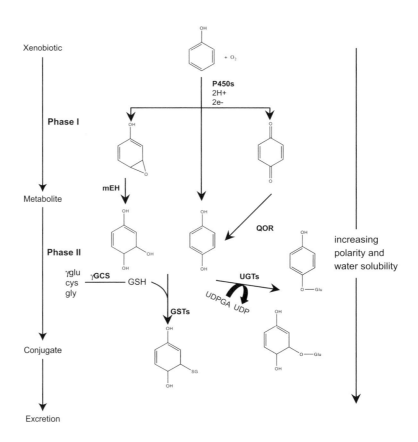

selected for over evolutionary time because they function in the normal biochemical processing of endogenous lipophilic chemicals (e.g., steroidal hormones) and because they reduce the toxicity and enhance the clearance of naturally occurring xenobiotics in the foods of herbivores and predators (Jondorf 1981). Many biotransformation pathways have been described (Scheline 1991), and the example shown in figure 9.3 illustrates only a few possible scenarios for biotransformation of a simple xenobiotic, phenol. This was a point well made in a study in which this compound was administered to 21 different mammalian species which showed different patterns of elimination (Baudinette et al. 1980). In some species, like the mouse (*Mus musculus*), nearly a third of the phenol was eliminated after phase I as an oxidized metabolite prior to conjugation, whereas in some species, like the rat (*Rattus norvegicus*), nearly all was eliminated after phase II in a conjugated form. All the species eliminated glucose-derived conjugates formed by UDP-glucuronosyltransferase enzymes (UGTs), but the proportion of phenol eliminated by this pathway ranged from nearly all (95% in koalas *Phascolarctos cinereus*) to very little (13% in the carnivorous marsupial

Dasyurus viverrinus), with no apparent pattern according to phylogeny or habitat. Thus, the generality of the two-phase process can overshadow a lot of variation among species in reliance on the various pathways. In humans, significant variation occurs even among individuals or different populations in biotransformation activity, and influences their relative sensitivity to drugs (Service 2005).

9.3.2 Phase I Reactions Tend to Increase the Reactivity and/or the Polarity of Xenobiotics

Typically, reactive groups such as -COOH, -OH, $-NH_2$, or SH are created by phase I reactions. The types of reactions include oxidation, hydrolysis, hydration, reduction, epoxidation, and dehalogenation (figure 9.4). Many of the enzyme catalysts are membrane-bound proteins located on smooth endoplasmic reticulum (SER) in many tissues, but especially those in more direct contact with the environment such as gills or lungs, gut, hepatopancreas (invertebrates), and liver (vertebrates) which receives blood directly from the gut via the hepatic portal vein. The activity of these enzymes is measured in extracted SER, which forms minute membrane vesicles called microsomes (Lake 1987) and which impart a common name given to these enzymes—microsomal monooxygenases (MOs) (methods box 9.1). This system of enzymes is referred to by two other names as well. One of the names is "mixed function oxidases" (MFOs), because oxidation reactions are particularly common and they are characterized by low substrate specificity. The enzymes are also called cytochrome P450s (CYP isoenzymes), named after the gene superfamily that initiates formation of many of the MOs. The P450 superfamily contains more than 100 genes and is widely found in plants, animals, and prokaryotes. There are other extramicrosomal phase I enzymes such as esterases, epoxide hydrolases, and nitroreductases (figure 9.4).

Variation among species in the concentration and activity of phase I enzymes has been explained in terms of varying ability or requirements for detoxication of xenobiotics. For example, the larvae of parsnip webworms (*Depressaria pastinacella*), a species which specializes on furanocoumarin-rich parsnip plants, have P450 MO activity against furanocoumarins that is 10–300 times higher than in other lepidopterans that have been studied (Berenbaum et al. 1992). MOs are generally higher in mammals and birds than in fish (figure 9.6), and this has been explained as a result of the lower dependence of fish on metabolic detoxication; thus there is less selection pressure for high biotransformation activity, because fish eliminate lipophilic compounds into water by passive diffusion across the gills (Jondorf 1981; Walker 1994). There is at least one other alternative explanation. Mammals and birds are endotherms and fish

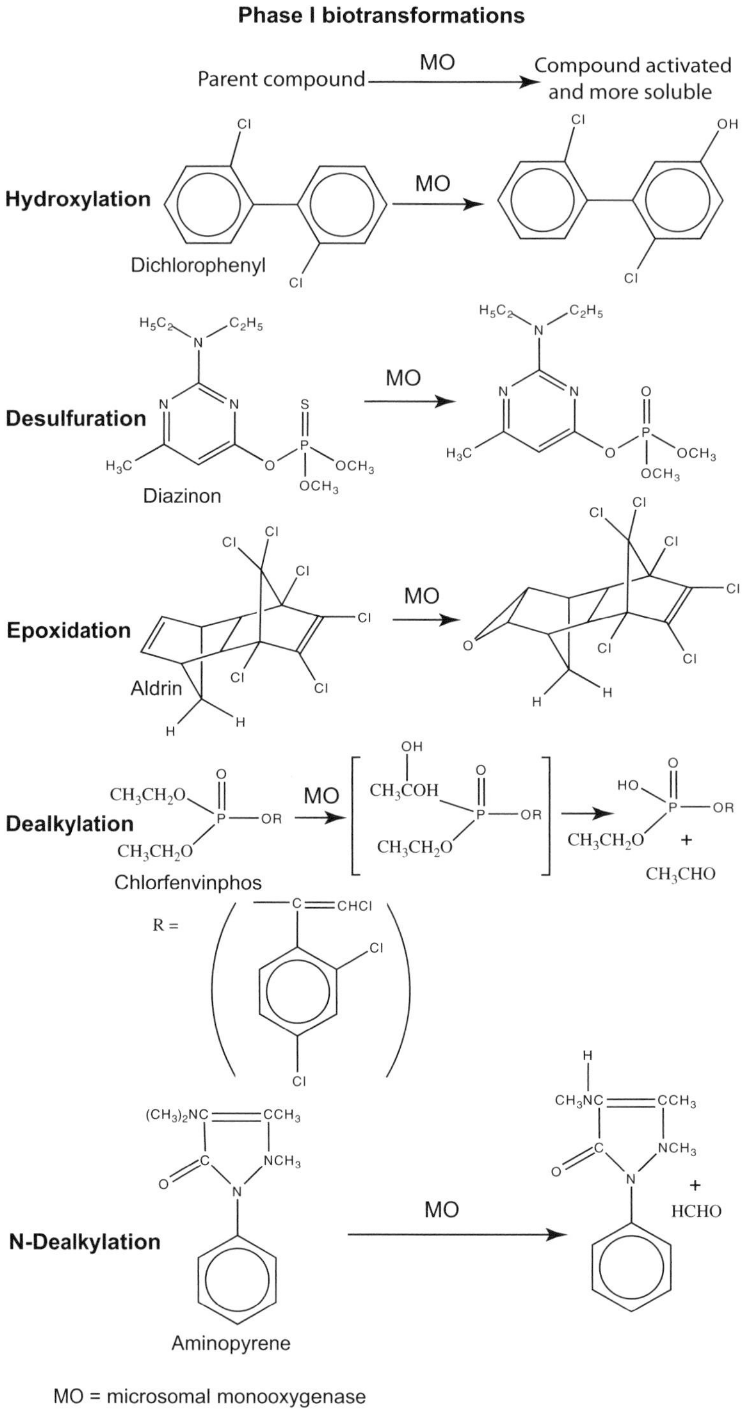

Figure 9.4. Phase I biotransformations, such as those shown here, activate the parent compound and make it slightly more water soluble. In these examples the parent compounds are manmade, but the same biotransformations act on naturally occurring toxins. MO = microsomal monoxygenase. (Adapted from Walker et al. 1996.)

Phase I biotransformations

MO = microsomal monooxygenase

Box 9.1. The Extraction, Measurement, and Nomenclature of Cytochrome P450 Monooxygenases

These enzymes are the most studied enzymes of phase I biotransformation and they are often used by ecologists to suggest an animal response to organic xenobiotics. Therefore, it is worth knowing a little about their nomenclature and how they are studied. To illustrate, lets follow the disposition of samples from leopard frogs, *Rana pipiens* (Huang et al. 1998, 2001).

Extraction

The activities of these enzymes can be measured in tissue homogenates but most studies isolate the particular cellular fraction where the activity resides, which is the endoplasmic reticulum (ER). Taking the frog liver tissue as an example, the liver was removed from the animal as soon as it was killed. The liver was minced and the cells were disrupted by careful grinding without generating excess heat, to preserve enzymatic activities. The broken fragments of ER membrane resealed to form small closed vesicles called microsomes. The microsomes were separated from all the other cellular fragments in the slurry through serial ultracentrifugation, in which the preparation and derivatives from it are rotated at increasingly higher speeds for longer periods of time. At relatively low speeds large fragments but not microsomes sediment to form a pellet which is discarded leaving behind a supernatant containing the microsomes. At even higher speeds for longer times (e.g., above 100,000 g for more than 1 h) the microsomes form a sediment, which is then isolated and frozen. Because isolated cytochrome P450 (CYP) enzymes are very sensitive to temperature, the above procedures need to be performed very carefully in icy conditions.

Measurement of Activity

An assay is performed on thawed microsomes and the rate of product formation is determined. In our frog example, the assay was the O-deethylation of the model substrate ethoxyresorufin by the enzyme ethoxyresorufin O-deethylase, whose activity is expressed by the acronym EROD. The formation of the reaction product was measured with a fluorescence spectrophotometer. Another common assay is the aryl hydrocarbon hydroxylase hydroxylation of benzo[*a*]pyrene, whose results are expressed as AHH activity. Substrates can be manmade or natural xenobiotics. Activity is typically expressed per unit of protein in the microsomal preparation (e.g., measured by the Lowry method). In the leopard frogs, EROD activity observed was about six times higher in frogs dosed with polychlorinated biphenyls compared with undosed controls. If activity is blocked by piperonyl

continued

continued

butoxide, a potent inhibitor of P450s, the activity can be more directly ascribed to monooxygenases (Snyder and Glendinning 1996). Alternatively, more direct tests for the enzymes can be made, as described next.

Direct Measurement of Expression of Protein Itself

Purification of some of the some of the particular enzymes has allowed production of polyclonal and monoclonal antibodies that can measure the enzyme directly using Western blot and enzyme-linked immunosorbent assays (ELISA). You can consult cell biology texts (e.g., Alberts et al. 1989) for methodological details. In our frog example, solubilized protein from the microsomes was electrophoresed through polyacrylamide gel and then transferred (or "blotted") to nylon paper. The nylon paper was exposed to a monoclonal antibody from a fish (scup; *Stenotomus chrysops*) against a specific monooxygenase called CYP1A that is known to catalyze EROD. The antibody (i.e., MAb 1-12-3) was coupled to a fluorescent dye, so the presence and relative amount of the protein in the original microsome sample could be determined (figure 9.5). Notice that the large difference in EROD activity (above) is paralleled by a large difference in amount of the protein CYP1A.

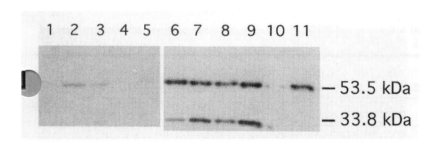

Figure 9.5. Immunoblots of liver microsomes from frogs exposed to PCB (lanes 6–9) show more staining for the monooxygenase CYP1A (~53–54 kDa) than microsomes from control frogs not exposed (lanes 1–5). Lanes 10 and 11 are standards which are microsomes with known amounts of CYP1A (respectively, 0.25 and 0.75 pmol). The bands at 33.8 kDa might reflect degraded CYP1A. (From Huang et al. 2001.)

One final interesting use of the antibody was that formalin-fixed slices of tissue from many organs of the frog were exposed to the antibody and analyzed immunohistochemically for the presence of CYP1A. The relative staining intensity in the tissues reflected relative amounts of the monooxygenase, providing data on where it occurred throughout the frog's body. Not surprisingly, tissues such as liver, intestine, stomach, and kidney expressed considerable CYP1A protein. A method like this may be useful to ecologists who can fix samples in formalin but cannot prepare and freeze microsomes under field conditions. The localization of CYP1A and other enzymes may help locate the sites and cell types where detoxification, bioactivation, and toxicity may occur. Retrospective studies using well-preserved historic samples may reveal whether organisms were

continued

continued

exposed to xenobiotics. In addition, cDNA probes have also been developed that can detect newly synthesized mRNA that codes for CYP1A (Haasch et al. 1989).

Nomenclature

There are many cytochrome P450 monooxygenases and their genes are grouped into gene families and subfamilies. The names of the genes and their products (mRNA and protein) are based on the root CYP (cytochrome P450), and given numbers and letters that designate the particular family (e.g., CYP1 vs CYP2), subfamily (e.g., CYP1A vs CYP1B), and gene (e.g., *CYP1A1* vs *CYP1A2*). The DNA name is italicized to distinguish it from the mRNA or protein. (from Newman 1998).

are ectotherms. These contrasting hypotheses might be evaluated in the future by including data from terrestrial ectothermic vertebrates such as reptiles. Another very intriguing pattern is that MO activities appear to be higher in omnivorous vertebrates that eat plant materials than in carnivorous vertebrates that eat mainly vertebrate flesh (Fossi et al. 1995) (figure 9.6).

While this seems consistent with the premise that there is an evolved match between many MOs and the presence of natural toxins in the diet, the issue deserves further study by careful comparison of the same enzyme systems in diverse vertebrates using appropriate phyogenetically informed comparative methods (see chapter 1).

Besides natural toxins, many anthropogenic chemicals can also be biotransformed by these enzyme systems, although some, such as chlorinated organic compounds, are relatively resistant. Sometimes, phase I biotransformation leads to an increase in toxicity, and some pesticides were synthesized with this effect in mind. For example, when pesticides such as malathion, chlorpyriphos, and diazinon undergo oxidative desulfuration (figure 9.4) the compounds formed are active inhibitors of acetylcholinesterase, which can thus disrupt proper nervous signal transmission across synapses. As another, natural, example, the poisoning of vertebrate herbivores by pyrrolizidine alkaloids from *Senecio* spp. (the weeds ragwort and groundsel) apparently occurs because hydrolysis of an ester group of the alkaloid creates a more toxic compound that binds to macromolecules like DNA. In this case, the relative toxicity of pyrrolizidine alkaloids in domestic herbivore species is positively correlated with the relative capacity of liver P450 enzymes (Cheeke 1994).

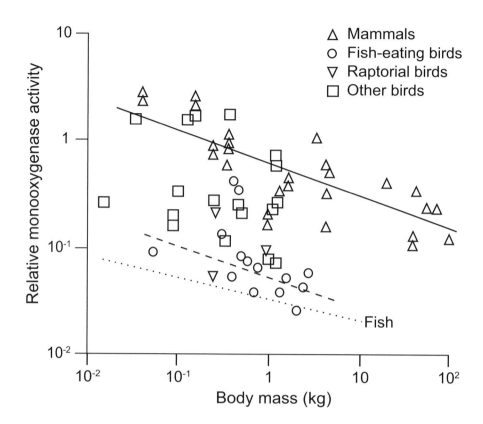

Figure 9.6. Mass-specific monooxygenase activity scales negatively with body mass and is higher in endothermic mammals and birds than in ectothermic fish. The data are activities of hepatic microsomal monooxygenases to a variety of substrates, expressed in relation to body mass. Each point represents one mammalian or avian species (males and females are sometimes entered separately). The dotted line shows average values in ectothermic fish in relation to the endothermic mammals (solid line) and fish-eating birds (dashed line). (Redrawn from Walker et al. 1996.)

MO activities are particularly low in molluscs, which has made them useful to humans in environmental monitoring programs such as the International Mussel Watch Programme. Because molluscs filter large quantities of water and particulate matter, but do not extensively biotransform and eliminate accompanying toxins, they are routinely collected and analyzed for anthropogenic chemicals that tend to bioaccumulate.

9.3.3 Phase II Reactions Conjugate Endogenous Substrates to Xenobiotics and Increase Their Solubility

The phase II reactions link (conjugate) the phase I metabolites to sugar derivatives, amino acids, peptides, and sulfate. Most of the conjugates are negatively charged (anions), which greatly increases water solubility and generally reduces toxicity. Important phase II enzymes include UDP-glucuronosyltransferase (UDP-GT), glutathion S-transferases (GSTs), and sulphotransferases (figure 9.7). The first enzyme, UDP-GT, is located in the endoplasmic reticulum, whereas the latter two enzyme systems are generally found in cytosol. The tissue distribution

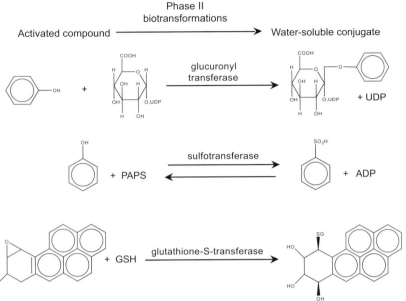

Figure 9.7. In phase II biotransformations, such as those shown here, endogenous substances such as glucuronic acid (top reaction), sulfate (middle reaction), and glutathione (bottom reaction) are conjugated to active sites on metabolites of xenobiotics, rendering the entire conjugated complex more water soluble. In the examples, glucuronic acid is supplied as uridine diphospfate (UDP)-glucuronic acid and sulfate is transferred from phosphoadenine phosphosulfate (PAPS). (Adapted from Walker et al. 1996.)

of phase II enzymes is similar to that of the phase I enzymes, which makes for an integrated system that adds or unmasks a functional group (phase I) and then conjugates this group with a phase II moiety. As was the case for phase I enzymes, variation among species in the concentration and activity of phase II enzymes has been explained in terms of varying requirements for detoxication of xenobiotics. For example, the glucuronosyltrasferase activity of dogs (canids) is high relative to that of cats (felids) (Walker 1994). Because the canids are more omnivorous than the generally carnivorous felids, this seems consistent with the premise that the enzymes' activity was selected for by the presence of natural toxins in the diet. If you are a pet owner the difference could be important, because it makes the cats more susceptible to toxic effects of salicylates like acetaminophen (Tylenol).

Phase II biotransformations are biosynthetic and utilize energy and material (figure 9.7). The UDP-GTs catalyze the conjugation of substrates (especially those with hydroxyls and amino groups) to glucuronic acid, a glucose derivative. It seems to be replaced in insects by conjugation to glucose (Brattsten 1979). GSTs catalyze the conjugation of substrates (especially epoxides produced by MOs) to glutathione, an endogenous tripeptide composed of the amino acids glutamate, cysteine, and glycine. Hydroxy or amino compounds can be conjugated with sulfate by sulfotransferases. The measurement of these conjugates in excreta offers a convenient method for estimating the energy and material detoxication costs of vertebrates and insects exposed to xenobiotics (below).

Another cost may be related to the fact that conjugation can also convert compounds that are neutral or weakly acidic into stronger organic acids that can have negative effects on metabolism (below). As with phase I processes, there can be considerable interspecific variation. As an example, the simple phenolic ring of benzoic acid is excreted mainly as a glycine conjugate by mammals, amphibia, fish and insects, as the ornithine conjugate by birds and reptiles, and as the arginine conjugate by arachnids and myriapoids (Harborne 1993).

9.3.4 Levels of Some Phase I and II Enzymes Are Adaptively Modulated

Some of the biotransformation enzymes are inducible by exposure to xenobiotics. This explains why exposure of animals to sublethal levels of toxicants may increase their tolerance to the chemicals (Snyder and Glendinning 1996), and it has stimulated the effort to determine the feasibility of using these enzymes in environmental monitoring.

The detoxication of the furanocoumarin xanthotoxin by the parsnip webworm, (*Depressaria pastinacella*) provides a particularly well-studied natural example. The species feeds only on three species in the plant family Apiacea, all of which contain xanthotoxin, a compound that crosslinks DNA in the presence of UV light and which is toxic to most insect herbivores but not to the parsnip webworm. In webworms, exposure to xanthotoxin increased by at least three times the midgut microsomal MOs that metabolize the allelochemical (Berenbaum and Zangerl 1994). Webworm larvae fed radiolabeled xanthotoxin metabolized 95% of the chemical (Nitao 1990). Notwithstanding this fine example, the importance of upward regulation of phase I and phase II enzymes in the adjustment of animals to novel plant foods has been relatively unstudied.

The mechanism of induction for one major MO in vertebrates, cytochrome P4501A1 (CYP1A1), has been extensively studied. Cells in tissues such as liver contain a soluble protein known as the Ah (aryl hydrocarbon) receptor, which binds to incoming inducing agents. The complex of receptor-inducing agent ultimately enters the nucleus, attaches to specific sites on DNA, and results in transcription of mRNA that codes for CYP1A1. The induction can be rapid and additional CYP1A1 can be detected in only a few hours after exposure. CYP1A1 levels have been observed to increase 10–100 times. A number of chemicals are inducing agents of CYP1A1, including polycyclic aromatic hydrocarbons (PAHs), and certain congeners of polychlorinated biphenyls (PCBs), polychlorinated dibenzo-*p*-dioxins (PCDDs), and polychlorinated dibenzofurans (PCDFs). The most potent inducers are chemicals that have a particular structure that interacts well with the Ah receptor, and they may or may not be effectively biotransformed by the CYP1A1 gene product that they induce. The various congeners of PCBs, PCDDs, and PCDFs have been ranked according to their relative potency for inducing CYP1A1, and researchers use the rankings to calculate indices of toxic equivalency of mixes relative to the most potent inducer, the dioxin 2,3,7,8-TCDD (tetrachlorodibenzo-p-dioxin). The indices are called toxic equivalency factors or concentrations (TEFs or TEQs) (Safe 1990).

The induction of CYP1A1 is used to monitor exposure of wild birds, mammals, and fish to industrial contaminants such as PAHs, PCBs, PCDDs, and PCDFs (Rattner et al. 1989; Whyte et al. 2000). Direct measurement of these chemicals may be difficult in water or tissues because of very low levels, and it is usually extremely expensive. But the assays for CYP1A1 mRNA (using Northern blots) or protein (using Western blots) or enzyme activity (measured in microsomal fractions; see box 9.1) are cheaper, and thus elevated levels in animal tissues may serve as an economical biomarker of exposure to the inducing organic contaminants. As an example, both contaminant levels (PCBs + PCDDs +PCDFs) and CYP1A1 activity (measured as EROD and PROD activity) were elevated in the progeny of common terns (*Sterna hirundo*) eating fish contaminated with the polyhalogenated hydrocarbons (figure 9.8). Researchers relying on this and other biomarkers need to be aware that biotransformation enzymes can also be induced by numerous other factors including temperature and endogenous hormone titers.

Exposure of mammals to various inducing agents can also result in elevation of phase II enzymes (Dearing et al. 2005). Studies of this sort in fish have

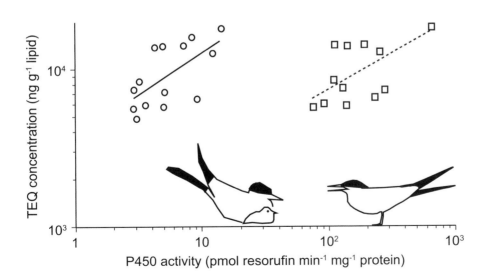

Figure 9.8. Cytochrome P450 monooxygenase activity is elevated in tern nestlings that are exposed to higher levels of PCBs, PCDDs and PCDFs. Eggs were collected from nesting colonies near the North Sea in the Netherlands, hatched in the laboratory, and then chemical residues and liver enzyme activities were measured in the nestlings. For the chemical mixes in the tissues, a toxic equivalency concentration (TEQ) was calculated (see text). In liver microsomes enzyme reaction rates using two substrates were measured: 7-ethoxyresorufin O-deethylase (EROD) and 7-pentoxyresorufin O-depentylase (PROD). Notice that for an approximately sixfold increase in exposure to contaminants there was approximately a tenfold increase in enzyme activity. (Redrawn from Bosveld et al. 1995.)

yielded mixed results. The Ah receptor appears to be lacking in marine invertebrates studied so far, and so treatment of molluscs and crustaceans has yielded
little difference in levels of phase I or phase II enzymes.

9.3.5 Symbiotic Microbes Can Be Agents of Biotransformation

There is a general notion that foregut fermenting herbivores are often more
resistant to plant toxins than hindgut fermenters because of the degradation or
inactivation of toxins in the foregut (Cheeke 1994). Although we do not know
of a rigorous survey demonstrating this, there are many studies that show that
specific microorganisms detoxify specific plant secondary metabolites (Cheeke
1994; Foley et al. 1999). We discuss this issue more extensively in chapter 6
(section 6.3.10). One of the most interesting examples is that of mimosine, a
toxic amino acid in a tropical legume (*Leucaena leucocephala*), which is degraded
by a specific rumen microorganism in tropical cattle in Indonesia. Mimosine
toxicity in cattle depends on whether the microorganism occurs in the rumen,
and once it was introduced into Australian cattle they were able to utilize the
legume without suffering toxicity (Cheeke 1994).

9.4 Elimination of Xenobiotics and Their Metabolites

9.4.1 The Kidney, Liver, and Gut Are Major Routes of Elimination

In vertebrates, many water-soluble parent xenobiotics and their metabolites
are excreted through the kidney, whose functional excretory unit is the
nephron (figure 9.9). Water-soluble compounds of sufficiently low molecular
weight are passively filtered from blood across the glomerular capillary wall,
which resembles an ultrafilter with a diameter of 75–100 Å. Compounds
bound to plasma protein (section 9.2.1) are too large to be filtered. Additionally,
organic acids and bases are selectively transported via mediated, energy-depend-
ent mechanisms in the kidney tubule (mainly proximal; see below). Further
modification of the filtrate in the tubules occurs by passive diffusion. For
example, lower pH in the distal tubule can cause protonation of organic acids
($A^- + H^+ \rightarrow AH$), and this more lipid soluble form can be reabsorbed passively
back into blood.

The hepatic lobule is the functional unit of the liver involved in excretion
(figure 9.10). Hepatocytes absorb xenobiotics and their metabolites across the
surface facing the blood, possibly biotransform them, and then secrete metabolites

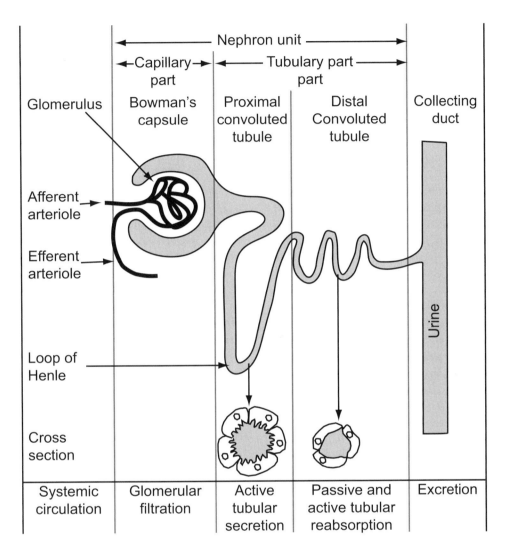

Figure 9.9. Schematic of the nephron, the basic structural/functional unit in the kidney of vertebrates. Xenobiotics that are not too large are filtered from blood in the glomerulus into the Bowman's capsule. As the filtrate proceeds through the proximal and distal convoluted tubules, the filtrate is further modified by reabsorption of material or secretion of other compounds including additional xenobiotics, or their metabolites. The modified filtrate enters the collecting duct, where additional water may be absorbed, and is conveyed to the bladder. The long loop of Henle is present in mammals and in some birds. (Modified from Ritschel 1999.)

back into blood or into the bile canaliculus across their other surface. The canaliculi drain into bile ductules and eventually into the bile duct which connects with the intestine. Hence, metabolites might be further biotransformed by intestinal microbes and either eliminated in feces or reabsorbed from the intestine and recirculated to the kidney or liver. Enterohepatic circulation, essentially multiple cycles of the xenobiotic and its metabolite(s) through the intestine and liver, increases persistence of a chemical in the body.

Whether excretion of a compound occurs via kidney or liver (into the gut) depends partly on molecular weight. Most excreted conjugates are organic anions, and in a number of mammalian species that have been studied (rodents, rabbit, human) if molecular weights are greater than 400 daltons then excretion

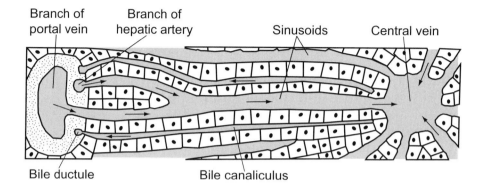

Branch of portal vein   Branch of hepatic artery   Sinusoids   Central vein

Bile ductule   Bile canaliculus

Figure 9.10. Schematic of the liver lobule, the basic structural/functional unit in the liver of vertebrates. Blood from the intestine carrying nutrients and xenobiotics enters via the portal vein and mixes in sinusoids with more oxygenated blood from the hepatic artery. Liver cells (hepatocytes) that line the sinusoids may absorb material from the blood, or secrete into it, across their apical membrane that faces the sinusoid. The basolateral membrane of each hepatocyte may secrete xenobiotics, or their metabolites, into the bile canaliculus. The canaliculus is drained by the bile ductule that conveys the material to the gall bladder and/or intestine. (Modified from Kappas and Alvares 1975.)

is mainly through the liver into feces and if less than 300 then excretion is mainly through the kidney into urine.

A number of classes of membrane transport proteins, sometimes called phase III mechanisms, mediate the elimination of xenobiotics from kidney tubule cells, hepatocytes, and intestinal cells. Experimental studies of membrane transport in both vertebrates and invertebrates often use p-aminohippurate (PAH) as a model organic anion (or weak organic acid that exists as an anion at physiological pH) and tetraethylammonium (TEA) as a model organic cation (or base). Net tubular secretion of PAH and TEA by the tubule cells of the vertebrate kidney is a mediated (i.e., saturable), energy-requiring process that appears to involve carrier proteins in the membranes facing both the peritubular fluid and the tubule lumen (Braun and Dantzler 1997). Similar mechanisms have been identified in invertebrate excretory organs such as Malpighian tubules of insects, antennal glands of crustaceans, and the kidney of snails (O'Donnell 1997).

Multidrug resistance proteins MRP2 mediate the ATP-dependent transport of amphipathic (i.e., containing both polar and nonpolar groups; chapter 2) anionic conjugates with glutathione, glucuronate, or sulfate. MRP2 is found in the apical membrane of the liver, proximal tubule, and intestinal cells that were identified above as important sites of phase I and phase II biotransformation and important routes of elimination (figure 9.11). Another type of transporter, called P-glycoprotein, is strictly speaking not a biotransformation mechanism because it can bind and transport lipophilic xenobiotics even prior to biotransformation. It can serve as a "first line of defense" against xenobiotics because P-glycoprotein in the apical membrane of intestinal cells can prevent toxic compounds from entering in the first place. Its mechanism depends on relatively nonspecific binding to lipophlic compounds as they diffuse through the membrane, and it uses ATP to pump them out. These proteins have been identified in many organisms from marine invertebrates to mammals (Epel 1998). Knowledge about these transport

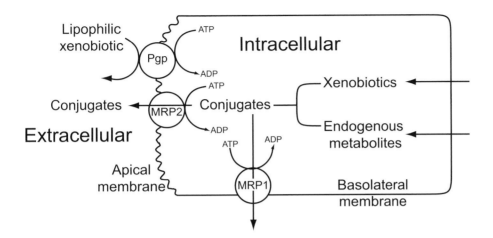

Figure 9.11. In intestinal and liver cells the elimination pathways for conjugated xenobiotics (or endogenous compounds such as bile salts) include multidrug resistance proteins in the apical membrane (MRP2). The apical membrane may also contain P-glycoprotein, which can prevent toxic compounds from entering in the first place (Epel 1998). These transporters (all indicated by circles embedded in membrane) utilize ATP to move substrates. (Adapted from Keppler et al. 1997.)

proteins has mainly accumulated in just the past decade, and there are many interesting discoveries yet to be made about variation in their function within and among species and their evolution (Green et al. 2005).

Another mode of elimination mediated via the gut occurs in some invertebrates. Metals accumulate in one of three types of granules in digestive or hepatopancreas cells and apparently are not remobilized (Walker et al. 1996). In spiders the digestive cells are sloughed and the metals are excreted subsequently in the feces. In isopods the elimination may not occur, as some of the hepatopancreas cells are permanent and so the metals accumulate throughout life, sometimes to very high levels (see 9.2.1 and 9.2.2).

9.4.2 Other Routes of Elimination

Routes of elimination associated with reproduction deserve mention because of their quantitative importance and utility in environmental monitoring. The vivi- or oviparous production of progeny, and their subsequent nurturing, may coincide with substantial elimination of xenobiotics. Lipophilic compounds especially pass into and accumulate in the developing fetus or egg yolk. For example, 20–30% or more of body burdens of PCBs or Hg are eliminated during egg laying in birds and amphibians. In one study on whales, it was estimated that 8% of PCBs ingested were exported to the offspring at birth, and a whopping 54% during lactation (Hickie et al. 1999).

Hair and feathers are sometimes important sites of elimination of heavy metals such as mercury. Levels in feathers correspond to levels in other body compartments such as blood (Bearhop et al. 2000). During molt, accumulated mercury is transferred from target organs into the growing feathers, and the body burden gradually falls. Feathers have become a favored tissue used in monitoring because they can be collected with little harm to the animal, represent the major route of elimination for some metals, such as mercury, and can contain 70% of the body burden. As an example of their use, the historical increase of mercury in the ecosphere due to increased human Hg inputs over the past 150 years (e.g., atmospheric Hg released from burning coal) is reflected in the temporal increase in mercury concentrations in feathers of Manx shearwaters (*Puffinus puffinus*) sampled in the seas near the United Kingdom (figure 9.12). Conceivably, data like these could also provide information about variation in exposure of birds to heavy metals across vast geographic areas. At one site in Trinidad, scarlet ibises unloaded so much mercury during molt at their roosts that they polluted the local environment (Klekowski et al. 1999)! They probably acquired the mercury during their annual migration to polluted wetlands in South America.

9.5 Costs of Xenobiotic Biotransformation and Elimination

Ingestion of xenobiotics is associated with losses of energy and material from animals. In some studies, addition of plant secondary metabolites (SMs) to the diet reduced the proportion of energy intake that could be allocated to production or respiration by as much as 30% (table 9.1). The exact mechanisms are in most cases unknown, but, as described below, the forms of the loss could include reduced net absorption of energy and material from the gut, increased excretion of energy and materials via urine, and increased heat production. The possible animal responses are compensatory increases in food intake or switching to alternative foods lacking xenobiotics. Otherwise, decreases in allocation to maintenance and production will occur (Sorensen et al. 2005).

9.5.1 Secondary Metabolites Can Reduce Digestibility of Carbohydrate, Protein, and Fiber

Secondary metabolites have been described that both increase and decrease digesta retention time (Cipollini 2000) and these might thus affect efficiency extracting nutrients, but we do not know of any examples. Examples of SMs that inhibit enzymatic and fermentative digestion are discussed below. Some SMs affect absorption as well. We have studied how some phenolics inhibit sugar absorption (Karasov et al. 1992; Skopec 2003). The classic examples are the flavonoids phlorezin and phloretin found in apple (*Malus*) bark and foliage, which specifically inhibit the apical membrane glucose transporter

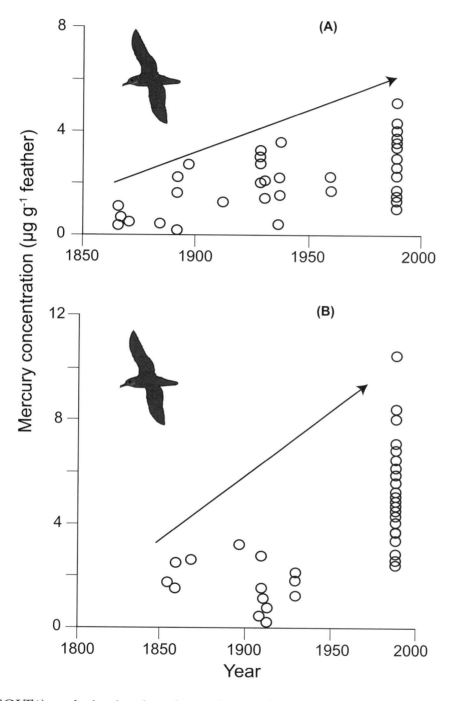

Figure 9.12. Exposure to mercury has increased recently, as indexed by mercury concentrations of feathers of Manx shearwaters sampled from (A) the southwest of, and (B) the northwest of Britain and Ireland. The lines with arrows are placed to show the long-term trend. (Redrawn from Thompson et al. 1992.)

(SGLT1) and the basolateral membrane glucose transporter (GLUT2), respectively. But we found that many other flavonoids depress glucose absorption efficiency in laboratory rats (figure 9.13). Many of these are found in fruits, but their effects on frugivores are unstudied. We suspect that SMs could have

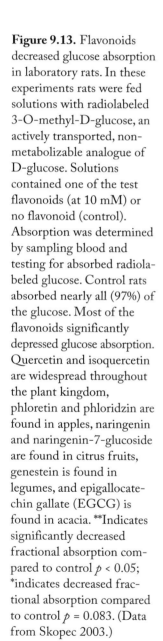

Figure 9.13. Flavonoids decreased glucose absorption in laboratory rats. In these experiments rats were fed solutions with radiolabeled 3-O-methyl-D-glucose, an actively transported, non-metabolizable analogue of D-glucose. Solutions contained one of the test flavonoids (at 10 mM) or no flavonoid (control). Absorption was determined by sampling blood and testing for absorbed radiolabeled glucose. Control rats absorbed nearly all (97%) of the glucose. Most of the flavonoids significantly depressed glucose absorption. Quercetin and isoquercetin are widespread throughout the plant kingdom, phloretin and phloridzin are found in apples, naringenin and naringenin-7-glucoside are found in citrus fruits, genestein is found in legumes, and epigallocatechin gallate (EGCG) is found in acacia. **Indicates significantly decreased fractional absorption compared to control $p < 0.05$; *indicates decreased fractional absorption compared to control $p = 0.083$. (Data from Skopec 2003.)

TABLE 9.1

Reduction of proportion of energy intake allocated to production or respiration on addition of plant secondary metabolites to the diet

Species	Diet + secondary metabolite	Reduction in energy (%)	Reference
Insects			
Manduca sexta	Semisynthetic diet + 0.5% nicotine	8	Schoonhoven and Meerman 1978
Danaus chrysippus	*Calotropis gigantea* + 0.5% caffeine	31	Muthukrishnan et al. 1979
Spodoptera littoralis	Semisynthetic diet + 0.01 M gallic acid	6	Mansour 1981
Birds			
Bonasa umbellus	*Quercus tremuloides* + 2.4% coniferyl benzoate	8	Guglielmo et al. 1996
Mammals			
Neotoma lepida	Semisynthetic diet + 3.7% resin from *Larrea tridentata*	6	Mangione et al. 2004

Source: Adapted from (Muthukrishnan and Pandian 1987).

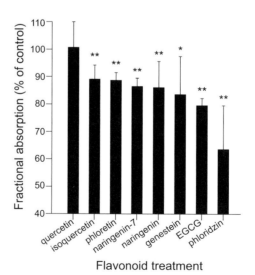

similar digestion altering effects in frugivorous bats and birds and, based on their behavioral sensitivity to their alterations in digestive efficiency, influence their fruit choice.

Tannins are water-soluble polyphenolic compounds with a molecular weight between 300 and 3000 Da, that interact with proteins and other macromolecules through H and hydrophobic bonds (chapter 2). Studying their effects on digestion is most straightforward using mammals because their feces are eliminated separately from urine. Urine might contain tannin components that were absorbed, or their biotransformation products, or energy- and N-containing products of postabsorptive processing. Even focusing just on mammals, however, it is apparent that the effects of tannins on animals are not general but depend on the particular tannin structure, concentration, and on particularities of consumer physiology (Zucker 1983; Bernays et al. 1989; Mole et al. 1993; Robbins et al. 1991; Ayres et al. 1997; Foley et al. 1999). Many ecologists we meet have the notion that tannins are general digestibility reducers, perhaps because it is known that in vitro both hydrolyzable and condensed tannins can bind enzymes and nutrients and decrease microbial fermentation rates (Bernays et al. 1989; Lowry et al. 1996) and hydrolyzable tannins can depress intestinal cell metabolism and cause intestinal lesions (Bernays et al. 1989; Karasov et al. 1992). But these very general effects in vitro do not translate to general effects on digestion in vivo.

Even if digestive enzymes are inhibited in vitro, the effects are potentially prevented or reversed in vivo by change in pH or by surfactants (detergents) such as bile acids or other tannin-binding material in the gut such as mucous (Bernays et al. 1989). Some mammals that commonly consume tannins secrete proline-rich (20–40% proline) proteins (PRPs) in their saliva that are thought to preferentially bind tannins (Hagerman and Butler 1981). The complexed tannins may escape both enzymatic and microbial degradation and be excreted in the feces, thus protecting the animal from either damage to the gut epithelium, true digestibility reduction, or toxicity (Austin et al. 1989). But this response leads to increased fecal loss of the energy and nitrogen in the tannin-protein complex and thus to a decline in apparent digestive efficiency. One of us (W.H.K.) collaborated with Skopec to study this response in laboratory rats whose production of salivary PRPs can be induced (Mehansho et al. 1992), so that the responses of noninduced and induced rats can be compared. When tannin-naive rats were first switched from a control diet to a diet with 3% pentagaloylglucose (a purified hydrolyzable tannin, radiolabeled) there was no significant decline in dry matter or nitrogen apparent digestibility (figure 9.14), and hence no direct effect on their digestion of the tannin at that concentration. After 10 days eating the diet, which is long enough for induction of salivary

Figure 9.14. Dietary tannin can reduce apparent digestibility in laboratory rats, not by disrupting digestive processes but instead by increasing endogenous fecal losses of matter and energy. Note that simply adding 3% pentagaloylglucose (a hydrolyzable tannin) to the diet did not depress apparent digestibility (compare "naive, 0% tannin" versus. "naive, 3% tannin"). After 10 days on the 3% tannin diet, which is long enough for induction of secretion of salivary proline-rich proteins (PRPs), which bind the tannin, there was in feces a ten fold increase in proline and a three fold increase in radiolabeled tannin, presumably from salivary PRP's complexed to the tannin, and coincident with that, a 10% decline in apparent dry matter digestibility and a 20% decline in apparent nitrogen digestibility (compare "naive" versus "acclimated, 3% tannin"). The benefit of secreting the PRPs is protection from effects of tannin, but the cost is the loss of material in the feces. Values shown are means ± standard error and the asterik denotes a significant decrease compared with controls (naive, 0% tannin). (Data from Skopec et al. 2004.)

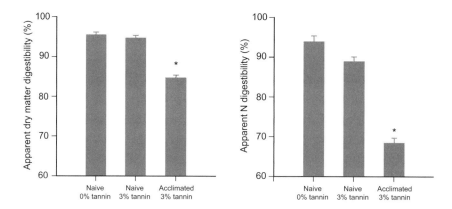

PRP production, there was a tenfold increase in proline in feces and a threefold increase in radiolabeled tannin, presumably from salivary PRPs complexed to the tannin. Coincident with this effect, there was a 10% decline in apparent dry matter digestibility and a 20% decline in apparent nitrogen digestibility (figure 9.13). The interpretation is that the tannin did not reduce digestive efficiency per se, but increased the fecal loss of endogenous material, especially PRPs, which has the effect of reducing the coefficient of apparent digestibility, especially of nitrogen. Relationships between dietary tannin content and reduction in apparent digestibility can be used to increase the accuracy of predictive equations of food digestibility based on food chemical composition (Hanley et al. 1992).

Potentially, tannins reduce cell wall digestibility by reducing rates of microbial fermentation (Bravo et al. 1994; Lowry et al. 1996). Grazing ruminant herbivores, which tend to have smaller parotid salivary glands than browsers (Kay et al. 1980), are more susceptible to this effect than browers because grazers may produce fewer salivary PRPs to deactivate tannins (Robbins et al. 1987). Indeed, this example of depression of cell wall digestibility in some ruminants may be one of the few examples where tannins truly act as digestibility reducers. When Bernays et al. (1989) reviewed impacts of tannins on insect herbivores, they concluded that there was no proven case of antidigestive effects of tannins in vivo in insects, and this is true today as far as we know. As they point out, "this is not to say that there are no negative effects—but they do not concern direct effects on digestion."

9.5.2 Impact of Xenobiotic Ingestion on Respiration

Maintaining and operating detoxication processes must cost metabolic energy, which if large enough might be measured as increased respiration. It has been estimated that the operating costs of MOs constituting phase I reactions (NADPH, etc.) are 20 J/mmol toxin (Brattsen 1988), which represents less than 1% of basal metabolism for a vertebrate. Nonetheless, at least one study demonstrated that absorption of xenobiotics is associated with an increase in resting metabolism. Phenolics infused in sheep led to increases in resting metabolism of 5% (Iason and Murray 1996). Some phenolic compounds act as uncouplers of oxidative phosphorylation, leading to greater oxygen consumption, and this effect might add to whatever detoxification costs exist. Another type of cost might be observed in animals fed SMs chronically. Resting heat production might increase if certain organs such as the liver are enlarged to enhance detoxication capacity, or if resources must be allocated to repair tissue damaged by absorbed SMs. These might be explanations why heat production increased 15% relative to controls in fasted voles that had been chronically fed the phenolic gallic acid (Thomas et al. 1988). Curiously, in another study on two species of South American rodents (*Phyllotis darwini* and *Octodon degus*), chronic feeding of tannic acid-laden diet had no significant effect on resting respiration rate, although maximal metabolic rate was elevated (Bozinovic and Novoa 1997).

9.5.3 Energy and Material Losses with Excreted Conjugated Xenobiotic Metabolites

As discussed above (section 9.3.3), many xenobiotics that are absorbed are ultimately excreted conjugated to sugar derivatives, amino acids, peptides, and sulfate. Theoretical calculations (Illius and Jessop 1995) indicate that the associated energy losses would reduce the metabolizability of food by about 5%, although empirical studies indicate that the effects can be higher in some animals. The highest losses have been observed in grouse, herbivorous birds, eating native vegetation (figure 9.15). The energy lost in glucuronides represented 5–15% of their metabolizable energy, compared with 3% or less in six mammalian species

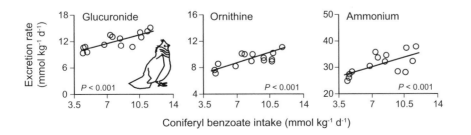

Figure 9.15. The excretion by ruffed grouse of biotransformation conjugates such as glucuronide and ornithine increased with increasing ingestion of the secondary metabolite coniferyl benzoate (CB) in their wild aspen diet. Ammonia excretion increased as well, perhaps a result of increasing catabolism of protein to produce bicarbonate (HCO_3^-) which can partially buffer organic acids such as glucuronic acid (see text). In this study, the intake of CB was manipulated by feeding grouse quaking aspen flower buds with increasing amounts of CB. (Replotted from Guglielmo et al. 1996.)

(Guglielmo et al. 1996). The generality of this difference, and reasons for it if it holds up, await further research.

Possibly even more important than these energy losses are costs measured in terms of nitrogen metabolism. Ruminant and other foregut fermenters absorb little dietary sugar, which is mostly microbially fermented proximal to the small intestine. Because fatty acids are a poor substrate for glucose production (chapter 7), ruminants must catabolize body protein to produce glucose for the conjugated glucuronides. Hence, high rates of xenobiotic absorption can lead to relatively high rates of protein catabolism and body mass loss (Illius and Jessop 1995). Also, protein may be catabolized to produce bicarbonate (HCO_3^-) which can partially buffer the organic acids formed during phase I and II detoxication (Foley et al. 1995). This helps explain why excretion of ammonia (a product of protein catabolism) increased with xenobiotic ingestion in figure 9.15. Finally, the increased loss of N in ammonia or amino acid and peptide conjugates (see figure 9.15) can potentially impose a negative N balance on plant diets with low N. In grouse, ornithine conjugation increased the minimum N requirement by 67–90%, depending on the rate of xenobiotic absorption (Guglielmo et al. 1996) (see chapter 11).

9.6 Modeling Approaches Can Integrate the Processes of Absorption, Distribution, and Elimination (Including Biotransformation and Excretion)

Imagine a simple experiment in which an animal is injected intravenously (IV) with a dose of a xenobiotic, and a body compartment such as blood is sampled. The xenobiotic's level in the blood falls over time as it distributes elsewhere in the body and is eliminated from the body (figure 9.16). If, in another experiment, the xenobiotic is also fed to the animal and some is absorbed, its level in blood will initially rise and then decline following a similar course. However, the postingestion rising phase might be missed if blood sampling does not begin early enough (figure 9.16).

The fraction of the xenobiotic that was absorbed, also called its bioavailability, can be calculated by comparing the oral and IV blood concentration-time curves. First, the two areas under the curve from time zero to infinity ($AUC_{0\rightarrow\infty}$) are calculated by the trapezoidal method. You may recall from calculus that this method subdivides the distance along the x axis into many subintervals of equal length which each become the bases of trapezoids whose vertical heights are the average distance from the x axis to the curve. The areas

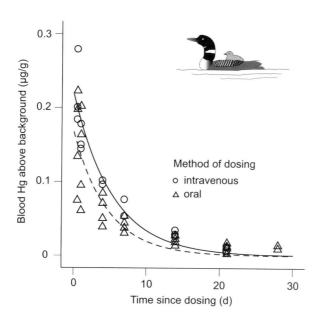

Figure 9.16. Absorption and elimination by common loon chicks of a dose of methylmercuric chloride. The blood Hg values are corrected to an equivalent dose administered. After either intravenous (IV) or oral administration, Hg appears in blood and then falls over time as it distributes elsewhere in the body and is eliminated. Notice that, for the equivalent doses used here, blood values following oral administration (triangles, dashed line) are slightly lower than following iv administration (circles, solid line). This is because less than 100% of the oral dose was absorbed. The amount absorbed, 83%, can be calculated by comparing the areas under the two curves (see text). (Data from Fournier et al. 2002.)

of all the trapezoids are summed and approximate the area under the curve. Then, the fraction absorbed (*F*) is calculated as follows:

$$F = \left(\frac{AUC_{0\to\infty}\ \text{oral}}{AUC_{0\to\infty}\ \text{IV}} \right) \left(\frac{\text{dose, IV}}{\text{dose, oral}} \right). \tag{9.1}$$

In the example in figure 9.16, in which young (35-day-old) common loons (*Gavia immer*) were injected and fed a dose of methyl mercury in a raw fish, the calculated bioavailability was 0.83, meaning 83% of the dose was absorbed. A strength of this method for measuring absorption, besides its simplicity, is that it relies on no assumptions regarding into how many compartments the methyl Hg may have distributed (figure 9.1) or the kinetics of its elimination. In this simple experiment, the relationship between postdose blood Hg level (above background) and time postdosing fit a monoexponential elimination model which corresponds to clearance from a single compartment:

$$C_i = \lambda_1 e^{-\alpha_1 t} \tag{9.2}$$

where C_t is the concentration of mercury in whole blood ($\mu g\ g^{-1}$) at time t, and t is time (days in this case) elapsed since the administration of the dose. The terms λ ($\mu g\ g^{-1}$) and α (day^{-1}) are unknown parameters that can be determined by nonlinear regression. Although the model assumes that there is a single compartment, there is no requirement that its identity be defined. In this single-compartment model, the variable α represents the fractional rate constant for excretion (r_e). Elimination rate can also be expressed as a biological half-life ($t_{1/2}$), calculated using the equation $t_{1/2} = \ln 2/r_e$. Sometimes elimination data such as these are tested for fit to a biexponential elimination model, which is consistent with clearance from two-compartments:

$$C_t = \lambda_1 e^{-\alpha_1 t} + \lambda_2 e^{-\alpha_2 t} \qquad (9.3)$$

The terms λ_1, λ_2 (μg g^{-1}) and α_1, α_2 (day^{-1}) are unknown parameters to be determined by nonlinear regression on the basis of experimental data. In our example in figure 9.16, the fit to the monoexponential model was superior, based on a statistical test that is very useful for comparing fits to different models (Motulsky and Ransnas 1987). Thus, applying simple models to the data from our simple experiment we have some insight into absorption, distribution, and elimination of methyl Hg in common loons.

Whereas this simple experiment involved a single dose of xenobiotic, wild animals are more likely to be exposed continually. Pharmacokinetic models can be extended to this situation as well. Consider the balance equation for an animal whose body mass (M) represents a single compartment into which the chemical can accumulate:

$$\text{rate of xenobiotic absorption} = \text{rate of loss} + \text{rate of storage.} \qquad (9.4)$$

Absorption can be expressed as the product of the chemical's concentration in the diet (C_f), the food intake rate (I), and the fractional absorption (F, often called bioavailability; equation 9.1). Rate of loss is determined by the product of the fractional rate constant for elimination, r_e, and the total amount of xenobiotic in the body (X). Rearranging to solve for rate of storage, we have

$$\text{rate of storage} = FIC_f - r_e X. \qquad (9.5)$$

We will divide through by body mass (M) in order to calculate the chemical's concentration (amount per unit mass C), which is most often measured by ecologists:

$$\text{rate of storage}/M = dC/dt = (FIC_f/M) - r_e C. \qquad (9.6)$$

This model indicates that, when input exceeds loss, the xenobiotic concentration in the body will rise, though at a decelerating rate because as C rises the amount eliminated per unit time, which is the product $r_e C$, rises. At some point a steady state is reached in which input = output (i.e., $dC/dt = 0$). Such a rise to a near steady-state situation is suggested for great skuas (*Catharacta skua*) continually ingesting methyl Hg (figure 9.17, left panel). For a one-compartment model, the concentration at steady state (C_{ss}) can be determined as

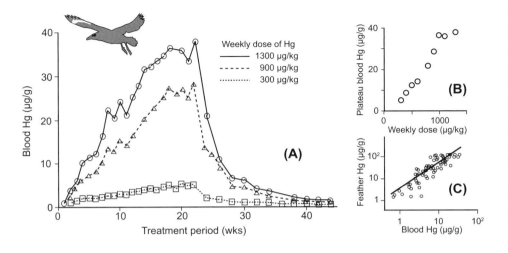

Figure 9.17. Patterns of Hg in great skuas (*Catharacta skua*) dosed with methylmercuric chloride. In the large figure (A), which plots data for three of the nine birds, the first dose was administered after blood sampling on week 0 and weekly dosing continued until week 21. Notice that at all doses blood Hg steadily rose but appeared to level off at 15-20 weeks, and then fell exponentially once dosing stopped. In the upper inset (B) the approximate plateau blood Hg level in all nine birds was linearly related to the Hg dose, as would be expected for first-order kinetics (see text). The lower inset (C) shows the linear relationship between blood Hg level and mercury in feathers grown at the same time. (Data replotted from Bearhop et al. 2000.)

$$C_{ss} = FIC_f/(Mr_e). \qquad (9.7)$$

As expected from equation 9.7, the highest blood Hg levels of skuas were linearly related to Hg dose. Much of the Hg was eliminated into feathers (see section 9.4.2), and there was also a linear relation between blood Hg level and feather Hg content (figure 9.17). Note that, for the skuas, once exposure was terminated by feeding a methyl-Hg-free food, depuration occurred, not unlike that which occurred in common loons (cf. figure 9.16 and left panel of figure 9.17).

Models such as these help us to think about how exposure to toxins might differ among species of different sizes, a little studied but important topic for ecological and evolutionary analyses as well as for veterinary treatment. The steady-state concentration in the body will vary according to how the parameters of I and r_e scale with body mass. In animals feeding to match energy expenditure $I \propto M^{3/4}$ (chapters 1 and 3), and in numerous (but not all) sets of pharmacokinetic data $r_e \propto M^{-1/4}$ (Riviere 1999). In this situation we therefore expect C_{ss} to be independent of body mass (i.e., $C_{ss} \propto M^{3/4}/(MM^{-1/4}) \propto M^0$). Previously, Freeland (1991) concluded that larger herbivores would have higher steady-state plasma concentrations than smaller herbivores when eating the same toxin-laden food, but we think that this conclusion is wrong in part because Freeland used a higher scaling factor for r_e. Freeland thought that larger herbivores select more diverse diets than smaller herbivores to avoid intoxication, but we suspect that the explanation probably lies elsewhere.

Simple compartment models with first-order kinetics (equations 9.2–9.7), called first-order because the kinetic rate constant is independent of concentration, are of great heuristic value even if they are only an approximation of what actually occurs. The approach is based on models in pharmacokinetics. For aquatic organisms, an additional expression is easily added to incorporate absorption and loss by diffusion across the body surface or gills. Sometimes a two-compartment system provides a better explanation than a one-compartment model, but the increase in complexity makes the approach more difficult. The field of toxicokinetics and modeling has blossomed and produced a wealth of different types of models that vary in their basic assumptions, structure, and complexity. Another class of models, called physiological (as opposed to compartmental) pharmacokinetic models, are based on blood flows through major tissues and organs. The modeling approach, be it compartmental or physiological, helps us predict the level of a chemical in an animal's body, given information about its level in food or in the environment. As discussed in the next section, an important application of these models has been to increase understanding of bioaccumulation and biomagnification in ecosystems and the prediction of risks to ecosystems. Models can also direct and focus research, as when we try to understand whether differences between groups in tolerance to chemicals are due to different levels of toxicant at the target site (because of differences in the toxicant's uptake and/or loss), or due to differences in the interaction of the chemical (or its metabolite) with the site of action.

9.7 Models Can Predict Bioaccumulation and Biomagnification in Ecosystems

Ecologists have found that sometimes the levels of contaminants in consumers are consistently higher than in their food. A term adopted to describe this situation is the bioaccumulation factor or *BAF*:

$$BAF = \frac{\text{concentration in whole organism}}{\text{concentration in food and/or water ingested}}. \tag{9.8}$$

(A related term, the bioconcentration factor (*BCF*), is applied when the denominator is simply concentration in water around the organism, and applies best to situations where contaminant uptake is across the body surface and not by ingestion.) *BAF*'s can be huge for some manmade liposoluble chemicals: for Baltic white eagles eating fish, *BAF* values were 4700 for DDE (a metabolite of DDT) and 3600 for PCBs (Paasivirta 1991).

If we return to equation 9.7 and reexpress it according to the quotient in equation 9.8, it becomes perfectly clear when bioaccumulation will occur:

$$BAF = \frac{C_{ss}}{C_f} = \frac{FI}{Mr_e}. \tag{9.9}$$

*BAF*s will be high when xenobiotic elimination (r_e in the denominator) is very low in relation to absorption (*FI* in the numerator). The chemicals already known to bioaccumulate in animals, like pp′ DDE, PCDDs, and PCBs have exceedingly low values of r_e, or, expressed in the inverse, very long half-lives in the body ($t_{1/2}$'s). For first-order kinetics, $t_{1/2} = \ln 2/r_e$. As examples, half-life for pp′ DDE in pigeons is 250 days (Walker 1990) and for TCDD in pheasants 378 days (Nosek et al. 1993). The major reason for the slow elimination is a very slow rate of biotransformation. Indeed, enzymatic reaction rates in vitro are sometimes (not always) correlated with whole-animal elimination rate constants (Walker 1980). The biotransformation mechanisms that have evolved in animals (section 9.3) are not effective against all anthropogenic compounds, especially those with high levels of halogenation (e.g., chlorinated hydrocarbons) that makes attack of C-H bonds difficult. Making the logical connection between biotransformation rates and bioaccumulation via equation 9.7 has led to the interesting proposal that in vitro measures of biotransformation, perhaps with microsomal preparations (box 9.1), could be used to predict relative *BCF*s for new chemicals, and thus predict their hazards (Walker 1990).

If you think about equation 9.9 you will realize that it might be used to make quantitative predictions of bioaccumulation throughout a food web within an ecosystem. Consider a trophic chain of prey consumed by a predator which, in turn, is also consumed by a predator. If we measured or estimated the variables in each species we could predict the elevation in body contaminant content up the food chain, starting with the contaminant concentration in the food of the first species. The methods are available to estimate the parameters, as shown in this section and in other chapters regarding energetic determinants of food intake (I). Combining this approach with methods described in chapter 8 for characterizing trophic position using stable isotopes, one predicts an ecosystem-wide correlation between contaminant level and ^{15}N enrichment, a proxy for trophic position (figure 9.18). In this example with toxaphene, the biomagnification, or increase in concentration over the entire food chain, is about 1000. Consideration of equations 9.6 and 9.7 leads to the conclusion that the magnitude of biomagnification will be positively related to the length of a food chain and a chemical's half-life in species along the food chain. What a terrific example of linkage between physiological ecology and an important ecosystem phenomenon!

9.8 Postingestional Effects of Xenobiotics on Feeding Behavior

Up to this point we have discussed many factors that determine the concentration of xenobiotics in an animal's body, such as absorption, distribution, elimination,

Figure 9.18. The biomagnification of the organochlorine pesticide toxaphene in subarctic lakes. The concentration of toxaphene in representatives of the biota increased with their higher trophic position, which was indexed by ^{15}N enrichment (see text and chapter 8). Regressions from biota of three different lakes are shown: Laberge (circles, solid line), Fox (triangles, dotted line), and Kusawa (square, dashed line). Each point is a mean value for a taxonomic group. The organisms analyzed included invertebrate species (filled symbols) from near the bottom of the food chain to vertebrate species (unfilled symbols; all fish) including top predatory fish such as lake trout. (Replotted from Kidd et al. 1995.)

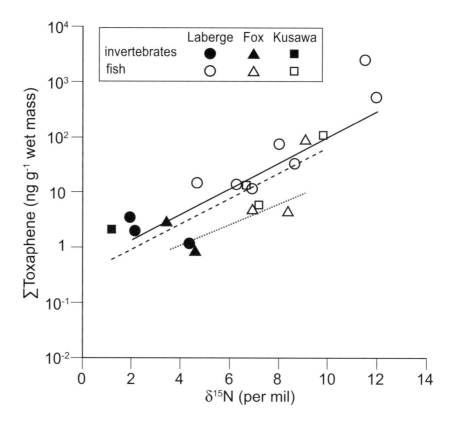

and even sometimes trophic position. But we have not discussed the role of the animal's own behavior in influencing toxin exposure. In chapter 2 we opined that SMs are so pervasive it is almost a certainty that any thorough analysis of a plant food, and maybe even any animal food, will find some present. It seems improbable that herbivores can avoid ingesting these chemicals. More likely, if animals can detect SMs perhaps they will regulate their intake to avoid harm if the SMs are too potent or too expensive to process. An integrated detection and regulatory response implies some kind of feedback mechanism, and in this final section we review evidence of regulation and the methods used to study possible feedback mechanisms.

9.8.1 Animals Sometimes Modulate Their Food Intake to Regulate Xenobiotic Ingestion

Let us return to the example of the ruffed grouse eating male aspen flower buds, described above. An important deterrent chemical in these buds is a phenolic ester called coniferyl benzoate. We know that the grouse can sense this chemical when it is present in the mouth, based on electrical signals along its trigeminal nerve which runs from the region of the oral cavity to the brain (Jakubas and Mason 1991). If grouse are provided with diets with added coniferyl benzoate, they sense it and on the first day they reduce their food intake. This occurs whether the addition is to a diet previously lacking coniferyl benzoate or to a diet with some lower level of the compound (figure 9.19). But within a day they typically habituate to the higher level of the chemical and their feeding rate returns to the level of grouse with no coniferyl benzoate in their diet. However, there is a limit to this response. Grouse eat normal amounts of food as long as it contains no more than about 4.5% coniferyl benzoate. Once the food contains more than this, they reduce daily food intake in response to progressive increases in coniferyl benzoate concentration (figure 9.19). If one calculates their daily intake of coniferyl benzoate on diets with more than 4.5%, it becomes apparent

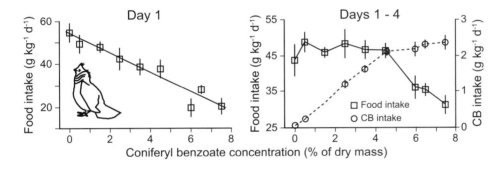

Figure 9.19. Ruffed grouse appear to regulate food intake according to coniferyl benzoate and not energy. If grouse are provided with diets with added coniferyl benzoate they sense it and on the first day they reduce their food intake ("day 1," left figure). But within a day they typically habituate to the higher level of the chemical and their feeding rate returns to the level of grouse with no coniferyl benzoate in their diet ("days 1–4," right figure). However, there is a limit to this response. Once the food contains more than 4.5% coniferyl benzoate, grouse reduce daily food intake with progressive increases in coniferyl benzoate concentration ("days 1–4"). It is daily coniferyl benzoate intake, and not daily food intake, that is held to a relatively constant level. (Replotted from Jakubas et al. 1993.)

that it is daily coniferyl benzoate intake, and not daily food intake, that is held to a relatively constant level. This is a very interesting exception to the general response of many animals to defend a relatively constant energy absorption rate (see chapter 3, section 3.4.1). Presumably, if the grouse ate more coniferyl benzoate they would suffer intoxication, and so eating too little food and losing weight is the lesser of two poor options. One interesting question is whether the level of maximum toxin intake corresponds to the grouse's maximum capacity to detoxify and eliminate coniferyl benzoate.

Apparent regulation like this has also been observed in mammals and insects. For example, when brushtail possums (*Trichosurus vulpecula*) were provided with food with increasing concentration of jensenone extracted from *Eucalyptus* their feeding rate on the diet progressively declined, whereas the jensenone intake reached a plateau (Stapley et al. 2000). Even increasing their energy needs by exposing them to cold could not stimulate the possums to increase their food intake if the associated jensenone intake would exceed the threshold value. Likewise, diluting a caffeine-containing diet of velvetbean caterpillars (*Anticarsia gemmatalis*) with water stimulated them to increase their fresh matter food intake to hold nutrient intake constant, but once they reached a threshold dose of caffeine (which is toxic) their intake began to decline and they grew slower (Slansky and Wheeler 1992). Insects, such as tobacco hornworms (*Manduca sexta*) and southern army worms (*Spodoptera eridania*) eating nicotine-laden food (Cresswell et al. 1992; Appel and Martin 1992), and parsnip webworms (*Depressaria pastinacella*) eating furanocoumarin-laden food (Berenbaum et al. 1989), also reduce their feeding rate as levels of the relevant toxic chemicals are increased in their diets. Notably, those insect species that evolved to become specialists on certain toxin-laden plants are not necessarily immune to the respective toxin(s). Pyrrolizidine alkaloids appear to be both nutritive and toxic to the ornate moth (*Utetheisa ornatirx*), and the larvae modulate their intake temporally to avoid toxicosis (Kelley et al. 2002). Eating less and surviving at even reduced growth rates to reproduce is apparently a better prospect than the consequences of direct toxicity. The methods of geometric analysis, introduced in chapter 3 (box 3.5), provide an experimental and theoretical framework for assessing such trade-offs (Simpson and Raubenheimer 2001).

The maximum tolerable toxin intake has tremendous predictive value, and it is unfortunate that it has been rarely determined in studies that demonstrate that animals are deterred by particular SMs. The maximum can serve as a constraint in a feeding model such as a linear programming model (Belovsky and Schmitz 1991; also see chapter 11, box 11.2). Also, it permits one to calculate the feeding rate of the animal as a function of toxin concentration, which should permit one to calculate reductions in body mass maintenance, growth, and so on.

What are the likely feedback mechanisms that permit an animal to achieve the kind of regulation indicated in figure 9.19, and how are they studied?

9.8.2 Conditioned Taste Aversions Are an Important Feedback Mechanism Regulating Toxin Intake

There are few examples where animals have evolved receptors that are highly specific for plant toxins. In fact, several tests for statistical relationships between toxicity and particular tastes or general palatability have been negative (Clark 1997). So, how do animals know to avoid toxins?

Chemesthesis is a term for the sensation of nonspecific stimuli that are often irritating or painful, mediated by specialized neurons called nociceptors (Clark 1997). An animal's immediate avoidance of a substance, not requiring any past experience or learning, might be mediated by such a mechanism. One of the best studied examples is a membrane channel protein that goes by the name of vanilloid receptor 1 (VR-1), which is expressed in sensory neurons in the mouth region of many mammals (Caterina et al. 2000). One noxious stimulus for VR-1 is capsaicin, a lipophilic alkaloid that is responsible for "hot" chili peppers' fiery taste, but temperature above 43°C and protons (acid) are also stimuli and so we must not assume that VR-1 evolved specifically to detect a plant SM. In any event, animals in which VR-1 is expressed (like you!) have a highly specific receptor for this alkaloid, which is found only in the genus of chilis (*Capsicum*), and those animals tend to be behaviorally deterred from eating fruit of this genus. Interestingly, the avian homologue of mammalian VR-1 does not interact with capsaicin (Clark 1997; Jordt and Julius 2002) which helps explain why some birds seem able to ingest chilis without compunction (Tewksbury and Nabhan 2001).

Most insects, mammals, and birds that have been studied also have neural elements that are sensitive to a range of chemicals that includes SMs and which are not chemesthetic (Chapman and Blaney 1979; Clark 1997). Animals are thought to regulate their intake of such SMs below toxic levels by learning to associate a distinctive taste with negative postingestive feedback from the toxin. When ingestion of a taste stimulus is paired with internal malaise, the animal remembers the taste and rejects its ingestion thereafter. This learning is referred to as conditioned taste aversion (CTA).

Conditioned taste aversions are demonstrated experimentally by feeding a food or flavor along with a dose of chemical that causes malaise, most often lithium chloride (LiCl). It is thought that the LiCl directly or indirectly stimulates the emetic system of the body, which resides in the medulla of the brain (Provenza et al. 1994). The CTA is then demonstrated by a subsequent reduction in intake of the food or flavor, even if the LiCl or other malaise-causing chemical is absent. Over time, the aversion may become extinct. The strength and duration of the aversion varies according to many factors. In domestic livestock, which have been intensively studied, greater stimulation of the emetic system with higher doses of LiCl strengthen the aversion, whereas long delays between ingestion and postingestion sensations weaken the aversion (Provenza et al. 1993) as do social factors such as observation of other animals eating the deterrent food. Some vertebrates may even limit their intake of a food that a member of their social group avoids (Thorhallsdottir et al. 1990). Besides mammals and birds, CTAs have been demonstrated in a wide variety of vertebrates including fish, amphibians, and reptiles as well as in some invertebrates such as snails and insects (Slansky and Wheeler 1992).

Foley and his colleagues have extensively studied the impact of SMs in *Eucalyptus* on Australian arboreal folivores. They examined the development

Figure 9.20. Conditioned flavor aversion to cineole in common ringtail possums. In the wild ringtails eat eucalyptus leaves that contain a potent toxin (jensenone) as well as a relatively nontoxic natural stimulus, 1,8-cineole (a monoterpene). When they were first captured (A; "at capture"), ringtails avoided consuming food with cineole (at 10.8% of dry mass here and elsewhere) even in a food lacking jensenone. But after 12 days to sample the cineole-containing food, they ate normal amounts, compared with controls (B; "acclimated"). In the third stage of the experiment cineole was paired with jensenone in the laboratory diet, and once again the ringtails exhibited an aversion to cineole-containing diet even when it lacked jensenone (C; "averted"). (Data from Lawler et al. 1999.)

of CTA using a natural stimulus (1,8-cineole, a monoterpene) and a natural toxin (jensenone, mentioned above in 9.7.1). When common ringtail possums (*Pseudocheirus peregrinus*) are captured and brought into captivity they are already averse to cineole, as shown by a large reduction in their daily intake of food to which the compound has been added (figure 9.20A). However, after 12 days of acclimation to diet with increasing amount of cineole (and no added toxin) they will tolerate as much as 50% more of the chemical as shown by significantly increased consumption of diet with 10.8% cineole (figure 9.20B). However, the aversion was effectively reestablished during a reconditioning period in which cineole was paired once again with the toxin jensenone, as shown by their low intake of diet with 10.8% cineole (and no added toxin) when tested again after postreconditioning (figure 9.20C). Foley and his colleagues collected nearly identical results in parallel studies with common brushtail possums (*Trichosurus vulpecula*) as well (Lawler et al. 1999).

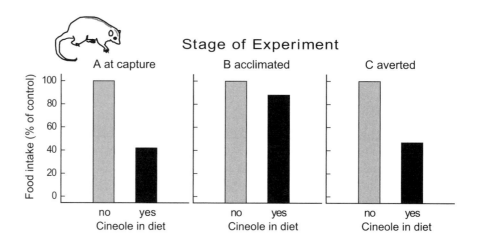

Relatively nontoxic cineole is an effective signal for these arboreal folivores because it has a strong taste and odor and its presence in *Eucalyptus* leaves is strongly correlated with toxic constituents like jensenone, so it can signal impending toxicosis. But ringtails and brushtails will respond to jensenone alone too. As mentioned above, when they were fed diets containing jensenone, both species regulated their food intake so that jensenone intake did not exceed an intake ceiling. The response to jensenone may be mediated by the same "emetic center" of the medulla that LiCl is thought to act upon. The emetic response is partly mediated by the neurotransmitter serotonin. When brushtail and ringtail possums were administered ondansetron, which is a selective antagonist of the serotonin receptors, they ate more jensenone than animals administered saline control solution (Lawler et al. 1998).

Our discussion has emphasized the importance of conditioned taste aversion to postingestive sensations mediated through the emetic center of the medulla. The effect of the antagonist ondansetron is a clue to us that the chemical may cause a sensation of nausea in the animal, which we cannot ask about its sensations. Can we conclude that if the deterrent effect of a SM is not affected by ondansetron then a postingestive effect is unlikely? As an illustration, Foley and his colleagues also studied the effects of another SM, the phenolic glycoside salicin (Pass and Foley 2000). As in other experiments, common brushtail possums regulated their intake of salicin so as not to exceed an intake ceiling. But ingestion of salicin did not seem to have any significant effects on nitrogen balance (no apparent detoxication cost), and administration of ondansetron did not increase their ingestion of salicin (no apparent emetic effect). This begged the question of whether a preingestive factor might be responsible, perhaps like trigeminal nerve stimulation as mentioned earlier regarding grouse. However, studies in human beings, whom we can ask about sensations, have shown that they will avoid foods containing chemicals that cause other negative consequences, such as cramps or headaches (Rozin 1988). You can see how challenging it can be to explain completely the mechanisms of deterrence by SMs.

Studies on conditioned taste aversions suggest several important points about plant SMs and feeding behavior. First, it is possible for one researcher to observe that a SM found in one species of plants deters feeding by an animal and for another researcher to observe no effect with another species of plant that also has the SM, because the chemical might not be a toxin in its own right but merely act as a cue to some other toxin that is present in the first case but not in the second case. Thus, the very widespread approach of correlating feeding behavior or food selection of wild animals with levels of plant SMs has low probability of leading to mechanistic understanding or predictive ability regarding the effect of SMs on feeding. Dose-response studies with isolated chemicals are more informative. Second, the behavioral studies with vertebrates suggest that aversive taste alone without toxic side effects will not powerfully inhibit feeding, because when the amount of nonaversive food is limited, hunger will stimulate the animal to sample aversive foods (Provenza 1996). This is why CTAs typically become extinct over time if not reinforced by some additional negative stimulus. Finally, the studies do suggest that if wild animals can detect SMs they will often regulate their intake to avoid harm if the SMs are too potent or too expensive to process. Foley et al. (1999) presented a mechanistic model based on pharmacokinetic principles like those that we have discussed showing how this might operate at the level of meal size and frequency. But they point out that there are no rigorous tests of this kind of approach. Also, regulation of intake to avoid SMs may not be as precise when animals feed on natural foods containing different amounts of SMs, and the ability of animals to form CTAs to SMs may be reduced when many food choices are available (Dearing et al. 2005), which is a typical situation for free-living animals.

9.9 Toxic Effects of Xenobiotics in Wild Animals

Whether an ingested xenobiotic will be toxic depends on the animal's exposure, which we have discussed at length, and on the interaction(s) of the xenobiotic with its site(s) of action, which leads to the expression of toxic effects. There are many comparative examples of two species ingesting the same putative toxin and one of them tolerating as much as a 1000 times more of the toxin than the other (Walker 1994). This begs the question whether the tolerant species has target site insensitivity or simply maintains much lower body concentration of the xenobiotic by virtue of very low absorption or very high elimination (consult equation 9.7). The issue has played a prominent role in discussions about why there are SMs in ripe fruits, because one hypothesis is that the same chemical might screen out seed predators and pulp thieves that are detrimental to seed dispersal but at the same time have little or no effect on seed dispersers (Cipollini 2000). Most tests so far have failed to provide good evidence of such "directed effects" (Cipollini 2000), although the example of chilis discussed above is an exception because mammalian seed predators are deterred whereas avian dispersers are not (Tewksbury and Nabhan 2001).

One group of animals in which to look for target site insensitivity is specialist species that have evolved to rely on a particular plant that contains putative toxins. However, those species are not necessarily immune to the respective toxin(s). Recall the parsnip webworms (*Depressaria pastinacella*) that eat furanocoumarin-laden food that few other lepidopterans can thrive on. They have P450 MO activity against furanocoumarins that is 10–300 times higher than in other lepidopterans that have been studied (Berenbaum et al. 1992). If this corresponds to differences in r_e in equation 9.7, then if all else is equal this leads to the initial expectation that steady state levels of furanocoumarins will be 10–300 times lower in their bodies. This may suffice to explain the webworms' tolerance, rather than reduced target site sensitivity. Indeed, the webworms eating furanocoumarin-laden food, and tobacco specialists such as tobacco hornworms (*Manduca sexta*) and southern army worms (*Spodoptera eridania*) eating nocotine-laden food (Cresswell et al. 1992; Appel and Martin 1992), reduce their feeding rate as levels of the respective toxic chemicals are increased in their diets, suggesting that they are not immune.

But there are examples of target site insensitivity. A natural example may be the monarch butterfly's apparent insensitivity to the cardenolides it stores in its body. These chemicals have an inhibitory effect on Na^+-K^+-ATPase and thus affect vertebrates whose cellular function is critically dependent on maintaining a cellular gradient for Na^+ with this enzyme. But phytophagous insects have K^+ as their major cation and Na plays a minor role, which in combination with ATPases that are relatively insensitive to cardenolides, makes them more tolerant of the compounds (Brattsten 1979). Walker (1994) has suggested that such qualitative differences in target site sensitivity will be more common between distant clades, such as vertebrates versus invertebrates, than between species within clades. Nonetheless, there are a few examples of apparent differences in target site insensitivity even within species. These have emerged from instances of selection for pesticide resistance. For example, some strains of houseflies and certain other insects developed resistance to DDT that is associated with insensitivity of the nervous system to the insecticide, which targets sodium channels in nerves (Walker 1994). Great progress has been made identifying the specific sodium channel gene mutations that cause the reduced neuronal sensitivity to DDT (Soderlund and Knipple 1999).

It is beyond the scope of this book to detail the mechanisms of toxicity of xenobiotics. Furthermore, it is not even clear how often animals suffer direct effects of natural toxins at specific cellular target sites because, as discussed in the previous section, if they can detect them they will often regulate their intake to avoid harm if the SMs are too potent or too expensive to process.

It has been argued that generalist herbivores maintain broad diet breadth partly because they must limit their intake of any single food to avoid toxicosis (Freeland and Janzen 1975). The method of linear programming (chapter 11, box 11.2) permits one to incorporate information on toxin constraints to describe the possible diet mixes that do not intoxicate (Belovsky and Schmitz 1991). But what happens if choices are restricted due to food limitation? There are a few examples where wild herbivores are apparently forced to rely on toxin-containing foods, and they apparently reduce their feeding rate and suffer mass loss. Desert woodrats in the Mojave Desert of North America suffer a seasonal bottleneck in food availability between summer and the onset of winter rains and new plant growth (Karasov 1989). During this period they are forced to rely on the leaves of creosote bush (*Larrea tridentata*) which are coated with a phenolic-containing resin. The woodrats can tolerate a limited amount of the resin (figure 9.21, left panel), and in the wild they limit their leaf intake to avoid outright toxicosis but as a consequence they suffer mass loss from energy imbalance and they risk death (figure 9.21, right panel). Snowshoe hares (*Lepus americanus*) in the boreal forest apparently provide another example of an herbivore that sometimes suffers a seasonal bottleneck in food availability and eats restricted amounts of toxin-laden browse, loses body mass, and risks death directly from starvation or due to increased risk of predation (Tahvanainen et al. 1991).

There are estimates that 1% of domesticated livestock in North America suffer losses due to toxicosis from plant toxins (deaths, abortions, birth defects,

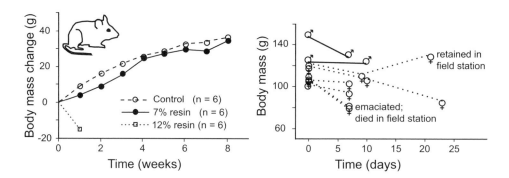

Figure 9.21. Desert woodrats cannot maintain themselves on a diet containing 12% resin of creosote bush (*Larrea tridentata*). The left-hand figure shows mean mass changes in groups of woodrats in the laboratory consuming chow with 0, 7%, or 12% resin (by dry mass) (redrawn from Meyer and Karasov 1991). The right-hand figure shows mass changes of individual males and females in the field during the winter when diets of free-living woodrats were nearly 100% creosote bush (redrawn from Karasov 1989). Notice that all wild woodrats that were recaptured lost mass, some to the point of emaciation. One female, which had lost mass when recaptured, was retained in the field station and gained mass when fed rodent chow and water.

organ toxicity, decreased performance, and photosensitization) (James et al. 1988). In almost all cases, poisoning occurred when the poisonous plant was the only forage available because of overgrazing, drought, snow and cold, or early green-up. Comparable estimates for wild herbivores are not available as far as we know, but the few documented cases of plant poisonings of wild mammalian herbivore populations occurred under similar kinds of circumstances (Meyer and Karasov 1991).

Such instances of outright toxicosis are probably rare in wild species. Most often there are multiple plant species from which to choose. In that case, the main effects of natural toxins are probably alterations in foraging behavior that affect foraging costs and risk of predation, and/or reduced productivity due to decreases in food intake and apparent digestibility or increases in energy losses associated with detoxication. Focused studies are much needed to evaluate these ideas. Though the extent of impacts due directly to toxicosis from natural xenobiotics in foods of wild animals is unknown and arguably small because of protective feedback mechanisms, this is not necessarily the case for manmade toxins. For manmade xenobiotics it is interesting to speculate whether the situation might be different. Have most animals had adequate time to evolve mechanisms of detection and avoidance? Perhaps some anthropogenic chemicals are so toxic, and so ubiquitous in foods through deliberate application or subsequent ecosystem transfer processes, that exposure and toxicosis become more likely.

9.10 Toxicogenomics: New Methodologies for the Integrative Study of Exposure, Postabsorptive Processing, and Toxicity in Animals Exposed to Natural and Manmade Toxins

It is a daunting task to achieve anything even approaching a comprehensive knowledge about interactions between animals and xenobiotic compounds.

To comprehend the challenge, recall three important features of ecotoxicology that we have reviewed. First, there are thousands of putative toxicants in the natural world (chapter 2) and made by humans. Second, when animals are exposed to them the xenobiotics are processed by scores of proteins grouped into at least three functional classes: those that biotransform xenobiotics directly by reactions such as oxidation, hydroxylation, and dealkylation (phase I enzymes), those that conjugate them to endogenous compounds to facilitate elimination (phase II enzymes), and those that directionally transport the biotransformed products into urine, from liver into bile, or from intestinal cells into the intestinal lumen (phase III enzymes/transporters). Third, there are myriad targets within animals that might be influenced by the parent xenobiotic or its metabolites. If you contemplate all the possible combinations of these features, and if you do not become too discouraged in doing so, you might wonder whether there are alternative study approaches that will advance knowledge rapidly, besides the study of one feature at a time.

The newly emerging field of toxicogenomics offers a promising alternative. This field elucidates the role that genes play in the biological responses of organisms exposed to environmental stressors. Although the field is too new to offer many illustrative examples, we will focus on one just to provide a stimulating peek at some of its potential.

In order to study more comprehensively the responses to a variety of chemicals Gerhold et al. (2001) exposed laboratory rats to one of four toxicants (3-methylcholanthrene (3MC), phenobarbital, dexamethasone, clofibrate) or to just the solution into which the toxicants were dissolved (called the vehicle control). The particular compounds were chosen because a lot is already known about them, but for our purposes you can just as well imagine that they are poorly understood plant secondary metabolites. The rats were dosed for three days and then their livers were removed to determine the overall response of this important detoxication organ. The researchers used DNA microarrays to test whether expression of specific genes was up- or down-regulated by exposure (box 9.2). The genes they focused on were 130 genes classified as xenobiotic-biotransforming genes, including 51 CYP and other phase I genes and also phase II biotransformation genes. You can see from the small subset of data displayed in table 9.2 that the compounds elicited different expression patterns. For example, 3MC and clofibrate for the most part induced different phase I and phase II enzymes. These different patterns were expected based on what was already known from many previous studies that focused on the compounds and genes one at a time, and so the results using the microarray essentially represent a validation of the microarray technology for measuring more comprehensively and economically the integrated response of the animal.

Box 9.2. Gene Expression Profiling Using cDNA Microarrays

Upon exposure to xenobiotics, a variety of genes, related or unrelated, might be induced or suppressed. The new technology called DNA microarrays or gene chips allows researchers to simultaneously screen tens of thousands of genes for regulatory changes after physiological perturbation. The general methods used by Gerhold et al. (2001) to study hepatic responses of rats following exposure to test xenobiotics are described in many sources, such as recent textbooks, web pages (Ballatori et al. 2003), and journal articles (e.g., Duggan et al. 1999). The fundamental principle upon which the method operates is the hybridization that occurs between nucleic acids (i.e., A-T and G-C for DNA; A-U and G-C for RNA). Gene-specific oligonucleotides are anchored to a matrix and can be used to detect complementary DNAs made from mRNA extracted from animal tissue. The major steps are (1) fabrication of a DNA-based microarray (a "gene chip") that contains probes corresponding to specific genes that are to be investigated; (2) isolation of RNA from tissue samples and the preparation of tagged cDNA or cRNA (copy RNA) representations of the original RNA pool, which are called targets; (3) exposure of the microarray to the tagged cDNA or cRNA so that hybridization can occur; (4) quantitation of the signal from hybridization; and (5) analysis of the data. Let us follow these steps in the study by Gerhold et al. (2001) (figure 9.22).

Figure 9.22. Schematic of how expression of RNA is measured using microarray technology (explanation in box 9.2). (Redrawn from http://www.cs.wustl.edu/~jbuhler/research/array/.)

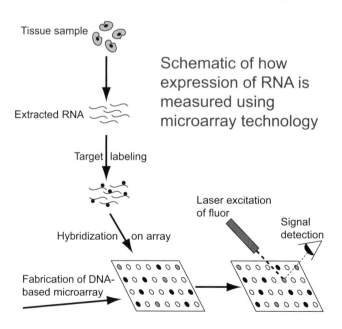

continued

continued

1. Fabrication of a DNA-based Microarray

The gene chip used by Gerhold et al. (2001) contained 1,443 genes and was manufactured commercially. Genes can be chosen from databases, are amplified by PCR (polymerase chain reaction), and after purification and quality control checks the clones that contain the genes are spotted onto a matrix made of membrane or glass. In most cases the DNA is anchored to the matrix by UV radiation.

2. Preparation of Targets from Tissue Samples

Gerhold et al. (2001) harvested liver from rats exposed to test solutions. Total RNA was extracted, which in some studies is then labeled, but Gerhold et al. (2001) went one step further and isolated mRNA specifically, using an additional procedure. The mRNA was "reverse transcribed" (using a reverse transcriptase) to produce a copy of the gene that would have been produced in the original liver cell. In some studies this complementary DNA, or cDNA, is labeled with a radioactive or fluorescent tag, but Gerhold et al. (2001) added a step. They transcribed the cDNA into copy RNA (cRNA) using RNA polymerase, and in this last transcription step they incorporated biotin-UTP and biotin-CTP into the cRNA. The biotin is a reporter molecule that identifies the presence of the cRNA after staining (see below). Some other studies use fluorescently labeled UTP and/or CTP, and for radioactive detection ^{33}P-CTP is used. The net result of all these steps was the creation of a pool of RNA from the original tissue sample amplified in quantity by 6–30 times and tagged with labels.

3. Hybridization

The tagged cRNA was suspended in a buffer solution and allowed to hybridize with the gene chip for 16 h at 45 ℃. Also during this step the chips were stained with streptavidin-R-phycoerythrin, a fluorescent chemical that binds biotin.

4. Signal Detection

The chips were scanned with a fluorescence detection instrument which first excites the fluor using a specific wavelength of light from a laser and then detects the emitted signal fluorescence from phycoerythrin. Because these experiments are carried out with a large excess of immobilized probe on the chip relative to the labeled target in the solution, the more of a specific mRNA in a tissue sample pool, the more hybridization occurs, and the stronger the resulting signal. Gerhold et al. (2001) used a type of microarray

continued

continued

design where a single sample is hybridized to each chip, and in our figure blackened cells represent high expression, cells filled in with gray represent low expression, and white cells represent no expression.

It is also common to cohybridize two samples on each chip. One sample is tagged with a fluorescent compound Cy3 while the other sample is tagged with another, Cy5. When the chip is read with a fluorescent detection instrument genes that are more highly expressed in the sample tagged with Cy3 will fluoresce red, while genes more highly expressed in the sample tagged with Cy5 will fluoresce green. If the gene is expressed equally in both samples it will fluoresce yellow. There are important nuances regarding interpretation of the fluorescent signals (Duggan et al. 1999), but for the sake of simplicity we will not discuss them.

5. Data Analysis

Data from each microarray were normalized to data from a single-vehicle microarray. mRNA expression for each gene was then reported as the relative units of fluorescence. By comparing the expression in samples from rats exposed to test chemicals to the expression in samples from control rats, the researchers could define instances of up- and down-regulation, or no change. The criterion used to define notable changes usually depends on the nature/characteristics of the physiological response. It is common that minimum ratios in the detection signals (test animal: control animal) ranging between 2 and 10 are adopted. Gerhold et al (2001) used a minimum ratio in the detection signal (test animal:control animal) of ≥ 2.

Another interesting finding came from the probes derived from genes involved in metabolic energy regulation. 3MC and clofibrate suppressed several genes that regulate metabolism of sugar (*PEPCK,* mediates gluconeogenesis) or lipids (*LCAT, APOA4, APOC1*) (table 9.2). As discussed by Gerhold et al. (2001), these regulatory events would be expected to cause decreased lipid turnover. Consistent with these events, 3MC was already known to cause lipid accumulation within liver cells, and clofibrate was already known to cause decreased serum lipid levels. Phenobarbital and dexamethasone induced some of these genes regulating lipid metabolism, which is consistent with other studies where they increased serum high-density lipoprotein levels. Findings such as these raise the prospect that microarray technology might also provide an initial screen for toxic effects of xenobiotics that might later be checked with more traditional dose-response studies.

TABLE 9.2
Up- and down-regulation of the expression of different genes
in rats exposed to different toxicants

Gene		Response to chemical (compared to control)			
Abbreviation	Function	Clofibrate	Dexa-methasone	Pheno-barbital	3-methyl-cholanthrene
Phase I					
CYP1A1	cytochrome P-450				+++
CYP1A2	cytochrome P-450				+++
CYP2A1	cytochrome P-450	+++			
CYP2B1	cytochrome P-450	+++	+	+++	
CYP2B2	cytochrome P-450	+++	+++	+++	
CYP4A1	cytochrome P-450	+++			
CYP4A2	cytochrome P-450	+++			
EHc	cytosolic epoxide hydrolase	+++			
EHm	mitochondrial epoxide hydrolase			+++	
Phase II					
UGT1A6	UDP-glucuronosyltransferase	+++	+++		+++
GSTA1	glutathione sulfotransferase	–	+++	+++	+++
GSTA2	glutathione sulfotransferase		+++	+++	+++
PST	phenol/aryl sulfotransferase	+	+		
Phase III					
NTCP	Na-taurocholate cotransporting polypeptide		–		
CMOAT	Canalicular multi-specific organic anion transporter	–		+++	
Metabolism-related genes					
PEPCK	Phosphoenolpyruvate carboxykinase C	– – –			– – –
LCAT	Lecithin:cholesterol acyltransferase	–			–

(*continued*)

Table 9.2 (*continued*)

Gene		Response to chemical (compared to control)			
Abbreviation	*Function*	*Clofibrate*	*Dexa-methasone*	*Pheno-barbital*	*3-methyl-cholanthrene*
Metabolism-related genes					
APOA1	Apolipoprotein AIV	–	+++	+++	
APOA4	Apolipoprotein AIV	– – –	+++	+++	–
APOC1	Apolipoprotein CI	–			–

Note: This table is a subset of a larger data set of gene regulation profiles in rats exposed for 3 days to 3-methylcholanthrene, clofibrate, dexamethasone, and Phenobarbital. Column 1 lists specific genes that were tested and column 2 provides a description of the kind of protein the gene codes for. In columns 3–6, underneath each chemical, is a designation for the change in expression, relative to control rats not exposed to the test chemicals. An increase of ≥ 2-fold relative to control is designated as +++, and smaller changes are designated as +. If the gene was repressed so that there was consistent expression of ≤ 1/2 of the control value, the designation is – – –, whereas consistent expression at values below but closer to control values is designated as –. Adapted from (Gerhold et al. 2001).

On the strength of successes like this researchers have increased the number of compounds assayed and the number of genes studied. One study exposed rats to 52 different compounds at two dose levels and then tested for expression of genes in liver using a library of about 25,000 genes (Waring et al. 2003). Scanning this many genes, of which the function of many is unknown, will facilitate the discovery of novel pathways involved in the response of animals to xenobiotics.

In summary, gene expression profiling has multiple potential uses in monitoring the responses of wildlife to xenobiotics (Ballatori et al. 2003), but more research is needed to determine whether this potential will be realized. Some must address major technical issues, such as whether we can effectively use other methods than DNA arrays (box 9.2) to study responses of animals whose genomes have not already been well characterized. One alternative is called differential display and suppression polymerase chain reaction subtractive hybridization (Ballatori et al. 2003). Another technical challenge has to do with how best to analyze the huge data sets generated from probing with hundreds to thousands of genes. With solutions to these problems in hand we can seek answers to the many interesting questions raised by the examples discussed above. Will particular chemicals cause specific signature patterns of gene expression changes that can be used as biomarkers of exposure? Can we link expression patterns to particular protective or detrimental effects? Can the patterns be used to identify common modes of action across species?

References

Alberts, B., D. Bray, J. Lewis, M. Raff, K. Roberts, and J. D. Watson. 1989. *Molecular biology of the cell*. Garland Publishing, New York.

Appel, H. M., and M. M. Martin. 1992. Significance of metabolic load in the evolution of host specificity of *Manduca sexta*. *Ecology* 73:216–228.

Austin, P. J., L. A. Suchar, C. T. Robbins, and A. E. Hagerman. 1989. Tannin-binding proteins in saliva of deer and their absence in saliva of sheep and cattle. *Journal of Chemical Ecology* 15:1335–1347.

Ayres, M. P., T. P. Clausen, S. F. MacLean, Jr., A. M. Redman, and P. B. Reichardt. 1997. Diversity of structure and antiherbivore activity in condensed tannins. *Ecology* 78:1696–1712.

Ballatori, N., J. L. Boiyer, and J. C. Rockett. 2003. Exploiting genome data to understand the function, regulation, and evolutionary origins of toxicologically relevant genes. *Enviromental Health Perspectives: Toxicogenomics* 111:61–65.

Baudinette, R. V., J. F. Wheldrake, S. Hewitt, and D. Hawke. 1980. The metabolism of [^{14}C] phenol by native Australian rodents and marsupials. *Australian Journal of Zoology* 28:511–520.

Bearhop, S., G. D. Ruxton, and R. W. Furness. 2000. Dynamics of mercury in blood and feathers of great skuas. *Environmental Toxicology and Chemistry* 19:1638–1643.

Belovsky, G. E., and O. J. Schmitz. 1991. Mammalian herbivore optimal foraging and the role of plant defenses. Pages 1-28 in R. T. Palo and C. T. Robbins (eds.), *Plant defenses against mammalian herbivory*. CRC Press, Boca Raton, Fl.

Berenbaum, M. R., M. B. Cohen, and M. A. Schuler. 1992. Cytochrome P450 monooxygenase genes in oligophagous Lepidoptera. Pages 114–124 in C. A. Mullin and J. G. Scott (eds.), *Molecular mechanisms of insecticide resistance.* American Chemical Society, Washington, D.C.

Berenbaum, M. R., and A. R. Zangerl. 1994. Costs of inducible defense: Protein limitation, growth, and detoxification in parsnip webworms. *Ecology* 75:2311–2317.

Berenbaum, M. R., A. R. Zangerl, and K. Lee. 1989. Chemical barriers to adaptation by a specialist herbivore. *Oecologia* 80:501–506.

Berger, B., R. Dallinger, and A. Thomaser. 1995. Quantification of metallothionein as a biomarker for cadmium exposure in terrestrial gastropods. *Environmental Toxicology and Chemistry* 14:781–791.

Bernays, E. A., G. C. Driver, and M. Bilgener. 1989. Herbivores and plant tannins. *Advances in Ecological Research* 19:263–302.

Bosveld, A.T.C., J. Gradener, A. J. Murk, A. Brouwer, M. v. Kampen, E.H.G. Evers, and M. Van den Berg. 1995. Effects of PCDDs, PCDFs and PCBs in common tern (*Sterna hirundo*) breeding in estuarine and coastal colonies in the Netherlands and Belgium. *Environmental Toxicology and Chemistry* 14:99–115.

Bozinovic, F., and F. Novoa. 1997. Metabolic costs of rodents feeding on plant chemical defenses: a comparison between an herbivore and an omnivore. *Comparative Biochemistry and Physiology A* 117:511–514.

Brattsen, L. B. 1988. Enzymic adaptations in leaf-feeding insects to host-plant allelochemicals. *Journal of Chemical Ecology* 14:1919–1939.

Brattsten, L. B. 1979. Biochemical defense mechanisms in herbivores against plant allelochemicals. Pages 199–270 in G. A. Rosenthal and D. H. Janzen (eds.), *Herbivores: Their interactions with plant secondary metabolites*. Academic Press, New York.

Braun, E. J., and W. H. Dantzler. 1997. Vertebrate renal system. Pages 481–576 in W. H. Dantzler (ed.), *Handbook of physiology. Section 13: Comparative physiology*. Oxford university Press, New York.

Bravo, L., R. Abia, M. A. Eastwood, and F. Saura-Calixto. 1994. Degradation of polyphenols (catechin and tannic acid) in the rat intestinal tract. Effect on colonic fermentation and faecal output. *British Journal of Nutrition* 71:933–946.

Brower, L. P., W. N. Ryerson, L. L. Coppinger, and S. C. Blazier. 1968. Ecological chemistry and the palatability spectrum of Monarch butterfly avian predators. *Science* 161:1349–1350.

Caterina, M. J., A. Leffler, A. B. Malmberg, W. J. Martin, J. Trafton, Z.K.R. Petersen, M. Koltzenburg, A. I. Basbaum, and D. Julius. 2000. Impaired nociception and pain sensation in mice lacking the capsaicin receptor. *Science* 288:306–313.

Chapman, R. F. and W. M. Blaney. 1979. How animals perceive secondary compounds. Pages 161–198 in G. A. Rosenthal and D. H. Janzen (eds.), *Herbivores: Their interaction with secondary plant metabolites*. Academic Press, New York.

Cheeke, P. R. 1994. A review of the functional and evolutionary roles of the liver in the detoxification of poisonous plants with special reference to pyrrolizidine alkaloids. *Veterinary and Human Toxicology* 36:240–247.

Cipollini, M. L. 2000. Secondary metabolties of vertebrate-dispersed fruits: Evidence for adaptive functions. *Revista Chilena de Historia Natural* 73:421–440.

Clark, L. 1997. Physiological, ecological, and evolutionary bases for the avoidance of chemical irritants by birds. Pages 1–37 in V. J. Nolan, E. D. Ketterson, and C. F. Thompson (eds.), *Current ornithology*, vol. 14. Plenum Press, New York.

Cresswell, J. E., S. Z. Merritt, and M. M. Martin. 1992. The effect of dietary nicotine on the allocation of assimilated food to energy metaboilism and growth in fourth-instar larvae of the southern armyworm, *Spodoptera eridania* (Lepidoptera: Noctuidae). *Oecologia* 89:449–453.

Dearing, M. D., W. J. Foley, and S. McLean. 2005. The influence of plant secondary metabolites on the nutritional ecology of herbivorous terrestrial vertebrates. *Annual Review of Ecology and Systematics* 36:169–189.

Deichmann, W. B., D. Henschler, B. Holmstedt, and G. Keil. 1986. What is there that is not poison? A study of the *Third Defense by Paracelsus. Archives of Toxicology* 58:207–213.

Duggan, D. J., M. Bittner, Y. Chen, P. Meltzer, and J. M. Trent. 1999. Expression profiling using cDNA microarrays. *Nature Genetics Supplement* 21:10–14.

Epel, D. 1998. Use of multidrug transporters as first lines of defense against toxins in aquatic organisms. *Comparative Biochemistry and Physiology* 120:23–28.

Foley, W. J., G. R. Iason, and C. McArthur. 1999. Role of plant secondary metabolites in the nutritional ecology of mammalian herbivores: how far have we come in

25 years? Pages 130–209 in H.-J.G. Jung and G. C. Fahey (eds.), *Nutritional ecology of herbivores*. American Society of Animal Science, Savoy, Il.

Foley, W. J., S. McLean, and S. J. Cork. 1995. Consequences of biotransformation of plant secondary metabolites on acid-base metabolism in mammals—a final common pathway? *Journal of Chemical Ecology* 21:721–743.

Fossi, M.C., A. Massi, L. Lari, L. Marsili, S. Focardi, C. Leonizio, and A. Rezoni. 1995. Interspecies differences in mixed function oxidase activity in birds: Relationships between feeding habits, detoxication activities and organochlorine accumulation. *Environmental Pollution* 90:15–24.

Fournier, F., W. H. Karasov, K. P. Kenow, M. W. Meyer, and R. K. Hines. 2002. The oral bioavailability and toxicokinetics of methylmercury in common loon (*Gavia immer*) chicks. *Comparative Biochemistry and Physiology A* 133:703–714.

Freeland, W. J. 1991. Plant secondary metabolites: biochemical coevolution with herbivores. Pages 61–81 in R. T. Palo and C. T. Robbins (eds.), *Plant defenses against mammalian herbivory*. CRC Press, Boca Raton, Fl.

Freeland, W. J., and D. H. Janzen. 1975. Strategies in herbivory by mammals: The role of plant secondary compounds. *American Naturalist* 108:269–289.

Gerhold, D., M. Ju, J. Xu, C. Austin, C. T. Caskey, and T. Rushmore. 2001. Monitoring expression of genes involved in drug metabolism and toxicology using DNA microarrays. *Physiological Research* 5:161–170.

Green, A. K., D. M. Barnes, and W. H. Karasov. 2005. A new method to measure intestinal activity of P-glycoprotein in avian and mammalian species. *Journal of Comparative Physiology* 175:57–66.

Guglielmo, C. G., W. H. Karasov, and W. J. Jakubas. 1996. Nutritional costs of a plant secondary metabolite explain selective foraging by ruffed grouse. *Ecology* 77:1103–1115.

Haasch, M. L., P. J. Wejksnora, J. J. Stegeman, and J. J. Lech. 1989. Cloned rainbow trout P 1 450 complementary DNA as a potential environmental monitor. *Toxicology and Applied Pharmacology* 98:362–368.

Hagerman, A. E., and L. G. Butler. 1981. Specificity of proantho-cyanidin-protein interactions. *Journal of Biological Chemistry* 256:4494–4497.

Hanley, T. A., C. T. Robbins, A. E. Hagerman, and C. McArthur. 1992. Predicting digestible protein and digestible dry matter in tannin-containing forages consumed by ruminants. *Ecology* 73:537–541.

Harborne, J. B. 1993. *Introduction to ecological biochemistry*. Academic Press, New York.

Hickie, B. E., D. Mackay, and J. d. Koning. 1999. Lifetime pharmacokietic model for hydrophobic contaminants in marine mammals. *Environmental Toxicology and Chemistry* 18:2622–2633.

Huang, Y. W., M. J. Melancon, R. E. Jung, and W. H. Karasov. 1998. Induction of cytochrome P450-associated monooxygenases in northern leopard frogs, *Rana pipiens* by 3,3′4,4′,5-pentachlorobiphenyl (PCB 126). *Environmental Toxicology and Chemistry* 17:1564–1569.

Huang, Y. W., J. J. Stegeman, B. R. Woodin, and W. H. Karasov. 2001. Immunohisto-chemical localization of cytochrome P450-associated monooxygenases induced by

3,3′,4,4′,5 pentachlorobiphenyl in multiple organs of leopard frog, *Rana pipiens. Environmental Toxicology and Chemistry* 21:191–197.

Iason, G. R., and A. H. Murray. 1996. The energy costs of ingestion of naturally occurring nontannin plant phenolics by sheep. *Physiological Zoology* 69:532–546.

Illius, A. W., and N. S. Jessop. 1995. Modeling metabolic costs of allelochemical ingestion by foraging herbivores. *Journal of Chemical Ecology* 21:693–719.

Jakubas, W. J., W. H. Karasov, and C. G. Guglielmo. 1993. Ruffed grouse tolerance and biotransformation of the plant secondary metabolite coniferyl benzoate. *Condor* 95:625–640.

Jakubas, W. J., and J. R. Mason. 1991. Role of avian trigeminal sensory system in detecting coniferyl benzoate, a plant allelochemical. *Journal of Chemical Ecology* 17:2213–2221.

James, P. S., M. H. Ralphs, and D. B. Nielsen. 1988. *The ecology and economic impact of poisonous plants.* Westview Press, Boulder, Colo.

Jondorf, W. R. 1981. Drug-metabolizing enzymes as evolutionary probes. *Drug Metabolism Reviews* 12:379–430.

Jordt, S. E., and D. Julius. 2002. Molecular basis for species-specific sensitivity to "hot" chili peppers. *Cell* 108:421–430.

Kappas, A., and A. P. Alvares. 1975. How the liver metabolizes foreign substances. *Scientific American* 232:22–31.

Karasov, W. H. 1989. Nutritional bottleneck in a herbivore, the desert wood rat (*Neotoma lepida*). *Physiological Zoology* 62:1351–1382.

Karasov, W. H., M. W. Meyer, and B. W. Darken. 1992. Tannic acid inhibition of amino acid and sugar absorption by mouse and vole intestine: Tests following acute and subchronic exposure. *Journal of Chemical Ecology* 18:719–736.

Kay, R. N. B., W. V. Engelhardt, and R. G. White. 1980. The digestive physiology of wild ruminants. Pages 743–761 in Y. Ruckebusch and P. Thivend (eds.), *Digestive physiology and metabolism in ruminants.* MTP Press, Lancaster, England.

Kelley, K. C., K. S. Johnson, and M. Murray. 2002. Temporal modulation of pyrrolizidine alkaloid intake and genetic variation in performance of *Utetheisa ornatrix* caterpillars. *Journal of Chemical Ecology* 28:669–685.

Keppler, D., I. Leier, and G. Jedlitschky. 1997. Transport of glutathione conjugates and glucuronides by the multidrug resistance proteins MRP1 and MRP2. *Biological Chemistry* 378:787–791.

Kidd, K. A., D. W. Schindler, D.C.G. Muir, W. L. Lockhart, and R. H. Hesslein. 1995. High concentrations of toxaphene in fishes from a subarctic lake. *Science* 269:240–242.

Klekowski, E. J., S. A. Temple, A. M. Siung-Chang, and K. Kumarsingh. 1999. An association of mangrove mutation, scarlet ibis, and mercury contamination in Trinidad, West Indies. *Environmental Pollution* 105:185–189.

Lake, B. G. 1987. Preparation and characterization of microsomal fractions for studies on xenobiotic metabolism. Pages 183–215 in K. Snell and B. Mullock (eds.), *Biochemical toxicology: A practical approach.* IRL Press, Oxford, U.K.

Lamb, J. G., J. S. Sorensen, and M. D. Dearing. 2001. Comparison of detoxification enzyme mRNAs in woodrats (*Neotoma lepida*) and laboratory rats. *Journal of Chemical Ecology* 27:845–857.

Lawler, I. R., W. J. Foley, G. J. Pass, and B. M. Eschler. 1998. Administration of a 5HT3 receptor antagonist increases the intake of diets containing *Eucalyptus* secondary metabolites by marsupials. *Journal of Comparative Physiology B* 168:611–618.

Lawler, I. R., J. Stapley, W. J. Foley, and B. M. Eschler. 1999. Ecological example of conditioned flavor aversion in plant-herbivore interactions: Effect of terpenes of *Eucalyptus* leaves on feeding by common ringtail and brushtail possums. *Journal of Chemical Ecology* 25:401–415.

Livingstone, D. R. 1991. Organic xenobiotic metabolism in marine invertebrates. Pages 45–185 in R. Gilles (ed.), *Advances in comparative and environmental physiology,* vol. 7. Springer-Verlag, Berlin.

Lowry, J. B., C. S. McSweeney, and B. Palmer. 1996. Changing perceptions of the effect of plant phenolics on nutrient supply in the ruminant. Australian Journal of Agricultural Research 47:829–842.

Mangione, A. M., M. D. Dearing, and W. H. Karasov. 2004. Creosote bush (*Larrea tridentata*) resin increases water demands and reduces energy availability in desert woodrats *(Neotoma lepida). Journal of Chemical Ecology* 30:1409–1429.

Mansour, M. H. 1981. Efficiency of two allelochemics on the conversion of ingested and digested food into the body tissues of *Spodoptera littoralis* (Biosd.) (Lepidoptera: Noctudidae). *Zeitschrift Angewandle Entomologic* 92:493–499.

Mehansho, H., T. M. Asquith, L. G. Butler, J. Rogler, and D. M. Carlson. 1992. Tannin-mediated induction of proline-rich protein synthesis. *Journal of Agricultural and Food Chemistry* 40:93–91.

Meyer, M. W., and W. H. Karasov. 1991. Chemical aspects of herbivory in arid and semiarid habitats. Pages 167–187 in R. T. Palo and C. T. Robbins (eds.), *Plant defenses against mammalian herbivory.* CRC Press, Boca Raton, Fl.

Mole, S., J. C. Rogler, and L. G. Butler. 1993. Growth reduction by dietary tannins: different effects due to different tannins. *Biochemical Systematics and Ecology* 21:667–677.

Motulsky, H. J., and L. A. Ransnas. 1987. Fitting curves to data using nonlinear regression: a practical and nonmathematical review. *Federation of American Societies of Experimental Biology Journal* 1:365–374.

Muthukrishnan, J., S. Mathavan, and K. Venkatasubbu. 1979. Effects of caffeine and theophylline on food utilization and emergence in *Danaus chrysippus* L. (Lepidoptera: Danidae). *Entomon* 4:307–312.

Muthukrishnan, J., and T. J. Pandian. 1987. Insecta. Pages 373-511 in T. J. Pandian and F. J. Vernberg (eds.), *Animal energetics,* vol. 1. Academic Press, San Diego, Calif.

Newman, M. C. 1998. *Fundamentals of ecotoxicology.* Sleeping Bear/Ann Arbor Press, Chelsea, Mich.

Nitao, J. K. 1990. Metabolism and excretion of the furanocoumarin xanthotoxin by parsnip webworm, *Depressaria pastinacella. Journal of Chemical Ecology* 16: 417–428.

Nosek, J. A., S. R. Craven, W. H. Karasov, and R. E. Peterson. 1993. 2,3,7,8 -tetrachlorodibenzo-p-dioxin in terrestrial environments: implications for resource management. *Wildlife Society Bulletin* 21:179–187.

O'Donnell, M. J. 1997. Mechanisms of excretion and ion transport in invertebrates. Pages 1207–1289 in W. H. Dantzler, ed., *Handbook of physiology. Section 13: Comparative physiology.* Oxford University Press, New York.

Paasivirta, J. 1991. *Chemical ecotoxicology.* Lewis, Chelsea, Mich.

Pass, G. J., and W. J. Foley. 2000. Plant secondary metabolites as mammalian feeding deterrents: separating the effects of the taste of salicin from its post-ingestive consequences in the common brushtail possum (*Trichosurus vulpecula*). *Journal of Comparative Physiology B* 170:185–192.

Provenza, F. D. 1996. Acquired aversions as the basis for varied diets of ruminants foraging on rangelands. *Journal of Animal Science* 74:2010–2020.

Provenza, F. D., J. V. Nolan, and J. J. Lynch. 1993. Temporal contiguity between food ingestion and toxicosis affects the acquisition of food aversions in sheep. *Applied Animal Behaviour Science* 38:269–281.

Provenza, F. D., L. Ortega-Reyes, C. V. Scott, J. J. Lynch, and E. A. Burritt. 1994. Anti-emetic drugs attenuate food aversions in sheep. *Journal of Animal Science* 72:1989–1994.

Rattner, B. A., D. J. Hoffman, and C. M. Marn. 1989. Use of mixed-function oxygenases to monitor contaminant exposure in wildlife. *Environmental Toxicology and Chemistry* 8:1093–1102.

Ritschel, W. A. 1999. *Handbook of basic pharmacokinetics.* American Pharmaceutical Association, Washington, D.C.

Riviere, J. E. 1999. *Comparative pharmacokinetics. Principles, techniques and applications.* Iowa State University Press, Ames, Iowa.

Robbins, C. T., A. E. Hagerman, P. J. Austin, C. McArthur, and T. A. Hanley. 1991. Variation in mammalian physiological responses to a condensed tannin and its ecological implications. *Journal of Mammalogy* 72:480–486.

Robbins, C. T., S. Mole, A. E. Hagerman, O. Hjeljord, D. L. Baker, C. C. Schwartz, and W. W. Mautz. 1987. Role of tannins in defending plants against ruminants: Reduction in dry matter digestion? *Ecology* 68:1606–1615.

Roesijadi, G. 1992. Metallothioneins in metal regulation and toxicity in aquatic animals. *Aquatic Toxicology* 22:81–114.

Rozin, P. 1988. Cultural approaches to human food preferences. Pages 137–153 in J. E. Morley, M. B. Sterman, and J. H. Walsh (eds.), *Nutritional modulation of neural function.* Academic Press, San Diego, Calif.

Safe, S. 1990. Polychlorinated biphenyls (PCBs), dibenzo-*p*-dioxins (PCDDs), dibenzofurans (PCDFs), and related compounds: Environmental and mechanistic considerations which support the development of toxic equivalence factors (TEFs). *Critical Reviews in Toxicology* 21:51–88.

Scheline, R. R. 1991. *CRC handbook of mammalian metabolism of plant compounds.* CRC Press, Boca Raton, Fl.

Schoonhoven, L. M., and J. Meerman. 1978. Metabolic cost of changes in diet and neutralization of allelochemicals. *Entomologia Experimentalis et Applicata* 24:689–693.

Service, R. F. 2005. Going from genome to pill. *Science* 308:1858–1860.

Simpson, S. J., and D. Raubenheimer. 2001. The geometric analysis of nutrient-allelochemical interactions: A case study using locusts. *Ecology* 82:422–439.

Skopec, M. M. 2003. Polyphenolics in the mammalian gut: Effects on glucose absorption and the efficacy of a salivary defense mechanism. Ph.D. dissertation, University of Wisconsin, Madison.

Skopec, M. M., A. E. Hagerman, and W. H. Karasov. 2004. Do salivary proline-rich proteins counteract dietary hydrolysable tannin in laboratory rats? *Journal of Chemical Ecology* 30:1679–1692.

Slansky, F., and G. S. Wheeler. 1992. Caterpillars' compensatory feeding response to diluted nutrients leads to toxic allelochemical dose. *Entomologia Experimentalis et Applicata* 65:171–186.

Snyder, M. J., and J. Glendinning. 1996. Causal connection between detoxification enzyme activity and consumption of a toxic plant compound. *Journal of Comparative Physiology A* 179:255–261.

Soderlund, D. M., and D. C. Knipple. 1999. Knockdown resistance to DDT and pyrethroids in the house fly (Diptera: Muscidae): From genetic trait to molecular mechanism. *Annals of the Entomological Society of America* 92:909–915.

Sorensen, J. S., J. D. McLister, and M. D. Dearing. 2005. Plant secondary metabolites compromise the energy budgets of specialist and generalist mammalian herbivores. *Ecology* 86:125–139.

Stapley, J., W. J. Foley, R. Cunningham, and B. Eschler. 2000. How well can common brushtail possums regulate their intake of *Eucalyptus* toxins? *Journal of Comparative Physiology B* 170:211–218.

Tahvanainen, J., P. Niemela, and H. Henttonen. 1991. Chemical aspects of herbivory in boreal forest-feeding by small rodents, hares, and cervids. Pages 115–131 in R. T. Palo and C. T. Robbins (eds.), *Plant defenses against mammalian herbivory*. CRC Press, Boca Raton, Fl.

Tewksbury, J. J., and G. P. Nabhan. 2001. Seed dispersal—Directed deterrence by capsaicin in chillies. *Nature* 412:403–404.

Thomas, D. W., C. Samson, and J. M. Bergeron. 1988. Metabolic costs associated with the ingestion of plant phenolics by *Microtus pennsylvanicus*. *Journal of Mammalogy* 69:512–515.

Thompson, D. R., R. W. Furness, and P. M. Walsh. 1992. Historical changes in mercury concentrations in the marine ecosystem of the north and north-east Atlantic ocean as indicated by seabird feathers. *Journal of Applied Ecology* 29:79–84.

Thorhallsdottir, A. G., F. D. Provenza, and D. F. Balph. 1990. Ability of lambs to learn about novel foods while observing or participating with social models. *Applied Animal Behaviour Science* 25:25–33.

Walker, C. H. 1980. Species variations in some hepatic microsomal enzymes that metabolize xenobiotics. Pages 113–164 in J. W. Brridges, L. F. Chasseaud, and G. G. Gordon (eds.), *Progress in drug metabolism*, vol. 5. John Wiley & Sons, New York.

Walker, C. H. 1990. Kinetic models to predict bioaccumulation of pollutants. *Functional Ecology* 4:295–301.

Walker, C. H. 1994. Comparative toxicology. Pages 193–218 in E. Hodgson and P. E. Levi (eds.), *Introduction to biochemical toxicology.* Appleton and Lange, Norwalk, Conn.

Walker, C. H., S. P. Hopkin, R. M. Sibly, and D. B. Peakall. 1996. *Principles of ecotoxicology.* Taylor and Francis, London.

Walker, C. H., and M.J.J. Ronis. 1989. The monooxygenases of birds, reptiles and amphibians. *Xenobiotica* 19:1111–1121.

Wallace, W. G., T.M.H. Brouwer, M. Brouwer, and G. R. Lopez. 2000. Alterations in prey capture and induction of metallothionein in grass shrimp fed cadmium-contaminated prey. *Environmental Toxicology and Chemistry* 19:962–971.

Waring, J. F., G. Cavet, R. A. Jolly, J. McDowell, H. Dai, R. Ciurlionis, C. Zhang, R. Stoughton, P. Lum, A. Ferguson, C. J. Roberts, and R. G. Ulrich. 2003. Development of a DNA microarray for toxicology based on hepatotoxin-regulated sequences. *Environmental Health Perspectives: Toxicogenomics* 111:53–60.

Whyte, J. J., R. E. Jung, C. J. Schmitt, and D. E. Tillitt. 2000. Ethoxyresorufin-*O*-deethylase (EROD) activity in fish as a biomarker of chemical exposure. *Critical Reviews in Toxicology* 30:347–370.

Zucker, W. V. 1983. Tannins: Does structure determine function? An ecological perspective. *American Naturalist* 121:335–365.

Limiting Nutrients

CHAPTER TEN

Ecological Stoichiometry

Until now, we have investigated how animals assimilate food and how they process it. This is important because what an animal eats matters not only to the animal itself. It also shapes the animal's role in its ecological community and determines its contribution to the flux of energy and materials in ecosystems. The study of foraging and of trophic relationships is at the center of ecology. Since Lindeman's (1942) paper on the trophic structure of communities, energy has been the currency of choice for ecologists. With few exceptions (Belovsky 1984), foraging/trophic ecology has focused on energy as "the" currency used by animals to make foraging decisions (Stephens and Krebs 1986). In chapter 1, we adopted an "energy-centric" view of life and throughout the book we have emphasized energy. For some time, a pair of binoculars (or a microscope) and a calorimetric bomb was all that was needed to understand how animals make feeding choices and to understand their ecological roles.

In our earlier chapter on food chemistry (Chapter 2), and in the next few chapters, we will focus our attention away from energy, and concentrate on the variation among foods in the ratios of carbon (which is a proxy for energy) to key nutrients such as water, nitrogen, and other minerals. Higher ratios in foods than in consumers can sometimes create relative nutrient scarcity. Such scarcity has been a powerful selective agent on animal physiology and behavior. In the next three chapters we explore how ecologists determine nutrient requirements, and how requirements can vary under different conditions. We approach the problem at a variety of levels of organization, from the individual organism to the consequences at the level of ecosystems. With information on nutrient requirements in hand we can assess whether needs match relative availability in the environment. Thus, we can examine how limiting nutrients are for organisms and what the effects of nutrient limitations are for animals and for the ecological roles that they play.

In this chapter we explore the consequences of adopting a thoroughly reductionist approach. We ask: How far we can get if we ignore the biochemical complexities of life and assume that animals simply acquire and process chemical elements? The energetic approach attempts to reduce a variety of processes to a single currency: the mighty Joule. Here we will examine the consequences of reducing organisms to entities that harvest, assimilate, and process chemical elements. In a pioneering paper, ecologist Bill Reiners (1986) proposed using the flow of different elements through ecosystems as a multidimensional way of studying questions that are refractory to the one-currency/energy approach. A handful of ecologists have taken Reiner's suggestion to heart and the vigorous field of ecological stoichiometry has been launched (Elser et al. 1996; Sterner and Elser 2002). This chapter is devoted to describing the benefits of considering the consequences of reducing organisms to their stoichiometry or elemental composition.

10.1 Ecological Stoichiometry: The Power of Elemental Analysis

Ecological stoichiometry considers ratios of key elements to analyze how the characteristics and activities of sources, producers, and consumers influence, and are influenced by, the ecosystems in which they are found. Ecological stoichiometry focuses on the relative elemental composition of the participants in ecological interactions. For reasons that will be outlined shortly, it emphasizes nitrogen (N), phosphorus (P), and carbon (C). Box 10.1 describes a CHN analyzer, one of the pieces of equipment used to determine the elemental composition of biological samples. Ecological stoichiometry's conceptual foundation is the application of the law of conservation of mass in the form of the balanced budgets described in chapter 1 to the flux of materials in ecosystems. Figure 10.2 shows examples of stoichiometric budgets in chemistry, biochemistry, and ecology. Equation C in this figure highlights the importance of considering the stoichiometric composition of prey and predators for ecosystem level processes.

10.1.1 Living Systems as "Molecules"

The last equation in figure 10.2 uses a remarkably useful analogy and one that is at the very foundation of ecological stoichiometry: that of living systems (construed broadly to include organisms and collections of organisms) as chemical compounds. Of course, organisms are neither molecules nor bags of individual atoms. We are astounding collections of innumerable macromolecules that interact

Box 10.1.

Ecological stoichiometry relies on the measurement of the elemental composition of organic compounds. Although automated reliable elemental analyzers (or CHN and CHNS analyzers) were developed only 30 years ago, they have rapidly become pivotal tools for ecologists. An elemental analyzer has a deceptively simple structure. Dry and ground samples are carefully weighed in a microbalance and packaged in small tin foil cups. The sample-containing cups are loaded into an autosampler that drops them into a quartz reactor. This reactor is placed within a furnace heated at 1050°C. The samples are "flash combusted" in a stream of helium that has been temporarily enriched with excess oxygen. The term flash combustion refers to a combustion in the presence of catalysts (tungstic oxide and copper) and aided by the exothermic oxidation of tin which increases the local temperature of the sample from 1050°C to from 1700 to 1800°C. The complete oxidation of an organic sample yields gaseous N_2, CO_2, H_2O, and SO_2, which are separated by a gas chromatography (GC) column. The amount of each gas is measured by a thermal conductivity detector (TCD) as it exits the GC. Each gas produces a peak in the signal of the detector. These peaks are then integrated by a computer workstation to calculate the percentage of N, C, H, and S in the sample (figure 10.1). The phosphorus content of a sample can be measured relatively simply using a colorimetric method (Woods et al. 2002).

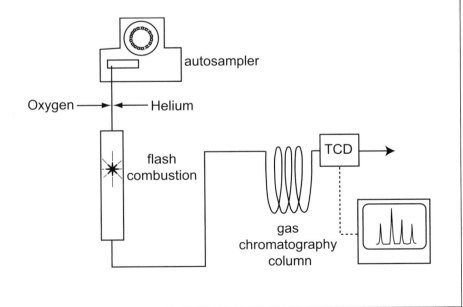

Figure 10.1. Diagram of a CHN analyzer.

Figure 10.2. Ecological stoichiometry applies mass-balance principles to the participants of ecological interactions. Mass balance must be met in simple inorganic chemical reactions (A), simple biochemical transformations (B), and complex ecological interactions (C). In the stoichiometry of a predator-prey interaction shown in equation C, a prey of a given nitrogen (N) and phosphorus (P) content is consumed by a predator of fixed composition, (x:y) to increase its biomass by a factor Q. If the predator's elemental composition differs from that of its prey (x:$y \neq a$:b), the composition of the predator's waste will have a composition that differs from that of the prey (a':$b' \neq a$:b). Thus, the disparity in the elemental ratio between predators and prey dictates the elemental composition, and thus the relative availability of different elements, of the recycled "waste" used by decomposers and autotrophs. (From Elser et al. 1996.)

Stoichiometry in chemistry

$$3CaCl_2 + 2Na_3PO_4 \rightleftarrows CaCO_3(PO_4)_2 + 6NaCL$$

Stoichiometry in biochemistry

$$C_6H_{12} + 6O_2 \rightleftarrows 6CO_2 + 6H_2O$$

Stoichiometry in ecology

$$(N_xP_y)_{predator} + (N_aP_b)_{prey} \rightleftarrows Q(N_xP_y)_{predator} + (N_{a'}P_{b'})_{waste}$$

with each other and with smaller molecules and free elements. The practitioners of biological stoichiometry do not question the importance of physiology and biochemistry. Rather, they attempt to use the metaphor of living systems as molecules in an attempt to construct a minimal, but more complete, ecological theory than the one based on a single currency (energy, or carbon, or nitrogen). A list of elements (or even of complex compounds) and their amounts does not reveal the architecture of an organism or a component of an ecosystem, but it is a useful starting point.

Let us push the analogy of organisms as molecules by taking a look at the approximate stoichiometric "formula" for a human being:

$$H_{375\times10^6}O_{132\times10^6}C_{85.7\times10^6}N_{6.43\times10^6}Ca_{1.5\times10^6}P_{1.02\times10^6}S_{206\times10^3}Na_{183\times10^3}K_{177\times10^3}$$
$$Cl_{127\times10^3}Mg_{40\times10^3}Si_{38.6\times10^3}Fe_{268\times10^3}Zn_{2.11\times10^3}Cu_{76}I_4Mn_{13}F_{13}CrSe_4Mo_3Co_1$$

This formula shows the relative, not the absolute, amounts of each element. It expresses all values relative to the value of the scarcest element (cobalt), which is assigned a value of 1 (Sterner and Elser 2002). Even a perfunctory look at it should justify ecological stoichiometry's attention to C, N, and P. Only 11 elements appear in all biological systems. In humans, these 11 elements constitute

TABLE 10.1

Percentages of four most common elements in humans

Element	Hydrogen	Oxygen	Carbon	Nitrogen	Others
Percentage	61.8	25.4	9.4	1.4	1.0

Source: (Frausto da Silva and Williams 1994).

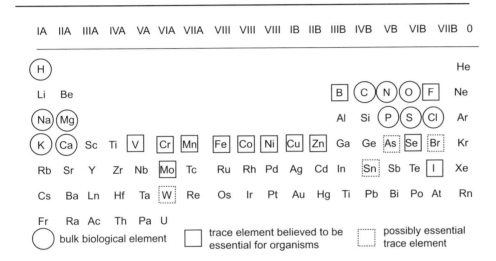

Figure 10.3. The chemistry of life is, to a very large extent, the chemistry of the lighter elements. Of the elements required by living organisms, all but molybdenum and iodine have atomic number less than 35. (After Frausto da Silva and Williams 1994.)

99.9% of the total number of atoms, but just four of these elements—carbon, nitrogen, hydrogen, and oxygen—make up 99% of this total (Frausto da Silva and Williams 1994; table 10.1). Oxygen and hydrogen are present in large amounts because living creatures contain large quantities of water. Carbon and nitrogen are next in abundance because, together with hydrogen and oxygen, they are the basic elements of organic structures and metabolites. The other seven elements (sodium, potassium, calcium, magnesium, phosphorus, sulfur, and chlorine) represent about 0.9% of the total number of atoms in the human body. As mentioned before, although phosphorus is not a member of the big four elements of life (C, H, O, and N), it is a focal element for biological stoichiometry. The reason is that it is a central, and sometimes limiting, constituent of the nucleic acids (DNA and RNA) that define living systems. Besides the 11 elements listed above, there are ten more that are required by most, but not all biological systems, and seven more that are required only by some animals or some plants (figure 10.3).

10.1.2 Redfield Ratios and the Stoichiometry of the Ocean

In the previous section, we portrayed the chemical formula of a human being. We are stoichiometrically quite homeostatic, and thus the elemental composition of humans probably varies relatively little among individuals. In the early 1930s, Alfred Redfield, an animal physiologist turned oceanographer, made a most astounding discovery. He found that we can associate an elemental ratio with a whole ecological realm! Redfield discovered that the atomic ratios between nitrogen, phosphorus, and carbon in marine plankton were very similar to their relative proportions in the open ocean. For every atom of phosphorus there are approximately 16 atoms of nitrogen and 106 atoms of organic carbon. The Redfield ratio is worth emphasizing in an equation:

$$C_{106}:N_{16}:P_1.$$

Here it is worth pointing out that in ecological stoichiometry the "stoichiometries" of interest are ratios of elements (C:N, N:P, C:P, and C:N:P). In ecological stoichiometry, these ratios are expressed not as mass, but as atomic/molar ratios.

Redfield found not only that these elemental ratios were similar between plankton and water, but also that they appeared to be remarkably congruent in numerous regions of the world's oceans. According to Falkowski (2000), ocean scientists have adopted Redfield ratios as canonical values comparable to Avogadro's number, the speed of light in a vacuum, and the Boltzmann constant. Falkowski explains the relative constancy of C:N:P ratios in the ocean in the following way: The organisms in plankton are functionally similar assemblages of biochemicals often encased in a shell formed from the most widely available ingredients. Most of the plankton eats other plankton with similar composition, and the outcome is that on average the N:P ratios of plankton are similar throughout the world (Falkowski 2004). When the dead planktonic organisms sink, bacteria consume them and oxidize the organic matter to form dissolved inorganic nutrients (CO_2, NO_3^-, and PO_4^{3-}). The ratio of these nutrients is similar to the C:N:P ratio in plankton. Falkowki's explanation works well at small scales, but does not explain the constancy across vast expanses of ocean unless the ocean behaves like a well-mixed vat. This seems to be the case: the residence time of N and P in the ocean ($\approx 10^4$ years) is long relative to the ocean's circulation time (10^3 years).

Falkowski's description is probably correct and explains (1) the similarity between plankton and the mineralized nutrients, and (2) the spatial and temporal constancy of these ratios. However, it does not explain why the ratio should be on average $C_{106}:N_{16}:P_1$. When phytoplankton are grown in culture in the laboratory N:P ratios can vary from 6:1 to 60:1. There is a lot of wiggle room for the stoichiometric ratios of marine life to vary, yet on average and at large scales, Redfield ratios seem to hold. Sterner and Elser (2002) give an account of the potential causes for the (in our view still mysterious), Redfield ratios in the ocean, and we refer you to them for details. Whatever the reasons are for the specific values of Redfield ratios, they suggest something quite striking: To a first approximation, the ocean is a vast biological vat in which enormous quantities of C, N, and P circulate both vertically and horizontally at very large scales and in relatively constant proportions.

Freshwater systems are different. Their stoichiometric characteristics are influenced very strongly by the geology of the sites in which they are situated, and by the characteristics of their watershed, which determines how big they are and how much material flows into them from terrestrial sources. Hence, their C:H:N ratios are very variable and this variation can have huge consequences for the food webs that inhabit them. It is only recently that Redfield's approach has been applied to terrestrial systems, and these show significant differences from marine ones. McGroddy et al. (2004) examined the C:N:P stoichiometry of forest foliage and soil litter throughout the world. They found global C:N:P ratios for foliage (1212:28:1) and litter (3007:45:1) that, when compared to the canonical oceanic 106:16:1 Redfield ratios, greatly favored C over N, and P, and slightly favored N over P. McGroddy and her collaborators also found variation among biomes and along latitudinal gradients. N:P ratios in foliage and litter seem to be twofold higher in tropical relative to temperate broadleaf forests, and conifer forests seem to have intermediate values. In subsequent sections we will explore in more detail the stoichiometric differences between the realms and their consequences, but before doing so we must explore the principles that guide inquiry in ecological stoichiometry.

10.1.3 The Four Founding Blocks of Ecological Stoichiometry

Ecological stoichiometry has its foundation on four simple but penetrating observations. *First*, there is a significant biochemical discontinuity between producers and consumers. Although the biomass of producers contains most of the substances essential for herbivore nutrition, the relative composition of these substances is often very different from that of herbivores, and it may also be different from their optimal food composition (see Sterner and Hessen 1994; and subsequent sections in this chapter). We can call this first observation the *consumer/producer stoichiometric dichotomy*. A *second* important difference between producers and consumers is the degree of constancy in body (and even cellular) elemental composition. In autotrophs, like algae and terrestrial plants, the relative content of carbon, nitrogen, and phosphorus can vary enormously. C:N:P ratios can vary greatly among plant parts, among plants of a single species, and among plant species (Nielsen et al. 1996). Animals, on the other hand, tend to regulate their elemental content much more strictly (Sterner 1995). To maintain a homeostatic body composition that differs from that of the producers that they consume, herbivores either have to be choosy in what they eat and/or they must retain and dispose of the elements that they get from plants differentially. This second observation can be called the *producer/ consumer homeostasis dichotomy. Third*, organisms with high production rates (as a result of either growth or reproduction) seem to have low C:P and N:P ratios (reviewed by Elser et al. 2000a). Elemental composition seems to shape and be shaped by an organism's life history. Following Elser et al. (2000a) we call this observation the *growth rate hypothesis*. The *fourth* and final observation is that factors leading to changes in the elemental composition of the consumer community can lead to changes in their stoichiometric ratios and thus to the composition of the waste products that they recycle. We can call this the *consumer-driven recycling principle*. The following sections describe these observations in more detail. The consumer/producer stoichiometric dichotomy (observation 2) is to a large extent a consequence of differences in elemental composition constancy ("homeostasis") between producers and consumers. Therefore, we deal with differences in homeostasis between produces and consumers (observation 2) first.

10.2 An Ecological Stoichiometry Primer

10.2.1 How Homeostatic Are Organisms? A Stoichiometric View

The concept of homeostasis has its origins in Claude Bernard's proposition that organisms often have two environments: the external environment and the internal environment (or *milieu intérieur*). Bernard was impressed with the constancy of the chemical composition and physical properties of the fluids that bathe animal cells (Bernard 1865). Years later, in the twentieth century, Cannon emphasized not only the relatively stable composition of body fluids, but also the coordinated regulatory mechanisms that achieve this constancy (Cannon 1932). They coined the term "homeostasis" both to refer to constancy in the internal environment and to describe the coordinated physiological processes that regulate it. Open almost any mammalian or human physiology textbook and you are likely to find reference to a homeostatic mechanism.

The concept of homeostasis is an important one in biology. Historian of science L. L. Langley (1973) stated: "Any paper (in biology) that is worth the paper that it is printed on, should clarify a homeostatic mechanism." Although Langley overstates the case, it is to the credit of the founders of ecological stoichiometry that they have brought homeostasis back to the forefront of ecological research. Homeostasis had been used in ecology before, but its use was associated with the notion of ecosystems as self-regulating, well-ordered, and stable systems. As this notion became first questioned, and then discredited, the term homeostasis was dropped from mainstream ecological thought (Worster 1994). In ecological stoichiometry, homeostasis has a limited definition. The term refers to the similarity between the elemental composition of organisms relative to the composition of the resources that they use. If the elemental composition of an organism reflects passively the composition of set of elements in its resources, then the organism is *nonhomeostatic*. In contrast, if an organism's elemental composition is completely independent of the resources that it consumes then the organism is *strictly homeostatic*. Because there are limits to the combinations of elements that can comprise a living organism, all organisms will exhibit some degree of homeostasis. This degree varies, however, and this variation has, as we will see, important ecological consequences.

How can we quantify "homeostasis" (in the limited stoichiometric sense of this chapter)? Figure 10.4 provides a possibility. In stoichiometric terms, nonhomesotatic organisms are what they eat. Therefore their stoichiometry reflects that of the resources that they use. Throughout this chapter, and following Sterner and Elser (2002), we use the word "stoichiometry" as shorthand for the composition of a mixture of elements. If we plot the stoichiometry of the resources on the x axis and that of the consumer organism on the y axis, then nonhomeostatic organisms fall on the $y = x$ line. If the stoichiometry of the organism falls on a line that passes through the origin but has a slope that is different from 1, the organisms are not what they eat, but they are still nonhomeostatic. Their elemental composition follows passively that of the environment (Sterner and Elser (2002) call this situation "constant proportionality"). Contrast these two situations with strictly homeostatic organisms, which keep their elemental composition independent despite changes in the chemical compositions of their resources. In these organisms the relationship between y and x in figure 10.4 is a horizontal line.

So far, we have represented graphically only two possible extremes. Sterner and Elser (2002) suggest that for many organisms that show partial homeostasis, the relationship between their stoichiometry and that of their resources can be modeled by power functions

$$y = cx^{1/H} \qquad (10.1)$$

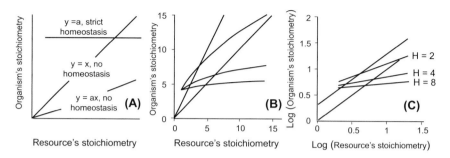

Figure 10.4. In ecological stoichiometry, the term homeostasis refers to the similarity between the elemental composition of organisms relative to the composition of the resources that they use. Thus, we can represent homeostasis graphically by plotting a metric of chemical composition (C:N, C:P, %P, . . .,etc.) of the organism in the y axis and that of its resources in the x axis (A). If the elemental composition of an organism lies along a line that passes through the origin, then the organism is nonhomeostatic. If an organism's elemental composition is completely independent of the resources that it consumes then the organism is strictly homeostatic and its composition can be represented by a horizontal line. For intermediate levels of stoichiometric homeostasis, the relationship between the organism's stoichiometry (y) and that of its chemical resources (x) can be represented by power functions (B), which can be linearized in log-log plots (C). The reciprocal of the log-log slope of this relationship (H) is an index of stoichiometric homeostasis. In the figure we varied H from 2 to 8.

which can be linearized using logarithms as

$$\log(y) = \log(c) + \frac{1}{H}\log(x). \tag{10.2}$$

The parameter H in equations 10.1 and 10.2 is an index of homeostasis. When H is large, the relationship between a consumer's stoichiometry and that of its resources has a shallow slope (figure 10.4). Indeed, when $H = \infty$, the organism shows strict homeostasis. If $H = 1$, the organism is nonhomeostatic. Figure 10.5 shows several examples of organisms with varying degrees of stoichiometric homeostasis.

It is worth emphasizing that the degree of homeostasis depends on both the element under consideration and the organism in question. Macroelements (e.g., C and N) tend to be regulated more homeostatic than trace elements (e.g., Se and Cu). It also appears that autotrophs (primary producers) are less stoichiometrically homeostatic than heterotrophs (see subsequent sections). Williams and Frausto da Silva (1996) and Sterner and Elser (2002)

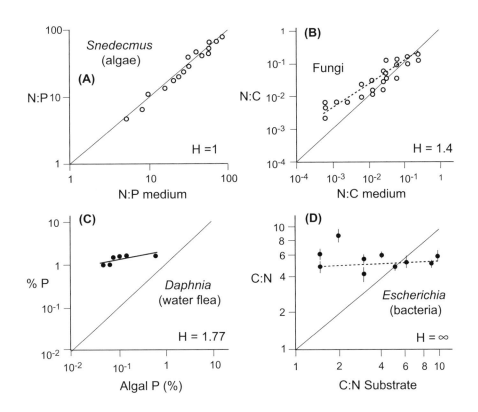

Figure 10.5. Examples of varying stoichiometric homeostasis in algae (*Scenedesmus*; A), fungi (B), a water flea (*Daphnia*; C), and bacteria (D). (Modified from Sterner and Elser 2002.)

present illuminating speculations on the potential evolutionary reasons for differences among groups of organisms in homeostasis. In this chapter, we will simply describe the ecological consequences of this variation.

10.2.2 Animals and Plants: Why Do They Differ in Stoichiometric Homeostasis

If you have been in a basic biology course, you have been exposed to the differences between eucaryotic animal and plant cells. Both types of cells have nuclei, mitochondria, ribosomes, and various organelles. Plant cells have chloroplasts and animal cells do not. Another important difference is the presence of cell walls (support structures) and a large central vacuole (a storage structure) in plant cells. This difference has major stoichiometric consequences. Because cell walls add carbon but little or no nitrogen and phosphorus, C:N and C:P ratios in terrestrial plants can be high and may be variable depending on the allocation of plants to structural elements. The large central vacuole is responsible for variation in the whole plant's stoichiometry because it allows plants to store both organic and inorganic compounds. Although the cytoplasm of plant cells can have a very constant chemical composition (i.e., it can be very stoichimetrically homeostatic), the whole plant cell can show large variation in chemical composition as a result of the variation in the content of its storage vacuoles. Leigh and Wyn-Jones (1985) appropriately call the cytoplasm of plant cells "selective" and the vacuole "promiscuous." The vacuole protects plants against the vagaries in the supply of nutrients and explains plants' potentially large stoichiometric variation.

There is one more important source of stoichiometric variation in higher plants. Plants can show dramatic stoichiometric variation among the organs (roots, stems, and leaves) of a single individual. Leaves, rootlets, and flowers are nutrient rich relative to woody stems. In apple trees, for example, the C:N ratio of the whole plant is about 165. However, the C:N ratio of various structures ranges from about 45 in leaves to 400 in the wood of the main stem (Kramer and Kozlowski 1979). Consider the stoichiometric implications of figure 10.6. The leaf to total mass for major plant types (and thus the stoichiometry of

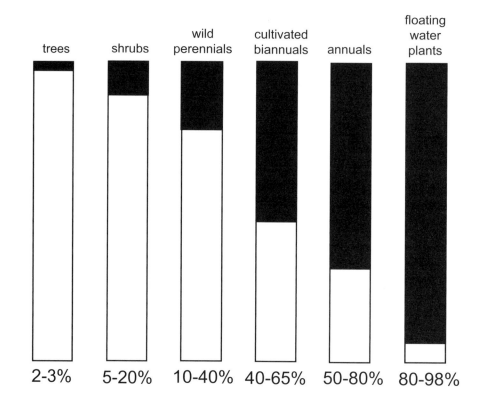

Figure 10.6. The leaf to total mass ratio (respectively, shaded to unshaded area) varies in major groups of plant life forms. In the figure, this ratio is expressed as the percentage of total dry mass. (After Körner 1993.)

trees   shrubs   wild perennials   cultivated biannuals   annuals   floating water plants

2-3%   5-20%   10-40%   40-65%   50-80%   80-98%

whole plants) varies 40-fold from deciduous trees to floating aquatic plants (Körner 1993). Enquist and Niklas (2002; Niklas and Ehnquist 2002) suggest that much of the variation in biomass allocation to leaves, stems, and roots can be explained by simple scaling power laws (see chapter 1). This observation, coupled to the seemingly consistent variation in stoichiometry among plant structures, suggests the possibility of a predictive theory linking allometry and ecological stoichiometry.

The many factors that impinge on how plants allocate nutrients among organs also have stoichiometric consequences. Assessing the effect of a variety of environmental conditions, such as nutrients, light levels, and CO_2 concentrations on the stoichiometry of plants is something that both algal and higher plant physiologists do routinely. The resulting vast literature documents extreme stoichiometric variation among autotrophs (reviewed by Sterner and Elser 2002). Although there is variation in stoichiometric homeostasis among plant species, plants are clearly less homeostatic than the consumers that eat them. The stoichiometric plasticity of plants leads to considerable variation in C:N:P composition at the base of food webs (see following section).

Although we believe that there is a clear difference in stoichiometric homeostasis between autotrophs and heterotrophs, it would be inappropriate to sweep under the rug the stoichiometric variation in heterotrophs. Like plants, animal cells can have storage structures and animals often have specialized storage tissues. Animals store lipids (sometimes in prodigious amounts; Hagen and Auen 2001) and vertebrates can use the skeleton as a repository for minerals. The size of these storage pools has direct stoichiometric consequences and the study of its variation is important and interesting. Indeed, we devote part of chapters 7 and 13 to the storage of lipids and a section in chapter 11 to bone as a potential mineral warehouse. Here we simply note that the effect of lipid is on C:N and C:P ratios (lipids lack N and P), whereas bone contains large amounts of P and therefore has the potential of changing the magnitude of C:P and N: P ratios. Furthermore, it may be inappropriate to lump all heterotrophs in a single group with homogeneous stoichiometric characteristics. Some bacteria, yeast, and fungi are ecologically important heterotrophs, and a variety of species in all these groups share with algae (and with autotrophic bacteria) the ability to accumulate and store phosphorus in the form of polyphosphate (Kulaev 1979). Some heterotrophic bacteria can accumulate large amounts of phosphorus. Polyphosphate can constitute 20% of the dry mass of the heterotrophic bacterium *Acinetobacter* (Deinema et al. 1980*)*. The capacity of some bacteria to accumulate phosphorus has been garnered by humans to clean up phosphate-rich wastewater (Grady and Filipe 2000).

10.2.3 Differences among the Realms: The Stoichiometry of Autotrophs and Heterotrophs in Aquatic and Terrestrial Systems

Having examined in a perfunctory fashion the differences in homeostasis between autotrophs and heterotrophs, we can now take a macroscopic look at the stoichiometric variation that one can find in nature. In a massive effort, a working group at the National Center for Ecological Analysis and Synthesis (NCEAS) compiled and analyzed the available data on the elemental composition of autotrophs and heterotrophs. The NCEAS ecological stoichiometry working group used only invertebrates (zooplankton and insects) in their survey of

Figure 10.7. In freshwater zooplankton and insects, C:N ratios are less variable than C:P and N:P ratios. This observation is consistent with the notion that P content is more variable than N content. Compare the distribution of C:N values with those of autotrophs in figure 10.8. (After Elser et al. 2000b.)

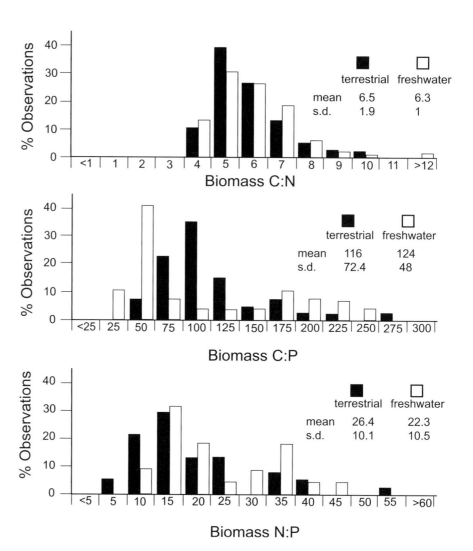

heterotrophic organisms. They found several remarkable results. First, they found that the means and variances in C:N, C:P, and N:P ratios were notably similar between zooplankton and terrestrial invertebrates (figure 10.7). This result is perhaps not surprising: all metazoans with relatively simple body plans should have similar stoichiometries and similar levels of homeostasis. This result does not mean that there is no meaningful stoichiometric variation among invertebrates. There is a clue to this variation, and its importance can be gleaned from the difference in the variation between C:N ratios and both C:P and N:P ratios. C:N ratios vary about twofold, whereas C:P and N:P ratios vary four- and fivefold (figure 10.8). The variation in P is a theme that we will return to in a subsequent section (*growth rate hypothesis*).

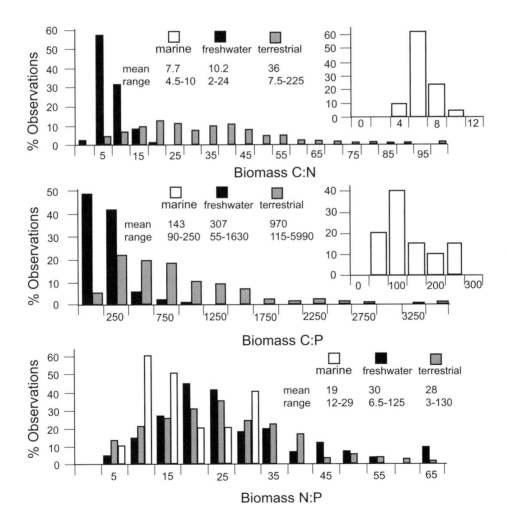

Figure 10.8. The stoichiometry of autotrophs in marine and freshwater ecosystems (estimated by the stoichiometry of seston) and of terrestrial ecosystems (estimated by the stoichiometry of foliage) varies significantly. C:N and C:P ratios are higher and more variable in terrestrial than in aquatic ecosystems. The variation in N:P ratios is not as dramatic.

As expected, and in contrast with heterotrophs, autotrophs show large stoichiometric differences among ecosystems and large variation within ecosystem types (figure 10.8). The producers at the base of marine pelagic systems have relatively homogeneous stoichiometries and provide consumers with food that is rich in nitrogen and phosphorus. The uniformity of marine ecosystems stands in contrast with the variation of freshwater lake ecosystems. The seston in lakes appear to be often poor in phosphorus. The producer base of terrestrial and aquatic ecosystems differ dramatically in C:nutrient ratios. Mean C:N and C:P ratios of terrestrial autotrophs are more than threefold higher than those of aquatic producers. Because the data for terrestrial autotrophs in figure 10.9 is for leaves, and a significant fraction of the terrestrial biomass is in woody tissue and bark, which have higher C:nutrient ratios, this comparison underestimates these ratios for terrestrial plants. In terrestrial ecosystems, a nonchoosy herbivore that attempted to feed on generalized plant biomass would obtain a lot less N and P than would a pelagic member of the zooplankton feeding on generalized seston. The similarity in N:P ratios between autotrophs in freshwater and terrestrial ecosystems is intriguing because it is often stated that primary production in lakes is limited by phosphorus (Schindler 1978), whereas primary production in terrestrial ecosystems is limited by nitrogen (Sprent 1987). Elser et al. (2000b) propose two alternatives: (1) the prevalence of nitrogen limitation is greater than previously thought, or (2) phosphorus limitation in terrestrial ecosystems is more common than previously acknowledged (see chapter 11).

The patterns shown in figures 10.7 and 10.8 must be considered with some care. By necessity, macroscopic patterns based on large data sets rely on assumptions. Some of the important assumptions of figure 10.9 are worth noting. The first one is that the stoichiometric ratios for "autotrophs" were measured in seston. Seston is a collective term for all particulate matter in free water. It includes both bioseston (phytoplankton, zooplankton, and heterotrophic bacteria) and tripton (nonliving particulate material). Although as argued by Elser et al. (2000b), seston may be dominated by phytoplankton, some of the variation in stoichiometry may be due to the presence of heterotrophic bacteria and tripton. This comment is not meant to criticize the useful work of the NCEAS ecological stoichiometry working group, but simply to point at the need for more data on the stoichiometry of the micro-organisms that probably really rule ecosystems. Biological oceanographer Lawrence Pomeroy has stated that "the marine food web consisting primarily of diatoms, copepods, and fishes is now generally seen as an excrescence from what is normally a microbial food web" (Pomeroy 2001). A similar comment could be applied to many freshwater and even to some terrestrial systems. Understanding the differences in stoichiometry among realms demands that we pay attention to microbes.

10.2.4 The Importance of C:N Ratios in Terrestrial Ecosystems

Ecological stoichiometry has been developed primarily by aquatic ecologists and applied the most to lakes. The application of stoichiometric approaches to these freshwater ecosystems has been fruitful and exciting. It has revealed tantalizing connections between nutrition, life history, and biogeochemistry. Stoichiometric analyses of terrestrial organisms and ecosystems are almost absent (see Markow et al. 1999 and Reich and Oleksyn 2004 for exception). At this point it is still unclear what the emphasis of ecological stoichiometry on N:P ratios that has been so successful in pelagic freshwater will reveal for terrestrial ecosystems (Sterner and Elser's 2002 opus magnum provides many hints). What is without doubt, is that C:N ratios are remarkably important in terrestrial ecosystems. The importance of C:N ratios is highlighted by the highly significant positive relationships between N and P content in leaf tissue found by Reich and Oleskyn (2004). Although there is a lot of residual variation in this relationship, nevertheless it appears that high N content implies high P in leaf tissues.

Although the N:P ratios of primary producers and consumers do not differ between freshwater and terrestrial ecosystems, the C:N ratios of autotrophs (primary producers) are much higher and much more variable in terrestrial than in aquatic ecosystems (Elser et al. 2000b, figure 10.8). This observation suggests that nitrogen might be more limiting for consumers on land than in water. Unlike plants, animals use structural materials based on various calcareous materials (bones, shells, and carapaces) and nitrogen (proteins and chitin in exoskeletons which is carbohydrate that contains nitrogen and is accompanied by protein, see chapter 2). Thus their C:N ratios are lower than those of plants, often by orders of magnitude (figures 10.7 and 10.8). Not only are the C:N ratios of plants high, but also the energy contained in plants' structural carbohydrates can be in a chemical form that often renders it inaccessible to assimilation by animals (Chapters 2 and 6). In general, it can be said that terrestrial herbivores are faced with the challenge of concentrating the nitrogen in plants to permit growth of their relatively low C:N tissues. This can be a daunting task. In a thoroughly documented book, T. C. R White (1993) provides example after example of animal populations that are limited not by the availability of energy, but by the scarcity of protein (or N).

Animals reduce the effective C:N ratio of their food by using a variety of nonexclusive behavioral and physiological stratagems (White 1993). Because these stratagems have been described in some detail before in chapters 3, 5 and 6, we will describe them here only briefly: (1) Herbivores can be selective. Plants must grow and reproduce, and growth is not well served by nitrogen-poor structural tissues. Herbivores often concentrate their feeding on protein-rich (and probably also phosphorus-rich) reproductive structures and actively growing parts. However, because plants often grow and reproduce in bursts, animals must synchronize their life histories to the plants' growth and reproduction cycles. (2) Herbivores can process food rapidly and assimilate food selectively. They can defecate the relatively indigestible cell wall carbohydrates intact, and skim the soluble cell contents which have lower C:N ratios. This strategy may require ingesting phenomenal amounts of food and thus may be accompanied by the high costs of having to feed incessantly and of having to maintain a large gut volume. (3) Herbivores can enlist the help of microorganisms (chapters 5 and 6). Gut microrganisms use the nitrogenated by-products of the herbivores' protein metabolism (urea and uric acid) and the carbon skeletons of plant's carbohydrates, to manufacture microbial protein. Microorganism-mediated nitrogen recycling has the consequence of improving the C:N ratio of the herbivore's ingested food. Finally, (4) many animals that rely on plants for energy, supplement their nitrogen budgets with animal protein. Terrestrial vertebrates that are often considered herbivores and that obtain most of their energy from food with dismally high C:N ratios (good examples are fruit and nectar), eat other animals to gather enough protein to grow and reproduce.

10.2.5 Global Change, C:N Ratios, and the Uncertain Future of Herbivore-Plant Interactions

The concentration of CO_2 is much higher now than it was before the industrial revolution, and it continues to rise. It is expected to roughly double from approximately 350 parts per million (ppm) in 2003 to 650 ppm by the mid twenty-first century. One of the uncontestable effects of increased atmospheric CO_2 on plants is an increase in the C:N ratios of their tissues (Curtis and Wang 1998). Loladze (2002) used a very simple mathematical model to predict that plant biomass C:nutrient ratios in general, not only C:N ratios, will increase. In short, the available evidence suggests that the nutritional quality of plants will decrease as atmospheric CO_2 increases. Bezemer and Jones (1998) and Watt et al. (1995) reviewed the effects of feeding plants grown under elevated CO_2 on insect herbivores. Their results suggest that the effects of CO_2 enhancement depend on insect guilds. Leaf chewers seem to eat more, grow more slowly, and take longer to develop when they eat plants grown in an elevated CO_2 atmosphere (Watt et al. 1995). In contrast, the effects of increased CO_2 on phloem feeders (e.g., aphids) can be positive (but whether the effect was positive, neutral, or negative, depends on species). It is difficult to predict from the results of these laboratory experiments the future of herbivore plant interactions in our CO_2-enriched future. However it seems likely that we will end up living in a world in which plant tissues are less nutritious, and some insect herbivores less abundant (Watt 1995).

Before abandoning the effect of CO_2 enrichment on herbivore-plant interactions we will dwell briefly on an observation from the previous paragraph. Leaf-chewing insects appear to compensate and eat more of plants with lower nitrogen content (and hence higher C:N ratios; Whittaker 2001). This result is often attributed to compensatory feeding, a topic that we discussed in detail in chapter 3. The implication seems to be that insects eat more to increase their nitrogen intake (figure 10.9). This belief is not justified and points to one of the dangers of relying solely on stoichiometric data to make inferences about ecological mechanisms. The compensatory response of insects to increased C:N ratios can be explained by (1) increased feeding to increase nitrogen intake, or, more parsimoniously, (2) as increased intake to compensate for reduced content

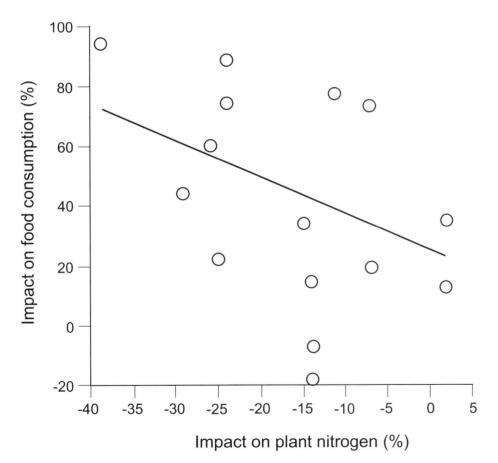

Figure 10.9. Folivorous insects increase their consumption as the C:N ratio of the plants that they consume increases as a result of elevated atmospheric CO_2. The change in consumption was measured as the % difference in intake between insects fed on plants grown under current CO_2 levels (350 ppm) and 700 ppm.

of digestible nutrients and energy. Increased C:N ratios are the potential result not only of reduced protein, but of increased structural (cell walls) and nonstructural (starch) carbohydrates that may be difficult to assimilate by insects, and of non-nitrogenous secondary compounds. Only a combination of detailed feeding experiments with controlled diets and measurements of macronutrient content in plants exposed to elevated CO_2 can distinguish between these two alternatives (chapter 3). Elemental analyses are powerful, but have severe limitations.

10.2.6 The Growth Rate Hypothesis

Recall from figure 10.7 that in invertebrates C:N ratios vary about twofold, whereas C:P and N:P ratios vary four- and fivefold (figure 10.7). What are the causes and ecological consequences of this variation in phosphorus content? Ecologists Jim Elser and Robert Sterner posed a bold hypothesis. They conjectured that high levels of cellular phosphorus are characteristic of productive cells, tissues, and organisms and hypothesized that these levels characterize organisms with high growth rates. They called this correlation between high relative P:N and P:C ratios and production rate the growth rate hypothesis (Elser et al. 2000a). Because this hypothesis can help to link processes at the cellular and organism level with ecosystem function, it is worth dwelling on it. Why should rapidly growing individuals show high relative phosphorus content? One reason is that, in order to grow rapidly, animals must contain high levels of messenger and ribosomal RNA, two central ingredients of the cellular machinery used to synthesize tissue protein. Because RNA has a lot of phosphorus relative to other cellular components (about 10% per unit weight), a high RNA content necessarily translates into a high relative phosphorus concentration (figure 10.10 and table 10.2). Aquaculturists and fish ecologists take advantage of the correlation between high RNA content and high production rates and use RNA:DNA ratios as indices of fish growth (Buckley and Caldarone 1999; box 10.2).

Ontogenetic variation in growth rate seems to be associated with high RNA content, and hence with low N:P ratios (Clemmensen 1994). Across species there seems to be a positive correlation between RNA content (and hence P:N ratios) and growth rate (reviewed by Elser et al. 2000a; figure 10.12). Elser et al. (2000a) hypothesize that variation in RNA content, and hence in growth/production rate, across species correlates with the molecular structure of the genes that regulate and code for the production of ribosomal RNA. They hypothesize that the length of the intergenic spacers and the number of copies of these genes underpins variation in the rate of ribosomal RNA synthesis, in

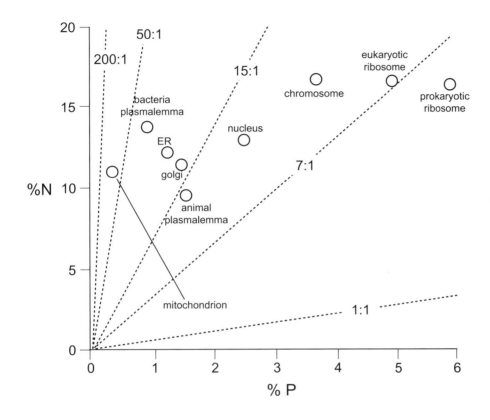

Figure 10.10. The nitrogen and phosphorus composition of organelles and cell structures varies significantly. Values along dotted lines have the same N:P ratio. Note that RNA-rich ribosomes have low N:P ratios. Hence, we expect cells and tissues with active protein synthesis, and hence with lots of ribosomes, to also have low N:P ratios. (After Elser et al. 1996.)

TABLE 10.2
Approximate stoichiometry of several important biochemicals

Compound	C(%)	N(%)	P(%)	C:N:P
Lipids				
Triacylglycerides (tripalmytin)	75	0	0	1:0:0
Phospholipids[a]	65	1.6	4.2	39:0.8:1
Carbohydrates (glucose and				
glucose polymers)	37	0	0	1:0:0
Chitin	41.4	7	0	5.1
Proteins	53	17	0	10.7:1:0
ATP	28	14	18	3.9:1.7:1
Nucleic acids	310.7	14.5	8.7	9.5:3.7:1

Note: Stoichiometric compositions are expressed both as percentages of mass and as atomic ratios. Because the biochemicals listed are not single compounds, the values in this table are averages. After Sterner and Elser (2002).

[a] The values in this row are the average for five common phospholipids (phosphatidylethanolamine, phosphatidylcholine, phosphatidylglycerol, phosphatidyldiglycerol, and sphyngomyelin).

Box 10.2.

The idea that RNA/DNA ratios can be used as indices of growth is based on two premises: (1) Growth is closely linked to protein synthesis, which is controlled by ribonucleic acid (RNA). Hence, one can assume that the quantity of RNA in a tissue will vary according to growth. (2) In contrast, desoxyribonucleic acid (DNA), the carrier of genetic information, is assumed to exist in a constant quantity in somatic cells. Hence DNA content should be an index of cell number. Bulow (1970) was the first to introduce the use of RNA/DNA ratios for assessing growth rate. Figure 10.11 shows one of his pioneering data sets. Although RNA/DNA ratios are a promising biochemical tool to assess growth rates in a variety of organisms, their use remains a matter of keen debate. RNA/DNA ratios are not magic bullets, they work very well in some species but not in others. In addition, results often depend on methodological nuances: methods differ in sensitivity and nucleic acid yield, generating discrepancies among studies. Bergerson (1997) summarizes the terms of this debate and Buckley and Caldarone (1999) summarize the analytical state of the art.

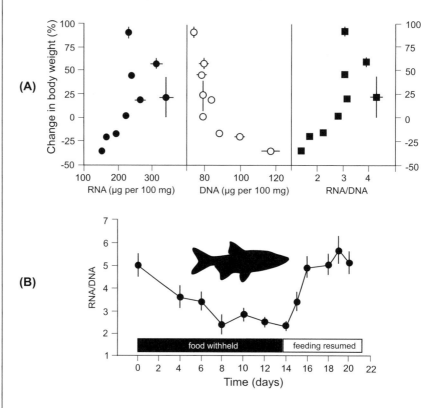

Figure 10.11. (A) In golden shiners (*Notemigonus crysoleucus*), growth rate increased with both RNA content of fat-free tissues and the ratio of RNA to DNA (points in A are means ± SD for 20 fish). In contrast, DNA content was negatively related to body mass changes. (B) RNA/DNA ratios were very responsive to physiological changes. They decreased rapidly in fasted fish and increased rapidly when fish were fed again. (After Bulow 1970.)

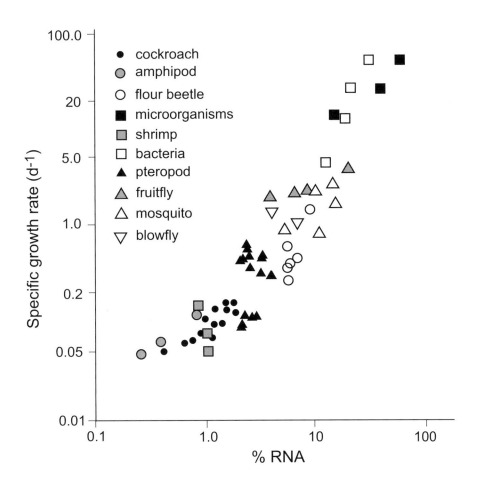

Figure 10.12. In a variety of arthropods and microorganisms RNA content is correlated with mass-specific growth rate (fraction of mass change per day). Note that arthropods and microorganisms seem to follow the same relationship. (After Sterner and Elser 2002.)

C:N:P ratios and, hence in growth and production rates. If their hypothesis is correct, ecological stoichiometry might have revealed unexpected connections between molecular biology, organismic function, and ecosystem processes (figure 10.13).

10.2.7 The Consumer-Driven Recycling Principle

Ecological stoichiometry explores the consequences of an old adage that declares "all waste is food." The material excreted by consumers end up making the raw materials used by both decomposers and consumers. Here we will use two examples to illustrate how differences in the composition of predator and prey determine the composition of the material wasted that then can become available for ecosystem recycling (Sterner 1990; and following sections). To illustrate this principle we will use an example of a study that had an important

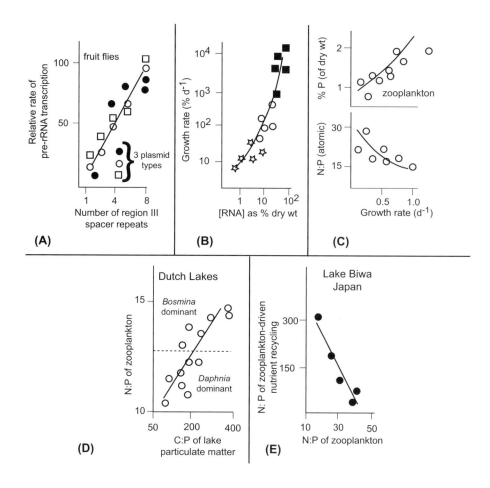

Figure 10.13. The growth rate hypothesis attempts to link the molecular genetics of growth with its ecological consequences. Elser et al. (2000a) hypothesize that selection for growth leads to changes in the frequency of genes associated with ribosomal genes. Specifically, they postulate a positive correlation between the number of repeats in the non-transcribed intergenic spacer and the rate at which rRNA is transcribed (A). Because increased growth is presumably accompanied by rapid protein synthesis (see box 10.2 and figure 10.12), RNA and growth rate must be positively correlated (B). Given that ribosomes are relatively rich in phosphorus (figure 10.10), population growth rates should increase with phosphorus content and decrease with N:P ratio. In ecosystems, food with unbalanced elemental composition can influence the community dominance of consumers. Lakes with P-rich seston should presumably be dominated by fast-growing *Daphnia* with low N:P ratios(C); in lakes with P-deficient seston, slow-growing (high N:P) copepods (like *Bosmina*) should dominate (D). Finally, differences in the somatic C:N:P of zooplankton can affect the excretion rate and the composition of the materials recycled and used by phytoplankton (E).

role in the history of ecological stoichiometry. Elser and his colleagues introduced piscivorous bass into a small lake (Tuesday Lake) in Michigan (Elser et al. 1988). The result was a drastic reduction in minnows that feed on zooplankton. *Daphnia* grew rapidly and came to dominate the zooplankton community, replacing slow-growing calanoid copepods. The result was phosphorus limitation in phytoplankton. In contrast, when they removed piscivores (Peter Lake), zooplankton became dominated by slow-growing consumers (calanoid copepods), and phytoplankton became nitrogen limited. A stoichiometric explanation for this unexpected result should be forming in your mind. Although at this point, the result should not be surprising, in 1988 it was. The introduction (or removal) of a top piscivore led to changes in the nutrient that limited phytoplankton growth.

What is the explanation? In Tuesday Lake, low minnow predation on zooplankton allowed the dominance of *Daphnia*. *Daphia* grow rapidly and have (as expected from the growth rate hypothesis) lower N:P ratios (N:P approximately 12:1) than slower-growing calanoid copepods (N:P greater than 30:1). Because, in order to grow rapidly, *Daphnia* retain P and excrete N, their dominance in the zooplankton community led to a high N:P recycling ratio and to phosphorus limitation in phytoplankton (Sterner et al. 1992). In contrast, in Peter Lake heavy minnow predation on *Daphnia* allowed the zooplankton to become dominated by calanoid copepods. Because the N:P ratio of the producer algal pool tends to be lower than that of those of copepods, the copepods retained nitrogen and disposed of phosphorus, and phytoplankton became nitrogen limited (figure 10.14).

In summary, the intensity of predation by fish in lakes can lead to changes in the stoichiometry of primary consumers and the composition of the waste products that they recycle, and hence to unexpected changes in the nutrients that limit the growth of phytoplankton. Elser and Urabe (1999) provide several more examples of the role of food-web stoichiometry in regulating phytoplankton community structure and function.

Our next example of the consumer-driven recycling principle is a simple, but informative, consequence of the contrasting C:N ratios in the food eaten by carnivores and herbivores. The stoichiometric considerations outlined in figure 10.2 suggest that the waste products of carnivores will contain high levels of the nutrients that are limiting for the growth of primary producers (nitrogen and phosphorus). In contrast, the carbon-rich waste products of many herbivores will end up mainly in the hyphae of decomposers. A good example of the potentially important role of stoichiometric recycling of consumer wastes for ecosystem function is the nitrogen and phosphorus subsidy of terrestrial

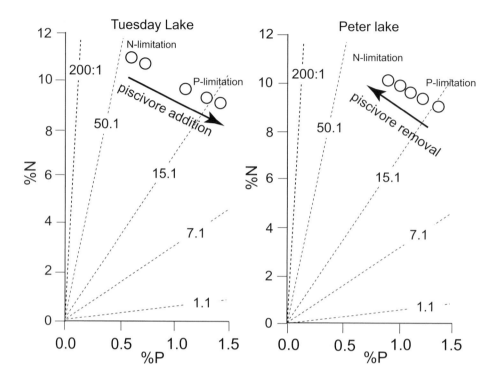

Figure 10.14. The experimental addition or removal of piscivorous predators to lakes illustrates how food web structure can determine whether primary producers are nitrogen or phosphorous limited. Elser et al. (1988) removed piscivorous predators from Peter Lake and added piscivorous predators to Tuesday Lake. After the removal of piscivores in Peter Lake, predation on *Daphnia* (a cladoceran with low N:P ratios) increased. Slow-growing copepods with a high N:P ratio came to predominate in the community of primary consumers. These copepods retained nitrogen and disposed of phosphorus, imposing a nitrogen limitation on phytoplankton. When piscivores were added to Tuesday Lake, low fish predation on zooplankton allowed the dominance of *Daphnia* with low N:P, an increase in the N:P ratio of the recycled product available to producers and hence to phosphorus limitation in phytoplankton. Each point in the stoichiometric diagrams represents a weekly measurement of the nitrogen and phosphorus composition of the zooplankton in Tuesday and Peter Lakes after each experimental manipulation. Values along dotted lines have the same N:P ratio. (Figure modified from Elser et al. 1996.)

productivity by seabird guano. In the Sea of Cortez, islands with seabird roosting colonies have N and P soil concentrations that are up to six times higher than those in islands without birds (Anderson and Polis 1999). Not only is the biomass and productivity of plants higher in bird than in nonbird islands, but also the plants in bird islands are significantly enriched in N and P and hence more nutritious. The consequence is that most consumers (rodents and beetles) have higher densities in bird islands (Sanchez-Piñero and Polis 2000). Stapp et al. (1999) hypothesize that several species of consumers would be unable to persist in some of these arid islands without the N and P subsidy brought by birds from marine ecosystems. As ecological stoichiometry suggests, the elemental ratios of what animals eat, and hence of what they retain and what they recycle, can have major ecological consequences.

10.3 Are Energy and Elements Two Independent Currencies?

Of course not: we established this dichotomy for didactic purposes and to stress the need to complement the energetic view of life with a stoichiometric perspective. In this section, we describe how these two perspectives can be linked. In chapter 1 we emphasized the integrative role that allometry and temperature play in both physiology and ecology. Here we describe a model that suggests that the same equations that are at the core of the metabolic theory of ecology may hold the key to bring together the energetic and stoichiometric perspectives. Our treatment follows the reasoning followed Gillooly and his collaborators, (2005). We will first derive the relationship between the concentration of RNA in a heterotrophic eukaryote and its body mass and then, from this relationship, we will find the relationship between total body phosphorus and body mass. The first part of the argument relies on the assumption that we can approximate the concentration of RNA in an organism by the concentration of ribosomal RNA. Because ribosomal RNA represents about 85% of total RNA in an organism, this is not the most risky or simplistic of all the assumptions that we will make!

The reasoning behind Gilooly et al.'s derivation depends on combining three ingredients to find out the relationship between the concentration of RNA in an organism ([RNA] in mg per gram of body mass) and body mass. The first ingredient is an equation from chapter 1 that gives the mass-specific aerobic metabolic rate (b) as a function of body mass (m_b) and body temperature (T) as

$$b \propto (m_b)^{-1/4} \, e^{-E_1/kT} = b_0(m_b)^{-1/4} e^{-E_i/kT},\qquad(10.3)$$

where E_i is the average activation energy for respiration, k is the Boltzmann constant, and b_0 is a normalization constant that differs among taxa. The second ingredient is an equation for the mass specific protein synthesis in an organism as a function of metabolic rate. We can approximate the rate of protein synthesis per unit mass (S in grams per gram) as the product of the average mass of an amino acid (M_{AA}) times the rate at which the organism processes energy in the form of ATPs to make peptide bonds:

$$S = \left(\frac{\text{energy invested in protein synthesis per unit time}}{\text{energy required per peptide bond}} \right) (M_{AA}). \quad (10.4)$$

Let us assume that each amino acid has a mass M_{AA}, that forming a peptide bond consumes four ATPs, each of which generates E_{ATP} Joules of energy, and that protein synthesis consumes a nearly constant fraction α of total respiration (α is about 0.2). Then,

$$S = \left(\frac{\alpha b}{4 E_{ATP}} \right) (M_{AA}) = \left(\frac{\alpha b_0 (m_b)^{-1/4} \, e^{-E_i / kT}}{4 E_{ATP}} \right) (M_{AA}). \quad (10.5)$$

The third ingredient requires expressing the mass-specific rate of protein synthesis as a function of the number of ribosomes per unit body mass times the rate of protein synthesis per ribosome:

$$S = (\text{number of ribosomes}) \times (\text{rate of protein synthesis per ribosome}) \quad (10.6)$$

Let us denote [RNA] as the concentration of ribosomal RNA (in mg of RNA per gram of tissue) and M_{RNA} the mass of a ribosome ($\approx 4 \times 10^{-21}$ mg in eukaryotes), then the number of ribosomes per gram of tissue equals [RNA]/M_{RNA}. The rate of protein synthesis per ribosome should equal the product of the average mass of an amino acid (M_{AA}), times the rate at which ribosomes make peptide bonds ($s_0 \approx 10^{11}$ peptide bonds per ribosome per second) times the temperature dependence of this process, which we can assume follows Arrhenius kinetics:

$$S = \left(\frac{[RNA]}{M_{RNA}} \right) \left(M_{AA} \, s_0 e^{-E_p / kT} \right). \quad (10.7)$$

We are almost there. We now combine equations 10.5 and 10.7 as

$$\left(\frac{\alpha b_0 (m_b)^{-1/4} \, e^{E_i / kT}}{E_{ATP}} \right) (M_{AA}) = \left(\frac{[RNA]}{M_{RNA}} \right) \left(M_{AA} \, s_0 e^{-E_p / kT} \right) \quad (10.8)$$

and obtain an expression for [RNA]:

$$[RNA] = \left(\frac{\alpha M_{RNA} b_0}{4 s_0 E_{ATP}} \right) (m_b)^{-1/4} e^{(E_p - E_i)/kT}. \quad (10.9)$$

This equation looks complicated but makes two simple and important predictions: (1) It predicts that the concentration of RNA in an organism will scale

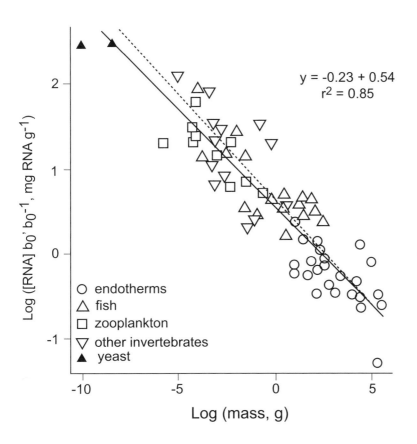

Figure 10.15. The concentration of RNA in heterotrophic eukaryotes is related allometrically to body mass to approximately the −1/4 power. Gillooly et al. (2005) divided [RNA] by the ratio between the standardization constant of the allometric relationship between metabolic rate and body mass for multicellular ectotherms (b_0') and the normalization constants for endotherms (3.90×10^8 W/g$^{3/4}$), for multicellular ectotherms (9.91×10^7 W/g$^{3/4}$), and for unicells (2.77×10^7 W/g$^{3/4}$). This standardization allows making an a priori prediction for the value of the intercept

as $\log \left(\dfrac{\alpha M_{\text{RNA}} b_0'}{4 s_0 \, E_{\text{ATP}}} \right) = 0.61$

or 4.07 mg of RNA per gram of dry body mass. Note that the value estimated in the regression (0.54 or 3.4 mg of RNA per gram of dry body mass) is very close to the value predicted a priori. The solid line represents the best fit to the data and the dashed line ($y = −0.25 + 0.61$) the one predicted a priori.

with body mass to the −1/4 power (figure 10.15), and (2) it predicts that [RNA] will be relatively independent of body temperature. This is because the activation energy of respiration (E_i) and that of protein synthesis (E_p) probably have similar values, and thus $E_p - E_i \approx 0$.

We can estimate the contribution of phosphorus (P) in RNA to whole-body P (P_{body}) by combining equation 10.9 with empirical data on the quantity of P in other parts of the organism, such as the skeleton (P_{other}):

$$[P_{\text{body}}] = [P_{\text{RNA}}] + [P_{\text{other}}] \qquad (10.10)$$

RNA contains a relatively constant fraction ($\approx 9\%$) of phosphorus by mass; if we call this fraction ω, then $[P_{\text{RNA}}] \approx \omega[\text{RNA}]$. The content of phosphorus in other structures is an allometric function of mass of the form $C(m_b)^\beta$. Therefore

$$\left[P_{\text{body}}\right] = \left[P_{\text{RNA}}\right] + \left[P_{\text{other}}\right] \approx \left(\frac{\alpha M_{\text{RNA}} b_0}{4 s_0 E_{\text{ATP}}} \right)(m_b)^{-1/4} + C(m_b)^\beta . \qquad (10.11)$$

Among vertebrates and invertebrates $[P_{other}]$ scales with body mass with an exponent b that is very close to 0 (i.e., phosphorus concentration represents a relatively constant fraction of body mass). Equation 10.11 implies that the relationship between $[P_{body}]$ and body mass will be nonlinear. It will decline with mass at small body masses, but the relationship will tend to become flat at high body masses (figure 10.16A). The percentage of phosphorus in RNA will be high in small animals and become small and relatively constant in large animals (figure 10.16B).

Note that the trend lines in figures 10.15 and 10.16 leave a lot of variation to be explained. Their axes are log-log and hence a range of variation of 1 unit implies variation of an order of magnitude. The model is useful not because it explains all variation in RNA and P content, but because it provides a baseline that permits us to assess the effects of other factors. It allows finding out the features of an organism that make it deviate from the RNA and P content that one would expect from its mass. The model is also an important tool in comparative tests of stoichiometric ideas. For example, a comparative test of the

Figure 10.16. (A) Although the % of P in RNA decreases allometrically with body mass (dashed line from equation 10.9), whole-body P depends only weakly on body mass. The solid curve was estimated from equation 10.11, assuming that $[P_{other}]$ for invertebrates is 0.6%. The data are for 169 adult insect species. (B) The percentage of P in RNA relative to total P ($100 \times [P_{RNA}]/[P_{body}]$) decreases with body mass. The dashed line for ectotherms and the solid line for endotherms were constructed from equation 10.11.

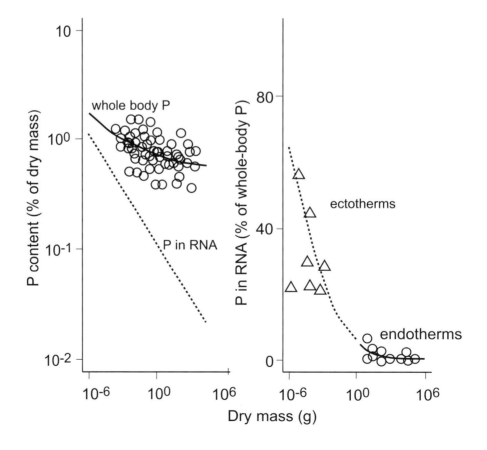

growth rate hypothesis requires that we first account for the effect that body mass has on both RNA and P content. At this point, you might want to reevaluate whether figure 10.12 provides unambiguous support for the growth rate hypothesis. How much of the variation in both growth and % RNA, and how much of their covariation, are explained by body mass rather than by the link between growth rate and RNA content? This question remains open.

We hope that you are surprised by the concordance between the model and the data, especially given the extraordinarily simplifying nature of the model's assumptions. This concordance strongly suggests that energy metabolism and stoichiometry are tightly linked. It indicates that the flux of energy has an influence on the currencies of ecological stoichiometry. Energy and materials are not alternative currencies, but complementary and closely connected ones. One of the important tasks of physiological ecologists is to establish the connections between the metabolic theory of ecology, with its emphasis on energy, and ecological stoichiometry, with its emphasis on materials.

References

Anderson, W. B., and G. A. Polis. 1999. Nutrient fluxes from water to land: seabirds affect plant nutrient status on Gulf of California Islands. *Oecologia* 118:324–3310.

Belovsky, G. E. 1984. Herbivore optimal foraging: a comparative test of three models. American Naturalist 124:97–115.

Bergerson, J. P. 1997. Nucleic acids in ichthyoplankton ecology: a review, with emphasis on recent advances for new perspectives. *Journal of Fish Biology* 51:284–3010.

Bernard, C. 1865. *Introduction à l'ètude de la mèdecine expèrimentale*. Bailliere, Paris.

Bezemer, T. M., and T. H. Jones. 1998. Plant-insect herbivore interactions in elevated atmospheric CO_2: Quantitative analyses and guild effects. *Oikos* 82:212–2210.

Buckley, L. E., and O.T.L. Caldarone. 1999. RNA-DNA ratios and other nucleic acid indicators for growth and condition of marine fishes. *Hydrobiologia* 401:265–277.

Bulow, F. J. 1970. RNA-DNA ratios as an indicators of recent growth rates of a fish. *Journal of the Fisheries Research Board of Canada* 27:2343–2349.

Cannon, W. B. 1932. *The wisdom of the body*. Norton, New York.

Clemmesen, C. 1994. The effect of food availability, age or size on the RNA/DNA ratio of individually measured herring larvae: Laboratory callibration. *Marine Biology* 118:337–3810.

Curtis, P. S., and X. Wang. 1998. A meta-analysis of elevated CO_2 effects on woody plant mass, form, and physiology. *Oecologia* 113:299–313.

Deinema, M. H., L.H.A. Habets, J. Scholten, E. Turkstra, and H. A. A. Webers. 1980. The accumulation of polyphosphate in Ecinetobacter spp. *FEMS Letters* 9:275–279.

Elser, J. J., M. M. Elser, N. A. McKay, and S. R. Carpenter. 1988. Zooplankton mediated transitions between N and P limited algal growth. *Limnology and Oceanography* 33:1–14.

Elser, J. J., D. R. Dobberfuhl, N. A. McKay, and J. H. Schampel. 1996. Organism size, life history, and N:P stoichiometry. *Bioscience* 46:674–684.

Elser, J. J., and J. Urabe. 1999. The stoichiometry of consumer-driven nutrient recycling: Theory, observations, and consequences. *Ecology* 80:735–751.

Elser, J. J., R. W. Sterner, E. Gorokhova, W. F. Fagan, T. A. Markow, J. B. Cotner, J. F. Harrsion, S. A. Hobbie, G. M. Odell, and L. J. Weider. 2000a. Biological stoichiometry from genes to ecosystems. *Ecology Letters* 3:540–550.

Elser, J. J., W. F. Fagan, R. F. Denno, D. R. Dobberfuhl, A. Folarin, A. Huberty, S. Interlandi, S. Kilham, E. McCauley, K. L. Schulz, E. H. Sieman, and R. W. Sterner. 2000b. Nutritional constraints in terrestrial and freshwater food webs. *Nature* 408:578–580.

Enquist, B. J., and K. J. Niklas. 2002. Global allocation patterns of biomass in seed plants. *Science* 295:1517–1520.

Falkowski, P. G. 2000. Rationalizing elemental ratios in unicellular algae. *Journal of Physiology* 36:3–6.

Falkowski, P. G., and C. S. Davis. 2004. Natural proportions. *Nature* 431:131.

Frausto da Silva, J.J.R., and R.J.P. Williams. 1994. *The biological chemistry of the elements.* Oxford University Press, Oxford.

Gillooly, J. F., A. P. Allen, J. H. Brown, J. J. Elser, C. Martínez del Rio, V. M. Savage, G. B. West, W. H. Woodruff, and H. A. Woods. 2005. The metabolic basis of whole-organism RNA and phosphorus content. *Proceeding of the National Academy of Science* 102:11923–11927.

Grady, C.P.L., and C.D.M. Filipe. 2000. Ecological engineering of bioreactors for wastewater treatment. *Water, Air, and Soil Pollution* 123:117–132

Hagen, W., and H. Auen. 2001. Seasonal adaptations and the role of lipids in oceanic zooplankton. *Zoology* 104:313–326.

Körner, C. 1993. Scaling from species to vegetation: the usefulness of functional groups. pages 117–140 in E. D. Schulze and H. Mooney (eds.), *Biodiversity and ecosystem function.* Springer-Verlag, Berlin.

Kramer, P. J., and T. T. Kozlowski. 1979. *Physiology of woody plants.* Academic Press, New York.

Kulaev, I. S. 1979. *The biochemistry of inorganic phosphates.* Wiley and Sons, New York.

Langley, L.L. 1973. *Homesotasis: Origins of the concept.* Dowden, Hutchinson and Ross, Stroudsburg, PA.

Leigh, R.A., and R. G. Wyn-Jones. 1985. Cellular compartmentalization in plant nutritions: The selective cytoplasm and the promiscuous vacuole. Pages 249–277 in B. Tinker and A. Läuchli (eds.), *Advances in plant nutrition.* Prager Scientific, New York.

Lindeman, R. L. 1942. The trophic dynamic aspects of ecology. *Ecology* 23:399–418.

Loladze, I. 2002. Rising atmospheric CO_2 and human nutrition: Toward globally imbalanced plant stoichiometry. *Trends in Ecology and Evolution* 17:457–461.

Markow, T. A., B. Raphael, D. Dobberfuhl, C. M Breitmeyer, J. J. Elser, and E. Pfeiler. 1999. Elemental stoichiometry of *Drosophila* and their hosts. *Functional Ecology* 13:78–84.

McGroddy, M., E., T. Daufresne, and L. O. Hedin. 2004. Scaling of C:N:P stoichiometry in forests worldwide: Implications of terrestrial Redfield-type ratios. *Ecology* 85:2390–2401.

Nielsen, S. L., S. Enríquez, C. M. Duarte, and K. Sand-Jensen. 1996. Scaling maximum growth rates across photosynthetic organisms. *Functional Ecology* 10:167–175.

Niklas, K. J., and N. J. Enquist. 2002. On the vegetative biomass partitioning of seed plant leaves, stems, and roots. *American Naturalist* 159:482–497.

Pomeroy, L. R. 2001. Microbial ecology of the oceans (book review). *Limnology and Oceanography* 2001:471–472.

Reich, P. B., and J. Oleskyn. 2004. Global patterns of plant and leaf N and P in relation to temperature and latitude. *Proceeding of the National Academy of Science* 101: 10849–10850.

Reiners, W. A. 1986. Complementry models for ecosystems. *American Naturalist* 127:59–73.

Sanchez-Piñero, F., and G. A. Polis. 2000. Bottom-up dynamics of allochtonous input: Direct and indirect effects of seabirds on islands. *Ecology* 81:3117–3131.

Schindler, D. W. 1978. Factors regulating phytoplankton production and standing crop in the world's freshwaters. *Limnology and Oceanography* 23:478–486.

Sprent, J. I. 1987. *The ecology of the nitrogen cycle.* Cambridge University Press, Cambridge.

Stapp, P., G. A. Polis, and F. Sánchez-Piñero. 1999. Stable isotopes reveal strong marine and El Niño effects on island food webs. *Nature* 401:467–469.

Stephens, D.W., and J.R. Krebs. 1986. *Foraging theory.* Princeton University Press, Princeton, N.J.

Sterner, R. W. 1990. The ratio of nitrogen to phosphorus re-supplied by herbivores: Zooplankton and the algal competitive arena. *American Naturalist* 136:209–229.

Sterner, R. W. 1995. Elemental stoichiometry of species in ecosystems. Pages 240–252 in C. G. Jones and J. H. Lawton (eds.), *Linking species and ecosystems.* Chapman and Hall, New York.

Sterner, R. W., J. J. Elser, and D. O. Hessen. 1992. Stoichiometric relationships among producers, consumers, and nutrient cycling in pelagic ecosystems. *Biogeochemistry* 17:49–67.

Sterner, R. W., and J. J. Elser, 2002. *Ecological stoichiometry: The biology of the elements from molecules to the atmosphere.* Princeton University Press, Princeton, N.J.

Sterner, R. W., and D. O. Hessen. 1994. Algal nutrient limitations and the nutrition of aquatic herbivores. *Annual Review of Ecology and Systematics* 25:1–29.

Watt, A. D., J. B. Whittaker, M. Docherty, G. Brooks, E. Lindsay, and D. T. Salt. 1995. The impact of elevated CO_2 on insect hervibores. pages 198–217 in R. Harrington and N. E. Stork (eds.), *Insects in a changing environment.* Academic Press, London.

White, T.C.R. 1993. The inadequate environment: *Nitrogen and the abundance of animals.* Springer-Verlag, New York.

Whittaker, J. B. 2001. Insects and plants in a changing atmosphere. *Journal of Ecology* 89:507–518.

Williams, R.J.P., and J.J.R. Frausto da Silva. 1996. *The natural selection of the chemical elements.* Oxford University Press, Oxford.

Woods, H. A., M. C. Perkins, J. J. Elser and J. F. Harrison. 2002. Absorption and storage of phosphorus by larval *Manduca sexta. Journal of Insect Physiology* 48:555–564.

Worster, D. 1994. *Nature's economy: A history of ecological ideas.* Cambridge University Press, Cambridge.

CHAPTER ELEVEN

Nitrogen and Mineral Requirements

11.1 Nitrogen Requirements and Limitation in Ecology

IN CHAPTER 10 WE introduced the very fundamental fact that C:nutrient ratios (e.g., C:N, C:P) of plants can be much higher than those of animals. This observation suggests a relative scarcity of certain nutrients. Furthermore, as detailed in chapters 2 and 6, some nutrients in food may not be in a form that can be incorporated into body tissue efficiently, as when there is a poor match between the amino acid profile of food and the requirements of animals. All this leads to the deduction that lack of specific nutrients probably looms large in the ecology of animals, especially herbivores. White (1993) has argued this point in the case of N. How much of specific nutrients animals need, how that compares with what is in their food, and what happens when there is a mismatch between need and availability are the subjects of this chapter.

11.1.1 Nitrogen Budgets and Minimum Requirements

For maintenance, animals must consume enough N to balance their losses through urine, feces, and shed epidermal tissue. Studies of N budgets of many kinds of animals have shown that this is not a major problem for animals that consume other animals (carnivores, insectivores) but it is a potential problem for animals that rely on plants (e.g., folivores, frugivores, granivores) that sometimes have quite low N content (chapter 2). For example, field studies of herbivores such as hispid cotton rats (*Sigmodon hispidus*) eating leaves in North American grasslands and green iguanas (*Iguana iguana*) eating leaves on Caribbean islands

Figure 11.1. Field studies of herbivores such as hispid cotton rats eating leaves in North American grasslands (left-hand figure; data from Randolph and Cameron 2001) and green iguanas eating leaves on Caribbean islands (right-hand figure; data from van Marken Lichtenbelt et al. 1993) indicate that for prolonged periods of the year the forages available have too little N to satisfy even maintenance requirements. Almost all the monocot and dicot food plants of cotton rats in winter and spring had less N than the maintenance requirements for cotton rats (indicated by the thick horizontal line), which are at least 15 mg/g (see text). In the study on green iguanas, the N content of food plants during the second and third months of the study (corresponding to December–January 1986) was below maintenance requirements (indicated by the thick horizontal line).

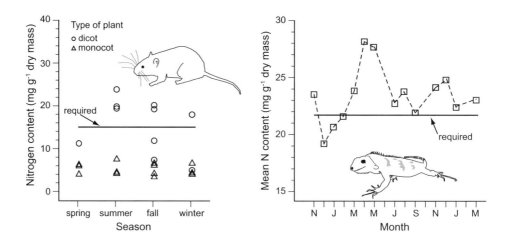

indicate that for long periods of the year the forages available have too little N to satisfy even maintenance requirements (figure 11.1). Most (approximately 85%) sugary fruits have too little N to satisfy minimum needs of frugivorous wood thrushes (*Hylocichla mustelina*) in northeastern forests of North America (Witmer 1998). Requirements including those for growth and reproduction would be even higher. How does one assemble a quantitative estimate of seasonal N requirements for comparison to available resources?

We can model the maintenance N requirement based on the sum of obligatory endogenous and other losses; this is called the factorial method. The requirement for digestible N must at least equal the minimal, constant losses in urine (EUN; box 11.1) and feces (metabolic fecal N, MFN). In individuals that are molting or shedding skin these venues of loss would also be included. The model includes the intake rate of dry matter (I) because the endogenous fecal losses of N (and possibly also of P and Ca; Murray 1995) are increased with increasing feeding rate (box 11.1). Also, the factorial model for N includes a variable reflecting the biological value of food's N (BV ranges from 0 to 1.0; see chapter 2 and box 11.1). The BV reflects how much of the food N is utilized in protein turnover, which depends on how much of the diet N is actually in the form of protein, and the amino acid makeup of that protein. Thus, the estimated required N per day equals (EUN + (MFN × I))/BV. Because most ecologists would find the required concentration in the diet to be a more useful expression than the daily rate required, the daily requirement itself can be normalized to (i.e., divided by) feeding rate to yield this expression for dietary N required, in mg g^{-1} dry matter:

$$\text{Diet N content required for maintenance} = \frac{(\text{EUN} + (\text{MFN} \times I))}{(\text{BV} \times I)} \quad (11.1)$$

Box 11.1. Measuring Endogenous Nitrogen Losses in Honey Possums

Endogenous urinary N (EUN) and metabolic fecal N (MFN) can be measured in animals eating a N-free diet, because in that case all N excreted in feces and urine is of endogenous origin. More commonly, however, they are measured indirectly as illustrated for honey possums (*Tarsipes rostratus*) below, using data from Bradshaw and Bradshaw (2001).

Honey possums (mean body mass = 11 g) were placed into special cages designed so that their urine and feces could be collected separately. They were fed four diets that differed in N content. After a period of habituation, daily measures were made of food consumed, feces produced, and urine produced. Subsamples of the food, and some or all of the feces and urine were dried (60°C) to determine the ratio of dry mass to wet mass to measure the N content (see also chapter 2). Rates of intake and excretion, which are typically positive functions of N intake (see figures below) were normalized to $(mass)^{3/4}$ because endogenous losses of N scale with body mass in this fashion in mammals, birds, and fish (Jobling 1993; Robbins 1993).

The obligatory fecal N loss, metabolic fecal N, is taken to be the *y* intercept of the linear regression of fecal N per unit of dry matter intake in the diet against diet N content (left-hand panel in figure 11.2). The regression was $y = 0.031x + 0.348$, $r^2 = 0.24$, $n = 20$, $P < 0.05$. The magnitude of MFN, 0.348 mg N g^{-1} food intake, is low compared to that of most animals (see figure 11.3), probably because of very low contribution of sloughed cells on a nonfibrous diet and low contribution of microbes in this nonfermenter. Using this value one can partition the fecal N loss and determine that about 80% was unabsorbed dietary N and 20% endogenous loss (Bradshaw and

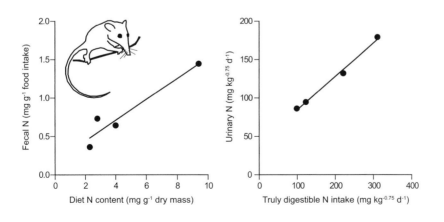

Figure 11.2. Fecal and urinary losses are positive functions of N intake. Shown are mean values for three to nine honey possums on each of four diets composed mainly of honey and pollen. (Redrawn from Bradshaw and Bradshaw 2001.)

continued

continued

Bradshaw 2001). Also, one can then estimate that the true digestibility of N, corrected for the endogenous loss, was about 76% (calculation not shown).

The obligatory urinary N loss, endogenous urinary N, is taken to be the y intercept of the linear regression of urinary N versus truly digested N (TDN = N intake − [fecal N loss + endogenous fecal loss]) (right-hand panel of figure 11.2). The regression was $y = 0.406x + 41.83$, $r^2 = 0.8$, $n = 20$, $p < 0.001$. The magnitude of EUN, 41.8 mg N $kg^{-3/4}$ d^{-1}, is relatively low, like most such measures in metatherians (see figure 11.3). This may relate to lower basal metabolic rate and lower protein turnover (Bradshaw and Bradshaw 2001). The slope of the relationship between urinary N and TDN, 0.406, is the urinary N, in excess of endogenous loss, per unit of food N. When the food N has low biological value (a poor mix of amino acids in relation to requirements) the slope is steep, and when the food N has high biological value (BV) the slope is flatter. In fact, the BV can be estimated as 1 − slope = 0.58.

This is a requirement for truly digestible N, and because most animals absorb at least 90% of their dietary N (see chapter 3, section 3.2.3) it should be adjusted upward, perhaps by dividing by 0.9. Also, many researchers report requirements or plant contents in terms of crude protein, which they typically calculate as 6.25 times the N value (see chapter 2, where we discouraged this practice).

One advantage of the modeling approach to estimate nutrient requirements is that the estimated requirement can be adjusted for different conditions. In the case of N requirement, for example, free-living animals might have higher feeding rates due to different environmental conditions, or different levels of activity, or because they are eating more of a diet diluted by high fiber content. Using equation 11.1, one can estimate the required diet N or protein under these different conditions (figure 11.3). By including in the numerator the daily rate at which N is deposited into new tissue during growth or reproduction, and increasing feeding rate in the denominator appropriately, one can adjust the requirement upward to estimate the requirement during production. Also, an ecological insight arises out of the modeling. When herbivores eat very low N diets they cannot necessarily compensate for low N content with very high intake rates. The metabolic fecal N loss, which increases with food intake rate, sets an absolute lower limit to the required minimum dietary N requirement. An animal cannot always "eat its way" out of trouble.

The model (figure 11.3) suggests that nonruminant eutherian herbivores eating an ideal N source (BV = 1) require at a minimum a plant N content of at

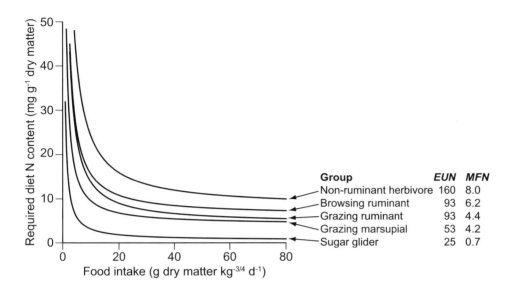

Figure 11.3. Minimum dietary nitrogen content necessary to meet maintenance requirements for various mammals. The curves were generated using equation 11.1 and values that are shown for endogenous urinary N (EUN; units of mg k$^{-0.75}$ d^{-1}) and metabolic fecal N (MFN; units of mg g^{-1} dry matter intake). Body masses used were 100 g for the non-ruminant eutherian, 100 kg for the browsing and grazing ruminants, 10 kg for the grazing marsupial, and 155 g for the sugar glider. This is a requirement for truly digestible N in dry food, and because most animals absorb at least 90% of their dietary N (see text) it might be adjusted upwards, perhaps by dividing by 0.9. (Adapted from Robbins 1993.)

least 11 mg N g^{-1} dry matter, and that some mammals can survive on lower dietary N contents than others due to differences in EUN and MFN. The differences in MFN can be largely understood on the basis of differences in dietary fiber content. A sap- and nectar-consuming sugar glider, for example, loses less fecal N than herbivores consuming high-fiber food, probably because its lower-roughage diet causes less abrasional loss of intestinal cells and microbes. The differences in EUN appear correlated with differences in overall energy expenditure, as marsupials have lower basal energy expenditures than eutherians. Some researchers have speculated that this reflects a causal relation between protein turnover and basal energy metabolism, but whether this correlation reflects such a cause-effect relationship remains a mystery. Hummingbirds provide a counter-example to this claim. These tiny birds have both prodigiously high metabolic rates for their mass and remarkably low nitrogen requirements (McWhorter et al. 2003). Another way the N requirement is minimized is by maximizing the biological value of food N, which is achieved by diet selection or through interaction with microbial symbionts, as discussed below.

The factorial method for estimating N requirement (equation 11.1) cannot easily be applied to animals that eliminate urine and feces mixed, so directly comparable data on birds and ectotherms are scarce, but let us consider them briefly:

Birds. The combined EUN + MFN losses have been measured in a number of granivorous/herbivorous avian species and the average, 270 mg kg$^{-3/4}$ d^{-1}, is intermediate between the losses in eutherian and marsupial mammals (Robbins 1993). As might be expected then, those birds' minimum requirements (430 mg N kg$^{-3/4}$ d^{-1}) are intermediate as well between those of eutherians

(580 mg N kg$^{-3/4}$ d^{-1}) and ruminants (356 mg N kg$^{-3/4}$ d^{-1}) (Robbins 1993). Fruit- and nectar-consuming birds appear to have combined EUN + MFN losses that are four to five times lower than, and lower requirements than, those of the granivore/herbivores studied so far, but the mechanistic basis for these lower requirements is unknown and, as described for hummingbirds, rather mysterious (Witmer 1998; Pryor et al. 2001).

Reptiles. In green iguanas, the MFN was similar to that in mammalian herbivores (Marken Lichtenbelt 1992). Minimum urinary N loss has been estimated in herbivorous chuckwalla lizards (45 mg kg$^{-3/4}$ d^{-1}; Nagy 1975) and in 14 fish species (mean = 51 mg kg$^{-3/4}$ d^{-1}; Bowen 1987; Jobling 1993). These values are lower than those observed in eutherian mammals (figure 11.3; Robbins 1993), perhaps because ectotherms have lower metabolic rates.

Insects. Minimal maintenance N requirements have been measured in only a few insects and ranged from 10 to 30 mg g^{-1} dry matter (Mattson 1980), which is similar to or higher than the estimate for nonruminants eutherians (above, and figure 11.3).

Animals have ameliorated the problem of nutritionally deficient diets by associating with microorganisms that can synthesize diet components absent in the animal's diet. In chapter 6 we discussed how the symbiotic interactions can improve N economy. Essentially, they convert dietary N of low biological value into bacterial protein that has a very high biological value, because microbes can synthesize some amino acids that are essential for vertebrates. At least three types of animals might then digest the microbial protein and absorb the amino acids: foregut fermenters (e.g., ruminants, macropod marsupials like kangaroos, and the hoatzin) that pass the microbes to the protein-digesting small intestine, hindgut fermenters that are capable also of hindgut protein breakdown and amino acid absorption (some avian herbivores), and hindgut fermenters that reingest the microbe-rich excreta (e.g., many vertebrate herbivores, termites, wood roaches are coprophageous). All these animals can potentially use their gut microbes to increase the biological value of the food N which then lowers the required dietary N content (see equation 11.1).

11.1.2 Nitrogen Requirements that Include Growth Costs

Often, especially for animals that grow more or less continuously, N requirements are determined in relation to growth. In Lepidopteran larvae, for example, growth rate is positively related to leaf N content (Slansky and Scriber 1985). A similar positive relation exists for growth rate and food N in fish (Fris and Horn 1993). Although these large data summaries over many species eating

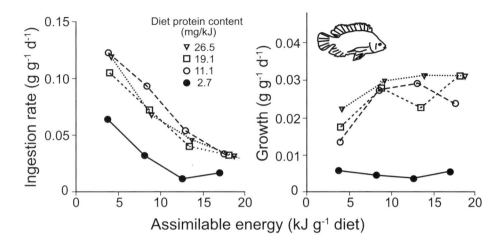

Figure 11.4. Tilapia fish feed to meet energy needs and grow more slowly when food N content is too low. In the left-hand figure food intake increased as the diet's assimilable energy content declined, and feeding rate was depressed on the diet with lowest protein. Notice in the right-hand figure that growth rate was proportional to diet protein content at the lowest diet energy level. Each point is a mean for several fish. (Redrawn from Bowen et al. 1995.)

many diets may suggest that growth rate is a continuously increasing function of dietary N intake, the situation may be more complex when focusing on an individual species. Consider the feeding and growth response of omnivorous tilapia fish (*Tilapia nilotica*) offered diets of differing N and energy content ad libitum (figure 11.4). Within each protein level, as metabolizable energy in the diet was reduced the fish compensated by eating more so that their metabolizable energy intake in kJ d^{-1} was independent of diet energy content (figure 11.4, left panel). This common response pattern in animals was discussed in chapter 3. Ingestion did not compensate for protein limitation and so protein intake was related to diet protein content. Growth rate was proportional to diet protein content only at the lowest diet energy level (figure 11.4, right panel). Thus, for diets with at least 8.4 kJ g^{-1} diet, growth rate was not a continuously increasing function of dietary protein but was relatively independent of it as long as diet protein content was ≥93.2 mg g^{-1} (i.e., 14.9 mg N g^{-1}).

This protein requirement for growth in tilapia is much lower than the requirement that is typically cited for maximum growth in fish, which is 300–400 mg g^{-1} diet (Bowen 1987). Most studies on fish have mainly focused on carnivores used in aquaculture, and their apparent high protein requirements for growth have led to the notion that plant-eating fish must consume animal material to obtain enough protein for growth. Wild herbivorous fish have rarely been studied and may be able to grow on much lower dietary protein contents. This is the case in insects where growth is supported on lower protein diets in plant-eating larvae compared with larvae eating animal tissue (Dadd 1985). The seaweed-eating fish *Cebidichthys violaceus* grew on diets with as little as 68–186 mg protein g^{-1} plant (Fris and Horn 1993). For comparison, dietary protein contents that yield maximum growth in rats and chickens are 200–220 mg protein g^{-1} diet (Bowen 1987), although those species will still grow, albeit more slowly, on diets with less protein. Notice, incidentally, that the requirements for maximal growth in the rat and chicken are at least double the requirement for maintenance N balance estimated in the previous section for nonruminant endotherms. In chapters 13 and 14 we will explore further the estimation of extra material costs for growth and reproduction.

Bowen et al. (1995) used their results on the effect of energy and N on growth of tilapia to develop a general model that predicts the growth of poikilotherms. They mapped the growth rates achieved on different combinations of assimilated energy and dietary protein content as isopleths on plots of energy assimilation rate (x axis) versus protein content (mg kJ^{-1}; y axis) (figure 11.5). An interesting

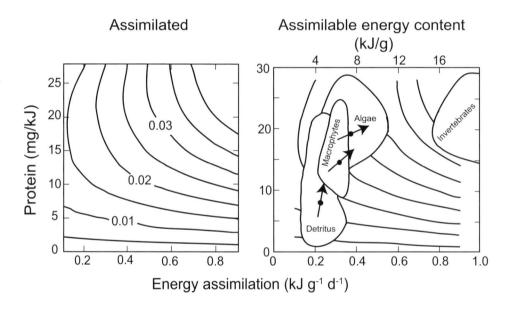

Figure 11.5. Predicted consumer growth rates based on experimental data from tilapia. In the left-hand figure, growth rates achieved on different combinations of assimilated energy and dietary protein content are plotted as isopleths on plots of energy assimilation rate (x axis) versus protein content (mg kJ^{-1}; y axis). In the right-hand figure, different food types with known energy and protein values are plotted onto the growth isopleths. For detritivores, the model predicts that increases in growth rate are achieved mainly by increases in protein content rather than energy (the vector for detritus is more vertical than horizontal). Protein and energy are equally important to herbivores feeding on macrophytes, and energy is somewhat more important for herbivores feeding on algae (the vector for algae is more horizontal than vertical). (Redrawn from Bowen et al. 1995.)

feature of the plot is its illustration of the circumstances when protein might limit growth. As growth isopleths tend to become parallel to the x axis, increases in dietary protein have the most positive effect on growth rate, but as isopleths tend to become parallel to the y axis increases in energy are relatively more important. Based on a survey of the energy and protein in different kinds of foods of aquatic poikilotherms, and assuming an intake rate of 5% of body mass d^{-1}, the authors mapped the different food types onto the growth isopleths. Not surprisingly they found that for invertebrate foods, which have relatively high protein and energy contents, growth rates of consumers will mainly be a function of energy intake.

The more interesting findings relate to plant foods and detritus, which provide much less energy and protein per gram of food consumed. Among these foods the relative importance of protein and energy as limiting substances varied. For detritivores, the model predicts that increases in growth rate are achieved mainly by increases in protein content rather than energy. Protein and energy are equally important to herbivores feeding on macrophytes, and energy is somewhat more important for herbivores feeding on algae. There are many interesting ecological implications of the plot (Bowen et al. 1995). We might expect consumers of detritus and macrophytes while feeding to be more selective about protein content than algal feeders.

11.1.3 Ingestion of Xenobiotics Can Increase the Dietary Requirement for Nitrogen

At least two defensive responses of animals to toxins and/or secondary metabolites (SMs) should, in theory, increase the N requirement. Some animals secrete proline-rich (20–40% proline) salivary proteins that are postulated to bind and inactivate dietary tannin, and the tannin-protein complexes are then excreted in feces (see section 9.5.1). The loss of N increases with dietary tannin content, and available data (Robbins et al. 1991; Hanley et al. 1992) suggest that for animals with maintenance protein requirements of 50–100 mg protein g^{-1} diet, like many vertebrate herbivores, the requirement would be increased by 10–20%. Because the salivary proteins are rich in proline, which can be synthesized by animals (chapter 2), much of the extra N can be supplied by any amino acid or even non-amino sources of N in the case of animals that improve their N economy with symbiotic microbes.

The second important detoxification response that should increase N requirement is the increased loss of N in ammonia or amino acid and peptide conjugates (see section 9.5.3). In grouse, for example, excretion of ornithine conjugates would increase minimum obligatory N losses by 67–90%, depending on the rate of xenobiotic absorption (Guglielmo et al. 1996), which should translate to similar large percentage increases in the minimum N required (equation 11.1).

We do not know of any N balance studies in animals that actually demonstrate that minimum N requirements are increased in animals due either to this mechanism or to increased salivary secretion of proline-rich proteins, although there are some strong hints of these effects. When frugivorous cedar waxwing birds (*Bombycilla cedrorum*) ate fruits of highbush cranberry (*Viburnum opulus*) they suffered higher than expected losses of N and were even further out of N balance than would be expected just on the basis of the fruit's very low N content (ca. 3.6 mg N g^{-1} dry mass; Witmer 2001). Witmer invoked (but did not measure) fruit SMs as the cause for the higher N losses suffered by the birds.

He observed that, in the wild, cedar waxwings tend to utilize *V. opulus* fruit only when a supplementary source of N is available. He observed the birds shuttling back and forth between bushes with fruit and catkin-laden cotton-wood trees (*Populus deltoides*). The pollen in these catkins is rich in protein, but catkins are only available in the early spring. In most years the ripe fruit of *V. opulus* remains uneaten all winter until the spring when cottonwoods bloom. Then hordes of waxwings rapidly strip the fruit from bushes.

11.1.4 Nitrogen Limitation in the Wild

Our survey of how N requirements are determined, and their general magnitude in relation to availability (as in figure 11.1, above) is consistent with the idea that N can often be limiting. As already introduced in chapters 6 and 10, animals respond to lower nitrogen availability with a variety of ploys involving behavioral changes, phenotypically plastic responses, and evolutionary adaptations. We will briefly discuss these ploys, which include (1) lowering N requirements, (2) processing N more efficiently, (3) consuming more N by eating more food, or (4) selecting food with higher N content.

(1) We know little about the mechanisms that account for lower endogenous urinary losses and thus lower maintenance requirements. These mechanisms are likely biochemical features integrated within metabolic pathways (e.g., efficient reamination of carbon skeletons with ammonia resulting from protein turnover). We suspect, but have little evidence for this suspicion, that these physiological mechanisms are not very adjustable over ecological time scales. Animals evolve them in response to diets with low N (nectar-feeding animals are a good example; McWhorter et al. 2003). On a different temporal scale, animals can match their N requirements to the availability of N in the environment. White (1993) provides many examples of species that synchronize their times of high N requirements (reproduction and raising of the young) with seasons of high N abundance, and examples of species that respond facultatively to temporal increases in N availability.

(2) The utilization of symbiotic microbes to improve N economy is a major strategy that has evolved multiple times among animals to cope with the universal problem of N limitation (White 1993). It is so important that we devoted a section in chapter 6 to it.

(3) It is controversial whether animals increase food intake on a N-poor diet in a compensatory fashion. We discussed in chapter 3 (section 3.4.1) how many animals regulate their food intake rate mainly to meet energy

requirements. Furthermore, an animal cannot eat enough food to meets its N requirement when the dietary N concentration drops below the level of the MFN losses (equation 11.1, above). It has been proposed, however, that some animals have evolved digestive and/or metabolic traits that make them routinely somewhat inefficient at extracting and/or retaining energy from their diet. These features force them to overingest foods and may end up having the net effect of permitting them to exist on a diet with less N (or some other critical nutrient). For example, Batzli and his colleagues studied the nutritional ecology of microtine rodents (voles, lemmings) in several different ecosystems, and they found that the apparent energy digestibilities of brown lemmings (*Lemmus sibericus*) on their native forages were particularly low (range 25–39%) compared with those of other microtines (Batzli and Cole 1979). Noting the relatively low nutrient contents of the lemmings' forages (Barkley et al. 1980), they hypothesized that low digestibility of energy might be an adaptation for exploiting poor quality forage. It is unknown how widespread these "adaptations for digestive inefficiency" are, and the specific traits that lead to inefficiency have been described only sketchily. A second alternative approach is to ingest and assimilate large amounts of low-N food and "burn off" the excess energy with a relatively high respiration rate rather than depositing excess energy as fat. In chapter 3 (section 3.4.1) we discussed evidence that some insects and fish use this tactic. Also, there is some evidence that captive frugivorous bats may do this (Delorme and Thomas 1999), although it is possible that the relatively high metabolic rates of captive fed bats are an artifact of captivity stress. A third alternative approach is exhibited by phloem-sucking insects such as aphids and scale insects, which ingest and assimilate prodigious amounts of sugar and then transform it and excrete it in the form of honeydew (White 1993; Ashford et al. 2000; Ashford et al. 1997). A variety of other organisms, including hundreds of species of insects in at least 49 families (Klingauf 1987), birds (Greenberg et al. 1993; Latta et al. 2001), and mammals (Carthew et al. 1999; Sharpe and Goldingay 1998) take advantage of this wasted energy. By one fanciful interpretation, the manna that saved the Israelites from starvation in the wilderness is a consequence of how some insects cope with a nitrogen-poor diet.

All of the above strategies may be insufficient and many animals often suffer from both acute and chronic N scarcity. Nitrogen limitation can have implications at multiple ecological scales. Growth rate and the time needed to reach reproductive maturity may be influenced by the N content of an herbivore's or omnivore's food, as described above for many insect larvae (Slansky and

Figure 11.6. Protein supplementation increased the growth rate and breeding intensity of free-living white-footed mice (*Peromyscus maniculatus*). (A) In 1997, a year of relatively lower growth in control populations, mice in the supplemented populations grew significantly faster than in the control populations. (B) In both years a significantly higher proportion of young-of-the-year females bred in protein-supplemented populations. (Data from McAdam and Millar 1999.)

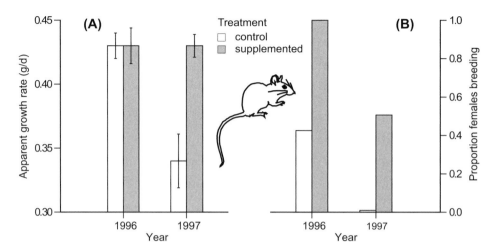

Scriber 1985) or demonstrated experimentally in some free-living vertebrates. For example, to test the hypotheses that growth and maturation of young-of-the-year omnivorous white-footed mice (*Peromyscus maniculatus*) were protein or energy limited, John Millar and his colleagues (Teferi and Millar 1993; McAdam and Millar 1999) supplemented populations with either a high-protein food or a low-protein, high-energy food. Protein supplementation increased growth rate and the proportion of young-of-the-year females that bred (figure 11.6). In contrast, over four seasons in the two studies, energy supplementation caused no significant increase in growth rate and only modest increases in maturation rate. In another example, red deer females that fed on vegetation that was naturally N fertilized with seabird guano had higher lifetime reproduction than females feeding on unfertilized vegetation, although both groups consumed similar amounts of forage (Iason et al. 1986). In this example the N supplementation occurred naturally, but this kind of situation is not unique (chapter 10). The impacts of N deposition by aggregations of birds (Bosman and Hockey 1986), mammals (Augustine and Frank 2001), and fish (Hilderbrand et al. 1999; Helfield and Naiman 2001) can have dramatic effects on plants and through them on the community structure and function of the animals that eat them.

If animals remain N limited in spite of the many traits that minimize loss and maximize intake and/or economy of N utilization, then a final strategy comes into play.

(4) Animals must carefully select plant foods with relatively higher N contents or even supplement their plant-based diets with animal food. Most nectar and fruit-eating birds, for example, must supplement the diet of their rapidly growing young with insects (Martinez del Rio 1994). The need to forage

selectively for N may influence habitat use. Hispid cotton rats (*Sigmodon hispidus*), are generalist herbivores that occur in grass-dominated habitats throughout the New World. Although the biomass of grasses (monocotyledons, or monocots) is several times greater than that of dicotyledonous plants (dicots) in such habitats, there is variation among vegetation patches. Some patches contain a mix of the two plant types and others are dominated entirely by grasses. Because grasses and dicots differ in N content, this spatial variation imposes a quandary to cotton rats. A diet of only monocots can meet their energy requirements, but the level of N in monocots (ca. 6 mg g^{-1} dry matter; figure 11.1) is too low to support maintenance, growth, and reproduction (above; Hellgren and Lochmiller 1997). A diet of dicots only meets both energy and N requirements (figure 11.1), but it takes too long to harvest (Randolph and Cameron 2001). Cotton rats sacrifice both energy maximization and time maximization and opt for a compromise: they eat a mixed diet of mono- and dicots. In the spring, when reproduction is most intense, their use of dicots peaks at 40–65% of the diet (Kincaid and Cameron 1982; Schetter et al. 1998). To eat a mixed diet, female cotton rats concentrate their activities in vegetation patches that contain mono- and dicot mixtures. When Eshelman and Cameron (1996) supplemented protein-poor monocot patches with a synthetic diet with high N (24 mg N g^{-1} dry matter), the cotton rats increased their utilization of those plots. Integrated studies such as these on the cotton rats, along with other examples described above, provide strong support for what has come to be called the "N limitation hypothesis": N is limiting in the habitats of many populations of wild herbivores (Mattson 1980; White 1993). By limiting the density of herbivores, N can have cascading effects on other trophic levels.

11.2 Mineral Requirements and Limitation in Ecology

So far in this chapter we have used N to highlight how nutrient requirements are determined, how they can vary under different conditions, and how one evaluates whether they might be scarce enough to influence food and habitat choice, survival, growth, or reproduction. Such a detailed presentation for all essential nutrients is not the intent of this book, and there are other sources that focus specifically on animal nutrition (Dadd 1985; Hume 1982; Klasing 1998; McDowell 1992; McDowell 1985; Robbins 1993; Van Soest 1994). Considering our goal of focusing primarily on ecological implications, we can utilize the vast amount of information on animal nutrition to identify those nutrients that are most or least likely to be limiting for free-living wild animals. For example,

TABLE 11.1
Elements and wildlife

Element	Dietary source	Deficiency?
Macroelements		
Calcium and phosphorus	Nuts, seeds, many arthropods, and animal flesh may have inadequate levels or poor Ca:P ratios; bones, plants, and snail or egg shells contain higher levels and can act as a supplementary source	Ca is perhaps the most often limiting mineral; Ca deficiency has been reported in some egg-laying birds and in wild carnivores eating only flesh, but most animals probably seek out Ca sources such as snail shells or bones (osteophagia), especially during reproduction, and thereby avoid deficiencies or imbalances
Sodium	Low levels in forage in areas where soil concentration is low; satisfactory levels in flesh; mineral licks or plants with higher Na levels sometimes sought	Unlikely in carnivores; herbivores may seek out special sources of sodium
Potassium	Present in adequate levels in most plant and animal tissue	Probably rare
Magnesium	Most food sources probably have adequate levels. Mg in lush young forage may have low bioavailability due to interactions with N, K⁺, long-chain fatty acids, organic acids.	Deficiency generally rare. If Mg bioavailability is low, sometimes bone resorption of Mg is not rapid enough to maintain serum and extracellular [Mg], causing an illness called grass tetany or grass staggers.
Trace elements		
Iron	High or adequate levels in animal tissue and most forage; lower levels in cereal grains and grasses grown in sandy soils	Rare
Iodine	Concentration in vegetation can be low in areas where soil concentration is low; e.g., northern half of U.S. from Great Lakes to Pacific NW	Goiter has been observed in free-living white-tailed deer, but is not often reported in wildlife

Copper	Present in almost all foods; concentration in forage generally reflects soil concentration	Deficiencies have been observed in moose populations in Alaska and bontebok grazing on a reserve in South Africa
Zinc	Animal tissue usually has adequate levels of Zn, which is available; in plants Zn can complex with phytate, or high Ca^{2+} can decrease availability	Deficiencies have not been reported
Selenium	In some large geographic regions many of the plant species have inadequate levels, e.g., most of Wisconsin to New England, Pacific NW	Deficiencies have been observed in mountain goats, pronghorn, woodchucks, various species of African antelope
	Some plant species accumulate toxic levels, e.g., some vetches (*Astragalus*), might cause high levels in adjacent forage plants	These plants are normally unpalatable to herbivores, but might be consumed under starvation conditions
Manganese	Present in seeds, forages, animal tissue	Deficiencies have not been reported
Molybdenum		Deficiencies have not been reported

Note: This list is not exhaustive. Some required elements are not listed, e.g., chlorine, sulfur, cobalt, fluoride, chromium. Adapted from Robbins (1993).

among the many macro- and trace elements that are essential to animals (defined in chapter 2), our comparison of the requirements from the literature (e.g., Robbins 1993) with the composition of various types of foods suggests that for many types of feeders, element requirements are satisfied incidentally as the animals consume enough food to meet energy needs (table 11.1). Analogously, vitamin needs are thought to be often met in the natural diet of free-living wildlife.

Most interesting are those situations in which foods normally eaten by a certain feeder may not satisfy requirements for particular nutrients. These are the nutrients that would therefore most likely influence diet selection, mortality, and/or productivity. We will explore some general issues in mineral nutrition by focusing on case studies for three such situations involving the macronutrients sodium (Na), calcium (Ca), and phosphorus (P). Recall that chapter 2 gave examples of limitations due to micronutrients.

11.2.1 Sodium May Be Limiting for Herbivores

In many parts of the world herbivores feed on diets with low Na concentrations. Na is easily leached from soil in ecosystems with moderate precipitation and is not generally accumulated by plants (Robbins 1993). Hence, the Na concentration of rainwater, soil, and plants tends to decrease from coastal to interior continental regions. For example, in the Jotunheimen mountain region in southern Norway (elevation 800–2300 m above sea level) all of the plant types surveyed ($n = 31$ species) in winter (May) and midsummer (July) had Na concentrations below 13 mmol kg^{-1} dry mass, or 300 mg kg^{-1} dry mass, except for fungi (figure 11.7). We will use this Na content as a benchmark because, as discussed below, it is probably the minimum Na content to support growth in a caribou (*Rangifer tarandus*). Sodium concentrations are even lower, ranging from 10 to 67 mg kg^{-1} dry mass, in the woody browse eaten by moose (*Alces alces*), white-tailed deer (*Odocoileus virginianus*), and other herbivores in eastern deciduous and boreal forests in North America that are even more isolated from oceanic aerosol inputs of Na than the forests of Scandanavia (Jordan 1987). This apparent mismatch between requirement and availability can influence feeding behavior, production, and probably movements of vertebrate herbivores. The issue of dietary Na is thought not to be so important for insects for whom Na is not an important osmolyte as it is in vertebrates (Dadd 1985), but there are interesting cases where Na apparently limits reproduction and insects seek it out (Smedley and Eisner 1996).

We are somewhat in the dark about the actual Na requirements of wildlife because there have been few thorough studies of requirement beyond those in

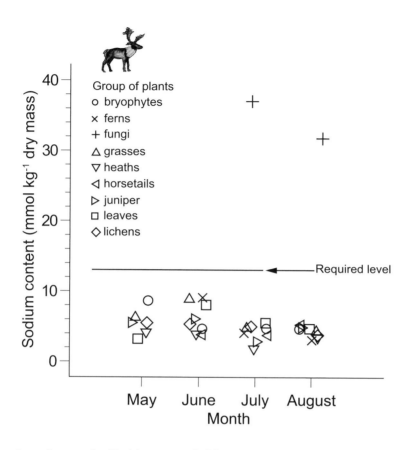

Figure 11.7. Most plants have too little sodium (Na) to meet requirements of free-living caribou in the Jotunheimen mountain region in southern Norway. The figure shows the mean values for each group of plants in comparison to the estimated requirement (shown as a horizontal solid line). Samples in May (month 5) represent winter samples. Notice that fungi have much higher Na contents. (Data from Staaland and Hove 2000.)

domesticated animals (Robbins 1993). The maintenance needs of white-tailed deer (*Odocoileus virginianus*, mean mass 51.4 kg), estimated from total collection trials on deer fed varying Na levels, were 8.75 mg kg$^{-3/4}$ d^{-1}, or 136 mg kg^{-1} dry mass of food (Hellgren and Pitts 1997). This requirement is of the same magnitude as the sodium requirement of voles (*Microtus pennsylvanicus*), which were in negative Na balance on a diets of 139 mg kg^{-1} dry mass of food (Christian et al. 1993). Metabolic fecal losses via gastrointestinal secretions, rather than urine losses, are the major determinant of Na maintenance requirements. Urine losses are greatly reduced during Na restriction through the action of aldosterone, a steroid hormone secreted by the adrenal cortex, which stimulates kidney Na absorption. In fact, Na-restricted mammalian and avian wildlife appear to excrete urine with negligible Na (Blair-West et al. 1968; Jordan 1987; figure 11.8).

Sodium requirement is increased during production. In our example with caribou (*Rangifer tarandus*, figure 11.7), the extra growth requirement was estimated based on 0.375 kg d^{-1} growth consisting of tissue with a Na content of 50 mmol (1150 mg) per kg (Staaland and Hove 2000). Because almost all dietary Na is absorbed, this translates to an additional dietary requirement of

Figure 11.8. Vertebrates can greatly decrease urinary Na loss when consuming low Na diets. (A; left-hand plot) is for white-tailed deer (*Odocoileus virginianus*) consuming diets of differing Na contents. Points are samples from individuals, and the solid line is a fit to the data (replotted from Hellgren and Pitts 1997). (B; right-hand plot) is for rufous hummingbirds (*Selasphorus rufus*) eating low Na nectar (replotted from Lotz and Martínez del Rio 2004). The four different kinds of points represent samples from four individuals, and the four curved solid lines are power equation fits to data for each individual. The straight dashed line in B represents the relationship between sodium intake and concentrations in excreta that would lead to neutral sodium balance (i.e., ingestion = excretion). The remarkable ability of the kidney to reabsorb Na is mediated by aldosterone, a steroid hormone secreted by the cortex of the adrenal gland. Aldosterone stimulates the kidney's distal tubule to reabsorb Na from the kidney filtrate.

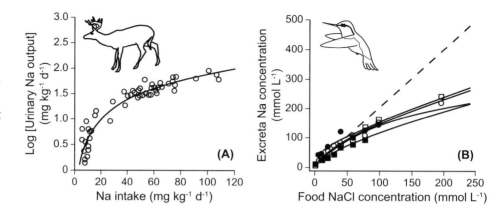

431 mg Na d^{-1}, which must be added to maintenance needs (139 mg d^{-1}, estimated from white-tailed deer). The sum of these two divided by daily food intake (1.86 kg d^{-1} for a 40 kg calf) yields 306 mg Na kg^{-1} dry mass of food required for growth. This estimated requirement is lower than the standard for domesticated and laboratory animals (500–2000 mg kg^{-1} dry mass; Robbins 1993), many of which have been selected for very high production rates. Na requirements are from 800 to 1000 mg kg^{-1} dry mass in nonlactating cattle and sheep and double that during lactation (McDowell 1992).

A huge range of plant species (≥ 85) and families contain plant secondary metabolites (PSMs) that can increase Na excretion from the kidney (Dearing et al. 2001) and thus potentially increase an herbivore's dietary requirement for Na. The compounds represent all the major classes such as terpenes, phenolics, and alkaloids, and their modes of action include specific depression of Na reabsorption by the ascending loop of kidney tubules (so-called "loop diuretics") or diuresis (increased urine volume) which might secondarily increase Na excretion. The ecological importance of these has been practically unexplored, although a number of studies have suggested that PSMs increased Na loss in vertebrate herbivores (reviewed by Robbins 1993; Foley et al. 1999). In many of those studies, however, it is difficult to determine whether the PSM directly increased Na loss, because the herbivores reduced their food intake and increased tissue catabolism to meet energy needs. Reduced intake and increased catabolism can lead to flushing of excess Na from the body. One way to control for this in future studies is to utilize an experimental design in which food intake of controls is paired to that of experimentals. Also, Foley et al. (1999) point out that future studies need to determine whether increased Na loss is an acute or chronic effect of PSMs.

In summary, only a few thorough studies of Na requirements of herbivores have been performed, and the possible interactive effects of SMs are also relatively unexplored. There are also few studies of whether the sometimes very high levels of K in plants might increase Na loss through diuresis (Robbins 1993). A few studies have taken advantage of ^{22}Na, which could be a powerful tool for studying Na dynamics in free-living herbivores (Green and Dunsmore 1978; Green et al. 1984). Na dynamics are studied using principles and methods similar to those used to study labeled water turnover, which we describe in the next chapter (box 12.1). Staaland and Hove (2000) injected ^{22}NaCl into caribou living at the site characterized in figure 11.7, and took blood samples after equilibration and then again months later. The measured fluxes in calves and females ranged from 8.4 to 12.7 mg kg$^{-3/4}$ d^{-1}, which is at or slightly above the minimum requirement estimated in deer species (above). Recall that the estimates from these studies only provide estimates of actual fluxes and not of minimal requirements. Although more studies of Na requirements and dynamics are clearly needed, all the estimates of Na requirements of wildlife, in comparison to the low Na content of plants in some ecosystems, underscore the real possibility of Na insufficiency for vertebrate herbivores in those ecosystems. Sodium is known to be a limiting factor for growth and survival of domesticated ruminants in many regions of the world. How do animals accommodate to life in a low-sodium world?

11.2.2 Sodium Hunger—A Classic Example of Behavioral Regulation

One of the fascinating aspects of Na nutrition is that many animals have an innate Na hunger. Rats and sheep made sodium deficient ingest salt (NaCl) offered to them within seconds (Schulkin 1991). Animals can even select Na-based salts from among a number of salt options (e.g., KCl, CaCl$_2$, MgCl$_2$), and in some species Na ingestion is commensurate with Na deficit. Specific hungers have been described for other nutrients such as P and perhaps Ca, protein, simple carbohydrates, and some vitamins (Schulkin 2001), but these hungers require some learning whereas the hunger for Na is innate (Schulkin 1991). Innate and learned specific hungers, and taste aversion learning (described in chapter 9), are instances of behavioral regulation, i.e., behavior serving the same end as physiology. They provide the basis for the ability of animals to select nutritionally balanced diets, which has been demonstrated in mammals, birds and reptiles. Richter was a major originator of concepts and research in this field and much of the progress in this field of psychobiology can be reviewed in the books by physiological psychologist Jay Schulkin (1991, 2001). Dethier's (1976) lovely book on hungry flies provides many examples of specific hungers in insects.

Many kinds of salt-directed behavior have been described in wild animals (Robbins 1993). Animals (and indeed humans) will travel long distances to natural salt deposits, they sometimes ingest mineral-rich soils, they lick urine-soaked ground, and they select Na-rich plants (Schulkin 2001). In some ecosystems in which plants are generally low in Na there may be a small number of plant species that have relatively high levels. Notice, for example, that in the Jotunheimen mountain region in southern Norway, fungi have Na contents that are from six to seven times higher than all the other plant types (figure 11.7), and caribou consume them when they are seasonally available (Staaland and Hove 2000). In northern forests, submerged and floating aquatic plant species have Na contents (dry mass basis) that are from 10 to 300 times higher than terrestrial species (Jordan 1987; Ohlson and Staaland 2001). In a classic study of optimal foraging Belovsky (1978) incorporated a moose's requirement for Na into a foraging model using the method of linear programming and predicted the optimal amounts of aquatic and terrestrial vegetation consumed (box 11.2).

These salt-directed behaviors are not necessarily only in reaction to salt depletion but can also be an anticipatory homeostatic response (Schulkin 2001). The consumption of salt in anticipation of greater need, say during pregnancy and lactation, may be an important response of animals living in a low-sodium world. Returning to our example of caribou for illustration, there may be specific seasons of relatively higher Na availability such as the summer when fungi or aquatic vegetation are available, or during a period when migrations bring them close to marine coastal areas. Consumption and storage of Na would be advantageous, and at least two storage sites have been proposed. In caribou, Staaland et al. (1982) estimated that half the Na in the body is in bone and 30–40% of that is exchangeable, making this a possible site of Na storage. Another 20–25% is in the rumen and it has been postulated that as the caribou becomes progressively Na depleted it relies on this as a reserve. Ruminants facing Na deprivation can substitute K for Na in their saliva, and caribou on low-Na forage exhibit rising ratios of K:Na in both saliva and rumen fluid.

In summary, caribou and many other herbivores live in a low-sodium world. In some habitats herbivores can get by because they minimize their losses and their innate specific hunger for Na motivates them to seek out a variety of potential supplements to their Na budget. For some, the problem of balancing their Na budget is solved on an annual basis by accumulating Na when it is available and then drawing on the reserve when it is in short supply. Besides impacting their feeding behavior and movements, the shortage of Na potentially influences herbivore population dynamics, a topic we will consider in a section below.

Box 11.2. Using Linear Programming to Determine Diets that Satisfy Nutrient Requirements

Belovsky (1978) championed the use of linear programming by using it to define what mixes of aquatic and terrestrial plants would satisfy requirements of moose. We will illustrate the approach using those data.

Step 1 is to define characteristics of the two plant types in relation to likely requirements or constraints that must be satisfied such as the need for energy, sodium, and avoidance of overfilling the gut:

Feature of plant	Terrestrial Plant	Aquatic Plant
Digestible energy content (kJ g^{-1} dry mass)	13.0	16
Sodium concentration (mg g^{-1} dry mass)	0.05	3
Volume (g wet mass g^{-1} dry mass)	4.04	20

Step 2 is to describe quantitatively the requirements and constraints for a 358 kg moose in terms of the amount per day consumed of terrestrial plants (I_{tp}, in g d^{-1}) and aquatic plants (I_{ap}, in g d^{-1}):

Requirement or constraint	Numerical value	Corresponding linear equation
Maintenance energy required, R	$R = 45{,}831$ kJ d^{-1}	$R \le 13\,I_{tp} + 16\,I_{ap}$
Sodium requirement, $I_{Na,min}$	$I_{Na,min} = 1{,}340$ mg d^{-1}	$I_{Na,min} \le 0.05\,I_{tp} + 3I_{ap}$
Rumen volumetric capacity, G_{max}	$G_{max} = 31{,}513$g d^{-1}	$G_{max} \ge 4.04\,I_{tp} + 20\,I_{ap}$

Belovsky (1978) also included a time constraint: moose could not spend more than 256 min d^{-1} foraging on land, which at a cropping rate of 0.065 min g^{-1} food translates to a maximum intake of 3938 g d^{-1}. Their cropping rate on aquatic plants was 0.05 min g^{-1} food.

Step 3 is to solve for the diets that satisfy all the requirements and constraints. Because we are dealing with only two food types this is easily accomplished by plotting the linear equations on an *x-y* graph (figure 11.9). Three-dimensional graphs are used when there are three food types, and if there are even more then it is easiest to use matrix algebra to find solutions.

continued

continued

continued

Figure 11.9. Linear programming of diet selection with constraints. Only the diets inside the small quadrilateral (gray area) satisfy all the requirements/constraints. Clearly, some aquatic plants must be consumed to meet Na requirements, but a diet of just aquatic plants cannot be consumed because it would exceed the rumen's daily capacity. A mixed diet is the solution, but which mix? An optimal diet can be identified once a goal or optimality criterion is defined. For example, if the moose's goal is to maximize its digestible energy intake, the optimal diet is located at the upper right-hand corner of the quadrilateral, which corresponds to a diet of about 17% aquatic plant material (by dry mass). Energy-maximizing moose should spend 296 min d^{-1} foraging on land and 39 min d^{-1} foraging in water.

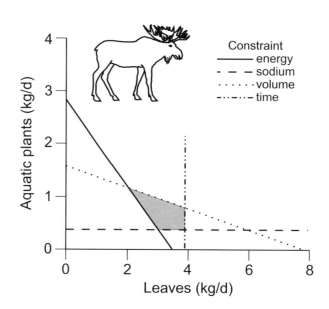

Linear programming is critically dependent on our understanding of the physiological requirements that are expressed in the linear equations, and there is debate about whether equations that have been adopted are biologically realistic and about other features of the approach (Owen-Smith 1996). But if parameter values are adequately defined the approach should be robust and arguably has been successful predicting diet choices of animals (Belovsky 1990, 1994).

11.2.3 Calcium Is Limiting for a Wide Variety of Animals

In the mid-1980s European ornithologists noticed a serious problem—the proportion of some forest passerine birds laying eggs with defective shells was on the increase in some forests. At first glance, the finding of eggs with thin shells that fail to hatch, or are abandoned, seemed reminiscent of poisoning by organochlorine compounds, but birds have generally recovered from this problem in Europe following the banning of DDT. Thanks to the thorough work of the Dutch biologist Jaap Graveland and his colleagues we now know that in some of these forests Ca availability limits breeding success of birds, and we can identify that humans had a hand in creating the problem by causing acid rain. Graveland and Van Der

Wal (1996) discovered that eggshell defects were inversely related to availability of snail shells, an important calcium source for egg laying (see below), and that density of snails was directly related to calcium content of the forest soil and litter (figure 11.10). Furthermore, they and others tested each of the suggestive correlations in figure 11.10 experimentally and showed that additions of calcium to the soil (Johannessen and Solhoy 2001), or reductions in the acidity that causes Ca leaching, increased the density of snails (Graveland and Van Der Wal 1996). When they supplemented the diet of free-living birds with a Ca source the prevalence of eggshell defects was reduced and the fraction of eggs that hatched increased (Graveland and Drent 1997). This fascinating causal chain of events hinges on the fact that birds and snails in these forests live on the cusp of Ca deficiency.

Many birds lay one egg per day, and the demands of producing a heavily calcified eggshell creates a daily calcium demand for reproduction much greater than in other vertebrates. Furthermore, the relatively low Ca contents of the normal foods of many birds, such as insects ($0.14–1.49$ mg g^{-1} dry mass; Studier and Sevick 1992; Graveland and Berends 1997) and nuts and seeds ($0.5–1.7$ mg g^{-1} dry mass; Graveland and van Gijzen 1994; Robbins 1993) cannot provide more than 25% of the Ca required for egg formation (Graveland and Berends 1997). The requirement can be estimated from the amount of Ca in the egg divided by the transfer efficiency from the parent's ingestion to the egg. The eggs of 18 g great tits (*Parus major*), for example, contain from 30 to 40 mg of Ca, and birds generally appear capable of transferring 40–70% of ingested Ca to their eggs (Graveland and Berends 1997). Thus, great tits laying an egg once a day require 43–100 mg Ca above daily maintenance needs. Considering their likely feeding rate of 7 g dry mass d^{-1} (Nagy 2001) and their prey Ca content (from above), one calculates that their prey provide at most 23% of the extra Ca requirement. Small passerines do not appear to store Ca prior to laying but rely on daily intake to meet their need (Krementz and Ankney 1995; Graveland and Berends

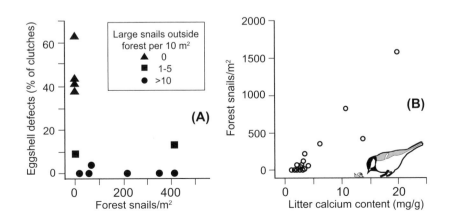

Figure 11.10. In some Dutch forests there was a high prevalence of defects in eggshells of passerine birds when snail density was very low and when there were no large snails outside the forest (A). Snail shells are an important calcium source for egg laying (see text). The availability of snails was directly related to calcium content of the forest soil and litter (B). (Redrawn from Graveland and Van Der Wal 1996.)

1997; Blum et al. 2001; box 11.3). It is not surprising, then, that when passerine birds are maintained on diets with moderate or low Ca, and are not provided a Ca supplement, they lay at least 50% fewer eggs many of which are defective (Holford and Roby 1993; Graveland and Berends 1997).

Normally, wild and captive passerine birds exhibit Ca hunger, that is, Ca-directed behaviors analogous to those described above for Na hunger. They search for Ca-rich material such as snail shells, bones (osteophagia), and calcareous grit, and perhaps particularly Ca-rich arthropod prey such as woodlice or millipedes which have 100 times more Ca than other prey (Grave-

Figure 11.11. Many non-passerine birds store calcium and phosphorus by depositing it within the marrow cavities of their bones. The calcium is held within an open, spongy medullary bone tissue that lines the walls of long bones—depicted as the stippled area in the figures of a chicken humerus. Specialized cells called osteoblasts accumulate and deposit calcium phosphate ($Ca_3(PO_4)_2$) and hydroxyapatite ($Ca_{10}(PO_4)_6(OH)_2$) crystals. When calcium and phosphorus demand is heavy because females are laying eggs, osteoblasts reabsorb the crystals and secrete calcium and potasium into the blood stream for use in the ovaries for the synthesis of egg components. Consequently, after egg laying there is a reduction in the amount of spongy tissue, as shown in right-hand figure. (Modified from Proctor and Lynch 1993.)

Box 11.3. Capital-or Income-Based Calcium Economy during Breeding?

In vertebrates, calcium and phosphorus are sequestered with calcium in the bone's minerals calcium phosphate ($Ca_3(PO_4)_2$) and hydroxyapatite ($Ca_{10}(PO_4)_6(OH)_2$). Although the pools of P and Ca in mineralized bone show slow turnover, these elements are not completely immobilized after being deposited. Throughout life the processes of reabsorption and mobilization are continuous and there is interchange of Ca and P between bone, blood, and other parts of the body (McDowell 1992). One major outcome is tight control of plasma Ca, which is important because Ca plays a big role in signaling in many tissues throughout the body. The entire process is regulated, with hormones calcitonin and parathyroid hormone playing big roles, even influencing the efficiency of Ca and P absorption from the diet.

The role of bone as a temporary reservoir for calcium has been well investigated in some vertebrates. Medullary bone deposited prior to egg laying in some birds and then mobilized to synthesize the calcite of egg shells is a good example (Dacke et al. 1993; figure 11.11).

The presence of medullary bone is clearly associated with reproduction in many species. However, its importance as a reservoir of calcium and

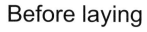

Before laying	After laying

continued

continued

phosphorus varies among bird taxa. Some species, such as chickens and quail, appear to accumulate calcium and phosphorus prior to egg laying. They depend, at least partially, on this stored "capital" to produce eggs. Pahl et al. (1997) used high-energy radiography to determine whether small songbirds used medullary bone seasonally. Wisely, one of the species they chose to study was the brown-headed cowbird (*Molothrus ater*). Because cowbird females can lay up to 50 eggs in a season, they are prime candidates for a species that could use somatic calcium and phosphorus stores. Pahl et al. (1997) found no evidence of medullary bone in any of the three songbird species that they examined. They hypothesized that songbirds base the reproduction of eggshells on their "income" of calcium and phosphorus over short times before and during the laying of a clutch. Species that do not rely on medullary bone may be especially sensitive to acid rain and chemical soil poisoning that decreases the availability of crucial minerals (Graveland and Drent 1997).

land and van Gijzen 1994). Captive great tits spent 20–40% of the daylight hours in such activities! The snails that birds and other animals search for may also be experiencing a Ca-poor world. The shell of land snails contains 97% (by mass) of calcium carbonate, and they lay eggs with Ca-rich shells, so they require relatively large amounts of Ca for growth and reproduction (Crowell 1973; Ireland 1991). One snail, *Achatina fulica*, apparently absorbed 48% of Ca in rabbit pellets (Ireland 1991), but laboratory studies have generally indicated that plant matter is not a sufficient source of Ca for growth and reproduction and that snails obtain additional calcium by ingesting soil (Crowell 1973; Wareborn 2002; Heller and Magaritz 1983; Gomot et al. 1989), from which they can apparently absorb 84% of soil calcium carbonate (Daouda and Haquou 1995). In locations on soils without calcium carbonate (e.g., bedrock of granite), or where soil is acidified, Ca may thus become limiting and densities of snails are relatively low. Availability of Ca may be a significant factor limiting the distribution of other kinds of crustaceans as well.

The problem of Ca insufficiency is not specific to birds and snails that produce Ca-rich shells. The reproduction of insectivorous bats and birds that rely on low Ca prey may be constrained by low Ca availability (Barclay 1994), exacerbated by effects of increased ecosystem acidity (Ormerod et al. 1991; St. Louis and Barlow 1993). Calcium requirements of growing mammals and birds range from

4 to 12 mg Ca g^{-1} diet (Robbins 1993), which is still quite a bit higher than the Ca content of nuts, seeds, and most invertebrate prey. Even maintenance needs may not be met by some of these foods. For example, inspect the plot of total Ca excretion of fox squirrels (*Sciurus niger*) feeding on nuts and seeds which all had less than 2.08 mg of Ca per gram of dry mass (figure 11.12). The endogenous metabolic losses can be estimated by the y intercept as 27 ± 5 mg kg^{-1} d^{-1}, although they are not necessarily fixed at this level. Endogenous fecal loss of calcium may be influenced by feeding rate, as was the case for N (Murray 1995). The slope in figure 11.12 is 0.545, which indicates a retention of dietary Ca of 45.5% (i.e., $1 - 0.545$), so the maintenance requirement is estimated as $27/0.455 = 59$ mg kg^{-1} d^{-1}. If the squirrels eat 30 g kg^{-1} d^{-1} food to meet energy needs (Havera and Smith 1979), then they need 2 mg Ca/g food, which is higher than the Ca content of all eight of the wild nuts fed in the study. Many seeds also have an excess of phosphorus (P), which can form insoluble complexes with Ca and reduce its absorption. Dietary Ca to P ratios of 1:1 to 2:1 are considered best for absorption and metabolism of Ca (Robbins 1993).

Carnivores are another group that may sometimes suffer from Ca insufficiency. Vertebrate flesh has very low Ca content and a low Ca:P ratio, and free-living foxes and vultures that routinely consumed vertebrate flesh without bones exhibited Ca deficiency syndromes such as osteoporosis (Robbins 1993). Herbivores seem less likely to suffer from inadequate Ca, although it happens. For example, in the studies of the nutritional ecology of cotton rats that were discussed above only one diet plant in wintertime, and none of the diet plants in springtime, had enough

Figure 11.12. Wild nuts and seeds fed to fox squirrels had inadequate calcium content because squirrels excreted more Ca than they consumed. The maintenance requirement can be estimated by dividing the apparent minimum endogenous loss of calcium, represented by the y intercept (27 mg kg^{-1} d^{-1}), by the proportion of ingested calcium that is retained, represented by 1 – slope ($1 - 0.545$). (Data from Havera and Smith 1979.)

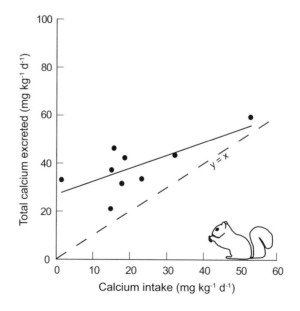

Ca to support growth (Randolph and Cameron 2001). In a different study, the grasses consumed by wild donkeys in Australia were extremely low in Ca (Freeland and Choquenot 1990) and the females showed signs of reduced bone density and ate soil, presumably to increase mineral intake (eating soil is called "pica"). Schulkin (2001) reviews examples of mammalian and reptilian herbivores that apparently exhibit a Ca hunger analogous to that described for passerine birds.

The problem of Ca insufficiency is thus relatively pervasive, and in some cases conservation schemes may address it. Researchers are testing whether animals living on low-Ca or acidic soils can benefit from soil additions of Ca (Ormerod and Rundle 1998; Johannessen and Solhoy 2001) and/or acid buffers such as lime (Graveland and Van Der Wal 1996) or from more direct Ca supplementation (Richardson et al. 1986). Carrion-feeding birds depend on mammalian carnivores to crack bones into pieces small enough to swallow. In areas of South Africa in which mammalian carnivores have been extirpated, carrion-feeding birds may suffer from Ca deficiencies. To alleviate them, wildlife managers provide cafeterias of cracked bones (Richardson et al. 1986).

11.2.4 Phosphorus Can Be Limiting, Especially in Aquatic Ecosystems

Phosphorus (P) limitation has been reported in a number of species of aquatic invertebrates, and there is a prevailing view that, in general, primary and secondary production are limited by P in freshwater ecosystems (Elser et al. 2000; Peters 1986). In low-P oligotrophic waters, autotrophs may have P contents too low to meet the needs of herbivores. In laboratory experiments the C:P ratios of alga grown in monoculture can be manipulated and the response of herbivores studied. Both planktonic and benthic herbivores grow at slower rates when fed algae with lower P (figure 11.13). The interpretation of such data is confounded by the fact that P-starved phytoplankton may also have lower concentration of essential biochemicals (such as polyunsaturated fatty acids, chapter 2) besides lower P content and may develop thicker cell walls. Carefully designed studies that test the effect of changing only algal P content concur that direct limitation of P to zooplankton is very important (Plath and Boersma 2001). A successful and widely applied model that links plankton growth to a limiting nutrient is the Droop equation (Sterner and Hessen 1994)

$$\mu = \hat{\mu}\left(q - \frac{q_{\min}}{q}\right), \tag{11.2}$$

Figure 11.13. Both planktonic *Daphnia magna* and benthic mayfly larvae (*Caenis* sp.) grew faster when fed algae with higher phosphorus (P) content relative to carbon (C). In each plot the growth rate on the lowest P:C ratio in the food was arbitrarily set at 1 and the growth rates on the other foods were expressed relative to that growth rate. The corresponding absolute growth rates at relative growth = 1 were 0.15 per day for Daphnia (DeMott et al. 1998) and 0.006 per day for mayflies (Frost and Elser 2002).

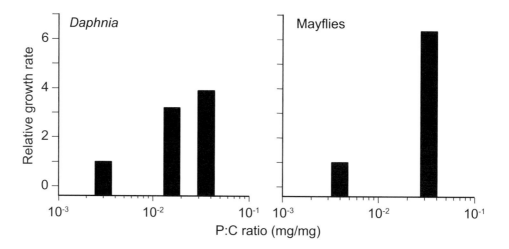

where μ is the specific growth rate per unit time, $\hat{\mu}$ is the maximum specific growth rate, q is the amount of limiting nutrient inside the plankton body, and q_{min} is the minimum required amount of the nutrient inside the plankton body.

As discussed in chapter 10, a fundamental problem in this situation is that animals may ingest less P and more energy (C) than required for maintenance and growth. Water fleas (*Daphnia* spp.), which appear to have higher P contents and thus higher P demands than other freshwater zooplankton (Hessen and Rukke 2000) may exhibit several compensatory adjustments to this challenge. First, on P-deficient diets they limit their P excretion (Gulati and DeMott 1997), but in strongly P-limited *Daphnia* there appears to be a low but consistent P excretion (DeMott et al. 1998) perhaps analogous to minimum endogenous metabolic losses. Second, studies with food dual labeled with radioactive ^{14}C and ^{32}P provided evidence that *Daphnia* reduce the assimilation of C across the gut wall while maintaining high P assimilation efficiency (DeMott et al. 1998). Direct uptake of P from water is negligible at naturally occurring concentrations (Sterner and Hessen 1994). Third, the dual-label studies provided evidence that *Daphnia* disposed of excess C through either increased respiration or C excretion during long-term feeding on highly P-deficient diets. This kind of digestive and metabolic dexterity has not been described in many animals and it will be important to confirm these findings and learn of more examples. These responses waste energy, and so growth efficiency (mass or joules added per unit mass or joules consumed) declines on low-P diets. A fourth possible response by *Daphnia* is that excess C might be stored as lipids, but so far there is little evidence for this possibility (Sterner and Hessen 1994), although there seem to be some interesting—and relatively complex—interactions between P limitation and fatty acid biochemistry in *Daphnia* (Gulati and DeMott 1997). The compensatory behavioral response of seeking out food items with higher P content is possible for some plankton like copepods (Sterner and Hessen 1994), but filter-feeding *Daphnia* do not have the possibility of selective feeding. At least one study suggested that feeding (filtering) activity increased with declining P content (Plath and Boersma 2001), but more studies have indicated decreased (Sterner and Hessen 1994) or constant (DeMott et al. 1998) feeding on P-limited diets.

A recent stochiometric analysis gives circumstantial evidence that P limitation may be as common and important in terrestrial as in aquatic ecosystems (see chapter 10; Elser et al. 2000). Terrestrial plants tend to have less P per unit energy (i.e., lower P:C ratio) than aquatic plants, a sign of differences in resource availability, but there is no major difference among the invertebrate herbivores of aquatic and terrestrial ecosystems in the demand for P as indexed by the P:C ratio of herbivores (Elser et al. 2000). Thus, the differential in P:C ratio across trophic levels may be less advantageous for invertebrates in terrestrial ecosystems. When *Manduca sexta* caterpillars were raised on natural and artificial diets with P concentrations that varied over an ecologically relevant range, increased dietary P significantly increased growth rate and body P content, and shortened the time to the final instar molt (Perkins et al. 2004). Thus, a variety of data suggest that dietary P may affect the growth rates and population dynamics of some insect herbivores.

For terrestrial vertebrates, we find it difficult to make a general statement about possible insufficiency. On the one hand, a comparison of the phosphorus requirement for growth of mammals and birds, from 3 to 6 mg P g^{-1} dry diet, and the typical P content of various food types (Robbins 1993) suggests that P deficiency might be uncommon. Also, Grasman and Hellgren (1993) suggested that P limitation might be unlikely except in very large mammals that produce very large antlers (which can contain 12% P and 24% Ca by mass) and in those that feed on grasses, which tend to have lower P content relative to forbs and browses. On the other hand, P limitation does occur among domestic ruminants under a variety of conditions. P deficiency in cattle has been reported in 46 countries and is the result of low P in plants (McDowell 1992). Globally, leaf P content declines toward the equator in five dominant plant groups (coniferous trees, and angiosperm grasses, herbs, shrubs, and trees (Reich and Oleksyn 2004). The P content in food plants appears to be inadequate to meet the needs of lactating cotton rats and wild donkeys at least in some habitats (Randolph and Cameron 2001; Freeland and Choquenot 1990).

Ungulate grazers in the Serengeti seem also to be P limited (see next section). As Sterner and Elser (2002) have pointed out, P may be limiting in many terrestrial ecosystems and for many terrestrial vertebrates. P limitation deserves as much attention on land as in water.

11.2.5 Mineral Limitation Influences Higher Levels of Ecological Organization

A variety of theoretical, correlative, and experimental studies point to the likely significance of mineral limitation at higher levels of ecological organization. We will finish up by organizing some examples, both new and some we have already cited, into several categories of ecological effects.

Mineral limitation may influence a primary population parameter such as birth rate. Calcium limitation of passerine reproduction was demonstrated experimentally in the wild (above). In a series of studies on microtine rodents (voles and lemmings) Batzli and his colleagues demonstrated that plant mineral contents at some seasons or places are too low to support growth and reproduction (Barkley et al. 1980; Batzli 1986). Their studies included an experimental laboratory study in which they supplemented apparently Na- and Ca-poor grass seeds with Na and Ca. Supplementation led to higher frequency of litter production, larger litter size, and higher survival and growth of young. McCreedy and Weeks (1993) tested the Na-limitation hypothesis by provisioning study plots with Na solutions, versus deionized water controls, and measuring growth of cottontail rabbits (*Sylvilagus floridanus*). The Na-"fertilized" plots were more intensively utilized by free-living rabbits than the controls, and females living in Na-supplemented plots grew larger.

In some marine systems mineral limitation influences, both indirectly and directly, a secondary population parameter such as population growth rate of heterotrophs. For example, iron (Fe) addition experiments seem to indicate that low Fe limits primary production in certain open ocean ecosystems (Downing et al. 1999). One consequence of Fe addition is higher grazing rates and higher rates of population increase among the grazing community that eats the autotrophs (Landry et al. 2000). In addition, the growth rates of heterotroph populations can directly respond positively to increases in the Fe content of their food (Chase and Price 1997). The positive populational responses of the heterotrophs can feed back negatively on the autotroph populations in such a way that the community structure, such as the number of autotroph size classes, is limited both from the bottom by micronutrient limitation and from the top by heterotroph grazing (Armstrong 1994).

Mineral limitation influences the distribution of herbivores across the landscape, as exemplified in the previous descriptions of snail diversity in relation to Ca, habitat selection by cotton rats, and movements of many herbivores in response to Na and Ca hunger. One of the most dramatic examples involves the mass seasonal movements of African migratory ungulates (McNaughton 1990; Murray 1995). Tropical grasses are generally of lower nutritional quality than temperate forages and it is well recognized that domesticated ungulates in the tropics are frequently mineral deficient unless supplemented. It is not surprising, therefore, that nutritional deficiencies have been documented in wild African ungulates (McNaughton and Georgiadis 1986). In the Serengeti ecosystem of eastern Africa, vast herds of millions of ungulates spend the dry season in subhumid northwestern grasslands but during the wet season they migrate to southeastern grasslands where calving occurs. Mean concentrations of many elements, including Na, Ca, and P, are higher in grasses in the southern regions where the plant community overlies more fertile volcanic soils (figure 11.14). P concentrations in particular (and maybe Na concentrations as well) are too low in the northern regions to support reproduction (figure 11.14). After reproduction at the end of the wet season, ungulates return to the northwest corner, which is often the only area supporting sufficient plant biomass during the dry season. Also, within each of these regions, and elsewhere in Africa, the herbivores were found to concentrate on sites with high nutrient contents (Weir 1972; Seagle and McNaughton 1992). Ungulates in other ecosystems, like Yellowstone in North America, may migrate for analogous reasons (Frank et al. 1998). McNaughton (1990) points out that findings such as these underscore how surveys of forage mineral properties might be useful guides in design of reserves and in ecosystem (habitat) rehabilitation.

An interesting feature of the interplay between minerals and the distribution of herbivores across the landscape is putative feedback loops by which grazers play a major role in ecosystem nutrient cycling and community structuring. Budgets of nutrient flow in ecosystems indicate that inputs by terrestrial herbivores from their excretion or carcasses can play important roles in long-term element balance, as exemplified in our previous discussion of N limitation. Also, the movements of herbivores coupled with the deposition of N and other elements influence spatial heterogeneity of soil properties and plants at many scales from individual plants to landscapes (McNaughton and Georgiadis 1986; Augustine and Frank 2001; Steinauer and Collins 2001). In analogous fashion, excretory patterns of pelagic invertebrates are known to influence mineral dynamics and community structure (see chapter 10 and references therein). All these examples reflect a growing appreciation for the importance of mineral limitation in ecology at many levels of organization.

Figure 11.14. When ungulates of the Serengeti ecosystem migrate to the southeast for calving, their food on the wet season range meets their nutrient requirements for lactation better than do the plants on the dry season range. The graphs show the mean element concentrations measured by two researchers, McNaughton (1990) and Murray (1995), in comparison with the estimated element requirement for lactation (from Murray 1995). Notice that both investigators found that P levels were too low on dry season range but sufficient on wet season range to support reproduction. McNaughton, but not Murray, also concluded that Na levels were too low on dry season range but sufficient on wet season range. The map was redrawn from Murray (1995).

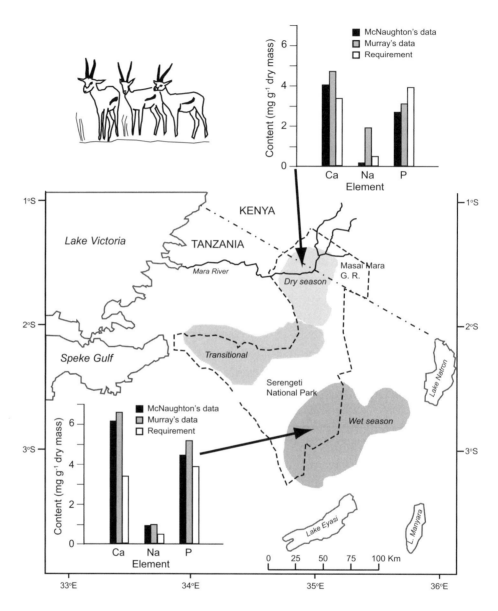

References

Armstrong, R. A. 1994. Grazing limitation and nutrient limitation in marine ecosystems: Steady state solutions of an ecosystem model with multiple food chains. *Limnology and Oceanography* 39:597–608.

Ashford, D. A., J. Pritchard, and A. E. Douglas. 1997. Honeydew sugars and osmoregulation in the pea aphid *Acyrthosiphon pisum*. *Journal of Experimental Biology* 200:2137–2143.

Ashford, D. A., W. A. Smith, and A. E. Douglas. 2000. Living on a high sugar diet: The fate of sucrose ingested by a phloem-feeding insect, the pea aphid *Acyrthosiphon pisum. Journal of Insect Physiology* 46:335–341.

Augustine, D. J., and D. A. Frank. 2001. Effects of migratory grazers on spatial heterogeneity of soil nitrogen properties in a grassland ecosystem. *Ecology* 82:3149–3162.

Barclay, R.M.R. 1994. Constraints on reproduction by flying vertebrates: energy and calcium. *American Naturalist* 144:1021–1031.

Barkley, S. A., G. O. Batzli, and B. D. Collier. 1980. Nutritional ecology of microtine rodents: a simulation model of mineral nutrition for brown lemmings. *Oikos* 34:103–114.

Batzli, G. O. 1986. Nutritional ecology of the California vole: effects of food quality on reproduction. *Ecology* 67:406–412.

Batzli, G. O., and F. R. Cole. 1979. Nutritional ecology of microtine rodents: Digestibility of forage. *Journal of Mammalogy* 60:740–750.

Belovsky, G. E. 1978. Diet optimization in a generalist herbivore. *Theoretical and Population Biology* 14:105–134.

Belovsky, G. E. 1990. How important are nutrient constraints in optimal foraging models or are spatial/temporal factors more important? Pages 255–278 in R. N. Hughes (ed.), *Behavioral mechanisms of food selection.* Springer-Verlag, Berlin.

Belovsky, G. E. 1994. How good must models and data be in ecology? *Oecologia* 100:475–480.

Blair-West, J. R., J. P. Coghlan, D. A. Denton, J. F. Nelson, E. Orchard, B. A. Scoggins, R. D. Wright, K. Myers, and C. L. Jungueira. 1968. Physiological, morphological, and behavioral adaptations to a sodium deficient environment by wild native Australian and introduced species of animals. *Nature* 217:922–928.

Blum, J. D., E. H. Taliaferro, and R. T. Holmes. 2001. Determining the sources of calcium for migratory songbirds using stable strontium isotopes. *Oecologia* 126:569–574.

Bosman, A. L., and P.A.R. Hockey. 1986. Seabird guano as a determinant of rocky intertidal community structure. *Marine Ecology Progress Series* 32:247–257.

Bowen, S. H. 1987. Dietary protein requirements of fishes—a reassessment. *Canadian Journal of Fisheries and Aquatic Sciences* 44:1995–2001.

Bowen, S. H., E. V. Lutz, and M. O. Ahlgren. 1995. Dietary protein and energy as determinants of food quality: trophic strategies compared. *Ecology* 76:899–907.

Bradshaw, F. J., and S. D. Bradshaw. 2001. Maintenance nitrogen requirement of an obligate nectarivore, the honey possum, *Tarsipes rostratus. Journal of Comparative Physiology B* 171:59–67.

Carthew, S. M., R. L. Goldingay, and D. L. Funnell. 1999. Feeding behaviour of the yellow-bellied glider (*Petaurus australis*) at the western edge of its range. *Wildlife Research* 26:199–208.

Chase, Z., and N. M. Price. 1997. Metabolic consequences of iron deficiency in heterotrophic marine protozoa. *Limnology and Oceanography* 42:1673–1684.

Christian, D. P., T. E. Manning, and C. J. Harth. 1993. Sodium and potassium balance of captive meadow voles (*Microtus pennsylvanicus*) fed laboratory chow and vegetation diets. *Comparative Biochemistry and Physiology A* 106A:571–579.

Crowell, H. H. 1973. Laboratory study of calcium requirements of the brown garden snail, *Helix aspersa* Muller. *Proceedings of the Malacological Society of London* 40:491–503.

Dacke, C. G., S. Arkle, D. J. Cook, I. M. Wormstone, M. Zadi, and Z. A. Bascal. 1993. Medullary bone and avian calcium regulation. *Journal of Experimental Biology* 184:63–88.

Dadd, R. H. 1985. Nutrition: organisms. Pages 313–390 in G. A. Kerkut and L. I. Gilbert (eds.), *Comprehensive insect physiology, biochemistry, and pharmacology*. Pergamon Press, Oxford.

Daouda, I., and A. H. Haquou. 1995. Effects of dietary calcium (source and way of presentation) on growth of the snail *Achatina achatina* L. *Cahiers Agricultures* 4:444–448.

Dearing, M. D., A. M. Mangione, and W. H. Karasov. 2001. Plant secondary compounds as diuretics: An overlooked consequence: *American Zoologist* 41:890–901.

Delorme, M., and D. W. Thomas. 1999. Comparative analysis of the digestive efficiency and nitrogen and energy requirements of the phyllostomid fruit-bat (*Artibeus jamaicensis*) and the pteropodid fruit-bat (*Rousettus aegyptiacus*). *Journal of Comparative Physiology B* 169:123–132.

DeMott, W. R., R. D. Gulati, and K. Siewertsen. 1998. Effects of phosphorus-deficient diets on the carbon and phosphorus balance of *Daphnia magna*. *Limnology and Oceanography* 43:1147–1161.

Downing, J. A., C. W. Osenberg, and O. Sarnelle. 1999. Meta-analysis of marine nutrient-enrichment experiments: Variation in the magnitude of nutrient limitation. *Ecology* 80:1157–1167.

Elser, J. J., W. F. Fagan, R. F. Denno, D. R. Doberfuhl, A. Folarin, A. Huberty, S. Interlandi, S. S. Kilham, E. McCauley, K. L. Schulz, E. H. Siemann, and R. W. Sterner. 2000. Nutritional constraints in terrestrial and freshwater food webs. *Nature* 408:578–580.

Eshelman, B. D., and G. N. Cameron. 1996. Experimentally induced habitat shifts by hispid cotton rats (*Sigmodon hispidus*): Response to protein supplementation. *Journal of Mammalogy* 77:232–239.

Fleming, R. H., S. C. Bishop, H. A. McCormack, D. K. Flock, and C. C. Whitehead. 1997. The heritability of bone characteristics affecting osteoporosis in laying hens. *British Poultry Science* 38:S21.

Foley, W. J., G. R. Iason, and C. McArthur. 1999. Role of plant secondary metabolites in the nutritional ecology of mammalian herbivores: how far have we come in 25 years? Pages 130–209 in H.-J.G. Jung and G. C. Fahey (eds.), *Nutritional ecology of herbivores*. American Society of Animal Science, Savoy, IU.

Frank, D. A., S. J. McNaughton, and B. F. Tracy. 1998. The ecology of the earth's grazing ecosystems: Profound functional similarities exist between the Serengeti and Yellowstone. *BioScience* 48:514–521.

Freeland, W. J., and D. Choquenot. 1990. Determinants of herbivore carrying capacity: Plants, nutrients, and *Equus asinus* in Northern Australia. *Ecology* 71:589–597.

Fris, M. B., and M. H. Horn. 1993. Effects of diets of different protein content on food consumption, gut retention, protein conversion, and growth of *Cebidichthys violaceus* (Girard), an herbivorous fish of temperate zone marine waters. *Journal of Experimental and Marine Biology and Ecology* 166:185–202.

Frost, P. C., and J. J. Elser. 2002. Growth responses of littoral mayflies to the phosphorus content of their food. *Ecology Letters* 5:232–240.

Gomot, A., L. Gomot, S. Boukraa, and S. Bruckert. 1989. Influence of soil on the growth of the land snail *Helix aspersa*. An experimental study of the absorption route for the stimulating factors. *Journal of Molluscan Studies* 55:1–7.

Grasman, B. T., and E. C. Hellgren. 1993. Phosphorus nutrition in white-tailed deer: nutrient balance, physiological responses, and antler growth. *Ecology* 74:2279–2296.

Graveland, J., and A. E. Berends. 1997. Timing of the calcium intake and effect of calcium deficiency on behaviour and egg laying in captive Great Tits, *Parus major*. *Physiological Zoology* 70:74–84.

Graveland, J., and R. H. Drent. 1997. Calcium availability limits breeding success of passerines on poor soils. *Journal of Animal Ecology* 66:279–288.

Graveland, J., and R. Van Der Wal. 1996. Decline in snail abundance due to soil acidification causes eggshell defects in forest passserines. *Oecologia* 105:351–360.

Graveland, J., and T. van Gijzen. 1994. Arthropods and seeds are not sufficient as calcium sources for shell formation and skeletal growth in passerines. *Ardea* 82:299–314.

Green, B., J. Anderson, and J. Whateley. 1984. Turnover of sodium and water by free-living rabbits, *Oryctolagus cuniculus*. *Australian Wildlife Research* 5:93–99.

Green, B., and J. D. Dunsmore. 1978. Turnover of tritiated water and [22]sodium in captive rabbits (*Oryctolagus cuniculus*). *Journal of Mammalogy* 59:12–17.

Greenberg, R., C. M. Caballero, and P. Bichier. 1993. Defense of homopteran honeydew by birds in the Mexican highlands and other warm temperate forests. *Oikos* 68:519–524.

Guglielmo, C. G., W. H. Karasov, and W. J. Jakubas. 1996. Nutritional costs of a plant secondary metabolite explain selective foraging by ruffed grouse. *Ecology* 77:1103–1115.

Gulati, R. D., and W. R. DeMott. 1997. The role of food quality for zooplankton: Remarks on the state-of-the-art, perspectives and priorities. *Freshwater Biology* 38:753–768.

Hanley, T. A., C. T. Robbins, A. E. Hagerman, and C. McArthur. 1992. Predicting digestible protein and digestible dry matter in tannin-containing forages consumed by ruminants. *Ecology* 73:537–541.

Havera, S. P., and K. E. Smith. 1979. A nutritional comparison of selected fox squirrel foods. *Journal of Wildlife Management* 43:691–704.

Helfield, J. M., and R. J. Naiman. 2001. Effects of salmon-derived nitrogen on riparian forest growth and implications for stream productivity. *Ecology* 82:2403–2409.

Heller, J., and M. Magaritz. 1983. From where do land snails obtain the chemicals to build their shells? *Journal of Molluscan Studies* 49:116–121.

Hellgren, E. C., and R. L. Lochmiller. 1997. Requirements for reproduction in the hispid cotton rat (*Sigmodon hispidus*): a re-evaluation. *Journal of Mammalogy* 78:691–694.

Hellgren, E. C., and W. J. Pitts. 1997. Sodium economy in White-tailed deer (*Odocoileus virginianus*). *Physiological Zoology* 70:547–555.

Hessen, D. O., and N. A. Rukke. 2000. The costs of moulting in *Daphnia*; mineral regulation of carbon budgets. *Freshwater Biology* 45:169–178.

Hilderbrand, G. V., T. A. Hanley, C. T. Robbins, and C. C. Schwartz. 1999. Role of brown bears (*Ursus arctos*) in the flow of marine nitrogen into a terrestrial ecosystem. *Oecologia* 121:546–550.

Holford, K. C., and D. D. Roby. 1993. Factors limiting fecundity of captive brown-headed cowbirds. *Condor* 95:536–545.

Hume, I. D. 1982. *Digestive physiology and nutrition of marsupials*. Cambridge University Press, Cambridge.

Iason, G. R., C. D. Duck, and T. H. Clutton-Brock. 1986. Grazing and reproductive success of red deer: The effect of local enrichment by gull colonies. *Journal of Animal Ecology* 55:507–515.

Ireland, M. P. 1991. The effect of dietary calcium on growth, shell thickness and tissue calcium distribution in the snail *Achatina fluica*. *Comparative Biochemistry and Physiology* 98:111–116.

Jobling, M. 1993. Bioenergetics: Feed intake and energy partitioning. Pages 1–44 in J. C. Rankin and F. B. Jensen (eds.), *Fish ecophysiology*. Chapman and Hall, London.

Johannessen, L. E., and T. Solhoy. 2001. Effects of experimentally increased calcium levels in the litter on terrestrial snail populations. *Pedobiologia* 45:234–242.

Jordan, P. A. 1987. Aquatic foraging and the sodium ecology of moose: A review. *Swedish Wildlife Research Supplement* 1:119–137.

Kincaid, W. B., and G. N. Cameron. 1982. Dietary variation in three sympatric rodents on the Texas coastal prairie. *Journal of Mammalogy* 63:668–672.

Klasing, K. 1998. *Comparative avian nutrition*. CABI Publishing, New York.

Klingauf, F. A. 1987. Feeding, adaptation, and excretion. Pages 225–253 in A. K. Minks and P. Harrewijn (eds.), *Aphids: Their biology, natural enemies, and control*, vol. 2A. Elsevier, Amsterdam.

Krementz, D. G., and C. D. Ankney. 1995. Changes in total body calcium and diet of breeding house sparrows. *Journal of Avian Biology* 26:162–167.

Landry, M. R., J. Constantinou, M. Latasa, S. L. Brown, R. R. Bidigare, and M. E. Ondrusek. 2000. Biological response to iron fertilization in the eastern equatorial Pacific (IronEx II). III. Dynamics of phytoplankton growth and microzooplankton grazing. *Marine Ecology Progress Series* 201:57–72.

Latta, S. C., H. A. Gamper, and J. R. Tietz. 2001. Revising the convergence hypothesis of avian use of honeydew: Evidence from Dominican subtropical dry forest. *Oikos* 93:250–259.

Lotz, C., and C. Martínez del Rio. 2004. The ability of rufous hummingbirds (*Selasphorus rufus*) to dilute and concentrate excreta. *Journal of Avian Biology* 35:54–62.

Marken Lichtenbelt, W.D.V. 1992. Digestion in an ectothermic herbivore, the green iguana (*Iguana iguana*): effect of food composition and body temperature. *Physiological Zoology* 65:649–673.

Martínez del Rio, C. 1994. Nutritional ecology in nectar- and fruit-eating volant vertebrates. Pages 103–127 in D. J. Chivers and P. Langer (eds.), *Food and form and function of the mammalian digestive tract*. Cambridge University Press, Cambridge.

Mattson, W. J. 1980. Herbivory in relation to plant nitrogen content. *Annual Review of Ecology and Systematics* 11:119–161.

McAdam, A. G., and J. S. Millar. 1999. Dietary protein constraint on age at maturity: An experimental test with wild deer mice. *Journal of Animal Ecology* 68:733–740.

McCreedy, C. D., and H. P. Weeks, Jr. 1993. Growth of cottontail rabbits (*Sylvilagus floridanus*) in response to ancillary sodium. *Journal of Mammalogy* 74:217–223.

McDowell, L. R. 1985. *Nutrition of grazing ruminants in warm climates*. Academic Press, Orlando, Fl.

McDowell, L. R. 1992. *Minerals in animal and human nutrition*. Academic Press, San Diego, Calif.

McNaughton, S. J. 1990. Mineral nutrition and seasonal movements of African migratory ungulates. *Nature* 345:613–615.

McNaughton, S. J., and N. J. Georgiadis. 1986. Ecology of African grazing and browsing mammals. *Annual Review of Ecology and Systematics* 17:39–65.

McWhorter, T. J., D. R. Powers, and C. Martinez del Rio. 2003. Are hummingbirds facultatively ammonotelic? Nitrogen excretion and requirements as a function of body size. *Physiological and Biochemical Zoology* 76:731–743.

Murray, M. G. 1995. Specific nutrient requirements and migraton of wildebeest. Pages 231–256 in A.R.E. Sinclair and P. Arcese (eds.), *Serengeti II: Dynamics, management, and conservation of an ecosystem*. University of Chicago Press, Chicago.

Nagy, K. A. 1975. Nitrogen requirement and its relation to dietary water and potassium content in the lizard *Sauromalus obesus*. *Journal of Comparative Physiology* 104:49–58.

Nagy, K. A. 2001. Food requirements of wild animals: Predictive equations for free-living mammals, reptiles, and birds. *Nutrition Abstracts and Reviews B* 71:21R–31R.

Ohlson, M., and H. Staaland. 2001. Mineral diversity in wild plants: Benefits and bane for moose. *Oikos* 94:442–454.

Ormerod, S. J., J. O'Halloran, S. D. Gribbin, and S. J. Tyler. 1991. The ecology of dippers *Cinclus cinclus* in relation to stream acidity in upland Wales: breeding performance, calcium physiology and nestling growth. *Journal of Applied Ecology* 28:419–433.

Ormerod, S. J., and S. D. Rundle. 1998. Effects of experimental acidification and liming on terrestrial invertebrates: Implications for calcium availability to vertebrates. *Environmental Pollution* 103:183–191.

Owen-Smith, N. 1996. Circularity in linear programming models of optimal diet. *Oecologia* 108:259–261.

Pahl, R., D. W. Winkler, J. Graveland, and B. W. Batterman. 1997. Songbirds do not create long-term stores of calcium in their legs prior to laying: results from high-resolution radiography. *Proceedings of the Royal Society (London) B* 264:239–244.

Perkins, M. C., H. A. Woods, J. F. Harrison, and J. J. Elser. 2004. Dietary phosphorus affects the growth of larval *Manduca sexta*. *Archives of Insect Biochemistry and Physiology* 55:153–168.

Peters, R. H. 1986. The role of prediction in limnology. *Limnology and Oceanography* 31:1143–1159.

Plath, K., and M. Boersma. 2001. Mineral limitation of zooplankton: Stochiometric constraints and optimal foraging. *Ecology* 82:1260–1269.

Proctor, N. S., and P. J. Lynch. 1993. *Manual of ornithology*. Yale University Press, New Haven, Ct.

Pryor, G. S., D. J. Levey, and E. S. Dierenfeld. 2001. Protein requirements of a specialized frugivore, Pesquet's parrot (*Psittrichas fulgidus*). *Auk* 118:1080–1088.

Randolph, J. C., and G. N. Cameron. 2001. Consequences of diet choice by a small generalist herbivore. *Ecological Monographs* 71:117–136.

Reich, P. B., and J. Oleksyn. 2004. Global patterns of plant leaf N and P in relation to temperature and latitude. *Proceedings of the National Academy of Sciences* 101:11001–11006.

Richardson, P.R.K., P. J. Mundy, and I. Plug. 1986. Bone crushing carnivores and their significance to osteodystrophy in Griffon vulture chicks. *Journal of Zoology* 200:23–43.

Robbins, C. T. 1993. *Wildlife feeding and nutrition*. Academic Press, San Diego.

Robbins, C. T., A. E. Hagerman, P. J. Austin, C. McArthur, and T. A. Hanley. 1991. Variation in mammalian physiological responses to a condensed tannin and its ecological implications. *Journal of Mammalogy* 72:480–486.

Schetter, T. A., R. L. Lochmiller, D. M. Leslie, Jr., D. M. Engle, and M. E. Payton. 1998. Examination of the nitrogen limitation hypothesis in non-cyclic populations of cotton rats (*Sigmodon hispidus*). *Journal of Animal Ecology* 67:705–721.

Schulkin, J. 1991. *Sodium hunger: The search for a salty taste*. Cambridge University Press, Cambridge.

Schulkin, J. 2001. *Calcium hunger: Behavioral and biological regulation*. Cambridge University Press, Cambridge.

Seagle, S. W., and S. J. McNaughton. 1992. Spatial variation in forage nutrient concentrations and the distribution of Serengeti grazing ungulates. *Landscape Ecology* 7:229–241.

Sharpe, D. J., and R. L. Goldingay. 1998. Feeding behaviour of the squirrel glider at Bungawalbin Nature Reserve, north-eastern New South Wales. *Wildlife Research* 25:243–254.

Slansky, F., and J. M. Scriber. 1985. Food consumption and utilization. Pages 87–163 in G. A. Kerkut and L. I. Gilbert (eds.), *Comprehensive insect physiology, biochemistry and pharmacology*, vol. 4. Pergamon Press, Oxford.

Smedley, S. R., and T. Eisner. 1996. Sodium: A male moth's gift to its offspring. *Proceedings of the National Academy of Sciences* 93:809–813.

St. Louis, V. L., and J. C. Barlow. 1993. The reproductive success of tree swallows nesting near experimentally acidified lakes in northwestern Ontario. *Canadian Journal of Zoology* 71:1090–1097.

Staaland, H., D. F. Holleman, J.R. Luick, and R. G. White. 1982. Exchangeable sodium pool size and turnover in relation to diet in reindeer. *Canadian Journal of Zoology* 60:603–610.

Staaland, H., and K. Hove. 2000. Seasonal changes in sodium metabolism in Reindeer (*Rangifer tarandus*) in an inland area of Norway. *Arctic, Antarctic, and Alpine Research* 32:286–294.

Steinauer, E. M., and S. L. Collins. 2001. Feedback loops in ecological hierarchies following urine deposition in tallgrass prairie. *Ecology* 82:1319–1329.

Sterner, R. W., and D. O. Hessen. 1994. Algal nutrient limitation and the nutrition of aquatic herbivores. *Annual Review of Ecology and Systematics* 25:1–29.

Studier, E. H., and S. H. Sevick. 1992. Live mass, water content, nitrogen and mineral levels in some insects from south-central Lower Michigan. *Comparative Biochemistry and Physiology A* 103:579–595.

Teferi, T., and J. S. Millar. 1993. Early maturation by northern *Peromyscus maniculatus*. *Canadian Journal of Zoology* 71:1743–1747.

van Marken Lichtenbelt, W. D., R. Wesselingh, J. T. Vogel, and K.B.M. Albers. 1993. Energy budgets in free-living green iguanas in a seasonal environment. *Ecology* 74:1157–1172.

Van Soest, P. J. 1994. *Nutritional ecology of the ruminant*. Cornell University Press, Ithaca, N.Y.

Wareborn, I. 2002. Reproduction of two species of land snails in relation to calcium salts in the foerna layer. *Malacologia* 18:177–180.

Weir, J. S. 1972. Spatial distribution of elephants in an African National Park in relation to environmental sodium. *Oikos* 23:1–13.

White, T.C.R. 1993. *The inadequate environment*. Springer-Verlag, Berlin.

Witmer, M. C. 1998. Ecological and evolutionary implications of energy and protein requirements of avian frugivores eating sugary diets. *Physiological Zoology* 71:599–610.

Witmer, M. C. 2001. Nutritional interactions and fruit removal: Cedar waxwing consumption of *Viburnum opulus* fruits in spring. *Ecology* 82:3120–3130.

CHAPTER TWELVE

Water Requirement and Water Flux

12.1 Water Budgets, Fluxes, and Requirements

ANIMALS IN TERRESTRIAL environments must produce through metabolism, or consume through drinking and eating, enough water to balance evaporative losses across their skin and lungs and the losses associated with excretion of feces and urine (figure 12.1). Studies of water budgets of many kinds of terrestrial animals have shown that this is not a major problem for animals that consume moist foods such as fruits or other animals. But it is a potential problem for animals that rely on drier foods such as desiccated leaves, like the black-tailed jackrabbit shown in figure 12.1, or for animals that rely on seeds and detritus. For these feeders, water may become a driving force influencing aspects of their behavior such as travel to water sources or diet selection, or even their populations' demographic features such as survivorship and growth rate. To emphasize this point in the ensuing discussion we will return repeatedly to the jackrabbit because, as you will see, things do not always turn out well for these denizens of deserts and semiarid environments. We do not turn to them because jackrabbits are the most desert adapted of species, which they are not, but because the relatively large number and varied studies on them permit one of the most synthetic and integrative views of the significance of water in an animal's ecology.

The challenges of balancing the water budget are most easily demonstrated by following the seasonal water relations of animals inhabiting arid environments (figure 12.2). An important tool ecologists have used in these studies is the measurement of total water flux in free-living animals using labeled water turnover (box 12.1). Although the method alone does not give any details about exchange through the various routes of gain and loss, the field studies are often

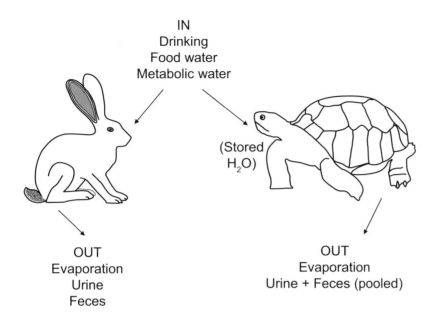

Figure 12.1. Avenues of water gain and loss in jackrabbit (*Lepus* sp.) and desert tortoise (*Gopherus agassizii*). Although water vapor is both taken up and lost, evaporation refers to the net loss. Urine and feces are pooled in the cloaca of reptiles but eliminated separately in most mammals. The tortoise may store dilute urine in its bladder for an extended time and, when dehydrated, reabsorb water until the urine becomes isosmotic with plasma.

complemented by more detailed laboratory studies. Nagy and his colleagues have championed this approach, and it is appropriate that the two examples in figure 12.2 are drawn from their work.

Water flux can vary tremendously over time because of large differences in the water content of food consumed and sometimes because of drinking. During seasons of new plant growth, plant eaters take in more water than the minimum they require because mainly they are eating to meet energy (not water) requirements (chapter 3), and growing plant tissue generally has relatively higher water content (figure 12.2). This is also the situation for animals that routinely consume moist foods. The excess water may be stored, as in desert tortoises that store dilute urine in their bladder (Nagy and Medica 1986). But many animals will more often balance output to input by increasing the water content of their excreta (urine and feces) and increasing their evaporative water loss rates, as we will describe for the jackrabbit. In this case, the measurement of total water flux conveys little information about minimum water requirement. Although hundreds of water flux measurements have been made in animals (Nagy and Peterson 1988), most were not made under water-restricted conditions and we might wonder about their utility. Comparisons of these fluxes across species and sizes of animals offer useful corrections for the effects of size on water relations (among different clades, scaling is proportional to $M^{0.64-0.96}$; Nagy and Peterson 1988) and also benchmarks for comparisons within and among clades. But even if one restricts attention to measurements

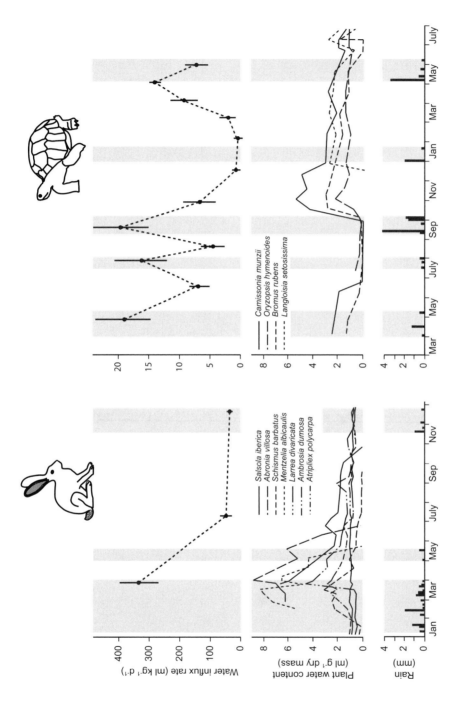

Figure 12.2. Water relations of free-living animals in the Mojave Desert of North America. Panels for black-tailed jackrabbit (*Lepus californicus*; left panels redrawn from Nagy et al. 1976) and desert tortoises (*Gopherus agassizii*; right panels redrawn from Nagy and Medica 1986) include information over 1–2 years on water influx, along with rainfall, and food plant water contents. Notice that for both species, there is tremendous variation over time in water availability (plant water content or rainfall) and hence in water influx. During some times of the year plant water contents were below 0.5 g H_2O g^{-1} dry mass (33% water) and there was no free-standing water. In both of these studies the researchers tried to correct water fluxes for differences in body size among individuals by dividing flux by body mass (hence, ml H_2O kg^{-1} d^{-1}), but other bases of normalization (e.g., ml $kg^{-3/4}$ d^{-1}) might be superior.

Box 12.1. Measuring Water (or Sodium) Flux
in Free-Living Animals

Rates of water (or sodium) influx and efflux in animals can be measured by administering isotopically labeled water (or sodium) and following the decline in specific activity of the isotope in the body pool over time. For water, typical isotopes are the stable isotope deuterium (^2H_2O) or the unstable and hence radioactive tritium (^3H_2O). Radioactive ^{22}Na is the Na isotope. For our description we will consider the case of water flux (Lifson and McClintock 1966), though similar principles apply for Na flux (e.g., see Green 1978).

In the case of water, the specific activity declines because of the loss of label from the animal through excretion and evaporation and dilution of the label from the input of unlabeled water via oxidative metabolism, eating, and drinking. After a dose of labeled water (d) is administered into the body water pool (W, in g H_2O) and distributes through it, the relationship between specific activity (above background, $^*H$) and time (t) fits a monoexponential elimination model, consistent with clearance from a single compartment:

$$^*H_t = [d / W]e^{-\alpha t} \qquad (12.1.1)$$

where $^*H_t$ is the specific activity in body water at time t, and t is time (days in this case) elapsed since the administration of the dose. The term α (day^{-1}) is the fractional rate of water turnover and can be determined from two (or more sampling) times. The variable α is also called the elimination rate constant. Most ecologists inject an animal with labeled water, wait for it to distribute within the animal's total body water pool (one hour in small animals to several hours in larger animals), take a blood (or urine) sample (time $t = 1$), and then release the animal. When the animal is recaptured later another blood (or urine) sample is taken (time $t = 2$). The specific activity $^*H$ in the blood water is measured, and α is calculated:

$$\alpha = [\ln{^*H_{t=1}} - \ln{^*H_{t=2}}] / t. \qquad (12.1.2)$$

Also, if the first sample is taken soon after the dose is administered and distributes through the body water, then the body water pool W can be computed by isotope dilution from the first equation as

$$W = d / {^*H_{t=1}}. \qquad (12.1.3)$$

Assuming that the behavior of labeled water matches that of unlabeled water, and that the pool size of body water did not change between the two

continued

continued

> sampling times (i.e., influx = efflux), the total water flux in g H_2O d^{-1} in the animal is calculated by multiplying α by the amount of body water:
>
> $$H_2O \text{ flux in g } H_2O \text{ } d^{-1} = \alpha W. \qquad (12.1.4)$$
>
> This is the simplest situation, and in a number of sources (Nagy and Costa 1980; Nagy 1983; Speakman 1997) there are modified equations for situations when influx and efflux are not equal and W changes over time, or for other types of deviations from the basic assumptions. Furthermore, there are many technical details having to do with distilling water from blood, tissue, or excretory samples, and measuring its specific activity. Though there are lots of nuances, the method is really very simple and generally accurate to at least ±10%.
>
> The use of isotopes to measure flux is popular for several reasons. With them, ecologists can gain information about the rate of resource utilization of free-living animals. Minimum requirements can be estimated under field conditions (see figure 12.3, below), and feeding rate can be estimated in some cases (Nagy 1975b). For example, if animals do not drink then water influx is due to food water intake (the product of feeding rate I and food water content w_{food}, in g H_2O g^{-1} dry matter) and water produced in metabolism (metabolic water production; described below). Rearranging to solve for feeding rate I, one calculates food intake:
>
> $$I = (H_2O \text{ flux in g } H_2O \text{ } d^{-1}$$
> $$- \text{ metabolic water produced in g } H_2O \text{ } d^{-1}) / w_{food}. \qquad (12.1.5)$$
>
> The method assumes that all the water in food is absorbed, which may not be the case always (McWhorter et al. 2003). Examples of this application are available for herbivores (Shoemaker et al. 1976), carnivores (Green 1978), insectivores (Anderson and Karasov 1981), nectarivores (van Helversen and Reyer 1984), and nursing mammals (Holleman et al. 1975). For the last application, corrections may be necessary for cycling of the isotope among mother and litter mates (Robbins 1993; Kretzmann et al. 1993; Scantlebury et al. 2000).

made only on free-living animals, one must not assume that the measured fluxes represent minimum requirements. Also, even if animals drink in the wild, this does not mean that they must drink.

Nutritionists determine the minimum requirement for a nutrient either by feeding it at different levels and monitoring animal performance as a function of intake, or by estimating it based on knowledge of minimum loss rates that must

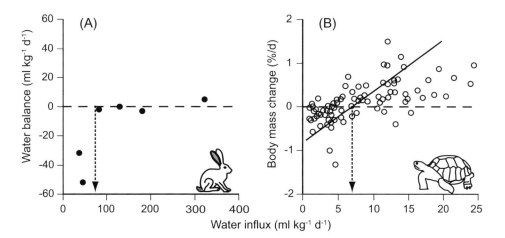

Figure 12.3. Estimating minimum water flux in free-living animals. In the experimental study with black-tailed jackrabbits (left-hand plot; replotted from Nagy et al. 1976) the animals were held over several weeks in outdoor enclosures and either ate natural vegetation with varying water content or were provided with dry alfalfa and water. Each filled circle represents the mean for 3 to 22 hares. In the correlational study with desert tortoises (replotted from Nagy and Medica 1986) tortoises were held in fenced study plots throughout the year and ate only natural vegetation. The minimum total water influx, indicated by a dotted line with an arrow, was defined as the influx where water balance (influx–efflux; left-hand plot) or body mass change (right-hand plot) was zero (indicated by a dashed line).

be replaced. We will illustrate the former method here and the latter method in a subsequent section. The minimum water requirement of the jackrabbits was measured under seminatural conditions by placing captives in outdoor enclosures with variable access to water and measuring the water influx rate that just matched water efflux (figure 12.3). With adequate acclimation time to minimize all routes of water loss, jackrabbits were losing about 100 ml kg^{-1} d^{-1} during the driest seasons, and so they required at least 85 ml kg^{-1} d^{-1} from food or drinking on top of the 15 ml kg^{-1} d^{-1} metabolic water that they produced from catabolizing their food for energy (figure 12.3A). In another approach, ecologists studying free-living animals sometimes regress body mass change against water influx in order to estimate minimum requirement (e.g., figure 12.3B). However, it is important to remember that there is no assurance in this case that intake of all other nutrients was adequate. If you now compare these estimated minimum total water fluxes of the jackrabbit (100 ml kg^{-1} d^{-1}) and desert tortoise (6.9 ml kg^{-1} d^{-1}) with their respective seasonal measures (figure 12.2) it is easy to see that for some of the year water influx is more than adequate, and to identify the time of year when water may have been limiting.

In light of these considerations, our focus will be to understand an animal's minimum requirement for water. A useful way to think about any nutrient requirement is to ask how much must be ingested to match obligatory losses. This is analogous to the factorial method we introduced earlier (chapter 11) for estimating N requirements (section 11.1.1). We will apply this focus in the following sections and draw some very general conclusions, although we admit that they are not universally accepted. One is that, even when accounting for water produced metabolically, only a few granivores (mainly small ones), but probably no herbivores, can survive without either some food water or drinking. Also, it will become apparent how intertwined water relations are with features of digestion and nutrition.

12.2 Avenues of Water Loss

12.2.1 Urinary and Fecal Water Loss Are Influenced by Food Intake and Digestion as well as by Kidney and Intestinal Water and Solute Absorption

In the steady state, the consumption of food is associated with the absorption of numerous ions that must be eliminated, along with metabolites from protein deamination such as ammonia, urea or uric acid, and xenobiotic metabolites. If these solutes are excreted through the kidney entirely dissolved in water, then the minimum urine water loss is (Schmidt-Nielsen 1964)

$$\text{minimum urine water loss in g d}^{-1}$$
$$\propto \frac{\text{mmol osmolytes excreted d}^{-1}}{\text{maximum urine concentration in mmol g}^{-1}\,H_2O}. \qquad (12.1)$$

The first term, mmol osmolytes excreted per day, is mainly the product of feeding rate and diet composition. Barring changes in osmolyte absorption efficiency, a higher feeding rate and/or a higher diet content of protein, ions, and xenobiotics will increase obligatory urinary water loss. Estimates of the osmolytes associated with different food types have been made based on food analysis and in balance trials with animals (e.g., Nagy et al. 1976; Nagy and Milton 1979). The data indicate that much of the osmolyte load (greater than 90% for animal prey) is likely related to food protein content. The total osmotic load in the context of equation 12.1 may not be a simple function of the addition of loads due to the protein metabolites and salt ions (Hill et al. 2004), which is why we include a proportionality rather than an equality in that equation. Although the protein metabolites from dietary protein typically dominate, some herbage can have high salt loads due to relatively high contents of primary minerals such as potassium, calcium, chloride, sodium, and magnesium. Some desert herbivores have been shown to consume parts of plants selectively to reduce salt intake (Kenagy 1972; Kam and Degen 1988).

The maximum urine osmolality varies among species. In the vertebrate kidney the loop of Henle is the feature that can produce a concentrated urine with a high osmolality, and most mammals and some birds, but no reptiles, fish, or amphibians possess these loops (see chapter 9, figure 9.9). A description of the fascinating way in which the loop achieves concentration of urine in conjunction with antidiuretic hormone-mediated changes in kidney-collecting-duct water permeability is available in most comparative physiology textbooks and so we forgo it here. The concentration ability is sometimes expressed relative to the osmolality of plasma (about 0.25 to 0.35 mmol g^{-1} H$_2$O in vertebrates) as a urine-to-plasma ratio (U:P), and so animals lacking loops of Henle have maximum ratios of 1, some birds have ratios of 2:1 to 3:1, and small desert rodents have maximum ratios of 20 to 30:1 (Degen 1997). It is not a perfect basis for comparison because some animals (e.g., mammals) regulate plasma osmolality, which is the denominator in the ratio, more tightly than other animals (e.g., birds; Williams and Tieleman 2001), and also because there are other mechanisms besides concentrating urine for reducing water lost per mole of N or all ions excreted.

Although most birds and all reptiles do not excrete a concentrated urine, many, including a few amphibians, minimize excretory water loss by excreting ions and uric acid in a slurry that contains a high ratio of excreted materials to water. The uric acid is filtered from plasma as well as excreted by the renal tubules and, because it has low aqueous solubility, it precipitates. In seemingly analogous fashion, some rodents (*Gerbillurus* spp.) of southern African arid regions excrete a precipitate of allantoin (another nitrogen excretory product) in their urine (Downs and Perrin 1991). A poorly understood but important feature of a uric acid precipitate is that it may contain significant quantities of inorganic cations, making it possible for these species to excrete more of such ions than could be accommodated by the maximum urine concentration (Braun and Dantzler 1997). The urine may be modified further once it enters the cloaca or following reflux from there to more proximal regions of the gastrointestinal tract such as the colon or cecum, but it cannot be made more concentrated (Withers 1992). Urine concentration is regulated in most vertebrates through the action of antidiuretic hormone (also called vasopressin), and vertebrates can also produce very dilute urine simply by filtering plasma, reabsorbing in the kidney tubules many of the ions in the filtrate, and excreting the watery residue. Thus, in the case of the jackrabbit, urine osmolyte concentration varied inversely through the seasons with water influx rate to a maximum about ten times that of plasma (figure 12.4). Salt glands represent another route of water loss associated with elimination of ions in some birds and reptiles (Peaker and Linzell 1975), although it generally contributes a smaller proportion of water loss than other routes. Water loss through the salt gland was 10% or less of the total water loss in several species of desert lizards (Nagy 1982). Water loss through sweating will be considered in the discussion below on evaporation.

Feces are not defecated without loss of some water. Higher feeding rates and/or less digestible food will increase obligatory fecal water loss. Fecal water

Figure 12.4. Black-tailed jackrabbits regulated primarily urinary and evaporative water loss to match their total rate of water influx and stay in balance. Water-restricted hares excreted relatively concentrated urine compared with hares ingesting excess water in moist food (upper left-hand panel; data from Nagy et al. 1976). Fecal water content (lower left-hand panel; data from Nagy et al. 1976) was low and fairly independent of total water influx. Without reducing their body temperatures, hares reduced evaporative water loss (right-hand figure) when they were chronically water restricted (dashed line) compared to when they had water ad libitum (solid line). (Data from Reese and Haines 1978.)

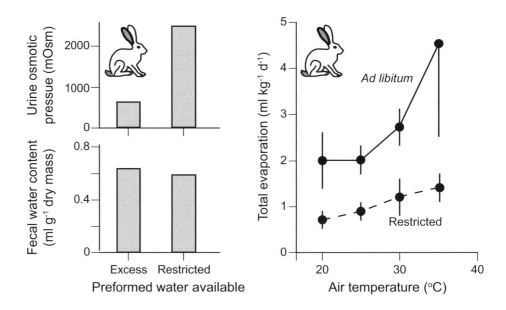

loss is relatively high in herbivores because they meet their energy requirements by consuming relatively large amounts of poorly digested dry matter with concomitant very high rates of fecal dry mass flux. The minimum fecal water loss is

$$\text{Minimum fecal water loss in ml/d} = I(1-D_m) \text{ (minimum g } H_2O \text{ g}^{-1} \text{ dry feces)} \qquad (12.2)$$

where I is the food intake in g d^{-1} and D_m is the diet dry matter digestibility (see chapter 3).

There are no active transport mechanisms for water known in animals. Water is absorbed in the digestive tract by osmosis in association with absorption of ions (Karasov and Hume 1997). Fecal water content may be regulated according to hydration state, but no vertebrates are reported to excrete feces drier than about 0.40 g H_2O g^{-1} dry matter (i.e., ca. 28% water; Withers 1992; Degen 1997; Tracy and Walsberg 2001b). In the case of the jackrabbit eating a natural diet, fecal water content was more or less invariant and relatively low, averaging around 0.6 g H_2O per gram of dry matter (i.e., about 38% water; figure 12.4). Terrestrial arthropods can do better, however. The insect Malpighian tubule can accumulate a very high osmotic concentration (up to 4.3 mmol g^{-1} H_2O) creating a large osmotic gradient for water absorption and thus reducing excretal water content to as low as 14% water (Withers 1992).

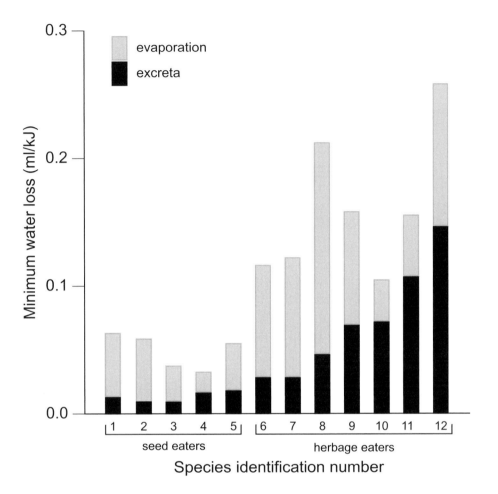

Figure 12.5. When animals eat herbage they generally lose more water per kJ metabolizable energy than when eating seeds. The solid black bars represent water losses in excreta (urine + feces) of animals provided with the minimum water necessary for water and mass balance, and thus they represent obligatory losses. The evaporative water loss (respiratory + cutaneous) in each trial is also shown, so that you can appreciate the relative magnitude of each major route of water loss. Species designations for seed eaters are as follows: 1 and 2 = larks (*Spizocorys starki, Eremopteryx verticalis*) (Willoughby 1968); 3 = Merriam's kangaroo rat (*Dipodomys merriami*) (Schmidt-Nielsen 1964); 4 = pocket mouse (*Perognathus parvus*) (Withers 1982), and 5=antelope ground squirrels (*Ammospermophilus leucurus*) (Karasov 1983). Herbivores: 6 = desert beetle. *Eleodes armata* (Cooper 1985); 7 = Arabian oryx (*Oryx beisa*) ((Taylor 1970, 1968); 8 = Grant's gazelle (Taylor 1968); 9 = chuckwalla lizard (*Sauromalus obesus*) (Nagy 1972, 1975a); 10 = desert woodrat (*Neotoma lepida*) (Karasov 1989), 11 = black-tailed jackrabbit (*Lepus californicus*) (Nagy et al. 1976), 12 = ostrich (*Struthio camelus*) (Withers 1983).

It is clear from this brief discussion of urinary and fecal water loss that consumption of food is associated with water loss. It should also be clear how these processing costs can be quantified by considering feeding rate, diet digestibility, and the ion and N load of the diet, using equations 12.1 and 12.2. Because of these nutritional factors, the minimum water lost per unit food consumed differs quite a bit according to diet (figure 12.5). In order to make these dietary comparisons clear, we have expressed the minimum urinary and fecal water lost relative to metabolizable energy for each food, because this mode of expression also makes it easier to compare among animals of different sizes. In addition, using this mode of expression contributes to the development of a conceptual framework for a large and ever growing body of field-collected data on water flux and metabolic rate of free-living animals, because the total water fluxes are sometimes expressed per kJ energy metabolized

(Nagy and Peterson 1988). The minimum processing cost in g H_2O kJ^{-1} is lowest for seeds because they are so digestible that relatively small amounts are consumed to meet energy requirements, feces production is low, and the contents and fluxes of ions and N are relatively low. The processing cost is generally higher for plant material (by an average of eight times for these examples) because to meet energy needs a herbivore must consume large amounts of poorly digestible dry matter with concomitant high rates of fecal dry mass flux and of ion and N flux. Of course, even a very high obligatory water loss through urine and feces may not be important if all the losses associated with food processing are insignificant compared to evaporative losses. But they are not. For the species summarized in figure 12.5, the losses through urine and feces were not small but accounted for 16–69% of total water loss. In some studies on insects consuming dry food, about 25% of total water loss was excretory (Edney 1977; Cooper 1985). Food processing losses are clearly important, especially for herbivores, but to round out our discussion we must consider evaporation, the other major route of loss, and then consider all the outputs in relation to inputs.

12.2.2 Evaporative Losses in the Total Water Budget

Evaporation occurs from wet surfaces (respiratory tract, eyes), or through porous ones like the integument, and from fluid excreted across the integument as sweat or from extrarenal salt glands. A variety of methods are used to measure evaporation (methods box 12.2). The basic guiding physical principle is that the rate of evaporation ($\dot{E}$; mass of H_2O time^{-1}) depends on the surface area over which evaporation occurs (A), the gradient in water vapor density between animal and environment, and the total resistance (θ_t) to evaporation (O'Connor and Spotila 1992):

$$\dot{E} = A((RH_{surf} \times \chi_{sat,T_{surf}}) - (RH_a \times \chi_{sat,T_a}))/\theta_t \qquad (12.3)$$

where $\chi_{sat,T_{surf}}$ is saturated water vapor density at tissue surface temperature (mass of H_2O per unit volume air), χ_{sat,T_a} is saturated water vapor density at air temperature, and RH is relative humidity (0 to 1) in environmental air (RH_a) or at the evaporating surface (RH_{surf}). Air is typically assumed to be saturated at the animal's evaporating surface ($RH_{surf} = 1$). The exact units of θ will depend on the particular time units in which $\dot{E}$ is measured (e.g., mg H_2O per sec, min, or h, per whole organism or unit area, etc.) and the particular units in which vapor density is expressed (e.g., mg H_2O per cm^3 or per liter air, etc.). The relationship

Box 12.2. Measuring Evaporative Water Loss

A simple way for a field ecologist to measure evaporative loss is by mass loss in a fasted animal. It works because fasted animals catabolizing mainly fat have a respiratory quotient ($RQ = CO_2$ production/O_2 consumption) of approximately 0.72, which means that the mass change due to respiratory losses of CO_2 is balanced by the mass of O_2 retained. Thus, any mass decrease observed is largely due to evaporation, or urinary and fecal loss, which are typically small in a fasted animal and are easily measured by placing the animal over mineral oil and weighing what is excreted.

Although this method is simple, the most common way evaporation is measured by ecophysiologists is in a metabolism chamber where the volumetric inflow and outflow of air, and its water content, is measured, and humidity, temperature, and animal activity level can be controlled. Some researchers have restrained animals in partitioned chambers, with the head snug in a latex sheet so that respiratory water loss could be separated from integumentary or cutanous water loss (e.g., Bernstein 1971; Wolf and Walsberg 1996). An alternative method is to use a mask to collect respiratory water loss (Tieleman and Williams 2002), which has the advantage of excluding evaporation from the eyes, which are wet surfaces, and the head's integument. In arthropods, to separate the relatively small respiratory H_2O loss via spiracles from the relatively large cuticular loss, researchers have used ultrasensitive mass measurements and infrared absorbance to detect H_2O vapor (Lighton 1994).

Evaporation is sometimes calculated by difference in a water budget. For example, if total flux is measured with labeled water (methods box 12.1) and urinary and fecal loss is measured gravimetrically, then an estimate of evaporation is readily available by subtraction. But this method can lead to an overestimate if vapor influx at high environmental humidity elevates the isotope turnover rate (Nagy and Costa 1980).

Finally, one can estimate total evaporative losses using allometric equations that summarize vast amounts of data across species in different clades over a range of body sizes (Calder 1984; Peters 1983; Schmidt-Nielsen 1984). Superior analyses are phylogenetically informed. More typically, the allometric approach is used in making comparisons, and can reveal evidence for evolutionary differences, such as the finding that desert-adapted birds tend to have lower total evaporative water loss rates than nondesert avian species (Williams 1996). Allometric analyses sometimes predict ecological patterns. For example, allometric comparison of evaporative water loss and metabolic water production from metabolism leads to the prediction that smaller granivorous species are more likely than larger species to survive without drinking (MacMillen and Hinds 1983; MacMillen 1990; Millen and Baudinette 1993).

Figure 12.6. The water vapor density of saturated air ($\chi_{\text{saturated}}$) is an increasing function of air temperature. This relationship can be approximated as follows (Daniels and Alberty 1975):

$$\chi_{\text{saturated}} = 9.16 \times 10^8 e^{-(5218/[273 + T])}$$

where $\chi_{\text{saturated}}$ has units of mg H_2O L^{-1} air and T is temperature in °C at the evaporating surface.

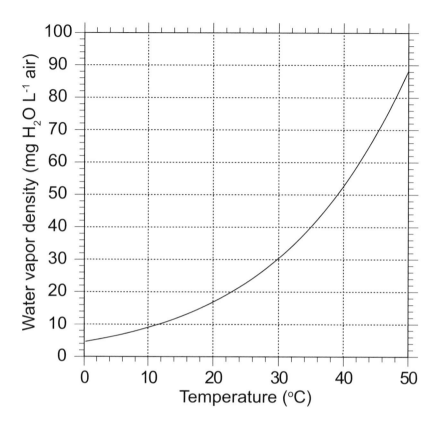

between χ_{sat} for saturated air at different temperatures is available in many sources in tabular form (List 1966) or figure form (figure 12.6), and can be approximated as follows (Daniels and Alberty 1975):

$$\chi_{\text{sat}} = (9.16 \times 10^8)e^{-(5218/[273 + T])}, \tag{12.4}$$

where χ_{sat} has units of mg H_2O L^{-1} air and T is temperature in °C at the evaporating surface.

At any given environmental temperature, total (cutaneous+respiratory) evaporative losses are a declining function of environmental vapor density ($= RH_a \times \chi_{sat,T_a}$) (figure 12.7; Welch 1980), as might be expected from equation 12.3. The effect of environmental air temperature on evaporative losses is a little more complex but in most regards it is still explicable in terms of equations 12.3 and 12.4. In most mammals, birds, reptiles, and arthropods at low to moderate environmental temperatures, evaporative loss (either total, cutaneous, or respiratory) is a slightly increasing function of T_a or independent of T_a, as for the species depicted in figures 12.7 and 12.8, which is discussed in the section "Cutaneous H_2O loss" below. But at high T_a total evaporative water loss typi-

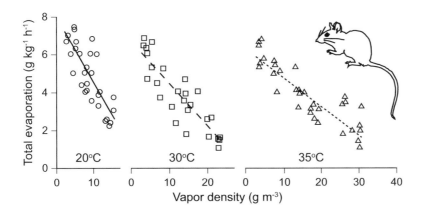

Figure 12.7. Total (cutaneous + respiratory) evaporative losses are a declining function of environmental vapor density in deer mice (*Peromyscus*), but different air temperatures (shown as different symbols) have relatively little effect. (Replotted from Welch 1980.)

cally increases sharply, which contributes to evaporative cooling. Any further analyses of these patterns are best performed by considering the two major routes of evaporative water loss separately. But it is apparent from consideration of these basic principles and patterns that a number of animal behaviors will reduce evaporative losses, including restriction of activity to cooler parts of the day, selection of microclimates that reduce thermal loading or have higher humidities, and storage of heat and thus tolerating a rise in body temperature, in lieu of immediate evaporative cooling. Stored heat can be dissipated later by conduction, convection, and radiation (see box 13.4 in chapter 13).

Respiratory H₂O loss (RWL) across lungs or trachea occurs in conjunction with O_2 consumption ($\dot{V}O_2$) and CO_2 production, although, when heat stressed, animals can sharply increase evaporative cooling by shallow panting (mammals) or gular fluttering (birds) without dramatically increasing gas exchange. Schmidt-Nielsen, in his seminal work on desert animals (1964), explained that variation in RWL ($\dot{E}_{respiratory}$, in mass of H_2O time^{-1}) might be explained in terms of the amount of water in exhaled air and changes in $\dot{V}O_2$ (volume O_2 time^{-1}) and oxygen extraction, which both determine ventilation rate in relation to metabolic demand for oxygen. We arranged these factors in a simple model:

$$E_{respiratory} = \dot{V}O_2 \times ((RH_{surf} \times \chi_{sat,T_{surf}}) - (RH_a \times \chi_{sat,T_a}))/(O_{2,in} - O_{2,out}) \quad (12.5)$$

where $(RH_{surf} \times \chi_{sat,T_{surf}}) - (RH_a \times \chi_{sat,T_a})$ is as above and $(O_{2,in} - O_{2,out})$ represents the difference in O_2 content (volume O_2 volume^{-1} air) of inspired and expired air (O_2 extraction). It is not uncommon for a water-restricted animal to reduce its evaporative water loss rate, as shown for the jackrabbits in figure 12.4. According to equation 12.5, animals can reduce respiratory loss by reducing energy expenditure, extracting more O_2 per breath, selecting a more humid environment, or reducing the temperature or relative humidity of expired air.

Only two vertebrate species are claimed to be able to desaturate exhaled air (camels (Schmidt-Nielsen et al. 1980) and ostriches (Withers et al. 1981)), but many species have a nasal countercurrent heat exchange mechanism to exhale air cooler than their body temperature, effectively recondensing and recovering some of the water contained in air expired from lungs (Schmidt-Nielsen et al. 1970; Withers 1992; Tieleman et al. 1999). This mechanism can be important for an endotherm because it can decouple $\dot{E}_{\mathrm{respiratory}}$ from $\dot{V}O_2$ so that the former is relatively constant with decreasing T_a while the latter increases.

In spite of this potential for the uncoupling of $\dot{E}_{\mathrm{respiratory}}$ and $\dot{V}O_2$, inspection of equation 12.5 makes it easy to see why some investigators express respiratory water loss per cm^3 O_2 consumed. According to Withers (1992), in vertebrate lungs this ratio can be reduced to around 0.25 mg H_2O cm^{-3} O_2. Insects can decrease this ratio even more (to about 0.11 mg H_2O cm^{-3} O_2) because their tracheoles can extract more O_2 from air. Another reason this ratio is useful for us is that, after converting $\dot{V}O_2$ to kJ d^{-1} (see chapter 1, box 1.1), it puts evaporative loss in the same units we used above for urinary and fecal water loss (i.e., g H_2O kJ^{-1}), which will come in handy when modeling the entire water budget below. Remember, however, that considering the various ways $\dot{E}_{\mathrm{respiratory}}$ can be decoupled from $\dot{V}O_2$, it is unlikely that this ratio is constant for a species. The lowest value of this $\dot{E}_{\mathrm{respiratory}}$ in vertebrates corresponds to 0.013 g H_2O kJ^{-1}. You can see that vertebrates exhibit total evaporative water loss values higher than this in balance trials over several days (figure 12.5). Evaporative water losses are higher partly due to the contribution of cutaneous water loss, which we will now consider.

Cutaneous evaporative H$_2$O loss (CWL). You may recognize that equation 12.3 is analogous to the first law of diffusion developed by the physiologist Adolph Fick (1829–1901), especially if one equates the resistance term θ_t with the quotient of the diffusion path length (cm) and the diffusion coefficient for water vapor in air (0.242 cm^2 sec^{-1}) (i.e., θ_t = diffusion path length/diffusion coefficient) (Monteith and Campbell 1980). Because cutaneous evaporation is actually a multistep process in which water or water vapor first crosses the cutaneous barrier(s) (e.g., skin, cuticle) which constitutes an internal resistance (θ_i) and then crosses the boundary layer of air (over the cutaneous barrier) which constitutes an external resistance (θ_e), the resistance term θ_t is actually the sum of two resistances to evaporation that are in series ($\theta_t = \theta_i + \theta_e$) (O'Connor and Spotila 1992). Typically, θ_t is calculated by measuring $\dot{E}$ across some or all of the integument under controlled conditions of relative humidity and air temperature (i.e., $RH_a \times \chi_{\mathrm{sat},T_a}$). For the pigeon depicted in figure 12.8, the CWL to dry air was only a slightly increasing function of air temperature (T_a) until it increased sharply above the pigeon's upper critical temperature (35°C) (Webster et al. 1985). This sharp increase does not occur in all species (e.g., Tieleman and Williams 2002). To calculate θ_t one must take into account that the temperature of the evaporating

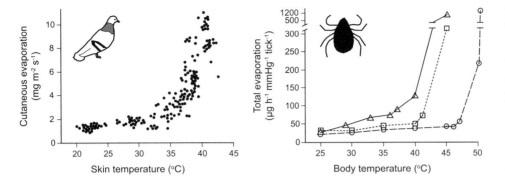

Figure 12.8. In many vertebrates and insects cutaneous (integumentary) water loss to dry air is only a slightly increasing function of temperature until it increases sharply at some transition temperature. In the pigeon (*Columba livia*) the transition occurs just above the bird's lower critical temperature (35°C) (replotted from Webster et al. 1985). In the rabbit tick (*Haemaphysalis leporispalustris*) the transition temperature varies according to developmental stage (unfilled triangles = engorged nymphs 1 day after drop off; unfilled squares = the same but 4 days after drop off; unfilled circles = the same but 15 days after drop off). The sharp transition in arthropods is generally associated with changes in the epicuticular/epidermal wax layer (replotted from Edney 1977). In endothermic vertebrates, the transition in total evaporative loss is associated with primarily respiratory evaporative cooling when air temperature > body temperature.

surface, the skin, was not constant but was positively related to T_a. This means that $\chi_{sat,T_{surf}}$ increased with T_a, and thus θ_t declined from around 160 to around 60 s cm^{-1} over the temperature range. This variation may be due to alterations in any of the serial resistances that make up the total resistance, such as resistance to water vapor diffusion across the stratum corneum of the skin, the feathers, or the boundary layer of unstirred air around the animal (Tracy 1982). The skin's resistance is most important for birds, and during heat stress birds can decrease this resistance (Hoffman and Walsberg 1999), and thus increase the CWL, by dilation of blood vessels in the dermal capillary bed (Ophir et al. 2002), effectively reducing the diffusion path length, and/or increasing the permeability of the skin to water vapor (Arieli et al. 1999). Also, in mammals that have sweat glands the cutaneous resistance to water loss may be short-circuited by secretion of water (along with NaCl) which then evaporates freely on the surface and cools the mammal. The boundary layer resistance to evaporation is most important for animals with wet skins such as amphibians, and there are biophysical models for calculating this resistance (Tracy 1976; O'Connor and Spotila 1992).

Across species, values for θ_t vary from below 5 s cm^{-1} for the integument of some amphibians and aquatic arthropods to as much as 4000–5000 s cm^{-1} for the cuticle for arid adapted arthropods (Withers 1992), and could decline under windy conditions due to diminution of the boundary layer. Some animals acclimating to dehydrating conditions increase their integumentary/cutaneous θ. For example, θ more than doubles when zebra finches (*Poephila guttata*) are water deprived, compared with hydrated controls (Menon et al. 1989). Such variation in θ within a species, or the marked variation among species, has generally been traced to differences in composition, quantity, and conformation of lipids in the stratum corneum of vertebrate skin or epicuticle of arthropods (Edney 1977; Hadley 1985; Gibbs 1998; Haugen et al. 2003). The waterproofing characteristics of these lipids are altered at high temperatures in some species, so the CWL may increase sharply at some critical temperature (Edney 1977) (figure 12.8).

12.3 The Dietary Requirement for Water

Having specified each of the obligatory losses of water, we can return to our question of how much must be ingested to match obligatory losses. It is useful to express the water budget algebraically and rearrange it slightly to highlight the most critical features. In developing a simple model we will use our handy notation of water flux per unit energy metabolized and we will rely on the following points:

(1) Requirement for water in food or drinking, in g d^{-1} = minimum fecal and urinary water loss + minimum evaporative loss − water produced in metabolism (figure 12.1). Water produced in metabolism is equal to metabolic rate in kJ d^{-1}, R, multiplied by the metabolic water produced per kJ metabolized, w_{met} (table 12.1). In analogous fashion, daily evaporative water lost can be expressed as the metabolic rate multiplied by the evaporative water lost per kJ metabolized, e (g kJ^{-1} metabolized).

(2) At steady state, minimum fecal and urinary loss is equal to the product $[R][l]$, where l is the minimum water lost per kJ metabolized (figure 12.5).

Thus, minimum daily preformed water requirement can now be expressed as

$$\text{minimum requirement in g d}^{-1} = R\,[l - w_{met} + e] \qquad (12.6)$$

This expression allows us to identify quickly that the only animals that might possibly (i.e., if e is not too large) maintain balance in the absence of drinking or food water are those for whom $l < w_{met}$, because otherwise the summation within the brackets is certainly more than zero. When values for l that were determined in balance trials (figure 12.5) are compared with w_{met} (table 12.1), it is apparent that

TABLE 12.1

Metabolic water = water formed when organic material oxidizes

e.g., $C_6H_{12}O_6 + 6O_2 \rightarrow 6CO_2 + 6H_2O$

(180g)   (192g)   (264g)   (108g)

Substrate	gH_2O formed per g substrate	gH_2O formed per KJ	$\mu L\,H_2O$ per mL O_2
Starch	0.56	0.032	0.62
Fat	1.07	0.026	0.53
Protein (Urea excreted)	0.39	0.021	0.43
(Uric acid exc.)	0.50	0.026	0.53

Source: From (Schmidt-Nielsen 1990).

granivores potentially can exist without drinking or food water but no vertebrate herbivore can do this. Water losses associated with processing food are just too high in herbivores. Water-conserving avian, mammalian, and insect granivores gain more water per kJ seed metabolized than they excrete in processing seeds, so if their evaporative losses per unit kJ are not too high this net gain can possibly balance evaporative losses and they can survive on dry seeds with no drinking water.

MacMillen and his colleagues have shown that small mammalian and avian granviores fit exactly this situation, especially as environmental temperature declines (MacMillen and Hinds 1983; MacMillen 1990). As temperature declines their metabolic rate rises, but recall from our earlier point that evaporative water loss changes little as temperature declines. Therefore, the ratio of evaporative loss per unit kJ declines. But, those small granivores that live in environments that are uniformly hot, even when active at night or resting in their burrows, probably fail to achieve water independence (Walsberg 2000). Also, the larger endothermic granivores, whose zones of thermal neutrality extend to quite low T_a, generally do not increase their metabolic rates enough to achieve water independence. As for insect granivores, there are interesting experimental studies that suggest that they increase their metabolic rates and hence water production during water restriction (Jindra and Sehnal 1990), but the hypothesis needs more study. Vertebrate (Hulbert and MacMillen 1988) and possibly invertebrate (Edney 1977) granivores achieve more positive water balance on carbohydrate-rich seeds that produce more water per kJ substrate (table 12.1), and when seeds are exposed to higher humidity and gain water hygroscopically (Frank 1988). However, maintaining water balance on dry seeds is marginal, and if the seeds have high salt or N content, which raises urinary water loss, or high indigestible fraction, which raises fecal water loss, then even small granivores will be out of water balance and lose body mass on dry seeds (Withers 1982). The most recent studies of diet and water relations of free-living Merriam's kangaroo rats (*Dipodomys merriami*), long considered the supreme example of water independence, actually cast doubt on whether they balance their water budget on an all-seed diet in the hottest, driest habitats of the Sonoran desert (Walsberg 2000). But, in the cooler Mojave desert they do maintain water and mass balance on an all seed diet during late summer and fall, which is the driest season of the year (Nagy and Gruchacz 1994).

Thus, granivores often, and herbivores always, require some water in addition to that formed metabolically, and if they cannot or do not drink then they must ingest water with their food. Equation 12.6 can be modified to calculate the required food water content, which is then easily measured in field-collected food samples. Divide both sides of equation 12.6 by feeding rate, which for

Figure 12.9. The required minimum water content of a food increases with its metabolizable energy content, the excretory water losses associated with processing it, and the animal's evaporative water loss. In this figure the predicted required minimum water contents for herbage (y axis) are shown as a function of food metabolizable energy content (x axis, in units of kJ g^{-1} dry mass) and minimal fecal and urinary water loss (variable l, in units of g H$_2$O/kJ metabolized). These predictions were made assuming the average evaporative loss found in seven species of arid-dwelling herbivores (e = 0.1 g H$_2$O/kJ metabolized, based on species listed in figure 12.5). Assumed values for l span the range from the lowest known value to higher values observed in arid-dwelling vertebrates (from figure 12.5). See text for full explanation.

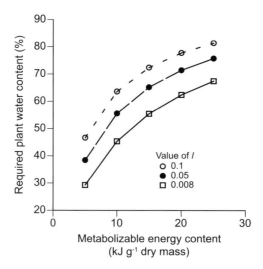

animals in steady state is given by equation 3.7 (in chapter 3) and after a tad of algebra you get

$$\text{required minimum H}_2\text{O content in g g}^{-1} = c(l - w_{met} + e) \qquad (12.7)$$

or,

$$\text{required minimum H}_2\text{O content in \% H}_2\text{O} = 100 \cdot \frac{c(l - w_{met} + e)}{1 + c(l - w_{met} + e)}$$

The new variable in the equation, c, is the food's metabolizable energy content in kJ g^{-1}. Based on equation 12.7, jackrabbits, our focal species in this section, apparently require 2.1 g H$_2$O g^{-1} dry matter, or 68% water (Nagy et al. 1976). The expression in equation 12.7 can lead us to a general prediction for all herbivores, based on the likely values of each of the parameters. Total evaporative loss (respiratory + cutaneous, or e) averaged 0.09 ± 0.04 g kJ^{-1} in the seven herbivores on restricted water represented in figure 12.5. Herbivores' values for c generally range from 5 to 15 kJ g^{-1} (chapter 3). Using these values, plus the likely values of l (0.05–0.15; figure 12.5) and w_{met} (Table 12.1), we find that it is improbable that any herbivore can get along with less than 0.5 g H$_2$O g^{-1} dry matter, or 33% water (figure 12.9).

The accomplished comparative physiologist C. R. Taylor thought that the African oryx (*Oryx beisa*), a desert antelope, might get along at this level (Taylor 1969), but this speculation has never been tested. Penned oryx in East Africa probably require a plant water content above 40% (King et al. 1975), and free-living Arabian oryx (*O. leucoryx*) in summertime consuming plants with 40% water lost, on average, 1.6% of their body mass each day (Williams et al. 2001). Another desert herbivore, the springbok antelope of the Kalahari Desert, apparently requires about 67% water (2 ml g^{-1} dry mass) in its food (Nagy and Knight 1994). Water balance studies of seven insect, mammalian, avian, and reptilian herbivores that are represented in figure 12.5 indicated an average required plant water content of around 50% water (i.e., 1.2 ± 0.2 [standard error] g H$_2$O g^{-1} dry food; range 0.61–1.84). Growth and/or reproduction would require higher plant water content, though this is also a little studied topic. For example, in fat sand rats (*Psammomys obesus*) fed saltbush *Atriplex halimus*, birth mass and growth rate of young were positively related to water content over a range of 60–85% (Kam and Degen 1994). The model (equation 12.7) also indicates that, if a plant has too little water in it, eating more of it will generally not help. Recall that feeding and metabolic rates were eliminated as variables in the algebraic rearrangement. Consistent with this theoretical finding, attempts to feed animals excess amounts of forage that is too dry did not rectify water imbalance (Nagy 1972; Scriber 1977).

If forage is too dry, animals must find a source of free water to drink. Locating surface water or digging are obvious solutions, but there are others. Drinking droplets of water from fog condensed on the body is important for desert reptiles and arthropods in several desert ecosystems (Louw 1972; Seely 1979; Cloudsley-Thompson 1991). Some desert arthropods absorb water from ingested damp soil (Cloudsley-Thompson 1991). No vertebrate can achieve net water absorption from unsaturated air, but a number of species of arachnids and insects do this nifty trick (Edney 1977).

Animals differ in how long they can tolerate desiccation. Generally, larger animals have longer desiccation endurance than smaller ones (Calder 1984), and ectotherms have an advantage over endotherms in how long they can tolerate dry conditions (Pough 1980). A simple model can explain both generalities. The body water content of animals scales isometrically with body size (i.e., $\propto M^{1.0}$) (Calder 1984). Assuming that animals tolerate a similar proportional loss of body water, e.g., equivalent to 15% of body mass, their desiccation endurance in days is estimated as $0.15M^{1.0}/$(daily rate of net water loss). The components of water flux scale with body mass to approximately the 3/4 power (Calder 1984). Therefore, desiccation endurance time will therefore scale with body mass to a power greater than 0:

$$\text{desiccation endurance} = \frac{\text{g of H}_2\text{O that can be lost}}{\text{g/d of net H}_2\text{O loss}}$$

$$\propto \frac{0.15M^{1.0}}{M^{3/4}} \propto 0.15M^{1/4}. \tag{12.8}$$

By this model, ectotherms have longer desiccation endurance than endotherms because, at any given size, they have much lower rates of net water loss (the denominator). Reptiles seem able to tolerate greater changes in the volumes of their body fluids and in the osmotic concentrations of the extracellular fluids than do birds and mammals (Minnich 1982), which further extends their endurance, but when conditions are extreme some species cease feeding and restrict activity (Cloudsley-Thompson 1991). Nagy (1988) compared herbivorous chuckwalla lizards (*Sauromalus obesus*) and desert tortoises (*Gopherus agassizii*) in the Mojave desert in these regards. Chuckwallas maintained more precise regulation of blood volume and osmotic pressure, which restricted them to eating green, moist vegetation in the spring and becoming relatively inactive during the late summer dry period. Tortoises tolerated wider swings in their osmotic and fluid balances. They stored some water when it was in excess, and during the summer and fall dry periods ate dry vegetation and drew on body water compartments.

Work by Nagy's student Brian Henen and other colleagues supports these findings in studies of desert tortoises at a number of desert sites over several years (Turner et al. 1986; Henen 1997; Henen et al. 1998). The studies on desert tortoises in many locations in the American southwest indicate that although desert tortoises in any particular season may not simultaneously be in positive balance for both water and energy, they balance things out on an annual basis and manage to reproduce most years, even drought years. Bradshaw (2003) raises the troubling question of whether the tortoises' features represent an evolved strategy for dealing with a harsh, unpredictable environment or reflect a highly stressed animal barely surviving in marginal habitats. He cites evidence that during the Pleistocene tortoises occupied more mesic habitats, and modern habitats have been altered due to increasing aridity and floral changes caused by human disturbances such as grazing and introduction of non-native species.

12.4 Ingestion of Xenobiotics Can Increase the Dietary Requirement for Water

Animals in arid lands just maintaining water balance on fairly dry foods (e.g., seeds, desiccated leaves) have another potential problem. The presence of secondary metabolites (SMs) in those foods can tip the balance and increase the minimum water requirement. As was described in chapter 9, absorption of SMs is typically followed by excretion of their water-soluble metabolites, and this addition to the diet osmolyte load will increase urinary water loss. Furthermore, many SMs act as diuretics; they increase urine flow directly via a variety of mechanisms such as decreasing the reabsorption of Na by the kidney, or by decreasing the production or blocking the effects of aldosterone, the hormone that regulates kidney water reabsorption (Dearing et al. 2001). The alkaloid caffeine from coffee and black tea is an example probably familiar to you, but diuretic SMs have been identified in other major classes of compounds as well (e.g., terpenoids, phenolics) and in over 100 species of plants from various families (Dearing et al. 2001).

Tests for diuretic effects in wild mammals and birds typically involve feeding diet with and without the putative diuretic and testing for increased water consumption, increased urine volume, and decreased urine concentration (Jakubas et al. 1993; Dearing et al. 2002). The presumption is that the SM increases urinary water loss and the animal compensates by increasing voluntary water intake. However, this conclusion is not very convincing, because it is possible that as a response to a different taste from the addition of the SM to the diet the animal consumes more water which in turn results in excretion of larger volume of dilute urine. The only way to discriminate between these alternatives is to control water availability experimentally. One of us (Karasov) in collaboration with Mangione and Dearing conducted such a study on desert woodrats (*Neotoma lepida*) (Mangione et al. 2004). When we added resin from creosote bush (*Larrea tridentata*) to the woodrats' chow diet, the rodents increased urine volume output, decreased urine concentration, and increased their voluntary water intake two to four times. What convinced us about the negative effect of creosote resin on the woodrats' water conserving ability was that when woodrats were water restricted, the minimal amount of water that they needed for body mass maintenance increased after resin was added to their diet. Inclusion of the resin increased the minimum water that the rats lost per kJ metabolized. Considering the prevalence of SMs with diuretic properties, it is reasonable to speculate that they often pose a challenge for herbivores and granivores living in arid environments.

12.5 Is Water Ecologically Limiting?

In some plant communities water content may be too low to support water balance in resident herbivores. Comparison of the required plant water content (0.5 to 1 g H_2O g^{-1} dry matter) with measured plant water contents in the Mojave Desert (figure 12.2) shows that in the late fall and early winter, before winter rains, nearly all plants available to herbivores were too dry. Though you may think that the Mojave Desert is an obvious place to look for water limitation, there are other examples in more moderate environments. In the seasonally semiarid environment of the green iguana (*Iguana iguana*) on Caribbean islands, no plant classes satisfied the lizard's water requirements during the four-month-long dry season (their diet water content was below 50%; Marken Lichtenbelt 1993). As a

Figure 12.10. Leaves with less than 60% water content were too dry to support normal growth of cecropia larvae. In this experiment cecropia larvae (*Hyalophora cecropia*) were held at 21.6 °C and relative humidities of 75% (upward-pointing triangles; designated as 75) and 35% (downward triangles, designated as 35). They were fed wild cherry leaves that were either hydrated by placing the petiole in water (designated with a filled triangle) or not (unfilled triangle). (Replotted from Scriber 1977.)

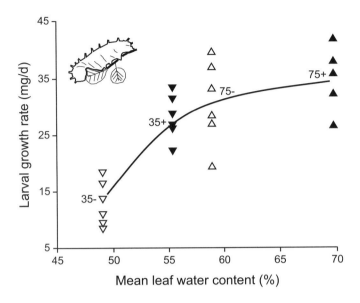

consequence, they lost mass in two study years (0.6–0.7% d^{-1}). The growth rates of insectivorous *Anolis* lizards on some of these Caribbean islands seem to be limited by water availability (Stamps and Tanaka 1981; Jenssen and Andrews 1984). In the deciduous forests of temperate North America (New York state) the water content of many of the wild cherry trees (*Prunus serotina*) eaten by cecropia moth (*Hyalophora cecropia*) larvae is too low (water content range 50–60%) to support normal growth (Scriber 1977; figure 12.10).

When food does not contain enough water animals must seek out local supplemental sources of water, migrate to an area with plants with adequate water, or suffer reduced growth and reproduction and consequently fitness, like the slower-growing cecropia moths feeding on low-water-content cherry leaves (figure 12.10). Let us consider examples of each type of response.

Invertebrate prey are the most likely source of supplemental water. Many vertebrate omnivores living in semiarid or arid habitats increase their consumption of insects during the dry season (Robbins 1993). The modeling approach permits a quantitative prediction of diet mixes that satisfy water requirements. For example, omnivorous antelope ground squirrels (*Ammospermophilus leucurus*) of the arid North American southwest are mainly granivorous during the dry season that occurs in fall and winter. But seeds are too dry to satisfy their water requirement (Karasov 1985). Based on the modeling approach described above, one of us (Karasov) predicted that ground squirrels require a dietary supplement of arthropods of about 20% of diet dry mass. It was satisfying to find that this amount was approximately what they ingested (box 12.3).

Box 12.3. Predicting Diet Mixes that Satisfy Water Requirements

Omnivorous antelope ground squirrels (*Ammospermophilus leucurus*) cannot survive on only dry seeds, so what mix of dry seeds and insects will simultaneously satisfy their energy and water needs? These diurnal rodents of the arid North American southwest remain active all year and maintain energy and water balance without drinking during the prolonged dry period in the Mojave Desert, which can last from May to the following December or January. Equations that predict food needs for energy (equation 3.7) and water needs (equation 12.6) can be solved simultaneously to provide a quantitative answer that can be tested against field observations.

$$[I_s][c_s] + [I_i][c_i] = R \tag{12.3.1}$$

$$[I_s][w_s] + [I_i][w_i] = R[1 - W_{met} + e] \tag{12.3.2}$$

The variables and most of their measured values are listed below, but the value for l requires some additional comment. Ground squirrels excreting maximally dry feces and maximally concentrated urine eliminate more water when feeding on insects ($l = 0.034$ g H_2O/kJ) than on seeds ($l = 0.019$) because insects contain more N to eliminate and are slightly less digestible (more feces per gram food) (Karasov 1983). To keep the problem as simple as possible, it can be solved for this range of values of l, or the mean of the two, although the model can be easily modified to provide a single exact solution. The solution, using the average l and calculated by the method of simultaneous equations in algebra, is for a diet containing 4 g seeds and 0.9 g insect (i.e., 18% insect by dry mass; 13–22% using the range of values of l). How does this compare with arthropod consumption by wild ground squirrels?

Analysis of stomach contents showed two peaks in arthropod consumption. One occurred in midsummer and coincided with the period when arthropod availability peaked, but the second occurred in fall and early winter when arthropods are rare and seeds and plants are too dry to provide supplemental water. Despite the relative scarcity of arthropods, ground squirrels increased their consumption of them during this dry period to a level of about 20% of dry matter consumed, similar to the prediction. As soon as new plant growth occurred following winter rains the ground squirrels switched to an entirely herbivorous diet that met all their nutrient requirements (Karasov 1985). The very same pattern was observed in a diet study on antelope ground squirrels at another Mojave Desert site as well (Bradley 1968).

The prediction of diets that satisfy multiple nutrient requirements by using multiple budget equations can be accurate. In another methods box

continued

continued

in chapter 11 (box 11.2) we provided another example using a slightly different, graphical technique.

I_s, I_i	feeding rate in g/d for seeds and insects, respectively
R	metabolic rate in kJ/d
c_s, c_i	metabolizable energy content of seeds and insects, respectively
w_s, w_i	water content in g H_2O/g dry mass of seeds and insects, respectively
l	minimum water lost per kJ metabolized
w_{met}	metabolic water produced per kJ metabolized
e	evaporative water loss per kJ metabolized

Measured values (Karasov 1985)
R = 90 kJ/d for 100 g ground squirrel
c_s = 18 kJ/g dry mass; c_i = 20 kJ/g dry mass
w_s = 0.1 g H_2O/g dry mass; w_i = 3 g/g dry mass
w_{met} = 0.029 g H_2O/kJ
e = 0.036 g/kJ

Their increased consumption of arthropods during the fall and winter was not simply a response to increased abundance of arthropods, which actually become scarcer at this cold time of year.

With some minor exceptions, stricter herbivores, like ungulates and the hares that we have discussed throughout this chapter, do not eat animal prey and therefore are very dependent on free water for drinking during their dry season (Robbins 1993). Western (1975) estimated that about two-thirds of the herbivore species in the Amboseli ecosystem of Kenya were obligate drinkers, and that their distributions during the dry season were related to rainfall and water availability. Hares and rabbits without access to free water require vegetation with water contents of at least 60% (above). During droughts they stop reproducing and their populations sometimes decline due to deaths or migration (Robbins 1993). In semiarid range in Wyoming in central North America, hares can breed repeatedly but they severely curtail reproduction when forage water content drops below 50% (Rogowitz 1992). In a correlative study, such as this, uncertainty remains about whether the reduction was mainly due to water shortage, especially because plant abundance and other features of plant quality like protein and fiber may change in concert with plant water content. We know of few experimental tests of the effect of water supplementation on demography

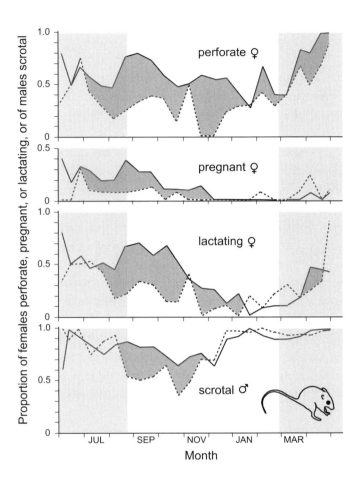

Figure 12.11. Experimental provisioning of drinking water extended the period of reproduction in the Namib desert rodent *Gerbillurus paeba*. On the experimental plot (solid line), compared with the control plot (no water provided, dashed line), significantly more females had bred (perforate vaginal opening, top figure) and were pregnant (second from top) and lactating (third from top) especially during the hot dry portion of the year from about September to February (indicated by the unshaded area). Proportionally more males on the experimental plot had scrotal (rather than abdominal) testes. These differences between the plots were statistically significant, and were also apparent in another omnivorous member of the rodent community (*Rhabdomys pumilio*) but not in another granivorous species that was studied (*Desmodillus auricularis*). (Replotted from Christian 1979.)

generally or in hares particularly, but in at least one rodent community water supplementation significantly extended the period of reproduction in some of the species (figure 12.11; Christian 1979). There was not a corresponding rise in number of rodents, but it is important to note that an increase in density is not a certain consequence of an increase in reproduction because of possible offsets due to increased mortality or emigration. In another experimental study (Hervert and Krausman 1986), the closure of water catchments forced mule deer (*Odocoileus hemionus*) does to travel outside their known home ranges through areas with little palatable forage seeking alternative surface water sources. The does drank once each day and then returned to their usual home ranges. At Hwange National Park in Zimbabwe, addition of water via boreholes resulted in an increase in animal numbers (elephant, buffalo, zebra), and the loss of some boreholes due to pump malfunction led to declines (Owen-Smith 1996).

Considering the sometimes important ecological effects of water shortage on survival and reproduction, the provisioning of artificial water resources may be considered an important management tool in wildlife conservation in arid regions (Robbins 1993; Owen-Smith 1996; Rosenstock et al. 1999; Rosenstock et al. 1999). In southern Africa, surface water availability forms the primary limitation on the distribution of large herbivore populations in semiarid environments (Owen-Smith 1996). Although herbivorous species in arid regions use artificial water sources and seem to rely on them (Robbins 1993), there are woefully few experimental tests of this dependence, leaving the issue unresolved (Broyles 1995): Witness the debate about water provisioning for desert bighorn sheep (*Ovis canadensis*) in North America (Broyles and Cutler 1999; Rosenstock et al. 2001; Broyles and Cutler 2001). Wildlife agencies maintain water catchment systems for bighorn sheep, but conservationists and some biologists question their utility. Most scientists would agree that careful study of animal responses to experimental manipulations of artificial water resources can help resolve such controversies. Wildlife ecologist (and physiological ecologist) Charles Robbins suggested that water budget approximations, like those described above, should guide the decision process and help identify when water is likely limiting (Robbins 1993). Even when a single species is shown to benefit from artificial water sources, these sources can have other effects on native flora and fauna in arid and semiarid zones which also need to be considered as management decisions are made (Owen-Smith 1996; James et al. 1999).

12.6 Testing the Evolutionary Match between Environmental Aridity and Water Relations

Physiological ecologists have been enormously successful describing how animals function in their natural environments and how they are adapted to them evolutionarily (Bennett 1987) or, expressed another way, in equilibrium with them (Feder 1987). Walsberg (2000) noted that the field began its rise to prominence with studies about desert animals by workers such as George A. Bartholomew, William R. Dawson, Bodil Schmidt-Nielsen, and Knut Schmidt-Nielsen. The behavioral and physiological features of the most desert-adapted species are recounted in a number of books, either general (Schmidt-Nielsen 1964) or with particular focus on invertebrates and reptiles (Edney 1977; Crawford 1981; Bradshaw 1986; Cloudsley-Thompson 1991), birds (Maclean 1996), and small mammals (Degen 1997). This huge compendium of knowledge about water relations of animals in relation to their physical environment represents one of the major products of physiological ecologists and accounts for a significant fraction of material in animal physiology textbooks (Withers 1992; Schmidt-Nielsen 1990; Randall et al. 1997; Willmer et al. 2000; McNab 2002). Indeed, more than a decade ago leaders pondered whether the field was a victim of its own success and where its future lay as a scientific enterprise (Feder et al. 1987). We will finish this chapter with a discussion of how some students of animal water relations have responded to this challenge.

Some researchers are testing more carefully the common assumption that particular physiological features represent genetic adaptations that evolved to increase dehydration resistance in response to increased aridity. Williams and Tieleman (2001) point out that a contrasting viewpoint about desert birds is that they do not have any particular physiological specialization(s) that permit them to occupy desert environments—what seems to be adaptive in birds to the desert environment is instead a generalized feature of birds. In one study they compared 12 species of larks (family Alaudidae) drawn from many different locations along a gradient of aridity (Tieleman et al. 2002). Their index of aridity, which incorporated information on average annual precipitation and maximum and minimum temperatures, was low in hot, dry deserts and high in cool, wet areas. After correcting for body mass differences, there were highly significant correlations between aridity and both basal metabolic rate and total evaporative water loss (TEWL), even when the analysis was phylogenetically informed. Thus, this result is very consistent with the idea that the differences indicate genetic changes brought about by natural selection. Low energy requirements and low TEWL may be adaptive in arid environments. However, besides genetics, the other determinants of phenotypic variation are development and environmental acclimatization (if in the field) or acclimation (if in the laboratory). The differences among the 12 lark species could be genetic, but they could also be irreversible differences that arose as larks at each site went through development under conditions of differing aridity or reversible differences that arose as larks responded to recent environmental aridity.

Researchers are teasing apart these determinants of phenotypic responses to aridity in studies of water relations of birds (MacMillen and Hinds 1998) and small mammals (Hewitt 1981; Buffenstein and Jarvis 1985). Tracy and Walsberg (2001b) compared populations of the Merriam's kangaroo rat (*D. merriami*) along an aridity gradient and found that individuals from the most arid locality were smaller and had lower whole-animal and mass-specific total evaporative water loss. The difference is ecologically important because evaporation is the single largest avenue of water loss in this species (Schmidt-Nielsen and Schmidt-Nielsen 1952). To test for developmental and environmental determinants, Tracy and Walsberg (2001a) captured pregnant females from the most arid site and from a mesic site and raised pups from the two lineages in either dry or wet conditions in the laboratory (figure 12.12). By comparing the progeny at sexual maturity (60 days of age) they tested for a developmental effect, and then, as a

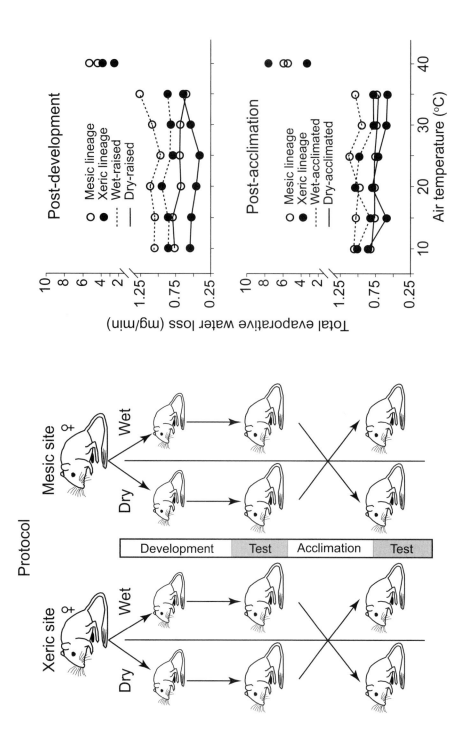

Figure 12.12. Testing for the genetic, developmental, and environmental determinants of variation in total evaporative water loss in Merriam's kangaroo rats (*Dipodomys merriami*). (Protocol) The pups of pregnant females from a xeric and mesic site were placed under one of two conditions in the laboratory ("Wet" or "Dry") until sexual maturity (60 days). Their evaporative water loss was measured, and then they were switched to opposite conditions, acclimated for 45 days, and retested. (Top right) In the first test, labeled "Post-development," total evaporative water loss was significantly lower in those raised under drier conditions, although differences among the lineages were still apparent. (Bottom right) In the second test, labeled "Post-acclimation" total evaporative water loss was significantly lower in kangaroo rats held under drier conditions, regardless of their development history. Although differences among the lineages were still apparent, the authors considered them of questionable biological significance, and they concluded that acclimation accounts for most of the observation and mutes earlier developmental and genetic differences in total evaporative water loss. The circles represent means (*n* = 8–12 pups in each case). (Replotted from Tracy and Walsberg 2001a.)

test for acclimation ability, they switched the kangaroo rats to the alternative laboratory environment and tested them again 45 days later. At first glance the findings seemed conflicting and confusing, because the environmental conditions altered the body mass of the individuals and the conclusions depended on whether the TEWL was expressed on a whole-animal or a per- gram animal basis. This kind of situation, which is not uncommon in physiological ecology, can be exasperating, but we agree with McNab's judgment that the whole-animal performance is probably most important ecologically or evolutionarily (McNab 1999). When the kangaroo rats were first tested at sexual maturity, whole-animal TEWL was significantly lower in those raised under xeric conditions, although differences among the lineages were still apparent (figure 12.12, top graph). But the apparent developmental differences were not fixed and were likely a reflection of acclimation. When tested a second time after acclimating to the alternative environment, the TEWL was significantly lower in kangaroo rats held under xeric conditions, independent of their development history (figure 12.12, bottom graph). Although differences among the lineages were still apparent, Tracy and Walsberg (2001a) considered them of questionable biological significance. Overall, they concluded that "acclimation mutes earlier developmental and genetic differences in total EWL."

The family of kangaroo rats (Heteromyidae), like the family of larks (above), shows considerable variation in features of water relations that matches up consistently with the physical environment of the respective species (French 1993). You should not conclude from our discussion that the apparent match between environmental aridity and animals' resistance to dehydration does not reflect genetic differences that arose from natural selection. Instead, take from this discussion an increased appreciation of the trifecta of responses of animals to environmental conditions (genetics, development, acclimatization) and some good examples of how to test most rigorously for their relative importance.

References

Anderson, R. A., and W. H. Karasov. 1981. Contrasts in energy intake and expenditure in sit-and-wait and widely foraging lizards. *Oecologia* 49:67–72.

Arieli, Y., N. Feinstein, P. Raber, M. Horowitz, and J. Marder. 1999. Heat stress induces ultrastructural changes in cutaneous capillary wall of heat-acclimated rock pigeon. *American Journal of Physiology* 277:R967–R974.

Bennett, A. F. 1987. Accomplishments of ecological physiology. Pages 1–8 in M. E. Feder, A. F. Bennett, W. W. Burggren, and R. B. Huey (eds.), *New directions in ecological physiology*. Cambridge University Press, Cambridge.

Bernstein, M. H. 1971. Cutaneous water loss in small birds. *Condor* 73:468–469.

Bradley, W. G. 1968. Food habits of the antelope ground squirrel in southern Nevada. *Journal of Mammalogy* 49:14–21.

Bradshaw, S. D. 1986. *Ecophysiology of desert reptiles.* Academic Press, Sydney.

Bradshaw, S. D. 2003. *Vertebrate ecophysiology.* Cambridge University Press, Cambridge.

Braun, E. J., and W. H. Dantzler. 1997. Vertebrate renal system. Pages 481–576 in W. H. Dantzler (ed.), *Handbook of physiology. Section 13: Comparative physiology.* Oxford University Press, New York.

Broyles, B., 1995. Desert wildlife water developments: Questioning use in the Southwest. *Wildlife Society Bulletin* 23:663–675.

Broyles, B., and T. L. Cutler. 1999. Effect of surface water on desert bighorn sheep in the Cabeza Prieta National Wildlife Refuge, southwestern Arizona. *Wildlife Society Bulletin* 27:1082–1088.

Broyles, B., and T. L. Cutler. 2001. Reply to Rosenstock et al. (2001) regarding effects of water on desert bighorn sheep at Cabeza Prieta National wildlife Refuge, Arizona. *Wildlife Society Bulletin* 29:738–743.

Buffenstein, R., and J. Jarvis. 1985. The effects of water stress on growth and renal performance of juvenile Namib rodents. *Journal of Arid Environments* 9:232–236.

Calder, W. A. 1984. *Size, function, and life history.* Harvard University Press, Cambridge, Mass.

Christian, D. P. 1979. Comparative demography of three Namib Desert rodents: Responses to the provision of supplementary water. *Journal of Mammalogy* 60:679–690.

Cloudsley-Thompson, J. L. 1991. *Ecophysiology of desert arthropods and reptiles.* Springer-Verlag, Berlin.

Cooper, P. D. 1985. Seasonal changes in water budgets in two free-ranging tenebrionid beetles, *Eleodes armata* and *Cryptoglossa verrucosa. Physiological Zoology* 58:458–472.

Crawford, C. S. 1981. *Biology of desert invertebrates.* Springer-Verlag, Berlin.

Daniels, F., and R. A. Alberty. 1975. *Physical chemistry.* Wiley, New York.

Dearing, M. D., A. M. Mangione, and W. H. Karasov. 2001. Plant secondary compounds as diuretics: An overlooked consequence. *American Zoologist* 41:890–901.

Dearing, M. D., A. M. Mangione, and W. H. Karasov. 2002. Ingestion of plant secondary compounds causes diuresis in desert herbivores. *Oecologia* 130:576–584.

Degen, A. A. 1997. *Ecophysiology of small desert mammals.* Springer-Verlag, New York.

Downs, C. T., and M. R. Perrin. 1991. Urinary concentrating ability of four *Gerbillurus spp.* of southern African arid regions. *Journal of Arid Environments* 20:71–82.

Edney, E. B. 1977. *Water balance in land arthropods.* Springer-Verlag, Berlin.

Feder, M. E. 1987. The analysis of physiological diversity: the prospects for pattern documentation and general questions in ecological physiology. Pages 38–70 in M. E. Feder, A. F. Bennett, W. W. Burggren, and R. B. Huey (eds.), *New directions in ecological physiology.* Cambridge University Press, Cambridge.

Feder, M. E., A. F. Bennett, W. W. Burggren, and R. B. Huey. 1987. *New directions in ecological physiology.* Cambridge University Press, Cambridge.

Frank, C. L. 1988. The relationship of water content, seed selection, and the water requirements of a heteromyid rodent. *Physiological Zoology* 61:527–534.

French, A. R. 1993. Physiological ecology of the Heteromyidae: economics of energy and water utilization. Pages 509–538 in H. H. Genoways and J. H. Brown (eds.), *Biology of the Heteromyidae.* American Society of Mammalogists, Stillwater, Okla.

Gibbs, A. G. 1998. Water-proofing properties of cuticular lipids. *American Zoologist* 38:471–482.

Green, B. 1978. Estimation of food consumption in the dingo, *Canis familiaris dingo*, by means of ^{22}Na turnover. *Ecology* 59:207–210.

Hadley, N. F. 1985. The adaptive role of lipids in biological systems. John Wiley, New York.

Haugen, M., J. B. Williams, P. Wertz, and B. I. Tieleman. 2003. Lipids of the stratum corneum vary with cutaneous water loss among larks along a temperature-moisture gradient. *Physiological and Biochemical Zoology* 76:907–917.

Henen, B. T. 1997. Seasonal and annual energy budgets of female desert tortoises (*Gopherus agassizii*). *Ecology* 78:283–296.

Henen, B. T., C. C. Peterson, I. R. Wallis, K. H. Berry, and K. A. Nagy. 1998. Effects of climactic variation on field metabolism and water relations of desert tortoises. *Oecologia* 117:365–373.

Hervert, J. J., and P. R. Krausman. 1986. Desert mule deer use of water developments in Arizona. *Journal of Wildlife Management* 50:670–676.

Hewitt, S. 1981. Plasticity of renal function in the Australian desert rodent *Notomys alexis*. *Comparative Biochemistry and Physiology A* 69:297–304.

Hill, R. W., G. A. Wyse, and M. Anderson. 2004. *Animal physiology*. Sinauer Associates. Sunderland, Mass.

Hoffman, T.C.M., and G. E. Walsberg. 1999. Inhibiting ventilatory evaporation produces an adaptive increase in cutaneous evaporation in mourning doves, *Zenaida macroura*. *Journal of Experimental Biology* 202:3021–3028.

Holleman, D. F., R. G. White, and J. R. Luick. 1975. New isotope methods for estimating milk intake and yield. *Journal of Dairy Science* 58:1814–1821.

Hulbert, A. J., and R. E. MacMillen. 1988. The influence of ambient temperature, seed composition and body size on water balance and seed selection in coexisting heteromyid rodents. *Oecologia* 75:521–526.

Jakubas, W. J., W. H. Karasov, and C. G. Guglielmo. 1993. Ruffed grouse tolerance and biotransformation of the plant secondary metabolite coniferyl benzoate. *Condor* 95:625–640.

James, C. D., J. Landsberg, and S. R. Morton. 1999. Provision of watering points in the Australian arid zone: a review of effects on biota. *Journal of Arid Environments* 41:87–121.

Jenssen, T. A., and R. M. Andrews. 1984. Seasonal growth rates in the Jamaican lizard, *Anolis opalinus*. *Journal of Herpetology* 18:338–341.

Jindra, M., and F. Sehnal. 1990. Linkage between diet humidity, metabolic water production and heat dissipation in the larvae of *Galleria mellonella*. *Insect Biochemistry* 20:389–395.

Kam, M., and A. A. Degen. 1988. Water, electrolyte, and nitrogen balances of fat sand rats (*Psammomys obesus*) when consuming the saltbush *Atriplex halimus*. *Journal of Zoology (London)* 215:453–462.

Kam, M., and A. A. Degen. 1994. Body mass at birth and growth rate of fat sand rat (*Psammomys obesus*) pups: effect of litter size and water content of *Atriplex halimus* consumed by pregnant and lactating females. *Functional Ecology* 8:351–357.

Karasov, W. H. 1983. Water flux and water requirement in free-living antelope ground squirrels, *Ammospermophilus leucurus*. *Physiological Zoology* 56:94–105.

Karasov, W. H. 1985. Nutrient constraints in the feeding ecology of an omnivore in a seasonal environment. *Oecologia* 66:280–290.

Karasov, W. H. 1989. Nutritional bottleneck in a herbivore, the desert wood rat (*Neotoma lepida*). *Physiological Zoology* 62:1351–1382.

Karasov, W. H., and I. D. Hume. 1997. Vertebrate gastrointestinal system. Pages 409–480 in W. Dantzler (ed.), *Handbook of comparative physiology*. American Physiological Society, Bethesda, Md.

Kenagy, G. J. 1972. Saltbush leaves: excision of hypersaline tissue by a kangaroo rat. *Science* 178:1094–1096.

King, J. M., G. P. Kingaby, J. G. Colvin, and B. R. Heath. 1975. Seasonal variation in water turnover by oryx and eland on the Galana Game Ranch Research Project. *East African Wildlife Journal* 13:287–296.

Kretzmann, M. B., D. P. Costa, and B. J. Le-Boeuf. 1993. Maternal energy investment in elephant seal pups: Evidence for sexual equality? *American Naturalist* 141:466–480.

Lifson, N., and R. McClintock. 1966. Theory of use of the turnover rates of body water for measuring energy and material balance. *Journal of Theoretical Biology* 12:46–74.

Lighton, J.R.B. 1994. Discontinuous ventilation in terrestrial insects. *Physiological Zoology* 67:142–162.

List, R. J. 1966. *Smithsonian meterological tables*. Publications 4014. Smithsonian Institution, Washington, D.C.

Louw, G. N. 1972. The role of advective fog in the water economy of certain Namib Desert animals. *Symposia of the Zoological Society of London* 31:297–314.

Maclean, G. L. 1996. *Ecophysiology of desert birds*. Springer-Verlag, Berlin.

MacMillen, R. E. 1990. Water economy of granivorous birds: A predictive model. *Condor* 92:379–392.

MacMillen, R. E., and R. V. Baudinette. 1993. Water economy of granivorous birds. *Functional Ecology* 7:704–712.

MacMillen, R. E., and D. S. Hinds. 1983. Water regulatory efficiency in heteromyid rodents: A model and its application. *Ecology* 64:152–164.

MacMillen, R. E., and D. S. Hinds. 1998. Water economy of granivorous birds: California House Finches. *Condor* 100:493–503.

Marken Lichtenbelt, W.D.V. 1993. Optimal foraging of a herbivorous lizard, the green iguana in a seasonal environment. *Oecologia* 95:246–256.

McNab, B. K. 1999. On the comparative ecological and evolutionary significance of total and mass-specific rates of metabolism. *Physiological and Biochemical Zoology* 72:642–644.

McNab, B. K. 2002. *The physiological ecology of vertebrates*. Comstock Publishing Associates, Ithaca, Nile.

McWhorter, T. J., C. Martinez del Rio, and B. Pinshow. 2003. Modulation of ingested water absorption by Palestine sunbirds: Evidence for adaptive regulation. *Journal of Experimental Biology* 206:659–666.

Menon, G. K., L. F. Baptista, B. E. Brown, and P. M. Elias. 1989. Avian epidermal differentiation II. Adaptive response of permeability barrier to water deprivation and replenishment. *Tissue and Cell* 21:83–92.

Minnich, J. E. 1982. The use of water. Pages 325–395 in C. Gans and F. H. Pough (eds.), *Biology of the reptilia* vol. 12. Academic Press, London.

Monteith, J. L., and G. S. Campbell. 1980. Diffusion of water vapour through integuments—potential confusion. *Journal of Thermal Biology* 5:7–9.

Nagy, K. A. 1972. Water and electrolyte budgets of a free-living desert lizard, *Sauromalus obesus*. *Journal of Comparative Physiology* 79:39–62.

Nagy, K. A. 1975a. Nitrogen requirement and its relation to dietary water and potassium content in the lizard *Sauromalus obesus*. *Journal of Comparative Physiology* 104:49–58.

Nagy, K. A. 1975b. Water and energy budgets of free-living animals: measurement using isotopically labeled water. Pages 227–245 in N. F. Hadley (ed.), *Environmental physiology of desert organisms*. Dowden, Hutchinson, and Ross, Stroudsburg.

Nagy, K. A. 1982. Field studies of water relations. Pages 483–501 in C. Gans and F. H. Pough (eds.), *Biology of the reptilia*, vol. 12. Academic Press, London.

Nagy, K. A. 1983. *The doubly labeled water method: A guide to its use*, Publication number 12(1417). Laboratory of Biomedical and Environmental Sciences, University of California, Los Angeles.

Nagy, K. A. 1988. Seasonal patterns of water and energy balance in desert vertebrates. *Journal of Arid Environments* 14:201–210.

Nagy, K. A., and D. P. Costa. 1980. Water flux in animals: analysis of potential errors in the tritiated water method. *American Journal of Physiology* 238:R454–R465.

Nagy, K. A., and M. J. Gruchacz. 1994. Seasonal water and energy metabolism of the desert-dwelling kangaroo rat (*Dipodomys merriami*). *Physiological Zoology* 67:1461–1478.

Nagy, K. A., and M. H. Knight. 1994. Energy, water, and food use by springbok antelope (*Antidorcas marsupialis*) in the Kalahari Desert. *Journal of Mammalogy* 75:860–872.

Nagy, K. A., and P. A. Medica. 1986. Physiological ecology of desert tortoises in southern Nevada. *Herpetologica* 42:73–92.

Nagy, K. A., and K. Milton. 1979. Aspects of dietary quality, nutrient assimilation and water balance in wild howler monkeys (*Alouatta palliata*). *Oecologia* 39:249–258.

Nagy, K. A., and C. C. Peterson. 1988. Scaling of water flux rate in animals. *University of California Publications in Zoology* 120:1–172.

Nagy, K. A., V. H. Shoemaker, and W. R. Costa. 1976. Water, electrolyte, and nitrogen budgets of jackrabbits (*Lepus californicus*) in the Mojave Desert. *Physiological Zoology* 49:351–363.

O'Connor, M. P., and J. R. Spotila. 1992. Consider a spherical lizard: animals, models, and approximations. *American Zoologist* 32:179–193.

Ophir, E., Y. Arieli, J. Marder, and M. Horowitz. 2002. Cutaneous blood flow in the pigeon *Columba livia*: its possible relevance to cutaneous water evaporation. *Journal of Experimental Biology* 205:2627–2636.

Owen-Smith, N. 1996. Ecological guidelines for waterpoints in extensive protective areas. *South African Journal of Wildlife Research* 26:107–113.

Peaker, M., and J. L. Linzell. 1975. *Salt glands in birds and reptiles.* Cambridge University Press, Cambridge.

Peters, R. H. 1983. *The ecological implications of body size.* Cambridge University Press, Cambridge.

Pough, F. H. 1980. The advantages of ectothermy for tetrapods. *American Naturalist* 115:92–112.

Randall, D., W. Burggren, and K. French. 1997. *Animal physiology: Mechanisms and adaptations.* W. H. Freeman and Co., New York.

Reese, J. B., and H. Haines. 1978. Effects of dehydration on metabolic rate and fluid distribution in the jackrabbit, *Lepus californicus. Physiological Zoology* 51:155–165.

Robbins, C. T. 1993. *Wildlife feeding and nutrition.* Academic Press, San Diego, Calif.

Rogowitz, G. L. 1992. Reproduction of white-tailed jackrabbits on semi-arid range. *Journal of Wildlife Management* 56:676–684.

Rosenstock, S. S., W. B. Ballard, and J. C. Jr. Devos. 1999. Benefits and impacts of wildlife water developments. *Journal of Range Management* 52:302–311.

Rosenstock, S. S., J. J. Hervert, V. C. Bleich, and P. R. Krausman. 2001. Muddying the water with poor science: A reply to Broyles and Cutler. *Wildlife Society Bulletin* 29:734–743.

Scantlebury, M., W. Hynds, D. Booles, and J. R. Speakman. 2000. Isotope recycling in lactating dogs (*Canis familiaris*). *American Journal of Physiology* 278:R669–R676.

Schmidt-Nielsen, K. 1964. *Desert animals.* Oxford University Press, New York.

Schmidt-Nielsen, K. 1984. *Scaling: Why is animal size so important?* Cambridge University Press, Cambridge.

Schmidt-Nielsen, K. 1990. *Animal physiology: Adaptation and environment.* Cambridge University Press, Cambridge.

Schmidt-Nielsen, K., F. R. Hainsworth, and D. E. Murrish. 1970. Countercurrent heat exchange in the respiratory passages: effects on water and heat balance. *Respiratory Physiology* 9:263–276.

Schmidt-Nielsen, K., R. C. Schroter, and A. Skolnik. 1980. Desaturation of the exhaled air in the camel. *Journal of Physiology* 305:74P–75P.

Schmidt-Nielsen, K. B., and B. Schmidt-Nielsen. 1952. Water metabolism of desert mammals. *Physiological Reviews* 32:135–166.

Scriber, J. M. 1977. Limiting effects of low leaf-water content on the nitrogen utilization, energy budget, and larval growth of *Hyalophora cecropia* (Lepidoptera: Saturniidae). *Oecologia* 28:269–287.

Seely, M. K. 1979. Irregular fog as a water source for desert dune beetles. *Oecologia* 42:213–228.

Shoemaker, V. H., K. A. Nagy, and W. R. Costa. 1976. Energy utilization and temperature regulation by jackrabbits (*Lepus californicus*) in the Mojave Desert. *Physiological Zoology* 49:364–375.

Speakman, J. R. 1997. *Doubly labelled water: Theory and practice.* Chapman & Hall, London.

Stamps, J., and S. Tanaka. 1981. The influence of food and water on growth rates in a tropical lizard (*Anolis aeneus*). *Ecology* 62:33–40.

Taylor, C. R. 1968. The minimum water requirements of some East African bovids. *Symposia of the Zoological Society of London* 21:195–206.

Taylor, C. R. 1969. The eland and the oryx. *Scientific American* 220:88–95.

Taylor, C. R. 1970. Strategies of temperature regulation: effect on evaporation in East African ungulates. *American Journal of Physiology* 219:1131–1135.

Tieleman, B. I., and J. B. Williams. 2002. Cutaneous and respiratory water loss in larks from arid and mesic environments. *Physiological and Biochemical Zoology* 75:590–599.

Tieleman, B. I., J. B. Williams, and P. Bloomer. 2002. Adaptation of metabolism and evaporative water loss along an aridity gradient. *Proceedings of the Royal Society (London) B* 270:207–214.

Tieleman, B. I., J. B. Williams, G. Michaeli, and B. Pinshow. 1999. The role of the nasal passages in the water economy of crested larks and desert larks. *Physiological and Biochemical Zoology* 72:219–226.

Tracy, C. R. 1976. A model of the dynamic exchanges of water and energy between a terrestrial amphibian and its environment. *Ecological Monographs* 46:293–326.

Tracy, C. R. 1982. Biophysical modeling in reptilian physiology and ecology. Pages 273–321 in C. Gans and F. H. Pough (eds.), *Biology of the reptilia* vol. 12. Academic Press, London.

Tracy, R. L., and G. E. Walsberg. 2001a. Developmental and acclimatory contributions to water loss in a desert rodent: Investigating the time course of adaptive change. *Journal of Comparative Physiology B* 171:669–679.

Tracy, R. L., and G. E. Walsberg. 2001b. Intraspecific variation in water loss in a desert rodent, *Dipodomys merriami*. *Ecology* 82:1130–1137.

Turner, F. B., P. Hayden, B. L. Burge, and J. B. Roberson. 1986. Egg production by the desert tortoise (*Gopherus agassizii*) in California. *Herpetologica* 42:93–104.

Van helversen, O., and H. U. Reyer. 1984. Nectar intake and energy expenditure in a flower visiting bat. *Oecologia* 63:178–184.

Walsberg, G. E. 2000. Small mammals in hot deserts: Some generalizations revisited. *BioScience* 50:109–120.

Webster, M. D., G. S. Campbell, and J. R. King. 1985. Cutaneous resistance to water-vapor diffusion in pigeons and the role of the plumage. *Physiological Zoology* 58:58–70.

Welch, W. R. 1980. Evaporative water loss from endotherms in thermally and hygrically complex environments: An empirical approach for interspecies comparsions. *Journal of Comparative Physiology* 139:135–143.

Western, D. 1975. Water availability and its influence on the structure and dynamics of a savannah large mammal community. *East African Wildlife Journal* 13:265–286.

Williams, J. B. 1996. A phylogenetic perspective of evaporative water loss in birds. *Auk* 113:457–472.

Williams, J. B., S. Ostrowski, E. Bedin, and K. Ismail. 2001. Seasonal variation in energy expenditure, water flux and food consumption of Arabian oryx *Oryx leucoryx*. *Journal of Experimental Biology* 204:2301–2311.

Williams, J. B., and B. I. Tieleman. 2001. Physiological ecology and behavior of desert birds. Pages 299–353 in V. Nolan and C. F. Thompson (eds.), *Current ornithology*, vol. 16. Kluwer Academic/Plenum Publishers, New York.

Willmer, P., G. Stone, and I. Johnston. 2000. *Environmental physiology of animals.* Blackwell Science, Oxford.

Willoughby, E. J. 1968. Water economy of the Stark's lark and grey-backed finch-lark from the Namib Desert of South West Africa. *Comparative Biochemistry and Physiology* 27:723–745.

Withers, P. C. 1982. Effect of diet and assimilation efficiency on water balance for two desert rodents. *Journal of Arid Environments* 5:375–384.

Withers, P. C. 1983. Energy, water, and solute balance of the ostrich *Struthio camelus. Physiological Zoology* 56:568–579.

Withers, P. C. 1992. *Comparative animal physiology.* Saunders College Publishing, Fort Worth, Tex.

Withers, P. C., W. R. Siegfried, and G. N. Louw. 1981. Desert ostrich exhales unsaturated air. *South African Journal of Science* 77:569–570.

Wolf, B. O., and G. E. Walsberg. 1996. Respiratory and cutaneous evaporative water loss at high environmental temperatures in a small bird. *Journal of Experimental Biology* 199:451–457.

Production in Budgets of Mass and Energy

CHAPTER THIRTEEN

Growth in Budgets of Mass and Energy

13.1 Overview of Chapters 13 and 14

ENERGY AND MATERIAL that are ingested but not lost to urine, feces, or heat can be allocated to production. Production includes growth, storage, or reproduction. We will discuss what determines its magnitude in this and the next chapter. Growth and development are critical for bringing animals to reproductive readiness. Storage enhances survival and provides resources for reproduction, which is the generation of progeny that transmit genes to later generations. We begin with a brief overview of two major approaches to the question of costs of production, one based on fluxes of mass and energy and one based more on demographic terms of survival, time to reproductive maturity, and fecundity. In keeping with the major themes of this book we will emphasize the first approach, but energy and mass flux inform the second approach as well. After these basics, we will discuss growth in particular. We will see that almost all animals share a basic pattern of growth that is probably traceable to fundamental allometries of energy gain and expenditure. Once we define the two components of the cost of growth we will see that growth is sometimes so costly that animals must make compensatory adjustments in other aspects of their biology. We will also consider the difficult problem of whether growth is maximized or regulated.

13.2 Two Approaches Are Used to Evaluate Costs of Production

Production is considered to be costly. As we will see, high production costs increase the food and foraging time requirement of animals, thereby forcing them sometimes to make compensatory adjustments in other components of their energy and time budgets. These production costs can be measured as increments in mass and energy, and so we will call this the bioenergetics approach. However, biologists also apply a life-history approach and measure the costs of production in demographic terms related to fitness, such as a decline in number or quality of young that are produced when resources are allocated to parental survival (maintenance) instead of production (Reznick 1985; Reznick 1992). The two approaches are complementary, and there are many ideas about how to synthesize them (Wootton 1998; Congdon et al. 1982; Owen-Smith 2002; Waldschmidt et al. 1987; Sinervo and Svensson 1998). Although it can be difficult to merge them, important insights arise from both approaches. In order to illustrate the two approaches, we will use both to answer a single question in a single "model" organism. The question is "What does it cost a female fish to reproduce?" This fish is the medaka (*Oryzias latipes*), a small fish native to Japan, Korea, and China that is widely used in genetic and embryological research in part due to its fecundity and short life cycle. The medaka has a breeding season of up to four months during which females lay eggs daily (Hirshfield 1980).

13.2.1 The Organizing Idea of the Bioenergetics Approach is the Budget of Energy and Mass

Hirshfield (1980) maintained medaka at three different temperatures (25, 27, and 29 °C), but we will focus for now on the results at 27 °C where production was highest. He fed them food at three levels and followed their allocation of energy for 25 d. In chapters 1 and 3 we introduced the simplest energy budget, which we present here rearranged algebraically to highlight this chapter's focus. The total production (P_{Total}, in J/d) for growth (P_G), storage (P_S), and reproduction (P_R) equals net energy absorbed from food minus energy allocated to respiration:

$$P_{\text{Total}} = P_G + P_S + P_R = (I - E) - R \qquad (13.1)$$

where I is consumption, E is excretory losses, and R is respiratory losses (all in units of J/d). P_{Total} is sometimes called "scope for growth," which is a misnomer because it represents energy available for storage and reproduction as well (Lucas 1996). P_R was calculated based on the production of eggs (mean = 5.4/d) and their energy content (4.64 J/egg), and ($P_G + P_S$) was calculated based on measured change in body energy content over 25 d. Respiration was not measured, but typically in ectotherm and endotherm energy budgets

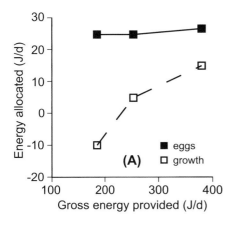

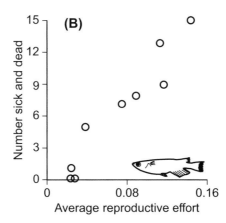

Figure 13.1. Two approaches to estimate the cost of egg production in medaka (*Orysias latipes*), a fish. (A) To support egg production (filled squares) as energy availability declined, medaka reduced growth rate and even drew on energy stored in tissue (unfilled squares), as indicated by negative values of energy allocated. (B) Medaka that allocated proportionally more energy to eggs (reproductive effort = egg energy/gross energy intake) exhibited higher disease and mortality. (Data replotted from Hirshfield 1980.)

R exceeds P_{Total} (Humphreys 1979). It was apparent from the negative value of $(P_G + P_S)$ at low feeding levels that the medaka kept up their constant egg production by allocating energy from their body tissue to reproduction when necessary (figure 13.1, left panel). Hirshfield (1980) expressed the reproductive effort of the medaka as a ratio of the energy allocated to reproduction relative to consumption (P_R/I). He found that the reproductive effort in his laboratory populations, which ranged from 0.07 to 0.14, was similar to that observed for free-living medaka in Japan (range 0.1–0.18). One question that cannot be addressed with the data was whether any of the respiratory energy was also expended for reproduction. In section 13.3.2 we will discuss this possibility.

Equation 13.1 offers an approach to questions of great interest and importance to ecologists. A fundamental one has to do with budget management. If it is selectively advantageous to increase P, is the entire budget expanded (increase I) or must there be compensatory adjustments, or tradeoffs (decrease R)? The dynamic energy budget approach favored by Kooijman, which is focused on fluxes of mass and energy, explicitly assumes that there are tradeoffs of some sort between maintenance, growth, development and reproduction (Kooijman 2000). Can we use the budget to calculate food consumption based on measurements of growth of wild animals and, if so, can we scale up from these budgets of individuals to consumption levels of entire populations in ecological communities? How will changing environmental conditions (e.g., temperature) influence each feature of the budget and thus the potential for population survival and growth? These are broad questions which we do not set out to answer for all animals, but we can sketch out the methods and principles with some case studies such as that for the medaka and others presented subsequently.

13.2.2 The Organizing Idea of Life-History Theory Is Trade-Offs
in Features of Demography

Life-history theory attempts to predict what pattern of resource allocation to maintenance, growth, or reproduction will be adaptive in particular environments. It assumes that allocation to one component reduces resources to another. Evidence for a trade-off in the medaka, for example, was that in the entire study there was a negative correlation between reproductive effort and growth (Hirshfield 1980). To make a prediction one must specify the demographic effects of a particular pattern in terms of survival, fecundity, and generation time. Allocation of resources thus involves trade-offs, and a core assumption is a trade-off between current reproduction and future expected reproductive output. Studies with the medaka provided evidence of two such tradeoffs (Hirshfield 1980). First, because medaka fecundity (number of eggs per day) was positively related to body size, the reduction in growth by current reproduction reduced future fecundity. Second, there was a positive correlation between reproductive effort and mortality, consistent with a tradeoff between reproduction and survival (figure 13.1 B). Hirshfield (1980) suggested that when reproduction drew on body tissue, as illustrated in figure 13.1A, mortality increased. Life-history theory based on demographic theory makes the following predictions: High, variable, or unpredictable adult mortality rates select for increased reproduction early in life. High, variable, or unpredictable juvenile mortality rates select for decreased reproduction and longer adult life. Thus, the energy an animal devotes to maintenance, growth, or reproduction depends on its probability of survival to future reproduction. Changes that reduce the value of juveniles and increase the value of adults will favor iteroparity (breeding multiple times) relative to semelparity (breeding only once) (Wootton 1998).

The study on medaka, as well as many others on fish (Wootton 1998) and reptiles (Congdon 1989; Congdon et al. 1982) illustrate the different but complementary aspects of the bioenergetic and life history approaches to questions about the cost of production. The study of bioenergetics can provide causal mechanisms for some of the correlations between reproduction, survival, and growth (Wootton 1985; Sinervo and Svensson 1998). As we will see below, a bioenergetics model is very useful in exploring the direct effects of current environmental factors such as temperature and food availability on life-history parameters such as growth rate. When differences are observed between populations of the same species in life-history features such as growth rate, the differences could result from such direct effects rather than from genetic differences between the populations.

Having introduced the general approaches to measuring costs of production, we will consider in more detail the bioenergetics of growth and reproduction (next chapter). Storage was discussed in chapter 7.

13.3 Energetics of Growth

13.3.1 Animals Share a Basic Pattern of Postnatal Growth that is Described in Growth Models

We will focus on postnatal growth (after birth or hatch) because it has more immediate ecological implications than embryonic growth, although the same bioenergetic principles apply to both. We will use a simple definition of growth as an increase in size, with increase in mass as a useful measure for most purposes. Development, which can proceed along with growth, refers to changes in tissue morphology and function. Maturation occurs as tissues approach more closely the adult form and function (Schew and Ricklefs 1998). The condition and functional capabilities of newly hatched or born animals varies tremendously, and the terms precocial and altricial are most widely used to describe the extremes of the range. Ricklefs (1983) and Hill (1992) describe the defining features for birds and mammals, but many of them can be applied generally. In our discussion precocial refers to neonates with well-developed functions (sensory, feeding, and mobility) and altricial refers to neonates with much less-developed functions. In general, altricial young depend on parents for nutrition, whereas precocial young feed themselves.

Growth data on animals are collected by measuring linear dimensions and/or mass as individuals age, repeatedly, or by compiling such data on individuals of known age in a population. Cumulative growth often has a logistic relationship with age (figure 13.2A), which means that the rate of size or mass increase is faster in younger animals and declines with age (figure 13.2B). This pattern is apparent in all species as far as we know, whether endothermic or ectothermic and whether altricial or precocial. One possible variation is that some animals, such as many fish (Wootton 1998) are said to have indeterminate growth, meaning that sexually

Figure 13.2. The same data on growth can be portrayed graphically in several ways. Plot A shows the typical logistic increase in mass with age. Plot B shows the rate of gain, which is actually the instantaneous slope from plot A. Plot C shows the rate of gain expressed as a % of mass per unit time.

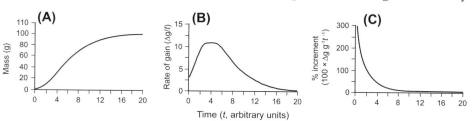

mature individuals do not have a characteristic adult size. This is in contrast to animals (e.g., many insects, birds, mammals) with a determinate growth pattern where growth stops abruptly when sexual maturity is reached. However, as we will discuss below, West et al. (2001) challenge this distinction between determinate and indeterminate growers and suggest that animals have fairly similar growth patterns when viewed from an appropriate, common time scale. Another variation is when reproductively active animals shrink in body size when food availability declines, as has been described for sea urchins *Diadema antillarum* (Levitan 1989), krill *Euphausia pacifica* (Marinovic and Mangel 1999), and marine iguanas *Amblyrhynchus cristatus* (Wikelski and Thom 2000). During shrinkage, when food is limited, somatic tissue is catabolized to meet energy needs. Because smaller individuals require less energy, shrinkage should, and in the case of marine iguanas does, increase survival.

The complexity of the growth rate pattern, with its changing rate over time, has stimulated much modeling work in order to derive a single or small number of parameters that express the growth rate and that can be used to make comparisons among populations or species. Furthermore, the generality of the growth pattern among animals has stimulated the search over decades for a biological theory and model that explains it. As a result, a plethora of growth models exist that vary from those that are purely descriptive and statistical to those that derive from fundamental physiological principles (see discussions by Ricklefs (1983), Blaxter (1989), Wootton (1998)). Many of the former type, such as the von Bertalanffy function (von Bertalanffy 1960), the Gompertz function, and the logistic equation are part of a family of curves known as the Richards function after its developer, plant biologist F. J. Richards (1959):

$$s_t = s_\infty (1 - b \exp^{-rt})^{-n} \text{ and } b = \frac{s_\infty^{-1/n} - s_0^{-1/n}}{s_\infty^{-1/n}} \tag{13.2}$$

in which

  n = a shape parameter related to the inflection point of the curve
  r = the growth rate constant with units of time^{-1}
  s_o = size at $t = 0$
  s_t = the size of an individual at time t, typically measured as mass
  s_∞ = asymptotic size

Changes in the shape parameter n modify the Richards function so that it incorporates many different shapes, and hence models, of growth curves (Ebert 1999; figure 13.3). Ebert (1999) provides a practical guide to the determination of some of the growth parameters in field studies when age is uncertain and when there are seasonal components such as lags. He also discusses how

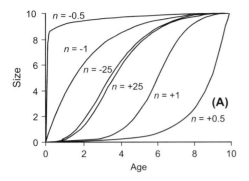
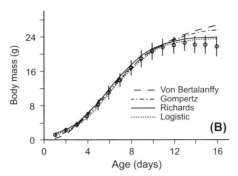

Figure 13.3. Examples of the Richards family of growth curves. (A; left-hand figure) These curves correspond to equation 13.2 with the specified shape parameter *n* selected so that growth is 99% of asymptotic value at age 10. The Bertalanffy curve has a value of *n* = −1 and the logistic *n* = +1. Curves for n = −25 and *n* = +25 converge on the Gompertz equation where the absolute value of *n* = infinity. (A) is redrawn from Ebert (1999), who points out that not all these possible forms are observed in organisms; most organisms seem to have absolute values of *n* no smaller than about 0.2. (B) Body mass growth of 35 tree swallow nestlings fit by Zach et al. (1984) to the four models, which all provide reasonably good fits until older ages. Circles are means and vertical bars are standard deviations. (Redrawn from Zach et al. 1984.)

growth-curve parameters can be combined with data on the size structure of a population to estimate survival rates.

Another class of growth models is based on physiological principles (Parks 1982; Baldwin and Sainz 1995). Some, like the model presented in box 13.1 that is derived from the allometric principles introduced in chapter 1, attempt to explain general features of growth in organisms. Other models can be used to predict growth of a specified animal under varying intake levels and environmental conditions (e.g., temperature). Several predictive models used in fish ecology developed by James Kitchell and his colleagues (Hanson et al. 1997) and Elliott (1994) are based on bioenergetic principles like those described in this chapter. In the study of wild fish populations, these models are embedded

---

Box 13.1. A General Model for Ontogenetic Growth

There is a long tradition of attempting to understand growth from principles of allometry (von Bertalanffy 1960; Reiss 1989). The following model is one of the most recent efforts (West 2001):

(1) Imagine the animal as composed of some total number of cells (N_c) which each have an average metabolic rate per cell (R_c) and mass per cell (M_c). Furthermore, imagine that it takes some amount of energy to create a new cell (J). These and other variables are listed and defined at the bottom of this box.

(2) During growth, some fraction of total metabolic rate (R) is due to maintenance metabolism (estimated as the product of $N_c R_c$) and some fraction to production of new tissue (estimated as $J(dN_c/dt)$). Hence, the total metabolic rate is the sum of the two components.

(3) At any time t the total body mass $M = M_c N_c$, so growth (change in mass per unit time) can be written as

continued

continued

$$dM/dt = (M_c/J)R - (R_c/J)M. \qquad (13.1.1)$$

We already know that total metabolic rate scales with (mass)$^{\sim 3/4}$. and that within a taxon can be estimated from an allometric equation such as $R = R_o M^{3/4}$. R_o is a constant for the given taxon. So, replace R in equation 13.1.1 with this expression:

$$dM/dt = (M_c/J)R_o M^{3/4} - (R_c/J)M. \qquad (13.1.2)$$

Now, let $a = (M_c/J)R_o$ and $b = (R_c/J)$, which gives a simplified equation:

$$dM/dt = aM^{3/4} - bM \qquad (13.1.3)$$

This equation predicts that as the animal grows, its growth rate (dM/dt) will decline with increasing size (M), ultimately reaching zero. When growth stops ($dM/dt = 0$), the final body size is the asymptotic maximum body size (M_∞), giving

$$0 = a\, M_\infty^{3/4} - b\, M_\infty,$$
$$a\, M_\infty^{3/4} = b\, M_\infty,$$
$$a/b = M_\infty / M_\infty^{3/4} = M_\infty^{1/4},$$
$$M_\infty = (a/b)^4 = (R_o M_c / R_c)^4. \qquad (13.1.4)$$

This model therefore predicts features of growth and asymptotic size on the basis of fundamental cellular properties not related directly to growth. A surprising prediction is that all animals will exhibit determinate growth, i.e., reach an asymptotic body mass M_∞. Yet growth studies show that many animals (especially ectotherms) are indeterminate growers and never reach M_∞. How is this rationalized? The authors (West et al. 2001) go on in their article to show that ectotherms grow relatively slowly and simply do not reach M_∞ over the course of our growth studies and even over the course of their normal lifetime. The authors show that, if growth of animals is scaled to dimensionless time, and size is expressed relative to the expected M_∞, then all animals actually fall on the same universal growth curve (figure 13.4).

R = average resting metabolic rate of the whole animal at time t
R_o = taxon-specific whole-animal metabolic rate per unit $m^{3/4}$
R_c = metabolic rate of a single cell
J = metabolic energy to create a single cell

continued

continued

M = total body mass
M_c = mass of a cell
M_o = total mass at birth
M_∞ = final asymptotic total body mass
N_c = total number of cells in the whole animal
φ = dimensionless mass ratio = $(M/M_\infty)^{1/4}$
t = time
τ = dimensionless time variable = $at/4\,M_\infty^{1/4} - \ln[1 - (M_o/M_\infty)^{1/4}]$

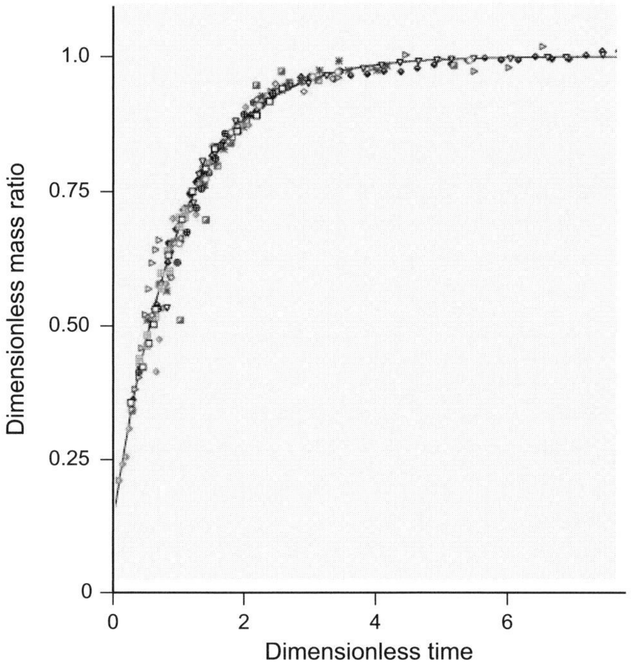

Figure 13.4. Universal growth curve. In this remarkable figure, West et al. (2001) plotted their values of the dimensionless mass ratio ($\varphi = (M/M_\infty)^{1/4}$) versus the dimensionless time variable ($\tau = (at/4M_\infty^{1/4}) - \ln [1 - (M_0/M_\infty)^{1/4}]$), for a wide variety of species that exhibit determinate and indeterminate growth. The species included six mammals (from shrew- to cowsized), three birds (robin, hen, and heron), three fish (guppy, salmon and cod), and one invertebrate (shrimp). (See the original publication for identity of specific symbols). When plotted in this way, the model predicts that growth curves for all organisms should fall on the same universal curve $1 - e^{-\tau}$ (shown as a solid line). (Modified from West et al. 2001.)

in individual-based models that predict growth rates in cohorts of fish of the same age (Hanson et al. 1997; Wootton 1998). We will elaborate on this approach in a section that follows.

The decades of scrutiny of growth patterns across hundreds of species has yielded some generalities about rates. If one plots the maximum growth rate (in g/d from figure 13.2B) observed from birth/hatch to maturity against the adult size (or asymptotic size), one finds that within a taxonomic clade growth rate scales approximately with $M^{3/4}$ across species (Case 1978; Peters 1973; Calder 1984; figure 13.5). Endotherms tend to have higher growth rates than

Figure 13.5. Growth rate (in g/d) scales with adult body mass to approximately the 3/4 power in endotherms and ectotherms. (A) Endothermic bird species (represented by the lines) tend to have higher growth rates than similar-sized ectothermic vertebrates (unfilled circles), and among birds the altricial species (solid line) tend to have higher growth rates than the precocial species (dashed line). (B) Mammals, which tend to have growth rates similar to those of precocial birds, have higher growth rates than invertebrates (unfilled circles). However, much of the difference disappears when one corrects for differences in body temperature. The dotted line represents the regression for the invertebrate data corrected to 38°C. (Figures replotted from Calow and Townsend 1981.)

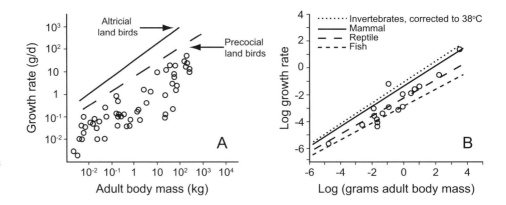

similar-sized ectotherms, although much of the difference disappears when one corrects for differences in body temperature (Calow and Townsend 1981; Gillooly et al. 2002; figure 13.5B). Finally, among birds, the species with altricial modes of development have higher growth rates than precocial species (Ricklefs 1968; figure 13.5A). The longstanding explanation for this is a bioenergetic one and was proposed first by William Dawson and his colleagues (Hill 1992). Precocial birds exhibit the endothermic metabolic pattern of relatively high stable body temperature (T_b) maintained by high rate of respiration at a much earlier age than altricial birds, which exhibit the ectothermic metabolic pattern at hatch. Thus, if both types have similar rates of energy ingestion, the precocial species will lose more energy in respiration whereas the altricial ones will allocate more to growth. Curiously precocial and altricial mammals do not show the same clear-cut difference in growth (Hill 1992).

13.3.2 The Cost of Growth Includes Energy and Mass Deposited in Tissues and Respiratory Costs

The rate energy or nutrient is deposited in an animal's tissues during growth are a function of its rates of mass gain and the energy or nutrient content of the gain. The body composition of animals changes during development and growth. As a general rule in vertebrates, during growth and development the dry mass per unit body mass increases as body water content declines, and the energy content per unit dry mass increases as fat:protein ratio of the dry mass increases (Robbins 1993; Jobling 1994; Weathers 1996). These changes occur earlier in precocial species than altricial species, and the water content of muscles or other tissues are often used as an index to extent of development or "maturity" (Starck and Ricklefs 1998). There are other variations that seem related to features of animals' life history such as higher fat accumulation in species that rely on unpredictable food sources, or higher fat accumulation in young animals that will likely experience negative energy budgets early in life because they live in very cold environments (e.g., marine mammals) or are ineffective feeders (Robbins 1993). We will not dwell on this issue except to emphasize that accurate determinations of costs requires knowledge about these changes in body composition.

Respiratory costs of production include the respiratory costs of digesting, absorbing and processing additional food consumed above maintenance, synthesizing new tissue, and perhaps the respiratory costs of foraging for extra food above maintenance. Peterson et al. (1999) measured these costs in free-living garter snakes (*Thamnophis sirtalis fitchi*), and the costs proved to be significant in several respects. In a population of growing juveniles and non-reproductive adults they used the doubly labeled water method to measure total respiratory energy expenditure (called field metabolic rate, or FMR; methods box 13.2) and they also measured rates of mass change. Not surprisingly, they

---

Box 13.2. The Doubly Labeled Water (DLW) Method for Measuring Respiration

Rates of CO_2 production can be measured in free-living animals by using this method, and then converted to rates of heat production using conversion factors for indirect calorimetry (see methods box 1.1). The technique has been validated in lots of captive animals (Nagy 1989; Speakman 1997), and is considered to have an accuracy of $\pm 5\%$ for terrestrial vertebrates in the field (Nagy et al. 1999). Why and how does it work? The method was invented by Nathan Lifson and his colleagues (Lifson et al. 1955; Lifson and McClintock 1966) after they made the key discovery that the isotopically labeled oxygen (*O) in body H_2O is in equilibrium with the oxygen in respiratory CO_2 (Lifson et al. 1949). The particular isotope used is often oxygen-18 (i.e., ^{18}O). The equilibrium occurs because of the carbonic anhydrase reaction in red blood cells and in the lung:

$$H_2{}^*O + CO_2 \xleftrightarrow{\text{carbonic anhydrase}} H_2C{}^*O_3$$
$$\text{and, then,} \quad H_2C{}^*O_3 \xleftrightarrow{\text{carbonic anhydrase}} H_2{}^*O + C{}^*O_2.$$

Therefore, once the body H_2O is enriched with *O, the specific activity of *O will decline for two reasons: (i) as $C{}^*O_2$ is expired, and (ii) as $H_2{}^*O$ is lost by evaporation, excretion and secretion, coupled with simultaneous

continued

continued

Figure 13.6. After body H_2O is isotopically enriched, the concentrations of the oxygen isotope (*O; filled circles) and the hydrogen isotope (*H; unfilled triangles) decrease exponentially towards natural abundance levels (called background). The rate constants for elimination of *O (r_{*O}) and *H (r_{*H}) are the slopes of the straight lines relating log[enrichment above background] to time. The fractional rate of decline of *O (r_{*O}) is faster than that of *H (r_{*H}) because *O is lost via both CO_2 and H_2O whereas *H is lost via H_2O alone.

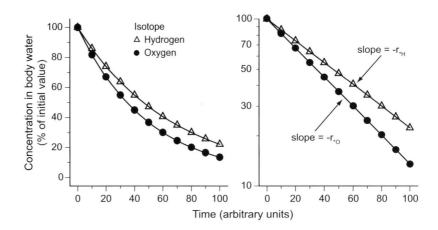

dilution by new, unlabeled H_2O consumed as food and drink or produced during oxidation of substrates. One can separate out the decline associated with H_2O by measuring simultaneously that effect alone. This is done by isotopically enriching body H_2O with labeled hydrogen (*H) and measuring the decline in its specific activity. The particular H isotope used is ^2H or ^3H (the latter is unstable and thus radioactive). Figure 13.6 shows the typical exponential pattern of decline, which appears linear on a semiarithmic log plot.

You can now understand that the rate of CO_2 production is a function of the difference $r_{*O} - r_{*H}$. The calculation of CO_2 production also requires, at a minimum, some knowledge of the pool size(s) for O (N_{*O}) and *H (they can differ) (Nagy 1980; Speakman 1997). If you assume that the "dilution spaces" of *O and *H are equal (let us call this quantity N), then the equation to estimate the rate of CO_2 production is: $V_{co_2} = (N/2)(r_{*O} - r_{*H})$. Note that we divide the size of the oxygen in body water pool by 2, because each molecule of CO_2 that is expired removes two oxygens.

When ecologists apply the method they follow several steps, each of which has several alternatives that are thoroughly discussed in a number of sources (Nagy 1983; Speakman 1997).

(1) *Inject a captured animal with a dose of water labeled with isotopes of *O and *H* — typical choices are ^{18}O, and either the stable isotope (^2H) or the unstable (radioactive) isotope (^3H). Each has virtues depending on size of the animal, cost, availability of permits for using radioisotopes, availability of analytical equipment.

(2) *Wait one to several hours for isotopes to equilibrate with body H_2O* — longer for larger body sizes and longer in ectotherms than endotherms.

continued

continued

> (3) *Take a blood (or urine or expired gas) sample and release the animal.*
> (4) *Recapture the animal days to weeks later for a second sample* — longer for larger body sizes and longer in ectotherms than endotherms.
> (5) *Analyze the specific activities of *O and *H in the H_2O from the blood (or in the urine or expired air)* — depending on the isotope used.
> (6) *Calculate CO_2 production and energy respired* — selection of equations depends on details of isotope kinetics such as the number of exchangeable pools and fractionation; conversion to energy respired depends on knowledge of the substrates oxidized.
>
> The DLW method has been shown to be relatively accurate in estimating the mean FMR of free-living animals. As examples of comparisons with alternative methods such as time-energy budget analysis or observation of metabolizable energy intake you can consult Masman et al. (1988) for a bird, Kenagy et al. (1989) for a mammal, and Marken Lichtenbelt (1991) for a reptile. Although Masman et al. (1988) showed very good correlation between DLW estimates and time-energy budget estimates of energy expenditure among individual kestrels (*Falco tinnunculus*), the DLW method's utility for studying variation among individuals within a population has been questioned. In a review, Speakman (1997) concluded that it might have limited ability to define the respiration rate of focal individuals. If this proves true, then this makes it more difficult to correlate respiration with factors that are expected to alter it such as environmental conditions or level of activity.

found that both FMR and growth rate increased with body size (figure 13.7A). In order to tease out the apparent respiratory cost of growth from data confounded by the effect of body size, they analyzed mass residuals of all the individuals, calculated from the regressions of FMR and growth rate on body mass. When they plotted the residual FMR versus the residual growth rate (figure 13.7 B), the slope represented the incremental increase in respiration per unit increment in growth, which was about 1.7–2.0 kJ g^{-1} wet mass gain, or 8.3 kJ g^{-1} dry mass gain because the ratio of wet mass to dry mass was 4.17 (Peterson et al. 1998). With this estimate in hand, they assembled a detailed energy budget of growing snakes and made two startling conclusions. First, the respiratory costs associated with growth were responsible for 58 to 69% of total respiratory costs in small, growing garter snakes. Most of the respiratory costs in these animals went into growing! Second, the respiration rate of these rapidly growing snakes was nearly the maximum respiration rate that was thought to

Figure 13.7. The respiratory cost of growth in free-living garter snakes, *Thamnophis sirtalis fitchi,* in northern California, USA, was estimated in a two-step process. In step 1, shown in the left-hand upper and lower plots A, growth rate and FMR (field metabolic rate, measured with doubly labeled water) were regressed against body size. In step 2, shown in the right-hand plots B, the residuals of FMR were regressed against the residuals of growth rate for all the snakes (top right figure; individual points not shown), and also for a subset of apparently efficient growers (bottom right figure). The slopes of the linear regressions (solid lines) in both cases are about 2 kJ/g of growth. Dotted lines show 95% confidence intervals. (Replotted from Peterson et al. 1999.)

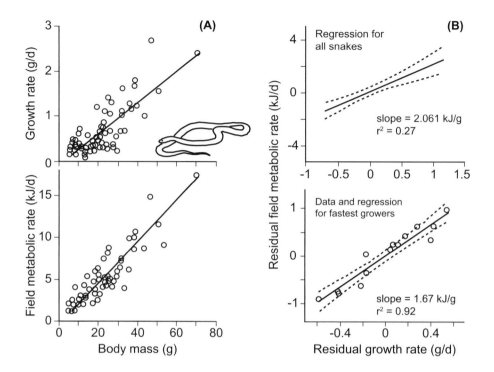

be achievable by these animals (Peterson et al. 1998). In order to trust this remarkable result and understand some of its ecological implications, we will consider briefly some other theoretical and empirical estimates of growth costs and their magnitude in some other animals.

13.3.3 What Is Responsible for Most of the Respiratory Cost of Growth?

Recall that animals continually undergo both catabolism (breakdown) and anabolism (synthesis, chapter 7). Growth occurs when anabolism exceeds catabolism. It is known from stochiometric calculations and bioenergetic measurements on unicellular organisms that synthesis requires energy (Wieser 1994) or, expressed another way, whenever chemical potential energy is transferred from one biomolecular form to another some heat is given off. In the dynamic energy budget approach favored by Kooijman, activities such as growth and reproduction incur overhead costs that are included in the animal's total respiration (Kooijman 2000). The release of heat associated with biochemical transformations was the basis for the heat increment of feeding (HIF) defined in chapter 7 and, in fact, the HIF contains an uncertain proportion of the respiratory cost of production (Parry 1983; Blaxter 1989; Wieser 1994).

In animal energetics, the relationship between the chemical potential energy transferred (P_G; e.g., growth energy stored in new tissue) and the heat released (R_G) is encompassed in the partial efficiency of production (Kleiber 1961), which in some texts is also denoted as K_3 to distinguish it from two other types of efficiencies that are sometimes calculated (box 13.3):

$$\text{Partial efficiency of production for growth} = \frac{P_G}{(P_G + R_G)}, \tag{13.3}$$

Box 13.3. Efficiencies in Nutritional Ecology

An efficiency is a ratio of output to input (chapter 1). For production efficiency, the numerator is the rate at which energy is allocated into new tissue, P (definition according to equation 13.1). The denominator may be the gross energy intake rate (I) as in equation 13.3.1, metabolizable energy intake ($I - E$) as in equation 13.3.2, or the sum of the production energy and the respiratory cost of production (also called the cost of biosynthesis) as in equation 13.3.3. In some texts (Wieser 1994; Wootton 1998) these different efficiencies are designated as K_1, K_2, and K_3, respectively:

$$\text{gross production efficiency} = P / I, \tag{13.3.1}$$

$$\text{net production efficiency} = P / (I - E) = P / (P + R), \tag{13.3.2}$$

$$\text{partial efficiency of production for growth} = \frac{P_G}{\left(P_G + R_G\right)}. \tag{13.3.3}$$

Why would ecologists be concerned with these efficiencies? For one thing, along with exploitation efficiency (the proportion of available food a population consumes) they influence the proportion of energy transferred between trophic levels and thus potentially the length of food chains (Ricklefs 1973). Their use can make the estimation of food intake or growth rate by wild animals more accurate. In production management situations they determine the economy of growing or producing materials from given amounts of food. To illustrate the latter two points, consider the typical utilization plot (as defined in box 3.3) between P_{total} and metabolizable energy intake, which we will call MEI (figure 13.8; modified from Wootton 1998).

At $MEI_{maintenance}$ just enough food is consumed to meet maintenance needs (i.e., $MEI_{maintenance} = R$). At higher or lower intakes the animal

continued

continued

Figure 13.8. The declining slope of the utilization plot (curved solid line), as metabolizable energy intake (MEI) increases, indicates diminishing production returns for progressive increases in food intake. The optimum intake that maximizes growth per unit intake is where the tangent originating at $P = 0$ intersects the utilization plot (dashed line).

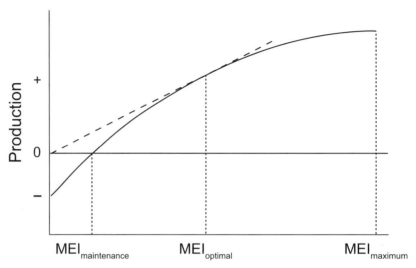

accumulates tissue energy ($P_{total} > 0$) or consumes it ($P_{total} < 0$). The utilization plot is not linear but the slope declines with increasing *MEI*. In domesticated mammals, birds, and fish the conversion of *MEI* into new tissue is less efficient at intakes above $MEI_{maintenance}$ than is the use of body tissue to provide energy for maintenance below $MEI_{maintenance}$ (Wieser 1994; Wootton 1998). The slope of the utilization plot (*dP/dMEI*) defines the partial efficiency of production (Kleiber 1961). You can see that an alternative estimate of partial efficiency is $P/(MEI - MEI_{maintenance})$. In nine species of endotherms, partial efficiencies averaged 0.66 ± 0.10 for $MEI > MEI_{maintenance}$ and 0.82 ± 0.08 for $MEI < MEI_{maintenance}$ (Blaxter 1989). Thus, if one knows the food intake rate or *MEI* of an animal in relation to its maintenance requirement, one can estimate its tissue energy gains or losses by dividing the difference in intake rate by the appropriate partial efficiency. Working in the other direction, if one knows maintenance energy needs and measures a growth rate, one can calculate more accurately the food intake rate.

Although partial efficiency of production is not constant but depends on level of feeding and the substrates synthesized or degraded (Blaxter 1989), Wieser (1994) advocated adoption of a "consensus" value of 0.75. Ricklefs (1974) also recommended using a value of 0.75, but Weathers (1996) reviewed growth studies in birds and considered that the value was too high. Using 0.75 as a guesstimate when no data are available is probably OK, but it is clearly better to make an accurate measurement.

Although partial efficiency of production is not constant but depends on level of feeding and the substrates synthesized or degraded (Blaxter 1989), Wieser (1994) advocated adopting a value of 0.75 for general "back of the envelope" use. Substituting 0.75 as a partial efficiency of production in equation 13.3 yields a respiratory cost of growth (R_G/P_G) 0.33 kJ heat per kJ deposited in new tissue, or 7.26 kJ heat per g dry mass gain assuming that most growing wild animals have 22 kJ g^{-1} dry mass (Wieser 1994). This value is remarkably similar, and hence lends credence, to the value derived in the analysis of growing garter snakes (above).

In a nutshell, a large but uncertain proportion of the respiratory cost of growth measured in the garter snakes and other animals can be accounted for by the heat increment of feeding associated with the extra food consumed to fuel the growth. If growth has been measured in an animal, be it endotherm or ectotherm, then the respiratory cost of growth can be approximated as the product of the energy gain and 0.33 J/J or of the dry mass gain and 7.3 kJ/g. However, because ectotherms have much lower maintenance respiration rates than endotherms, it follows that the relative contribution of the respiratory cost of growth to total respiration will be higher in ectotherms Parry (1983). Furthermore, because animals have highest mass specific growth rates when they are youngest (figure 13.2), it follows that the relative contribution of the respiratory cost of growth to total respiration will be higher in young, small animals Parry (1983). These considerations render plausible the first startling conclusion of Peterson et al. (1999) on energetics of growing garter snakes, that respiratory costs associated with growth were responsible for the majority of total respiratory costs. However, the impact of growth costs was not nearly so great in energy budgets of other growing reptiles studied with doubly labeled water (Nagy 2000), perhaps because growth rates were not so high. Peterson et al. (1999) may have amplified the growth rate of juvenile garter snakes by capturing them and holding them without food (but with water) for 3 to 21 d before they were injected with DLW and released. This kind of accelerated growth following food deprivation is called catch-up growth and is discussed below (section 13.4.1).

13.3.4 Do Animals Need to Make Compensatory Adjustments when Growth Cost Is High?

The other remarkable finding of Peterson et al. (1998) was that when the FMR of the free-living snakes were compared with laboratory measurements of maximum respiration during exercise they were "virtually indistinguishable"! The

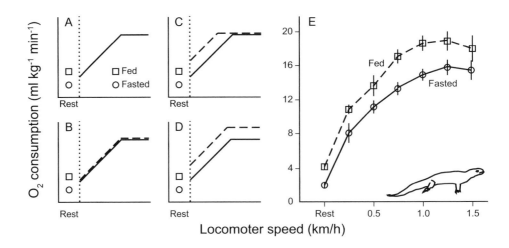

Figure 13.9. In a test of whether the maximum rate of respiration R_{max} can be increased, the hypothetical and observed patterns of respiration rate during rest and exercise (terrestrial locomotion) in fasted and postprandial (after a meal) monitor lizards (*Varanus exanthematicus*) were compared. As background, panel A shows the oxygen consumption rate in fasted lizards resting (to the left of the vertical dotted line) versus speed of walking (to the right of the dotted line). The typical pattern observed in many animals is an approximately linear increase with speed up to an asymptotic maximum rate of oxygen consumption, R_{max}. Panels B and C represent predictions of a model of respiration that assume that expansion in R_{max} is not possible, with priority then given either to exercise (B) or to food processing (C). Fasted lizards are represented by unfilled circles, solid lines, and fed lizards are represented by unfilled squares, dashed lines. Panel D represents the prediction of a model that assumes that expansion in R_{max} is possible. Measurements shown in panel E unequivocally supported the latter prediction (D). The vertical lines through each point are standard errors of the mean. (Redrawn from Bennett and Hicks 2001.)

authors pointed out that "obviously snakes could not have been exercising at their maximal rates . . . for 24 h/d for 2 wk." The high rate of expenditure was probably due to the heat increment of feeding inclusive of growth costs. Respiration rates after a meal ("postprandial") that are as high as respiration during activity have also been measured in other snakes (Secor et al. 2000), in fish (Priede 1985; Korsmeyer and Dewar 2001), and in nestling altricial birds (Chappell and Bachman 1998). The comparison raises an ecologically important question—is there a conflict between digestion and activity? As for most animals engaged in terrestrial locomotion, the respiration rate of moving snakes increases with velocity in a fairly linear fashion to a maximal aerobic speed at which a maximal rate of respiration (R_{max}) is achieved (figure 13.9A). Because the parameter measured is O_2 consumption, this R_{max} is typically called $\dot{V}O_{2(max)}$. If this maximum were set by the ability of the respiratory organs (e.g., gills, lungs) or blood to capture and transport O_2 (Weibel 2000), then there would be a fundamental conflict if the heat increment of feeding inclusive of costs of growth were added to locomotion expenditures and the total exceeded $\dot{V}O_{2(max)}$. Unless the total respiration budget can be expanded, animals must limit feeding and production to permit activity, or limit activity after feeding. For growth in particular, the high respiratory costs of young, growing animals may necessitate trade-offs within the total energy budget (Wieser 1994; Rombough 1994; Nagy 2000).

The total respired energy (R_{Total}) can be broken down to include expenditures for growth (R_G) and reproduction (R_R) plus the minimal expenditures for maintenance (R_S, which is basal or standard respiration rate), for activity (R_A), for thermoregulation (R_T; endotherms only), and for the post absorptive processing of nutrients (R_F) to support the latter three components:

$$R_{Total} = R_S + R_G + R_R + R_A + R_T + R_F. \qquad (13.4)$$

Korsmeyer and Dewar (2001), reviewing studies on tuna and other fish species, wrote that "All fishes have a potential power budgeting problem in that the combined capacities for [each of the components] will likely exceed maximum aerobic capacity." Thus, researchers have suggested that fish and aquatic invertebrates resolve the conflict by either slowing down feeding and growth and extending it over a longer period to reduce its instantaneous cost, or by compromising activity in favor of digestion (Priede 1985; Korsmeyer and Dewar 2001; Owen 2001; Whiteley et al. 2001). Bennett and Hicks (2001) came to an opposite conclusion based on their studies of reptiles. Using monitor lizards (*Varanus exanthematicus*), they tested the predictions of contrasting models of respiration that assume expansion in R_{max} is not possible, with priority then given either to exercise (figure 13.9B) or to food processing (figure 13.9C), or expansion in R_{max} is possible (figure 13.9D). Their results (figure 13.9E) unequivocally supported the latter prediction (figure 13.9D). Unlike the situation suspected for aquatic species, neither exercise nor nutrient processing was compromised at the expense of the other.

There is room for much more research on energy budgets of growing animals, and the findings have many interesting and provocative implications in ecology, behavior, and physiology. Here are a few. For ecology, field studies of neonate reptile time-activity budgets are needed to determine how they sometimes appear to achieve unexpectedly low field metabolic rates while simultaneously growing rapidly (Nagy 2000). For behavior, detailed energy budgets of altricial nestling birds are needed to understand whether begging behavior is "costly" which is assumed in many evolutionary models (Johnstone and Godfray 2002). On the one hand, begging behavior apparently can be energetically costly as it can reduce the growth rate of birds (Kilner 2001; Rodriguez-Girones et al. (2001), but on the other hand the actual rates and durations of energy expenditure seem too low to explain the effect (Bachman and Chappell 1988; Chappell and Bachman 1998; Chappell and Bachman 2002). For physiology, as pointed out by Bennett and Hicks (2001), the finding that the total budget for respiration can be expanded, as in the monitor lizards, is a significant caveat to the use of exercise alone to define $V_{O_{2(max)}}$, and a significant challenge to the hypothesis of symmorphosis, which proposes that organisms are designed without excess capacity (chapter 3, box 3.4). And, as pointed out by Wieser (1994), the apparent fixed size of the respiration budget of growing larval fish suggests that the net growth efficiency (equation 13.2) may change during development, but how?

13.4 Rates of Growth

Why do some animals grow faster than others? Students seeking answers to this longstanding question need to be clear about what they are comparing and whether they seek proximate or ultimate answers. When researchers use field data on growth rates they risk confounding environmental effects on growth, such as differences in food availability or microclimate, with the intrinsic growth rate characteristic for a species (Arendt 1997). When comparing differences in intrinsic growth rate, one can ask about the physiological mechanisms by which more rapid growth is accomplished or about the selective value of different rates of growth (Case 1978). Remain aware of these important distinctions in the following discussion of growth rates.

13.4.1 Most Animals Seem to Have Intrinsic Growth Rates that Are Genetically Programmed

Besides the large scale differences in growth rate due to allometry, metabolic pattern (endothermy versus ectothermy), and developmental mode (altricial versus precocial) that we discussed above (section 13.3.1), there can still be substantial differences in growth rate intraspecifically between populations and between closely related species. The Atlantic silverside *Menidia menidia* provides an example of population differences in intrinsic growth rate. The range of this species extends along the eastern seaboard of North America from Gulf of St. Lawrence to northern Florida, and so the species experiences a gradient in mean water temperature and also differences in length of the growing season (Conover and Present 1990). Individuals from northern populations, relative to those in southern populations, hatch later in spring and experience a shorter growth season, yet they gain mass 2–3 times faster (Present and Conover 1992) and so by the end of the growing season their size is similar to that of individuals in the southern populations. The differences in growth cannot be simply explained by differences in food availability or temperature. Under controlled conditions larvae from the northern populations grow faster than larvae from southern populations at both low and high temperatures when food is in excess (Conover and Present 1990). A genetic basis for the difference was demonstrated by using larvae from laboratory stocks reared in a common environment for several generations.

Results such as these suggest that normal growth rate is not at the physiological maximum but rather that it is more regulated, and operates in response to a growth program that is perhaps optimal for the animals' particular environment. A reasonable adaptive interpretation for the faster intrinsic growth of northern *Menidia*, for example, is that the high growth rate in the northern population compensates for the shorter growing season and allows the young to reach a size that improves their chances of surviving their first winter (Conover and Present 1990; Wootton 1998). But this begs the question of why individuals in the southern population do not grow as fast as possible. Why are they holding back? Case (1978) and Arendt (1997) have raised this as a major question about intrinsic growth rates: "Why are most growth rates not maximized (i.e., what are the likely trade-offs)"? Life history theory suggests that low growth rate is favored when juvenile mortality is low relative to adult mortality (Case 1978). If very rapid growth is achieved by behaviors (e.g., increased activity) that increase risk of predation, then a program of restrained feeding and growth may have higher fitness than a program that maximizes short-term growth rate (Houston and McNamara 1990).

A related issue is how animals respond to ephemeral periods of poor food conditions. If growth is limited by food supply, for example, can animals adjust by showing a marked growth spurt when supply is no longer limited? This type of growth is usually referred to as catch-up or compensatory growth (Goss 1978, figure 13.10). Such accelerated growth following a period of food restriction

Figure 13.10. After a period of limited growth, perhaps due to food limitation, animals may or may not exhibit catch-up growth. In the left-hand figure, the normal growth curve is represented as a dotted line, and the solid line connecting filled circles shows catch-up growth after a period of 2 time units of no growth, perhaps due to food limitation (gray-filled circles). The dashed line connecting unfilled squares shows a different possible growth pattern, in the absence of catch-up growth following a period of food limitation (gray-filled squares). Still another no-catch-up alternative might be achievement of normal asymptotic mass at a much later age. The right-hand figure shows a key test for catch-up growth, which would be higher than normal growth rate (i.e., mass/time). As before, the normal growth rate is represented as a dotted line.

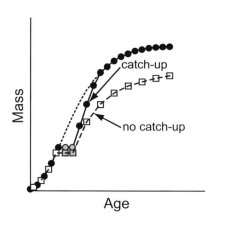
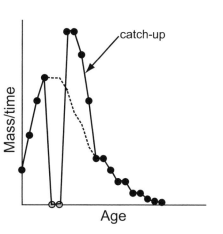

would be another kind of evidence that the normal growth rate is not at the physiological maximum but rather is more regulated (Schew and Ricklefs 1998). Understanding whether catch-up growth occurs is important for predicting growth responses to changing ecological conditions and the impacts of ecological food limitation (Metcalfe and Monaghan 2001). Reports of catch-up growth exist for species in numerous taxa including invertebrates (Calow and Woolhead 1977), amphibians (Alford and Harris 1988), fish (Dobson and Holmes 1984; Jobling 1994; Wootton 1998), and mammals (Wilson and Osbourn 1960). Tests with altricial bird species indicate limited ability to exhibit catch-up growth, but it apparently occurs in some avian species with precocial young (Schew and Ricklefs 1998; Lepczyk and Karasov 2000). Blaxter (1989) points out that in a variety of species exhibiting accelerated mass gain, the tissue composition of the gain is not normal; water and protein contents are relatively higher and fat content is lower relatively lower than in the normal growth of an animal of the same mass.

High responses to artificial selection for growth rate have been found in invertebrates (Bayne et al. 1999), fish (Wootton 1998), birds (Jackson and Diamond 1996), and mammals (Blaxter 1989), which supports the view that there is a genetic component (program) that determines growth patterns. These responses to artificial selection suggest that natural selection for similar traits can occur in natural populations. Studies with lines of animals selected for rapid growth offer answers to key questions about intrinsic growth rates, such as "What are the mechanism(s) by which some are faster than others," and "Why not grow as fast as possible?" As for the first question, the mechanisms by which individuals from genetically selected lines grow faster may include hyperphagia, changes in components of the respiration budget (equation 13.3), or even improvements in the net efficiency of growth. In the fish *Menidia menidia*, genetically based natural latitudinal differences in growth rate were due to differences in feeding rate and possibly in net growth efficiency (Present and Conover 1992). Similar mechanisms are observed in species that have been subject to artificial selection. For example, in quail (*Coturnix;* Lilja et al. 1985) and chickens (*Gallus;* Jackson and Diamond 1996), hyperphagia is one of the major changes that results from artificial selection for more rapid growth, and it is associated with an increase in the relative size of the digestive organs. Rock oysters (*Saccostrea commercialis*) selected for rapid growth feed at up to twice the rate and with greater net growth efficiency than nonselected oysters, but have similar resting rates of respiration (Bayne et al. 1999; figure 13.11). How can net growth efficiency be increased? Several studies (reviewed by (Bayne et al. 1999; Bayne and Hawkins 1997) suggest that reduced rates of protein synthesis and degradation (turnover), but with synthesis still exceeding degradation, can yield high growth rate with reduced energetic cost.

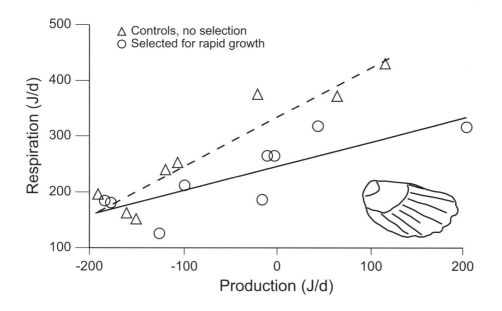

Figure 13.11. Oysters (*Saccostrea commercialis*) that were selected for rapid growth seem to have higher net growth efficiency. This is suggested from the lower respiratory energy losses of oysters selected for rapid growth (unfilled circles, solid line), relative to non-selected oysters (unfilled triangles, dashed line), at production rates above zero. The gentler slope for oysters selected for rapid growth is consistent with the idea that they have a higher partial efficiency of production (see methods box 13.3 for further discussion of production efficiencies). (Figure replotted from Bayne et al. 1999.)

Presumably there are biochemical trade-offs with this strategy because otherwise all animals would engage in them, but what are they? While no one has a certain answer to this question, studies with animals selected for rapid growth do suggest some physiological reasons why growth may not always be maximized (table 13.1). In some species selected for rapid growth there are apparent trade-offs in rate of tissue maturation, and increases in developmental errors, susceptibility to disease, and other health problems, and overall lower reproductive success. Another approach to assess trade-offs of rapid growth might be to administer key hormone(s) that control growth (e.g., growth hormone) and study the effects (Ketterson and Nolan 1994; Ketterson et al. 1996), reasoning that the alterations in hormone level(s) mimic the phenotypic expression of certain genes. This approach, called "phenotypic engineering" by Ketterson and Nolan (1994), was also called "phenotypic manipulation" by Sinervo and Svensson (1998), who presented some case studies of its application and discussed the strengths and weaknesses of this approach relative to those of selection studies.

The number of studies that investigate how physiology translates into demographic costs in life history is relatively small. For researchers interested in questions about evolution of growth, selection studies and/or phenotypic manipulation usefully complement other field and laboratory studies that seek to understand the relative contribution of intrinsic and extrinsic constraints on growth. In some cases the information will have implications for conservation. We are in the midst of enormous global experiments on how our activities affect life history features of aquatic organisms and on the consequences of such

TABLE 13.1

Some physiological reasons why growth is not always maximized, from studies with animals selected for rapid growth

Putative trade-off	Type of study in which putative trade-off was demonstrated
Between growth and differentiation	Japanese quail selected for rapid growth have lower skeletal muscle capacity for thermogenesis (Shea et al. 1995).
Between growth and "quality control" during development	Within a line of rats selected for rapid growth, more developmental asymmetries in morphology were observed, indicative of developmental errors (Leamy and Atchley 1985)
Between growth and defense	Strains of mice selected for rapid growth had higher susceptibility to external parasites than nonselected strains (Bradford and Famula 1984). Strains of salmon selected for rapid growth tended to have higher susceptibility to bacterial infection than nonselected strains (Smoker 1986). Selection of turkeys for fast growth rate was accompanied by a reduction in specific immune responses (Bayyari et al. 1997).
Between growth and survival	Strains of mice selected for rapid growth had shorter life spans and more tumors than nonselected strains (Eklund and Bradford 1977). Coronary arterial lesions were less abundant and less advanced in a cultured strain of Atlantic salmon that grew more slowly than another cultured, fast growing strain (Saunders et al. 1992).
Between growth and reproduction	Embryo mortality was higher in eggs of quail selected for rapid growth compared with nonselected lines (Peebles and Marks 1991). But, cattle selected for higher growth had similar reproductive output as control lines (Mercadante et al. 2003)

Source: Adapted from (Arendt 1997).

changes (Wootton 1998; Zimmer 2003). When we select or genetically engineer for rapidly growing organisms in aquaculture, and they escape to the wild, what will be the consequences for related wild populations with which they genetically mix? If the tradeoffs suggested in table 13.1 are general and strong, perhaps the effects will be minimal because individuals with rapid growth traits will be strongly selected against in the wild. In the meanwhile, however, the high abundance and body size of escapees might mask habitat degradation, enhance predator populations, and encourage fishery exploitation to increase to a level beyond what the wild population can support (Myers et al. 2004).

13.4.2 A Species' Realized Growth Rate May be Greatly Influenced by the Physical Environment

Temperature, moisture, and nutrients are key environmental factors that could potentially exert influences on growth rates and life histories of animals. A deviation from the normal growth pattern is an example of phenotypic plasticity. Two types are often distinguished, called imposed and induced variation by Schew and Ricklefs (1998) or phenotypic modulation and developmental conversion, respectively, by Smith-Gill (1983).

- Imposed variation is a direct result of environmental change on growth, as when too little energy is available for the normal allocation of energy to new tissue.
- Induced variation results from an active response by a neonate's genetic program to a change in the environment. For example environmental cues might be used to initiate an alternative growth developmental program (Smith-Gill 1983).

The key difference is whether there are one or more growth/developmental programs. Schew and Ricklefs (1998) point out that it is often difficult in practice to determine when phenotypic changes are imposed versus induced.

The following examples of phenotypic plasticity in growth correspond most closely to the situation of imposed variation, but we will also discuss below induced variation. Because in previous sections of this book we have provided examples of nutritional limitation of growth (by water, N, and minerals), we will focus here on energy. As reviewed by Porter and Tracy (1983) and Porter et al. (1994, 2000), progressive increases in sophistication of biophysical models of animals and the size of the empirical data base on animal energetics has led to substantial current ability to accurately model mass and energy fluxes of animals under different environmental conditions. We will present two of the most basic, important points and forgo some of the most technical details.

13.4.3 Discretionary Energy for Growth and Production Varies with Temperature

A simple graphical model makes it is easy to understand why the energy available for growth will often be temperature dependent (figure 13.12). The net absorbed energy available for allocation to maintenance respiration (R_S) and growth (P_G) depends on temperature, at least in ectotherms, because of temperature-dependent variation in intake and digestion (chapter 4). Endotherms, with relatively constant body temperatures, are considered to have relatively temperature-independent maximum rates of metabolizable energy intake, perhaps close to values listed in table 3.1, or lower and more variable if

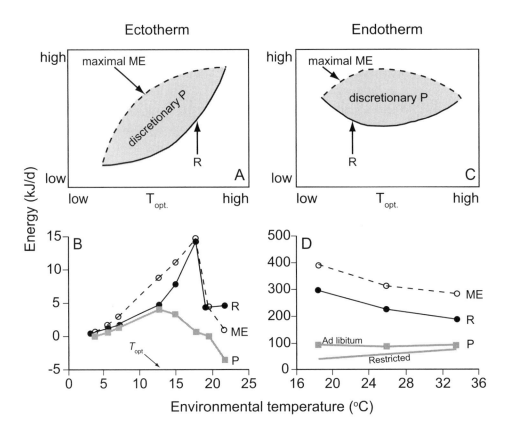

Figure 13.12. A simple graphical model predicts that the discretionary energy available for production (sometimes called "scope for growth") will often be temperature-dependent. (A) For ectotherms (top, left), the presumed relationship between temperature and maximal metabolizable energy (ME; $= I - E$ in equation 13.1) is positive, as is energy expended in respiration (R). The difference between absorbed and expended energy is discretionary energy for production (P). (B, bottom left) Energy flow in brown trout (from Elliott 1976) corresponding to the model. Notice that the temperature that corresponds to maximum discretionary energy (optimum for production, $T_{opt.}$, is below the temperature of maximum ME. (C; top right) In endotherms (top, right) maximal ME is assumed to be depressed at very high and low environmental temperatures, and energy expended in maintenance respiration increases with declining temperature and may also increase with high temperature during evaporative cooling. (D; bottom right) Energy flow in a poultry chick (from Olson et al. 1972) corresponded to the model when ME was restricted to 300 kJ/d, but when chicks were allowed to feed ad libitum they maintained P constant with declining temperature by increasing ME (unfilled circles, dashed line). (Panels A and C modified from Porter et al. 2000.)

their appetite and foraging behavior varies with temperature. The respired energy expended for maintenance is also temperature dependent in all animals, although the particular pattern differs between ecto- and endotherms (chapter 1).

The difference between the two curves of intake and expenditure is an oval area representing the discretionary energy for growth or other production (P).

Curves such as these have been observed in many species of ectotherms (e.g., Elliott 1976; Foyle et al. 1989; Jonsson et al. 2001) and a few species of endotherms (e.g., Olson et al. 1972; Smith and Teeter 1993; Mandal et al. 1999 (figure 13.12). This model, which predicts growth as a function of temperature, has been applied extensively for aquatic species because it is so straightforward to characterize the effective environmental temperature, which is simply the water temperature. One of the most extensive studies were those of J. Elliott during the 1970s on the brown trout (*Salmo trutta*), which are summarized in his book (Elliott 1994). Notice that the temperature at which specific growth rate was maximized was lower than that at which the rate of food consumption was maximized, because expenditure for maintenance respiration rose rapidly at higher temperatures (figure 13.12B). Elliott found that the temperature at which specific growth rate was maximized was dependent on feeding level, declining when food was limiting. Although this inserts some uncertainty into ecological predictions, it turns out that this curve predicts very well the growth rate in the wild. For example, differences in the mean growth rates of trout populations in most locations are explained by differences in temperature regimes (Edwards et al. 1979; Elliott 1994). Kitchell developed a similar model for yellow perch (*Perca flavescens*) (Kitchell et al. 1977) which proved equally successful and has been adapted as a computer package applicable to many aquatic species (Hanson et al. 1997). Most impressively, the model has been usefully applied in many different kinds of ecological studies and management scenarios (Hewett and Johnson 1992; Wootton 1998) (table 13.2).

13.4.4 Environmental Conditions Influence Energy Gain because They Constrain Activity Time Budgets

More time available for feeding increases daily food intake and net absorption from food in both ectotherms (Adolph and Porter 1993; Sinervo 1990; Sinervo and Adolph 1994) and some endotherms (Kvist and Lindstrom 2000). In terrestrial ecosystems the available activity time for feeding is sometimes limited because of adverse environmental conditions that jeopardize heat balance. Small animals especially must stay in or near heat balance, because they have low heat capacities (low thermal inertia), and imbalances can cause rising or falling body temperature (T_b) and bring them to critical limits rapidly. Small animals can utilize heterogeneous thermal environments by shuttling, for example, between shaded microenvironments that cool them and unshaded microenvironments that heat them, but shuttling can reduce time for foraging

TABLE 13.2
Interesting applications of bioenergetics models in aquatic ecology

Ecological application	References
Predicting impact of predation:	
A model was used to predict accurately that stocking Lake Michigan with predatory salmonid fish would result in a decline in primary prey species alewife (*Alosa pseudoharengus*)	(Stewart et al. 1981; Kitchell and Crowder 1986)
Explaining environmental impacts:	
In a pond receiving heated effluent, a direct effect of temperature was ruled out as the cause of the decline in condition of largemouth bass; reduced prey availability was the likely explanation	(Rice et al. 1983)
Explaining geographic range limits:	
In snow crabs (*Chionoecetes opilio*) the discretionary energy for production is available over a very limited temperature range, which limits them energetically to cold water	(Foyle et al. 1989)
Defining habitat requirements:	
Models of production versus intake and temperature are used in habitat indices for salmonids	(Elliott 1994)
Predicting impacts of global warming:	
A model was used to examine how climate warming would affect consumption and growth of fish species in the Great Lakes	(Hill and Magnuson 1990)
Modeling ecosystem transfer of contaminants:	
Bioenergetics model of alewife (*Alosa pseudoharengus*) was used to model contaminant levels in invertebrate prey and ecosystem biomagnification	(Jackson and Carpenter 2003)
Managing fisheries:	
Model was used to determine which species, what sizes, and at what times to stock pike in reservoirs with different thermal regimes	(Bevelhimer et al. 1985)

Source: Adapted from Hewett and Johnson (1992); Wootton (1998).

if not all the microenvironments offer the same opportunities for feeding. Although these principles are well established and demonstrated for terrestrial ectotherms (Porter et al. 1973; Porter and Tracy 1983; Adolph and Porter 1993), they apply as well to many small endotherms like desert ground squirrels (Chappell and Bartholomew 1981) and desert birds (Wolf and Walsberg 1996; Walsberg 1993). In fact, even endotherms as large as moose, which have very large heat capacities making them much less sensitive to heat imbalances, sometimes reduce feeding opportunities by seeking particular microclimates for thermoregulation (Schwab and Pitt 1991; Demarchi and Bunnell 1995).

The effect of environmental conditions on T_b is a complex function of heat exchange by conduction, convection, radiation, and evaporation. It can be modeled as a sum of inputs and outputs as we have done for other types of mass and energy flow. A useful term that incorporates many of these fluxes is operative environmental temperature (T_e) (methods box 13.4). This measure of

Box 13.4.

Thermal energy is exchanged between the surfaces of animals and their environment by radiation, conduction, convection, and when water evaporates or condenses. Figures 13.13 and 13.14 help explain two major reasons why it is complicated to predict these exchanges for free-living terrestrial animals.

First, the thermal environment can be very complex and heterogeneous over space and time. In the example in figure 13.13, snakes sitting on the

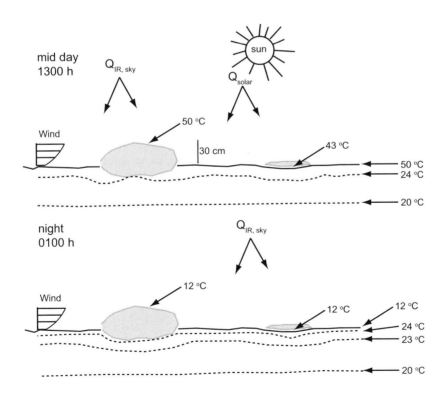

Figure 13.13. The thermal environment can be very complex and heterogeneous over space and time. These figures illustrate the temperatures in microhabitats for snakes (on top and beneath thick and thin rocks, on or under soil) at midday (1300 h; upper figure) and at night (0100 h; lower figure). Notice that wind velocity, which affects convective heat exchange, increases with height above the ground, and soil temperature, which influences conductive heat exchange, decreases with soil (burrow) depth. See the text for definition of terms. (Adapted from Huey et al. 1989.)

continued

continued

continued

Figure 13.14. Models of varying complexity are used to characterize heat exchange between animals and their environments and to estimate body temperature and respiration (Porter, et al. 2000). See the text for definition of terms. In this very simple model, the animal's shape is assumed to be spheroid, endogenous heat production from respiration is uniform through the body, and many details are unspecified, such as how radiant energy absorbed at the fur/feather–air interface is transmitted to the animal. (Adapted from Porter et al. 2000.)

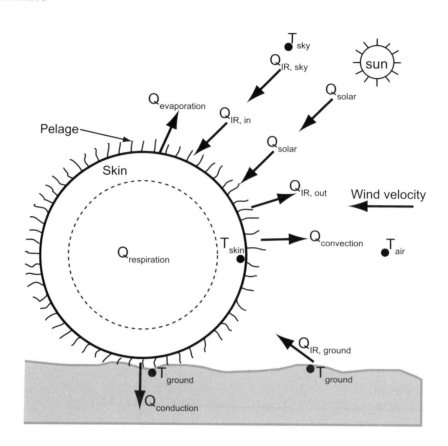

soil surface, climbing above it, or descending beneath it experienced different soil and air temperatures and wind speed, which influence conduction and convection. Snakes sitting in and out of shade or underground experienced different radiant fluxes from the environment.

Second, animals have features that influence heat exchange in complex ways. Body size and shape, and the presence and color of scales, hair, and feathers influence the avenues of heat flux (Q) and complicate the definition of the animal's temperature (T) at different points on the body. The model in figure 13.14 is very simple, and other models take into account heat flow across layers of the body with different properties, such as the interfaces between the body core and the skin, skin and pelage, and pelage and the environment (Porter et al. 2000).

Ecologists interested in climactic constraints on time and energy budgets have taken two approaches to deal with the complexity. One is to measure as much as possible — the environmental variables (e.g., air temperature, wind

continued

continued

speed, short- and long-wave radiation) and the crucial animal properties (e.g., body size, shape, orientation, pelage characteristics, surface reflectivity to radiant fluxes) — and then construct a fairly complete heat budget (Porter and Gates 1969; Campbell 1977). The basic skeleton looks like this:

$$Q_{solar} + Q_{IR,in} + Q_{respiration} = Q_{evaporation} + Q_{IR,out} + Q_{convection}$$
$$+ Q_{conduction} + Q_{storage} \qquad (13.4.1)$$

where all fluxes Q are in W/animal and where Q_{solar} is radiant heat gain from direct sunlight, $Q_{IR,in}$ is radiant heat gain from infrared radiation, $Q_{respiration}$ is heat generated by respiration, $Q_{evaporation}$ is heat lost by evaporation, and $Q_{convection}$ and $Q_{conduction}$ are net heat fluxes by convection and conduction, respectively. If the other fluxes do not balance (i.e., sum to zero) then there is heat storage or loss ($Q_{storage}$) from the body, which corresponds to a rising or falling body temperature (T_b). Each flux in this equation is itself described by an equation with parameters that include environmental conditions and features of the animals. Each of the submodels can vary in detail according to the assumptions and simplifications made. With this approach an ecologist might determine the respiration rate or T_b of an animal in specified environmental conditions, or what microclimate(s) should be selected to maintain respiration rate or T_b within certain limits. This approach requires many kinds of environmental sensors (thermometer, anemometer, radiometer, spectral refractometer) and facility with mathematics.

There is another approach for ecologists that either lack the requisite material and skill for constructing complete heat budgets or do not desire to do so. In this approach the thermal features of complex natural environments are incorporated within a single temperature index, called operative environmental temperature T_e. T_e equals the sum of air temperature (T_a) and a temperature increment or decrement, ΔT_R, that subsumes radiative, conductive, and convective factors, but not respiration or evaporation:

$$T_e = T_a + \Delta T_R. \qquad (13.4.2)$$

T_e is the effective temperature of the environment for a specific animal at a specific place and time, combining conduction, convection, and radiation, or expressed another way, it is the equilibrium temperature of an animal in the absence of heat flux from respiration or evaporation/condensation, e.g., a typical terrestrial ecotherm (Bakken 1976). The theoretical rationale for T_e, and its calculation from the complete heat budget above, have been described (Bakken 1976; Campbell 1977). What makes it most valuable is

continued

continued

that it can be measured using a suitable model or taxidermic mount of the animal as a T_e thermometer. A range of devices have been used as T_e thermometers by inserting thermocouples into hollow metal casts of the animal painted to match the animal's radiant energy absorptivity across the solar spectrum (290 – 2600 nm), into freeze-dried animals, and into closed cylinders constructed of plastic or metal and appropriately painted (Walsberg and Wolf 1996a). Color matching can be validated through measurements using a spectral refractometer. Additional validation is important, and this can be achieved by placing animals and models in the same environment and comparing their equilibrium temperatures (Grant and Dunham 1988; Walsberg and Wolf 1996a). Bakken (1992) provides examples of the myriad uses in ecology of T_e thermometers, including mapping the spatial pattern of the thermal environment, testing computed heat budget models, quantifying the benefits of thermoregulatory adjustments, and expressing environmental conditions as a single-number index for multivariate studies.

Analogous physical models are sometimes made for endotherms by constructing a heated taxidermic mount, calibrating it in the laboratory for the range of conditions to be encountered, and then placing it in the environment to determine the power requirement for maintaining a particular T_b (Bakken 1976). These mounts also require validation, and several should be used because there is variation in their response (Walsberg and Wolf 1996b).

environmental conditions has made it easier for ecologists to map the spatial pattern of the thermal environment, quantify the benefits of behavioral or physiological thermoregulatory adjustments, and express environmental conditions as a single-number index for multivariate studies (Bakken 1992). Let us consider its application to the question of population differences in growth rate.

Populations inhabiting sites with different environmental conditions will potentially have different daily and seasonal activity periods and hence different potential daily energy intakes and growth rates. Dunham and his colleagues studied energetics of three populations of the insectivorous desert lizard *Sceloporus merriami* along an altitudinal gradient in Texas, United States (Grant and Dunham 1988; Dunham et al. 1989; Grant and Dunham 1990; Beaupre et al. 1993; Beaupre and Dunham 1995). Growth rates were significantly higher at the mid-elevation site, and they wondered whether thermal constraints on time-energy budgets might explain the differences among sites, because there were no significant differences in feeding strikes per h active. They measured T_e's of various microclimates using painted, hollow-body, copper models of the lizard that accurately estimated microclimate temperatures. By placing the models randomly at each site they established two key points. First, at the high, elevation site T_e remained cool enough for lizards to be active all day on the surface (about 13.2 h/d) whereas at the lower elevations T_e's at midday were too hot for surface activity (figure 13.15). At mid elevation the lizards had to retreat to cooler microclimates amid rocks or underground for nearly 5 h/d and at the low-elevation site it was too hot for foraging for up to 9 h/d. This could explain why

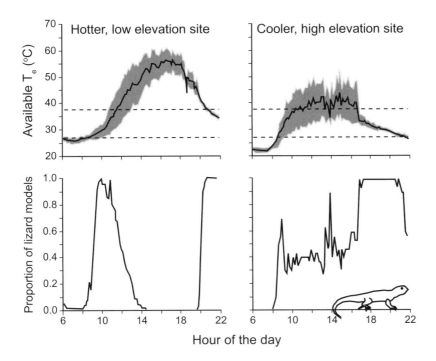

Figure 13.15. Operative temperatures (T_e's) in the habitat of *Sceloporus merriami,* measured with painted, hollow-body copper models of the lizard, can be used to determine thermal constraints on activity time. Top figures: Notice the higher T_e's midday at the low-elevation site, relative to the high-elevation site. Within each of these top figures, the dark lines represent the average of the T_e's measured by many models placed in each habitat, and the gray-shaded regions enclose ±1 standard deviation. The dashed lines forming the interval 27–38°C approximate the range of body temperatures the lizards tolerate. Lower figures: The lower plots show the fraction of all measured T_e's within that interval of 27– 38°C. Low fractions indicate relatively more thermal constraint on lizard activity because few microhabitats are cool enough. The authors assumed that a fraction of 0.1 represented the minimum frequency of favorable thermal microclimates for activity. Lizards at the high-elevation site could be active at the surface all day, but at low elevation the available activity time was curtailed. Activity was also curtailed at a mid-elevation site (not shown). At 2130 activity ended at both sites, and lizards sought out retreats in rock crevices that were cooler at the high-elevation site (24.4°C) than at the mid (30.3°C) or the low-elevation site (31.5°C). (Redrawn from Grant and Dunham 1990.)

lizards grew faster at mid compared with low elevation. Second, during the inactive period high-elevation lizards spent the night in rock crevices at around 24°C, which severely limited their digestion rates relative to lizards at lower elevations that spent the night in crevices at 30–32°C (Beaupre et al. 1993). Thus, lizards at mid elevation grew fastest most likely because they were not as time limited as low-elevation lizards due to thermal constraints, and because they were not process limited like lizards at high elevation.

A number of more recent studies have confirmed that thermal constraints acting through both time and process limitation can explain variation in growth rate among populations of *Sceloporus* lizards, although these studies are also finding evidence of genetic differences in intrinsic growth rate as well (Sinervo 1990; Sinervo and Adolph 1994; Angilletta 2001; Niewiarowski 2001). Nonetheless, relatively simple models of temperature-dependent energy budgets can do a good, although not perfect, job of predicting lizard production in the wild, as was the case for fish (above). For example, among most populations of *Sceloporus undulatus* differences in annual egg masses and fecundities, and in age but not size at maturity, are explained by differences in

---

Box 13.5. Thermal Time

It is well documented in ectothermic animals that growth rate and development time depend on temperature (Atkinson 1994). The development rate of leopard frogs (*Rana pipiens*) from fertilization to beginning of first cleavage *versus* temperature shows the typical pattern (figure 13.16A; redrawn from Gilbert and Raworth 1996). There is a lower threshold or base temperature (T_{base}) of around 5°C below which no growth or development occurs, but above T_{base} the rate of growth or development (time $^{-1}$) increases up to a maximum, levels off, and then finally declines (discussed also in section 6.3.3). The value of T_{base} is often derived by reverse extrapolation of the fairly linear part of the relationship to the x-intercept where rate = zero. We say "fairly linear" because the expected relationship for processes depending on enzyme reaction rates is an exponential increase with temperature T (chapter 1), which is apparent in this figure prior to the leveling off. But if we accept the linear approximation, then the relationship between rate and temperature can be described by a simple model

$$\text{rate} = (1/U)(T - T_{base}) \tag{13.5.1}$$

where U, which is the inverse of the slope, is a thermal constant for development that has units of degree-days. U has also been called a thermal

continued

continued

Figure 13.16.

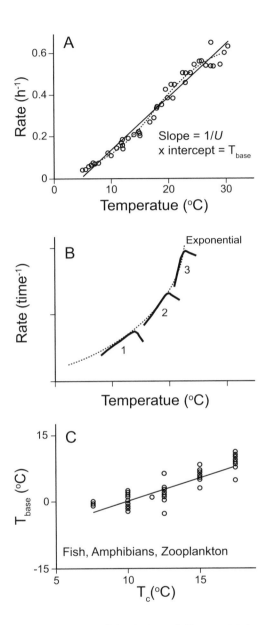

time requirement, a heat sum (Trudgill and Perry 1994) and a sum of effective temperatures (Honek 1999). It is a rather nifty approximation, because once it is measured for a population one can predict the rate of growth or development as a function of ecologically relevant temperatures, constant or not. If an animal has a particular developmental process

continued

continued

to complete and environmental temperature(s) is/are known, then the necessary time can be predicted from the inverse of the above relationship. Because of its utility, this model has been applied widely to understand development time in animals and plants.

The parameters U and T_{base} can vary among populations within a species (Trudgill and Perry 1994) and can change when animals are exposed experimentally for many generations to different temperature regimes (Huey et al. 1991). Among similar species, U and T_{base} are often negatively correlated, U increases with body size (Trudgill and Perry 1994; Honek 1999), and T_{base} is often positively correlated with natural environmental temperature (Honek 1999). It turns out that all of these patterns among species are somewhat predictable from allometric principles like those outlined in box 13.1 and the basic biochemical kinetics of metabolism. This was shown by Charnov and Gillooly (2003), who modified the growth model presented in box 13.1 by incorporating a temperature effect (growth rate $\propto M^{3/4} e^{-Ei/kT}$; see also chapter 1) and then applied a linear approximation. We can easily appreciate most of their findings by foregoing the formal derivation and inspecting their simple graphical model (figure 13.16B, redrawn from Charnov and Gillooly 2003). It shows that even though body-mass-corrected rate increases exponentially with temperature, a given population or species occupies only a certain interval along the curve (three intervals are shown), resulting in an approximately linear relation for each. If you extrapolate each back to its respective T_{base}, and if you consider how the slopes (which are the inverse of U) are changing, then you realize that larger values of U (lower slopes) will occur with lower values of T_{base}. This explains the often -observed negative correlation between U and T_{base}, referred to above. Also, if for each population/species you define the midpoint of its temperature interval as its mean temperature of development (T_c), you can see that there will be a correlation between T_{base} and T_c, as shown for fish, amphibians, and zooplankton in the bottom figure of 13.16 (redrawn from Charnov and Gillooly 2003). In the figure, the x axis = T_c and the y axis = T_{base}. This explains the often-observed positive correlation between T_{base} and natural environmental temperature, referred to above. Charnov and Gillooly (2003) noticed that T_{base} is about 10°C below T_c, which prompted them to call it the "10°C rule."

temperature regimes (Adolph and Porter 1993; Adolph and Porter 1996). This approach integrating time and temperature is analogous to degree-day measures used for insects and marine invertebrates (methods box 13.5) (Taylor 1981; Kingsolver 1989; Roff 1992). Ever more sophisticated models are being introduced that integrate the heat budget with mass budgets for nutrients and water to predict population dynamics (Grant and Porter 1992; Porter et al. 1994, 2000).

13.5 Growth in Relation to Life History Transitions

The picture that emerged in the last section is that in many species growth proceeds in relation to a program, within constraints set by environmental conditions such as temperature or food or other resource availability. Expressed in terms of an energy budget (equation 13.1), we might say that temperature and food availability affect P_{Total}, and growth then reflects the proportional allocation of P_{Total} to growth versus storage, and reproduction. Life histories have been described as heritable sets of rules that determine such age-specific allocations (Dunham et al. 1989). Allocation in the dynamic energy budget approach favored by Kooijman is described by the "κ rule," according to which an age-independent proportion κ of energy utilized is spent on maintenance plus growth, and the remaining portion 1 − κ on development plus reproduction (Kooijman 2000). Many animals seem to have fairly distinct ontogenetic transitions that reflect changes in the allocation, such as metamorphosis or reproductive maturation, and these can be characterized by the age and size at which they occur. For many animals, the combinations of age and size at which life history transitions occur can change in response to environmental conditions. Risk of predation, for example, has been shown to speed up growth and/or development in both ectothermic vertebrates (Wilbur and Collins 1973; Werner 1986; Reznick 1996) and invertebrates (Crowl and Covich 1990; Ball and Baker 1996), and the speed-up reduced their subsequent risk relative to individuals that did not speed up. Considering our emphasis so far on food availability and temperature as key environmental features, we will focus on these as factors causing change in life history transitions.

13.5.1 In Most Ectotherms, Decreased Food Intake Results in Slower Growth and Delayed Maturation at a Smaller Size

This pattern has been observed in studies with both invertebrates (e.g., Twombly 1996; Hentschel and Emlet 2000) and vertebrates (e.g., Reznick 1990; Leips and Travis 1994; Morey and Reznick 2000), although not universally (O'Laughlin and Harris 2000). For example, in a study of spadefoot toads, the pattern was very apparent in one species (*Scaphiopus intermontanus*) but not the other (*S. couchii*) (figure 13.17). Because of the generally-held perception that for most species decreased food increases age at maturity or reproduction (Day and Rowe 2002), researchers have sought a general model to explain the pattern. A feature that could explain it is a developmental threshold, such as a minimum size or condition that must be attained before

Figure 13.17 Decreased food results in slower growth and delayed maturation at a smaller size in many, although not all, ectotherms. In these plots of size versus age in two species of spadefoot toads (*Scaphiopus spp.*), the asymptotes denote when metamorphosis occurred. In both species decreased food resulted in slower growth, and age at metamorphosis was delayed in *S. intermontanus* (right-hand figure), but not in S. *couchii*. (Replotted from Morey and Reznick 2000.)

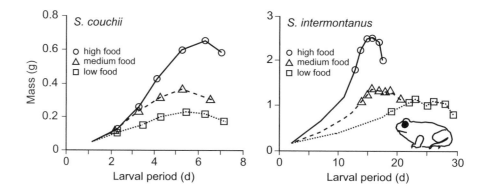

the life history transition can occur (Day and Rowe 2002). Wilbur and Collins (1973), studying amphibian metamorphosis, linked the idea of a threshold to knowledge about endocrinological conditions that were necessary for amphibian metamorphosis to occur. Day and Rowe (2002) present a short history of many studies testing elements of the Wilbur-Collins model, and how the results of many studies can be explained by a theoretical model incorporating a developmental threshold. A pattern of later maturation at a smaller size when food is limiting, like that in figure 13.17, is a pattern of phenotypic plasticity, and can be generated from several models. For example, it can be generated from a model that assumes that maturation begins at a fixed critical body size and proceeds for a fixed amount of time, or from models that assume that the critical size itself, or the time from initiation to completion of maturation, respond to food availability in an adaptive fashion (Reznick 1990; Morey and Reznick 2000). The important distinction between these two kinds of models is that in the former model, phenotypic plasticity is simply the result of a fixed pattern of development in different environments (imposed variation *sensu* Schew and Ricklefs (1998) whereas in the latter model the phenotypic plasticity is adaptive (Reznick 1990) (induced variation *sensu* Schew and Ricklefs 1998). Reznick's studies of amphibians and fish are lucid and powerful expositions of the models and the very careful studies that define the nature of the thresholds that are necessary to discriminate among these models. His studies provide strong evidence that growth and development programs are not necessarily fixed. The implication for animal energy budgets, then, is that the rules for allocation of energy are not necessarily fixed but may respond to food availability. The physiological links between food availability and different patterns of allocation are poorly understood but important to understand if we are to predict animal responses to changing environments.

13.5.2 In Most Ectotherms, Decreased Temperature Results in Slower Growth and Delayed Maturation at a Larger Size

Atkinson (1994) reviewed over 100 studies of the effects of rearing temperature on age and size at maturity in ectotherms. Not surprisingly he found that lower temperature led to lower rates of growth, and lower growth rate resulted in later maturation as was the case for limited food. However, whereas life history analysis generally suggests that it is adaptive for adults to emerge smaller if reared in conditions that slow down juvenile growth (Atkinson and Sibly 1997), in 84% of the species reared in the cold, maturation occurred at a larger rather than smaller size. The species included a variety of invertebrates (e.g., insects, crustaceans, mollusks) and vertebrates but the studies were heavily skewed toward insects. The observation has no generally accepted explanation, and a number of models have been proposed (Berrigan and Charnov 1994; Atkinson and Sibly 1997), including two that seek an explanation by considering the general effect of temperature on reaction rates (chapter 1). One type of model, originating back to von Bertalanffy (1960), proposes differential effects of temperature on anabolic versus catabolic reactions (Atkinson and Sibly 1996). Another type of model proposes differential effects of temperature on processes supporting growth versus those supporting differentiation (Gibert and De Jong 2001; van der Have and De Jong 1996; Gibert and De Jong 2001). Resolution awaits future work, and there are valid concerns about the generality of the phenomenon and its occurrence in ecological situations (Bernardo and Reagan-Wallin 2002).

Putting these findings together with those on the effect of food limitation, the most general statement that can be made is that conditions that slow down growth in ectotherms, either food limitation or low temperature, typically cause delayed maturation. This phenotypic plasticity may be the result of a fixed program of development that is expressed in different environments or the result of an adaptively flexible program.

13.5.3 Growth and Maturation Responses of Endotherms to Food Limitation Varies among Species

Some neonate birds and mammals respond to poor feeding conditions by delaying maturation. For examples, Schew and Ricklefs(1998) review a number of studies on avian species that delay feather growth or tissue maturation during a period in which growth in mass is delayed by food limitation. They

Figure 13.18. Growth and maturation responses of endotherms to food limitation varies. (A) In one pattern (left-hand figure), maturation is delayed during a period in which growth in mass is delayed by food limitation (shown by gray-filled circles). Thus, the normal relationship between growth and maturation is maintained, although there is a delay in fledging or weaning. (B) In another pattern (right-hand figure), development proceeds at a fixed rate even if growth is slowed during food limitation (grey-filled circles). For animals that exhibit this pattern, unless they later exhibit catch-up growth, their growth may continue to lag behind maturation, and if growth does not reach adult size by maturation they may suffer permanent stunting. Do not confuse this figure with Fig. 13.10, which plots mass (or growth) vs. time. (Modified from Schew and Ricklefs 1998.)

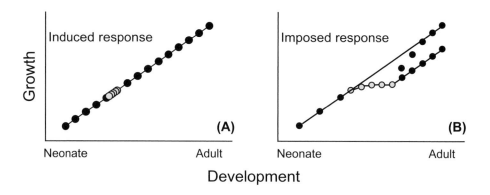

viewed this as beneficial because it maintains the normal relationship between growth and maturation, although it delays fledging or weaning (figure 13.18A). But there are other avian and mammalian species that develop at a fixed rate even if their growth is slowed during food limitation. Unless they later exhibit catch-up growth, their growth may continue to lag behind maturation, and if growth does not reach adult size by maturation they may suffer permanent stunting (figure 13.18B). These different response patterns to food limitation can even occur within the same species, depending on the age at which the limitation is imposed (Schew and Ricklefs 1998). There is much room for more research on this topic, especially if we are to understand all this variation within the context of life-history evolution. We have such limited ability to predict a *priori* how growth and development of an endothermic species will respond to food limitation. If it is important to know this in the context of a species conservation scheme, the best advice is to measure it in an experimental study.

References

Adolph, S. C., and W. P. Porter. 1993. Temperature, activity, and lizard life histories. *American Naturalist* 142:273–295.

Adolph, S. C., and W. P. Porter. 1996. Growth, seasonality, and lizard life histories: Age and size at maturity. *Oikos* 77:267–278.

Alford, R. A., and R. N. Harris. 1988. Effects of larval growth history on anuran metamorphosis. *American Naturalist* 131:91–106.

Angilletta, M. J., Jr. 2001. Thermal and physiological constraints on energy assimilation in a widespread lizard (*Sceloporus undulatus*) *Ecology* 82:3044–3056.

Arendt, J. D. 1997. Adaptive intrinsic growth rates: an integration across taxa. *Quarterly Review of Biology* 72:149–177.

Atkinson, D. 1994. Temperature and organism size — a biological law for ectotherms? *Advances in Ecological Research* 25:1–58.

Atkinson, D., and R. M. Sibly. 1996. On the solutions to a major life-history puzzle. *Oikos* 77:359–365.

Atkinson, D., and R. M. Sibly. 1997. Why are organisms usually bigger in colder environments? Making sense of a life history puzzle. *Trends in Ecology and Evolution* 12:235–239.

Bachman, G. C., and M. A. Chappell. 1988. The energetic cost of begging behaviour in nestling house wrens. *Animal Behaviour* 55:1607–1618.

Bakken, G. S. 1976. A heat transfer analysis of animals: Unifying concepts and the application of metabolism chamber data to field ecology. *Journal of Theoretical Biology* 60:337–384.

Bakken, G. S. 1992. Measurement and application of operative and standard operative temperatures in ecology. *American Zoologist* 32:194–216.

Baldwin, R. L., and R. D. Sainz. 1995. Energy partitioning and modeling in animal nutrition. *Annual Review of Nutrition* 15:191–211.

Ball, S. L., and R. L. Baker. 1996. Predator-induced life history changes: Antipredator behavior costs or facultative life history shifts? *Ecology* 77:1116–1124.

Bayne, B. L., and A. J. S. Hawkins. 1997. Protein metabolism, the costs of growth, and genomic heterozygosity: experiments with the mussel *Mytilus galloprovincialis* Lmk. *Physiological Zoology* 70:391–402.

Bayne, B. L., S. Svennsson, and J. A. Nell. 1999. The physiological basis for faster growth in the Sydney rock oyster, *Saccostrea commercialis*. Biological Bulletin (*Woods Hole*) 197:377–387.

Bayyari, G. R., W. E. Huff, N. C. Rath, J. M. Balog, L. A. Newberry, J. D. Villines, J. K. Dkeeles, N. B. Anthony, and K. E. Nestor. 1997. Effect of the genetic selection of turkeys for increased body weight and egg production on immune and physiological responses. *Poultry Science* 76:289–296.

Beaupre, S. J., and A. E. Dunham. 1995. A comparison of ratio-based and covariance analyses of a nutritional data set. *Functional Ecology* 9:876–880.

Beaupre, S. J., A. E. Dunham, and K. L. Overall. 1993. The effects of consumption rate and temperature on apparent digestibility coefficient, urate production, metabolizable energy coefficient and passage time in canyon lizards (*Sceloporus merriami*) from two populations. *Functional Ecology* 7:273–280.

Bennett, A. F., and J. W. Hicks. 2001. Postprandial exercise: prioritization or additivity of the metabolic responses? *Journal of Experimental Biology* 204:2127–2132.

Bernardo, J., and N. L. Reagan-Wallin. 2002. Plethodontid salamanders do not conform to "general rules" for ectotherm life histories: Insights from allocation models about why simple models do not make accurate predictions. *Oikos* 97:398–414.

Berrigan, D., and E. L. Charnov. 1994. Reaction norms for age and size at maturity in response to temperature: A puzzle for life historians. *Oikos* 70:474–478.

Bevelhimer, M. S., R. A. Stein, and R. F. Carline. 1985. Assessing significance of physiological differences among three esocids with a bioenergetics model. *Canadian Journal of Fisheries and Aquatic Sciences* 42:57–69.

Blaxter, K. 1989. *Energy metabolism in animals and man.* Cambridge University Press, Cambridge.

Bradford, G. E., and T. R. Famula. 1984. Evidence for a major gene for rapid postweaning growth in mice. *Genetical Research* 44:293–308.

Calder, W. A. 1984. *Size, function, and life history.* Harvard University Press, Cambridge.

Calow, P., and C. R. Townsend. 1981. Resource utilization in growth. Pages 220–244 in C. R. Townsend and P. Calow (eds.), *Physiological ecology: An evolutionary approach to resource use.* Sinauer Associates, Sunderland, Mass.

Calow, P., and A. S. Woolhead. 1977. The relationship between ration, reproductive effort and age-specific mortality in the evolution of life-history strategies—some observations on freshwater triclads. *Journal of Animal Ecology* 46:765–781.

Campbell, G. S. 1977. *An introduction to environmental physics.* Springer, New York.

Case, T. J. 1978. On the evolution and adaptive significance of postnatal growth rates in the terrestrial vertebrates. *Quarterly Review of Biology* 53:243–282.

Chappell, M. A., and G. C. Bachman. 1998. Exercise capacity of house wren nestlings: Begging chicks are not working as hard as they can. *Auk* 115:863–870.

Chappell, M. A., and G. C. Bachman. 2002. Energetic costs of begging behavior. Pages 143–162 in J. Wright and M. L. Leonard (eds.), *The evolution of begging: Competition, cooperation, and communication.* Kluwer Academic, Dordrecht.

Chappell, M. A., and G. A. Bartholomew. 1981. Activity and thermoregulation of the antelope ground squirrel *Ammospermophilus leucurus* in winter and summer. *Physiological Zoology* 54:215–223.

Charnov, E. L. and J. F. Gillooly. 2003. Thermal time: Body size, food quality and the 10°C rule. *Evolutionary Ecology Research* 5:43–51.

Congdon, J. D. 1989. Proximate and evolutionary constraints on energy relations of reptiles. *Physiological Zoology* 62:356–373.

Congdon, J. D., A. E. Dunham, and D. W. Tinkle. 1982. Energy budgets and life histories of reptiles. Pages 233–271 in C. Gans and F. H. Pough (eds.), *Biologia of the reptilia,.* vol. 13. Academic Press, New York.

Conover, D. O., and T. M. C. Present. 1990. Countergradient variation in growth rate: Compensation for length of growing season among Atlantic silversides from different latitudes. *Oecologia* 83:316–324.

Crowl, T. A., and A. P. Covich. 1990. Predator-induced life history changes in freshwater snail. *Science* 247:949–951.

Day, T., and L. Rowe. 2002. Developmental thresholds and the evolution of reaction norms for age and size at life history transitions. *American Naturalist* 159:338–350.

Demarchi, M. W., and F. L. Bunnell. 1995. Forest cover selection and activity of cow moose in summer. *Acta Theriologica* 40:23–36.

Dobson, S. H., and R. M. Holmes. 1984. Compensatory growth in the rainbow trout, *Salmo gairdneri* Richardson. *Journal of Fish Biology* 25:649–656.

Dunham, A. E., B. W. Grant, and K. L. Overall. 1989. Interfaces between biophysical and physiological ecology and the population ecology of terrestrial vertebrate ectotherms. *Physiological Zoology* 62:335–355.

Ebert, T. A. 1999. *Plant and animal populations.* Academic Press, San Diego.

Edwards, R. W., J. W. Densem, and P. A. Russell. 1979. An assessment of the importance of temperature as a factor controlling the growth rate of brown trout in streams. *Journal of Animal Ecology* 48:501–507.

Eklund, J., and G. E. Bradford. 1977. Longevity and lifetime body weight in mice selected for rapid growth. *Nature* 265:48–49.

Elliott, J. M. 1976. The energetics of feeding, metabolism and growth of brown trout (*Salmo trutta* L.) in relation to body weight, water temperature, and ration size. *Journal of Animal Ecology* 45:923–948.

Elliott, J. M. 1994. *Quantitative ecology and the brown trout.* Oxford University Press, Oxford.

Foyle, T. P., R. K. O'dor, and R. W. Elner. 1989. Energetically defining the thermal limits of the snow crab. *Journal of Experimental Biology* 145:371–393.

Gibert, P., and G. De Jong. 2001. Temperature dependence of development rate and adult size in *Drosophila* species: Biophysical parameters. *Journal of Evolutionary Biology* 14:267–276.

Gilbert, N., and D. A. Raworth. 1996. Insects and temperature—a general theory. *Canadian Entomologist* 128:1–13.

Gillooly, J. F., E. L. Charnov, G. B. West, V. M. Savage, and J. H. Brown. 2002. Effects of size and temperature on developmental time. *Nature* 417:70–73.

Goss, R. J. 1978. *The physiology of growth.* Academic Press, New York.

Grant, B. W., and A. E. Dunham. 1988. Thermally imposed time constraints on the activity of the desert lizard *Sceloporus merriami*. *Ecology* 69:167–176.

Grant, B. W., and A. E. Dunham. 1990. Elevational covariation in environmental constraints and life histories of the desert lizard *Sceloporus merriami*. *Ecology* 71:1765–1776.

Grant, B. W., and W. P. Porter. 1992. Modeling global macroclimatic constraints on ectotherm energy budgets. *American Zoologist* 32:154–178.

Hanson, P. C., T. B. Johnson, D. E. Schindler, and J. F. Kitchell. 1997. Fish Bioenergetics 3.0 for Windows. University of Wisconsin Sea Grant Institute, Madison, WI.

Hentschel, B. T., and R. B. Emlet. 2000. Metamorphosis of barnacle nauplii: Effects of food variability and a comparison with amphibian models. *Ecology* 81:3495–3508.

Hewett, S. W., and B. L. Johnson. 1992. Fish Bioenergetics Model 2. University of Wisconsin Sea Grant Institute, Madison.

Huey, R. B., C. R. Peterson, S. J. Arnold, and W. P. Porter. 1989. Hot rocks and not-so-hot rocks: Retreat-site selection by garter snakes and its thermal consequences. *Ecology* 70:931–944.

Hill, D. K., and J. J. Magnuson. 1990. Potential effects of global warming on the growth and prey consumption of Great Lakes fishes. *Transactions of the American Fisheries Society* 119:265–275.

Hill, R. W. 1992. The altricial/precocial contrast in the thermal relations and energetics of small mammals. Pages 122–159 in T. E. Tomasi and T. S. Horton (eds.), *Mammalian energetics.* Comstock Publishing Associates, Ithaca, N.Y.

Hirshfield, M. F. 1980. An experimental analysis of reproductive effort and cost in the Japanese medaka, *Oryzias latipes. Ecology* 61:282–292.

Honek, A. 1999. Constraints on thermal requirements for insect development. *Entomological Sciences* 2:615–621.

Houston, A. I., and J. M. McNamara. 1990. The effect of environmental variability on growth. *Oikos* 59:15–20.

Huey, R. B., L. Partridge, and K. Fowler. 1991. Thermal sensitivity of *Drosophila melanogaster* responds rapidly to laboratory natural selection. *Evolution* 45:751–756.

Humphreys, W. F. 1979. Production and respiration in animal populations. *Journal of Animal Ecology* 48:427–453.

Jackson, L. J., and S. R. Carpenter. 2003. PCB concentrations of Lake Michigan invertebrates: Reconstruction based on PCB concentrations of alewives (*Alosa pseudoharengus*) and their bioenergetics. *Journal of Great Lakes Research* 21:112–120.

Jackson, S., and J. Diamond. 1996. Metabolic and digestive responses to artificial selection in chickens. *Evolution* 50:1638–1650.

Jobling, M. 1994. *Fish bioenergetics.* Chapman & Hall, London.

Johnstone, R. A., and H.C.J. Godfray. 2002. Models of begging as a signal of need. Pages 1–20 in J. Wright and M. Leonard (eds.), *The evolution of begging.* Kluwer Academic, Dordrecht.

Jonsson, B., T. Forseth, A. J. Jensen, and T. F. Naesje. 2001. Thermal performance of juvenile Atlantic salmon, *Salmo salar L. Functional Ecology* 15:701–711.

Kenagy, G. J., S. M. Sharbaugh, and K. A. Nagy. 1989. Annual cycle of energy and time expenditure in a golden-mantled ground squirrel population. *Oecologia* 78:269–282.

Ketterson, E. D., V. J. Nolan, M. J. Cawthorn, P. G. Parker, and C. Ziegenfus. 1996. Phenotypic engineering—using hormones to explore the mechanistic and functional bases of phenotypic variation in nature. *Ibis* 138:70–86.

Ketterson, E. D., and V. J. Nolan. 1994. Hormones and life histories: An integrative approach. *American Naturalist* 140:S33–S62.

Kilner, R. M. 2001. A growth cost of begging in captive canary chicks. *Proceedings of the National Academy of Sciences,* 98:11394–11398.

Kingsolver, J. G. 1989. Weather and the population dynamics of insects: integrating physiology and population ecology. *Physiological Zoology* 62:314–334.

Kitchell, J. F., and L. B. Crowder. 1986. Predator-prey interactions in Lake Michigan: Model predictions and recent dynamics. *Environmental Biology of Fishes* 16:205–211.

Kitchell, J. F., W. E. Stewart, and D. Weininger. 1977. Applications of a bioenergetics model to perch (*Perca flavescens*) and walleye (*Stizostedion vitreum*). *Journal of the Fisheries Research Board of Canada* 34:1922–1935.

Kleiber, M. 1961. *The fire of life.* John Wiley, New York.

Kooijman, S. A. L. M. 2000. *Dynamic energy and mass budgets in biological systems.* Cambridge University Press, Cambridge.

Korsmeyer, K. E., and H. Dewar. 2001. Tuna metabolism and energetics. Pages 35–78 in B. A. Block and E. D. Stevens (eds.), *Tuna: Physiology, ecology, and evolution.* Academic Press, San Diego, Calif.

Kvist, A., and A. Lindstrom. 2000. Maximum daily energy intake: It takes time to lift the metabolic ceiling. *Physiological and Biochemical Zoology* 73:30–36.

Leamy, L., and W. Atchley. 1985. Directional selection and developmental stability: Evidence from fluctuating asymmetry of morphometric characters in rats. *Growth* 49:8–18.

Leips, J., and J. Travis. 1994. Metamorphic responses to changing food levels in 2 species of hylid frogs. *Ecology* 75:1345–1356.

Lepczyk, C. A., and W. H. Karasov. 2000. The effect of ephemeral food restrictions on the growth of house sparrows (*Passer domesticus*). *Auk* 117:164–174.

Levitan, D. R. 1989. Density-dependent size regulation in *Diadema antillarum*: Effects on fecundity and survivorship. *Ecology* 70:1414–1424.

Lifson, N., G. B. Gordon, and R. McClintock. 1955. Measurement of total carbon dioxide production by means of $D_2^{18}O$. *Journal of Applied Physiology* 7:704–710.

Lifson, N., G. B. Gordon, M. B. Visscher, and A. O. Nier. 1949. The fate of utilised molecular oxygen and the source of the oxygen of respiratory carbon dioxide, studied with the aid of heavy oxygen. *Journal of Biological Chemistry* 180:803–811.

Lifson, N., and R. McClintock. 1966. Theory of use of the turnover rates of body water for measuring energy and material balance. *Journal of Theoretical Biology* 12:46–74.

Lucas, A. 1996. *Bioenergetics of aquatic animals.* Taylor & Francis, London.

Mandal, A. B., G. S. Bisht, H. Singh, and N. N. Pathak. 1999. Effect of environmental temperature on growth performance and nutrient utilization in growing guineakeets *(Numida meleagris). Tropical Agriculture* 76:74–77.

Marinovic, B., and M. Mangel. 1999. Krill can shrink as an ecological adaptation to temporarily unfavourable environments. *Ecology Letters* 2:338–343.

Marken Lichtenbelt, W.D.V. 1991. Energetics of the green iguana *Iguana iguana* in a semi-arid environment. Ph.D. dissertation, University of Groningen.

Masman, D., S. Daan, and H.J.A. Beldhuis. 1988. Ecological energetics of the kestrel: Daily energy expenditure throughout the year based on time-energy budget, food intake and doubly labeled water methods. *Ardea* 76:64–81.

Meijer, T., D. Masman, and S. Daan. 1989. Energetics of reproduction in female kestrels. *Auk* 106:549–559.

Mercadante, M.E.Z., I. U. Packer, A. G. Razook, J. N. S. G. Cyrillo, and L. A. Figueiredo. 2003. Direct and correlated responses to selection for yearling weight on reproductive performance of Nelore cows. *Journal of Animal Science* 81:376–384.

Metcalfe, N. B., and P. Monaghan. 2001. Compensation for a bad start: Grow now, pay later? *Trends in Ecology and Evolution* 16:254–260.

Morey, S., and D. Reznick. 2000. A comparative analysis of plasticity in larval development in three species of spadefoot toads. *Ecology* 81:1736–1749.

Myers, R. A., S. A. Levin, R. Lande, F. C. James, W. W. Murdoch, and R. T. Paine. 2004. Hatcheries and endangered salmon. *Science* 303:1980.

Nagy, K. A. 1980. CO_2 production in animals: analysis of potential errors in the doubly-labeled water method. *American Journal of Physiology* 238:R466–R473.

Nagy, K. A. 1983. The doubly labeled water method: A guide to its use. Publication Number 12(1417). Laboratory of Biomedical and Environmental Sciences, University of California, Los Angeles.

Nagy, K. A. 1989. Doubly-labeled water studies of vertebrate physiological ecology. Pages 270–287 in P. W. Rundel, J. R. Ehleringer, and K. A. Nagy (eds.), *Stable isotopes in ecological research.* Springer-Verlag, New York.

Nagy, K. A. 2000. Energy costs of growth in neonate reptiles. *Herpetological Monographs* 14:378–387.

Nagy, K. A., I. A. Girard, and T. K. Brown. 1999. Energetics of free-ranging mammals, reptiles, and birds. *Annual Review of Nutrition* 19:247–277.

Niewiarowski, P. H. 2001. Energy budgets, growth rates, and thermal constraints: Toward an integrative approach to the study of life-history variation. *American Naturalist* 157:421–433.

O'Laughlin, B. E., and R. N. Harris. 2000. Models of metamorphic timing: an experimental evaluation with the pond dwelling salamander *Hemidactylium scutatum* (Caudata: Plethodontidae). *Oecologia* 124:343–350.

Olson, D. W., M. L. Sunde, and H. R. Bird. 1972. The effect of temperature on metabolizable energy determination and utilization by the growing chick. *Poultry Science* 51:1915–1922.

Owen-Smith, R. N. 2002. *Adaptive herbivore ecology: From resources to populations in variable environments.* Cambridge University Press, Cambridge.

Owen, S. F. 2001. Meeting energy budgets by modulation of behaviour and physiology in the eel *(Anguilla anguilla L.). Comparative Biochemistry and Physiology A* 128:631–644.

Parks, J. R. 1982. *A theory of feeding and growth of animals.* Springer-Verlag, Berlin.

Parry, G. D. 1983. The influence of the cost of growth on ectotherm metabolism. *Journal of Theoretical Biology* 101:453–477.

Peebles, E. D., and H. L. Marks. 1991. Effects of selection for growth and selection diet on eggshell quality and embryonic development in Japanese quail. *Poultry Science* 70:1474–1480.

Peters, R. H. 1983. *The ecological implications of body size.* Cambridge University Press, Cambridge.

Peterson, C. C., B. M. Walton, and A. F. Bennett. 1998. Intrapopulation variation in ecological energetics of the garter snake *Thamnophis sirtalis*, with analysis of the precision of doubly labeled water measurements. *Physiological Zoology* 71:333–349.

Peterson, C. C., B. M. Walton, and A. F. Bennett. 1999. Metabolic costs of growth in free-living Garter Snakes and the energy budgets of ectotherms. *Functional Ecology* 13:500–507.

Porter, W. P., S. Budaraju, W. E. Stewart, and N. Ramankutty. 2000. Calculating climate effects on birds and mammals: Impacts on biodiversity, conservation, population parameters, and global community structure. *American Zoologist* 40:597–630.

Porter, W. P. and D. M. Gates. 1969. Thermodynamic equilibria of animals with environment. *Ecological Monographs* 39:245–270.

Porter, W. P., J. W. Mitchell, W. A. Beckman, and C. B. De Witt. 1973. Behavioral implications of mechanistic ecology. Thermal and behavioral modeling of desert ectotherms and their microenvironment. *Oecologia* 13:1–54.

Porter, W. P., J. C. Munger, W. E. Stewart, S. Budaraju, and J. Jaeger. 1994. Endotherm energetics: from a scalable individual-based model to ecological applications. *Australian Journal of Zoology* 42:125–162.

Porter, W. P. and C. R. Tracy. 1983. Biophysical analyses of energetics, time-space utilization, and distributional limits. Pages 55–83 in R. B. Huey, E. R. Pianka, and T. W. Schoener (eds.) *Lizard ecology: Studies of a model organism.* Harvard University Press, Cambridge, Mass.

Present, T.M.C., and D. O. Conover. 1992. Physiological basis of latitudinal growth differences in *Menidia menidia:* Variation in consumption or efficiency? *Functional Ecology* 6:23–31.

Priede, I. G. 1985. Metabolic scope in fishes. Pages 33–64 in P. Tytler and P. Calow (eds.), *Fish energetics: New perspectives.* Croom Helm, Beckenham, U.K.

Reiss, M. J. 1989. *The allometry of growth and reproduction.* Cambridge University Press, Cambridge.

Reznick, D. 1985. Costs of reproduction: An evaluation of the empirical evidence. *Oikos* 44:257–267.

Reznick, D. 1992. Measuring the costs of reproduction. *Trends in Ecology and Evolution* 7:42–45.

Reznick, D. 1996. Life history evolution in guppies: A model system for the empirical study of adaptation. *Netherlands Journal of Zoology* 46:172–190.

Reznick, D. N. 1990. Plasticity in age and size at maturity in male guppies (*Poecilia reticulata*): An experimental evaluation of alternative models of development. *Journal of Evolutionary Biology* 3:185–203.

Rice, J. A., J. E. Breck, S. M. Bartell, and J. F. Kitchell. 1983. Evaluating the constraints of temperature, activity, and consumption on growth of largemouth bass. *Environmental Biology of Fishes* 9:263–275.

Richards, F. J. 1959. A flexible growth function for empirical use. *Journal of Experimental Botany* 10:290–300.

Ricklefs, R. E. 1968. Patterns of growth in birds. *Ibis* 110:419–431.

Ricklefs, R. E. 1973. *Ecology.* Chiron Press, Newton, Mass.

Ricklefs, R. E. 1974. Energetics of reproduction in birds. Pages 152–292 in R. A. Paynter, Jr. (ed.), *Avian energetics*. Nutthal Ornithological Society, Cambridge Mass.

Ricklefs, R. E. 1983. Avian postnatal development. Pages 1–83 in D. S. Farmer and J. R. King (eds.), *Avian biology*, vol. VII. Academic Press, New York.

Robbins, C. T. 1993. *Wildlife feeding and nutrition*. Academic Press, San Diego, Calif.

Rodriguez-Girones, M. A., J. M. Zuniga, and T. Redondo. 2001. Effects of begging on growth rates of nestling chicks. *Behavioral ecology* 12:269–274.

Roff, D. A. 1992. *The evolution of life histories*. Chapman and Hall, New York.

Rombough, P. J. 1994. Energy partitioning during fish development: additive or compensatory allocation of energy to support growth? *Functional Ecology* 8:178–186.

Saunders, R. L., A. P. Farrell, and D. E. Knox. 1992. Progression of coronary arterial lesions in Atlantic salmon (*Salmo salar*) as a function of growth rate. *Canadian Journal of Fisheries and Aquatic Sciences* 49:878–884.

Schew, W. A., and R. E. Ricklefs. 1998. Developmental plasticity. Pages 288–304 in J. M. Starck and R. E. Ricklefs (eds.), *Avian growth and development: Evolution within the altricial-precocial spectrum*. Oxford University Press, Oxford.

Schwab, F. E., and M. D. Pitt. 1991. Moose selection of canopy cover types related to operative temperature, forage, and snow depth. *Canadian Journal of Zoology* 69:3071–3077.

Secor, S. M., J. W. Hicks, and A. F. Bennett. 2000. Ventilatory and cardiovascular responses of pythons (*Python molurus*) to exercise and digestion. *Journal of Experimental Biology* 203:2447–2454.

Shea, R. E., I. H. Choi, and R. E. Ricklefs. 1995. Growth rate and function of skeletal muscles in Japanese quail selected for 4-week body mass. *Physiological Zoology* 68:1045–1076.

Sinervo, B., 1990. Evolution of thermal physiology and growth rate between populations of the western fence lizard *(Sceloporus occidentalis)*. *Oecologia* 83:228–237.

Sinervo, B., and S. C. Adolph. 1994. Growth plasticity and thermal opportunity in *Sceloporus lizards*. *Ecology* 75:776–790.

Sinervo, B., and E. Svensson. 1998. Mechanistic and selective causes of life history trade-offs and plasticity. *Oikos* 83:432–442.

Smith-Gill, S. J. 1983. Developmental plasticity: Developmental conversion versus phenotypic modulation. *American Zoologist* 23:47–55.

Smith, M. O. and R. G. Teeter. 1993. Effects of feed intake and environmental temperature on chick growth and development. *Journal of Agricultural Science* 121:421–425.

Smoker, W. W. 1986. Variability of embryo development rate, fry growth, and disease susceptibility in hatchery stocks of chum salmon. *Aquaculture* 57:219–226.

Speakman, J. R. 1997. *Doubly labelled water: Theory and practice*. Chapman & Hall, London.

Starck, J. M., and R. E. Ricklefs. 1998. *Avian growth and development*. Oxford University Press, New York.

Stewart, D. J., J. F. Kitchell, and L. B. Crowder. 1981. Forage fishes and their salmonid predators in Lake Michigan. *Transactions of the American Fisheries Society* 110:751–763.

Taylor, F. 1981. Ecology and evolution of physiological time in insects. *American Naturalist* 117:1–23.

Trudgill, D. L., and J. N. Perry. 1994. Thermal time and ecological strategies — a unifying hypothesis. *Annals of Applied Biology* 125:521–532.

Twombly, S. 1996. Timing of metamorphosis in a freshwater crustacean: Comparison with anuran models. *Ecology* 77:1855–1866.

van der Have, T. M., and G. De Jong. 1996. Adult size in ectotherms: Temperature effects on growth and differentiation. *Journal of Theoretical Biology* 183:329–340.

von Bertalanffy, L. 1960. Principles and theory of growth. Pages 137–259 in W. W. Nowinski (ed.), *Fundamental aspects of normal and malignant growth.* Elsevier, Amsterdam.

Waldschmidt, S. R., S. M. Jones, and W. P. Porter. 1987. Reptilia. Pages 553–619 in T. J. Pandian and F. J. Vernberg, eds., *Animal energetics* vol. 2: *Bivalvia through reptila.* Academic Press, San Diego, Calif.

Walsberg, G. E. 1993. Thermal consequences of diurnal microhabitat selection in a small bird. *Ornis Scandinavica* 24:174–182.

Walsberg, G. E., and B. O. Wolf. 1996a. A test of the accuracy of operative temperature thermometers for studies of small ectotherms. *Journal of Thermal Biology* 21:275–281.

Walsberg, G. E., and B. O. Wolf. 1996b. An appraisal of operative temperature mounts as tools for studies of ecological energetics. *Physiological Zoology* 69:658–681.

Weathers, W. W. 1996. Energetics of postnatal growth. Pages 461–496 in C. Carey, (ed.), *Avian energetics and nutritional ecology.* Chapman and Hall, New York.

Weibel, E. R. 2000. *Symmorphosis: on form and function shaping life.* Harvard University Press, Cambridge, Mass.

Werner, E. E. 1986. Amphibian metamorphosis: Growth rate, predation risk, and the optimal size at transformation. *American Naturalist* 128:319–341.

West, G. B., J.H. Brown, and B.J. Enquist. 2001. A general model for ontogenetic growth. *Nature* 413:628–631.

Whiteley, N. M., R. F. Robertson, J. Meagor, A. J. El Haj, and E. W. Taylor. 2001. Protein synthesis and specific dynamic action in crustaceans: Effects of temperature. *Comparative Biochemistry and Physiology A* 128:595–606.

Wieser, W. 1994. Cost of growth in cells and organisms: General rules and compartive aspects. *Biological Review* 68:1–33.

Wikelski, M., and C. Thom. 2000. Marine iguanas shrink to survive El Nino. *Nature* 403:37.

Wilbur, H. M. and J. P. Collins. 1973. Ecological aspects of amphibian metamorphosis. *Science* 182:1305–1314.

Wilson, P. N., and D. F. Osbourn. 1960. Compensatory growth after undernutrition in mammals and birds. *Biological Reviews (Cambridge Philosophical Society)* 35:324–363.

Wolf, B. O., and G. E. Walsberg. 1996. Thermal effects of radiation and wind on a small bird and implications for microsite selection. *Ecology* 77:2228–2236.

Wootton, R. J. 1985. Energetics of reproduction. Pages 231–254 in P. Tytler and P. Calow (eds.), *Fish energetics: New perspectives.* Johns Hopkins University Press, Baltimore.

Wootton, R. J. 1998. *Ecology of teleost fishes.* Kluwer Academic, Dordrecht.

Zimmer, C. 2003. Rapid evolution can foil even the best-laid plans. *Science* 300:895.

CHAPTER FOURTEEN

Reproduction in Budgets of Mass and Energy

Our DISCUSSION OF reproduction follows a similar format to that used in our discussion of growth in chapter 13. We will discuss component costs of reproduction in animals and assess their relative magnitude in several ways. Reproductive costs, like growth costs, can be high, which has necessitated the synchronization of these life-cycle features with periods of environmental resource availability. We therefore include a brief overview of the timing of production in animals and how changes in resource availability influence production. We also discuss how mass and energy budgets are influenced by temperature. This theme is important because understanding it provides a powerful tool to predict the impacts of a changing world on animals.

14.1 Allocation to Reproduction: Trade-Off with Development and Effects of Body Size

One of the key aspects of maturation is the age when animals begin reproduction, and this may or may not be after they have completed growth. As discussed in chapter 13, one possible variation is that some animals, often called determinate growers (e.g., many insects, birds, mammals), typically reproduce only after attaining nearly asymptotic size, whereas other animals, such as many fish species, called indeterminate growers (Wootton 1998), reach their age for first reproduction before reaching asymptotic size. West (2001) pointed out that indeterminate growers die before they reach asymptotic size and that their fundamental difference compared with determinate growers is that they begin to allocate energy to reproduction before growth is complete, which reduces the growth rate. This trade-off idea recognizes that there are material costs of reproduction, and implies that there is a fundamental upper limit to how much budgets can be expanded to accommodate them.

In order to explore these and other issues relating to costs of reproduction in budgets of mass and energy, we will refer repeatedly to one long-term, intensive, and creative study on the cost of reproduction in kestrels (*Falco tinnunculus*) by Daan and his colleagues in the Netherlands, but we will supplement that presentation with additional examples from other taxa. As a shorthand notation, we will often express the energy that an adult female can devote to reproduction as some multiple of her basal or maintenance (nonreproductive) energy needs. In doing so, we are making an assumption that, across species of different sizes within a taxon, the energy devoted per unit time to reproduction scales with body mass in the same way as basal or maintenance energy needs. There are both theoretical and empirical reasons for this assumption. The theoretical argument is that the energy surplus that can be channeled to growth or reproduction is the difference between assimilated energy ($I - E$) and maintenance respiration ($R_{maintenance}$). Because each of these scales approximately with $M^{3/4}$ (chapters 1 and 3), their difference will scale in a similar fashion (Reiss 1989). The empirical support for the assumption comes from allometric analyses of the energy invested in reproduction per unit time by females in numerous taxa. Reiss (1989) reviewed a large number of studies of vertebrates and invertebrates, ectotherms and endotherms, and found that mass exponents ranged from 0.51 to 0.95. Even larger surveys of energy flow through populations have shown that assimilation, respiration, and production scale with body mass in a similar fashion and that ratios of production to assimilation are independent of body mass (Humphreys 1979).

14.2 Approaches for Measuring Costs of Reproduction

14.2.1 Costs in Chemical Potential Energy

Animals incur costs during various defined stages of reproduction, such as territorial defense, nest building, mating, egg production, gestation, and postnatal parental care. The simplest cost to measure is that associated with energy or mass of nutrients in reproductive tissues, because this mainly involves carefully measuring tissue masses and their energy or nutrient contents. The relevant tissues include material built up in reproductive tissues (gonads and accessories) and retained in the body, which might also be counted as tissue growth, as well as external products of reproduction such as eggs or live-born young produced by females, which will be discussed below.

We will use our focal species, the kestrel (body mass 300 g), to illustrate all these points. First, let us learn some basics about the kestrel's reproductive cycle. Reproduction occurs once per year between March and August in the Netherlands (Masman et al. 1988). Kestrels begin mating in March. Clutches are initiated over the course of two months beginning in April, and decline in size from six eggs for clutches initiated early to four eggs for clutches initiated late. Eggs are laid every other day and so a clutch of five eggs is laid in approximately nine days (Meijer and Drent 1999). Eggs hatch after 29 days of incubation, and young fledge when they are about 30 days old. From about two weeks before clutch initiation until the young are ten days old, the male provides all food to the female, and both parents provide food to the young for 5 to 30 days after fledging.

Each kestrel egg has about 100 kJ, and thus peak energy deposition rate for a clutch of five eggs is around 56 kJ/d ($5 \times 100/9$ days; Meijer et al. 1989). This is 41% of the bird's basal metabolic rate (BMR), which is 138 kJ/d. Each egg has a mass of 21 g, which is about 7% of the female's body mass. Therefore the female experiences day-to-day fluctuations in body mass while she is laying (figure 14.1). Notice in figure 14.1 that the female builds up considerable body mass prior to laying. This mass is made up mostly by the oviduct (7 g) and reserves (42 g). Almost all of this mass of potential reserves remains intact at the end of egg laying, and so most of the energy for egg laying must come from food delivered by the male. The reserves are used during the early nestling period when the

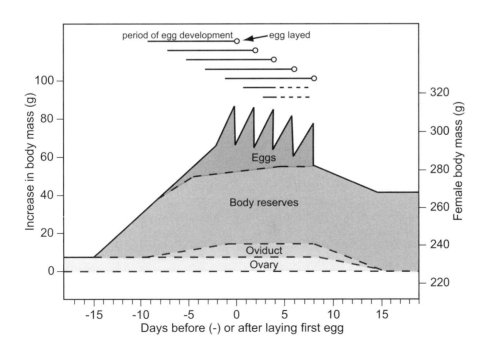

Figure 14.1. Female kestrels (*Falco tinnunculus*) build up body reserves before egg laying begins, and then during egg laying their day-to-day changes in body mass reflect the two day cycle from egg laying to deposition to egg laying. According to this model, the resources for egg laying do not come from the body reserves but from intake by the female. The reserves are used later, during the feeding of nestlings. (Redrawn from Meijer et al. 1989.)

females have high respiration rate due to high activity costs in feeding young (Meijer and Drent 1999). Masman et al. (1986) calculated the metabolizable energy delivered to females by males during egg laying based on field observations, and found that it was 371 kJ/d. Thus, peak energy deposition directly into eggs represented 15% of the female's metabolizable energy. The remaining 85% was used for respiration. The allocation to eggs can be twice as high in other avian species, such as waterfowl, that produce higher-energy-content eggs more frequently (table 14.1). Analogous measures made in many other kinds of animals show that one- to two-thirds of metabolizable energy can be allocated to reproductive products by females (table 14.1). How much of their respired energy represents additional reproductive costs will be discussed below.

In male kestrels, and in vertebrate males generally, the tissue costs of reproduction (gonads and sperm) are relatively low (e.g., Ricklefs 1974; Walsberg 1983; Wootton 1998). But there are interesting cases among other animals of

TABLE 14.1
Examples of allocation of metabolizable energy into reproductive products

Animal	Product	*Percentage allocated* *Mean + SD* *(range)*	Notes
Birds	Eggs (peak production)	$13 \pm 9\%$[b] (4 to 28%)	12 species reviewed by Meijer and Drent (1999)
Altricial eutherian mammals	Milk (peak production)	47 to 62%[c]	Based on allometric relationships summarized by Robbins (1993)
Lizards	Eggs	$25 \pm 3\%$ (16 to 27%)	Eight species reviewed by Anderson and Karasov (1988)
Fish	Eggs (assumed)[a]	$40 \pm 6\%$	15 species reviewed by Brett and Groves (1979)
Insects	Eggs	$34 \pm 19\%$ (5 to 68%)	17 species reviewed by Slansky and Scriber (1985)

[a] Reproductive energetics have been studied in only a few fish species (Elliott 1994; Wootton 1998), but growth energetics were summarized by Brett and Groves (1979). According to Jobling (1994), as most fish age the growth energy is reallocated to egg production, and growth is reduced.
[b] Based on measured energy in clutches and laying intervals, and daily respiration of free-living birds measured with doubly labeled water or estimated.
[c] Estimated for mammals weighing 0.1–100 kg as $100 \times P_{milk}/(P_{milk} + R)$, where P_{milk} is peak energy per day in milk from an allometric equation, and R is field metabolic rate, measured with doubly labeled water, also estimated from an allometric equation.

high costs to males. For example, in the bean weevil (*Bruchidius dorsalis*) males transfer about 7% of their body mass to females during copulation. Female weevils may copulate ten times and gather enough material from males to pay most of the nutritional cost of egg production (Takakura 1999).

Postnatal provisioning of energy and material can be extremely important, as in production of milk in mammals and crop milk in some birds (including pigeons, some penguins, and flamingos; Fisher 1972). There have been many measurements of the composition of milk in mammalian species (Robbins 1993), but we know of only a few such studies in doves and pigeons (Vandeputte 1980). Rates of transfer of material and energy in mammals have been estimated by pairing the measures of milk composition with either suckling rates, changes in mass of suckling young and mother that include corrections for fecal and urine excretion, or isotopic measures of water influx that include corrections for metabolic water production and possible recycling (see box 12.1; Gittleman and Thompson 1988; Robbins 1993). Rates of material and energy transfer rise during lactation to a peak and then fall during weaning. Among mammalian species the peak energy transfer rate scales with body mass to the approximately 3/4 power (Robbins 1993). Because basal metabolic rate (BMR) scales to the same exponent, lactational energy transfer is expressed as a multiple of basal metabolic rate, and in some species is two to three times the BMR.

Reproductive costs can sometimes be measured by altering the load on the parent and measuring its feeding response. This works fairly well in many small mammals, partly because most female small mammals meet the energy demands of reproduction primarily through an increase in food intake (Thompson 1992). Recall, for example, the experiment described in chapter 4 in which litter size was manipulated in mice and the females' food intake increased. Although mice seemed to compensate adequately, many studies with mammals have shown that the increase in parental intake or allocation does not keep pace with increasing litter size, and so offspring size and growth rate typically decrease (Sikes and Ylonen 1998). Analogous studies with doves, which feed their young crop milk, have shown that expanding the brood from the normal two up to three results in a slowing of growth rate and production of smaller young (Westmoreland and Best 1987; Blockstein 1989). Adjusting brood size experimentally is an example of phenotypic manipulation, which is a useful method for evaluating the fitness of alternative hypothetical natural histories or of investigating how opposing selection pressures influence life history trade-offs (Sinervo and Svensson 1998; see section 13.4.1). In another example of phenotypic manipulation, Landwer (1994) performed yolkectomy surgery on tree lizards (*Urosaurus ornatus*) to reduce egg number and hence production by 50% during vitellogenesis. Yolkectomized females grew more and survived better compared to intact controls, which demonstrated the expected cost of reproduction in both energetic and demographic terms. Researchers have also used this kind of manipulation to study the evolutionary trade-off between offspring size and offspring number (box 14.1).

Box 14.1. Allometric Engineering: The Experimental Manipulation of Offspring Body Size

Consider a female's typical clutch or litter size, and then ask "why not produce a little more?" The famous ornithologist David Lack argued that if total allocation is limited, then the selective advantage of producing many offspring from more, but smaller eggs, is balanced by the selective advantage of larger young born from larger eggs (Lack 1954). Thus, natural selection should favor a compromise between offspring quantity and quality. Allometric engineering is an experimental method that can be used to evaluate some of the underlying core assumptions of this hypothesis, such as whether offspring size and number are inversely related, or whether offspring quality is related to egg size. It includes a variety of techniques that change the number of offspring, or increase or decrease their size. These include the manipulation of hormones that determine the timing of the maturation and release of eggs in the ovary, the removal of immature eggs ("follicle ablation"), and the reduction in the amount of yolk in each egg ("yolkectomy").

As an example of hormonal manipulation to test the first assumption that offspring size and number are inversely related, Oksanen et al. (2002) administered gonadotropin hormones (follicle stimulating hormone, FSH, and lueteinizing hormone, LH) to enlarge litter size in bank voles (*Clethrionomys glareolus*) that were maintained under natural conditions in outdoor enclosures. Treated females, compared with controls, produced significantly larger litters of significantly smaller offspring. Likewise, when Sinervo and Licht administered FSH to side-blotched lizards (*Uta stansburiana*), mean clutch size was significantly increased and mean egg mass was significantly decreased, compared with controls (Sinervo and Licht 1991).

Sinervo and colleagues (1992) used side-blotched lizards to test the second assumption of Lack's conjecture. Is offspring quality related to egg size?

continued

continued

Sinervo et al. (1992) produced miniaturized and gigantisized hatchling lizards using methods such as ablation of some follicles which results in others becoming larger (Sinervo and Licht 1991), and yolk removal from freshly laid eggs (Sinervo 1990). Then they released them at field sites and monitored their survival. Offspring quality, indexed by survival (l) to one month, was sometimes but not always positively related to egg size (which was the prediction). Larger offspring did not invariably survive best, survival also depended on sex, season, and site. Fecundity (m; number of offspring that survived to one month) was negatively related to egg size, reflecting the inverse relation between clutch size and egg size (above). The predicted optimal egg size was the size at which the product $l \times m$ was maximal. This optimal size was indeed intermediate, as predicted (reflecting counteracting selection forces), and the predicted optimal egg size was higher for second and later clutches than for first clutches, which is the empirical observation as well.

Allometric engineering has been accomplished in other organisms besides voles and lizards. Some of the first studies of this sort were performed with echinoderms (Sinervo and McEdward 1988). Experimental manipulations with echinoderm eggs have demonstrated that larger larvae develop faster (McEdward 1996) and survive longer (Emlet and Hoegh-Guldberg 1997). The approach has also been used to understand the basis for differences in traits between populations. For example, among populations of fence lizards (*Sceloporus occidentalis*) there were differences in a number of life-history traits such as a trade-off between clutch and egg size, differences in egg incubation time and hatchling size (both correlated with egg size), and differences in the hatchlings' mass-specific growth rate, sprint speed, and stamina (Sinervo 1990). Because the latter three traits are influenced by the hatchlings' sizes, a point we emphasized in chapter 1, it is difficult to determine whether the population differences in all the traits resulted mainly from the differences in egg and hatchling size, or whether there might be other selective forces as well. By miniaturizing hatchlings from a larger-sized population, Sinervo and Huey found that differences between populations in sprint speed could be explained simply by population differences in egg/body size, but this factor alone could not explain all the difference in stamina, growth rate, or incubation time and so presumably those differences involved other evolved factors as well (Sinervo and Huey 1990). Thus, besides facilitating the study of offspring size/offspring number trade-offs, allometric engineering can permit inferences about the proximate cause(s) of trait evolution and an experimental test of the proximate influence of body size on a trait (Sinervo and Huey 1990).

14.2.2 Respiratory Costs of Reproduction

There are at least three kinds of respiratory cost of reproduction. One is an additional respiratory cost of production, which, although complicated to determine exactly, might be approximated as the product of the new tissue energy times 0.33 J/J (see section 13.3.3). Another is the basal respiration rate of reproductive tissues in the parent, such as the placenta, which represents 7% of the mass of the gravid uterus in mammals (80% is the fetuses; Blaxter 1989). In a few cases, direct measurements of the expected increase in respiration of reproductive tissues have been made by measurement of arteriovenous differences in oxygen content of blood and its rate of flow (Blaxter 1989), but more often researchers estimate this cost by comparing resting respiration rates of reproducing and nonreproducing animals. For example, the resting respiration rates of lactating golden-mantled ground squirrels (*Spermophilus saturatus*) increased 20% above that of nonreproductive individuals, but all this increase could be accounted for by increased specific dynamic effect associated with increased feeding rate (Kenagy et al. 1989). Some studies with reproducing insects, amphibians, and lizards have provided more dramatic examples of respiratory costs. In three species of insects reviewed by Slansky and Scriber (1985), respiration rates of mated females were 28–86% higher than in nonmated females. In gravid viviparous (live-bearing) lizards, resting respiration was reported to increase from 21% (Beuchat and Vleck 1990) to 67% (Guillette 1982), compared with nonreproductive individuals, and in the oviparous (egg-laying) amphibian *Ambystoma texanum* the resting respiration of presumed postabsorptive gravid females was reported to be a whopping 100% higher than that in postgravid females, after correcting for differences in carcass mass (Finkler and Cullum 2002). In the oviparous lizard *Sceoporus undulates*, Angilletta and Sears (2000) also estimated that the respiration rate of gravid females, adjusted for the respiration rate of their clutch, was 100% higher than in nonreproductives. The factor(s) accounting for large increases such as these are unknown, and it is difficult to evaluate this respiratory cost without concurrent measures of rates of energy transfer to new tissues including the mother's reproductive tissue as well as to the developing embryos and without more information about the duration of elevated respiration rate. In European starlings, whose respiration was measured throughout egg production (Vezina and Williams 2002), the increased respiratory costs can apparently be accounted for using the measured rate of energy transfer to oviduct and yolking follicles and the assumed value of 0.33 J respired per J transferred (see section 13.3.3, above).

The other type of respiratory cost of reproduction results from changes in the time/energy budgets of the parent(s), and these costs are also sometimes complicated to measure. One way to measure behavioral costs associated with territorial defense or parental care is to compare energy flow in reproducing individuals with that in nonreproductives. For example, respiration in free-living adult kestrels, estimated with either doubly labeled water or time-activity budgets, peaked during feeding of nestlings at levels 24% higher (females) to 70% higher (males) than in nonreproductive individuals, largely due to increased activity and flight costs (Masman et al. 1988). As another example, Ricklefs and Williams (1984) found that the field metabolic rate (FMR, measured with doubly labeled water) of adult European starlings (*Sturnus vulgaris*) increased during the nesting period to a level 86% higher than the FMR of nonreproductive birds. These results seem correct because we anticipate that the parents will spend extra time in flight or foraging in order to collect food for the nestlings. But, does it follow that if we observe no difference then the energetic cost is nil? The answer is no, because estimating costs as difference between energy flow of reproducers and nonproducers will result in error if there are compensatory changes occurring within the budget of the reproducer. Gittleman and Thompson (1988) suggested that the better estimates come from a combination of caloric consumption, respirometry, and time budgets of behavior. Examples of the application of this behavioral energetics approach to estimating reproduction costs in aquatic and terrestrial ectotherms and endotherms can be found in Aspey and Lustick (1983).

One of the most thorough studies of this sort in mammals was that by Kenagy et al. (1990) on free-living golden-mantled ground squirrels. They found that females at peak lactation had field metabolic rates insignificantly higher than nonlactating individuals, and among lactating individuals the FMR increased with litter size only very slightly. Finding no major compensatory adjustments in time/energy budgets, they concluded that direct energy investments in reproductive tissues (progeny and milk) accounted for the major energetic cost of reproduction in this species. But such a conclusion would not apply to female kestrels, for which peak energy deposition into eggs raised energy flow less than 20% per day as compared with a 24% increase in respiration during feeding of nestlings (above).

Another interesting approach to measure behavioral costs is to manipulate the behavior and then measure cost. For example, Marler et al. (1995) implanted free-living male mountain spiny lizards (*Sceloporus jarrovi*) with testosterone implants to increase their territorial defense, and measured their field metabolic rates with doubly labeled water. Lizards with implants had a mean FMR 31% higher than did sham controls, which had implants lacking crystalline testosterone. Marler et al. (1995) also showed that the testosterone implant did not increase resting respiration. Thus, the elevated FMR was most likely due to behavioral changes in the field, and, indeed, the researchers had previously shown that the testosterone-implanted males had longer daily activity periods. The testosterone-implanted males also had lower survivorship, compared with controls, which can also be explained by longer activity periods which presumably led to greater predation risk.

In another type of phenotypic manipulation, researchers have manipulated the clutch or brood size of birds to test for an impact on parental energy expenditure of increased or decreased activity in feeding different numbers of nestlings. Dijkstra et al. (1990) and Deerenberg et al. (1995) performed this test on kestrels and found that each additional nestling increased parental energy expenditure, measured with time-energy budgets and doubly labeled water, by 12 to 24 kJ/d, or about a 3–6% increase in daily respiration for each extra youngster to feed (figure 14.2). In a thoughtful review, Williams and Vezina (2001) noted that in nine studies of this sort only three found a positive relationship between respiration measured with doubly labeled water (DLW) and brood size. There is considerable variation among estimates of respiration for individuals within each of the studies, but it does not typically correlate with variation in reproductive effort. There are two different types of explanations for this. One explanation might be that measurement errors in the DLW method are too large and samples sizes have been too small to detect the patterns. Another explanation is that individuals operate within relatively fixed budgets that differ among individuals, and the individuals make compensatory adjustments as necessary within their budgets. This calls into question whether respiration is a good measure of reproductive effort, and suggests to some that a better approach for studying costs is to modify expenditure and look for effects on reproduction (Williams and Vezina 2001).

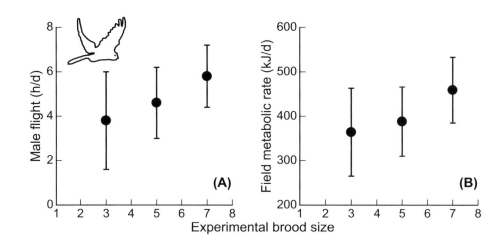

Figure 14.2. In kestrels (*Falco tinnunculus*), the time males spend hunting per day (figure A) and their daily energy expenditure (figure B) increased when brood size was experimentally changed from the natural mean of 5 young/brood. Data are from Dijkstra et al. (1990) and Deerenberg et al. (1995).

14.2.3 Meeting the Total Energetic Costs of Reproduction

Our discussion so far leads to the expectation that the energy and material costs of reproduction may have several components (P and R) and may vary over the stages of reproduction. For some groups of animals there have been enough studies to make some general statements about the overall magnitude of the energy costs of reproduction, which bears on questions about resource demand and time budgeting. The summaries in table 14.1 indicate that many animals double their daily food intake if they meet the increased costs of egg or milk production in this way rather than by drawing on reserves. Small mammals have been one of the best-studied groups in this regard (Thompson 1992). Most studies have shown that energy use during reproduction increases to a peak during lactation that is six to seven times basal metabolic rate (BMR), or two to three times maintenance needs, assuming that $R_{maintenance}$ is two to three times the BMR (figure 14.3). Averaged over the entire reproductive period, many female mammals approximately double their feeding rate during reproduction (Gittleman and Thompson 1988). Some comparisons of mated and unmated female insects show similar increments in feeding rate (Slansky and Scriber 1985). Consideration of such large increases in daily food intake also helps explain the importance of the upward adjustments in gut size of reproducing animals that were discussed in chapter 4.

But many animals reduce the daily demand for food during reproduction by storing energy and material in advance and then reallocating it during egg

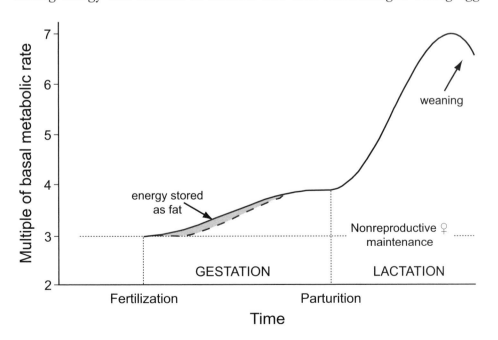

Figure 14.3. Typical pattern of metabolizable energy intake for small mammals during gestation and lactation. According to the model, nonreproductive females typically have energy intake and expenditures of around 3 times basal metabolic rate, and this increases <33% during gestation but >100% at peak lactation. The model shows that during gestation the intake may exceed the sum of respiration + allocation to reproductive tissue, with the difference being stored as fat. This fat store may or may not be used during lactation. (Redrawn from Thompson 1992.)

production. The percentages of energy for egg production that are drawn from endogenous reserves can vary widely within some taxa. Looking back to the groups summarized in table 14.1, for example, the ranges were from 0 to 100% reliance on reserves in birds (Meijer and Drent 1999), 0 to 90% in lizards (Anderson and Karasov 1988; Karasov and Anderson 1998), and 0 to 100% in insects (Slansky and Scriber 1985). The majority of fish species are iteroparous (they breed multiple times) rather than semelparous (breed once; Jobling 1994). Because the cost of reproduction is considerable, many iteroparous fish species show distinct patterns of energy storage and later depletion linked to cycles of reproduction. Fish researchers and managers that want to predict reproductive output often monitor the pattern of energy gain and loss by calculating a body condition index, which is mass/a(length)^b (where $b \sim 3$) (see box 1.4 in chapter 1 and section 7.5 in chapter 7).

14.3 Material Costs of Reproduction

Ecologists' intense focus on energy causes them sometimes to lose sight of other important factors. In previous chapters we described examples of limitation to reproduction by water and elements, such as calcium. In chapter 10 we discussed the fundamental fact that the C:N ratios of plants are much higher than those of animals, which raises the question of how animals that rely on plant-based diets reproduce themselves on a N-poor diet. We will illustrate the problem of meeting both energy and material needs during reproduction using data on N and energy from well-studied groups like birds and mammals, but we suspect that the problem is general across other groups and types of material.

The total N requirement during egg production in birds is 1.7 to 3.4 times higher than the maintenance N requirement, depending on factors such as egg mass, egg N content, and laying frequency (Robbins 1993). This may not represent any particular challenge if a nonreproductive bird is already surviving on a N-adequate diet and increases its food intake by a similar amount to meet its energy needs. In that case, the increased need for N is met coincidentally. However, our previous analysis of energy needs during reproduction suggests that birds do not need to increase their intake to this extent to meet the energy demands of egg production (e.g., table 14.1). How much more N do they require in their diet, and does reproduction necessitate a shift to an entirely different type of food (e.g., from plant to animal matter)? We explore this in the following analysis.

Budgets of mass and energy are the starting points for evaluating potential energy-material mismatches during production (table 14.2). Taking the American coot (*Fulica americana*) as an example, notice that the total N demands

TABLE 14.2
Increase in nitrogen requirements during reproduction

Status	Metabolizable energy (kJ/d)	N requirement (mg/d)	mg N per metabolizable kJ
American coot during egg-laying			
(A) Maintenance, nonreproductive	600[a]	256[c]	0.43
(B) Production in eggs	95.4[a]	625[a]	
(C) Production respiration	31.8[b]		
Total requirement A + B + C	727	881	1.21
White-tailed deer during peak lactation			
(A) Maintenance, nonreproductive	16,010[c]	10,100[c]	0.63
(B) Production in milk, peak lactation	8,513[c]	14,500[d]	
(C) Production respiration	2,837[b]		
Total requirement A + B + C	27,360	24,600	0.90

Note: The nitrogen requirement increases more than the energy requirement, which increases the dietary requirement for mg N per metabolizable kJ of dietary energy. Compare the estimated energy and N required when not reproducing (A; Maintenance, nonreproductive) with that when reproducing (A+B+C) in two endothermic vertebrates, the American coot (*Fulica americana*, body mass = 0.5 kg) and the white-tailed deer (*Odocoileus virginianus*, body mass = 70 kg).
[a] Meijer and Drent (1999).
[b] Assumes respiratory cost of production as the product of production energy times 0.33 J/J.
[c] Robbins (1993).
[d] Peak milk production for a 70-kg white-tailed deer, 1.26 kg/d; milk's N content, 11.5 g/kg (Hanwell and Peaker 1977).

during reproduction increase 3.4 times compared with the nonreproductive situation, which is a much larger increase than the analogous increase in need for metabolizable energy (1.2 times). Consequently, the ratio of total N:metabolizable energy demands during egg laying nearly triples, compared with the nonreproductive situation. This increase is consistent with the increase in dietary N requirement for laying in birds such as pheasants, quail, turkeys, and grouse (Robbins 1993). Although a three-fold increase sounds dramatic and suggests that a major diet shift is necessary during reproduction, the next step is to compare these ratios with the foods available to the coot. Plant foods consumed by birds have ratios of N to metabolizable energy that range from about

0.6 to 2.5 mg N/kJ (Karasov 1990), and so it is entirely possible that the coot can reproduce on a plant-based diet if it consumes plants with high enough N content. If it does not, then there are three alternatives. First, it can store ahead of time some reserve N and draw on the reserve during egg laying. Coots are known to do this, as body N content increases after they arrive on their breeding grounds and then decreases during egg laying, by an amount equivalent to about a quarter of the N in the clutch (Alisauskas and Ankney 1985). Second, they can supplement their diet with animal matter which, compared with plant materials, has a much higher ratio of N:metabolizable kJ (about 4.5 for arthropods and 6.9 for vertebrate flesh; Karasov (1990). Searching for scarcer foods with high concentrations of particular nutrients takes time. So some birds rely on fat stores to supply some of their energy needs so they can devote time to search for N (Krapu 1981) or Ca (see chapter 11). There is a third alternative, which is to eat more of the same diet and dispense somehow with excess energy (deposit it in fat or increase respiration). We have argued elsewhere in this book that there is not good evidence for this alternative in vertebrates.

Similar analyses for a dozen well-studied avian species (Meijer and Drent 1999) and one mammalian example (table 14.2) yield similar conclusions. Carnivores, like the kestrel, and insectivores consume food containing excess N relative to metabolizable energy, and easily meet N requirements for reproduction in the course of meeting energy requirements. In the white-tailed deer, as was the case for all the birds, the increase in requirement above maintenance is larger for N (2.4 times) than for metabolizable energy (1.7 times). This may necessitate a diet shift as the dietary N content per unit metabolizable energy must increase about 50%. However, the ratios in deer are well within the range of plant materials, and a female should be able to meet N requirements by selective feeding on more N-rich species or by consuming new plant growth with relatively higher N content, whose peak abundance coincides with the season of peak lactation—perhaps not coincidentally. Indeed, the relatively high energy and material needs of reproduction underscore the importance of timing reproduction in relation to environmental patterns of resource abundance.

When calculating material budgets during reproduction one must sometimes take account of recovery of material by the mother. In some species of mammals, for example, some water exported during lactation may be recovered by reingestion of the urine, feces, or saliva of the young (Baverstock and Green 1975; Friedman and Bruno 1976; Scantlebury et al. 2000). Similar recovery could occur for elements, which could be important if they were limiting. If recovery were complete, the minimum material cost during reproduction, at least up until the point the young begin feeding on their own, could be estimated by the mass gains of the young.

14.4 Nutritional Control of Reproduction

If the energy and material resources necessary for reproduction are not freely available then the timing and/or magnitude of reproduction may be altered or the fitness of the parent may be reduced as it struggles to gather scarce resources. The topic of controls over timing of reproduction has been written about extensively, and our main goal here is to provide a few illustrative examples reflecting the pervasive importance of nutrition.

14.4.1 The Importance of the Match between Supply and Demand

At the most general level, the availability and quality of food required for successful reproduction is usually the most important ultimate factor determining the timing of breeding. Ultimate factors refer to explanations that invoke evolutionary mechanisms such as natural selection. Timing is one key feature of a species' reproductive "strategy," and natural selection favors the reproductive strategy which, in a given environment, is most likely to yield the largest number of young that will survive to breed and/or is most likely to support the survival of parents until they breed again. The importance of the match between food availability and a species' need for food during reproduction was wonderfully illustrated in a recent study of seasonally breeding birds called blue tits (*Parus caeruleus*; Thomas et al. 2001). Donald Thomas, Jacques Blondel and their colleagues studied the energetics of these birds in habitats in Europe where they were breeding at the height of abundance of their main prey, caterpillars, and in habitats where the blue tits were breeding about two weeks earlier than when peak caterpillar abundance occurred. Thomas et al. called the former birds "matched" and the latter birds "mismatched" to seasonally varying food supply (figure 14.4). They found that parents mismatched to their food supply spent significantly more energy (measured using doubly labeled water) feeding nestlings than those matched, and that the nestlings of mismatched parents were significantly smaller than those of matched parents, presumably because they were provisioned with less food. Mismatched parents may have been working at levels at or beyond their sustainable limit, and when the researchers inspected data over several years they found that proportionally fewer parents persisted in the mismatched populations, reflecting a difference in survival. Besides nicely illustrating our point about the importance of temporal matching between reproduction and food supply, the example begs the question of how well animals will adjust phenotypically or genetically if climate warming or habitat disturbances

Figure 14.4. The energetics, survival, and future breeding success of adult blue tit (*Parus caeruleus*) parents differed at sites that were matched or mismatched with regard to food supply and demand. Panel A shows that at mismatched sites the timing of breeding (indicated by the horizontal black bar) preceded the interannual mean date in peak caterpillar abundance (indicated by the vertical black arrow), whereas at matched sites (panel B) the dates correspond. Panel C shows that the field metabolic rate, measured with doubly labeled water, was significantly higher in the most mismatched individuals, compared with those that bred at matched sites. The metabolic rates are expressed as the multiples of basal metabolic rate (BMR) during the daily activity period. When parents bred at matched sites they survived better (panel D), and a larger proportion of them bred for more than one year (panel E). (Data and redrawn panels A and B from Thomas et al. 2001.)

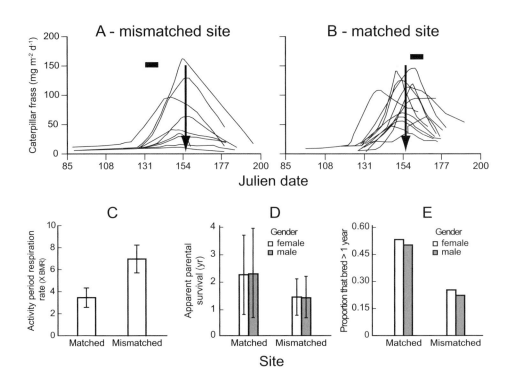

create such mismatches between peak food supply and reproductive demands of wildlife.

The reproductive responses of animals to changes in nutritional resources depend partly on the kind of environment in which they evolved. Ecologists commonly (and a bit sketchily) characterize environments according to the temporal availability of their food resources: (1) relatively unchanging environments where food is constantly available and reproduction is continuously possible, (2) environments where conditions favorable for reproduction are not continuous but occur predictably, and (3) environments where food availability is erratic and less predictable. Although these environments are characterized according to their temporal pattern of food availability, food is not necessarily the proximate factor which directly affects reproductive performance. Proximate factors refer to physiological and behavioral mechanisms. Such mechanisms can act as inducing factors that start and end reproduction, or as phasing factors that entrain endogenous rhythms of reproductive activity. Day length, as illustrated below, is an important non-nutritional proximate factor that may help insure that reproduction coincides with the favorable nutritional conditions necessary to support reproduction. With this as background, let us consider some examples where changing levels of nutritional resources probably act as a proximate factor affecting the timing and magnitude of reproduction. Although we illustrate with vertebrate examples, examples of all of these responses patterns are seen in invertebrates as well (Heatwole 1996).

14.4.2 Nutrition May Act Secondarily to Photoperiod as a Proximate Control to Spring and Summer Reproduction in Relatively Predictable Environments

Conditions favorable for breeding occur predictably in temperate regions, such as the Netherlands, and arctic regions. For the kestrel, abundance of its mammalian prey rises as its young are fledging, which is one of the most energy demanding times because parents are collecting food to feed their young and to meet their own expenditures, which are relatively high at that time (Masman et al. 1988). But the kestrels' breeding begins months earlier, and clearly the rise in prey abundance is not the immediate cue. In this situation, long-term predictive cues such as photoperiod are particularly useful for timing the initiation of reproduction (Gwinner 1986; Hahn et al. 1997). Photoperiod can act as the driver of the reproductive schedule, or it can act by synchronizing (entraining) an animal's endogenous circannual (approximately 1 year) rhythm. The photoperiodic responsiveness of kestrels was demonstrated when Meijer (1989) exposed captives to long days (Light:Dark (LD) 17.5:6.5 and LD 13:11) from December 1 onward and found that they initiated breeding earlier. To be sure that the bird has an endogenous circannual rhythm, researchers would have to hold them in very constant conditions with no time cues, such as changing daylength, and determine over several years whether they initiate breeding on an approximately 12-month schedule. Not surprisingly, such studies have been done on relatively few species because of the time costs.

You can consult the review by Hahn et al. (1997) for more information and for an entry into the extensive literature about principles of photoperiodic control of reproduction in animals, but the point that we want to emphasize is that even when control of reproductive timing is dominated by these mechanisms, the timing may be fine-tuned according to nutritional conditions. For example, in a 16-year study of snowshoe hares (*Lepus americanus*) in North America, hares commenced mating every year within a three-week period centered around April 15 (Cary and Keith 1979). Breeding occurred later when food (wild browse) was limiting, and, in an experimental study, Lloyd Keith and his colleagues showed that wild hares bred earlier when supplemented with food (Vaughan and Keith 1981). Thus, even though mating in hares is under control of an endogenous rhythm cued by lengthening daylength (Davis and Meyer 1972), it is fine-tuned by nutritional status. Mammalian herbivores with longer gestation times than hares, like white-tailed deer (*Odocoileus virginianus*), begin breeding earlier than hares according to the cue of shortening, fall days, but the exact time of breeding is also advanced or delayed according to food intake level (Verme 1965; Kirkpatrick 1988). In analogous fashion, experimental food supplementation in free-living kestrels has been shown to advance laying date by up to three weeks (Meijer et al. 1988). Indeed, in 20 field tests in which various avian species were supplemented with food, laying date was significantly advanced in 75% of the species (Meijer and Drent 1999). When food acts as a supplementary cue to photoperiodic control it can accelerate or inhibit the rate of reproductive development.

14.4.3 Nutritional Factors Can Be Primary Inducing Factors in Environments where Reproduction Can Potentially Occur more Continuously or more Unpredictably than in Highly Seasonal Environments

In some environments daylength is not a reliable cue for reproduction because food availability is independent of changes in daylength. In arid central Australia, for example, conditions suitable for reproduction of many desert birds occur for brief periods after rainfall. At least one species, the zebra finch (*Taeniopygia guttata*), has the potential to breed year round, based on observed breeding in every month (although not every year; Zann et al. 1995), but mainly breeds following irregular rains. Vleck and Priedkalns (1985) studied zebra finch reproductive biology and found that severely dehydrated males had gonads with sperm-producing activity, but the gonads were relatively small compared with those in hydrated controls. Once the dehydrated birds were provided with water ad libitum their testis size increased within three weeks. The onset of gonadal activity and growth in males is called recrudescence, which means breaking out into renewed activity. Vleck pointed out that gonadal recrudescence in rehydrated zebra finches is fast. In other avian species that breed seasonally, gonadal recrudescence in males can take over two months. Vleck also found that availability of green grass or high humidity, which add dietary water or reduce evaporative water loss, respectively, facilitated gonadal recovery. Other factors are also important in cueing male zebra finch breeding, such as interaction with females or grass seed availability (Vleck and Priedkalns 1985; Zann et al. 1995). Vleck concluded that water stress inhibited reproduction and that subsequent rehydration released the inhibition. This explanation also seems consistent with two examples that we described in chapter 12 of water influencing reproduction in free-living mammalian herbivores.

There are some very interesting examples in which animals' reproductive timing is influenced primarily by secondary metabolites (SMs) in their foods. One of our favorites regards a vole (*Microtus montanus*) in the intermountain region of western North America, whose controls for reproduction were revealed in a series of fascinating studies by Norm Negus, Pat Berger, and their colleagues. When nonbreeding female voles in captivity were maintained on a nutritious diet plus water and were also provided with a tiny amount of sprouted wheat, they went into estrous and became receptive for mating (Sanders et al. 1981). Those not provided with sprouts remained anestrous. The trivial mass of sprouts was of no consequence to the voles' energy or balance for major nutrients, but the researchers isolated a plant SM that was responsible for inducing their reproduction (Sanders et al. 1981). The compound, which goes by the mouthful 6-methoxybenzoxazlinone (or 6-MBOA for short), is structurally similar to compounds like serotonin, which is an endogenous neuroactive substance. Apparently, this naturally occurring substance in some germinating plants interacts with the vole's neuroendocrine mechanisms that control reproduction. Presumably, there was selection on voles' responsiveness to this SM because it results in vole reproduction synchronizing with periods of new plant growth. In an experimental field study in winter, Negus et al. were able to induce reproduction in otherwise nonbreeding voles by supplementing with oats coated with solvent containing 6-MBOA, whereas no reproduction occurred on control plots supplemented with oats coated with solvent alone (Berger et al. 1981). There are interesting examples of other SMs that share chemical features with vertebrate hormones and disrupt reproduction in mammals (Labov 1977) and birds (Leopold 1976).

14.4.4 Nutrition Influences the Magnitude of Reproduction

In many studies where researchers provided supplementary food to birds to test whether breeding date was advanced, they also checked whether clutch or egg size was increased. In almost all cases, food supplementation did not increase either clutch or egg size (Meijer and Drent 1999). But, even if extra nutrition does not affect the magnitude of allocation within individual birds, it can still have a population-level effect by influencing the proportion of adults that come into reproductive readiness, or by influencing growth rate and the time to reach sexual maturity. Also, among herbivorous or granivorous avian species, variation in diet quality (e.g., N content) does influence laying rate, egg size, and egg hatchability (e.g., see Sharp and Moss 1981; Robbins 1993; and section 13.8).

There is more evidence for a quantitative effect of nutrition on magnitude of allocation in individuals of other groups. A general review of the effect of nutrition on domesticated mammals can be found in Blaxter (1989). Kirkpatrick reviewed many examples of wild mammalian species in which the number of young born was positively related to level of nutrition, presumably because of an increase in ovulation rate (Kirkpatrick 1988). Managers of domestic livestock such as sheep have long relied on such an effect, called "flushing," in which they provide a higher level of nutrition just before breeding and thereby increase the proportion of females bearing twins. Kirkpatrick also reviewed instances in which better nutrition in wild mammals increased the proportion of adult females breeding, advanced the age of sexual maturity, and increased the mass and/or survival of young *in utero* or just at birth. These reproductive responses of wild mammals to variation in dietary energy supply are similar to responses seen in domesticated mammals (Blaxter 1989).

In ectotherms there is often a positive relationship between amount of food consumed and number or mass of eggs laid. Slansky and Scriber (1985) provide a number of examples for insects. For a vertebrate ectotherm example, let us consider fish, in which feeding level can affect fecundity in at least two ways. First, there is a size-dependent effect because in many fish species fecundity is positively related to body size. Growth rate and hence size are positively influenced by feeding level (section 13.5.1). There can also be a size-independent effect in which fecundity increases with feeding rate for a fish of some specified size. Both effects are apparent in laboratory studies with iteroparous fish such as three-spined sticklebacks, cichlids, and guppies (Wootton 1998). Batch fecundity, the number of eggs/spawning, is mainly a positive function of fish size. The frequency of spawning, however, is positively related to feeding level once the effect of fish size is removed. Also, the poorer the food supply, the lower the proportion of the population that proceeds with gonadal development and comes into spawning condition (Jobling 1994). There are, of course, exceptions to these generalities. In rainbow trout, for example, fecundity is reduced when feeding level is reduced, even after correcting for the effects of differences of fish size on egg production (Jobling 1994).

Patterns such as these are apparent for reptiles as well, with clutch mass generally increasing with body size, and frequency of laying increasing with food intake rate (Anderson and Karasov 1988). Some lizard species will lay multiple clutches per year when food availability is high but only one clutch/yr, or even zero, if food is more limiting.

14.5 Putting Energy and Material Costs of Reproduction in Perspective

The discussion so far, which focused on the energy and material costs of reproduction, has emphasized (1) how high the costs can be, (2) that high cost has necessitated the synchronization of reproductive expenditures with the natural pattern of environmental resource availability, and (3) that significant variation in resource availability relative to the costs of reproduction can lead to variation in reproductive allocation. The study that we described above about reproduction by blue tits that were temporally matched versus unmatched to resource availability encapsulates these points and also brings us back to the point made at the outset of this chapter about alternative ways to measure costs. When reproduction was mismatched to resource availability, blue tit parents expended more energy and provided fewer prey to their young, which are energy and material costs, and the parents exhibited lower survivorship and their young were smaller, which are fitness costs. This observational study, therefore, seems to show the reduction in parental reproductive value due to reproductive effort that is predicted by life-history theory (Williams 1966). Experimental studies provide more rigorous tests, and we will return to our focal species, the kestrel, for a final example of such a study.

As we described above, Daan and his colleagues have performed manipulative studies in which kestrel broods were enlarged, and they found that parents increased their hunting activity and their daily energy expenditure, but they lost mass and their nestlings grew slower (Dijkstra et al. 1990; Deerenberg et al. 1995). Data such as these suggest that birds raising their normal brood sizes work at near maximal capacity, and even though they might be able to raise extra nestlings, the higher level of work possibly compromises survival and future reproductive success. Indeed, Daan and his collaborators analyzed survival of kestrels over the short term (Dijkstra et al. 1990) and longer term (Daan et al. 1996) and found that adult survival was negatively related with manipulated brood size. Much of the extra mortality occurred in the winter following the summertime experiment. The exact mechanism is unknown. Daan et al. (1996) argued that the long delay argues against the mechanism being increased predation or disease risk during the episode of parental care. They speculated that the manipulated parents' higher metabolic expenditures or loss of body mass influenced the rate of senescence or longer-term immune function. Sinervo and Svensson (1998) and Ricklefs and Wikelski (2002) have suggested complementary and additional mechanisms. It is feasible to test such hypotheses. For example, in another manipulative study, Lemon (1993) found that captive zebra finches that were forced to work harder during reproduction also lost weight, waited longer to reproduce again, and had lower survival. With a tractable experimental animal such as the zebra finch, future researchers could reveal the mechanism(s) by which the work of reproduction compromises future reproduction and survival.

We hope that our focus on costs of production has underscored the utility of retaining a focus on both types of costs—energy and material costs as well as demographic/fitness costs. In some cases there are causal connections between the two types of costs. Also, when differences are observed in life-history features between populations of the same species, a bioenergetic model of resource allocation can be used as a hypothesis against which to compare an alternative hypothesis of genetic adaptation. Recall, for example, Elliott's (1994) conclusion that life-history differences between dozens of trout populations that were thought to be "evolutionary" (i.e., genetic) were satisfactorily explained as being due to effects of environmental temperature on energy budgets.

We find compelling reasons for quantifying energy and material costs even in the absence of simultaneous evaluation of fitness costs. First, improved understanding of principles of mass and energy exchange increase explanatory power across the board in ecology. The energy cost of reproduction, and limits on parents' abilities to allocate energy to reproduction, are thought to have profound effects on behavior, social structure, group size, mating/breeding strategy, etc. that so many scientific peers study (Thompson 1992). Second, we gain a quantitative measure of resource demand by animals on their environment, which has a variety of uses in ecology, including management of natural resources. Individual budgets of energy and mass can be used to assemble budgets for entire populations (Lucas 1996). If energy and material are limiting for a particular species, we may gain some insights into whether and how the balance of energy and material gains and losses translate into demographic costs that ultimately constrain reproduction (Sinervo and Svensson 1998). Finally, this kind of understanding is necessary to predict responses of animals to human-induced changes in habitat or in global climate (Porter et al., 2000, 1994). The ability of ecologists to contribute to solutions to these problems depends partly on their understanding of the principles of energy and material flow in animals.

References

Alisauskas, R. T., and C. D. Ankney. 1985. Nutrient reserves and the energetics of reproduction in American coots. *Auk* 102:133–145.

Anderson, R. A., and W. H. Karasov. 1988. Energetics of the lizard *Cnemidophorus tigris* and life history consequences of food-acquisition mode. *Ecological Monographs* 58:79–110.

Angilletta, M. J., Jr., and M. W. Sears. 2000. The metabolic cost of reproduction in an oviparous lizard. *Functional Ecology* 14:39–45.

Aspey, W. P., and S. I. Lustick. 1983. *Behavioral energetics: The cost of survival in vertebrates.* Ohio State University Press, Columbus.

Baverstock, P., and B. Green. 1975. Water recycling in lactation. *Science* 187:657–658.

Berger, P. J., N. C. Negus, E. H. Sanders, and P. D. Gardner. 1981. Chemical triggering of reproduction in *Microtus montanus*. *Science* 214:69–70.

Beuchat, C. A., and D. Vleck. 1990. Metabolic consequences of viviparity in a lizard, *Sceloporus jarrovi*. *Physiological Zoology* 63:555–570.

Blaxter, K. 1989. *Energy metabolism in animals and man.* Cambridge University Press, Cambridge.

Blockstein, D. E. 1989. Crop milk and clutch size in mourning doves. *Wilson Bulletin* 101:11–25.

Brett, J. R., and T.D.D. Groves. 1979. Physiological energetics. Pages 279–352 in W. S. Hoar, D. J. Randall, and J. R. Brett (eds.), *Fish physiology.* Academic Press, London.

Cary, J. R., and L. B. Keith. 1979. Reproductive change in the 10-year cycle of snowshoe hares. *Canadian Journal of Zoology* 57:375–390.

Daan, S., C. Deerenberg, and C. Dijkstra. 1996. Increased daily work precipitates natural death in the kestrel. *Journal of Animal Ecology* 65:539–544.

Davis, G. J., and R. K. Meyer. 1972. The effect of daylength on pituitary FSH and LH and gonadal development of snowshoe hares. *Biology of Reproduction* 6:264–269.

Deerenberg, C., I. Pen, C. Dijkstra, B. J. Arkies, G. H. Visser, and S. Daan. 1995. Parental energy expenditure in relation to manipulated brood size in the European kestrel *Falco tinnunculus*. *Zoology—Jena* 99:39–48.

Dijkstra, C., A. Bult, S. Bijlsma, S. Daan, T. Meijer, and M. Tijlstra. 1990. Brood size manipulations in the kestrel (*Falco tinnunculus*): Effects on offspring and parent survival. *Journal of Animal Ecology* 59:269–286.

Elliott, J. M. 1994. *Quantitative ecology and the brown trout.* Oxford University Press, Oxford.

Emlet, R. B., and O. Hoegh-Guldberg. 1997. Effects of egg size on postlarval performance: Experimental evidence from a sea urchin. *Evolution* 51:141–152.

Finkler, M. S., and K. A. Cullum. 2002. Sex-related differences in metabolic rate and energy reserves in spring-breeding small-mouthed salamanders (*Ambystoma texanum*). *Copeia* 2002:824–829.

Fisher, H. 1972. The nutrition of birds. Pages 431–469 in D. S. Farner, J. R. King, and K. C. Parkes (eds.), *Avian Biology*. Academic Press, London.

Friedman, M. I., and J. P. Bruno. 1976. Water recycling during lactation. *Science* 19:409–410.

Gittleman, J. L., and S. D. Thompson. 1988. Energy allocation in mammalian reproduction. *American Zoologist* 28:863–875.

Guillette, L. J., Jr. 1982. Effects of gravidity on the metabolism of the reproductively bimodal lizard, *Sceloporus aeneus*. *Journal of Experimental Zoology* 223:33–36.

Gwinner, E. 1986. *Circannual rhythms. Endogenous annual clocks in the organization of seasonal processes*. Springer-Verlag, Berlin.

Hahn, T. P., T. Boswell, J. C. Wingfield, and G. F. Ball. 1997. Temporal flexibility in avian reproduction. Pages 39–80 in V. Nolan, Jr., E. D. Ketterson, and C. F. Thompson (eds.), *Current ornithology*, vol. 14. Plenum Press, New York.

Hanwell, A., and M. Peaker. 1977. Physiological effects of lactation on the mother. *Symposia of the Zoological Society of London* 41:297–312.

Heatwole, H. 1996. *Energetics of desert invertebrates*. Springer-Verlag, Berlin.

Humphreys, W. F. 1979. Production and respiration in animal populations. *Journal of Animal Ecology* 48:427–453.

Jobling, M. 1994. *Fish bioenergetics*. Chapman & Hall, London.

Karasov, W. H. 1990. Digestion in birds: Chemical and physiological determinants and ecological implications. *Studies in Avian Biology* 13:391–415.

Karasov, W. H., and R. A. Anderson. 1998. Correlates of average daily metabolism of field-active zebra-tailed lizards (*Callisaurus draconoides*). *Physiological Zoology* 71:93–105.

Kenagy, G. J., D. Masman, S. M. Sharbaugh, and K. A. Nagy. 1990. Energy expenditure during lactation in relation to litter size in free-living golden-mantled ground squirrels. *Journal of Animal Ecology* 59:73–88.

Kenagy, G. J., R. D. Stevenson, and D. Masman. 1989. Energy requirements for lactation and postnatal growth in captive golden-mantled ground squirrels. *Physiological Zoology* 62:487–470.

Kirkpatrick, R. C. 1988. *Comparative influence of nutrition on reproduction and survival of wild birds and mammals—an overview*. Caesar Kleberg Wildlife Research Institute at Texas A&I University, Kingsville, Tex.

Krapu, G. L. 1981. The role of nutrient reserves in Mallard reproduction. *Auk* 98:29–38.

Labov, J. B. 1977. Phytoestrogens and mammalian reproduction. *Comparative Biochemistry and Physiology A* 57:3–9.

Lack, D. 1954. *The natural regulation of animal numbers*. Clarendon, Oxford.

Landwer, A. J. 1994. Manipulation of egg production reveals costs of reproduction in the tree lizard (*Urosaurus ornatus*). *Oecologia* 100:243–249.

Lemon, W. C. 1993. The energetics of lifetime reproductive success in the zebra finch *Taeniopygia guttata*. *Physiological Zoology* 66:946–963.

Leopold, A. S. 1976. Phytoestrogens: Adverse effects on reproduction in California quail. *Science* 191:98–100.

Lucas, A. 1996. *Bioenergetics of aquatic animals.* Taylor & Francis, London.

Marler, C. A., G. E. Walsberg, M. L. White, and M. Moore. 1995. Increased energy expenditure due to increased territorial defense in male lizards after phenotypic manipulation. *Behavioral Ecology and Sociobiology* 37:225–231.

Masman, D., S. Daan, and H. J. A. Beldhuis. 1988. Ecological energetics of the kestrel: Daily energy expenditure throughout the year based on time-energy budget, food intake and doubly labeled water methods. *Ardea* 76:64–81.

Masman, D., M. Gordijn, S. Daan, and C. Dijkstra. 1986. Ecological energetics of the kestrel: Field estimates of energy intake througout the year. *Ardea* 74:24–39.

McEdward, L. R. 1996. Experimental manipulation of parental investment in echinoid echinoderms. *American Zoologist* 36:169–179.

Meijer, T. 1989. Photoperiodic control of reproduction and molt in the kestrel, *Falco tinnunculus. Journal of Biological Rhythms* 4:351–364.

Meijer, T., S. Daan, and C. Dijkstra. 1988. Female condition and reproductive decisions: The effect of food manipulations in free-living and captive kestrels. *Ardea* 76:141–154.

Meijer, T., and R. Drent. 1999. Re-examination of the capital and income dichotomy in breeding birds. *Ibis* 141:399–414.

Meijer, T., D. Masman, and S. Daan. 1989. Energetics of reproduction in female kestrels. *Auk* 106:549–559.

Oksanen, T. A., E. Koskela, and T. Mappes. 2002. Hormonal manipulation of offspring number: Maternal effort and reproductive costs. *Evolution* 56:1530–1537.

Porter, W. P., S. Budaraju, W. E. Stewart, and N. Ramankutty. 2000. Calculating climate effects on birds and mammals: Impacts on biodiversity, conservation, population parameters, and global community structure. *American Zoologist* 40:597–630.

Porter, W. P., J. C. Munger, W. E. Stewart, S. Budaraju, and J. Jaeger. 1994. Endotherm energetics: from a scalable individual-based model to ecological applications. *Australian Journal of Zoology* 42:125–162.

Reiss, M. J. 1989. *The allometry of growth and reproduction.* Cambridge University Press, Cambridge.

Ricklefs, R. E. 1974. Energetics of reproduction in birds. Pages 152–292 in R. A. Paynter, Jr. (ed.), *Avian energetics.* Nutthal Ornithological Society, Cambridge, Mass.

Ricklefs, R. E., and M. Wikelski. 2002. The physiology/life history nexus. *Trends in Ecology and Evolution* 17:462–468.

Ricklefs, R. E., and J. B. Williams. 1984. Daily energy expenditure and water turnover rate of adult European starlings (*Sturnus vulgaris*) during the nesting cycle. *Auk* 101:707–716.

Robbins, C. T. 1993. *Wildlife feeding and nutrition.* Academic Press, San Diego, Calif.

Sanders, E. H., P. D. Gardner, P. J. Berger, and N. C. Negus. 1981. 6-Methoxybenzoxazolinone: A plant derivative that stimulates reproduction in *Microtus montanus. Science* 214:67–69.

Scantlebury, M., W. Hynds, D. Booles, and J. R. Speakman. 2000. Isotope recycling in lactating dogs (*Canis familiaris*). *American Journal of Physiology* 278:R669–R676.

Sharp, P. J., and R. Moss. 1981. A comparison of the responses of captive willow ptarmigan (*Lagopus lagopus lagopus*), red grouse (*Lagopus lagopus scoticus*) and hybrids to increasing day length with observations on the modifying effects of nutrition on red grouse. *General and Comparative Endocrinology* 45:181–188.

Sikes, R. S., and H. Ylonen. 1998. Considerations of optimal litter size in mammals. *Oikos* 83:452–465.

Sinervo, B. 1990. The evolution of maternal investment in lizards: an experimental and comparative analysis of egg size and its effect on offspring performance. *Evolution* 44:279–294.

Sinervo, B., P. Doughty, R. B. Huey, and K. Zamudio. 1992. Allometric eingineering—a causal analysis of natural selection on offspring size. *Science* 258:192–1930.

Sinervo, B., and R. B. Huey. 1990. Allometric engineering—an experimental test of the causes of interpopulational differences in performance. *Science* 248:1106–1109.

Sinervo, B., and P. Licht. 1991. Hormonal and physiological control of clutch size, egg size and egg shape in side-blotched lizards (*Uta stansbrugiana*): Constraints on the evolution of lizard life histories. *Journal of Experimental Zoology* 257:252–264.

Sinervo, B., and L. R. McEdward. 1988. Developmental consequences of an evolutionary change in egg size: An experimental test. *Evolution* 42:885–899.

Sinervo, B., and E. Svensson. 1998. Mechanistic and selective causes of life history trade-offs and plasticity. *Oikos* 83:432–442.

Slansky, F., and J. M. Scriber. 1985. Food consumption and utilization. Pages 87–163 in G. A. Kerkut and L. I. Gilbert (eds.), *Comprehensive insect physiology, biochemistry and pharmacology*, vol. 4. Pergamon Press, Oxford.

Takakura, K. I. 1999. Active female courtship behavior and male nutritional contribution to female fecundity in *Bruchidius dorsalis* (Fahraeus) (Coleoptera: Bruchidae). *Researches on Population Ecology—Tokyo* 41:269–273.

Thomas, D. H., J. Blondel, P. Perret, M. M. Lambrechts, and J. R. Speakman. 2001. Energetic and fitness costs of mismatching resource supply and demand in seasonally breeding birds. *Science* 291:2598–2600.

Thompson, S. D. 1992. Gestation and lactation in small mammals: Basal metabolic rate and the limits of energy use. Pages 213–259 in T. E. Tomasi and T. S. Horton (eds.), *Mammalian energetics*. Comstock Publishing Company, Ithaca, N.Y.

Vandeputte, P. J. 1980. Feeding, growth and metabolism of the pigeon, *Columba livia domestica*: Duration and role of crop milk feeding. *Journal of Comparative Physiology B* 135:97–100.

Vaughan, M. R., and L. B. Keith. 1981. Demographic response of experimental snowshoe hare populations to overwinter food shortage. *Journal of Wildlife Management* 45:354–380.

Verme, L. J. 1965. Reproduction studies on penned white-tailed deer. *Journal of Wildlife Management* 29:74–79.

Vezina, F., and T. D. Williams. 2002. Metabolic costs of egg production in the European starling (*Sturnus vulgaris*). *Physiological and Biochemical Zoology* 75:377–385.

Vleck, C. M., and J. Priedkalns. 1985. Reproduction in zebra finches: Hormone levels and effect of dehydration. *Condor* 87:37–46.

Walsberg, G. E. 1983. Avian ecological energetics. Pages 161–220 in D. S. Farner and J. R. King (eds.), *Avian biology*, vol. VII. Academic Press, New York.

West, G. B. 2001. A general model for ontogenetic growth. *Nature* 413:628–631.

Westmoreland, D., and L. B. Best. 1987. What limits mourning doves to a clutch of two eggs? *Condor* 89:486–493.

Williams, G. C. 1966. Natural selection, the costs of reproduction, and a refinement of Lack's principle. *American Naturalist* 100:687–690.

Williams, T. D., and F. Vezina. 2001. Reproductive energy expenditure, intraspecific variation and fitness in birds. Pages 355–406 in V. J. Nolan and C. S. Thompson (eds.), *Current ornithology*, vol. 16. Kluwer Academic/Plenum Publishers, New York.

Wootton, R. J. 1998. *Ecology of teleost fishes*. Kluwer Academic, Dordrecht.

Zann, R. A., S. R. Morton, K. R. Jones, and N. T. Burley. 1995. The timing of breeding by zebra finches in relation to rainfall in central Australia. *Emu* 95:208–222.

INDEX

hypothesis, 370, 372–375. *See also* biochemical indices, of nutritional status in ruminant herbivores
rumination, 357

salicin, 515
sea urchin (*Diadema antillarum*), 652
secondary metabolites (SMs), 103–107, 177, 479, 501, 510, 512, 587, 628–629, 715; effect of on feeding/taste, 513–516; in fruit, 516; inactive forms, 106–107; phytochemical assays, 105. *See also* xenobiotic biotransformation, costs of, and secondary metabolites (SMs)
sediments, 309
selenium (Se), 99–103; selenium deficiency (white muscle disease/nutritional muscular dystrophy), 101–102; selenosis ("blind staggers"), 102–103
serial endosymbiotic theory (SET), 241
seston, 260, 550
shrimp: brine shrimp (*Artemia salina*), 482; grass shrimp (*Palaemonetes*), 482
silverfish (order Thysanura), 313
Sitka spruce (*Picea sitchensis*), 468
smooth endoplasmic reticulum (SER), 485
snakes: boa, 415; Burmese python, 414; python, 415
sodium (Na), 584–587, 614; measuring sodium flux, 611–612; sodium hunger, 587–588, 600
Soxhlet apparatus, 106
spectrophotometry, 174–175
spiders (arachnids), 492, 497
stable isotope mass spectrometry, 436–438
starches, 65–66
starvation: biochemistry of, 417–420; biochemistry of fasting-adapted

species, 420–421; starvation-predation risk hypothesis, 411
stoichiometry. *See* ecological stoichiometry
streptomycin, 239
strontium, 464–465
sulfide, 272, 275, 278. *See also* hydrogen sulfide
sulfur (S), 275–76, 289, 537
sulphotransferases, 490
suspended particulate matter (SPM), 260
Symbiodinium zooxanthellae (*Symbiodinium A, B, C, D,* and *E*), 262–64
symbionts, 246; and the nitrogen economy of termites, 326–28; stacking of endocytobiotic symbionts, 319–20; *Streblomastix*, 319; *Trichomitopsis*, 319
symbioses, 238, 248; and the chimerical nature of organisms, 242–243; contributions of symbionts to the nutrition of their hosts, 276–277; and endocytobiotic symbionts, 319–320; insect-bacteria symbioses, 289–290; photosynthetic symbioses, 244, 246, 248, 251; possession of dual symbionts, 283; protection of symbionts, 248; symbiotic microorganisms, 284. *See also* deep sea symbioses
"symmorphosis," 137–138
Synergistes jonesii, 368
"syntrophs," 309

tannins, 103–105, 501–502
temperature, 30–32; and the fluidity of biological membranes, 85; operative environmental temperature, 677
termites, 293–94, 314–315; differences from leaf-cutting ants, 332–333; digestive processes in, 315, 317; and endogenous cellulases, 317–318; and the *Enterococcus* bacteria, 322;